Neutrino Physics: From Discovery to Cosmic Neutrinos

Neutrino Physics: From Discovery to Cosmic Neutrinos

Edited by:

Stefano Spezia

ARCLER PRESS

www.arclerpress.com

Neutrino Physics: From Discovery to Cosmic Neutrinos

Stefano Spezia

Arcler Press
224 Shoreacres Road
Burlington, ON L7L 2H2
Canada
www.arclerpress.com
Email: orders@arclereducation.com

ISBN: 978-1-77407-767-2 (Hardcover)

DECLARATION

Some content or chapters in this book are open access copyright free published research work, which is published under Creative Commons License and are indicated with the citation. We are thankful to the publishers and authors of the content and chapters as without them this book wouldn't have been possible.

ABOUT THE EDITOR

Stefano Spezia is Ph.D. holder in Applied Physics at the University of Palermo since April 2012. His major research experience is in noise-induced effects in nonlinear systems, especially in the fields of modeling of complex biological systems and simulation of semiconductor spintronic devices. Associate member of the Italian Physical Society and European Physical Society.

TABLE OF CONTENTS

LIST OF CONTRIBUTORS

G. Barenboim
Universitat de València – CSIC, València, Spain

Surender Verma
Department of Physics and Astronomical Science, Central University of Himachal Pradesh, Dharamshala 176215, India

Dimitrios K. Papoulias
AHEP Group, Institut de Física Corpuscular – CSIC/Universitat de València, Valencia, Spain

Theocharis S. Kosmas
Division of Theoretical Physics, University of Ioannina, Ioannina, Greece

Yoshitaka Kuno
Department of Physics, Osaka University, Osaka, Japan

Ilaria Brivio
Niels Bohr International Academy and Discovery Centre, Niels Bohr Institute, University of Copenhagen, Blegdamsvej 17, DK-2100 Copenhagen, Denmark

Michael Trott
Niels Bohr International Academy and Discovery Centre, Niels Bohr Institute, University of Copenhagen, Blegdamsvej 17, DK-2100 Copenhagen, Denmark

Peter Ballett
Institute for Particle Physics Phenomenology, Department of Physics, Durham University, South Road, Durham DH1 3LE, United Kingdom

Matheus Hostert
Institute for Particle Physics Phenomenology, Department of Physics, Durham University, South Road, Durham DH1 3LE, United Kingdom

Silvia Pascoli
Institute for Particle Physics Phenomenology, Department of Physics, Durham University, South Road, Durham DH1 3LE, United Kingdom

M. Aker
Institute for Nuclear Physics (IKP), Karlsruhe Institute of Technology (KIT), Hermann-von-Helmholtz-Platz 1, 76344 Eggenstein-Leopoldshafen, Germany
Institute for Technical Physics (ITEP), Karlsruhe Institute of Technology (KIT), Hermann-von-Helmholtz-Platz 1, 76344 Eggenstein-Leopoldshafen, Germany

K. Altenmüller
Max-Planck-Institut für Physik, Föhringer Ring 6, 80805 München, Germany
Technische Universität München, James-Franck-Straße 1, 85748 Garching, Germany
IRFU (DPhP & APC), CEA, Universit´e Paris-Saclay, 91191 Gif-sur-Yvette, France

M. Arenz
Helmholtz-Institut für Strahlen- und Kernphysik, Rheinische Friedrich-Wilhelms-Universität Bonn, Nussallee 14-16, 53115 Bonn, Germany

M. Babutzka
Institute of Experimental Particle Physics (ETP), Karlsruhe Institute of Technology (KIT), Wolfgang-Gaede-Straße 1, 76131 Karlsruhe, Germany

J. Barrett
Laboratory for Nuclear Science, Massachusetts Institute of Technology, 77 Massachusetts Avenue, Cambridge, Massachusetts 02139, USA

S. Bauer
Institut für Kernphysik, Westfälische Wilhelms-Universität Münster, Wilhelm-Klemm-Straße 9, 48149 Münster, Germany

M. Beck
Institut für Kernphysik, Westfälische Wilhelms-Universität Münster, Wilhelm-Klemm-Straße 9, 48149 Münster, Germany
Institut für Physik, Johannes-Gutenberg-Universität Mainz, 55099 Mainz, Germany

A. Beglarian
Institute for Data Processing and Electronics (IPE), Karlsruhe Institute of Technology (KIT), Hermann-von-Helmholtz-Platz 1, 76344 Eggenstein-Leopoldshafen, Germany

J. Behrens
Institute of Experimental Particle Physics (ETP), Karlsruhe Institute of Technology (KIT), Wolfgang-Gaede-Straße 1, 76131 Karlsruhe, Germany
Institute for Nuclear Physics (IKP), Karlsruhe Institute of Technology (KIT), Hermann-von-Helmholtz-Platz 1, 76344 Eggenstein-Leopoldshafen, Germany
Institut für Kernphysik, Westfälische Wilhelms-Universität Münster, Wilhelm-Klemm-Straße 9, 48149 Münster, Germany

T. Bergmann
Max-Planck-Institut für Physik, Föhringer Ring 6, 80805 München, Germany
Technische Universität München, James-Franck-Straße 1, 85748 Garching, Germany
Institute for Data Processing and Electronics (IPE), Karlsruhe Institute of Technology (KIT), Hermann-von-Helmholtz-Platz 1, 76344 Eggenstein-Leopoldshafen, Germany

U. Besserer
Institute for Nuclear Physics (IKP), Karlsruhe Institute of Technology (KIT), Hermann-von-Helmholtz-Platz 1, 76344 Eggenstein-Leopoldshafen, Germany
Institute for Technical Physics (ITEP), Karlsruhe Institute of Technology (KIT), Hermann-von-Helmholtz-Platz 1, 76344 Eggenstein-Leopoldshafen, Germany

K. Blaum
Max-Planck-Institut für Kernphysik, Saupfercheckweg 1, 69117 Heidelberg, Germany

F. Block
Institute of Experimental Particle Physics (ETP), Karlsruhe Institute of Technology (KIT), Wolfgang-Gaede-Straße 1, 76131 Karlsruhe, Germany

S. Bobien
Institute for Technical Physics (ITEP), Karlsruhe Institute of Technology (KIT), Hermann-von-Helmholtz-Platz 1, 76344 Eggenstein-Leopoldshafen, Germany

K. Bokeloh
Institut für Kernphysik, Westfälische Wilhelms-Universität Münster, Wilhelm-Klemm-Straße 9, 48149 Münster, Germany

J. Bonn
Institut für Physik, Johannes-Gutenberg-Universität Mainz, 55099 Mainz, Germany

B. Bornschein
Institute for Nuclear Physics (IKP), Karlsruhe Institute of Technology (KIT), Hermann-von-Helmholtz-Platz 1, 76344 Eggenstein-Leopoldshafen, Germany
Institute for Technical Physics (ITEP), Karlsruhe Institute of Technology (KIT), Hermann-von-Helmholtz-Platz 1, 76344 Eggenstein-Leopoldshafen, Germany

L. Bornschein
Institute for Nuclear Physics (IKP), Karlsruhe Institute of Technology (KIT), Hermann-von-Helmholtz-Platz 1, 76344 Eggenstein-Leopoldshafen, Germany

H. Bouquet
Institute for Data Processing and Electronics (IPE), Karlsruhe Institute of Technology (KIT), Hermann-von-Helmholtz-Platz 1, 76344 Eggenstein-Leopoldshafen, Germany

T. Brunst
Max-Planck-Institut für Physik, Föhringer Ring 6, 80805 München, Germany
Technische Universität München, James-Franck-Straße 1, 85748 Garching, Germany

T. S. Caldwell
Department of Physics and Astronomy, University of North Carolina, Chapel Hill, North Carolina 27599, USA
Triangle Universities Nuclear Laboratory, Durham, North Carolina 27708, USA

L. La Cascio
Institute of Experimental Particle Physics (ETP), Karlsruhe Institute of Technology (KIT), Wolfgang-Gaede-Straße 1, 76131 Karlsruhe, Germany

S. Chilingaryan
Institute for Data Processing and Electronics (IPE), Karlsruhe Institute of Technology (KIT), Hermann-von-Helmholtz-Platz 1, 76344 Eggenstein-Leopoldshafen, Germany

W. Choi
Institute of Experimental Particle Physics (ETP), Karlsruhe Institute of Technology (KIT), Wolfgang-Gaede-Straße 1, 76131 Karlsruhe, Germany

T. J. Corona
Department of Physics and Astronomy, University of North Carolina, Chapel Hill, North Carolina 27599, USA
Triangle Universities Nuclear Laboratory, Durham, North Carolina 27708, USA
Laboratory for Nuclear Science, Massachusetts Institute of Technology, 77 Massachusetts Avenue, Cambridge, Massachusetts 02139, USA

K. Debowski
Department of Physics, Faculty of Mathematics and Natural Sciences, University of Wuppertal, Gaußstraße 20, 42119 Wuppertal, Germany
Institute of Experimental Particle Physics (ETP), Karlsruhe Institute of Technology (KIT), Wolfgang-Gaede-Straße 1, 76131 Karlsruhe, Germany

M. Deffert
Institute of Experimental Particle Physics (ETP), Karlsruhe Institute of Technology (KIT), Wolfgang-Gaede-Straße 1, 76131 Karlsruhe, Germany

M. Descher
Institute of Experimental Particle Physics (ETP), Karlsruhe Institute of Technology (KIT), Wolfgang-Gaede-Straße 1, 76131 Karlsruhe, Germany

P. J. Doe
Center for Experimental Nuclear Physics and Astrophysics, and Department of Physics, University of Washington, Seattle, Washington 98195, USA

O. Dragoun
Nuclear Physics Institute of the CAS, v. v. i., CZ-250 68 Řež, Czech Republic

G. Drexlin
Institute of Experimental Particle Physics (ETP), Karlsruhe Institute of Technology (KIT), Wolfgang-Gaede-Straße 1, 76131 Karlsruhe, Germany
Institute for Nuclear Physics (IKP), Karlsruhe Institute of Technology (KIT), Hermann-von-Helmholtz-Platz 1, 76344 Eggenstein-Leopoldshafen, Germany

J. A. Dunmore
Center for Experimental Nuclear Physics and Astrophysics, and Department of Physics, University of Washington, Seattle, Washington 98195, USA

S. Dyba
Institut für Kernphysik, Westfälische Wilhelms-Universität Münster, Wilhelm-Klemm-Straße 9, 48149 Münster, Germany

F. Edzards
Max-Planck-Institut für Physik, Föhringer Ring 6, 80805 München, Germany
Technische Universität München, James-Franck-Straße 1, 85748 Garching, Germany

L. Eisenblätte
Institute for Data Processing and Electronics (IPE), Karlsruhe Institute of Technology (KIT), Hermann-von-Helmholtz-Platz 1, 76344 Eggenstein-Leopoldshafen, Germany

K. Eite
Institute for Nuclear Physics (IKP), Karlsruhe Institute of Technology (KIT), Hermann-von-Helmholtz-Platz 1, 76344 Eggenstein-Leopoldshafen, Germany

E. Ellinger
Department of Physics, Faculty of Mathematics and Natural Sciences, University of Wuppertal, Gaußstraße 20, 42119 Wuppertal, Germany

R. Engel
Institute for Nuclear Physics (IKP), Karlsruhe Institute of Technology (KIT), Hermann-von-Helmholtz-Platz 1, 76344 Eggenstein-Leopoldshafen, Germany
Institute of Experimental Particle Physics (ETP), Karlsruhe Institute of Technology (KIT), Wolfgang-Gaede-Straße 1, 76131 Karlsruhe, Germany

S. Enomoto
Center for Experimental Nuclear Physics and Astrophysics, and Department of Physics, University of Washington, Seattle, Washington 98195, USA

M. Erhard
Institute of Experimental Particle Physics (ETP), Karlsruhe Institute of Technology (KIT), Wolfgang-Gaede-Straße 1, 76131 Karlsruhe, Germany

D. Eversheim
Helmholtz-Institut für Strahlen- und Kernphysik, Rheinische Friedrich-Wilhelms-Universität Bonn, Nussallee 14-16, 53115 Bonn, Germany

M. Fedkevych
Institut für Kernphysik, Westfälische Wilhelms-Universität Münster, Wilhelm-Klemm-Straße 9, 48149 Münster, Germany

A. Felden
Institute for Nuclear Physics (IKP), Karlsruhe Institute of Technology (KIT), Hermann-von-Helmholtz-Platz 1, 76344 Eggenstein-Leopoldshafen, Germany

S. Fischer
Institute for Nuclear Physics (IKP), Karlsruhe Institute of Technology (KIT), Hermann-von-Helmholtz-Platz 1, 76344 Eggenstein-Leopoldshafen, Germany
Institute for Technical Physics (ITEP), Karlsruhe Institute of Technology (KIT), Hermann-von-Helmholtz-Platz 1, 76344 Eggenstein-Leopoldshafen, Germany

B. Flatt
Institut für Physik, Johannes-Gutenberg-Universität Mainz, 55099 Mainz, Germany

J. A. Formaggio
Laboratory for Nuclear Science, Massachusetts Institute of Technology, 77 Massachusetts Avenue, Cambridge, Massachusetts 02139, USA

F. M. Fränkle
Institute for Nuclear Physics (IKP), Karlsruhe Institute of Technology (KIT), Hermann-von-Helmholtz-Platz 1, 76344 Eggenstein-Leopoldshafen, Germany
Department of Physics and Astronomy, University of North Carolina, Chapel Hill, North Carolina 27599, USA
Triangle Universities Nuclear Laboratory, Durham, North Carolina 27708, USA

G. B. Franklin
Department of Physics, Carnegie Mellon University, Pittsburgh, Pennsylvania 15213, USA

H. Frankrone
Institute for Data Processing and Electronics (IPE), Karlsruhe Institute of Technology (KIT), Hermann-von-Helmholtz-Platz 1, 76344 Eggenstein-Leopoldshafen, Germany

F. Friedel
Institute of Experimental Particle Physics (ETP), Karlsruhe Institute of Technology (KIT), Wolfgang-Gaede-Straße 1, 76131 Karlsruhe, Germany

D. Fuchs
Max-Planck-Institut für Physik, Föhringer Ring 6, 80805 München, Germany
Technische Universität München, James-Franck-Straße 1, 85748 Garching, Germany

A. Fulst
Institut für Kernphysik, Westfälische Wilhelms-Universität Münster, Wilhelm-Klemm-Straße 9, 48149 Münster, Germany

D. Furse
Laboratory for Nuclear Science, Massachusetts Institute of Technology, 77 Massachusetts Avenue, Cambridge, Massachusetts 02139, USA

K. Gauda
Institut für Kernphysik, Westfälische Wilhelms-Universität Münster, Wilhelm-Klemm-Straße 9, 48149 Münster, Germany

H. Gemmeke
Institute for Data Processing and Electronics (IPE), Karlsruhe Institute of Technology (KIT), Hermann-von-Helmholtz-Platz 1, 76344 Eggenstein-Leopoldshafen, Germany

W. Gil
Institute for Nuclear Physics (IKP), Karlsruhe Institute of Technology (KIT), Hermann-von-Helmholtz-Platz 1, 76344 Eggenstein-Leopoldshafen, Germany

F. Glück
Institute for Nuclear Physics (IKP), Karlsruhe Institute of Technology (KIT), Hermann-von-Helmholtz-Platz 1, 76344 Eggenstein-Leopoldshafen, Germany

S. Görhardt
Institute for Nuclear Physics (IKP), Karlsruhe Institute of Technology (KIT), Hermann-von-Helmholtz-Platz 1, 76344 Eggenstein-Leopoldshafen, Germany

S. Groh
Institute of Experimental Particle Physics (ETP), Karlsruhe Institute of Technology (KIT), Wolfgang-Gaede-Straße 1, 76131 Karlsruhe, Germany

S. Grohmann
Institute for Technical Physics (ITEP), Karlsruhe Institute of Technology (KIT), Hermann-von-Helmholtz-Platz 1, 76344 Eggenstein-Leopoldshafen, Germany

R. Grössle
Institute for Nuclear Physics (IKP), Karlsruhe Institute of Technology (KIT), Hermann-von-Helmholtz-Platz 1, 76344 Eggenstein-Leopoldshafen, Germany
Institute for Technical Physics (ITEP), Karlsruhe Institute of Technology (KIT), Hermann-von-Helmholtz-Platz 1, 76344 Eggenstein-Leopoldshafen, Germany

R. Gumbsheimer
Institute for Nuclear Physics (IKP), Karlsruhe Institute of Technology (KIT), Hermann-von-Helmholtz-Platz 1, 76344 Eggenstein-Leopoldshafen, Germany

M. Ha Minh
Max-Planck-Institut für Physik, Föhringer Ring 6, 80805 München, Germany
Technische Universität München, James-Franck-Straße 1, 85748 Garching, Germany

M. Hackenjos
Institute for Nuclear Physics (IKP), Karlsruhe Institute of Technology (KIT), Hermann-von-Helmholtz-Platz 1, 76344 Eggenstein-Leopoldshafen, Germany
Institute for Technical Physics (ITEP), Karlsruhe Institute of Technology (KIT), Hermann-von-Helmholtz-Platz 1, 76344 Eggenstein-Leopoldshafen, Germany
Institute of Experimental Particle Physics (ETP), Karlsruhe Institute of Technology (KIT), Wolfgang-Gaede-Straße 1, 76131 Karlsruhe, Germany

V. Hannen
Institut für Kernphysik, Westfälische Wilhelms-Universität Münster, Wilhelm-Klemm-Straße 9, 48149 Münster, Germany

F. Harms
Institute of Experimental Particle Physics (ETP), Karlsruhe Institute of Technology (KIT), Wolfgang-Gaede-Straße 1, 76131 Karlsruhe, Germany

J. Hartmann
Institute for Data Processing and Electronics (IPE), Karlsruhe Institute of Technology (KIT), Hermann-von-Helmholtz-Platz 1, 76344 Eggenstein-Leopoldshafen, Germany

N. Haußmann
Department of Physics, Faculty of Mathematics and Natural Sciences, University of Wuppertal, Gaußstraße 20, 42119 Wuppertal, Germany

F. Heizmann
Institute of Experimental Particle Physics (ETP), Karlsruhe Institute of Technology

(KIT), Wolfgang-Gaede-Straße 1, 76131 Karlsruhe, Germany

K. Helbing
Department of Physics, Faculty of Mathematics and Natural Sciences, University of Wuppertal, Gaußstraße 20, 42119 Wuppertal, Germany

S. Hickford
Institute for Nuclear Physics (IKP), Karlsruhe Institute of Technology (KIT), Hermann-von-Helmholtz-Platz 1, 76344 Eggenstein-Leopoldshafen, Germany
Department of Physics, Faculty of Mathematics and Natural Sciences, University of Wuppertal, Gaußstraße 20, 42119 Wuppertal, Germany

D. Hilk
Institute of Experimental Particle Physics (ETP), Karlsruhe Institute of Technology (KIT), Wolfgang-Gaede-Straße 1, 76131 Karlsruhe, Germany

B. Hillen
Institut für Kernphysik, Westfälische Wilhelms-Universität Münster, Wilhelm-Klemm-Straße 9, 48149 Münster, Germany

D. Hillesheimer
Institute for Nuclear Physics (IKP), Karlsruhe Institute of Technology (KIT), Hermann-von-Helmholtz-Platz 1, 76344 Eggenstein-Leopoldshafen, Germany
Institute for Technical Physics (ITEP), Karlsruhe Institute of Technology (KIT), Hermann-von-Helmholtz-Platz 1, 76344 Eggenstein-Leopoldshafen, Germany

D. Hinz
Institute for Nuclear Physics (IKP), Karlsruhe Institute of Technology (KIT), Hermann-von-Helmholtz-Platz 1, 76344 Eggenstein-Leopoldshafen, Germany

T. Höhn
Institute for Nuclear Physics (IKP), Karlsruhe Institute of Technology (KIT), Hermann-von-Helmholtz-Platz 1, 76344 Eggenstein-Leopoldshafen, Germany

B. Holzapfel
Institute for Technical Physics (ITEP), Karlsruhe Institute of Technology (KIT), Hermann-von-Helmholtz-Platz 1, 76344 Eggenstein-Leopoldshafen, Germany

S. Holzmann
Institute for Technical Physics (ITEP), Karlsruhe Institute of Technology (KIT), Hermann-von-Helmholtz-Platz 1, 76344 Eggenstein-Leopoldshafen, Germany

T. Houdy
Max-Planck-Institut für Physik, Föhringer Ring 6, 80805 München, Germany

Technische Universität München, James-Franck-Straße 1, 85748 Garching, Germany

M. A. Howe
Department of Physics and Astronomy, University of North Carolina, Chapel Hill, North Carolina 27599, USA
Triangle Universities Nuclear Laboratory, Durham, North Carolina 27708, USA

A. Huber
Institute of Experimental Particle Physics (ETP), Karlsruhe Institute of Technology (KIT), Wolfgang-Gaede-Straße 1, 76131 Karlsruhe, Germany

T. M. James
Institute for Technical Physics (ITEP), Karlsruhe Institute of Technology (KIT), Hermann-von-Helmholtz-Platz 1, 76344 Eggenstein-Leopoldshafen, Germany

A. Jansen
Institute for Nuclear Physics (IKP), Karlsruhe Institute of Technology (KIT), Hermann-von-Helmholtz-Platz 1, 76344 Eggenstein-Leopoldshafen, Germany

A. Kaboth
Laboratory for Nuclear Science, Massachusetts Institute of Technology, 77 Massachusetts Avenue, Cambridge, Massachusetts 02139, USA

C. Karl
Max-Planck-Institut für Physik, Föhringer Ring 6, 80805 München, Germany
Technische Universität München, James-Franck-Straße 1, 85748 Garching, Germany

O. Kazachenko,[21]
[21]Institute for Nuclear Research of Russian Academy of Sciences, 60th October Anniversary Prospect 7a, 117312 Moscow, Russia

J. Kellerer
Institute of Experimental Particle Physics (ETP), Karlsruhe Institute of Technology (KIT), Wolfgang-Gaede-Straße 1, 76131 Karlsruhe, Germany

N. Kernert
Institute for Nuclear Physics (IKP), Karlsruhe Institute of Technology (KIT), Hermann-von-Helmholtz-Platz 1, 76344 Eggenstein-Leopoldshafen, Germany

L. Kippenbrock
Center for Experimental Nuclear Physics and Astrophysics, and Department of Physics, University of Washington, Seattle, Washington 98195, USA

M. Kleesiek

Institute of Experimental Particle Physics (ETP), Karlsruhe Institute of Technology (KIT), Wolfgang-Gaede-Straße 1, 76131 Karlsruhe, Germany

M. Klein
Institute for Nuclear Physics (IKP), Karlsruhe Institute of Technology (KIT), Hermann-von-Helmholtz-Platz 1, 76344 Eggenstein-Leopoldshafen, Germany
Institute of Experimental Particle Physics (ETP), Karlsruhe Institute of Technology (KIT), Wolfgang-Gaede-Straße 1, 76131 Karlsruhe, Germany

C. Köhler
Max-Planck-Institut für Physik, Föhringer Ring 6, 80805 München, Germany
Technische Universität München, James-Franck-Straße 1, 85748 Garching, Germany

L. Köllenberger
Institute for Nuclear Physics (IKP), Karlsruhe Institute of Technology (KIT), Hermann-von-Helmholtz-Platz 1, 76344 Eggenstein-Leopoldshafen, Germany

A. Kopmann
Institute for Data Processing and Electronics (IPE), Karlsruhe Institute of Technology (KIT), Hermann-von-Helmholtz-Platz 1, 76344 Eggenstein-Leopoldshafen, Germany

M. Korzeczek
Institute of Experimental Particle Physics (ETP), Karlsruhe Institute of Technology (KIT), Wolfgang-Gaede-Straße 1, 76131 Karlsruhe, Germany

A. Kosmider
Institute for Nuclear Physics (IKP), Karlsruhe Institute of Technology (KIT), Hermann-von-Helmholtz-Platz 1, 76344 Eggenstein-Leopoldshafen, Germany

A. Kovalík
Nuclear Physics Institute of the CAS, v. v. i., CZ-250 68 Řež, Czech Republic

B. Krasch
Institute for Nuclear Physics (IKP), Karlsruhe Institute of Technology (KIT), Hermann-von-Helmholtz-Platz 1, 76344 Eggenstein-Leopoldshafen, Germany
Institute for Technical Physics (ITEP), Karlsruhe Institute of Technology (KIT), Hermann-von-Helmholtz-Platz 1, 76344 Eggenstein-Leopoldshafen, Germany

M. Kraus
Institute of Experimental Particle Physics (ETP), Karlsruhe Institute of Technology (KIT), Wolfgang-Gaede-Straße 1, 76131 Karlsruhe, Germany

H. Krause
Institute for Nuclear Physics (IKP), Karlsruhe Institute of Technology (KIT), Hermann-

von-Helmholtz-Platz 1, 76344 Eggenstein-Leopoldshafen, Germany

L. Kuckert
Institute for Nuclear Physics (IKP), Karlsruhe Institute of Technology (KIT), Hermann-von-Helmholtz-Platz 1, 76344 Eggenstein-Leopoldshafen, Germany

B. Kuffner
Institute for Nuclear Physics (IKP), Karlsruhe Institute of Technology (KIT), Hermann-von-Helmholtz-Platz 1, 76344 Eggenstein-Leopoldshafen, Germany

N. Kunka
Institute for Data Processing and Electronics (IPE), Karlsruhe Institute of Technology (KIT), Hermann-von-Helmholtz-Platz 1, 76344 Eggenstein-Leopoldshafen, Germany

T. Lasserre
IRFU (DPhP & APC), CEA, Universit´e Paris-Saclay, 91191 Gif-sur-Yvette, France
Technische Universität München, James-Franck-Straße 1, 85748 Garching, Germany
Max-Planck-Institut für Physik, Föhringer Ring 6, 80805 München, Germany

T. L. Le
Institute for Nuclear Physics (IKP), Karlsruhe Institute of Technology (KIT), Hermann-von-Helmholtz-Platz 1, 76344 Eggenstein-Leopoldshafen, Germany
Institute for Technical Physics (ITEP), Karlsruhe Institute of Technology (KIT), Hermann-von-Helmholtz-Platz 1, 76344 Eggenstein-Leopoldshafen, Germany

O. Lebeda
Nuclear Physics Institute of the CAS, v. v. i., CZ-250 68 Řež, Czech Republic

M. Leber
Center for Experimental Nuclear Physics and Astrophysics, and Department of Physics, University of Washington, Seattle, Washington 98195, USA

B. Lehnert
Institute for Nuclear and Particle Astrophysics and Nuclear Science Division, Lawrence Berkeley National Laboratory, Berkeley, California 94720, USA

J. Letnev
University of Applied Sciences (HFD) Fulda, Leipziger Straße 123, 36037 Fulda, Germany

F. Leven
Institute of Experimental Particle Physics (ETP), Karlsruhe Institute of Technology (KIT), Wolfgang-Gaede-Straße 1, 76131 Karlsruhe, Germany

S. Lichter
Institute for Nuclear Physics (IKP), Karlsruhe Institute of Technology (KIT), Hermann-von-Helmholtz-Platz 1, 76344 Eggenstein-Leopoldshafen, Germany

V. M. Lobashev
Institute for Nuclear Research of Russian Academy of Sciences, 60th October Anniversary Prospect 7a, 117312 Moscow, Russia

A. Lokhov
Institut für Kernphysik, Westfälische Wilhelms-Universität Münster, Wilhelm-Klemm-Straße 9, 48149 Münster, Germany
Institute for Nuclear Research of Russian Academy of Sciences, 60th October Anniversary Prospect 7a, 117312 Moscow, Russia

M. Machatschek
Institute of Experimental Particle Physics (ETP), Karlsruhe Institute of Technology (KIT), Wolfgang-Gaede-Straße 1, 76131 Karlsruhe, Germany

E. Malcherek
Institute for Nuclear Physics (IKP), Karlsruhe Institute of Technology (KIT), Hermann-von-Helmholtz-Platz 1, 76344 Eggenstein-Leopoldshafen, Germany

K. Müller
Institute for Nuclear Physics (IKP), Karlsruhe Institute of Technology (KIT), Hermann-von-Helmholtz-Platz 1, 76344 Eggenstein-Leopoldshafen, Germany

M. Mark
Institute for Nuclear Physics (IKP), Karlsruhe Institute of Technology (KIT), Hermann-von-Helmholtz-Platz 1, 76344 Eggenstein-Leopoldshafen, Germany

A. Marsteller
Institute for Nuclear Physics (IKP), Karlsruhe Institute of Technology (KIT), Hermann-von-Helmholtz-Platz 1, 76344 Eggenstein-Leopoldshafen, Germany
Institute for Technical Physics (ITEP), Karlsruhe Institute of Technology (KIT), Hermann-von-Helmholtz-Platz 1, 76344 Eggenstein-Leopoldshafen, Germany

E. L. Martin
Department of Physics and Astronomy, University of North Carolina, Chapel Hill, North Carolina 27599, USA
Triangle Universities Nuclear Laboratory, Durham, North Carolina 27708, USA
Center for Experimental Nuclear Physics and Astrophysics, and Department of Physics, University of Washington, Seattle, Washington 98195, USA

C. Melzer
Institute for Nuclear Physics (IKP), Karlsruhe Institute of Technology (KIT), Hermann-von-Helmholtz-Platz 1, 76344 Eggenstein-Leopoldshafen, Germany
Institute for Technical Physics (ITEP), Karlsruhe Institute of Technology (KIT), Hermann-von-Helmholtz-Platz 1, 76344 Eggenstein-Leopoldshafen, Germany

A. Menshikov
Institute for Data Processing and Electronics (IPE), Karlsruhe Institute of Technology (KIT), Hermann-von-Helmholtz-Platz 1, 76344 Eggenstein-Leopoldshafen, Germany

S. Mertens
Max-Planck-Institut für Physik, Föhringer Ring 6, 80805 München, Germany
Technische Universität München, James-Franck-Straße 1, 85748 Garching, Germany
Institute for Nuclear and Particle Astrophysics and Nuclear Science Division, Lawrence Berkeley National Laboratory, Berkeley, California 94720, USA
Institute for Nuclear Physics (IKP), Karlsruhe Institute of Technology (KIT), Hermann-von-Helmholtz-Platz 1, 76344 Eggenstein-Leopoldshafen, Germany

L. I. Minter
Center for Experimental Nuclear Physics and Astrophysics, and Department of Physics, University of Washington, Seattle, Washington 98195, USA

S. Mirz
Institute for Nuclear Physics (IKP), Karlsruhe Institute of Technology (KIT), Hermann-von-Helmholtz-Platz 1, 76344 Eggenstein-Leopoldshafen, Germany
Institute for Technical Physics (ITEP), Karlsruhe Institute of Technology (KIT), Hermann-von-Helmholtz-Platz 1, 76344 Eggenstein-Leopoldshafen, Germany

B. Monreal
Department of Physics, Case Western Reserve University, Cleveland, Ohio 44106, USA

P. I. Morales Guzmán
Max-Planck-Institut für Physik, Föhringer Ring 6, 80805 München, Germany
Technische Universität München, James-Franck-Straße 1, 85748 Garching, Germany

K. Müller
Institute for Nuclear Physics (IKP), Karlsruhe Institute of Technology (KIT), Hermann-von-Helmholtz-Platz 1, 76344 Eggenstein-Leopoldshafen, Germany

U. Naumann
Department of Physics, Faculty of Mathematics and Natural Sciences, University of Wuppertal, Gaußstraße 20, 42119 Wuppertal, Germany

W. Ndeke
Institut für Physik, Humboldt-Universität zu Berlin, Newtonstraße 15, 12489 Berlin, Germany

H. Neumann
Institute for Technical Physics (ITEP), Karlsruhe Institute of Technology (KIT), Hermann-von-Helmholtz-Platz 1, 76344 Eggenstein-Leopoldshafen, Germany

S. Niemes
Institute for Nuclear Physics (IKP), Karlsruhe Institute of Technology (KIT), Hermann-von-Helmholtz-Platz 1, 76344 Eggenstein-Leopoldshafen, Germany
Institute for Technical Physics (ITEP), Karlsruhe Institute of Technology (KIT), Hermann-von-Helmholtz-Platz 1, 76344 Eggenstein-Leopoldshafen, Germany

M. Noe
Institute for Technical Physics (ITEP), Karlsruhe Institute of Technology (KIT), Hermann-von-Helmholtz-Platz 1, 76344 Eggenstein-Leopoldshafen, Germany

N. S. Oblath
Laboratory for Nuclear Science, Massachusetts Institute of Technology, 77 Massachusetts Avenue, Cambridge, Massachusetts 02139, USA

H.-W. Ortjohann
Institut für Kernphysik, Westfälische Wilhelms-Universität Münster, Wilhelm-Klemm-Straße 9, 48149 Münster, Germany

A. Osipowicz
University of Applied Sciences (HFD) Fulda, Leipziger Straße 123, 36037 Fulda, Germany

B. Ostrick
Institut für Kernphysik, Westfälische Wilhelms-Universität Münster, Wilhelm-Klemm-Straße 9, 48149 Münster, Germany

E. Otten
Institut für Physik, Johannes-Gutenberg-Universität Mainz, 55099 Mainz, Germany

D. S. Parno
Department of Physics, Carnegie Mellon University, Pittsburgh, Pennsylvania 15213, USA
Center for Experimental Nuclear Physics and Astrophysics, and Department of Physics, University of Washington, Seattle, Washington 98195, USA

D. G. Phillips II
Department of Physics and Astronomy, University of North Carolina, Chapel Hill, North Carolina 27599, USA
Triangle Universities Nuclear Laboratory, Durham, North Carolina 27708, USA

P. Plischke
Institute for Nuclear Physics (IKP), Karlsruhe Institute of Technology (KIT), Hermann-von-Helmholtz-Platz 1, 76344 Eggenstein-Leopoldshafen, Germany

A. Pollithy
Max-Planck-Institut für Physik, Föhringer Ring 6, 80805 München, Germany
Technische Universität München, James-Franck-Straße 1, 85748 Garching, Germany

A. W. P. Poon
Institute for Nuclear and Particle Astrophysics and Nuclear Science Division, Lawrence Berkeley National Laboratory, Berkeley, California 94720, USA

J. Pouryamout
Department of Physics, Faculty of Mathematics and Natural Sciences, University of Wuppertal, Gaußstraße 20, 42119 Wuppertal, Germany

M. Prall
Institut für Kernphysik, Westfälische Wilhelms-Universität Münster, Wilhelm-Klemm-Straße 9, 48149 Münster, Germany

F. Priester
Institute for Nuclear Physics (IKP), Karlsruhe Institute of Technology (KIT), Hermann-von-Helmholtz-Platz 1, 76344 Eggenstein-Leopoldshafen, Germany
Institute for Technical Physics (ITEP), Karlsruhe Institute of Technology (KIT), Hermann-von-Helmholtz-Platz 1, 76344 Eggenstein-Leopoldshafen, Germany

M. Röllig
Institute for Nuclear Physics (IKP), Karlsruhe Institute of Technology (KIT), Hermann-von-Helmholtz-Platz 1, 76344 Eggenstein-Leopoldshafen, Germany
Institute for Technical Physics (ITEP), Karlsruhe Institute of Technology (KIT), Hermann-von-Helmholtz-Platz 1, 76344 Eggenstein-Leopoldshafen, Germany

C. Röttele
Institute for Nuclear Physics (IKP), Karlsruhe Institute of Technology (KIT), Hermann-von-Helmholtz-Platz 1, 76344 Eggenstein-Leopoldshafen, Germany
Institute of Experimental Particle Physics (ETP), Karlsruhe Institute of Technology (KIT), Wolfgang-Gaede-Straße 1, 76131 Karlsruhe, Germany
Institute for Technical Physics (ITEP), Karlsruhe Institute of Technology (KIT), Hermann-von-Helmholtz-Platz 1, 76344 Eggenstein-Leopoldshafen, Germany

P. C.-O. Ranitzsch
Institut für Kernphysik, Westfälische Wilhelms-Universität Münster, Wilhelm-Klemm-Straße 9, 48149 Münster, Germany

O. Rest
Institut für Kernphysik, Westfälische Wilhelms-Universität Münster, Wilhelm-Klemm-Straße 9, 48149 Münster, Germany

R. Rinderspacher
Institute for Nuclear Physics (IKP), Karlsruhe Institute of Technology (KIT), Hermann-von-Helmholtz-Platz 1, 76344 Eggenstein-Leopoldshafen, Germany

R. G. H. Robertson
Center for Experimental Nuclear Physics and Astrophysics, and Department of Physics, University of Washington, Seattle, Washington 98195, USA

C. Rodenbeck
Institut für Kernphysik, Westfälische Wilhelms-Universität Münster, Wilhelm-Klemm-Straße 9, 48149 Münster, Germany

P. Rohr
Institute for Data Processing and Electronics (IPE), Karlsruhe Institute of Technology (KIT), Hermann-von-Helmholtz-Platz 1, 76344 Eggenstein-Leopoldshafen, Germany

Ch. Roll
Institut für Physik, Humboldt-Universität zu Berlin, Newtonstraße 15, 12489 Berlin, Germany

S. Rupp
Institute for Nuclear Physics (IKP), Karlsruhe Institute of Technology (KIT), Hermann-von-Helmholtz-Platz 1, 76344 Eggenstein-Leopoldshafen, Germany
Institute for Technical Physics (ITEP), Karlsruhe Institute of Technology (KIT), Hermann-von-Helmholtz-Platz 1, 76344 Eggenstein-Leopoldshafen, Germany
Institute of Experimental Particle Physics (ETP), Karlsruhe Institute of Technology (KIT), Wolfgang-Gaede-Straße 1, 76131 Karlsruhe, Germany

M. Ryšavý
Nuclear Physics Institute of the CAS, v. v. i., CZ-250 68 Řež, Czech Republic

R. Sack
Institut für Kernphysik, Westfälische Wilhelms-Universität Münster, Wilhelm-Klemm-Straße 9, 48149 Münster, Germany

A. Saenz
Institut für Physik, Humboldt-Universität zu Berlin, Newtonstraße 15, 12489 Berlin, Germany

P. Schäfer
Institute for Nuclear Physics (IKP), Karlsruhe Institute of Technology (KIT), Hermann-von-Helmholtz-Platz 1, 76344 Eggenstein-Leopoldshafen, Germany
Institute for Technical Physics (ITEP), Karlsruhe Institute of Technology (KIT), Hermann-von-Helmholtz-Platz 1, 76344 Eggenstein-Leopoldshafen, Germany

L. Schimpf
Institute of Experimental Particle Physics (ETP), Karlsruhe Institute of Technology (KIT), Wolfgang-Gaede-Straße 1, 76131 Karlsruhe, Germany

K. Schlösser
Institute for Nuclear Physics (IKP), Karlsruhe Institute of Technology (KIT), Hermann-von-Helmholtz-Platz 1, 76344 Eggenstein-Leopoldshafen, Germany

M. Schlösser
Institute for Nuclear Physics (IKP), Karlsruhe Institute of Technology (KIT), Hermann-von-Helmholtz-Platz 1, 76344 Eggenstein-Leopoldshafen, Germany
Institute for Technical Physics (ITEP), Karlsruhe Institute of Technology (KIT), Hermann-von-Helmholtz-Platz 1, 76344 Eggenstein-Leopoldshafen, Germany

L. Schlüter
Max-Planck-Institut für Physik, Föhringer Ring 6, 80805 München, Germany
Technische Universität München, James-Franck-Straße 1, 85748 Garching, Germany

H. Schön
Institute for Technical Physics (ITEP), Karlsruhe Institute of Technology (KIT), Hermann-von-Helmholtz-Platz 1, 76344 Eggenstein-Leopoldshafen, Germany

K. Schönung,[11,1,2]
[11]Max-Planck-Institut für Kernphysik, Saupfercheckweg 1, 69117 Heidelberg, Germany
[12]Institut für Physik, Johannes-Gutenberg-Universität Mainz, 55099 Mainz, Germany

M. Schrank
Institute for Nuclear Physics (IKP), Karlsruhe Institute of Technology (KIT), Hermann-von-Helmholtz-Platz 1, 76344 Eggenstein-Leopoldshafen, Germany

B. Schulz
Institut für Physik, Humboldt-Universität zu Berlin, Newtonstraße 15, 12489 Berlin, Germany

J. Schwarz
Institute for Nuclear Physics (IKP), Karlsruhe Institute of Technology (KIT), Hermann-von-Helmholtz-Platz 1, 76344 Eggenstein-Leopoldshafen, Germany

H. Seitz-Moskaliuk
Institute of Experimental Particle Physics (ETP), Karlsruhe Institute of Technology (KIT), Wolfgang-Gaede-Straße 1, 76131 Karlsruhe, Germany

W. Seller
University of Applied Sciences (HFD) Fulda, Leipziger Straße 123, 36037 Fulda, Germany

V. Sibille
Laboratory for Nuclear Science, Massachusetts Institute of Technology, 77 Massachusetts Avenue, Cambridge, Massachusetts 02139, USA

D. Siegmann
Max-Planck-Institut für Physik, Föhringer Ring 6, 80805 München, Germany
Technische Universität München, James-Franck-Straße 1, 85748 Garching, Germany

A. Skasyrskaya
Institute for Nuclear Research of Russian Academy of Sciences, 60th October Anniversary Prospect 7a, 117312 Moscow, Russia

M. Slezák
Max-Planck-Institut für Physik, Föhringer Ring 6, 80805 München, Germany
Nuclear Physics Institute of the CAS, v. v. i., CZ-250 68 Řež, Czech Republic

A. Špalek
Nuclear Physics Institute of the CAS, v. v. i., CZ-250 68 Řež, Czech Republic

F. Spanier
Institute for Nuclear Physics (IKP), Karlsruhe Institute of Technology (KIT), Hermann-von-Helmholtz-Platz 1, 76344 Eggenstein-Leopoldshafen, Germany

M. Steidl
Institute for Nuclear Physics (IKP), Karlsruhe Institute of Technology (KIT), Hermann-von-Helmholtz-Platz 1, 76344 Eggenstein-Leopoldshafen, Germany

N. Steinbrink
Institut für Kernphysik, Westfälische Wilhelms-Universität Münster, Wilhelm-Klemm-Straße 9, 48149 Münster, Germany

M. Sturm
Institute for Nuclear Physics (IKP), Karlsruhe Institute of Technology (KIT), Hermann-von-Helmholtz-Platz 1, 76344 Eggenstein-Leopoldshafen, Germany
Institute for Technical Physics (ITEP), Karlsruhe Institute of Technology (KIT), Hermann-von-Helmholtz-Platz 1, 76344 Eggenstein-Leopoldshafen, Germany

M. Suesser
Institute for Technical Physics (ITEP), Karlsruhe Institute of Technology (KIT), Hermann-von-Helmholtz-Platz 1, 76344 Eggenstein-Leopoldshafen, Germany

M. Sun
Center for Experimental Nuclear Physics and Astrophysics, and Department of Physics, University of Washington, Seattle, Washington 98195, USA

D. Tcherniakhovski
Institute for Data Processing and Electronics (IPE), Karlsruhe Institute of Technology (KIT), Hermann-von-Helmholtz-Platz 1, 76344 Eggenstein-Leopoldshafen, Germany

H. H. Telle
Departamento de Química Física Aplicada, Universidad Autonoma de Madrid, Campus de Cantoblanco, 28049 Madrid, Spain

T. Thümmler
Institute for Nuclear Physics (IKP), Karlsruhe Institute of Technology (KIT), Hermann-von-Helmholtz-Platz 1, 76344 Eggenstein-Leopoldshafen, Germany
Institut für Kernphysik, Westfälische Wilhelms-Universität Münster, Wilhelm-Klemm-Straße 9, 48149 Münster, Germany

L. A. Thorne
Department of Physics, Carnegie Mellon University, Pittsburgh, Pennsylvania 15213, USA

N. Titov
Institute for Nuclear Research of Russian Academy of Sciences, 60th October Anniversary Prospect 7a, 117312 Moscow, Russia

I. Tkachev
Institute for Nuclear Research of Russian Academy of Sciences, 60th October Anniversary Prospect 7a, 117312 Moscow, Russia

N. Trost
Institute for Nuclear Physics (IKP), Karlsruhe Institute of Technology (KIT), Hermann-von-Helmholtz-Platz 1, 76344 Eggenstein-Leopoldshafen, Germany

K. Urban
Max-Planck-Institut für Physik, Föhringer Ring 6, 80805 München, Germany
Technische Universität München, James-Franck-Straße 1, 85748 Garching, Germany

D. V´enos
Nuclear Physics Institute of the CAS, v. v. i., CZ-250 68 Řež, Czech Republic

K. Valerius
Institute for Nuclear Physics (IKP), Karlsruhe Institute of Technology (KIT), Hermann-von-Helmholtz-Platz 1, 76344 Eggenstein-Leopoldshafen, Germany
Institut für Kernphysik, Westfälische Wilhelms-Universität Münster, Wilhelm-Klemm-Straße 9, 48149 Münster, Germany

B. A. VanDevender
Center for Experimental Nuclear Physics and Astrophysics, and Department of Physics, University of Washington, Seattle, Washington 98195, USA

R. Vianden
Helmholtz-Institut für Strahlen- und Kernphysik, Rheinische Friedrich-Wilhelms-Universität Bonn, Nussallee 14-16, 53115 Bonn, Germany

A. P. Vizcaya Hernández
Department of Physics, Carnegie Mellon University, Pittsburgh, Pennsylvania 15213, USA

B. L. Wall
Center for Experimental Nuclear Physics and Astrophysics, and Department of Physics, University of Washington, Seattle, Washington 98195, USA

S. Wüstling
Institute for Data Processing and Electronics (IPE), Karlsruhe Institute of Technology (KIT), Hermann-von-Helmholtz-Platz 1, 76344 Eggenstein-Leopoldshafen, Germany

M. Weber
Institute for Data Processing and Electronics (IPE), Karlsruhe Institute of Technology (KIT), Hermann-von-Helmholtz-Platz 1, 76344 Eggenstein-Leopoldshafen, Germany

C. Weinheimer
Institut für Kernphysik, Westfälische Wilhelms-Universität Münster, Wilhelm-Klemm-Straße 9, 48149 Münster, Germany

C. Weiss
Project, Process, and Quality Management (PPQ), Karlsruhe Institute of Technology

(KIT), Hermann-von-Helmholtz-Platz 1, 76344 Eggenstein-Leopoldshafen, Germany

S. Welte
Institute for Nuclear Physics (IKP), Karlsruhe Institute of Technology (KIT), Hermann-von-Helmholtz-Platz 1, 76344 Eggenstein-Leopoldshafen, Germany
Institute for Technical Physics (ITEP), Karlsruhe Institute of Technology (KIT), Hermann-von-Helmholtz-Platz 1, 76344 Eggenstein-Leopoldshafen, Germany

J. Wendel
Institute for Nuclear Physics (IKP), Karlsruhe Institute of Technology (KIT), Hermann-von-Helmholtz-Platz 1, 76344 Eggenstein-Leopoldshafen, Germany
Institute for Technical Physics (ITEP), Karlsruhe Institute of Technology (KIT), Hermann-von-Helmholtz-Platz 1, 76344 Eggenstein-Leopoldshafen, Germany

K. J. Wierman
Department of Physics and Astronomy, University of North Carolina, Chapel Hill, North Carolina 27599, USA
Triangle Universities Nuclear Laboratory, Durham, North Carolina 27708, USA

J. F. Wilkerson
Department of Physics and Astronomy, University of North Carolina, Chapel Hill, North Carolina 27599, USA
Triangle Universities Nuclear Laboratory, Durham, North Carolina 27708, USA

J. Wolf
Institute of Experimental Particle Physics (ETP), Karlsruhe Institute of Technology (KIT), Wolfgang-Gaede-Straße 1, 76131 Karlsruhe, Germany

W. Xu
Laboratory for Nuclear Science, Massachusetts Institute of Technology, 77 Massachusetts Avenue, Cambridge, Massachusetts 02139, USA

Y.-R. Yen
Department of Physics, Carnegie Mellon University, Pittsburgh, Pennsylvania 15213, USA

M. Zacher
Institut für Kernphysik, Westfälische Wilhelms-Universität Münster, Wilhelm-Klemm-Straße 9, 48149 Münster, Germany

S. Zadorozhny
Institute for Nuclear Research of Russian Academy of Sciences, 60th October Anniversary Prospect 7a, 117312 Moscow, Russia

M. Zboril
Institut für Kernphysik, Westfälische Wilhelms-Universität Münster, Wilhelm-Klemm-Straße 9, 48149 Münster, Germany
Nuclear Physics Institute of the CAS, v. v. i., CZ-250 68 Řež, Czech Republic

G. Zeller
Institute for Nuclear Physics (IKP), Karlsruhe Institute of Technology (KIT), Hermann-von-Helmholtz-Platz 1, 76344 Eggenstein-Leopoldshafen, Germany
Institute for Technical Physics (ITEP), Karlsruhe Institute of Technology (KIT), Hermann-von-Helmholtz-Platz 1, 76344 Eggenstein-Leopoldshafen, Germany

Lena Funcke
Perimeter Institute for Theoretical Physics, 31 Caroline Street North, Waterloo, Ontario N2L 2Y5, Canada

Georg Raffelt
Max-Planck-Institut für Physik (Werner-Heisenberg-Institut), Föhringer Ring 6, 80805 München, Germany

Edoardo Vitagliano
Max-Planck-Institut für Physik (Werner-Heisenberg-Institut), Föhringer Ring 6, 80805 München, Germany

G. Anton
Erlangen Centre for Astroparticle Physics, Friedrich-Alexander-University Erlangen-Nürnberg, Erlangen 91058, Germany

I. Badhrees
Physics Department, Carleton University, Ottawa, Ontario K1S 5B6, Canada

P. S. Barbeau
Department of Physics, Duke University, and Triangle Universities Nuclear Laboratory, Durham, North Carolina 27708, USA

D. Beck
Physics Department, University of Illinois, Urbana-Champaign, Illinois 61801, USA

V. Belov
Institute for Theoretical and Experimental Physics named by A.I. Alikhanov of National Research Center “Kurchatov Institute,” 117218 Moscow, Russia

T. Bhatta
Department of Physics, University of South Dakota, Vermillion, South Dakota 57069, USA

M. Breidenbach
SLAC National Accelerator Laboratory, Menlo Park, California 94025, USA

T. Brunner
Physics Department, McGill University, Montreal H3A 2T8, Quebec, Canada
TRIUMF, Vancouver, British Columbia V6T 2A3, Canada

G. F. Cao
Institute of High Energy Physics, Beijing 100049, China

W. R. Cen
Institute of High Energy Physics, Beijing 100049, China

C. Chambers
Physics Department, Colorado State University, Fort Collins, Colorado 80523, USA

B. Cleveland
Department of Physics, Laurentian University, Sudbury, Ontario P3E 2C6, Canada

M. Coon
Physics Department, University of Illinois, Urbana-Champaign, Illinois 61801, USA

A. Craycraft
Physics Department, Colorado State University, Fort Collins, Colorado 80523, USA

T. Daniels
Department of Physics and Physical Oceanography, University of North Carolina at Wilmington, Wilmington, North Carolina 28403, USA

M. Danilov
Institute for Theoretical and Experimental Physics named by A.I. Alikhanov of National Research Center "Kurchatov Institute," 117218 Moscow, Russia

L. Darroch
Physics Department, McGill University, Montreal H3A 2T8, Quebec, Canada

S. J. Daugherty
Physics Department and CEEM, Indiana University, Bloomington, Indiana 47405, USA

J. Davis
SLAC National Accelerator Laboratory, Menlo Park, California 94025, USA

S. Delaquis
SLAC National Accelerator Laboratory, Menlo Park, California 94025, USA

A. Der Mesrobian-Kabakian
Department of Physics, Laurentian University, Sudbury, Ontario P3E 2C6, Canada

R. DeVoe
Physics Department, Stanford University, Stanford, California 94305, USA

J. Dilling
TRIUMF, Vancouver, British Columbia V6T 2A3, Canada

A. Dolgolenko
Institute for Theoretical and Experimental Physics named by A.I. Alikhanov of National Research Center "Kurchatov Institute," 117218 Moscow, Russia

M.J. Dolinski
Department of Physics, Drexel University, Philadelphia, Pennsylvania 19104, USA

J. Echevers
Physics Department, University of Illinois, Urbana-Champaign, Illinois 61801, USA

W. Fairbank, Jr.
Physics Department, Colorado State University, Fort Collins, Colorado 80523, USA

D. Fairbank
Physics Department, Colorado State University, Fort Collins, Colorado 80523, USA

J. Farine
Department of Physics, Laurentian University, Sudbury, Ontario P3E 2C6, Canada

S. Feyzbakhsh
Amherst Center for Fundamental Interactions and Physics Department, University of Massachusetts, Amherst, Massachusetts 01003, USA

P. Fierlinger
Physik Department and Excellence Cluster Universe, Technische Universität München, Garching 80805, Germany

D. Fudenberg
Physics Department, Stanford University, Stanford, California 94305, USA

P. Gautam
Department of Physics, Drexel University, Philadelphia, Pennsylvania 19104, USA

R. Gornea
Physics Department, Carleton University, Ottawa, Ontario K1S 5B6, Canada
TRIUMF, Vancouver, British Columbia V6T 2A3, Canada

G. Gratta
Physics Department, Stanford University, Stanford, California 94305, USA

C. Hall
Physics Department, University of Maryland, College Park, Maryland 20742, USA

E. V. Hansen
Department of Physics, Drexel University, Philadelphia, Pennsylvania 19104, USA

J. Hoessl
Erlangen Centre for Astroparticle Physics, Friedrich-Alexander-University Erlangen-Nürnberg, Erlangen 91058, Germany

P. Hufschmidt
Erlangen Centre for Astroparticle Physics, Friedrich-Alexander-University Erlangen-Nürnberg, Erlangen 91058, Germany

M. Hughes
Department of Physics and Astronomy, University of Alabama, Tuscaloosa, Alabama 35487, USA

A. Iverson
Physics Department, Colorado State University, Fort Collins, Colorado 80523, USA

A. Jamil
Wright Laboratory, Department of Physics, Yale University, New Haven, Connecticut 06511, USA

C. Jessiman
Physics Department, Carleton University, Ottawa, Ontario K1S 5B6, Canada

M. J. Jewell
Physics Department, Stanford University, Stanford, California 94305, USA

A. Johnson
SLAC National Accelerator Laboratory, Menlo Park, California 94025, USA

A. Karelin
Institute for Theoretical and Experimental Physics named by A.I. Alikhanov of National Research Center "Kurchatov Institute," 117218 Moscow, Russia

L. J. Kaufman
SLAC National Accelerator Laboratory, Menlo Park, California 94025, USA

T. Koffas
Physics Department, Carleton University, Ottawa, Ontario K1S 5B6, Canada

R. Krücken
TRIUMF, Vancouver, British Columbia V6T 2A3, Canada

A. Kuchenkov
Institute for Theoretical and Experimental Physics named by A.I. Alikhanov of National Research Center "Kurchatov Institute," 117218 Moscow, Russia

K. S. Kumar
Department of Physics and Astronomy, Stony Brook University, SUNY, Stony Brook, New York 11794, USA

Y. Lan
TRIUMF, Vancouver, British Columbia V6T 2A3, Canada

A. Larson
Department of Physics, University of South Dakota, Vermillion, South Dakota 57069, USA

B. G. Lenardo
Physics Department, Stanford University, Stanford, California 94305, USA

D. S. Leonard
IBS Center for Underground Physics, Daejeon 34126, Korea

G. S. Li
Physics Department, Stanford University, Stanford, California 94305, USA

S. Li
Physics Department, University of Illinois, Urbana-Champaign, Illinois 61801, USA

Z. Li
Wright Laboratory, Department of Physics, Yale University, New Haven, Connecticut 06511, USA

C. Licciardi
Department of Physics, Laurentian University, Sudbury, Ontario P3E 2C6, Canada

Y. H. Lin
Department of Physics, Drexel University, Philadelphia, Pennsylvania 19104, USA

R. MacLellan
Department of Physics, University of South Dakota, Vermillion, South Dakota 57069, USA

T. McElroy
Physics Department, McGill University, Montreal H3A 2T8, Quebec, Canada

T. Michel
Erlangen Centre for Astroparticle Physics, Friedrich-Alexander-University Erlangen-Nürnberg, Erlangen 91058, Germany

B. Mong
SLAC National Accelerator Laboratory, Menlo Park, California 94025, USA

D. C. Moore
Wright Laboratory, Department of Physics, Yale University, New Haven, Connecticut 06511, USA

K. Murray
Physics Department, McGill University, Montreal H3A 2T8, Quebec, Canada

O. Njoya
Department of Physics and Astronomy, Stony Brook University, SUNY, Stony Brook, New York 11794, USA

O. Nusair
Department of Physics and Astronomy, University of Alabama, Tuscaloosa, Alabama 35487, USA

A. Odian
SLAC National Accelerator Laboratory, Menlo Park, California 94025, USA

I. Ostrovskiy
Department of Physics and Astronomy, University of Alabama, Tuscaloosa, Alabama 35487, USA

A. Piepke
Department of Physics and Astronomy, University of Alabama, Tuscaloosa, Alabama 35487, USA

A. Pocar
Amherst Center for Fundamental Interactions and Physics Department, University of Massachusetts, Amherst, Massachusetts 01003, USA

F. Retière
TRIUMF, Vancouver, British Columbia V6T 2A3, Canada

A. L. Robinson
Department of Physics, Laurentian University, Sudbury, Ontario P3E 2C6, Canada

P. C. Rowson
SLAC National Accelerator Laboratory, Menlo Park, California 94025, USA

D. Ruddell
Department of Physics and Physical Oceanography, University of North Carolina at Wilmington, Wilmington, North Carolina 28403, USA

J. Runge
Department of Physics, Duke University, and Triangle Universities Nuclear Laboratory, Durham, North Carolina 27708, USA

S. Schmidt
Erlangen Centre for Astroparticle Physics, Friedrich-Alexander-University Erlangen-Nürnberg, Erlangen 91058, Germany

D. Sinclair
Physics Department, Carleton University, Ottawa, Ontario K1S 5B6, Canada
TRIUMF, Vancouver, British Columbia V6T 2A3, Canada

A. K. Soma
Department of Physics and Astronomy, University of Alabama, Tuscaloosa, Alabama 35487, USA

V. Stekhanov
Institute for Theoretical and Experimental Physics named by A.I. Alikhanov of National Research Center "Kurchatov Institute," 117218 Moscow, Russia

M. Tarka
Amherst Center for Fundamental Interactions and Physics Department, University of Massachusetts, Amherst, Massachusetts 01003, USA

J. Todd
Physics Department, Colorado State University, Fort Collins, Colorado 80523, USA

T. Tolba
Institute of High Energy Physics, Beijing 100049, China

T. I. Totev
Physics Department, McGill University, Montreal H3A 2T8, Quebec, Canada

B. Veenstra
Physics Department, Carleton University, Ottawa, Ontario K1S 5B6, Canada

V. Veeraraghavan
Department of Physics and Astronomy, University of Alabama, Tuscaloosa, Alabama 35487, USA

P. Vogel
Kellogg Lab, Caltech, Pasadena, California 91125, USA

J.-L. Vuilleumier
LHEP, Albert Einstein Center, University of Bern, Bern CH-3012, Switzerland

M. Wagenpfeil
Erlangen Centre for Astroparticle Physics, Friedrich-Alexander-University Erlangen-Nürnberg, Erlangen 91058, Germany

J. Watkins
Physics Department, Carleton University, Ottawa, Ontario K1S 5B6, Canada

M. Weber
Physics Department, Stanford University, Stanford, California 94305, USA

L. J. Wen
Institute of High Energy Physics, Beijing 100049, China

U. Wichoski
Department of Physics, Laurentian University, Sudbury, Ontario P3E 2C6, Canada

G. Wrede
Erlangen Centre for Astroparticle Physics, Friedrich-Alexander-University Erlangen-Nürnberg, Erlangen 91058, Germany

S. X. Wu
Physics Department, Stanford University, Stanford, California 94305, USA

Q. Xia
Wright Laboratory, Department of Physics, Yale University, New Haven, Connecticut

06511, USA

D. R. Yahne
Physics Department, Colorado State University, Fort Collins, Colorado 80523, USA

L. Yang
Physics Department, University of Illinois, Urbana-Champaign, Illinois 61801, USA

Y.-R. Yen
Department of Physics, Drexel University, Philadelphia, Pennsylvania 19104, USA

O. Ya. Zeldovich
Institute for Theoretical and Experimental Physics named by A.I. Alikhanov of National Research Center "Kurchatov Institute," 117218 Moscow, Russia

T. Ziegler
Erlangen Centre for Astroparticle Physics, Friedrich-Alexander-University Erlangen-Nürnberg, Erlangen 91058, Germany

R. Arnold
IPHC, ULP, CNRS/IN2P3, 67037 Strasbourg, France

C. Augier
LAL, Université Paris-Sud, CNRS/IN2P3, Université Paris-Saclay, 91405 Orsay, France

A. S. Barabash
NRC "Kurchatov Institute", ITEP, 117218 Moscow, Russia

A. Basharina-Freshville
UCL, London WC1E 6BT, UK

S. Blondel
LAL, Université Paris-Sud, CNRS/IN2P3, Université Paris-Saclay, 91405 Orsay, France

S. Blot
University of Manchester, Manchester M13 9PL, UK

M. Bongrand
LAL, Université Paris-Sud, CNRS/IN2P3, Université Paris-Saclay, 91405 Orsay, France

D. Boursette
LAL, Université Paris-Sud, CNRS/IN2P3, Université Paris-Saclay, 91405 Orsay, France

V. Brudanin
JINR, 141980 Dubna, Russia
National Research Nuclear University MEPhI, 115409 Moscow, Russia

J. Busto
Aix Marseille Université, CNRS, CPPM, 13288 Marseille, France

A. J. Caffrey
Idaho National Laboratory, Idaho Falls, ID 83415, USA

S. Calvez
LAL, Université Paris-Sud, CNRS/IN2P3, Université Paris-Saclay, 91405 Orsay, France

M. Cascella
UCL, London WC1E 6BT, UK

C. Cerna
CENBG, Université de Bordeaux, CNRS/IN2P3, 33175 Gradignan, France

J. P. Cesar
University of Texas at Austin, Austin, TX 78712, USA

A. Chapon
LPC Caen, ENSICAEN, Université de Caen, CNRS/IN2P3, 14050 Caen, France

E. Chauveau
CENBG, Université de Bordeaux, CNRS/IN2P3, 33175 Gradignan, France

A. Chopra
UCL, London WC1E 6BT, UK

L. Dawson
UCL, London WC1E 6BT, UK

D. Duchesneau
LAPP, Université de Savoie, CNRS/IN2P3, 74941 Annecy-le-Vieux, France

D. Durand
LPC Caen, ENSICAEN, Université de Caen, CNRS/IN2P3, 14050 Caen, France

R. Dvornický
JINR, 141980 Dubna, Russia
FMFI, Comenius University, 842 48 Bratislava, Slovakia

V. Egorov
JINR, 141980 Dubna, Russia

G. Eurin
LAL, Université Paris-Sud, CNRS/IN2P3, Université Paris-Saclay, 91405 Orsay, France
UCL, London WC1E 6BT, UK

J. J. Evans
University of Manchester, Manchester M13 9PL, UK

L. Fajt
Institute of Experimental and Applied Physics, Czech Technical University in Prague, 12800 Prague, Czech Republic

D. Filosofov
JINR, 141980 Dubna, Russia

R. Flack
UCL, London WC1E 6BT, UK

X. Garrido
LAL, Université Paris-Sud, CNRS/IN2P3, Université Paris-Saclay, 91405 Orsay, France

C. Girard-Carillo
LAL, Université Paris-Sud, CNRS/IN2P3, Université Paris-Saclay, 91405 Orsay, France

H. Gómez
LAL, Université Paris-Sud, CNRS/IN2P3, Université Paris-Saclay, 91405 Orsay, France

B. Guillon
LPC Caen, ENSICAEN, Université de Caen, CNRS/IN2P3, 14050 Caen, France

P. Guzowski
University of Manchester, Manchester M13 9PL, UK

R. Hodák[1]
UCL, London WC1E 6BT, UK

A. Huber
CENBG, Université de Bordeaux, CNRS/IN2P3, 33175 Gradignan, France

P. Hubert
CENBG, Université de Bordeaux, CNRS/IN2P3, 33175 Gradignan, France

C. Hugon
CENBG, Université de Bordeaux, CNRS/IN2P3, 33175 Gradignan, France

S. Jullian
LAL, Université Paris-Sud, CNRS/IN2P3, Université Paris-Saclay, 91405 Orsay, France

O. Kochetov
JINR, 141980 Dubna, Russia

S. I. Konovalov
NRC "Kurchatov Institute", ITEP, 117218 Moscow, Russia

V. Kovalenko
JINR, 141980 Dubna, Russia

D. Lalanne
LAL, Université Paris-Sud, CNRS/IN2P3, Université Paris-Saclay, 91405 Orsay, France

K. Lang
University of Texas at Austin, Austin, TX 78712, USA

Y. Lemière
LPC Caen, ENSICAEN, Université de Caen, CNRS/IN2P3, 14050 Caen, France

T. Le Noblet
LAPP, Université de Savoie, CNRS/IN2P3, 74941 Annecy-le-Vieux, France

Z. Liptak
University of Texas at Austin, Austin, TX 78712, USA

X. R. Liu
UCL, London WC1E 6BT, UK

1

P. Loaiza
LAL, Université Paris-Sud, CNRS/IN2P3, Université Paris-Saclay, 91405 Orsay, France

G. Lutter
CENBG, Université de Bordeaux, CNRS/IN2P3, 33175 Gradignan, France

M. Macko
CENBG, Université de Bordeaux, CNRS/IN2P3, 33175 Gradignan, France
Institute of Experimental and Applied Physics, Czech Technical University in Prague, 12800 Prague, Czech Republic

C. Macolino
LAL, Université Paris-Sud, CNRS/IN2P3, Université Paris-Saclay, 91405 Orsay, France

F. Mamedov
Institute of Experimental and Applied Physics, Czech Technical University in Prague, 12800 Prague, Czech Republic

C. Marque
CENBG, Université de Bordeaux, CNRS/IN2P3, 33175 Gradignan, France

F. Mauger
LPC Caen, ENSICAEN, Université de Caen, CNRS/IN2P3, 14050 Caen, France

A. Minotti
LAPP, Université de Savoie, CNRS/IN2P3, 74941 Annecy-le-Vieux, France

B. Morgan
University of Warwick, Coventry CV4 7AL, UK

J. Mott
UCL, London WC1E 6BT, UK
Present Address: Boston University, Boston, MA 02215, USA

I. Nemchenok
JINR, 141980 Dubna, Russia

M. Nomachi
Osaka University, 1-1 Machikaneyama Toyonaka, Osaka 560-0043, Japan

F. Nova
University of Texas at Austin, Austin, TX 78712, USA

F. Nowacki
IPHC, ULP, CNRS/IN2P3, 67037 Strasbourg, France

H. Ohsumi
Saga University, Saga 840-8502, Japan

G. Oliviéro
LPC Caen, ENSICAEN, Université de Caen, CNRS/IN2P3, 14050 Caen, France

R. B. Pahlka
University of Texas at Austin, Austin, TX 78712, USA

C. Patrick
UCL, London WC1E 6BT, UK

F. Perrot
CENBG, Université de Bordeaux, CNRS/IN2P3, 33175 Gradignan, France

A. Pin
CENBG, Université de Bordeaux, CNRS/IN2P3, 33175 Gradignan, France

F. Piquemal
CENBG, Université de Bordeaux, CNRS/IN2P3, 33175 Gradignan, France
Laboratoire Souterrain de Modane, 73500 Modane, France

P. Povinec
FMFI, Comenius University, 842 48 Bratislava, Slovakia

P. Pˇridal
Institute of Experimental and Applied Physics, Czech Technical University in Prague, 12800 Prague, Czech Republic

Y. A. Ramachers
University of Warwick, Coventry CV4 7AL, UK

A. Remoto
LAPP, Université de Savoie, CNRS/IN2P3, 74941 Annecy-le-Vieux, France

J. L. Reyss
LSCE, CNRS, 91190 Gif-sur-Yvette, France

C. L. Riddle
Idaho National Laboratory, Idaho Falls, ID 83415, USA

E. Rukhadze
Institute of Experimental and Applied Physics, Czech Technical University in Prague, 12800 Prague, Czech Republic

R. Saakyan
UCL, London WC1E 6BT, UK

A. Salamatin
JINR, 141980 Dubna, Russia

R. Salazar[11],
[11] University of Texas at Austin, Austin, TX 78712, USA

X. Sarazin
LAL, Université Paris-Sud, CNRS/IN2P3, Université Paris-Saclay, 91405 Orsay, France

J. Sedgbeer
Imperial College London, London SW7 2AZ, UK

Yu. Shitov
JINR, 141980 Dubna, Russia

L. Simard
LAL, Université Paris-Sud, CNRS/IN2P3, Université Paris-Saclay, 91405 Orsay, France
Institut Universitaire de France, 75005 Paris, France

F. Šimkovic
JINR, 141980 Dubna, Russia
FMFI, Comenius University, 842 48 Bratislava, Slovakia

A. Smetana
Institute of Experimental and Applied Physics, Czech Technical University in Prague, 12800 Prague, Czech Republic

K. Smolek
Institute of Experimental and Applied Physics, Czech Technical University in Prague, 12800 Prague, Czech Republic

A. Smolnikov
JINR, 141980 Dubna, Russia

S. Söldner-Rembold
University of Manchester, Manchester M13 9PL, UK

B. Soulé
CENBG, Université de Bordeaux, CNRS/IN2P3, 33175 Gradignan, France

I. Štekl
Institute of Experimental and Applied Physics, Czech Technical University in Prague, 12800 Prague, Czech Republic

J. Suhonen
Jyväskylä University, 40351 Jyvaskyla, Finland

C. S. Sutton
MHC, South Hadley, MA 01075, USA

G. Szklarz
LAL, Université Paris-Sud, CNRS/IN2P3, Université Paris-Saclay, 91405 Orsay, France

H. Tedjditi
[8]Aix Marseille Université, CNRS, CPPM, 13288 Marseille, France

J. Thomas
UCL, London WC1E 6BT, UK

V. Timkin
JINR, 141980 Dubna, Russia

S. Torre
UCL, London WC1E 6BT, UK

Vl. I. Tretyak
Institute for Nuclear Research, Kiev 03028, Ukraine

V. I. Tretyak
JINR, 141980 Dubna, Russia

V. I. Umatov
NRC "Kurchatov Institute", ITEP, 117218 Moscow, Russia

I. Vanushin
NRC "Kurchatov Institute", ITEP, 117218 Moscow, Russia

C. Vilela
UCL, London WC1E 6BT, UK

V. Vorobel
Faculty of Mathematics and Physics, Charles University in Prague, 12116 Prague, Czech Republic

D. Waters
UCL, London WC1E 6BT, UK

F. Xie
UCL, London WC1E 6BT, UK

A. Žukauskas
Faculty of Mathematics and Physics, Charles University in Prague, 12116 Prague, Czech Republic

Gaetana Anamiati
AHEP Group, Instituto de Física Corpuscular, C.S.I.C./Universitat de València Edifício de Institutos de Paterna, Apartado 22085, E-46071 València, Spain

Renato M. Fonseca
AHEP Group, Instituto de Física Corpuscular, C.S.I.C./Universitat de València Edifício de Institutos de Paterna, Apartado 22085, E-46071 València, Spain

Martin Hirsch
AHEP Group, Instituto de Física Corpuscular, C.S.I.C./Universitat de València Edifício de Institutos de Paterna, Apartado 22085, E-46071 València, Spain

Mauro Mezzetto
INFN sez. di Padova, Via F. Marzolo 8, I-35131 Padova, Italy

Francesco Terranova
Department of Physics, Universe of Milano-Bicocca, Piazza della Scienza 3, I-20126 Milano, Italy
INFN sez. di Milano-Bicocca, Piazza della Scienza 3, I-20126 Milano, Italy

Patrick D. Bolton
Department of Physics and Astronomy, University College London, Gower Street, London WC1E 6BT, United Kingdom

Frank F. Deppisch
Department of Physics and Astronomy, University College London, Gower Street, London WC1E 6BT, United Kingdom

Institut für Hochenergiephysik, Österreichische Akademie der Wissenschaften, Nikolsdorfer Gasse 18, 1050 Wien, Austria

C. Rott
Department of Physics, Sungkyunkwan University, Suwon, 440-746, South Korea

A. Taketa
Earthquake Research Institute, University of Tokyo, 1-1-1 Yayoi, Bunkyo-ku, Tokyo, Japan

D. Bose
Department of Physics, Sungkyunkwan University, Suwon, 440-746, South Korea

Kimihiro Okumura
Research Center for Cosmic Neutrino (RCCN), Institute for Cosmic Ray Research (ICRR), University of Tokyo, 5-1-5 Kashiwa-no-ha, Kashiwa, Chiba, Japan

Morihiro Honda
Institute for Cosmic Ray Research, the University of Tokyo, 5-1-5 Kashiwa-no-ha, Kashiwa, Chiba 277-8582, Japan

Kevin J. Kelly
Theory Department, Fermi National Accelerator Laboratory, Batavia, Illinois 60510, USA

Pedro A. N. Machado
Theory Department, Fermi National Accelerator Laboratory, Batavia, Illinois 60510, USA

Ivan Martinez-Soler
Theory Department, Fermi National Accelerator Laboratory, Batavia, Illinois 60510, USA
Department of Physics & Astronomy, Northwestern University, Evanston, Illinois 60208, USA
Colegio de Física Fundamental e Interdisciplinaria de las Américas (COFI), 254 Norzagaray street, San Juan, Puerto Rico 00901, USA

Stephen J. Parke
Theory Department, Fermi National Accelerator Laboratory, Batavia, Illinois 60510, USA

Yuber F. Perez-Gonzalez
Theory Department, Fermi National Accelerator Laboratory, Batavia, Illinois 60510, USA

Department of Physics & Astronomy, Northwestern University, Evanston, Illinois 60208, USA
Colegio de Física Fundamental e Interdisciplinaria de las Américas (COFI), 254 Norzagaray street, San Juan, Puerto Rico 00901, USA

Francesco Capozzi
Center for Cosmology and AstroParticle Physics (CCAPP), Ohio State University, Columbus, Ohio 43210, USA
Department of Physics, Ohio State University, Columbus, Ohio 43210, USA
Max-Planck-Institut für Physik (Werner-Heisenberg-Institut), 80805 München, Germany

Shirley Weishi Li
Center for Cosmology and AstroParticle Physics (CCAPP), Ohio State University, Columbus, Ohio 43210, USA
Department of Physics, Ohio State University, Columbus, Ohio 43210, USA
SLAC National Accelerator Laboratory, Menlo Park, California 94025, USA

Guanying Zhu
Center for Cosmology and AstroParticle Physics (CCAPP), Ohio State University, Columbus, Ohio 43210, USA
Department of Physics, Ohio State University, Columbus, Ohio 43210, USA

John F. Beacom
Center for Cosmology and AstroParticle Physics (CCAPP), Ohio State University, Columbus, Ohio 43210, USA
Department of Physics, Ohio State University, Columbus, Ohio 43210, USA
Department of Astronomy, Ohio State University, Columbus, Ohio 43210, USA

Carlos A. Argüelles
Massachusetts Institute of Technology, Cambridge, Massachusetts 02139, USA
Wisconsin IceCube Particle Astrophysics Center, Madison, Wisconsin 53703, USA

Vedran Brdar
PRISMA Cluster of Excellence, 55099 Mainz, Germany and Mainz Institute for Theoretical Physics, Johannes Gutenberg-Universität Mainz, 55099 Mainz, Germany

Joachim Kopp
PRISMA Cluster of Excellence, 55099 Mainz, Germany and Mainz Institute for Theoretical Physics, Johannes Gutenberg-Universität Mainz, 55099 Mainz, Germany

Francis Halzen
Department of Physics & Wisconsin IceCube Particle Astrophysics Center, UW-Madison, Madison, WI, United States

Ali Kheirandish
Department of Physics & Wisconsin IceCube Particle Astrophysics Center, UW-Madison, Madison, WI, United States

Markus Ahlers
Wisconsin IceCube Particle Astrophysics Center (WIPAC) and Department of Physics, University of Wisconsin-Madison, Madison, WI 53706, USA

Francis Halzen
Wisconsin IceCube Particle Astrophysics Center (WIPAC) and Department of Physics, University of Wisconsin-Madison, Madison, WI 53706, USA

Jeffrey M. Berryman
Center for Neutrino Physics, Department of Physics, Virginia Tech, Blacksburg, Virginia 24061, USA

M. Blennow
Departamento de Física Teórica and Instituto de Física Teórica, IFT-UAM/CSIC, Universidad Autónoma de Madrid, Cantoblanco, 28049 Madrid, Spain
Department of Physics, School of Engineering Sciences, KTH Royal Institute of Technology, AlbaNova University Center, Roslagstullsbacken 21, 106 91 Stockholm, Sweden

E. Fernandez-Martinez
Departamento de Física Teórica and Instituto de Física Teórica, IFT-UAM/CSIC, Universidad Autónoma de Madrid, Cantoblanco, 28049 Madrid, Spain

A. Olivares-Del Campo
Institute for Particle Physics Phenomenology, Department of Physics, Durham University, South Road, Durham DH1 3LE, UK

S. Pascoli
Institute for Particle Physics Phenomenology, Department of Physics, Durham University, South Road, Durham DH1 3LE, UK

S. Rosauro-Alcaraz
Departamento de Física Teórica and Instituto de Física Teórica, IFT-UAM/CSIC, Universidad Autónoma de Madrid, Cantoblanco, 28049 Madrid, Spain

A. V. Titov
Institute for Particle Physics Phenomenology, Department of Physics, Durham University, South Road, Durham DH1 3LE, UK

LIST OF ABBREVIATIONS

BCS	Bardeen-Cooper-Schrieffer
BAO	Baryon acoustic oscillation
BAU	Baryon Symmetry of the Universe
BPs	Benchmark points
BBN	Big Bang nucleosynthesis
BDT	Boosted decision tree
CKM	Cabibbo-Kobayashi-Maskawa
CKM	Cabibbo–Kobayashi–Maskawa
CCDs	Charged coupled devices
CC	Charged Current
CCQE	Charged Current Quasi-Elastic scatterings
CCM	Coherent Captain-Mills
CENNS	Coherent elastic neutrino-nucleus scattering
CW	Coleman-Weinberg
CMB	Core mantle boundary
CMB	Cosmic microwave background
DNN	Deep neural network
DUNE	Deep Underground Neutrino Experiment
DSM	Deformed Shell-Model
DANSS	Detector of Anti Neutrino based on Solid Scintillator
DOMs	Digital optical modules
EM	Electromagnetic
ECC	Emulsion Cloud Chamber
EBL	Extragalactic background light
FR	Fanaroff–Riley
FD	Far detector
FC	Feldman and Cousins
FV	Fiducial volume

FSD	Final-state distribution
FCNC	Flavor changing neutral-current
GUTs	Grand Unified Theories
HESE	High-energy starting events
ISS	Inverse Seesaw
JUNO	Jiangmen Underground Neutrino Observatory
LAAPDs	Large area avalanche photodiodes
LMA	Large Mixing Angle
LSS	Large scale structures
LNFV	Lepton flavor and number violating
LFV	Lepton flavor violation
LNC	Lepton number conserving
LNV	Lepton number violating
LSS	Linear Seesaw
LLR	Log-likelihood ratio
LBL	Long-baseline
LANSCE	Los Alamos Neutron Science Center
ML	Maximum-likelihood
MSW	Mikheyev-Smirnov-Wolfenstein
MC	Monte Carlo
NERSC	National Energy Research Scientific Computing Center
NC	Neutral-current
NLO	Next to Leading Order
NEG	Non evaporable getter
NO	Normal ordering
NME	Nuclear Matrix Element
NPP	Nuclear power plants
PPB	Part-per-billion
PMT	Photomultiplier tubes
PVA	Polyvinyl alcohol
PDF	Probability density functions
QFT	Quantum field theory
QM	Quantum mechanics
QD	Quasi-Dirac

QRPA	Quasi-particle random phase approximation
ROI	Region of interest
RGE	Renormalization Group Equation
SSD	Single state dominance
SMEFT	Standard Model Effective Field Theory
SSM	Standard Solar Model
SNO	Sudbury Neutrino Observatory
TOF	Time of flight
TPC	Time projection chamber
TMP	Turbo molecular pump
VEV	Vacuum Expectation Value
WIPP	Waste Isolation Pilot Plant

PREFACE

Neutrino physics began two decades ago when nuclear beta decay experiments discovered the continuous spectrum of beta rays posing a serious problem for the theorists. From those experiments, we have witnessed a revolution in the field, i.e., the observation that neutrinos have nonzero masses and can change from one flavor to another. Recently, neutrino's studies are multidisciplinary, involving the interplay of particle and nuclear physics, astrophysics, and cosmology with the investigation of neutrinos of different origins: solar, reactor, accelerator, atmospheric, cosmogenic, etc.

This book begins with an introductory Section 1 which reviews the properties of neutrinos in and beyond the Standard Model (SM), giving an overview of recent progress in the theoretical and phenomenological studies of neutrino masses, flavor mixing, CP violation, and the neutrino-nucleus interactions.

Section 2 focuses on the seesaw model, where the electroweak scale, the Higgs mass, and the neutrino masses are generated simultaneously, on a model that invokes the existence of a dark neutrino sector and new hidden gauge symmetry. Moreover, it discusses of neutrino mass measurement made in the recent experiment KATRIN, and of the possibility of distinguishing Majorana from Dirac neutrinos, assuming a gravitational origin of neutrino masses.

Section 3 deals with direct neutrino mass searches, double beta decay, and neutrino oscillations. It focuses on the search for neutrinoless double-beta decay by using a deep neural network with the complete EXO-200 dataset, and on the measurement of the half-life of the two neutrinos double decay by using the full dataset of the NEMO-3 experiment. The last part of the Section 3 dedicates only to neutrino oscillations by showing the results of experiments designed to measure the remaining unknown neutrino oscillation parameters via the oscillation of accelerator neutrinos over long distances, and then, by discussing the phenomenology of quasi-Dirac neutrino oscillations, the effect of lepton number violating nonstandard interactions, and by demonstrating the utility of large neutrino detectors for improving our understanding of the Earth's inner chemical composition.

Section 4 presents the comparison of the measurements of atmospheric neutrino flux by several experiments with the theoretical predictions, and the study of the correlation of atmospheric muons and neutrinos, at the aim to reduce the calculation error of the atmospheric neutrino flux. Moreover, it proposes the use of liquid argon chambers for estimating how the future Deep Underground Neutrino Experiment (DUNE) will be able to measure sub-GeV atmospheric neutrinos and extract complementary information to the accelerator neutrino program. Lastly, Section 4 treats the solar neutrinos, discussing the great potential of DUNE in delivering world-leading results.

Section 5 begins with discussing the production of sterile neutrinos in supernovae calculating their hypothetical flux over the keV-MeV mass range, and it continues with analyzing the status of the cosmic neutrino flux measurements by the IceCube experiment.

Finally, the last Section 6 focuses on neutrinos in cosmology. In particular, it explores the complementarity between the experiments of terrestrial neutrino oscillation and the astrophysical measurements in searching for the existence of sterile neutrinos. Moreover, it discusses the possibility that dark matter mainly interacts with neutrinos.

SECTION 1
INTRODUCTION TO NEUTRINO PHYSICS

CHAPTER

1

Neutrino Physics

G. Barenboim

Universitat de València – CSIC, València, Spain

ABSTRACT

The Standard Model has been incredibly successful in predicting the outcome of almost all the experiments done up so far. In it, neutrinos are mass-less. However, in recent years we have accumulated evidence pointing to tiny masses for the neutrinos (as compared to the charged leptons). These masses allow neutrinos to change their flavour and oscillate. In these lectures I review the properties of neutrinos in and beyond the Standard Model.

INTRODUCTION

The last decade witnessed a revolution in neutrino physics. It has been observed that neutrinos have nonzero masses, and that leptons mix. This fact was proven by the observation that neutrinos can change from one type, or "flavour", to another. Almost all the knowledge we have gathered about

Citation: Barenboim, G. "Neutrino Physics" Proceedings of the 2012 School of High-Energy Physics, La Pommeraye, Anjou, France, 6 – 19 June 2012, edited by C. Grojean and M. Mulders, CERN–2014–008 (CERN, Geneva, 2014). http://dx.doi.org/10.5170/CERN-2014-008.161

neutrinos, is only fifteen years old. But before diving into the recent "news" about neutrinos, lets find out how neutrinos were born.

The '20s witnessed the assassination of many sacred cows, and physics was no exception, one of physic's most holly principles, the conservation of energy, appeared not to hold within the subatomic world. For some radioactive nuclei, it seemed that part of its energy just disappear, leaving no footprint of its existence.

In 1920, in a letter to a conference, Pauli wrote, "Dear radioactive Ladies and Gentlemen, ... as a desperate remedy to save the principle of energy conservation in beta decay, ... I propose the idea of a neutral particle of spin half". Pauli postulated that the energy loss was taken off by a new particle, whose properties were such that it would not yet be seen: it had no electric charge and rarely interacted with matter at all. This way, the neutrino was born into the world of particle physics.

Soon afterwards, Fermi wrote the four-Fermi Hamiltonian for beta decay using the neutrino, electron, neutron and proton. A new field came to existence: the field of weak interactions. And two decades after Pauli's letter, Cowan and Reines finally observed anti-neutrinos emitted by a nuclear reactor. As more and more particles were discovered in the following years and observed to participate in weak processes, weak interactions got legitimacy as a new force of nature and the neutrino became a key ingredient of this interactions.

Further experiments over the course of the next 30 years showed us that there were three kinds, or "flavours" of neutrinos (electron neutrinos (νe), muon neutrinos (νμ) and tau neutrinos (ντ)) and that, as far as we could tell, had no mass at all. The neutrino saga might have stop there, but new experiments in solar physics taught us that the neutrino story was just beginning....

Within the Standard Model, neutrinos have zero mass and therefore interact diagonally in flavour space,

$$\begin{array}{lcl} W^+ \longrightarrow e^+ + \nu_e & ; & Z \longrightarrow \nu_e + \bar{\nu}_e \\ W^+ \longrightarrow \mu^+ + \nu_\mu & ; & Z \longrightarrow \nu_\mu + \bar{\nu}_\mu \\ W^+ \longrightarrow \tau^+ + \nu_\tau & ; & Z \longrightarrow \nu_\tau + \bar{\nu}_\tau \end{array} \quad (1)$$

Since they are mass-less, they move at the speed of light and therefore their flavour remains the same from production up to detection. It is obvious then, that at least as flavour is concerned, zero mas neutrinos are almost not interesting as compared to quarks.

On the other hand, neutrinos masses different from zero, mean that there are three neutrino mass eigenstates ν_i, i = 1, 2, . . ., each with a mass mi . The meaning of leptonic mixing can be understood by analysing the leptonic decays, $W^+ \longrightarrow \nu_i + \overline{\ell_\alpha}$ of the charged W boson. Where, α = e, μ, or τ , and l_e represents the electron, l_μ the muon, or l_τ the tau. We refer to particle l_α as the charged lepton of flavour α. Mixing essentially means that when the W^+ decays to a given flavour of charged lepton $\overline{\ell_\alpha}$, the neutrino that comes along is not always the same mass eigenstate ν_i . Any of the different ν_i can show up. The amplitude for the decay of a W^+ to a specific combination $\overline{\ell_\alpha} + \nu_i$ is designated by $U^*_{\alpha i}$. The neutrino that is emitted in W^+ decay along with the given charged lepton $\overline{\ell_\alpha}$ is then

$$|\nu_\alpha >= \sum_i U^*_{\alpha i} \, |\nu_i > \qquad (2)$$

This particular combination of mass eigenstates is the neutrino of flavour αThe quantities $U_{\alpha i}$ can be collected in a unitary matrix (analogue to the CKM matrix of the quark sector) known as the leptonic mixing matrix [1]. The unitarity of U guarantees that every time a neutrino of flavour α interacts in a detector and produces a charged lepton, such a charged lepton will be always `α, the charged lepton with flavour α. That is, a ν_e indefectibly creates an e, a ν_μ a μ, and a ν_τ a τ .

The relation (2), describing a neutrino of definite flavour as a linear combination of mass eigenstates, may be inverted to describe each mass eigenstate ν_i as a linear combination of flavours:

$$|\nu_i >= \sum_\alpha U_{\alpha i} \, |\nu_\alpha > \qquad (3)$$

The α-flavour "content" (or fraction) of ν_i is clearly $|U_{\alpha i}|^2$. When a vi interacts and generates a charged lepton, this α-flavour fraction becomes the probability that the emerging charged lepton be of flavour α.

NEUTRINO OSCILLATIONS IN VACUUM

A standard neutrino flavour transition, or "oscillation", can be understood as follows. A neutrino is produced by a source together with a charged lepton $\overline{\ell_\alpha}$ of flavour α. Therefore, at the production point, the neutrino is a ν_α. Then, after birth, the neutrino travels a distance L until it is detected. There, it is where it reaches a target with which it interacts and produces another charged lepton l_β of flavour β. Thus, at the interaction point, the neutrino is a ν_β. If β 6= α (for example, if l_α is a μ but l_β is a τ), then, during its trip from the source to the detection point, the neutrino has transitioned from a ν_α into a ν_β.

This morphing of neutrino flavour, $\nu_\alpha \longrightarrow \nu_\beta$, is a text-book example of a quantum-mechanical effect.

Because, as described by Eq. (2), a ν_α is really a coherent superposition of mass eigenstates ν_i, the neutrino that propagates since it is created until it interacts, can be any one of the νi's, therefore we must add the contributions of all the different ν_i coherently. Then, the transition amplitude, $\text{Amp}(\nu_\alpha \longrightarrow \nu_\beta)$ contains a share of each ν_i and it is a product of three factors. The first one is the amplitude for the neutrino born at the production point in combination with an $\overline{\ell_\alpha}$ to be, specifically, a ν_i. As we have mentioned already, this amplitude is given by $U^*_{\alpha i}$. The second factor is the amplitude for the ν_i created by the source to propagate until it reaches the detector. We will call this factor Prop(ν_i) and will find out its value later. The third factor is the amplitude for the charged lepton produced by the interaction of the ν_i with the detector to be, specifically, an l_β. As the Hamiltonian that describes the interaction of neutrinos, charged leptons and W bosons is hermitian , it ensues that if $\text{Amp}(W \longrightarrow \overline{\ell_\alpha}\nu_i) = U^*_{\alpha i}$. Therefore, the third and last factor in the ν_i contribution is $U_{\beta i}$, and

$$\text{Amp}(\nu_\alpha \longrightarrow \nu_\beta) = \sum U^*_{\alpha i}\ \text{Prop}(\nu_i)\ U_{\beta i}\ . \qquad (4)$$

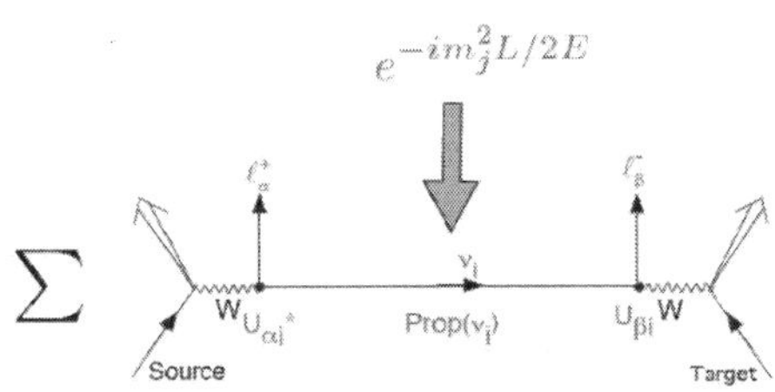

Figure 1: Neutrino flavour change (oscillation) in vacuum.

It still remains to be established the value of Prop(ν_i). To determine it, we'd better study the νi in its rest frame. We will label the time in that system τ_i. If ν_i does have a rest mass m_i, then in this frame its state vector satisfies the good old Schrödinger equation

$$i\frac{\partial}{\partial\tau_i}|\nu_i(\tau_i)> = m_i|\nu_i(\tau_i)> \qquad (5)$$

whose solution is given clearly by

$$|\nu_i(\tau_i)> = e^{-im_i\tau_i}|\nu_i(0)> \quad (6)$$

Then, the amplitude for the mass eigenstate ν_i to propagate for a time τ_i, is simply the amplitude $<\nu_i(0)|\nu_i(\tau_i)>$ for observing the original ν_i $|\nu_i(0)>$ after some time as the evoluted state $|\nu_i(\tau_i)>$, i.e. $\exp[-im_i\tau_i]$. Thus Prop(νi) is only this amplitude where we have used that the time taken by νi to travel from the neutrino source to the detector is τi , the proper time.

Nevertheless, if we want Prop(ν_i) to be of any use to us, we must write it first in terms of variables in the laboratory system. The natural choice is obviously the laboratory-frame distance, L, that the neutrino covers between the source and the detector, and the laboratory-frame time t, that slips away during the journey. The distance L is set by the experimentalists through the selection of the place of settlement of the source and that of the detector. Likewise, the value of the time t is selected by the experimentalists through their election for the time at which the neutrino is created and that when it is detected. Thus, L and t are chosen (hopefully carefully) by the experiment design, and are the same for all the ν_i components of the beam. Different ν_i do travel through an identical distance L, in an identical time t.

We still need two other laboratory-frame variables, they are the laboratory-frame energy E_i and momentum pi of the neutrino mass eigenstate ν_i. With the four lab-frame variable and using Lorentz invariance, we can obtain the phase $m_i\tau_i$ in the ν_i propagator Prop(ν_i) we have been looking for, which (expressed in terms of laboratory frame variables) is given by

$$m_i\tau_i = E_it - p_iL \quad (7)$$

At this point however one may argue that, in real life, neutrino sources are basically constant in time, and that the time t that slips away since the neutrino is produced till it dies in the detector is actually not measured. This argument is right. In real life, an experiment averages over the time t used by the neutrino to complete its journey. However, lets consider that two constituents of the neutrino beam, the first one with energy E_1 and the second one with energy E_2 (both measured in the lab frame), contribute coherently to the neutrino signal produced in the detector.

Now, if we call t to the the time used by the neutrino to cover the distance separating the production and detection points, then by the time the constituent whose energy is Ej (j = 1, 2) arrives to the detector, it has raised a phase factor $\exp[-iE_jt]$. Therefore, we will have an interference between the

E_1 and E_2 beam participants that will include a phase factor $\exp[-i(E_1 - E_2) t]$. When averaged over the non-observed travel time t, this factor goes away, except when E2 = E1. Therefore, only components of the neutrino beam that share the same energy contribute coherently to the neutrino oscillation signal [2,3]. Specifically, only the different mass eigenstate constituents of a beam that have the same energy contribute coherently to the oscillation signal.

A mass eigenstate νi , with mass mi , and energy E, has a momentum pi given by

$$p_i = \sqrt{E^2 - m_i^2} \cong E - \frac{m_i^2}{2E} \quad (8)$$

Where, we have used that as the masses of the neutrinos are miserably small, $m_i^2 \ll E^2$ for any energy E attainable at a realistic experiment. From Eqs. (7) and (8), we see that at energy E the phase $m_i\tau_i$ appearing in Prop(νi) takes the value

$$m_i \tau_i \cong E(t - L) + \frac{m_i^2}{2E} L \quad (9)$$

As the phase E(t − L) appears in all the interfering terms it will eventually disappear when calculating the transition amplitude. Thus, we can get rid of it already now and use

$$\text{Prop}(\nu_i) = \exp[-im_i^2 \frac{L}{2E}] \quad (10)$$

Applying this result, we can obtain from Eq. (4) that the amplitude for a neutrino born as a ν_α to be detected as a ν_β after covering a distance L through vacuum with energy E yields

$$\text{Amp}(\nu_\alpha \longrightarrow \nu_\beta) = \sum_i U_{\alpha i}^* e^{-im_i^2 \frac{L}{2E}} U_{\beta i} \quad (11)$$

The expression above is valid for an arbitrary number of neutrino flavours and mass eigenstates. The probability $P(\nu_\alpha \longrightarrow \nu_\beta)$ for $\nu_\alpha \longrightarrow \nu_\beta$ can be found by squaring it, giving

$$\begin{aligned} \mathrm{P}(\nu_\alpha \longrightarrow \nu_\beta) &= |\mathrm{Amp}(\nu_\alpha \longrightarrow \nu_\beta)|^2 \\ &= \delta_{\alpha\beta} - 4\sum_{i>j} \Re(U_{\alpha i}^* U_{\beta i} U_{\alpha j} U_{\beta j}^*) \sin^2\left(\Delta m_{ij}^2 \frac{L}{4E}\right) \\ &\quad + 2\sum_{i>j} \Im(U_{\alpha i}^* U_{\beta i} U_{\alpha j} U_{\beta j}^*) \sin\left(\Delta m_{ij}^2 \frac{L}{2E}\right) , \end{aligned} \quad (12)$$

with

$$\Delta m_{ij}^2 \equiv m_i^2 - m_j^2 \quad (13)$$

In order to get Eq. (12) we have used that the mixing matrix U is unitary.

The oscillation probability $\mathrm{P}(\nu_\alpha \longrightarrow \nu_\beta)$ we have just obtained corresponds to that of a neutrino, and not to an antineutrino, as we have used that the oscillating neutrino was produced along with a charged antilepton , and gives birth to a charged lepton ` once it reaches the detector. The corresponding probability $\mathrm{P}(\overline{\nu_\alpha} \longrightarrow \overline{\nu_\beta})$ for an antineutrino oscillation can be obtained from $\mathrm{P}(\nu_\alpha \longrightarrow \nu_\beta)$ taking advantage of the fact that the two transitions $\overline{\nu_\alpha} \longrightarrow \overline{\nu_\beta}$ and $\nu_\beta \longrightarrow \nu_\alpha$ are CPT conjugated processes. Thus, assuming that neutrino interactions respect CPT [4],

$$\mathrm{P}(\overline{\nu_\alpha} \longrightarrow \overline{\nu_\beta}) = \mathrm{P}(\nu_\beta \longrightarrow \nu_\alpha) \quad (14)$$

Then, from Eq. (12) we obtain that

$$\mathrm{P}(\nu_\beta \longrightarrow \nu_\alpha;\ U) = \mathrm{P}(\nu_\alpha \longrightarrow \nu_\beta;\ U^*) \quad (15)$$

Therefore, if CPT is a good symmetry (with respect to neutrino interactions), Eq. (12) tells us that

$$\begin{aligned} \mathrm{P}(\overset{(-)}{\nu_\alpha} \longrightarrow \overset{(-)}{\nu_\beta}) &= \delta_{\alpha\beta} - 4\sum_{i>j} \Re(U_{\alpha i}^* U_{\beta i} U_{\alpha j} U_{\beta j}^*) \sin^2\left(\Delta m_{ij}^2 \frac{L}{4E}\right) \\ &\quad \underset{(-)}{+} 2\sum_{i>j} \Im(U_{\alpha i}^* U_{\beta i} U_{\alpha j} U_{\beta j}^*) \sin\left(\Delta m_{ij}^2 \frac{L}{2E}\right) \end{aligned} \quad (16)$$

These expressions make it clear that if the mixing matrix U is complex, $\mathrm{P}(\overline{\nu_\alpha} \longrightarrow \overline{\nu_\beta})$and $\mathrm{P}(\nu_\alpha \longrightarrow \nu_\beta)$will not be identical, in general. As $\overline{\nu_\alpha} \longrightarrow \overline{\nu_\beta}$ and $\nu_\alpha \longrightarrow \nu_\beta$are CP conjugated processes, $\mathrm{P}(\overline{\nu_\alpha} \longrightarrow \overline{\nu_\beta}) \neq \mathrm{P}(\nu_\alpha \longrightarrow \nu_\beta)$would be an evidence of CP violation in neutrino oscillations. So far, CP violation has been observed only in the quark sector, so its measurement in neutrino oscillations would be quite exciting.

So far, we have been working in natural units, if we return now the ħ's and c factor (we have happily left out) into the oscillation probability we find that

$$\sin^2\left(\Delta m^2_{ij}\frac{L}{4E}\right) \longrightarrow \sin^2\left(\Delta m^2_{ij}c^4\frac{L}{4\hbar cE}\right) \tag{17}$$

Having done that, it is easy and instructive to explore the semi-classical limit, ħ→ 0. In this limit the oscillation length goes to zero (the oscillation phase goes to infinity) and the oscillations are averaged out. The same happens if we let the mass difference Δm^2 become large. This is exactly what happens in the quark sector (and the reason why we never study quark oscillations despite knowing that mass eigenstates do not coincide with flavour eigenstates).

In terms of real life units (which are not "natural" units), the oscillation phase is given by

$$\Delta m^2_{ij}\frac{L}{4E} = 1.27\,\Delta m^2_{ij}(\mathrm{eV}^2)\frac{L\,(\mathrm{km})}{E\,(\mathrm{GeV})} \tag{18}$$

then, since $\sin^2[1.27\,\Delta m^2_{ij}(\mathrm{eV}^2)L\,(\mathrm{km})/E\,(\mathrm{GeV})]$ can be experimentally observed only if its argument is of order unity or larger, an experimental set-up with a baseline L (km) and an energy E (GeV) is sensitive to neutrino mass squared differences $\Delta m^2_{ij}(\mathrm{eV}^2)$ larger that or equal to ~ [L(km)/E (GeV]$^{-1}$. For example, an experiment with L ~ 10^4 km, roughly the size of Earth's diameter, and E ~ 1 GeV is sensitive to Δm^2_{ij} down to ~10^{-4} eV^2. This fact makes it clear that neutrino oscillation experiments can test super tiny neutrino masses. It does so by exhibiting quantum mechanical interferences between amplitudes whose relative phases are proportional to these super tiny neutrino mass squared differences, which can be transformed into sizeable effects by choosing an L/E large enough.

But let's keep analysing the oscillation probability and see whether we can learn more about neutrino oscillations by studying its expression.

It is clear from $\mathrm{P}(\overset{(-)}{\nu_\alpha} \longrightarrow \overset{(-)}{\nu_\beta})$ that if neutrinos have zero mass, in such a way that all $\Delta m^2_{ij} = 0$, then, $\mathrm{P}(\overset{(-)}{\nu_\alpha} \longrightarrow \overset{(-)}{\nu_\beta}) = \delta_{\alpha\beta}$. Therefore, the experimental observation that neutrinos can morph from one flavour to a different one indicates that neutrinos are not only massive but also that their masses are not degenerate. Actually, it was precisely this evidence the one that led to the conclusion that neutrinos are massive.

However, every neutrino oscillation seen so far has involved at some point neutrinos that travel through matter. But the expression we derived is valid only for flavour change in vacuum, and does not take into account any interaction between the neutrinos and the matter traversed between their source and their detector. Thus, one might ask whether flavour-changing interactions between neutrinos and matter are indeed responsible of the observed flavour changes, and not neutrino masses. Regarding this question, two points can be made. First, although it is true that the Standard Model of elementary particle physics contains only mass-less neutrinos, it provides an amazingly well corroborated description of neutrino interactions, and this description clearly establishes that neutrino interactions with matter do not change flavour. Second, for at least some of the observed flavour changes, matter effects are expected to be miserably small, and there is solid evidence that in these cases, the flavour transition probability depends on L and E through the combination L/E, as anticipated by the oscillation hypothesis. Modulo a constant, L/E is precisely the proper time that goes by in the rest frame of the neutrino as it covers a distance L possessing an energy E. Thus, these flavour transitions behave as if they were a progression of the neutrino itself over time, rather than a result of interaction with matter.

Now, suppose the leptonic mixings were trivial. This would mean that in the decay $W^{+}\longrightarrow \overline{\ell_\alpha} + \nu_i$, which as we established has an amplitude $U^*_{\alpha i}$, the emerging charged antilepton `α of flavour α comes along always with the same neutrino mass eigenstate νi . That is, if $U^*_{\alpha i} \neq 0$, then Uαj becomes zero for all $j \neq i$. Therefore, from Eq. (16) it is clear that, $P(\overset{(-)}{\nu_\alpha} \longrightarrow \overset{(-)}{\nu_\beta}) = \delta_{\alpha\beta}$. Thus, the observation that neutrinos can change flavour indicates mixing.

Then, there are basically two ways to detect neutrino flavour change. The first one is to observe, in a beam of neutrinos which are all created of the same flavour, say α, some appearance of neutrinos of a new flavour β that is different from the flavour α we started with. This is what is called an appearance experiment. The second way is to start with a beam of identical ναs, whose flux is known, and observe that this known να flux is depleted. This is called a disappearance experiment.

As Eq. (16) shows, the transition probability in vacuum does not only depend on L/E but also oscillates with it. It is because of this fact that neutrino flavour transitions are named "neutrino oscillations". Now notice also that neutrino transition probabilities depend only on neutrino squared-mass splittings, and not on the individual squared neutrino masses themselves. Thus, oscillation experiments can only measure the neutrino squared-mass spectral pattern, but not its absolute scale, i.e. the distance above zero the entire pattern lies.

It is clear that neutrino transitions cannot modify the total flux in a neutrino beam, but simply alter its distribution between the different flavours. Actually, from Eq. (16) and the unitarity of the U matrix, it follows that

$$\sum_{\beta} \mathrm{P}(\overset{(-)}{\nu_\alpha} \longrightarrow \overset{(-)}{\nu_\beta}) = 1 \ , \tag{19}$$

where the sum runs over all flavours β, including the original flavour α. Eq. (19) makes it transparent that the probability that a neutrino morphs its flavour, added to the probability that it does not do so, is one. Ergo, flavour transitions do not change the total flux. Nevertheless, some of the flavours β 6= α into which a neutrino can oscillate into may be sterile flavours; that is, flavours that do not take part in weak interactions and therefore escape detection. If any of the original (active) neutrino flux turns into sterile, then an experiment measuring the total active neutrino flux—that is, the sum of the ν_e, ν_μ, and ν_τ fluxes—will find it to be less than the original flux. In the experiments performed up today, no flux was ever missed.

In the literature, description of neutrino oscillation normally assume that the different mass eigenstates νi that contribute coherently to a beam share the same momentum, rather than the same energy as we have argued they must have. While the supposition of equal momentum is technically wrong, it is an inoffensive mistake, since, as can easily be shown [5], it conveys to the same oscillation probabilities as we have found.

A relevant and interesting case of the (not that simple) formula for $\mathrm{P}(\overline{\nu_\alpha} \longrightarrow \overline{\nu_\beta})$ is the case where only two flavours participate in the oscillation. The only-two-neutrino scenario is a rather rigorous description of a vast number of experiments. Lets assume then, that only two mass eigenstates, which we will name ν1 and ν2, and two corresponding flavour states, which we will name νμ and ντ , are relevant.

There is then only one squared-mass splitting, $m_2^2 - m_1^2 \equiv \Delta m^2$. Even more, neglecting phase factors that can be proven to have no effect on oscillation, the mixing matrix U takes the simple form

$$\begin{pmatrix} \nu_\mu \\ \nu_\tau \end{pmatrix} = \begin{pmatrix} \cos\theta & \sin\theta \\ -\sin\theta & \cos\theta \end{pmatrix} \begin{pmatrix} \nu_1 \\ \nu_2 \end{pmatrix} \tag{20}$$

The U of Eq. (20) is just a 2×2 rotation matrix, and the rotation angle θ within it is referred to as the mixing angle. Inserting the U of Eq. (20) and the unique Δm^2 into the general expression for $P(\overline{\nu_\alpha} \longrightarrow \overline{\nu_\beta})$, Eq. (16), we immediately find that, for $\beta \neq \alpha$, when only two neutrinos are relevant,

$$P(\overline{\nu_\alpha} \longrightarrow \overline{\nu_\beta}) = \sin^2 2\theta \sin^2\left(\frac{\Delta m^2 L}{4E}\right) \tag{21}$$

Moreover, the survival probability, i.e. the probability that the neutrino remains with the same flavour its was created with is, as usual, unity minus the probability that it changes flavour.

NEUTRINO OSCILLATIONS IN MATTER

When we create a beam of neutrinos on earth through an accelerator and send it up to thousand kilometres away to a meet detector, the beam does not move through vacuum, but through matter, earth matter. The beam of neutrinos then scatters (coherently forward) from particles it meets along the way. Such a scattering can have a large effect on the transition probabilities. We will assume that neutrino interactions with matter are flavour conserving, as described by the Standard Model. Then a neutrino in matter have two possibilities to enjoy coherent forward scattering from matter particles . First, if it is an electron neutrino, νe—and only in this case—can exchange a W boson with an electron. Neutrinoelectron coherent forward scattering via W exchange opens up an extra interaction potential energy VW suffered exclusively by electron neutrinos. Obviously, this additional weak interaction energy has to be proportional to G_F, the Fermi coupling constant. In addition, the interaction energy coming from νe − e scattering grows with N_e, the number of electrons per unit volume. From the Standard Model, we find that

$$V_W = +\sqrt{2}\, G_F\, N_e \tag{22}$$

clearly, this interaction energy affects also antineutrinos (in a opposite way though), it changes sign if we replace the νe by $\overline{\nu_e}$.

The second interaction corresponds to the case where a neutrino in matter exchanges a Z boson with an matter electron, proton, or neutron. The Standard Model teaches us that weak interactions are flavour blind. Every flavour of neutrino enjoys them, and the amplitude for this Z exchange is always the same. It also teaches us that, at zero momentum transfer, the Z couplings to electrons and protons have equal strength and opposite sign. Therefore, counting on the fact that the matter through which our neutrino moves is electrically neutral (it contains equal number of electrons and protons), the electron and proton contribution to coherent forward neutrino scattering through Z exchange will add up to zero. Then, the effect of the Z exchange contribution to the interaction potential energy V^Z will be equal to all flavors and will depends exclusively on N_n, the number density of neutrons. From the Standard Model, we find that

$$V_Z = -\frac{\sqrt{2}}{2} G_F N_n \quad (23)$$

as was the case before, for VW , this contribution will flip sign if we replace the neutrinos by antineutrinos.

But we already learnt that Standard Model interactions do not change neutrino flavour. Therefore, unless non-Standard-Model flavour changing interactions play a role, neutrino flavour transitions or neutrino oscillations points also to neutrino mass and mixing even when neutrinos are propagating through matter.

Neutrino propagation in matter is easy to understand when analyzed through a time dependent Schrödinger equation in the laboratory frame

$$i\frac{\partial}{\partial t}|\nu(t) > = \mathcal{H}|\nu(t) > \quad (24)$$

where, |ν(t)> is a multicomponent neutrino vector state, in which each neutrino flavour corresponds to one component. In the same way, the Hamiltonian H is a matrix in flavour space. To make our lives easy, lets analyze the case where only two neutrino flavours are relevant, say ν_e and ν_μ. Then

$$|\nu(t) > = \begin{pmatrix} f_e(t) \\ f_\mu(t) \end{pmatrix} \quad (25)$$

where $fe(t)^2$ is the fraction of the neutrino that is a νe at time t, and similarly for fμ(t). Analogously, H is a 2×2 matrix in $\nu_e - \nu_\mu$ space

It will prove to be clarifying to work out the two flavour case in vacuum first, and add matter effects afterwards. Using Eq. (2) for $|\nu_\alpha >$ as a linear combination of mass eigenstates, we can see that the $\nu_\alpha - \nu_\beta$ matrix element of the Hamiltonian in vacuum, HV_{ac}, can be written as

$$< \nu_\alpha|\mathcal{H}_{\text{Vac}}|\nu_\beta > \; = \; < \sum_i U^*_{\alpha i}\nu_i|\mathcal{H}_{\text{Vac}}|\sum_j U^*_{\beta j}\nu_j >$$
$$= \sum_j U_{\alpha j}U^*_{\beta j}\sqrt{p^2+m_j^2} \; . \tag{26}$$

where we are supposing that neutrino belongs to a beam where all its mass components (the mass eigenstates) share the same definite momentum p. (As we have already mentioned, this supposition is technically wrong, however it leads anyway to the right oscillation probability.) In the second line of Eq. (26), we have used that

$$\mathcal{H}_{\text{Vac}}|\nu_j >= E_j|\nu_j > \tag{27}$$

with $E_j = \sqrt{p^2+m_j^2}$ the energy of the mass eigenstate νj with momentum p, and the fact that the mass eigenstates of the Hermitian Hamiltonian HVac constitute a basis and therefore are orthogonal.

As we have already mentioned, neutrino oscillations are the archetype quantum interference phenomenon, where only the relative phases of the interfering states play a role. Therefore, only the relative energies of these states, which set their relative phases, are relevant. As a consequence, if it proves to be convenient, we can feel free to happily remove from the Hamiltonian H any contribution proportional to the identity matrix I. As we have said, this substraction will leave unaffected the differences between the eigenvalues of H, and therefore will leave unaffected the prediction of H for flavour transitions.

It goes without saying that as in this case only two neutrinos are relevant, there are only two mass eigenstates, ν_1 and ν_2, and only one mass splitting $\Delta m^2 \equiv m_2^2 - m_1^2$, and therefore the U matrix is given by Eq. (20). Inserting this matrix into Eq. (26), and applying the high momentum approximation $(p^2+m_j^2)^{1/2} \cong p + m_j^2/2p$, and removing from HVac a term proportional to the the identity matrix (a removal we know is going to be harmless), we get

$$\mathcal{H}_{\text{Vac}} = \frac{\Delta m^2}{4E}\begin{pmatrix} -\cos 2\theta & \sin 2\theta \\ \sin 2\theta & \cos 2\theta \end{pmatrix} \tag{28}$$

To write this expression, we have used that p ~= E, where E is the average energy of the neutrino mass eigenstates in our neutrino beam of ultra high momentum p.

It is not difficult to corroborate that the Hamiltonian H_{Vac} of Eq. (28) for the two neutrino scenario would give an identical oscillation probability , Eq. (21), as the one we have already obtained in a different way. For example, lets have a look at the oscillation probability for the process $\nu_e \longrightarrow \nu_\mu$. From Eq. (20) it is clear that in terms of the mixing angle, the electron neutrino state composition is

$$|\nu_e> = \quad |\nu_1> \cos\theta + |\nu_2> \sin\theta \ , \tag{29}$$

while that of the muon neutrino is given by

$$|\nu_\mu> = -|\nu_1> \sin\theta + |\nu_2> \cos\theta \tag{30}$$

Now, the eigenvalues of H_{Vac}, Eq.25, read

$$\lambda_1 = -\frac{\Delta m^2}{4E} \ , \ \lambda_2 = +\frac{\Delta m^2}{4E} \tag{31}$$

The eigenvectors of this Hamiltonian, $|\nu_1>$ and $|\nu_2>$, can also be written in terms of $|\nu_e>$ and $|\nu_\mu>$ by means of Eqs. (29) and (30). Therefore, with H, its vacuum expression, H_{Vac} of Eq. (28), the Schrödinger equation of Eq. (24) tells us that if at time t = 0 we begin from a $|\nu_e>$, then after some time t this $|\nu_e>$ will progress into the state

$$|\nu(t)> = |\nu_1> e^{+i\frac{\Delta m^2}{4E}t} \cos\theta + |\nu_2> e^{-i\frac{\Delta m^2}{4E}t} \sin\theta \tag{32}$$

Thus, the probability $P(\nu_e \longrightarrow \nu_\mu)$ that this evoluted neutrino be detected as a different flavour ν_μ, from Eqs. (30) and (32), is given by,

$$\begin{aligned} P(\nu_e \longrightarrow \nu_\mu) &= |<\nu_\mu|\nu(t)>|^2 \\ &= |\sin\theta\cos\theta(-e^{i\frac{\Delta m^2}{4E}t} + e^{-i\frac{\Delta m^2}{4E}t})|^2 \\ &= \sin^2 2\theta \sin^2\left(\Delta m^2 \frac{L}{4E}\right) . \end{aligned} \tag{33}$$

Where we have substituted the time t travelled by our highly relativistic state by the distance L it has covered. The flavour transition or oscillation probability of Eq. (33), as expected, is exactly the same we have found before, Eq. (21).

We can now move on to analyze neutrino propagation in matter. In this case, the 2×2 vacuum Hamiltonian H_{Vac} receives two additional contributions and becomes HM, which is given by

$$\mathcal{H}_M = \mathcal{H}_{\text{Vac}} + V_W \begin{pmatrix} 1 & 0 \\ 0 & 0 \end{pmatrix} + V_Z \begin{pmatrix} 1 & 0 \\ 0 & 1 \end{pmatrix}$$

(34)

In the new Hamiltonian, the first additional contribution corresponds to the interaction potential due to W exchange, Eq. (22). As this interaction is suffered only by νe, this contribution is different from zero only in the upper left, νe − νe, element of HM. The second additional contribution, the last term of Eq. (34) comes from the interaction potential due to Z exchange, Eq. (23). Since this interaction is flavour blind, it affects every neutrino flavour in the same way, its contribution to HM is proportional to the identity matrix, and can be safely neglected. Then

$$\mathcal{H}_M = \mathcal{H}_{\text{Vac}} + \frac{V_W}{2} + \frac{V_W}{2} \begin{pmatrix} 1 & 0 \\ 0 & -1 \end{pmatrix}$$

(35)

where (for reasons that arc going to become clear later) we have divided the W-exchange contribution into two pieces, one a multiple of the identity matrix (that we will disregard in the next step) and, a piece that it is not a multiple of the identity matrix. Disregarding the first piece as promised, we have from Eqs. (28) and (35)

$$\mathcal{H}_M = \frac{\Delta m^2}{4E} \begin{pmatrix} -(\cos 2\theta - A) & \sin 2\theta \\ \sin 2\theta & (\cos 2\theta - A) \end{pmatrix}$$

(36)

where

$$A \equiv \frac{V_W/2}{\Delta m^2/4E} = \frac{2\sqrt{2} G_F N_e E}{\Delta m^2}$$

(37)

Clearly, A shows the size (the importance) of the matter effects as compared to the neutrino squared-mass splitting and signal the situations when they become important.

Now, if we define

$$\Delta m_M^2 \equiv \Delta m^2 \sqrt{\sin^2 2\theta + (\cos 2\theta - A)^2}$$

(38)

and

$$\sin^2 2\theta^M \equiv \frac{\sin^2 2\theta}{\sin^2 2\theta + (\cos 2\theta - A)^2} \tag{39}$$

H_M can be given as

$$\mathcal{H}_M = \frac{\Delta m^2_M}{4E} \begin{pmatrix} -\cos 2\theta^M & \sin 2\theta^M \\ \sin 2\theta^M & \cos 2\theta^M \end{pmatrix} \tag{40}$$

That is, the Hamiltonian in matter, H_M, becomes formally identical to its vacuum counterpart, H_{Vac}, Eq. (28), except that the vacuum parameters Δm2 and θ are now given by the matter ones, Δm^2_M and θ^M, respectively.

Obviously, the eigenstates of H_M are not identical to their vacuum counterparts. The splitting between the squared masses of the matter eigenstates is not the same as the vacuum splitting Δm^2, and the same happens with the mixing angle, the mixing in matter—the angle that rotates from the ν_e, ν_μ basis, to the mass basis—is different from the vacuum mixing angle θ. Clearly however, all the physics of neutrino propagation in matter is controlled by the matter Hamiltonian H_M. However, according to Eq. (40), at least at the formal level, H_M has the same functional dependence on the matter parameters Δm^2_M and θ^M in as the vacuum Hamiltonian H_{Vac}, Eq. (28), on the vacuum ones, Δm^2 and θ. Therefore, Δm^2_M corresponds to the effective splitting between the squared masses of the eigenstates in matter, and θ^M corresponds to the effective mixing angle in matter.

In a typical experimental set-up where the neutrino beam is generated by an accelerator and sent away to a detector that is, say, several hundred, or even thousand kilometers away, it traverses through earth matter, but only superficially, it does not get deep into the earth. The matter density met by such a beam en voyage can be taken to be approximately constant. Therefore, the electron density Ne is also constant, and the same happens with the parameter A, and the matter Hamiltonian H_M. They all become approximately position independent, and therefore quite analogue to the vacuum Hamiltonian H_{Vac}, which was absolutely position independent. Comparing Eqs. (40) and (28), we can immediately conclude that since H_{Vac} gives rise to the vacuum oscillation probability $P(\nu_e \longrightarrow \nu_\mu)$ of Eq. (33), H_M must give rise to a matter oscillation probability of the form

$$P_M(\nu_e \longrightarrow \nu_\mu) = \sin^2 2\theta^M \sin^2\left(\Delta m^2_M \frac{L}{4E}\right) \tag{41}$$

That is, the oscillation probability in matter (formally) is the same as in vacuum, except that the vacuum parameters θ and Δm^2 are replaced by their matter counterparts.

In theory, judging simply its potential, matter effects can have very drastic effects. From Eq. (39) for the effective mixing angle in matter, θM, we can appreciate that even when the vacuum mixing angle θ is incredible small, say, $\sin^2 2\theta = 10^{-4}$, if we get to have A ~= cos 2θ, then $\sin^2 2\theta^M$ can be brutally enhanced as compared to its vacuum value and can even reach its maximum possible value, one. This brutal enhancement of a tiny mixing angle in vacuum up to a sizeable one in matter is the "resonant" version of the Mikheyev-Smirnov-Wolfenstein effect [6–9]. In the beginning of solar neutrino experiments, people entertained the idea that this brutal enhancement was actually taking place inside the sun. Nonetheless, as we will see soon the solar neutrino mixing angle is quite sizeable (~ 34°) already in vacuum [10]. Then, although matter effects on the sun are important and they do enhance the solar mixing angle, unfortunately they are not as drastic as we once dreamt.

EVIDENCE FOR NEUTRINO OSCILLATIONS

Atmospheric and Accelerator Neutrinos

Almost fifteen year have elapsed since we were presented convincing evidence of neutrino masses and mixings, and since then, the evidence has only grown. SuperKamiokande (SK) was the first experiment to present compelling evidence of ν_μ disappearance in their atmospheric neutrino fluxes, see [12] . In Fig. 2 the zenith angle (the angle subtended with the horizontal) dependence of the multi-GeV ν_μ sample is shown together with the disappearance as a function of L/E plot. These data fit amazingly well the naive two component neutrino hypothesis with

$$\Delta m^2_{\text{atm}} = 2 - 3 \times 10^{-3}\text{eV}^2 \quad \text{and} \quad \sin^2\theta_{\text{atm}} = 0.50 \pm 0.15 \tag{42}$$

Roughly speaking SK corresponds to an L/E for oscillations of 500 km/GeV and almost maximal mixing (the mass eigenstates are nearly even admixtures of muon and tau neutrinos). No signal of an involvement of the third flavour, ν_e is found so the assumption is that atmospheric neutrino disappearance is basically $\nu_\mu \longrightarrow \nu_\tau$.

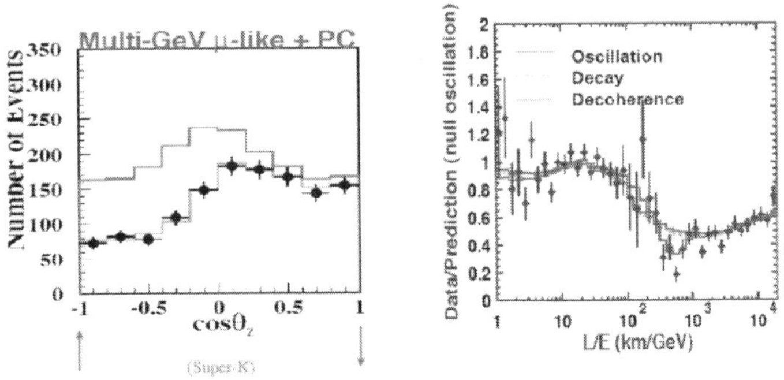

Figure 2: Superkamiokande's evidence for neutrino oscillations both in the zenith angle and L/E plots.

After atmospheric neutrino oscillations were established, two new experiments were built, sending (man-made) beams of ν_μ neutrinos to detectors located at large distances: the K2K experiment [13, 14], sends neutrinos from the KEK accelerator complex to the old SK mine, with a baseline of 120 km while the MINOS experiment [15], sends its beam from Fermilab, near Chicago, to the Soudan mine in Minnesota, a baseline of 735 km. Both experiments have seen evidence for νμ disappearance consistent with the one found by SK. The results of both are summarised in Fig. 3.

Reactor and Solar Neutrinos

The KamLAND reactor experiment, an antineutrino disappearance experiment, receiving neutrinos from sixteen different reactors, at distances ranging from hundred to thousand kilometers, with an average baseline of 180 km and neutrinos of a few ev, [16, 17], has seen evidence of neutrino oscillations . Such evidence was collected not only at a different L/E than the atmospheric and accelerator experiments but also consists on oscillations involving electron neutrinos, νe, the ones which were not involved before. These oscillations have also been seen for neutrinos coming from the sun (the sun produces only electron neutrinos). However,in order to compare the two experiments we should assume that neutrinos (solar) and antineutrinos (reactor) behave in the same way, i.e. assume CPT conservation. The best fit values in the two neutrino scenario for the KamLAND experiment are

$$\Delta m^2_\odot = 8.0 \pm 0.4 \times 10^{-5}\text{eV}^2 \quad \text{and} \quad \sin^2\theta_\odot = 0.31 \pm 0.03 \tag{43}$$

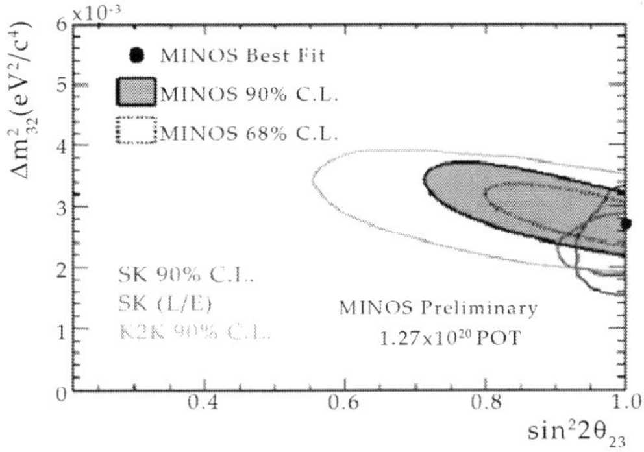

Figure 3: Allowed regions in the Δm^2_{atm} vs $\sin^2\theta_{atm}$ plane for MINOS data as well as for K2K data and two of the SK analyses. MINOS's best fit point is at $\sin^2\theta_{atm} = 1$ and $\Delta m^2_{atm} = 2.7 \times 10^{-3}eV^2$.

In this case, the L/E involved is 15 km/MeV which is more than an order of magnitude larger than the atmospheric scale and the mixing angle, although large, is clearly not maximal.

Fig. 4 shows the disappearance probability for the $\bar{\nu}_e$ for KamLAND as well as several older reactor experiments with shorter baselines.The second panel depicts the flavour content of the ^{8}Boron solar neutrino flux (with GeV energies) measured by SNO, [18], and SK, [19]. The reactor outcome can be explained in terms of two flavour oscillations in vacuum, given that the fit to the disappearance probability, is appropriately averaged over E and L.

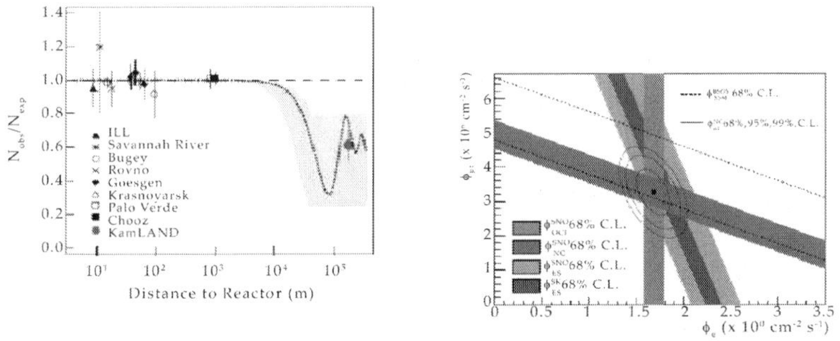

Figure 4: Disappearance of the $\bar{\nu}_e$ observed by reactor experiments as a function of distance from the reactor. The flavour content of the ^{8}Boron solar neutrinos for the various reactions for SNO and SK. CC: ν_e+d $\longrightarrow$ e^-+p+p, NC:ν_x + d $^-\!\!\longrightarrow \nu_x$ + p + n and ES: $\nu_\alpha + e^- \longrightarrow \nu_\alpha + e^-$.

The analysis of neutrinos coming from the sun is slightly more sophisticated because it should include the matter effects that the neutrinos suffer since they are born (at the centre of the sun) until they leave it, which are important at least for the ^{8}Boron neutrinos. The pp and ^{7}Be neutrinos are less energetic and therefore are not significantly altered by the presence of matter and leave the sun as if it where ethereal. ^{8}Boron neutrinos on the other hand, leave the sun strongly affected by the presence of matter and this is evidenced by the fact that they leave the sun as the ν_2 mass eigenstate and therefore do not undergo oscillations. This difference is, as mentioned, due mainly to their differences at birth. While pp (^{7}Be) neutrinos are created with an average energy of 0.2 MeV (0.9 MeV), ^{8}B are born with 10 MeV and as we have seen the impact of matter effects grows with the energy of the neutrino.

However, we should stress that we do not really see solar neutrino oscillations. To trace the oscillation pattern, we need a kinematic phase of order one. In the case of neutrinos coming from the sun the kinematic phase is

$$\Delta_\odot = \frac{\Delta m_\odot^2 L}{4E} = 10^{7\pm1}. \tag{44}$$

Therefore, solar neutrinos behave as "effectively incoherent" mass eigenstates once they leave the sun, and remains being so once they reach the earth. Consequently the ν_e survival probability is given by

$$\langle P_{ee} \rangle = f_1 \cos^2 \theta_\odot + f_2 \sin^2 \theta_\odot \tag{45}$$

where f_1 is the ν_1 content or fraction of ν_μ and f_2 is the ν_2 content of ν_μ and therefore both fractions satisfy

$$f_1 + f_2 = 1. \tag{46}$$

However, as we have mentioned, pp and ^{7}Be solar neutrinos are not affected by the solar matter and oscillate as in vacuum and thus, in their case $f_1 \approx \cos^2 \theta_\odot = 0.69$ and $f_2 \approx \sin^2 \theta_\odot = 0.31$. In the ^{8}B a neutrino case, however, matter effects are important and the corresponding fractions are substantially altered, see Fig. 5.

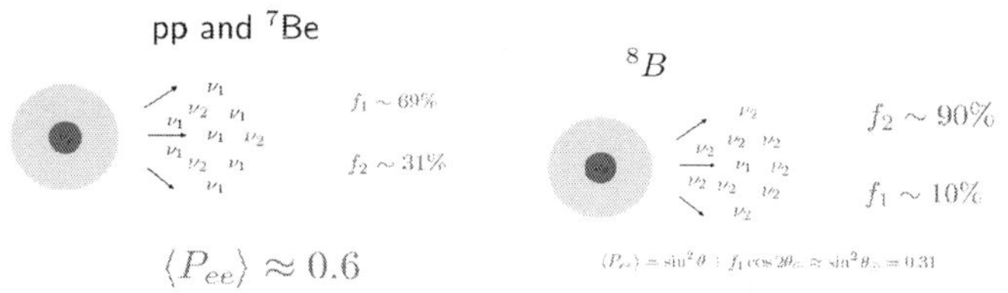

Figure 5: The sun produces ν_e in the core but once they exit the sun thinking about them in the mass eigenstate basis is useful. The fraction of ν_1 and ν_2 is energy dependent above 1 MeV and has a dramatic effect on the ^{8}Boron solar neutrinos, as first observed by Davis.

In a two neutrino scenario, the day-time CC/NC measured by SNO, which is roughly identical to the day-time average νe survival probability, <Pee>, reads

$$\left.\frac{CC}{NC}\right|_{\text{day}} = \langle P_{ee} \rangle = f_1 \cos^2\theta_\odot + f_2 \sin^2\theta_\odot, \tag{47}$$

where f_1 and $f_2 = 1 - f_1$ are the ν_1 and ν_2 contents of the muon neutrino, respectively, averaged over the ^{8}B neutrino energy spectrum appropriately weighted with the charged current current cross section.

Therefore, the ν1 fraction (or how much f2 differs from 100%) is given by

$$f_1 = \frac{\left(\left.\frac{CC}{NC}\right|_{\text{day}} - \sin^2\theta_\odot\right)}{\cos 2\theta_\odot} = \frac{(0.347 - 0.311)}{0.378} \approx 10\% \tag{48}$$

where the central values of the last SNO analysis, [18], were used. As there are strong correlations between the uncertainties of the CC/NC ratio and $\sin^2\theta_\odot$ it is not obvious how to estimate the uncertainty on f_1 from their analysis. Note, that if the fraction of ν_2 were 100%, then $\left.\frac{CC}{NC}\right|_{\text{day}} = \sin^2\theta_\odot$.

Using the analytical analysis of the Mikheyev-Smirnov-Wolfenstein (MSW) effect, provided in [20], one can obtain the mass eigenstate fractions, which are given by

$$f_2 = 1 - f_1 = \langle \sin^2\theta_\odot^M + P_x \cos 2\theta_\odot^M \rangle_{^8\text{B}}, \tag{49}$$

with $\theta_\odot^M$ being the mixing angle as given at the νe production point and Px is the probability of the neutrino to hop from one mass eigenstate to the second one during the Mikheyev-Smirnov resonance crossing. The average $<\ldots_{8B}$ is over the electron density of the ^{8}B νe production region in the centre of the Sun as given by the Solar Standard Model and the energy spectrum of ^{8}B neutrinos appropriately weighted with SNO's charged current cross section. All in all, the ^{8}B energy weighted average content of ν_2's measured by SNO is

$$f_2 = 91 \pm 2\% \text{ at the } 95\ \% \text{ C.L.} \quad (50)$$

Therefore, it is obvious that the ^{8}B solar neutrinos are the purest mass eigenstate neutrino beam known so far and SK super famous picture of the sun taken (underground) with neutrinos is made with approximately 90% of ν_2.

On March 8, 2012 a newly built reactor neutrino experiment, the Daya Bay experiment, located in China, announced the measurement of the third mixing angle [11], the only one which was still missing and found it to be

$$\sin^2(2\theta_{12}) = 0.092 \pm 0.017 \quad (51)$$

The fact that this angle, although smaller that the other two, is still sizeable opens the door to a new generation of neutrino experiments aiming to answer the open questions in the field.

ν STANDARD MODEL

Now that we have understood the physics behind neutrinos oscillations and have learnt the experimental evidence about the parameters driving this oscillations, we can move ahead and construct the Neutrino Standard Model:

– it consists of three light (mi < 1 eV) neutrinos, i.e. it involves only two mass differences $\Delta m^2_{\text{atm}} \approx 2.5 \times 10^{-3}\text{eV}^2$ and $\Delta m^2_{\text{solar}} \approx 8.0 \times 10^{-5}\text{eV}^2$.

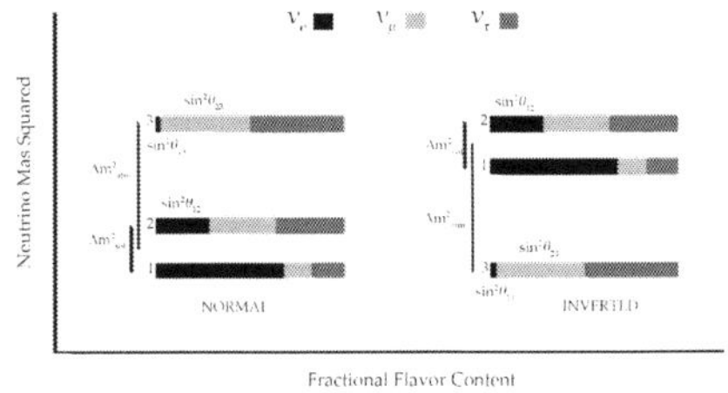

Figure 6: Flavour content of the three neutrino mass eigenstates (not including the dependence on the cosine of the CP violating phase δ).If CPT is conserved, the flavour content must be the same for neutrinos and anti-neutrinos. Notice that oscillation experiments cannot tell us how far above zero the entire spectrum lies.

– so far we have not seen any experimental indication (or need) for additional neutrinos. As we have measured long time ago the invisible width of the Z boson and found it to be 3, if new neutrinos are going to be incorporated into the model, they cannot couple to the Z boson, they cannot enjoy weak interactions, so we call them sterile. However, as sterile neutrinos have not been seen, and are not needed, our Neutrino Standard Model will contain only the three active flavours: e, μ and τ.

– the unitary mixing matrix, called the PMNS matrix, which describes the relation between flavour eigenstates and mass eigenstates, comprises three mixing angles (the so called solar mixing angle:θ_{12}, the atmospheric mixing angle θ_{23}, and the recently measured reactor mixing angle θ_{13}) , one Dirac phase (δ) and possibly two Majorana phases (α, β) and is given by

$$| \nu_\alpha \rangle = U_{\alpha i} \, | \nu_i \rangle$$

$$U_{\alpha i} = \begin{pmatrix} 1 & & \\ & c_{23} & s_{23} \\ & -s_{23} & c_{23} \end{pmatrix} \begin{pmatrix} c_{13} & & s_{13}e^{-i\delta} \\ & 1 & \\ -s_{13}e^{i\delta} & & c_{13} \end{pmatrix} \begin{pmatrix} c_{12} & s_{12} & \\ -s_{12} & c_{12} & \\ & & 1 \end{pmatrix} \begin{pmatrix} 1 & & \\ & e^{i\alpha} & \\ & & e^{i\beta} \end{pmatrix}$$

where sij = sin θ_{ij} and cij = cos θ_{ij}. Thanks to the hierarchy in mass differences (and to a less extent the smallness of the reactor mixing angle) we are allowed to identify the (23) label in the three neutrino scenario with the atmospheric δm^2_{atm} we obtained in the two neutrino scenario, identically the (12) label can be assimilated to the solar $\delta m^2_{\odot}$. The (13) sector drives the ν_e flavour oscillations at the atmospheric scale, and the depletion in reactor neutrino fluxes see [21]. Therefore,

$$\sin^2\theta_{12} = 0.31 \pm 0.03 \quad ; \quad \sin^2\theta_{23}) = 0.50 \pm 0.15 \quad ; \quad \sin^2\theta_{13} = 0.15 \pm 0.03$$

and the mass splittings are

$$| \Delta m^2_{32} | = 2.7 \pm 0.4 \times 10^{-3} \mathrm{eV}^2 \quad \text{and} \quad \Delta m^2_{21} = +8.0 \pm 0.4 \times 10^{-5} \mathrm{eV}^2.$$

These mixing angles and mass splittings are summarised in Fig. 6.

– The absolute mass scale of the neutrinos, or the mass of the lightest neutrino is not know yet, but the heaviest one must be lighter than about .5 eV.

–As transition or survival probabilities depend on the combination $U^*_{\alpha i} U_{\beta i}$ no trace of the Majorana phases could appear on oscillation phenomena, however they can have observable effects in those processes where the Majorana character of the neutrino is essential for the process to happen, like neutrino-less double beta decay.

NEUTRINO MASS AND CHARACTER

Absolute neutrino mass

The absolute mass scale of the neutrino cannot be obtained in oscillation experiments, however this does not mean we cannot have it. Direct experiments like tritium beta decay, or neutrinoless double beta decay and indirect ones, like cosmological observations, have potential to feed us the information on the absolute scale of neutrino mass, we so desperately need. The Katrin tritium beta decay experiment, [22], has sensitivity down to 200 meV for the "mass" of ν_e defined as

$$m_{\nu_e} = | U_{e1} |^2 m_1 + | U_{e2} |^2 m_2 + | U_{e3} |^2 m_3. \tag{52}$$

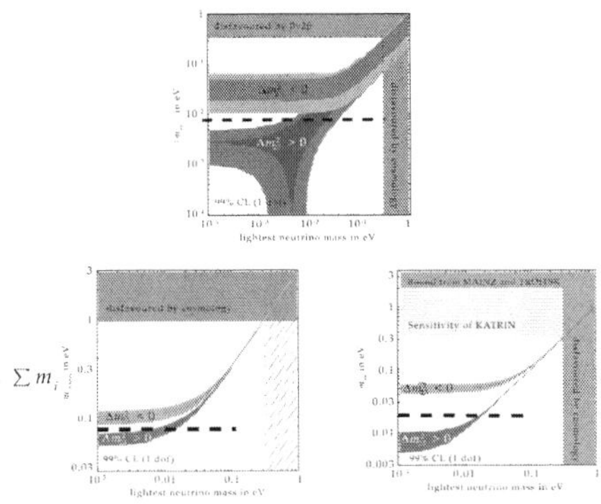

Figure 7: The effective mass measured in double β decay, in cosmology and in Tritium β decay versus the mass of the lightest neutrino. Below the dashed lines, only the normal hierarchy is allowed.

Neutrino-less double beta decay experiments, see [23] for a review, do not measure the absolute mass of the neutrino directly but a combination of neutrino masses and mixings,

$$m_{\beta\beta} = | \sum m_i U_{ei}^2 | = | m_a c_{13}^2 c_{12}^2 + m_2 c_{13}^2 s_{12}^2 e^{2i\alpha} + m_3 s_{13}^2 e^{2i\beta} |, \tag{53}$$

where it is understood that neutrinos are taken to be Majorana particles. The new generation of experiments seeks to reach below 10 meV for $m_{\beta\beta}$ in double beta decay.

Cosmological probes measure the sum of the neutrino masses

$$m_{\text{cosmo}} = \sum_i m_i. \tag{54}$$

If $\sum m_i \approx 50\,\text{eV}$, the energy balance of the universe saturates the bound coming from its critical density. The current limit, [24], is a few % of this number, ~ 1 eV. These bounds are model dependent but they do all give numbers of the same order of magnitude. However, given the systematic uncertainties characteristic of cosmology, a solid limit of less that 100 meV seems way too aggressive.

Fig. 7 shows the allowed parameter space for the neutrino masses (as a function of the absolute scale) for both the normal and inverted hierarchy.

Majorana vs Dirac

A fermion mass is nothing but a coupling between a left handed state and a right handed one. Thus, if we examine a massive fermion at rest, then one can regards this state as a linear combination of two massless particles, one right handed and one left handed particle. If the particle we are examining is electrically charged, like an electron, both particles, the left handed as well as the right handed must have the same charge (we want the mass term to be electrically neutral). This is a Dirac mass term. However, for a neutral particle, like a sterile neutrino, a new possibility opens up, the left handed particle can be coupled to the right handed anti-particle, (a term which would have a net charge, if the fields are not absolutely and totally neutral) this is a Majorana mass term. Thus a neutral particle does have two ways of getting a mass term, a la Dirac or a la Majorana, and in principle can have them both, as shown :

		CPT:		
Left Chiral	ν_L	$\Leftrightarrow$	$\bar{\nu}_R$	
	$\Updownarrow$		$\Updownarrow$	Dirac Masses
Right Chiral	ν_R	$\Leftrightarrow$	$\bar{\nu}_L$	
		Majorana Masses		

In the case of a neutrino, the left chiral field couples to SU(2) × U(1) implying that a Majorana mass term is forbidden by gauge symmetry. However, the right chiral field carries no quantum numbers, is totally and absolutely neutral. Then, the Majorana mass term is unprotected by any symmetry and it is expected to be very large, of the order of the largest scale in the theory. On the other hand, Dirac mass terms are expected to be of the order of the electroweak scale times a Yukawa coupling, giving a mass of the order of magnitude of the charged lepton or quark masses. Putting all the pieces together, the mass matrix for the neutrinos results as in Fig. 8.

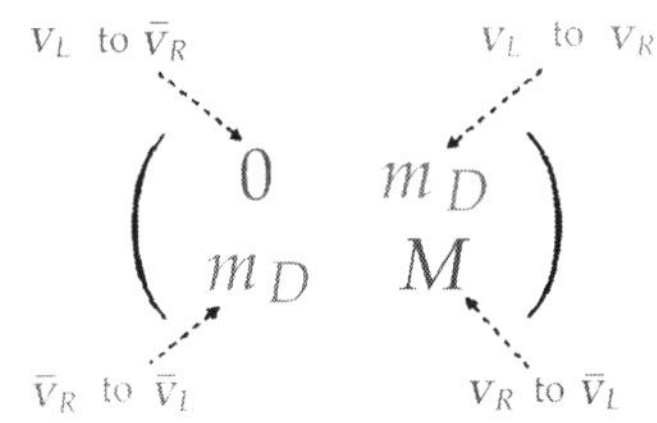

Figure 8: The neutrino mass matrix with the various right to left couplings, MD is the Dirac mass terms while 0 and M are Majorana masses for the charged and uncharged (under SU(2) × U(1)) chiral components.

To get the mass eigenstates we need to diagonalise the neutrino mass matrix. By doing so, one is left with two Majorana neutrinos, one super-heavy Majorana neutrino with mass $\simeq M$ and one light Majorana neutrino with mass m^2_D/M, i.e. one mass goes up while the other sinks, this is what we call the seesaw mechanism, [25–27]. The light neutrino(s) is(are) the one(s) observed in current experiments (its mass differences) while the heavy neutrino(s) are not accessible to current experiments and could be responsible for explaining the baryon asymmetry of the universe through the generation of a lepton asymmetry at very high energy scales since its decays can in principle be CP violating (they depend on the two Majorana phases on the PNMS matrix which are invisible for oscillations).

If neutrinos are Majorana particles lepton number is no longer a good quantum number and a plethora of new processes forbidden by lepton number conservation can take place, it is not only neutrinoless double beta decay. For example, a muon neutrino can produce a positively charged muon. However, this process and any processes of this kind, would be suppressed by $(m\nu/E)^2$ which is tiny, 10^{-20}, and therefore, although they are technically allowed, are experimentally unobservable.

CONCLUSIONS

The experimental observations of neutrino oscillations, meaning that neutrinos have mass and mix, answered questions that had endured since the establishment of the Standard Model. As those veils have disappeared, new questions open up and challenge our understanding:

- what is the nature of the neutrino? are they Majorana or Dirac? are neutrinos totally neutral?

- is the spectrum normal or inverted?
- is CP violated (is sin δ 6= 0)?
- which is the absolute mass scale of the neutrinos?
- are there new interactions?
- can neutrinos violate CPT [28]?
- are these intriguing signals in short baseline reactor neutrino experiments (the missing fluxes) a real effect? Do they imply the existence of sterile neutrinos?

We would like to answer these questions. For doing it, we are doing right now, and we plan to do new experiments. These experiments will, for sure bring some answers and clearly open new, pressing questions. Only one thing is clear. Our journey into the neutrino world is just beginning.

ACKNOWLEDGEMENTS

I would like to thank the students and the organisers of the European School on HEP for giving me the opportunity to present these lectures in such a wonderful atmosphere. I did enjoy each day of the school enormously.

REFERENCES

1. Z. Maki, M. Nakagawa and S. Sakata, Prog. Theor. Phys. 28, 870 (1962).
2. H. J. Lipkin, Phys. Lett. B 579, 355 (2004) [arXiv:hep-ph/0304187].
3. L. Stodolsky, Phys. Rev. D 58, 036006 (1998) [arXiv:hep-ph/9802387].
4. An analysis of CPT violation in the neutrino sector can be found in: G. Barenboim and J. D. Lykken, Phys. Lett. B 554 (2003) 73 [arXiv:hep-ph/0210411] ; G. Barenboim, J. F. Beacom, L. Borissov and B. Kayser, Phys. Lett. B 537 (2002) 227 [arXiv:hep-ph/0203261].
5. See for example, Neutrino mass, by B.Kayser, in D. E. Groom et al. [Particle Data Group], Eur. Phys. J. C 15, 1 (2000).
6. L. Wolfenstein, Phys. Rev. D 17, 2369 (1978).
7. S. P. Mikheev and A. Y. Smirnov, "Resonance enhancement of oscillations in matter and solar neutrino Sov. J. Nucl. Phys. 42, 913 (1985) [Yad. Fiz. 42, 1441 (1985)].
8. S. P. Mikheev and A. Y. Smirnov, "Neutrino oscillations in a variable-density medium and ν− bursts due Sov. Phys. JETP 64, 4 (1986) [Zh. Eksp. Teor. Fiz. 91, 7 (1986)] [arXiv:0706.0454 [hep-ph]].
9. S. P. Mikheev and A. Y. Smirnov, "Resonant amplification of neutrino oscillations in matter and solar Nuovo Cim. C 9, 17 (1986).
10. N. Tolich [SNO Collaboration], J. Phys. Conf. Ser. 375 (2012) 042049.
11. F. P. An et al. [DAYA-BAY Collaboration], Phys. Rev. Lett. 108 (2012) 171803 [arXiv:1203.1669 [hep-ex]].
12. Y. Ashie et al. [Super-Kamiokande Collaboration], "A measurement of atmospheric neutrino oscillation parameters by Phys. Rev. D 71, 112005 (2005) [arXiv:hep-ex/0501064]; Y. Takeuchi [SuperKamiokande Collaboration], arXiv:1112.3425 [hep-ex].
13. E. Aliu et al. [K2K Collaboration], "Evidence for muon neutrino oscillation in an accelerator-based Phys. Rev. Lett. 94, 081802 (2005) [arXiv:hep-ex/0411038].
14. M. H. Ahn et al. [K2K Collaboration], Phys. Rev. Lett. 90, 041801 (2003) [arXiv:hep-ex/0212007]; M. H. Ahn et al. [K2K Collaboration], Phys. Rev. D 74 (2006) 072003 [hep-ex/0606032].
15. D. Naples [MINOS Collaboration], J. Phys. Conf. Ser. 375 (2012) 042073.
16. K. Eguchi et al. [KamLAND Collaboration], "First results from

KamLAND: Evidence for reactor anti-neutrino Phys. Rev. Lett. 90, 021802 (2003) [arXiv:hep-ex/0212021].

17. T. Araki et al. [KamLAND Collaboration], "Measurement of neutrino oscillation with KamLAND: Evidence of spectral Phys. Rev. Lett. 94, 081801 (2005) [arXiv:hep-ex/0406035].
18. B. Aharmim et al. [SNO Collaboration], "Electron energy spectra, fluxes, and day-night asymmetries of B-8 solar Phys. Rev. C 72, 055502 (2005) [arXiv:nucl-ex/0502021].
19. S. Fukuda et al. [Super-Kamiokande Collaboration], "Determination of solar neutrino oscillation parameters using 1496 days of Phys. Lett. B 539, 179 (2002) [arXiv:hep-ex/0205075].
20. S. J. Parke and T. P. Walker, Phys. Rev. Lett. 57, 2322 (1986) [Erratum-ibid. 57, 3124 (1986)].
21. M. Apollonio et al. [CHOOZ Collaboration], Phys. Lett. B 466, 415 (1999) [arXiv:hepex/9907037].
22. A. Osipowicz et al. [KATRIN Collaboration], [arXiv:hep-ex/0109033.]
23. S. R. Elliott and P. Vogel, Ann. Rev. Nucl. Part. Sci. 52, 115 (2002) [arXiv:hep-ph/0202264].
24. D. N. Spergel et al. [WMAP Collaboration], "Wilkinson Microwave Anisotropy Probe (WMAP) three year results: Astrophys. J. Suppl. 170, 377 (2007) [arXiv:astro-ph/0603449].
25. M. Gell-Mann, P. Ramond and R. Slansky, in Supergravity, edited by P.van Nieuwenhuizen and D. Freedman, (North-Holland,1979), p.315.
26. R. N. Mohapatra and G. Senjanovic, Phys. Rev. Lett. 44, 912 (1980).
27. M. Fukugita and T. Yanagida, Phys. Lett. B 174, 45 (1986).
28. G. Barenboim, L. Borissov, J. D. Lykken and A. Y. Smirnov, JHEP 0210 (2002) 001 [hepph/0108199]. G. Barenboim and J. D. Lykken, Phys. Rev. D 80 (2009) 113008 [arXiv:0908.2993 [hep-ph]].

CHAPTER

2

Theoretical and Phenomenological Status of Neutrino Physics: A Brief Review

Surender Verma

Department of Physics and Astronomical Science, Central University of Himachal Pradesh, Dharamshala 176215, India

ABSTRACT

We present an overview of recent progress in the theoretical and phenomenological studies of neutrino masses, lepton avor mixing, and CP violation. Firstly, We discuss the status of neutrino mass with in the Standard Model (SM) of particle physics. Then the possible ways in which neutrino mass terms can be included in the SM are discussed. The inclusion of new physics beyond the SM inevitably brings new parameters which are not

Citation: Surender Verma. "Theoretical and Phenomenological Status of Neutrino Physics: A Brief Review" Hindawi Publishing Corporation Advances in High Energy Physics Volume 2015, Article ID 385968, 15 pages http://dx.doi.org/10.1155/2015/385968

constrained by the present experimental data on neutrino masses and mixing angles and, thus, are free parameters of the theory. We, also, discuss various theoretically motivated phenomenological approaches which can be used to reduce the number of free parameters and, thus, provide an excellent tool to understand the underlying physics of neutrino masses and mixings. Current experimental constraints on the neutrino mass spectrum and the lepton avor mixing parameters, including the recent observation of nonzero θ_{13}, have been summarized. Finally, We discuss the renewed interest in the possible existence of one or more sterile neutrinos and their phenomenology.

INTRODUCTION

Neutrino physics dates back to early years of the twentieth century when nuclear beta decay experiments [1, 2] discovered the continuous spectrum of beta rays which posed a serious problem for theoretical interpretation. There were two possible ways to justify the observed continuous beta ray spectrum: (i) energy conservation does not hold in the nucleus or, alternatively, (ii) a neutral particle that carries missing energy is emitted together with electrons.The second viewpoint was advocated by Pauli in a public letter to "Gruppe der Radioaktiven" at the Gauvereins Tagung in Tubingen written in December 1930 [3] as a "verzweifelter ausweg" (desperate wayout) to save the law of conservation of energy in nuclear beta decay process. According to Pauli's conjecture, the postulated particle should be electrically neutral with a mass much smaller ($0.01m_p$ at most) than the proton. The particle was supposed to have spin 1/2 and to respect the Pauli exclusion principle. A pioneering experiment by Ray Davis and his team needed 30 years to catch 2000 solar electron neutrinos in a mine in south Dakota. His count was much lower than predicted by theoretical calculations. The disappearance of neutrinos seems to be tied to a rather mind boggling behavior of these elusive particles: neutrino oscillations. As neutrinos travel through matter and space, they transform from one type into another, either appearing as electron neutrino, muon neutrino, or tau neutrino. It took almost 26 years for the experimental detection of the neutrinos and, thus, validate the Pauli neutrino hypothesis. The nuclear beta decay experiments, thus, gave us glimpses of the intricate working of nature. Since then, after decades of painstaking experimental and theoretical work, neutrinos have become an essential part of the quantum description of the fundamental particles and forces, namely, the Standard Model (SM) of particle physics.

Neutrino physics has played a crucial role in particle physics since the birth of the theory of weak interaction. The advances in the field requiring the detection of such an elusivc particle have been characterized by long time scales until the observation of Super-Kamiokande (SK) experiment in 1998 which triggered an impressive acceleration and interest in the field. Questions about neutrino oscillations and a fourth neutrino are only some of many neutrino puzzles yet to be solved. Exactly how heavy is a single neutrino? What do neutrinos tell us about the origin of mass? Are neutrinos connected to extra dimensions? Do neutrinos violate the matterantimatter symmetry or other fundamental symmetries of the Universe? Are neutrinos Dirac or Majorana type?, are some of the puzzles, yet to be answered, in neutrino physics. There are various neutrino experiments primarily aiming to confront these questions. Turning these theoretical ideas and possibilities into reality will require years of research and development to overcome technological challenges nothing new for neutrino physicists, who have seen that the need for patience when unraveling the mysteries of the Universe is their goal. At present, neutrino physics is passing through a phase of spectacular advancements on both theoretical and experimental fronts [4, 5]. Various neutrino experiments have conclusively demonstrated that neutrinos not only have mass but they also mix and the observed pattern of masses and mixings is quite different from that of the quark sector.

A BIRD EYE VIEW OF THE STANDARD MODEL OF PARTICLE PHYSICS

The Standard Model (SM) of particle physics is one of the most successful models in modern physics. It is a mathematical description of interactions of the fundamental building blocks of nature, that is, quarks and leptons. Moreover, it is a quantum description of the three out of four fundamental interactions (SM does not include gravity) mediated by the exchange of particles: the electromagnetic (EM) interaction is mediated by the exchange of photons, the strong interaction by the exchange of gluons, and weak interaction by the exchange of massive vector bosons.

In technical terms, the SM is based on the gauge group

$$G_{\mathrm{SM}} = SU(3)_C \times SU(2)_L \times U(1)_Y \tag{1}$$

with three fermion generations. Each generation consists of five different representations of this gauge group:

$$\left(1,2,-\frac{1}{2}\right),\left(3,2,\frac{1}{6}\right),(1,1,-1),\left(3,1,\frac{2}{3}\right),\left(3,1,-\frac{1}{3}\right), \tag{2}$$

where the numbers in the parenthesis represent the corresponding charges under the gauge group. In this notation, the electric charge is given by

$$Q_{EM} = T_{L3} + Y. \tag{3}$$

The field and matter content of the SM required to describe the observed elementary particle interactions are given in Tables 1 and 2, respectively. In addition, the SM contains a single Higgs boson doublet ϕ with charges (1, 2, 1/2) whose vacuum expectation value (vev)

$$\langle\phi\rangle = \begin{pmatrix} 0 \\ \frac{v}{\sqrt{2}} \end{pmatrix} \tag{4}$$

breaks the gauge symmetry of the SM

$$SU(3)_C \times SU(2)_L \times U(1)_Y \longrightarrow SU(3)_C \times U(1)_Y \,. \tag{5}$$

This is the only piece of the SM which is not confirmed experimentally. Indeed, the search for the Higgs boson remains one of the premier challenges for the present and future high energy experiments. It can be seen from Table 2 that neutrinos are fermions that have neither strong nor electromagnetic interactions; that is, they are singlets under the symmetry $S(3)_C \times U(1)_Y$. The SM has three active neutrinos accompanying the charged leptons e, μ, and τ and there are weak charged current (CC) interactions between the neutrinos and the corresponding charged leptons given by

$$-\mathcal{L}_{CC} = \frac{g}{\sqrt{2}} \sum_l \bar{\nu}_{Ll} \gamma^{\mu} l_L^{-} W_{\mu}^{+} + h.c.. \tag{6}$$

In addition, SM neutrinos have, also, neutral current (NC) interactions:

$$-\mathcal{L}_{NC} = \frac{g}{2\cos 2\theta_W} \sum_l \bar{\nu}_{Ll} \gamma^{\mu} \nu_{Ll} Z_{\mu}^{o}. \tag{7}$$

Equations (6) and (7) describe, within the SM, all the neutrino interactions. In the SM, fermion masses arise from the Yukawa interactions which couple a right-handed fermion with its left-handed doublet and the Higgs field in the following manner:

$$-\mathscr{L}_{\text{Yukawa}} = Y_{ij}^{d}\overline{Q}_{Li}\phi D_{Rj} + Y_{ij}^{u}\overline{Q}_{Li}\tilde{\phi}U_{Rj} + Y_{ij}^{l}\bar{l}_{Li}\phi E_{Rj} + h.c., \tag{8}$$

(where $\tilde{\phi} = i\tau_2\phi^*$) which after spontaneous symmetry breaking lead to charged fermion masses:

$$m_{ij}^{f} = Y_{ij}^{f}\frac{v}{\sqrt{2}}. \tag{9}$$

In the SM neutrinos are massless. This is due to the fact that, in the SM, neutrinos are described by left-handed chiral fields ν_{eL}, $\nu_{\mu L}$, and $\nu_{\tau L}$ only. Since corresponding right-handed fields ν_{eR}, $\nu_{\mu R}$, and $\nu_{\tau R}$ do not exist in the SM, a Dirac mass term

$$\mathscr{L}^{D} = \sum_{\alpha,\beta=e,\mu,\tau} \overline{\nu}_{\alpha L} M_{\alpha\beta}^{D} \nu_{\beta R} + h.c. \tag{10}$$

is precluded. An important feature of the SM relevant for neutrino mass is the fact that the SM with the gauge symmetry (equation (1)) and the particle content of Table 2 has an accidental global symmetry

$$G_{\text{SM}}^{\text{global}} = U(1)_B \times U(1)_{L_e} \times U(1)_{L_\mu} \times U(1)_{L_\tau}, \tag{11}$$

where $U(1)_B$ is the baryon number symmetry and $U(1)_{Le,L\mu,L\tau}$ are the three lepton flavor symmetries with total lepton number given by $L=L_e + L_\mu + L_\tau$. It is an accidental symmetry in the sense that we do not impose it. It is a consequence of the gauge symmetry and the representations of the physical states. In principle, neutrino mass can arise from loop corrections. In the SM, however, this cannot happen because the only possible neutrino mass term that can be constructed with the SM fields is the bilinear $l_L l_L^C$ which violates the total lepton number symmetry by two units. But as mentioned above, total lepton number is a global symmetry of the SM and therefore L-violating terms cannot be induced by loop corrections. Thus, the field content and the gauge symmetries of the SM preclude the existence of neutrino mass. This prediction of the SM is in contradiction with the experimental evidence for neutrino masses and mixings [6–12]. Therefore, it is necessary to extend the SM to accommodate nonzero neutrino masses.

Table 1: The gauge field content of the SM

	Related symmetry	Fields
Electroweak bosons	$U(1)_Y$	B_μ
	$SU(2)_L$	W_μ^i $(i = 1, 2, 3)$
Gluons	$SU(3)_C$	V_μ^j $(j = 1, \ldots, 8)$

Table 2: Particle content of the SM

L_L (1, 2, −1/2)	Q_L (3, 2, 1/6)	E_R (1, 1, −1)	U_R (3, 1, 2/3)	D_R (3, 1, −1/3)
$\begin{pmatrix} \nu_e \\ e \end{pmatrix}_L$	$\begin{pmatrix} u \\ d \end{pmatrix}_L$	e_R	u_R	d_R
$\begin{pmatrix} \nu_\mu \\ \mu \end{pmatrix}_L$	$\begin{pmatrix} c \\ s \end{pmatrix}_L$	μ_R	c_R	s_R
$\begin{pmatrix} \nu_\tau \\ \tau \end{pmatrix}_L$	$\begin{pmatrix} t \\ b \end{pmatrix}_L$	τ_R	t_R	b_R

The simplest possible extension of the SM is the introduction of the three right-handed neutrino fields ν_{eR}, $\nu_{\mu R}$, and $\nu_{\tau R}$ which are singlets under the gauge symmetry of the SM. In this way the neutrino fields become similar to the other massive fermion fields which have both left-handed and right-handed components. The Dirac mass term for the neutrino fields can be generated in a manner similar to Dirac masses for charged leptons and quarks. In addition, the Majorana mass term for the right-handed neutrinos can be written as

$$\mathcal{L}_R^M = \frac{1}{2} \sum_{\alpha,\beta=e,\mu,\tau} \overline{\nu_{\alpha L}^C} \left(M_R^M\right)_{\alpha\beta} \nu_{\beta R} + h.c. \tag{12}$$

which is invariant under the gauge symmetries of the SM and is, hence, allowed. The fundamental difference between a Majorana fermion and a Dirac fermion is that, for a Majorana fermion, the particle and antiparticle states coincide. In other words, charge conjugation does not have any effect on a Majorana fermion field. So $(\nu_{\alpha L})^C = \nu_{\alpha R}^C$ for $\alpha = e, \mu, \tau$. The introduction of right-handed neutrino fields leads to new physics, that is, Majorana neutrino masses and the existence of processes which violate lepton number by two units ($|\Delta L| = 2$). In general, in a model with left-handed and right-handed neutrino fields, the neutrino mass term is given by

$$\begin{aligned} \mathcal{L}^{D+M} &= \mathcal{L}^D + \mathcal{L}_R^M \\ &= \frac{1}{2} \begin{pmatrix} \overline{\nu}_L & \overline{\nu_L^C} \end{pmatrix} \begin{pmatrix} 0 & M^D \\ \left(M^D\right)^T & M_R^M \end{pmatrix} \begin{pmatrix} \nu_R^C \\ \nu_R \end{pmatrix} + h.c.. \end{aligned} \tag{13}$$

The Dirac mass matrix M^D is generated by the Higgs mechanism of the SM and its elements are proportional to the VEV of the Higgs doublet

$V_{SM} = (\sqrt{2}G_F)^{-1/2} = 246$ GeV, where G_F is the Fermi constant. Thus, the magnitudes of the elements of M^D are protected by the gauge symmetries of the SM. On the other hand, the Majorana mass term for the neutrino fields is invariant under the gauge symmetries of the SM; the elements of M_R^M are not protected by the SM gauge symmetries. In other words, the elements of M_R^M can have arbitrary large values. If M_R^M is generated by the Higgs mechanism at a high energy scale of new physics beyond the SM, the elements of M_R^M are expected to be of the order of such high energy scale which could be as high as the grand unification scale of about 1015 GeV. For $M_R^M \gg M^D$ the neutrino mass matrix

$$M = \begin{pmatrix} 0 & M^D \\ (M^D)^T & M_R^M \end{pmatrix} \tag{14}$$

can be diagonalized to yield the effective light Majorana neutrino mass matrix given by

$$m_{\text{eff}} = -M^D \left(M_R^M\right)^{-1} \left(M^D\right)^T \tag{15}$$

and a heavy 3×3 mass matrix

$$M_{\text{heavy}} \approx M_R^M. \tag{16}$$

Therefore, there are three heavy neutrinos decoupled from the low energy physics and three light neutrinos whose masses are suppressed with respect to the elements of Dirac mass matrix M^D by the small matrix factor $(M_R^M)^{-1}(M^D)^T$ which is known as seesaw mechanism [13–16].

THE GENERAL STRUCTURE OF LEPTON MIXING MATRIX

In any gauge theory, in order to identify physical particles, one must diagonalize all relevant mass matrices which typically result from gauge symmetry breaking. Mechanisms giving mass to neutrinos generally imply the need for new interactions whose Yukawa couplings will coexist with that of the charged leptons. The lepton mixing matrix, U, follows from a mismatch between the diagonalization of the charged lepton mass matrix and that of the neutrino mass matrix. This is similar to the way the Cabibbo-Kobayashi-Maskawa (CKM) matrix arises in the quark sector from a mismatch between up- and down-type Yukawa couplings. Hence, like quarks, massive neutrinos will generally mix. The means that the flavor states ν_α, $\alpha = e, \mu, \tau$ do not coincide with the mass eigenstates ν_i, $i = 1, 2, 3$. The flavor states are combinations of the mass eigenstates

$$\nu_\alpha = U_{\alpha i}\nu_i, \tag{17}$$

where the mixing parameters $U_{\alpha i}$ form the PMNS mixing matrix, U_{PMNS}. In the mass basis, leptonic charged current interaction is given by

$$-\mathscr{L}_{\text{CC}} = \frac{g}{\sqrt{2}}\left(\bar{e}_L \;\; \bar{\mu}_L \;\; \bar{\tau}_L\right)\gamma^\mu U \begin{pmatrix} \nu_1 \\ \nu_2 \\ \nu_3 \end{pmatrix} W_\mu^- + h.c., \tag{18}$$

where (ν_1, ν_2, ν_3) and (e, μ, τ) are neutrino mass eigenstates and charged lepton mass eigenstates, respectively. U is a 3×3 unitary lepton mixing matrix with $UU^\dagger = 1$.The mass term is a sum of Lorentz invariant products of left-handed and righthanded components of the fields. The charge conjugate fields $\nu_L^C = C\overline{\nu}_L^{\,T}$ and $\nu_R^C = C\overline{\nu}_R^{\,T}$ are right-handed and left-handed, respectively. Here, C is the unitary charge conjugation matrix having $C\gamma_\mu^T C^{-1} = -\gamma_\mu$, $C^T = -C$. . Using these we can find that $\overline{\nu_L^C} = -\nu_L^T C^{-1}$ and $\overline{\nu_R^C} = -\nu_R^T C^{-1}$. Taking into account that ν_L^C is the right-handed component one can easily write the mass term for neutrinos. In the basis where both charged lepton and the neutrino mass matrices are nondiagonal, the mass term for the leptons can be written as

$$\mathscr{L}_m = \left[\nu_L^T C^{-1} M_\nu \nu_L + \overline{l_L} M_l l_R + h.c.\right]. \tag{19}$$

The charged lepton mass matrix "M_l" can be diagonalized by the matrices V_l and V_R^l:

$$V^{l\dagger} M_l V_R^l = \begin{pmatrix} m_e & 0 & 0 \\ 0 & m_\mu & 0 \\ 0 & 0 & m_\tau \end{pmatrix}. \tag{20}$$

The complex symmetric Majorana neutrino mass matrix M^ν can be diagonalized by an orthogonal matrix V^ν :

$$V^{\nu}M_{\nu}V^{\nu T} = \begin{pmatrix} m_{\nu_e} & 0 & 0 \\ 0 & m_{\nu_\mu} & 0 \\ 0 & 0 & m_{\nu_\tau} \end{pmatrix}. \quad (21)$$

The lepton mixing matrix U_{PMNS} is given by

$$(U_{\text{PMNS}})_{ij} = V_{ik}^{l\dagger} V_{kj}^{\nu}. \quad (22)$$

For Majorana (Dirac) neutrinos contains (for generations) a total of [] real parameters out of which are angles and [] can be interpreted as physical phases. In particular, if there are only three Majorana neutrinos, is a unitary matrix analogous to the CKM matrix for quarks [17] and depends upon six real parameters which include three mixing angles and three phases. It can be conveniently parameterized as

$$U_{\text{PMNS}} = \begin{pmatrix} c_{12}c_{13} & s_{12}c_{13} & s_{13}e^{-i\delta} \\ -s_{12}c_{23} - c_{12}s_{13}s_{23}e^{i\delta} & c_{12}c_{23} - s_{12}s_{13}s_{23}e^{i\delta} & c_{13}s_{23} \\ s_{12}s_{23} - c_{12}s_{13}c_{23}e^{i\delta} & -c_{12}s_{23} - s_{12}s_{13}c_{23}e^{i\delta} & c_{13}c_{23} \end{pmatrix} \cdot \begin{pmatrix} 1 & 0 & 0 \\ 0 & e^{i\alpha} & 0 \\ 0 & 0 & e^{i(\beta+\delta)} \end{pmatrix}, \quad (23)$$

where $c_{ij} \equiv \cos\theta_{ij}$ and $s_{ij} = \sin\theta_{ij}$. The phases δ and (α, β) are Dirac-type and Majorana-type CP violating phases, respectively.

NEUTRINO OSCILLATIONS

Neutrino oscillations are a manifestation of leptonic mixing. The first hint that neutrinos have mass came from the pioneering Homestake experiment [18] performed by the American scientist Raymond Davis Jr. for the detection of solar neutrinos. In this experiment, it was found that only about one-third of the number of neutrinos predicted by standard solar model (SSM) were reaching the detector on the earth. This result puzzled both the solar and neutrino physicists. However, Mikheyev and Smirnov [19] put forward the idea, proposed previously by Wolfenstein [20], that the solar neutrinos might be changing into something else. Only electron neutrinos are emitted by the Sun and they could be converting into muon and tau neutrinos which were not being detected by the experiment. This effect of neutrino interconversion over time from one kind to another, namely, "neutrino oscillations," was first proposed by Gribov and Pontecorvo [21]. The precise mechanism for "solar neutrino oscillations" proposed by Mikheyev, Smirnov, and Wolfenstein (MSW) involved the resonant enhancement of neutrino oscillations due to matter effects. Just as light passing through the matter slows down, which is equivalent to the photon gaining a small effective mass, neutrinos passing through matter also get slowed down and gain a small effective mass. The effective neutrino mass is largest when matter density is highest which in the case of solar neutrinos is in the core of the Sun. In particular, electron neutrinos generated in the core of the Sun will be subject to such matter effects. This can result in a resonant enhancement of solar neutrino oscillations which is known as the MSW effect [20].

The idea of neutrino oscillations gained support from the Japanese experiment Super-Kamiokande [22, 23] in which similar deficit was observed in the atmospheric neutrino flux. The atmospheric neutrinos are produced when cosmic rays, composed mainly of protons, interact with the upper atmosphere. Most of the neutrinos pass through the earth unhindered; Super-Kamiokande was able to detect muon neutrinos coming from above as well as below and found that while the expected number of muons was coming from above only about a half of the expected number came from below. These results were attributed to the muon neutrinos coming from below having oscillated into tau neutrinos over an oscillation length L equal to the diameter of the earth. However, a number of muon neutrinos coming from above have negligible oscillation length and have no time to oscillate yielding the expected number of muon neutrinos from above. More recently, the Sudbury Neutrino Observatory (SNO) [24, 25] in Canada has confirmed the solar neutrino oscillations. There are various other neutrino experiments which not only have confirmed the neutrino oscillation phenomenon but also have measured the neutrino oscillation parameters with increased accuracies [26]. In fact, through neutrino oscillations, quantum mechanics allowed us to probe the smallest masses measured so far. More importantly, the results of neutrino experiments are, thus, providing the motivation for probing physics beyond the SM.

Neutrino Oscillations in Vacuum

Neutrino oscillations are quantum mechanical phenomenon in which a neutrino born with flavor α changes to a different flavor β while propagating in vacuum or in matter. Observation of neutrino oscillations in various neutrino experiments has shown that there is a mismatch between the flavor and mass eigenstates of neutrinos. Neutrinos are produced and detected via weak interactions. The neutrino state created in the decay $W^+ \to l_\alpha^+ + \nu$ is given by

$$|\nu_\alpha\rangle = \sum_{i=1}^{n} U_{\alpha i}^* |\nu_i\rangle . \quad (24)$$

This superposition of neutrino mass eigenstates produced in association with the charged lepton of flavor α is the state we refer to as the neutrino of flavor α. It is evident from (24) that the probability of finding a neutrino created in a given flavor to be in the same state (or any other state) oscillates with time. The time evolution of a neutrino produced in a given flavor at $x=t=0$ is given by

$$|\nu_\alpha(t)\rangle = \sum_{i=1}^{n} U_{\alpha i}^* |\nu_i(t)\rangle = \sum_{i=1}^{n} U_{\alpha i}^* e^{-iE_i t} |\nu_i(0)\rangle . \quad (25)$$

The neutrino oscillation probability that is, the probability of transformation from one flavor eigenstate $|\nu_\alpha\rangle$ to another flavor eigenstate $|\nu_\beta\rangle$ is given by

$$P(\nu_\alpha \longrightarrow \nu_\beta) = P_{\alpha\beta} = |\langle \nu_\beta \mid \nu_\alpha(t)\rangle|^2$$

$$= \left| \sum_{i=1}^{n}\sum_{j=1}^{n} U_{\alpha i}^* U_{\beta j} |\nu_j\rangle\, \nu_i(t) \right|^2 . \quad (26)$$

For ultrarelativistic neutrinos, one can assume $p_i \simeq p_j \equiv p = E$ so that

$$E_i = \sqrt{p_i^2 + m_i^2} \simeq p + \frac{m_i^2}{2E}, \quad (27)$$

where E_i and m_i are the energy and mass of the neutrino mass eigenstate ν_i. Using (27) along with the orthogonality relation $\langle \nu_j \mid \nu_i\rangle = \delta_{ij}$ we obtain the transition probability $P_{\alpha\beta}$ given by

$$P_{\alpha\beta} = \delta_{\alpha\beta} - 4\sum_{i>j} \Re\left(U_{\alpha i}^* U_{\beta i} U_{\alpha j} U_{\beta j}^*\right) \sin^2\left[1.27\Delta m_{ij}^2 \left(\frac{L}{E}\right)\right]$$

$$+ 2\sum_{i>j} \Im\left(U_{\alpha i}^* U_{\beta i} U_{\alpha j} U_{\beta j}^*\right) \sin\left[2.54\Delta m_{ij}^2 \left(\frac{L}{E}\right)\right], \quad (28)$$

where $\Delta m_{ij}^2 \equiv m_i^2 - m_j^2$ is in eV^2, L is in km, and E is in GeV. Thus, the essential ingredients for neutrino oscillations to occur are twofold: one is that neutrinos must not only be massive but also nondegenerate in mass and the second is that mass eigenstates of the neutrinos must be different from the flavor eigenstates. Assuming that *CPT* invariance holds

$$P\left(\bar{\nu}_\alpha \longrightarrow \bar{\nu}_\beta\right) = P\left(\nu_\beta \longrightarrow \nu_\alpha\right). \tag{29}$$

But from (28), we see that

$$P\left(\nu_\beta \longrightarrow \nu_\alpha; U\right) = P\left(\nu_\alpha \longrightarrow \nu_\beta; U^*\right). \tag{30}$$

Thus when CPT holds,

$$P\left(\bar{\nu}_\alpha \longrightarrow \bar{\nu}_\beta; U\right) = P\left(\nu_\alpha \longrightarrow \nu_\beta; U^*\right); \tag{31}$$

that is, the probability for oscillation of an antineutrino is the same as that for a neutrino, but the mixing matrix U is replaced by its complex conjugate. Thus, if U is not real, the neutrino and antineutrino oscillation probabilities can differ by having opposite sign in the last term in (28). When *CPT* holds any difference in these probabilities, this will indicate violation of *CP* invariance. For the case of two generations of neutrinos, the above analysis becomes quite simplified without any loss of physical understanding. The mixing matrix depends only on one parameter θ (known as mixing angle) and is given by

$$U = \begin{pmatrix} \cos\theta & \sin\theta \\ -\sin\theta & \cos\theta \end{pmatrix} \tag{32}$$

and there is a single mass-squared difference Δm^2 . With this form of U (28) becomes

$$P_{\alpha\beta} = \sin^2 2\theta \sin^2\left(1.27\Delta m^2 \frac{L}{E}\right). \tag{33}$$

Neutrino Oscillations in Matter

The propagation of neutrinos becomes significantly modified in the presence of matter due to their interactions with the matter composed of electrons, protons, and neutrons. Moreover, the coherent forward elastic scattering amplitudes are not the same for all neutrino flavors, that is, ν_e, ν_μ, and ν_τ. The electron neutrinos ν_e have additional contribution due to their charged current (CC) interactions with matter which are mediated by $W\pm$ bosons. On the basis of this fact, Mikheyev and Smirnov [19], following the work of Wolfenstein [20], showed that an exciting phenomenon occurs when neutrinos travel through dense matter. In the two-flavor case, if neutrinos v_e and v_μ travel through the Sun, the propagation of v_1 and v_2 will be modified due to the different interactions of ve and v_μ with electrons. The mixing angle in the presence of matter is given by

$$\sin^2 2\theta_m = \frac{\sin^2 2\theta}{(\cos 2\theta - L/L_m)^2 + \sin^2 2\theta}, \quad (34)$$

where L_m, a length characteristic of motion in matter of density N_e, is given by

$$L_m = \frac{2\pi}{\sqrt{2}G_F N_e}, \quad (35)$$

where G_F is the Fermi constant. Mikheyev and Smirnov [19] pointed out that(34) has a resonant character. If the neutrinos travel through a varying matter density, one may have

$$\frac{L}{L_m} = \cos 2\theta \quad (36)$$

so that

$$\sin^2 2\theta_m = 1, \quad (37)$$

resulting in maximal mixing and the survival probability [27], to a very good approximation, can be written as

$$P_{\alpha\alpha} = \frac{1}{2} + \frac{1}{2}\cos 2\theta \cos 2\theta_m. \quad \textbf{(38)}$$

RECENT DEVELOPMENTS IN NEUTRINO PHENOMENOLOGY AND THE CURRENT STATUS OF NEUTRINO MASSES AND MIXINGS

Spectacular advancements on both the theoretical and the experimental fronts have made possible the measurements of neutrino oscillation parameters. Recent major progress in neutrino physics, in particular and particle physics, in general, is related to studies of the neutrino masses and mixings. The first phase of these studies is essentially over and the main results from this phase are summarized in the following:(1)discovery of nonzero neutrino mass;(2)determination of the dominant structure of the lepton mixing, that is, discovery of two large (solar mixing angle θ_{12} and atmospheric mixing angle θ_{23}) and one small (reactor) mixing angles;(3)establishing strong differences in the quark and lepton mass spectra and their mixing patterns.

In a sense, we have now a "standard model of neutrinos" that can be formulated in the following way:(1)there are only three types of light neutrinos;(2)their interactions are described by the standard electroweak theory;(3)masses and mixings are generated in vacuum and they originate from some high energy (short range) physics at the electroweak and/or higher scales.

However, there exist many open questions about massive neutrinos and lepton flavor mixing. The following are examples.(1)Are massive neutrinos Dirac or Majorana particles? If massive neutrinos are Dirac particles, they can be distinguished from their antiparticles. By definition, a Majorana neutrino is identical to its antiparticle.

It is possible to identify the Majorana nature of massive neutrinos through the observation of the neutrinoless double beta decay of some even-even nuclei in which the total lepton number is not conserved.(2) Why are neutrino masses so tiny? The fact that the masses of neutrinos are considerably smaller than the masses of charged leptons or quarks is a big puzzle in particle physics.

Although, a lot of theoretical models about neutrino masses have been proposed at either low or high energy scales, none of them has proved to be very successful and conceivable.(3)What is the absolute scale of neutrino masses?

It is very important to know the absolute values of three neutrino masses because they are fundamental parameters of flavor physics. The mass scale of neutrinos is likely to indicate the energy scale of new physics responsible for the generation of neutrino masses and lepton flavor mixing(4) Why are the mixing angles $\theta 12$ and θ_{23} so large while θ_{13} is, relatively, small? The bilarge neutrino mixing pattern is, also, a mystery to theorists because it is "anomalously" different from the familiar trismall quark mixing pattern. (5) Is there leptonic CP violation? A necessary condition for the existence of CP and T violation in normal neutrino oscillations is $\theta_{13} = \not{0}$. As CP violation has been discovered in the quark sector there is no reason why CP should be conserved in the lepton sector. (6) Can the leptonic CP violating phases be determined? If lepton flavor mixing is correlated with CP violation, one has to determine the relevant CP violating phases through various possible experiments. The neutrinoless double beta decay and long-baseline appearance neutrino oscillations are expected to be sensitive to the Majorana and Dirac-type *CP* violating phases, respectively.

The immediate goal is to search for new physics beyond the Standard Model. The next phase of studies will be associated with new generation of neutrino experiments.

The main objectives of this new phase include the determination of the absolute scale of neutrino mass and subdominant structures of mixing matrix, namely, deviation of the θ_{23} from maximal value and the *CP* violation phase(s).

The objectives, also, include identification of neutrino mass hierarchy and precision measurements of already known parameters. For the case of three generations, there are four mixing parameters, namely, the three mixing angles and the Dirac-type phase which will manifest in the neutrino oscillation experiments. The neutrino oscillation experiments are sensitive to only two mass squared differences, all the three mixing angles, and the Dirac-type *CP* phase. We, already, have fairly good knowledge of the two mass squared differences and all the three mixing angles. Results from solar neutrino experiments [28–32] which have been collecting data for more than four decades have now culminated in choosing the large mixing angle (LMA) solution as a solution of the solar neutrino problem.

The latest addition to this huge repertoire of experimental data are the results from the Borexino experiment [33] which are consistent with the LMA solution. This conclusion from the solar neutrino experiments has been investigated independently by the KamLAND reactor antineutrino experiment [26, 34–36] and a combined analysis of the solar and KamLAND data gives $\Delta m^2_{21} = 7.6 \times 10^{-5}$ eV2 and $\sin^2 2\theta_{12} = 0.32$ [12, 36–41]. The other masssquared difference Δm^2_{32} and mixing angle θ_{23} are now pretty well determined by the zenith angle dependent atmospheric ν_μ data from Super-Kamiokande [22, 23, 32, 42] and the long baseline experiments K2K [43, 44] and MINOS [45]. The combined data from the atmospheric and long baseline experiments have pinned down $|\Delta m^2_{32}| = 2.4 \times 10^{-3}$ eV2 and $\sin^2 2\theta_{12}$ = 1. With the observation of nonzero θ_{13} [46–53], we now have complete knowledge of all the three mixing angles θ_{12}, θ_{23}, and θ_{13}. However, we do not have any information on Dirac-type *CP* violating phase δ and it is fully unconstrained as yet. In Table 3, we have shown the present status of the neutrino oscillation parameters [54].

Table 3: Global analysis of solar, atmospheric, reactor, and accelerator neutrino data in the framework of three-neutrino oscillations. Here, $\Delta m_{3l}^2 \equiv \Delta m_{31}^2$ for NH and $\Delta m_{3l}^2 \equiv \Delta m_{32}^2$ for IH

Parameter	Normal hierarchy (NH)		Inverted hierarchy (IH)	
	Best fit $\pm 1\sigma$	3σ range	Best fit $\pm 1\sigma$	3σ range
$\sin^2\theta_{12}$	$0.304^{+0.013}_{-0.012}$	0.270–0.344	$0.304^{+0.013}_{-0.012}$	0.270–0.344
$\sin^2\theta_{23}$	$0.452^{+0.052}_{-0.028}$	0.382–0.643	$0.579^{+0.025}_{-0.037}$	0.389–0.644
$\sin^2\theta_{13}$	$0.0218^{+0.0010}_{-0.0010}$	0.0186–0.0250	$0.0218^{+0.0010}_{-0.0010}$	0.0188–0.0251
$\frac{\Delta m_{21}^2}{10^{-5}eV^2}$	$7.50^{+0.19}_{-0.17}$	7.02–8.09	$7.50^{+0.19}_{-0.17}$	7.02–8.09
$\frac{\Delta m_{3l}^2}{10^{-3}eV^2}$	$+2.457^{+0.047}_{-0.047}$	+2.317–+2.607	$-2.449^{+0.048}_{-0.047}$	−2.590–−2.307

NEUTRINO PHYSICS AND CP VIOLATION

Symmetries play an important role in physics since they describe the invariance of physical phenomena under fundamental transformations. The symmetries P, C, and T correspond to the invariance of the Lagrangian of the field theory under these discrete operations. Parity (P) and time reversal (T) are space-time symmetries. The parity operation reverses all spatial coordinates ($r \to -r$) and, in consequence, the directions of all momentum vectors are reversed while leaving spins and angular momenta unchanged. The time reversal operation T corresponds to the replacement of T by $-T$ so that the directions of momenta and spins are reversed. All complex numbers such as coupling constants are replaced by their complex conjugates under the T operation. The operation of charge conjugation "C" is to interchange the role of the annihilation and creation operators of the particle with those of the antiparticle. The C operation flips the signs of the internal charges such as baryon number, electric charge, strangeness, and lepton number.

After observation of parity violation in weak interactions in 1956, the discovery of CP violation in the decay of neutral kaons in 1964 was the major breakthrough in our understanding of the real world. Since then considerable experimental and theoretical work has been devoted to understanding its occurrence and origin. In the SM, the possibility of CP violation arises from the presence, in the Lagrangian, of complex Yukawa couplings between fermions and the Higgs boson. In the quark sector, weak couplings are described by the Cabibbo-Kobayashi-Maskawa (CKM) mixing matrix [17, 55] given by

$$V_{\mathrm{CKM}} = \begin{pmatrix} c_{12}^q c_{13}^q & s_{12}^q c_{13}^q & s_{13}^q e^{-i\delta^q} \\ -s_{12}^q c_{23}^q - c_{12}^q s_{23}^q s_{13}^q e^{i\delta^q} & c_{12}^q c_{23}^q - s_{12}^q s_{23}^q s_{13}^q e^{i\delta^q} & s_{23}^q c_{13}^q \\ s_{12}^q s_{23}^q - c_{12}^q c_{23}^q s_{13}^q e^{i\delta^q} & -c_{12}^q s_{23}^q - s_{12}^q c_{23}^q s_{13}^q e^{i\delta^q} & c_{23}^q c_{13}^q \end{pmatrix}, \quad (39)$$

where $c_{ij}^{q} = \cos\theta_{ij}^{q}, s_{ij}^{q} = \sin\theta_{ij}^{q}, \theta_{ij}^{q}$ are quark mixing angles and δ^{q} is Dirac-type CP violating phase.

Neutrinos were introduced into the SM as massless fermions for which no gauge invariant renormalizable mass term can be constructed. Therefore, there is neither mixing nor CP violation in the leptonic sector and δ^{q} is the only source of CP violation in the SM. Moreover, the amount of CP violation predicted by the SM is far too small to yield the observed ratio of baryons to photons in the Universe [56–58]. The experimental evidence for neutrino masses and mixings provided an unambiguous signal of new physics where new

$$L^{\text{phys}} = \nu_L^T C d \nu_L + \bar{l}_L d_l l_R + \frac{g}{\sqrt{2}} \bar{l}_{iL} \gamma^{\mu} \left(U_{ij}\right)_{\text{PMNS}} \nu_{jL} W_{\mu}^{-}$$

$$+ h.c., \qquad (40)$$

where d and dl are diagonal, real, and positive matrices. The 3×3 unitary matrix U_{PMNS} is, in general, parameterized by six phases and three mixing angles. Three of these phases can be factored out and rotated away through the redefinition of the charged leptons l_L. These phases, thus, appearing in dl can be eliminated by the simultaneous redefinition of fields l_R. Another two of the six phases of U can be factored out to the right; however, in this case these two phases are physical because rotating them away from U corresponds to transferring them to the mass term of the light neutrinos. This is an important difference from the quark sector resulting from the fact that in the seesaw framework neutrinos have Majorana masses and are Majorana particles unlike quarks. These factorizable phases which cannot be removed from the theory are called Majorana phases. As a result, in the seesaw framework, there are additional sources of CP violation beyond the Kobayashi Maskawa mechanism of the quark sector. In the leptonic sector, there is one Dirac-type CP violating phase δ and two Majorana-type CP violating phases α, β. The PMNS matrix can be parameterized as

$$U_{\text{PMNS}} = \begin{pmatrix} c_{12}c_{13} & s_{12}c_{13} & s_{13}e^{-i\delta} \\ -s_{12}c_{23} - c_{12}s_{13}s_{23}e^{i\delta} & c_{12}c_{23} - s_{12}s_{13}s_{23}e^{i\delta} & c_{13}s_{23} \\ s_{12}s_{23} - c_{12}s_{13}c_{23}e^{i\delta} & -c_{12}s_{23} - s_{12}s_{13}c_{23}e^{i\delta} & c_{13}c_{23} \end{pmatrix} \cdot \begin{pmatrix} 1 & 0 & 0 \\ 0 & e^{i\alpha} & 0 \\ 0 & 0 & e^{i(\beta+\delta)} \end{pmatrix}, \quad (41)$$

where $c_{ij} = \cos\theta_{ij}$, $s_{ij} = \sin\theta_{ij}$, θ_{ij} are neutrino mixing angles. Leptonic *CP* violation at low energies can be observed through neutrino oscillations which are sensitive to the Dirac-type *CP* phase δ but insensitive to the Majorana-type *CP* violating phases. The Dirac-type *CP* violation can be related to the low energy weak basis (WB) invariant [59]

$$I_1 = \text{Tr}\left[H_\nu, H_l\right]^3, \quad (42)$$

where $H_\nu = M_\nu M_\nu^\dagger$ and $H_l = M_l M_l^\dagger$ with M_ν and M_l being neutrino and charged lepton mass matrices, respectively. In the basis where charged lepton mass matrix is diagonal I_1 can be written as

$$I_1 = -6i\left(m_\mu^2 - m_e^2\right)\left(m_\tau^2 - m_\mu^2\right)\left(m_\tau^2 - m_e^2\right) \cdot \Im\left[(H_\nu)_{12}(H_\nu)_{23}(H_\nu)_{31}\right]. \quad (43)$$

In terms of neutrino masses and mixings, I_1 is given by

$$I_1 = 6i\left(m_\mu^2 - m_e^2\right)\left(m_\tau^2 - m_\mu^2\right)\left(m_\tau^2 - m_e^2\right)\left(m_3^2 - m_1^2\right) \cdot \left(m_2^2 - m_1^2\right)\left(m_3^2 - m_2^2\right)J_{CP}, \quad (44)$$

where $J_{CP} = \Im[U_{11}U_{22}U_{12}^*U_{21}^*]$ determines the magnitude of *CP* violation in neutrino oscillations. Using (42)

$$J_{CP} = \frac{1}{8}\sin 2\theta_{12}\sin 2\theta_{13}\sin 2\theta_{23}\cos\theta_{13}\sin\delta. \quad (45)$$

Thus, a nonzero value of θ_{13} is an encouraging sign for the prospects of measuring low energy leptonic *CP* violation mediated through a Dirac-type *CP* violating phase. In the quark sector, *JCP* is of the order of 10_{-5} [60] whereas in the leptonic sector, with θ_{13} equal to the present experimental value, J_{CP} can be of the order of 10^{-2} which is about three orders of magnitude larger than that in the quark sector. Apart from the WB invariant I_1 there are two more WB invariants I_2, I_3 [61] related to the Majorana nature of massive neutrinos which are sensitive to the Majorana-type CP violating phases α, β and are given by

$$I_2 = \Im \operatorname{Tr}\left[H_l M_\nu M_\nu^* M_\nu H_l^* M_\nu^*\right],$$

$$I_3 = \Im \operatorname{Det}\left[M_\nu^* H_l M_\nu, H_l^*\right]. \qquad (46)$$

A necessary and sufficient condition for low energy CP invariance in the leptonic sector is that these three WB invariants should be identically zero [62, 63].

PHENOMENOLOGY OF NEUTRINO MASS MATRIX

In view of the recent experimental and theoretical developments in neutrino physics, it is quite clear that there exists physics beyond the SM. Almost every extension of the SM has additional sources of CP violation. In addition, there is a great puzzle in cosmology known as the baryon symmetry of the Universe (BAU) [64]. CP violation has been discussed phenomenologically since the sixties. Gauge theories, naturally, account for its presence but leave its ultimate origin unanswered. Understanding the origin of CP violation from first principles constitutes one of the central challenges of theoretical elementary particle physics. With the historic discovery of neutrino oscillations, the issue of leptonic CP violation occupied the centre stage of the particle and nuclear physics. The existence of CP violation in the lepton sector is expected in gauge theories of neutrino mass. The main difference with respect to CP violation in the quark sector is the appearance of new phases associated with the Majorana nature of neutrinos and/or with the admixture of $SU(3) \times SU(2) \times U(1)$ singlet leptons in the charged-current weak interaction.The latter also leads to the effective violation of unitarity in the lepton mixing matrix describing neutrino oscillations. Irrespective of the underlying origin of neutrino mass, Majorana phases are a generic feature of gauge theories that account for the smallness of neutrino mass through the feebleness of lepton number violation.

Within the SM of particle physics, Yukawa couplings are free parameters which must span several orders of magnitude to accommodate hierarchical spectrum of fermion masses and mixings. The fermion masses and the mixings are derived from these Yukawa couplings. Also, radiative corrections can obscure the underlying structure. In the absence of complete knowledge of parameters appearing in the fermion mass matrices it is impossible, without some additional assumptions, to determine all the elements of the Yukawa coupling matrices for quarks and leptons. Therefore, without a significant breakthrough in the theoretical understanding of the fermion flavors, the phenomenological approaches are bound to play a crucial role in interpreting new experimental data on quark and lepton mixings. These approaches are expected to provide useful hints towards unraveling the dynamics of fermion mass generation, CP violation, and identification of possible underlying symmetries of the lepton flavors from which realistic models of lepton mass generation and flavor mixing could, hopefully, be constructed. Even though the grand unification on its own does not shed any light on the flavor problem, the grand unified theories (GUTs) provide the optimal framework in which possible solutions to the flavor problem could be embedded. This is because the GUTs predict definite group theoretical relations between the fermion mass matrices. For this purpose, it is useful to find out possible leading order forms of the neutrino mass matrix. Such forms of neutrino mass matrix provide useful hints for model building which will eventually shed important light on the dynamics of lepton mass generation and flavor mixing. The first step in this direction is the reconstruction of the neutrino mass matrix in the flavor basis. However, the reconstruction results in a large variety of possible structures of mass matrices depending strongly on the mass scale, mass hierarchy, and the Majorana phases. There are several classes of models/Ansätze for Yukawa couplings. An incomplete list is as follows:(1)approximate flavor symmetries [65–68] in which the entries in the Yukawa matrices are small parameters by which the flavor symmetries are broken;(2)Fritzsch Ansätze and/or GUT inspired models [69–74] in which some entries in the Yukawa mass matrices are assumed to be zero (e.g., by discrete flavor symmetry) and others may be related by some GUT relation;(3)flavor democracy models [75, 76] in which all the entries in the Yukawa matrices are equal (i.e., no flavor symmetry) and hierarchy comes from diagonalization and renormalization group equation (RGE) running;(4)string inspired models, composite models, and so forth.

In the context of type-I seesaw mechanism with three generations of right-handed neutrinos, which are singlets of $S(2)$, after spontaneous symmetry breaking, the leptonic mass terms are given by

$$\mathscr{L} = -\left[\overline{\nu_L^0} M_D \nu_R^0 + \nu_R^{0T} C M_R \nu_R^0 + \overline{l_L^0} M_l l_R^0\right] + h.c., \tag{47}$$

where M_R, M_D, and M_l are right-handed Majorana neutrino, Dirac neutrino, and charged lepton mass matrices, respectively. The complex symmetric neutrino mass matrix, M_v, is given by type I seesaw mechanism (equation (15)) [13– 16]. All the information about lepton masses and mixings is encoded in the hermitian charged lepton mass matrix Ml and the complex symmetric neutrino mass matrix M_v.

In the basis where charged lepton mass matrix "M_l" is diagonal, the neutrino mass matrix "M_v" is diagonalized by a complex unitary matrix V_v

$$M_\nu = V_\nu M_\nu^{\text{diag}} V_\nu^T, \qquad (48)$$

Where

$$M_\nu^{\text{diag}} = \begin{pmatrix} m_1 & 0 & 0 \\ 0 & m_2 & 0 \\ 0 & 0 & m_3 \end{pmatrix} \qquad (49)$$

is the diagonal neutrino mass matrix. In the standard, Particle Data Group (PDG), parameterization of the neutrino mixing matrix, V_v, is given by

$$V_\nu \equiv UP = \begin{pmatrix} c_{12}c_{13} & s_{12}c_{13} & s_{13}e^{-i\delta} \\ -s_{12}c_{23} - c_{12}s_{13}s_{23}e^{i\delta} & c_{12}c_{23} - s_{12}s_{13}s_{23}e^{i\delta} & c_{13}s_{23} \\ s_{12}s_{23} - c_{12}s_{13}c_{23}e^{i\delta} & -c_{12}s_{23} - s_{12}s_{13}c_{23}e^{i\delta} & c_{13}c_{23} \end{pmatrix} \cdot \begin{pmatrix} 1 & 0 & 0 \\ 0 & e^{i\alpha} & 0 \\ 0 & 0 & e^{i(\beta+\delta)} \end{pmatrix}, \qquad (50)$$

where $s_{ij} = \sin\theta_{ij}$, $c_{ij} = \cos\theta_{ij}$, and P is the Majorana phase matrix. In addition to the Majorana phases α, β, the phase matrix contains Dirac-type phase δ just to make (1, 1) element of M_v (mee) independent of δ. The neutrino mass matrix "M_v" can be written as

$$M_\nu = UPM_\nu^{\text{diag}} P^T U^T. \qquad (51)$$

In terms of the neutrino masses, mixing angles and CP violating phases, the neutrino mass matrix "M_v" is given by

$$M_\nu = \begin{pmatrix} m_{ee} & m_{e\mu} & m_{e\tau} \\ m_{e\mu} & m_{\mu\mu} & m_{\mu\tau} \\ m_{e\tau} & m_{\mu\tau} & m_{\tau\tau} \end{pmatrix}, \quad (52)$$

where

$$m_{ee} = c_{12}^2 c_{13}^2 m_1 + c_{13}^2 e^{2i\alpha} m_2 s_{12}^2 + e^{2i\beta} m_3 s_{13}^2,$$

$$m_{e\mu} = c_{13}\left(e^{i(2\beta+\delta)} m_3 s_{13} s_{23} - c_{12} m_1 \left(c_{23} s_{12} + c_{12} e^{i\delta} s_{13} s_{23}\right) + e^{2i\alpha} m_2 s_{12}\left(c_{12} c_{23} - e^{i\delta} s_{12} s_{13} s_{23}\right)\right),$$

$$m_{e\tau} = c_{13}\left(c_{23} e^{i(2\beta+\delta)} m_3 s_{13} - e^{2i\alpha} m_2 s_{12}\left(c_{23} e^{i\delta} s_{12} s_{13} + c_{12} s_{23}\right) + c_{12} m_1\left(-c_{12} c_{23} e^{i\delta} s_{13} + s_{12} s_{23}\right)\right),$$

$$m_{\mu\mu} = c_{13}^2 e^{2i(\beta+\delta)} m_3 s_{23}^2 + m_1\left(c_{23} s_{12} + c_{12} e^{i\delta} s_{13} s_{23}\right)^2 + e^{2i\alpha} m_2\left(c_{12} c_{23} - e^{i\delta} s_{12} s_{13} s_{23}\right)^2,$$

$$m_{\mu\tau} = c_{13}^2 c_{23} e^{2i(\beta+\delta)} m_3 s_{23} + m_1\left(c_{12} c_{23} e^{i\delta} s_{13} - s_{12} s_{23}\right) \cdot \left(c_{23} s_{12} + c_{12} e^{i\delta} s_{13} s_{23}\right) - e^{2i\alpha} m_2\left(c_{23} e^{i\delta} s_{12} s_{13} + c_{12} s_{23}\right) \cdot \left(c_{12} c_{23} - e^{i\delta} s_{12} s_{13} s_{23}\right),$$

$$m_{\tau\tau} = c_{13}^2 c_{23}^2 e^{2i(\beta+\delta)} m_3 + e^{2i\alpha} m_2\left(c_{23} e^{i\delta} s_{12} s_{13} + c_{12} s_{23}\right)^2 + m_1\left(c_{12} c_{23} e^{i\delta} s_{13} - s_{12} s_{23}\right)^2. \quad (53)$$

It can be seen from (53) and (54) that the neutrino mass matrix contains nine physical parameters, namely, the three neutrino mass eigenvalues (m_1, m_2, and m_3), three mixing angles (θ_{12}, θ_{23}, and θ_{13}), and three CP violating phases δ, α, and β. Out of these nine free parameters of the theory not all are measured experimentally (Table 3). In particular, the CP violating phases are completely unconstrained. Thus, it is impossible to fully reconstruct the neutrino mass matrix from the results of feasible experiments and, thus, one cannot be definite about the origin and magnitude of CP violation in the leptonic sector. However, as a first step, it is worthwhile to investigate the neutrino mass matrix in the presence of some theoretically motivated inputs imposed on M_{ν} so as to reduce the number of free parameters of the theory and induce relationships amongst them. The possible forms of these theoretically motivated inputs are constrained by the current experimental data on neutrino masses and mixings and include the imposition of "texture zeros" [77–83], "hybrid textures" [84–86] in the neutrino mass matrix, requirement of zero determinant [87, 88], and zero trace condition [89] to name just a few.

In the texture zero Ansätze, it is assumed that some of the elements of neutrino mass matrix are anomalously small as compared to the other elements and can be, effectively, replaced by zeros. This is the simplest procedure to reduce the number of free parameters in the mass matrix. The texture Ansätze has been quite successful in the quark sector. A phenomenologically favored texture of quark mass matrix has been presented earlier by Gupta and Rajpoot [90–93]. In the spirit of quark-lepton similarity, the same texture has been prescribed for the charged lepton and the Dirac neutrino mass matrices [94]. The same texture for the right-handed neutrino mass matrix in the seesaw mechanism might follow from universal flavor symmetry hidden in a more fundamental theory of mass generation. Thus, texture zeros in different positions of the neutrino mass matrix, in particular and fermion mass matrices, in general could be the consequence of an underlying symmetry [95, 96]. Such universal textures of fermion mass matrices can, theoretically, be obtained in the context of GUTs based on SO(10) [97]. Moreover, neutrino mass matrices with texture zeros have important implications for leptogenesis [98].

Some attempts aimed at understanding the pattern of quark/lepton masses and mixings by introducing Abelian or non-Abelian flavor symmetries, naturally, lead to texture zeros in the mass matrices. However, the current low energy data are consistent only with a limited number of texture zero schemes of the neutrino mass matrix. They can, also, be realized within the framework of the seesaw mechanism. Furthermore, these textures can also be derived from a simple flavor group A_4 or its Z_3 subgroup [99]. The current neutrino oscillation data disallow all neutrino mass matrices with three or more texture zeros in the flavor basis. Out of fifteen possible neutrino mass matrices with two texture zeros, only seven are compatible with the current data on neutrino masses and mixings. Some sets of these texture zeros can be obtained by suitable weak basis (WB) transformations and have no physical meaning as such. However, a large class of sets of leptonic texture zeros considered in the literature imply the vanishing of certain CP-odd WB invariants and one can, thus, recognize a lepton flavor model in which the texture zeros are not explicitly present but correspond to a particular texture structure in a certain WB. The presence of texture zeros, in general, leads to a decrease in the number of independent CP violating phases. A particular texture zero structure gives rise to definite relationships between different CP violating phases [100, 101]. Such exact relations in closed form were obtained in [79]. Correlations between Dirac- and Majorana-typc CP violating phases for a particular texture zero scheme were studied in detail in [100,101]. It is, therefore, important to examine the interrelationships between the CP odd WB invariants which are required to vanish as a necessary and sufficient condition for CP conservation.

Moreover, a unified description of flavor physics and CP violation in the quark and lepton sectors is absolutely necessary. This can be achieved by constructing a low energy effective theory with the SM and some discrete non-Abelian family symmetry and, subsequently, embedding this theory into grand unified theory (GUT) models like $S(10)$ [97]. For this reason, the discrete symmetry will have to be a subgroup of $S(3)$ or $SU(3)$ [102]. The search for an adequate discrete symmetry has mainly focussed on the minimal subgroups of these groups with at least one singlet and one doublet irreducible representation to accommodate the fermions belonging to each generation. One such subgroup is the quaternion group $Q8$ [85] which not only accommodates the three generations of fermions but also explains the rather large difference between the values of 2-3 mixings in the quark and lepton sectors. Quaternion symmetry like some other discrete symmetries leads to nontrivial relationships amongst the nonzero mass matrix elements which underscores the inadequacy of texture zero analyses [84– 86] in isolation. Such textures which have equalities between different elements along with the vanishing of some elements of the mass matrix have been referred to as "hybrid textures" in the literature. Frigerio and Smirnov [86] presented a comprehensive analysis of the hybrid textures along with other possibilities for the neutrino mass matrix. The discrete quaternion groups have been extensively applied to flavor physics. It is, therefore, extremely important to investigate the connection between the textures of fermion mass matrices and the observables of flavor mixing. In particular, it is important to subject models based on Q_8 symmetry to the test of a viable leptogenesis to explain the baryon asymmetry of the Universe (BAU).

The phenomenological analyses of the neutrino mass matrix with "texture zeros" or "hybrid textures" are expected to provide useful hints towards unraveling the dynamics of fermion mass generation, *CP* violation, and identification of possible underlying symmetries of the lepton flavors from which realistic models of lepton mass generation and flavor mixing could be, hopefully, constructed. Moreover, an interesting cosmological implication of leptonic *CP* violation is that it opens an attractive possibility of accounting for the matter-antimatter asymmetry in the Universe within the socalled thermal leptogenesis mechanism.

STERILE NEUTRINOS AND ASSOCIATED PHENOMENOLOGY

The recent agreement of MiniBooNE antineutrino data [103] with the short-baseline $\bar{\nu}_\mu \to \bar{\nu}_e$ oscillation signal observed by LSND experiment [104] has opened an intense theoretical and experimental activity aimed at the clarification of the explanation of these observations in a framework compatible with the data of other neutrino oscillation experiments. The results of solar, atmospheric, and long-baseline neutrino oscillation experiments led us to the current standard threeneutrino mixing paradigm, in which the three active neutrinos ν_e, ν_μ, ν_τ are superpositions of three massive neutrinos]1,]2,]3 with masses m_1, m_2, m_3, respectively. The measured solar and atmospheric mass-squared dofferences are given in Table 3. The completeness of this paradigm has been challenged by the LSND [104] and MiniBooNE observations of short-baseline $\bar{\nu}_\mu \to \bar{\nu}_e$ transitions at different values of distance (L) and energy (E) but approximately at the same L/E. Since the distance and energy dependences of neutrino oscillations occur through this ratio, the agreement of MiniBooNE and LSND signals raised interest in the possibility of existence of one or more mass squared differences larger than about 0.5 eV, which is much larger than the values of Δm^2_{21} and Δm^2_{31}. Hence, we are led to the extension of threeneutrino mixing with the introduction of one or more sterile neutrinos which do not have weak interactions and do not contribute to the invisible width of the Z boson [105].

The possible existence of sterile neutrinos is an exciting possibility which could open a powerful window to investigate physics beyond the SM of particle physics. Schemes of neutrino mixing with sterile neutrinos have been studied extensively in the literature ([106] and the references therein) with more emphasis on the schemes with one or two sterile neutrinos. Sterile neutrinos, if they exist, would lead to rich experimental phenomena. Their admixture with active neutrinos would modify the neutrino flavor mixing and lead to the observable effects in neutrino oscillation experiments. Moreover, they could interact with gauge bosons, resulting in significant corrections to certain nonoscillation processes, for example, in the neutrinoless double beta decay ($0\nu\beta\beta$) amplitude [107] or in the beta decay spectra such as in KATRIN experiment [108, 109].

In the presence of $n_s = n-3$ sterile neutrinos, the neutrino mass matrix is an $n\times n$ matrix M] which can be diagonalized by an $n\times n$ unitary matrix U. The neutrino flavor eigenstates] (for $f = e, \mu, \tau, s_1, s_2,\ldots,s_{n-3}$) are then related to their mass eigenstates $\nu_i\,(\nu_1, \nu_2, \nu_3, \nu_{s_1}, \nu_{s_2}, \ldots)$ (for $i = 1, 2, 3, \ldots, n$) via

$$\nu_f = \sum_{i=1}^{n} U_{fi}\nu_i. \tag{54}$$

In general, for n massive families including $ns = n-3 \neq 0$ massive sterile neutrinos, one has $n-1 = n_s + 2$ Majorana phases, $3(n-2) = 3(n_s+1)$ mixing angles, and $2n-5 = 2n_s+1$ Dirac phases. The number of angles and Dirac phases is less than the naive $(1/2)(n-1)$ angles and $(1/2)(n-1)(n-2)$ phases because the $(1/2)n_s(n_s-1)$ rotations between sterile states are unphysical. For illustration, in the case of only one sterile neutrino, U is typically parameterized by

$$U = R_{34}\widetilde{R_{24}}\widetilde{R_{14}}R_{23}\widetilde{R_{13}}R_{12}P, \tag{55}$$

where the matrices are rotations in space; that is,

$$R_{34} = \begin{pmatrix} 1 & 0 & 0 & 0 \\ 0 & 1 & 0 & 0 \\ 0 & 0 & c_{34} & s_{34} \\ 0 & 0 & -s_{34} & c_{34} \end{pmatrix} \tag{56}$$

or

$$\widetilde{R_{14}} = \begin{pmatrix} c_{14} & 0 & 0 & s_{14}e^{-i\delta_{14}} \\ 0 & 1 & 0 & 0 \\ 0 & 0 & 1 & 0 \\ -s_{14}e^{i\delta_{14}} & 0 & 0 & c_{14} \end{pmatrix}, \quad (57)$$

where $s_{ij} = \sin\theta_{ij}$, $c_{ij} = \cos\theta_{ij}$. The diagonal P matrix contains the three Majorana phases α, β, and γ:

$$P = \text{diag}\left(1, e^{i\alpha/2}, e^{i(\beta/2+\delta_{13})}, e^{i(\gamma/2+\delta_{14})}\right). \quad (58)$$

Note that there are in total three Dirac CP-violating phases δ_{ij}. The above definition of P is constructed in such a way that only Majorana phases show up in the effective mass governing neutrino-less double beta decay. Similarly, one can parameterize the mixing matrix for 2 sterile neutrinos as

$$U = \widetilde{R_{25}}R_{34}R_{25}\widetilde{R_{24}}R_{23}\widetilde{R_{15}}\widetilde{R_{14}}\widetilde{R_{13}}R_{12}P, \quad (59)$$

where

$$P = \text{diag}\left(1, e^{i\alpha/2}, e^{i(\beta/2+\delta_{13})}, e^{i(\gamma/2+\delta_{14})} e^{i(\phi/2+\delta_{15})}\right). \quad (60)$$

Neutrino-less double beta decay ($0\nu\beta\beta$) is the only realistic test of lepton number violation. While there are several mechanisms to mediate the process (e.g., heavy neutrinos, right-handed currents, or SUSY particles), light Majorana neutrino exchange is presumably the best motivated scenario. In the presence of one and two sterile neutrinos, the effective neutrino mass in ($0\nu\beta\beta$) is, respectively, given by

$$\langle m_{ee}\rangle_1 = \left|c_{12}^2c_{13}^2c_{14}^2m_1 + s_{12}^2c_{13}^2c_{14}^2m_2e^{i\alpha} + s_{13}^2c_{14}^2m_3e^{i\beta} + s_{14}^2m_4e^{i\gamma}\right|, \quad \textbf{(61)}$$

$$\langle m_{ee}\rangle_2 = \left|c_{12}^2c_{13}^2c_{14}^2c_{15}^2m_1 + s_{12}^2c_{13}^2c_{14}^2c_{15}^2m_2e^{i\alpha} + s_{13}^2c_{14}^2c_{15}^2m_3e^{i\beta} + s_{14}^2c_{15}^2m_4e^{i\gamma} + s_{15}^2m_5e^{i\phi}\right|. \quad \textbf{(62)}$$

CONCLUSIONS

In the absence of significant breakthrough in our understanding of the parameters of neutrino mass matrix, the phenomenological approaches such as texture zeros, hybrid textures, zero determinant, and zero trace condition, to name a few, are bound to play important role in our understanding of neutrino masses and mixings. This in turn is intimately related to the possible observation CP violation in the leptonic sector. In this direction, recent observation of nonzero θ_{13} has very important implication for CP violation in the leptonic sector as it requires a nonzero value of $\theta 13$. The texture zeros are not weak basis (WB) invariants. This means that a given set of texture zeros which arise in a certain WB may not be present at all or may appear in different entries in another WB. A large class of sets of leptonic texture zeros considered in the literature imply the vanishing of certain C-odd weak basis invariants.Thus, we can recognize a lepton mass model in which the texture zeros are not explicitly present and which corresponds to a particular texture scheme in a certain WB. The relevance of CP-odd WB invariants in the analysis of the texture zero Ansatze is¨ due to the fact that texture zeros lead to a decrease in the number of the independent CP violating phases. A minimum number of CP-odd WB invariants can be found which will all vanish for the CP invariant mass matrices as a necessary and sufficient condition. Despite the success of SM threeneutrino oscillations in explaining the results of solar, atmospheric, reactor, and accelerator experiments by two large and one small mixing angles and two distinct mass squared splittings, MiniBooNE and LSND experiments are hinting to the existence of extra mass squared splittings and therefore the mixing of the three SM active neutrinos with extra sterile neutrinos. The existence of sterile neutrinos greatly influencess the phenomenology of the neutrino mass matrix as it would change the neutrino oscillation probabilities. Furthermore, there would exist new sources of CP violation affecting the amplitude of neutrinoless double beta decay (0v$\beta\beta$). In view of the recent enthusiastic results obtained from the experiments and the future experiments which are at the planning stage, the future of the neutrino physics is very buoyant and the next step is the observation of CP violation in the leptonic sector which is very important for the particle and astroparticle points of view.

ACKNOWLEDGMENTS

The author acknowledges the financial support provided by University Grants Commission (UGC)-Basic Science Research (BSR), Government of India vide Grant no. F.20-2/2013(BSR).

REFERENCES

1. L. Meitner and O. Hahn, "Über die Verteilung der Beta-Strahlen auf die einzelnen Produkte des aktiven Niederschlags des Thoriums," *Physikalische Zeitschrift*, vol. 13, p. 390, 1912.
2. J. Chadwick, "Distribution in intensity in the magnetic spectrum of the β-rays of radium," *Verhandlungen der Deutschen Physikalischen Gesellschaft*, vol. 16, pp. 383–391, 1914.
3. W. Pauli, "Dear radioactive ladies and gentlemen," in *Proceedings of the Tubingen conference*, December 1930.
4. S. F. King, A. Merle, S. Morisi, Y. Shimizu, and M. Tanimoto, "Neutrino mass and mixing: from theory to experiment," *New Journal of Physics*, vol. 16, Article ID 045018, 2014.
5. S. M. Bilenky, "Neutrino in standard model and beyond," http://arxiv.org/abs/1501.00232.
6. S. M. Bilenky and B. Pontecorvo, "Lepton mixing and neutrino oscillations," *Physics Reports*, vol. 41, no. 4, pp. 225–261, 1978.
7. S. M. Bilenky and S. T. Petcov, "Massive neutrinos and neutrino oscillations," *Reviews of Modern Physics*, vol. 59, no. 3, pp. 671–754, 1987.
8. S. M. Bilenky, C. Giunti, and W. Grimus, "Phenomenology of neutrino oscillations," *Progress in Particle and Nuclear Physics*, vol. 43, no. 1, pp. 1–86, 1999.
9. S. M. Bilenky, C. Giunti, J. A. Grifols, and E. Massó, "Absolute values of neutrino masses: status and prospects," *Physics Reports*, vol. 379, no. 2, pp. 69–148, 2003.
10. M. Gonzalez-Garcia and Y. Nir, "Neutrino masses and mixing: evidence and implications," *Reviews of Modern Physics*, vol. 75, no. 2, pp. 345–402, 2003.
11. M. Maltoni, T. Schwetz, M. Tórtola, and J. W. F. Valle, "Status of global fits to neutrino oscillations," *New Journal of Physics*, vol. 6, p. 122, 2004.
12. G. L. Fogli, E. Lisi, A. Marrone, and A. Palazzo, "Global analysis of three-flavor neutrino masses and mixings," *Progress in Particle and Nuclear Physics*, vol. 57, no. 2, pp. 742–795, 2006.
13. T. Yanagida, in *Proceedings of the Workshop on the Baryon Number of the Universe and Unified Theories*, Tsukuba, Japan, February 1979.

14. P. Minkowski, "μ→eγ at a rate of one out of 109 muon decays?" *Physics Letters B*, vol. 67, no. 4, pp. 421–428, 1977.
15. M. Gell-Mann, P. Ramond, and R. Slansky, *Supergravity*, North-Holland, Amsterdam, The Netherlands, 1979.
16. R. N. Mohapatra and G. Senjanović, "Neutrino mass and spontaneous parity nonconservation," *Physical Review Letters*, vol. 44, no. 14, pp. 912–915, 1980.
17. M. Kobayashi and T. Maskawa, "CP-violation in the renormalizable theory of weak interaction," *Progress of Theoretical Physics*, vol. 49, no. 2, pp. 652–657, 1973.
18. R. Davis, D. S. Harmer, and K. C. Hoffman, "Search for neutrinos from the sun," *Physical Review Letters*, vol. 20, no. 21, pp. 1205–1209, 1968.
19. S. P. Mikheyev and A. Y. Smirnov, "Resonant amplification of oscillations in matter and solar-neutrino spectroscopy," *Il Nuovo Cimento C*, vol. 9, no. 1, pp. 17–26, 1986.
20. L. Wolfenstein, "Neutrino oscillations in matter," *Physical Review D*, vol. 17, no. 9, pp. 2369–2374, 1978.
21. V. Gribov and B. Pontecorvo, "Neutrino astronomy and lepton charge," *Physics Letters B*, vol. 28, no. 7, pp. 493–496, 1969.
22. Y. Fukuda, T. Hayakawaa, E. Ichiharaa et al., "Evidence for oscillation of atmospheric neutrinos," *Physical Review Letters*, vol. 81, no. 8, pp. 1562–1567, 1998.
23. Y. Fukuda, T. Hayakawa, E. Ichihara et al., "Study of the atmospheric neutrino ux in the multi-GeV energy range," *Physics Letters B*, vol. 436, pp. 33–41, 1998.
24. Q. R. Ahmad, R. C. Allen, T. C. Andersen et al., "Measurement of the rate of νe+d→p+p+e- interactions produced by ^{8}B solar neutrinos at the sudbury neutrino observatory," *Physical Review Letters*, vol. 87, no. 7, Article ID 071301, 6 pages, 2001.
25. Q. R. Ahmad, R. C. Allen, T. C. Andersen et al., "Direct evidence for neutrino flavor transformation from neutral-current interactions in the sudbury neutrino observatory," *Physical Review Letters*, vol. 89, Article ID 011301, 2002.
26. K. Eguchi, S. Enomoto, K. Furuno et al., "First results from KamLAND: evidence for reactor antineutrino disappearance," *Physical Review Letters*, vol. 90, Article ID 021802, 2003.

27. S. J. Parke, "Nonadiabatic level crossing in resonant neutrino oscillations," *Physical Review Letters*, vol. 57, no. 10, pp. 1275–1278, 1986.
28. B. T. Cleveland, T. Daily, R. Davis Jr. et al., "Measurement of the solar electron neutrino flux with the homestake chlorine detector," *The Astrophysical Journal*, vol. 496, p. 505, 1998.
29. J. N. Abdurashitov, E. P. Veretenkin, V. M. Vermul et al., "Solar neutrino flux measurements by the Soviet-American gallium experiment (SAGE) for half the 22-year solar cycle," *Journal of Experimental and Theoretical Physics*, vol. 95, no. 2, pp. 181–193, 2002.
30. J. N. Abdurashitov, "Solar neutrino flux measurements by the Soviet-American gallium experiment (SAGE) for half the 22-year solar cycle," *Zhurnal Eksperimental'noi i Teoreticheskoi Fiziki*, vol. 122, pp. 211–226, 2002.
31. W. Hampel, J. Handt, G. Heusser et al., "GALLEX solar neutrino observations: results for GALLEX IV," *Physics Letters B*, vol. 447, no. 1-2, pp. 127–133, 1999.
32. S. Fukuda, Y. Fukudaa, M. Ishitsuka et al., "Determination of solar neutrino oscillation parameters using 1496 days of Super-Kamiokande-I data," *Physics Letters B*, vol. 539, no. 3-4, pp. 179–187, 2002.
33. C. Arpesella, G. Bellini, J. Benziger et al., "First real time detection of ^{7}Be solar neutrinos by Borexino," *Physics Letters B*, vol. 658, no. 4, pp. 101–108, 2008.
34. T. Araki, K. Eguchi, S. Enomoto et al., "Measurement of neutrino oscillation with KamLAND: evidence of spectral distortion," *Physical Review Letters*, vol. 94, no. 8, Article ID 081801, 5 pages, 2005.
35. S. Abe, T. Ebihara, S. Enomoto et al., "Precision measurement of neutrino oscillation parameters with KamLAND," *Physical Review Letters*, vol. 100, no. 22, Article ID 221803, 5 pages, 2008.
36. I. Shimizu, in *Proceedings of the International Conference on Topics in Astroparticle and Underground Physics (TAUP'07)*, 2007.
37. M. C. Gonzalez-Garcia and M. Maltoni, "Phenomenology with massive neutrinos," *Physics Reports*, vol. 460, no. 1–3, pp. 1–129, 2008.
38. M. Maltoni, T. Schwetz, M. A. Tortola, and J. W. F. Valle, "Status of global fits to neutrino oscillations," *New Journal of Physics*, vol. 6, no. 1, article 122, 2004.
39. S. Goswami, "Neutrino oscillations and masses," *International Journal*

of Modern Physics A, vol. 21, pp. 1901–1913, 2006.

40. A. Bandyopadhyay, S. Choubey, S. Goswami, S. T. Petcov, and D. P. Roy, "Update of the solar neutrino oscillation analysis with the 766 Ty KamLAND spectrum," *Physics Letters B*, vol. 608, no. 1-2, pp. 115–129, 2005.
41. G. L. Fogli, E. Lisi, A. Marrone et al., "Observables sensitive to absolute neutrino masses: a reappraisal after WMAP 3-year and first MINOS results," *Physical Review D*, vol. 75, Article ID 053001, 2007.
42. Y. Ashie, J. Hosaka, K. Ishihara et al., "Measurement of atmospheric neutrino oscillation parameters by Super-Kamiokande I," *Physical Review D*, vol. 71, Article ID 112005, 2005.
43. M. H. Ahn, S. Aoki, H. Bhang et al., "Indications of neutrino oscillation in a 250 km long-baseline experiment," *Physical Review Letters*, vol. 90, no. 4, Article ID 041801, 5 pages, 2003.
44. E. Aliu, S. Andringa, S. Aoki et al., "Evidence for muon neutrino oscillation in an accelerator-based experiment," *Physical Review Letters*, vol. 94, no. 8, Article ID 081802, p. 5, 2005.
45. D. G. Michael, P. Adamson, T. Alexopoulos et al., "Observation of muon neutrino disappearance with the MINOS detectors in the NuMI neutrino beam," *Physical Review Letters*, vol. 97, Article ID 191801, 2006.
46. J. I. Crespo-Anadon, "Double Chooz: latest results," http://arxiv.org/abs/1412.3698.
47. Y. Abe, C. Aberle, J. C. dos Anjos et al., "Reactor ν-e disappearance in the double chooz experiment," *Physical Review D*, vol. 86, no. 5, Article ID 052008, 21 pages, 2012.
48. Y. Abe, C. Aberle, J. C. dos Anjos et al., "First measurement of 13 from delayed neutron capture on hydrogen in the Double Chooz experiment," *Physics Letters B*, vol. 723, no. 1–3, pp. 66–70, 2013.
49. Y. Abe, J. C. dos Anjose, J. C. Barriere et al., "Background-independent measurement of θ13 in Double Chooz," *Physical Letters B*, vol. 735, pp. 51–56, 2014.
50. Y. Abe, J. C. dos Anjos, J. C. Barriere et al., "Ortho-positronium observation in the Double Chooz Experiment," *Journal of High Energy Physics*, vol. 2014, article 32, 2014.
51. F. P. An, A. B. Balantekin, H. R. Band et al., "Spectral measurement of electron antineutrino oscillation amplitude and frequency at Daya

Bay," *Physical Review Letters*, vol. 112, Article ID 061801, 2014.

52. P. Adamson, I. Anghel, C. Backhouse et al., "Electron neutrino and antineutrino appearance in the full MINOS data sample," *Physical Review Letters*, vol. 110, no. 17, Article ID 171801, 2013.
53. K. Abe, J. Adam, H. Aihara et al., "Observation of electron neutrino appearance in a muon neutrino beam," *Physical Review Letters*, vol. 112, Article ID 061802, 2014.
54. M. C. Gonzalez-Garcia, M. Maltoni, and T. Schwetz, "Updated fit to three neutrino mixing: status of leptonic CP violation," *Journal of High Energy Physics*, vol. 2014, article 52, 2014.
55. N. Cabibbo, "Unitary symmetry and leptonic decays," *Physical Review Letters*, vol. 10, pp. 531–533, 1963.
56. A. H. Guth and P. J. Steinhardt, "The inflationary universe," *Scientific American*, vol. 250, no. 5, pp. 90–102, 1984.
57. L. Fromme, S. J. Huber, and M. Seniuch, "Baryogenesis in the two-Higgs doublet model," *Journal of High Energy Physics*, vol. 2006, no. 11, p. 038, 2006.
58. K. Kajantie, M. Laine, K. Rummukainen, and M. E. Shaposhnikov, "The electroweak phase transition: a non-perturbative analysis," *Nuclear Physics B*, vol. 466, no. 1-2, pp. 189–258, 1996.
59. C. Jarlskog, "Commutator of the quark mass matrices in the standard electroweak model and a measure of maximal CP nonconservation," *Physical Review Letters*, vol. 55, no. 10, pp. 1039–1042, 1985.
60. C. Amsler, M. Doser, M. Antonelli et al., "Review of particle physics," *Physics Letters B*, vol. 667, no. 1–5, pp. 1–6, 2008.
61. G. C. Branco, L. Lavoura, and M. N. Rebelo, "Majorana neutrinos and CP violation in the leptonic sector," *Physics Letters B*, vol. 180, no. 3, pp. 264–268, 1986.
62. G. C. Branco and L. Lavoura, "On the addition of vector-like quarks to the standard model," *Nuclear Physics B*, vol. 278, pp. 738–754, 1986.
63. H. K. Dreiner, J. S. Kim, O. Lebedev, and M. Thormeier, "Supersymmetric Jarlskog invariants: the neutrino sector," *Physical Review D: Particles, Fields, Gravitation and Cosmology*, vol. 76, no. 1, Article ID 015006, 2007.
64. A. D. Sakharov, "Violation of CP invariance, C asymmetry, and baryon

asymmetry of the universe," *ZhETF Pis ma Redaktsiiu*, vol. 5, p. 32, 1967.

65. C. D. Froggat and H. B. Nielsen, "Hierarchy of quark masses, cabibbo angles and *CP* violation," *Nuclear Physics B*, vol. 147, no. 3-4, pp. 277–298, 1979.
66. C. D. Froggatt and H. B. Nielsen, "Statistical analysis of quark and lepton masses," *Nuclear Physics B*, vol. 164, pp. 114–140, 1980.
67. A. Antaramian, L. J. Hall, and A. Rašin, "Flavor-changing interactions mediated by scalars at the weak scale," *Physical Review Letters*, vol. 69, no. 13, pp. 1871–1873, 1992.
68. S. F. King and C. Luhn, "Neutrino mass and mixing with discrete symmetry," *Reports on Progress in Physics*, vol. 76, no. 5, Article ID 056201, 2013.
69. H. Fritzsch, "Weak-interaction mixing in the six-quark theory," *Physics Letters B*, vol. 73, no. 3, pp. 317–322, 1978.
70. H. Fritzsch, "Quark masses and flavor mixing," *Nuclear Physics B*, vol. 155, pp. 189–207, 1979.
71. S. Dimopoulos, "Natural generation of fermion masses," *Physics Letters B*, vol. 129, no. 6, pp. 417–428, 1983.
72. T. P. Chang and M. Sher, "Mass-matrix ansatz and flavor nonconservation in models with multiple Higgs doublets," *Physical Review D*, vol. 35, no. 11, pp. 3484–3491, 1987.
73. H. Georgi and C. Jarlskog, "A new lepton-quark mass relation in a unified theory," *Physics Letters B*, vol. 86, no. 3-4, pp. 297–300, 1979.
74. S. Dimopoulos, L. J. Hall, and S. Raby, "Predictive framework for fermion masses in supersymmetric theories," *Physical Review Letters*, vol. 68, no. 13, pp. 1984–1987, 1992.
75. H. Harari, H. Haut, and J. Weyers, "Quark masses and cabibbo angles," *Physics Letters B*, vol. 78, no. 4, pp. 459–461, 1978.
76. H. Fritzsch and J. Plankl, "Flavour democracy and the lepton-quark hierarchy," *Physics Letters, Section B: Nuclear, Elementary Particle and High-Energy Physics*, vol. 237, no. 3-4, pp. 451–456, 1990.
77. P. H. Frampton, S. L. Glashow, and D. Marfatia, "Zeroes of the neutrino mass matrix," *Physics Letters, Section B: Nuclear, Elementary Particle and High-Energy Physics*, vol. 536, no. 1-2, pp. 79–82, 2002.
78. B. R. Desai, D. P. Roy, and A. R. Vaucher, "Three-neutrino mass

matrices with two texture zeros," *Modern Physics Letters A*, vol. 18, no. 20, pp. 1355–1365, 2003.

79. Z.-Z. Xing, "Texture zeros and majorana phases of the neutrino mass matrix," *Physics Letters B*, vol. 530, no. 1–4, pp. 159–166, 2002.
80. W. Guo and Z.-Z. Xing, "Implications of the KamLAND measurement on the lepton flavor mixing matrix and the neutrino mass matrix," *Physical Review D*, vol. 67, Article ID 053002, 2003.
81. A. Merle and W. Rodejohann, "Elements of the neutrino mass matrix: allowed ranges and implications of texture zeros," *Physical Review D*, vol. 73, no. 7, Article ID 073012, 11 pages, 2006.
82. S. Dev and S. Kumar, "Neutrino parameter space for a vanishing *ee* element in the neutrino mass matrix," *Modern Physics Letters A*, vol. 22, p. 1401, 2007.
83. M. Honda, S. Kaneko, and M. Tanimoto, "Prediction of U e3 in neutrino mass matrix with two zeros," *Journal of High Energy Physics*, vol. 2003, no. 9, article 28, 2003.
84. S. Kaneko, H. Sawanaka, and M. Tanimoto, "Hybrid textures of neutrinos," *Journal of High Energy Physics*, vol. 2005, no. 8, article 73, 2005.
85. M. Frigerio, S. Kaneko, E. Ma, and M. Tanimoto, "Quaternion family symmetry of quarks and leptons," *Physical Review D*, vol. 71, Article ID 011901, 2005.
86. M. Frigerio and A. Y. Smirnov, "Neutrino mass matrix: inverted hierarchy and CP violation," *Physical Review D*, vol. 67, Article ID 013007, 2003.
87. G. C. Branco, R. González Felipe, F. R. Joaquim, and T. Yanagida, "Removing ambiguities in the neutrino mass matrix," *Physics Letters B*, vol. 562, no. 3-4, pp. 265–272, 2003.
88. B. C. Chauhan, J. Pulido, and M. Picariello, "Neutrino mass matrices with vanishing determinant," *Physical Review D—Particles, Fields, Gravitation and Cosmology*, vol. 73, no. 5, Article ID 053003, 2006.
89. X.-G. He and A. Zee, "Neutrino masses with a 'zero sum' condition: mv1+mv2+mv3 = 0," *Physical Review D*, vol. 68, no. 3, Article ID 037302, p. 4, 2003.
90. S. N. Gupta and S. Rajpoot, *Quark Mass Matrices and the Top Quark Model*, Wayne State University, September 1990.

91. S. N. Gupta and J. M. Johnson, "Quark mixing and CP violation," *Physical Review D*, vol. 44, no. 7, pp. 2110–2113, 1991.
92. S. Rajpoot, "Evading the top mass bound in the fritzsch scheme," *Modern Physics Letters A*, vol. 7, no. 4, pp. 309–313, 1992.
93. H. Fritzsch and Z.-Z. Xing, "A symmetry pattern of maximal *CP* violation and a determination of the unitarity triangle," *Physics Letters B*, vol. 353, pp. 114–118, 1995.
94. G. Ahuja, S. Kumar, M. Randhawa, M. Gupta, and S. Dev, "Texture 4 zero Fritzsch-like lepton mass matrices," *Physical Review D: Particles, Fields, Gravitation and Cosmology*, vol. 76, no. 1, Article ID 013006, 2007.
95. P. H. Frampton, M. C. Oh, and T. Yoshikawa, "Majorana mass zeros from Higgs triplet vacuum expectation values without a Majoron problem," *Physical Review D*, vol. 66, Article ID 033007, 2002.
96. W. Grimus and L. Lavoura, "On a model with two zeros in the neutrino mass matrix," *Journal of Physics G: Nuclear and Particle Physics*, vol. 31, no. 7, pp. 693–702, 2005.
97. H. S. Goh, R. N. Mohapatra, and S.-P. Ng, "Minimal supersymmetric SO(10) model and predictions for neutrino mixings and leptonic CP violation," *Physical Review D*, vol. 68, no. 11, Article ID 115008, 2003.
98. S. Kaneko, M. Katsumata, and M. Tanimoto, "Leptogenesis in neutrino textures with two zeros," *Journal of High Energy Physics*, vol. 2003, article 025, 2003.
99. M. Hirsch, A. S. Joshipura, S. Kaneko, and J. W. F. Valle, "Predictive flavor symmetries of the neutrino mass matrix," *Physical Review Letters*, vol. 99, no. 15, Article ID 151802, 2007.
100. S. Dev, S. Kumar, S. Verma, and S. Gupta, "Phenomenological implications of a class of neutrino mass matrices," *Nuclear Physics B*, vol. 784, no. 1-2, pp. 103–117, 2007.
101. S. Dev, S. Kumar, S. Verma, and S. Gupta, "Phenomenology of two-texture zero neutrino mass matrices," *Physical Review D*, vol. 76, no. 1, Article ID 013002, 15 pages, 2007.
102. I. de Medeiros, S. F. Varzielas, and G. G. Ross, "Tri-bimaximal neutrino mixing from discrete subgroups of SU(3) and SO(3) family symmetry," *Physics Letters B*, vol. 644, no. 2-3, pp. 153–157, 2007.
103. A. A. Aguilar-Arevalo, C. E. Anderson, S. J. Brice et al., "Event excess in the MiniBooNE search for $\bar{\nu}_\mu \rightarrow \bar{\nu}_e$ Oscillations," *Physical Review*

Letters, vol. 105, Article ID 181801, 2010.

104. A. Aguilar, L. B. Auerbach, R. L. Burman et al., "Evidence for neutrino oscillations from the observation of ν¯e appearance in a ν¯μ beam," *Physical Review D*, vol. 64, Article ID 112007, 2001.
105. Z. Hou, R. Keisler, L. Knox, M. Millea, and C. Reichardt, "How massless neutrinos affect the cosmic microwave background damping tail," *Physical Review D*, vol. 87, no. 8, Article ID 083008, 11 pages, 2013.
106. M. Lavedor and C. Guinti, "Phenomenology of neutrino oscillations and mixing," *Acta Physica Polonica B*, vol. 44, no. 11, pp. 2323–2330, 2013.
107. S. M. Bilenky, S. Pascoli, and S. Petcov, "Majorana neutrinos, neutrino mass spectrum, CP violation, and neutrinoless double β decay. II. Mixing of four neutrinos," *Physical Review D*, vol. 64, Article ID 113003, 2001.
108. A. S. Riis and S. Hannestad, "Detecting sterile neutrinos with KATRIN like experiments," *Journal of Cosmology and Astroparticle Physics*, vol. 2011, article 011, 2011.
109. J. A. Barrett and J. Formaggio, "Resolving the reactor neutrino anomaly with the KATRIN neutrino experiment," *Physics Letters B*, vol. 706, no. 1, pp. 68–71, 2011.

CHAPTER

3

Recent Probes of Standard and Non-standard Neutrino Physics with Nuclei

Dimitrios K. Papoulias[1], Theocharis S. Kosmas[2] and Yoshitaka Kuno[3]

[1]AHEP Group, Institut de Física Corpuscular – CSIC/Universitat de València, Valencia, Spain

[2]Division of Theoretical Physics, University of Ioannina, Ioannina, Greece

[3]Department of Physics, Osaka University, Osaka, Japan

ABSTRACT

We review standard and non-standard neutrino physics probes that are based on nuclear measurements. We pay special attention on the discussion of prospects to extract new physics at prominent rare event measurements

Citation: Papoulias DK, Kosmas TS and Kuno Y (2019) "Recent Probes of Standard and Non-standard Neutrino Physics With Nuclei". Front. Phys. 7:191. https://doi.org/10.3389/fphy.2019.00191

looking for neutrino-nucleus scattering, such as the coherent elastic neutrino-nucleus scattering (CEνNS) that may involve lepton flavor violation (LFV) in neutral-currents (NC). For the latter processes several appreciably sensitive experiments are currently pursued or have been planed to operate in the near future, like the COHERENT, CONUS, CONNIE, MINER, TEXONO, RED100, νGEN, Ricochet, NUCLEUS, etc. We provide a thorough discussion on phenomenological and theoretical studies, in particular those referring to the nuclear physics aspects in order to provide accurate predictions for the relevant experiments. Motivated by the recent discovery of CEνNS at the COHERENT experiment and the active experimental efforts for a new measurement at reactor-based experiments, we summarize the current status of the constraints as well as the future sensitivities on nuclear and electroweak physics parameters, non-standard interactions, electromagnetic neutrino properties, sterile neutrinos and simplified scenarios with novel vector Z' or scalar ϕ mediators. Indirect and direct connections of CEνNS with astrophysics, direct Dark Matter detection and charge lepton flavor violating processes are also discussed.

Keywords: coherent elastic neutrino-nucleus scattering (CENNS), non-standard interactions, electromagnetic neutrino properties, sterile neutrinos, novel mediators

INTRODUCTION

During the last few decades, intense research effort has been devoted to multidisciplinary neutrino searches involving physics within and beyond the standard model (SM) in the theory, phenomenology and experiments that drops in the interplay of particle, nuclear physics, astrophysics and cosmology.

Astrophysical and laboratory searches [1] offer unique opportunities to probe great challenges in modern-day physics such as the underlying physics of the fundamental electroweak interactions within and beyond the SM [2, 3] in the neutral and charged-current sector of semi-leptonic neutrino-nucleus processes [4–6]. To meet the sufficient energy and flux requirements, the relevant studies consider different low-energy neutrino sources including (i) Supernova (SN) neutrinos, (ii) accelerator neutrinos (from pion decay at rest, π-DAR) and (iii) reactor neutrinos, while interesting proposals aiming to use ^{51}Cr and beta-beam neutrino sources have appeared recently. The detection mechanism of low-energy neutrino interactions with nucleons and nuclei is experimentally hard and limited by the tiny nuclear recoils produced by the scattering process. To this purpose, the nuclear detector materials are carefully selected to fulfill the requirement of achieving a-few-keV or sub-keV threshold capabilities. The detectors developed are based on cutting edge technologies such as scintillating crystals (CsI[Na], NaI[Tl]), p-type point-contact (PPC) germanium detectors, single-phase or double liquid noble gases (LAr, LXe) charged coupled devices (CCDs), cryogenic bolometers, etc.

The neutral-current coherent elastic neutrino nucleus scattering (CEνNS) was proposed about four decades ago [7–9], while it was experimentally confirmed in 2017 by the COHERENT Collaboration [10] at the Spallation Neutron Source, in good agreement with the SM expectation. The observation of CEνNS has opened up a new era, triggering numerous theoretical studies to interpret the available data [11] in a wide spectrum of new physics opportunities, with phenomenological impact on astroparticle physics, neutrino oscillations, dark matter (DM) detection, etc. (see [12] for various applications). In particular, the recent works have concentrated on non-standard interactions (NSI) [13–19], electromagnetic properties [20–23], sterile neutrinos [24–26], CP-violation [27] and novel mediators [28–31]. Nuclear and atomic effects are explored in Cadeddu et al. [32], Ciuffoli et al. [33], Huang and Chen [34], Aristizabal Sierra et al. [35], Papoulias et al. [36], Arcadi et al. [37], and Cadeddu et al. [38] which may have direct implications to the neutrino-floor [39–41] and to DM searches [42–44]. Being a rapidly developing field, there are several experimental programs aiming to observe CEνNS in the near future, such as the TEXONO [45], CONNIE [46], MINER [47], νGEN [48], CONUS [49], Ricochet [50], and NUCLEUS [51].

Future CEνNS measurements have good prospects to shed light on the exotic neutrino-nucleus interactions expected in the context of models describing flavor changing neutral-current (FCNC) processes [52] as well as to subleading NSI oscillation effects [53–57] and various open issues in nuclear astrophysics [58, 59]. The main goal of this review article is to provide an up-to-date status of the conventional and exotic neutrino physics probes of CEνNS and to summarize the necessary aspects for the interpretation of the experimental data. We focus on the theoretical modeling, calculations and analysis of the data that are relevant at the time of writing and we mainly concentrate on the theoretical and phenomenological physics aspects. For a recent review on the experimental advances of CEνNS, see Akimov et al. [60].

This review article has been organized as follows: section 2 provides the theoretical treatment of low-energy neutrino-nucleus processes for both coherent and incoherent channels and its connection to the more general lepton-nucleus case with a particular emphasis on the nuclear physics aspects. Section 3 presents the current status of constraints on SM and exotic physics parameters resulted from the analysis of the COHERENT data and discusses the projected sensitivities from future CEνNS measurements at π-DAR and reactor facilities. In section 4 we briefly summarize the most important connections of CEνNS with DM searches, charged lepton flavor violation (cLFV) and astrophysics. Finally, the main conclusions are given in section 5.

THEORETICAL STUDY OF NEUTRINO-NUCLEUS INTERACTION

At low- and intermediate-energies, the neutrino being a key input to understand open issues in physics within and beyond the SM (see below), necessitated a generation of neutrino experiments for exploring neutrino scattering processes with nucleons and nuclei for both charged-current (inelastic) and neutral-current (coherent elastic and incoherent scattering) processes. Theoretically, the neutral-current neutrino-nucleus scattering we are interested here, is a well studied process for both coherent [61] an incoherent channels: [62, 63]. The accurate evaluation of the required transition matrix elements describing the various interaction channels of the electroweak processes between an initial and a final (many-body) nuclear state, is obtained on the basis of reliable nuclear wavefunctions. From a nuclear theory point of view, such results have been obtained by paying special attention on the accurate contruction of the nuclear ground state in the framework of the quasi-particle random phase approximation (QRPA), using schematic Skyrme [64] or realistic Bonn C-D pairing interactions [65]. Focusing on the latter method, the authors of Papoulias and Kosmas [61] solved iteratively the Bardeen-Cooper-Schrieffer (BCS) equations, achieving a high reproducibility of the available nuclear charge-density-distribution experimental data [66].

Coherent and Incoherent Neutrino-Nucleus Cross Sections

In the Donnelly-Walecka theory [67] all semi-leptonic nuclear processes at low and intermediate energies may be described by an effective interaction Hamiltonian through the leptonic j_{μ}^{lept} and hadronic $J\mu$ current densities,

$$\hat{H}_{eff} = \frac{G}{\sqrt{2}} \int \hat{\ell}_{\mu}^{lept}(x) \hat{\mathcal{J}}^{\mu}(x)\, d^3x, \tag{1}$$

where $G = G_F$ is the Fermi coupling constant for neutral-current processes and $G = G_F \cos\theta_c$ (θ_c is the Cabbibo angle) for charged-current processes. For partial scattering rates, the evaluation of the transition amplitudes $\langle f | \hat{H}_{eff} | i \rangle$ are treated via a multipole decomposition analysis of the hadronic current (see the Appendix). Then, for a given set of an initial $|J_i\rangle$ and a final $|J_f\rangle$ nuclear state, the double differential SM cross section becomes [68]

$$\frac{d^2\sigma_{i\to f}}{d\Omega\, d\omega} = \frac{G^2}{\pi} F(Z, \varepsilon_f) \frac{|\mathbf{k}_f| \varepsilon_f}{(2J_i + 1)} \left(\sum_{J=0}^{\infty} \sigma_{CL}^J + \sum_{J=1}^{\infty} \sigma_T^J \right) \tag{2}$$

with ε_f ($|\mathbf{k}_f|$) denoting the final energy (momentum) of the outgoing lepton, while $\omega = \varepsilon_i - \varepsilon_f$ stands for the excitation energy of the nucleus where ε_i is the initial lepton energy. For charged-current processes, the Fermi function $F(Z, \varepsilon_f)$, takes into account the final state interaction of the outgoing charged particle, while for neutral-current processes such as coherent and incoherent neutrino-nucleus scattering it is $F(Z, \varepsilon_f) = 1$.

The individual cross sections in Equation (2) receive contributions from the so-called Coulomb $\hat{\mathcal{M}}$, longitudinal $\hat{\mathcal{L}}$, transverse electric $\hat{\mathcal{T}}^{el}$ and transverse magnetic $\hat{\mathcal{T}}^{mag}$ operators for both vector and axial vector components (see the Appendix). The cross sections σ_{CL}^J and σ_T^J are expressed in terms of the reduced matrix elements of the eight basic irreducible tensor operators [67]

$$\begin{aligned} \sigma_{CL}^J = {} & (1 + a\cos\theta) |\langle J_f || \hat{\mathcal{M}}_J(\kappa) || J_i \rangle|^2 \\ & + (1 + a\cos\theta - 2b\sin^2\theta) |\langle J_f || \hat{\mathcal{L}}_J(\kappa) || J_i \rangle|^2 \\ & + \left[\frac{\omega}{\kappa}(1 + a\cos\theta) + d \right] 2\Re e |\langle J_f || \hat{\mathcal{L}}_J(\kappa) || J_i \rangle || \langle J_f || \hat{\mathcal{M}}_J(\kappa) || J_i \rangle|^*, \end{aligned} \tag{3}$$

$$\begin{aligned} \sigma_T^J = {} & (1 - a\cos\theta + b\sin^2\theta) \Big[|\langle J_f || \hat{\mathcal{T}}_J^{mag}(\kappa) || J_i \rangle|^2 \\ & + |\langle J_f || \hat{\mathcal{T}}_J^{el}(\kappa) || J_i \rangle|^2 \Big] \\ & \mp \left[\frac{(\varepsilon_i + \varepsilon_f)}{\kappa}(1 - a\cos\theta) - d \right] \\ & 2\Re e |\langle J_f || \hat{\mathcal{T}}_J^{mag}(\kappa) || J_i \rangle || \langle J_f || \hat{\mathcal{T}}_J^{el}(\kappa) || J_i \rangle|^*. \end{aligned} \tag{4}$$

Here, the + (−) sign refers to neutrino (antineutrino) scattering and θ represents the scattering angle, while the parameters a, b, d are expressed as

$$a = \frac{|\mathbf{k}_f|}{\varepsilon_f} = \sqrt{1 - \left(\frac{m_f}{\varepsilon_f}\right)^2}, \quad b = \frac{\varepsilon_i \varepsilon_f a^2}{\kappa^2}, \quad d = \frac{m_f^2}{\kappa \varepsilon_f}. \tag{5}$$

The 4-momentum transfer is trivially obtained from the kinematics of the process and in natural units reads

$$q^2 \equiv q_\mu q^\mu = q_0^2 - \mathbf{q}^2, \tag{6}$$

while for later convenience the magnitude of the 3-momentum transfer is defined as

$$\kappa = |\mathbf{q}| \equiv |\vec{q}| = \left[\omega^2 + 2\varepsilon_i \varepsilon_f (1 - a\cos\theta) - m_f^2\right]^{1/2} \tag{7}$$

For sufficiently small momentum transfer, i.e., $q \leq 1/R$ where R is the inverse nuclear radius[1], CEνNS dominates (see Figure 1). In this case, only ground state to ground state (g.s. → g.s.) transitions occur and lead to the following simplifications: the kinematics of the reaction imply $m_f = 0$ and $|\mathbf{k}_f| = \varepsilon_f$ so that $a = 1$ and $d = 0$, while the momentum transfer can be cast in terms of the incoming neutrino energy E_ν in the simple form

$$Q^2 = -q^2 = 4E_\nu^2 \sin^2\frac{\theta}{2}, \tag{8}$$

where the usual notation $\varepsilon_i = \varepsilon_f \equiv E_\nu$ has been adopted. Note also that the excitation energy in this case is $\omega = 0$ and $\kappa = \sqrt{-q^2} = \sqrt{Q^2}$, while angular momentum conservation implies that for CEνNS processes the only non-vanishing operator is the Coulomb, $T_1^0 \equiv \hat{\mathcal{M}}_0^0$ (see the Appendix for the definition of the operators T_i^J). Then the corresponding differential cross section is further simplified and takes the form

$$\left(\frac{d\sigma}{d\cos\theta}\right)_{\rm SM} = \frac{G_F^2}{2\pi} E_\nu^2 (1 + \cos\theta) \, |\langle g.s. || \hat{\mathcal{M}}_0^0(Q) || g.s. \rangle|^2, \tag{9}$$

where the matrix element for g.s. → g.s. transitions is explicitly written in terms of the nuclear form factors for protons $F_p(Q^2)$ and neutrons $F_n(Q^2)$, as

$$\langle g.s. || \hat{\mathcal{M}}_0^0(Q) || g.s. \rangle = \frac{1}{2}\left[\left(1 - 4\sin^2\theta_W\right) Z F_p(Q^2) - N F_n(Q^2)\right] \tag{10}$$

At CEνNS experiments the detection mechanism is sensitive to the tiny nuclear recoils generated in the aftermath of the scattering process. It is therefore reasonable to express the differential cross section with respect to the nuclear recoil energy T_N, which in the low energy approximation $T_N \ll E_\nu$, reads

$$\left(\frac{d\sigma}{dT_N}\right)_{SM} = \frac{G_F^2 M}{4\pi}\left(1 - \frac{M\,T_N}{2E_\nu^2}\right) |\langle g.s.||\hat{\mathcal{M}}_0^0(Q)||g.s.\rangle|^2 , \quad (11)$$

where $T_N=Q^2/2M$ and M is the mass of the nuclear isotope. The calculations of Papoulias and Kosmas [61] involved the BCS form factors for protons (neutrons)

$$F_{N_n} = \frac{1}{N_n}\sum_j \sqrt{2j+1}\langle g.s.||j_0(\kappa r)||g.s.\rangle \left(v^j_{p(n)}\right)^2 , \quad (12)$$

with N_n = Z or N and vjp(n)vp(n)j represents the occupation probability amplitude of the j-th single nucleon level.

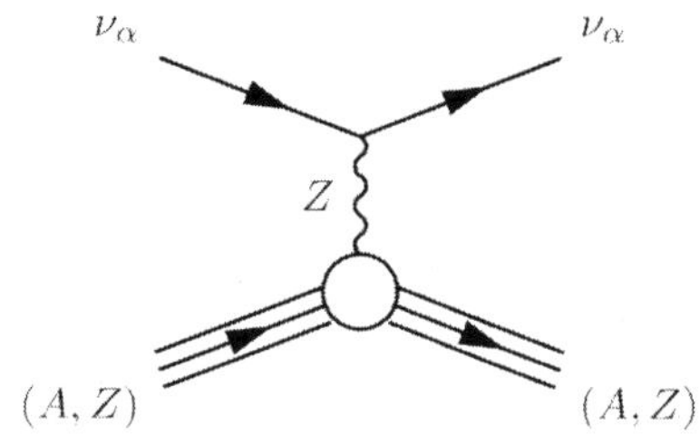

Figure 1. Feynman diagram illustrating the tree-level SM CEvNS process.

The method described above, involves realistic nuclear structure calculations making it more reliable compared to the use of phenomenological form factors, especially for accelerator neutrino sources (see the discussion in section 2.2). For the reader's convenience, Equation (11) is also expressed through the vector weak nuclear charge $Q^V{}_W$ in the approximation of equal proton and neutron form factors, as [69]

$$\left(\frac{d\sigma}{dT_N}\right)_{SM} = \frac{G_F^2 M}{\pi}(Q_W^V)^2\left(1 - \frac{M\,T_N}{2E_\nu^2}\right) F(Q)^2 , \quad (13)$$

where the vector QVWQWV weak charge is given by [70]

$$Q_W^V = \left[2(g_u^L + g_u^R) + (g_d^L + g_d^R)\right] Z + \left[(g_u^L + g_u^R) + 2(g_d^L + g_d^R)\right] N , \quad (14)$$

with the left- and right-handed couplings of u and d quarks to the Z-boson being

$$g_u^L = \rho_{\nu N}^{NC}\left(\frac{1}{2} - \frac{2}{3}\hat{\kappa}_{\nu N}\hat{s}_Z^2\right) + \lambda^{u,L},$$
$$g_d^L = \rho_{\nu N}^{NC}\left(-\frac{1}{2} + \frac{1}{3}\hat{\kappa}_{\nu N}\hat{s}_Z^2\right) + \lambda^{d,L},$$
$$g_u^R = \rho_{\nu N}^{NC}\left(-\frac{2}{3}\hat{\kappa}_{\nu N}\hat{s}_Z^2\right) + \lambda^{u,R},$$
$$g_d^R = \rho_{\nu N}^{NC}\left(\frac{1}{3}\hat{\kappa}_{\nu N}\hat{s}_Z^2\right) + \lambda^{d,R}. \tag{15}$$

The latter expressions include the radiative corrections from the PDG [71]: $\rho_{\nu N}^{NC} = 1.0082$, $\hat{\kappa}_{\nu N} = 0.9972$, $\lambda^{u,L} = -0.0031$, $\lambda^{d,L} = -0.0025$ and $\lambda^{d,R} = 2\lambda^{u,R} = 3.7 \times 10^{-5}$ while concerning the weak mixing-angle the adopted value is $\hat{s}_Z^2 \equiv \sin^2\theta_W = 0.2382$. Regarding the incoherent neutrino-nucleus cross section and for the sake of completeness we note that apart from the Donnelly-Walecka method given in Equation (2) a usefull formalism has been recently given in Bednyakov and Naumov [62].

The differential cross sections $d\sigma/dT_N$ and $d\sigma/d\cos\theta$ are shown in the upper left and upper right panel of Figure 2, from where it can be seen that large differences appear if the form factor dependence is neglected. On the other hand at low neutrino energies, i.e., $E_\nu \leq 20$MeV (relevant for reactor and solar neutrinos), the agreement of these two approximations is rather good. It is worth mentioning that forward scattering ($\theta = 0$) leads to maximum $d\sigma/d\cos\theta$, as well as that for this particular case the form factor is by definition equal to unity due to the zero momentum transfer, see Equation (8). Finally the bottom panel illustrates a comparison of the CEvNS cross section by incorporating the nuclear form factors and by assuming $F = 1$.

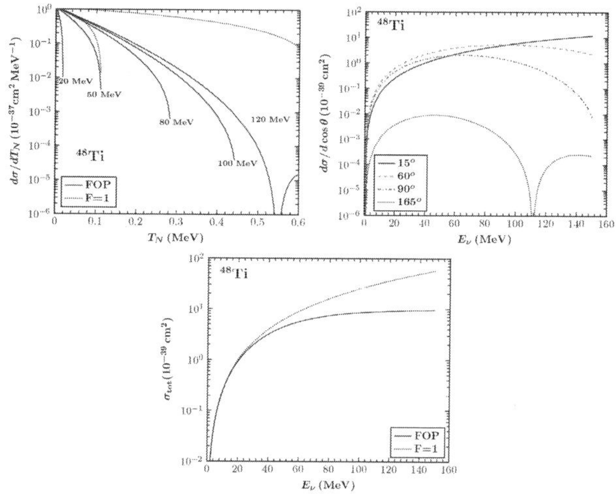

Figure 2. (Upper Left) The differential cross section $d\sigma/dT_N$ in terms of the nuclear recoil energy T_N for different incident neutrino energies. The results are compared to the case of point-like nucleus (F = 1) for $E_v = 50$ and $E_v = 120$ MeV. **(Upper Right)** The differential cross section $d\sigma/d\cos\theta$ as a function of the incoming neutrino energy E_v, for different scattering angles (for backward scattering the cross section vanishes). **(Bottom)** The CEvNS cross section σ_{tot} as a function of the neutrino energy. An asymptotic behavior is found at neutrino energies $E_v \geq 80$ MeV or higher. Taken from Papoulias [72].

Theoretical Methods for Obtaining the Nuclear Form Factors

Electron scattering data provide high precision measurements of the proton charge density distribution [73]. The absence of similar data for neutron densities, restricts us to rely on the approximation of $\rho_p(\mathbf{r}) = \rho_n(\mathbf{r})$ and thus assume $F_p(Q) = F_n(Q) \equiv F(Q)$ (see Equation 13). In the context of nuclear theory, it is possible to treat separately the proton and neutron nuclear form factors by employing non-trivial techniques. The most reliable methods for this purpose include the large-scale Shell-Model [74, 75], the QRPA [76], Microscopic Quasiparticle Phonon Model (MQPM) [77], the deformed Shell-Model (DSM) method [39] and others. Recently, crucial information on important nuclear parameters has been extracted from the analysis of the recent COHERENT data in Cadeddu et al. [32], Ciuffoli et al. [33], Huang and Chen [34], and Papoulias [78].

The point-nucleon charge density distribution $\rho(\mathbf{r})$, is defined as the expectation value of the density operator [79]

$$\hat{\rho}(\mathbf{r}) = \sum_{j=1}^{A} \frac{1}{2}(1 \pm \tau_{3j})\delta(\mathbf{r} - \mathbf{r}_j), \tag{16}$$

where the + (−) sign refers to the point-proton (neutron) charge density distribution. Assuming the nuclear ground state to be approximately described by a Slater determinant constructed from single-particle wavefunctions, the distributions of Equation (16) are given by summing in quadrature the point-nucleon wavefunctions. According to Kosmas and Vergados [79], for closed (sub)shell nuclei the charge density distribution is assumed to be spherically symmetric while the interesting radial component ($r = |\mathbf{r}|$) of the proton charge density distribution, $\rho_p(r)$, can be cast in the form

$$\rho_p(r) = \frac{1}{4\pi} \sum_{\substack{(n,l)j \\ \text{occupied}}} (2j+1)|R_{nlj}(r)|^2, \tag{17}$$

where $R_{nlj}(r)$ denotes the radial component of the single-particle wavefunction with quantum numbers n, l, and j. The nuclear form factor depends on the three momentum transfer squared $\mathbf{q}^2 \equiv |\mathbf{q}|^2$ and can be obtained via a Fourier transformation

$$F_{p\,(n)}(\mathbf{q}^2) = \frac{4\pi}{N_n} \int \rho_{p\,(n)}(r) j_0(|\mathbf{q}|r)\, r^2\, dr, \quad N_n = Z \text{ or } N \tag{18}$$

where $j_0(x) = \sin x/x$ denotes the zero-order Spherical Bessel function of first kind. The nuclear form factors lead to a suppression of the CEνNS cross section and subsequently to a suppression of the expected event rates (see [32] for a comparison with the COHERENT data). The uncertainties of the nuclear form factors are explored in Aristizabal Sierra et al. [35] where it is pointed out that studies looking for physics beyond the SM can be seriously affected by the uncertainty of the neutron form factor [80]. It is therefore important to treat with special care the nuclear form factors since new physics could be claimed or missed, if their uncertainties are not properly taken into account. In addition to the form factors obtained in the framework of the nuclear BCS method of Equation (12), below we present a summary of various form factor approximations widely considered in the literature.

(i) Form factors from available electron-scattering experimental data

The proton nuclear form factors $F_p(\mathbf{q}^2)$, may be evaluated through a model independent analysis (e.g., by employing a Fourier-Bessel expansion)

of the electron scattering data [66], having however the disadvantage of assuming $F_p(\mathbf{q}^2) = F_n(\mathbf{q}^2)$.

(ii) Fractional occupation probabilities (FOP) in a simple Shell-Model

For Harmonic Oscillator (h.o.) wavefunctions the nuclear form factor $F_p(\mathbf{q}^2)$ for protons can be expressed in polynomial form [81, 82]

$$F_p(\mathbf{q}^2) = \frac{1}{Z} e^{-(|\mathbf{q}|\, b)^2/4} \Phi\left(|\mathbf{q}|\, b, Z\right), \quad \Phi\left(|\mathbf{q}|\, b, Z\right) = \sum_{\lambda=0}^{N_{\max}} \theta_\lambda (|\mathbf{q}|\, b)^{2\lambda}, \tag{19}$$

with $N_{max} = (2n + l)_{max}$ denoting for the number of quanta of the highest occupied proton (neutron) level. In a similar manner, the radial nuclear charge density distribution $\rho_p(r)$ is written in terms of the polynomials $\Pi(r/b, Z)$ in the following compact form [81, 82]

$$\rho_p(r) = \frac{1}{\pi^{3/2} b^3} e^{-(r/b)^2} \Pi\left(\frac{r}{b}, Z\right), \qquad \Pi\left(\chi, Z\right) = \sum_{\lambda=0}^{N_{\max}} f_\lambda \chi^{2\lambda}, \tag{20}$$

with the definition $\chi = r/b$ (b stands for the h.o. size parameter). The explicit expressions for calculating the coefficients θ_λ and f_λ are given in the Appendix.

The occupation probabilities entering Equations (18) and (19) are assumed equal to unity (zero) for the states below (above) the Fermi surface, e.g., the filling numbers of the states for closed (sub)shell nuclei are those predicted by the simple Shell-Model. Going one step further, Kosmas and Vergados [79] introduced depletion/occupation numbers to describe the occupation probabilities of the surface levels, which satisfy the relation

$$\sum_{\substack{(n,l)j \\ \text{all}}} \alpha_{nlj}(2j+1) = N_n\,. \tag{21}$$

In this framework, there is a number of active surface nucleons (above or below the Fermi level) with non-vanishing occupation probability $0 \leq \alpha_{nlj} \leq 1$ and a number of core levels with $\alpha_{nlj} = 1$. The parameters are properly adjusted so that a high reproducibility of the experimental data is achieved [66]. By introducing four parameters α_i, i = 1, 2, 3, 4 in Equation (21) the polynomial $\Pi(\chi, Z)$ of Equation (20) reads

$$\Pi(\chi, Z, \alpha_i) = \Pi(\chi, Z_2)\frac{\alpha_1}{Z_1 - Z_2} + \Pi(\chi, Z_1)\left[\frac{\alpha_2}{Z_c - Z_1} - \frac{\alpha_1}{Z_1 - Z_2}\right]$$

$$+\Pi(\chi, Z_c)\left[\frac{Z'-Z}{Z'-Z_c} - \frac{\alpha_2}{Z_c - Z_1} - \frac{\alpha_3}{Z'-Z_c}\right]$$
$$+\Pi(\chi, Z')\left[\frac{Z-Z_c}{Z'-Z_c} + \frac{\alpha_3}{Z'-Z_c} - \frac{\alpha_4}{Z''-Z'}\right]$$
$$+\Pi(\chi, Z'')\left[\frac{\alpha_4}{Z''-Z'} - \frac{\lambda}{Z'''-Z''}\right]$$
$$+\Pi(\chi, Z''')\frac{\lambda}{Z'''-Z''},$$

with $\lambda = \alpha_1 + \alpha_2 - \alpha_3 - \alpha_4$ (see [61] for the fitted values). (22)

with $\lambda = \alpha_1 + \alpha_2 - \alpha_3 - \alpha_4$ (see [61] for the fitted values).

(iii) Use of effective expressions for the nuclear form factors

Besides calculations in the spirit of a nuclear structure model, a reliable description of the nuclear form factors (at least for low-energy reactor and solar neutrinos) may be obtained through the use of phenomenological approximations of the charge density distribution. The Helm-type density distribution is a convolution of a uniform nucleonic density with cut-off radius R_0 (accounting for the interior density) with a Gaussian falloff with folding width s (surface thickness). The corresponding Helm form factor takes the analytical form as [83]

$$F_{\text{Helm}}(Q^2) = 3\frac{j_1(QR_0)}{qR_0}e^{-(Qs)^2/2}, \tag{23}$$

where $j_1(x) = \frac{\sin x}{x^2} - \frac{\cos x}{x}$ is the 1st-order Spherical Bessel function. The first three moments can be analytically expressed as [84]

$$\langle R_n^2\rangle = \frac{3}{5}R_0^2 + 3s^2$$
$$\langle R_n^4\rangle = \frac{3}{7}R_0^4 + 6R_0^2s^2 + 15s^4$$
$$\langle R_n^6\rangle = \frac{1}{3}R_0^6 + 9R_0^4s^2 + 63R_0^2s^4 + 105s^6. \tag{24}$$

The surface thickness parameter is fixed to s = 0.9 by fitting muon spectroscopy data [85], having also the advantage of improving the matching between the Helm and the symmetrized Fermi (SF) distributions [86]. The SF approximation follows from a Woods-Saxon charge density distribution and is expressed through the half density radius c and the diffuseness parameter a, as [87]

$$F_{\text{SF}}(Q^2) = \frac{3}{Qc\left[(Qc)^2 + (\pi Qa)^2\right]}\left[\frac{\pi Qa}{\sinh(\pi Qa)}\right]\left[\frac{\pi Qa\sin(Qc)}{\tanh(\pi Qa)} - Qc\cos(Qc)\right], \tag{25}$$

with

$$c = 1.23A^{1/3} - 0.60\,(\mathrm{fm}), \quad a = 0.52\,(\mathrm{fm}), \tag{26}$$

while the surface thickness is written as t = 4aln 3 [32]. The corresponding first three moments of the SF form factor read [84]

$$\begin{aligned}
\langle R_n^2 \rangle &= \frac{3}{5}c^2 + \frac{7}{5}(\pi a)^2 \\
\langle R_n^4 \rangle &= \frac{3}{7}c^4 + \frac{18}{7}(\pi a)^2 c^2 + \frac{31}{7}(\pi a)^4 \\
\langle R_n^6 \rangle &= \frac{1}{3}c^6 + \frac{11}{3}(\pi a)^2 c^4 + \frac{239}{15}(\pi a)^4 c^2 + \frac{127}{5}(\pi a)^6 .
\end{aligned} \tag{27}$$

The Klein-Nystrand (KN) distribution is obtained from the convolution of a Yukawa potential with range $a_k = 0.7$ fm over a Woods-Saxon distribution (hard sphere with radius R_A). The resulting KN form factor reads [88]

$$F_{\mathrm{KN}} = 3\frac{j_1(QR_A)}{QR_A}\left[1 + (Qa_k)^2\right]^{-1}, \tag{28}$$

and is adopted by the COHERENT Collaboration, while in this case root mean square (rms) radius reads

$$\langle R_n^2 \rangle_{\mathrm{KN}} = 3/5R_A^2 + 6a_k^2 \tag{29}$$

Figure 3 presents the charge density distribution in the top panel and the corresponding nuclear form factors for ^{40}Ar (interesting for LAr CEνNS detectors) and ^{48}Ti (interesting for $\mu^- \rightarrow e^-$ conversion in nuclei) in the lower panel, while the results are compared for the various methods used. A comparison of the form factors for ^{127}I and ^{133}Cs that are of interest for COHERENT, evaluated with the DSM method (not covered here), with those of the Helm, SF and KN parametrizations, is given in Papoulias et al. [36]. By incorporating realistic nuclear structure calculations on the basis of the BCS method, the SM CEνNS cross section is given in Figure 4 for a set of different isotopes throughout the periodic table. For heavier isotopes the form factor suppression is more pronounced and therefore the cross sections flatten more quickly, since the nuclear effects become significant even at low-energies.

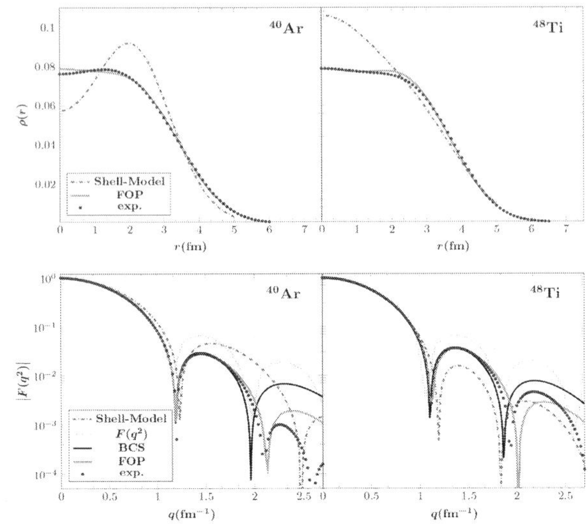

Figure 3. Charge density distribution **(Top)** and nuclear form factor **(Bottom)**. The results refer to the ^{40}Ar and ^{48}Ti isotopes and are compared for different nuclear methods. Figure adapted from Papoulias and Kosmas [61] under the terms of the Creative Commons Attribution 4.0 International license.

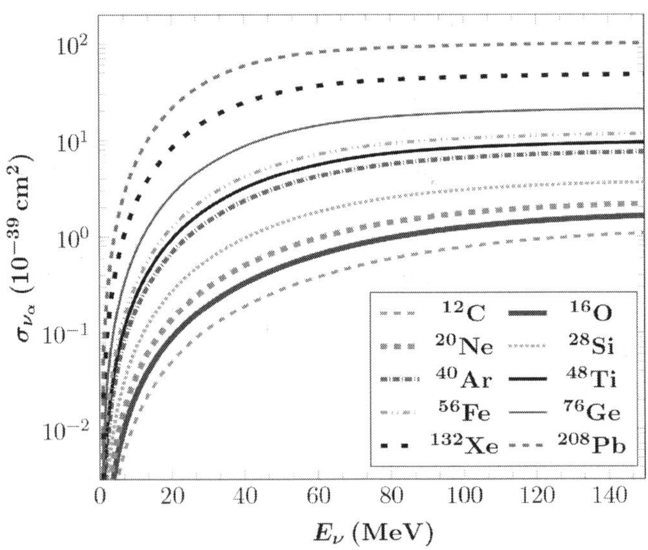

Figure 4. Integrated CEvNS cross sections σνα(ν̄ α)(Eν)σνα(ν̄α)(Eν) for a set of nuclear targets ranging from light to heavy isotopes. Figure adapted from Papoulias and Kosmas [61] under the terms of the Creative Commons Attribution 4.0 International license.

CONSTRAINTS WITHIN AND BEYOND THE SM FROM CENNS

The observation of CEνNS by the COHERENT experiment with a π-DAR neutrino source is a portal to new physics triggering a considerable number of phenomenological studies at low-energies. New constraints have been put on neutrino, electroweak and nuclear physics parameters, that we devote an effort to summarize below.

The experimental confirmation of CEνNS has also prompted a great rush in the experimental physics community and several projects are aiming to measure CEνNS using reactor neutrinos from nuclear power plants (NPP). It should be stressed that given the large potential of improvement in detector technology and control of systematics, it is feasible to further explore the low-energy and precision neutrino frontier.

The CONUS experiment is currently running at the Brokdorf NPP (Germany) and has already released preliminary results while the COHERENT experiment has released new results from the engineering run with a LAr detector [89].

Moreover, a number of prominent experiments are in preparation such as: the MINER experiment at the TRIGA Nuclear Science Center at Texas A&M University (USA), the CONNIE project at the Angra NPP (Brazil), the NUCLEUS and Ricochet experiments at the Chooz NPP (France)[2], the TEXONO program at the Kuo-Sheng NPP (Taiwan), the νGEN and RED100 experiments at the Kalinin NPP (Russia), the Coherent Captain-Mills (CCM) project at Los Alamos Neutron Science Center (LANSCE) as well as new proposals for a CEνNS measurement by employing a ^{51}Cr source [91] and new possibilities in China[3] (an exhaustive review of the CEνNS experimental developments is given in [60]).

Finally, it has been recently discussed the possibility of measuring CEνNS at the European Spallation Source (ESS) [92].

Table 1 lists a summary of the current and future experimental projects, while Figure 5 demonstrates the differential event rate for the various target nuclei at π-DAR [see [100]] and at the various reactor CEνNS experiments neglecting detector efficiencies and quenching factors (QF).

Table 1. Current and future experimental proposals for CEvNS searches.

Experiment	T_{th}	Baseline (m)	Target	Mass (kg)	Technology	Source
COHERENT [93]	6.5 keV	19.3	CsI[Na]	14.57	Scintillating crystal	π-DAR SNS
	5 keV	22	Ge	10	HPGe PPC	
	20 keV	29	LAr	2×10^3	Single phase	
	13 keV	28	NaI[Tl]	185*/3388	Scintillating crystal	
CCM [94]	10–20 keV	20–40	LAr	10^4	Scintillation	π-DAR Lujan
CONUS [95]	300 eV	17	Ge	4	HPGe	NPP 3.9 GW
MINER [47]	10 eV	1	Ge/Si	30	cryogenic	NPP 1 MW
CONNIE [96]	28 eV	30	Si	1	Si CCDs	NPP 3.8 GW
Ricochet [50]	50–100 eV	<10	Ge/Zn	10	Ge, Zn bolometers	NPP 8.54 GW
NUCLEUS [97]	20 eV	<10	$CaWO_4$ Al_2O_3	10^{-3}	Cryogenic $CaWO_4$ Al_2O_3 calorimeter array	NPP 8.54 GW
RED100 [98]	500 eV	19	Xe	100	LXe dual phase	NPP 3 GW
νGEN	350 eV	10	Ge	4 × 0.4	Ge PPC	NPP 3 GW
TEXONO [99]	150–200 eV	28	Ge	1	p-PCGe	NPP 2 × 2.9 GW

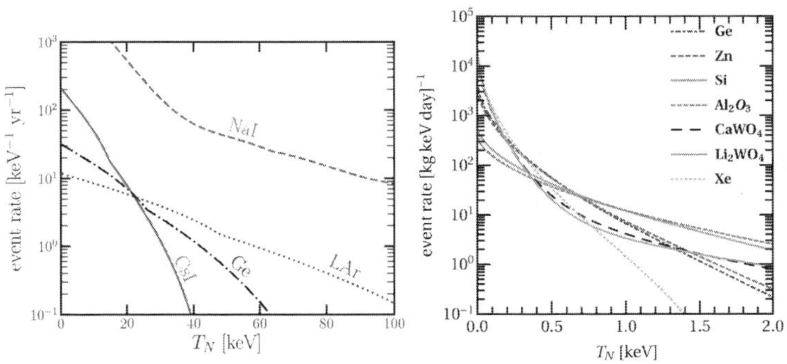

Figure 5. Expected CEvNS event rate for the different detector subsystems of the COHERENT experiment **(Left)** and for the different target nuclei relevant to reactor based experiments **(Right)**. For the case of COHERENT the results are shown according to the setups of Table 1, while for reactor based experiments the calculation assumes 1 kg of each target located at 20 m from a 4 GW reactor NPP. The impact of QF and efficiency is ignored.

In reality however, these experiments are sensitive to an ionization energy (e.g., electron equivalent energy eV_{ee}) since a large portion of the nuclear recoil energy eV_{nr} is lost to heat (conversion to phonons). The energy discrepancy has to be determined experimentally and is taken into account in terms of the QF. The latter quantity is crucial for such processes and depends on the nuclear recoil energy as well as on the target nucleus in question. Theoretically it follows the empirical form arising from the Lindhard theory [86]

$$Q(T_N) = \frac{\kappa g(\gamma)}{1+\kappa g(\gamma)} \qquad (30)$$

with $g(\gamma)=3\gamma^{0.15}+0.7\gamma^{0.6}+\gamma$ and $\gamma=11.5T_N(\text{keV})Z-7/3$, $\kappa=0.133Z^{2/3}A^{-1/2}$. The left and right panels of Figure 6 quantify the effect of the QF in the case of CEνNS .

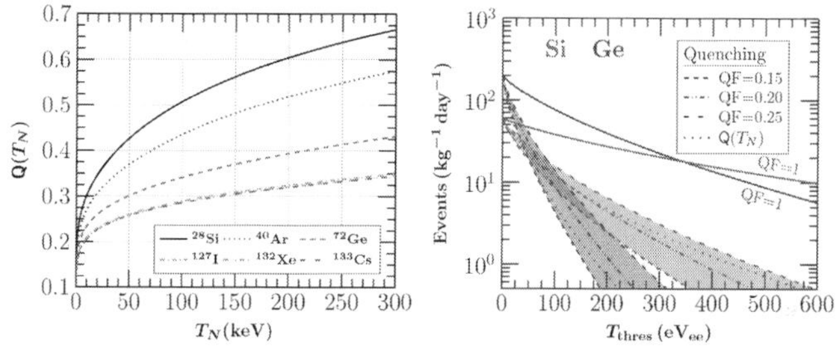

Figure 6. (Left) The QF as a function of the nuclear recoil energy. **(Right)** The impact of the QF on the expected number of events at a reactor based experiment assuming a typical neutrino flux of $10^{13}cm^{-2}s^{-1}$ for CEνNS off Silicon and Germanium detectors. The figure in the left panel has been adapted from Miranda et al. [22] under the terms of the Creative Commons Attribution 4.0 International license.

Electroweak and Nuclear Physics

The left panel of Figure 7 shows the expected number of events at the CsI[Na] detector of COHERENT and gives a comparison with the experimental data, from where it can be seen that a good agreement is reached. In Papoulias and Kosmas [21] the authors analyzed the CEνNS data and obtained a low-energy determination of the weak mixing angle, as illustrated in the right panel of Figure 7. The obtained constraint at 90% C.L. reads [21]

$$\sin^2\theta_W = 0.197^{+0.128}_{-0.080}\,. \qquad (31)$$

An interesting analysis combining atomic parity violating (APV) and CEνNS data was performed in Cadeddu and Dordei [101], while the prospects regarding the future reactor-based CEνNS experiments such as those presented in Table 1, have been extracted in Cañas et al. [102]. On the other hand, an improved determination of the CsI[Na] quenching factor can in principle lead to a significantly better agreement between the experimental results and the theoretical simulations [103], as well as to an improved sensitivity on the weak mixing angle [78].

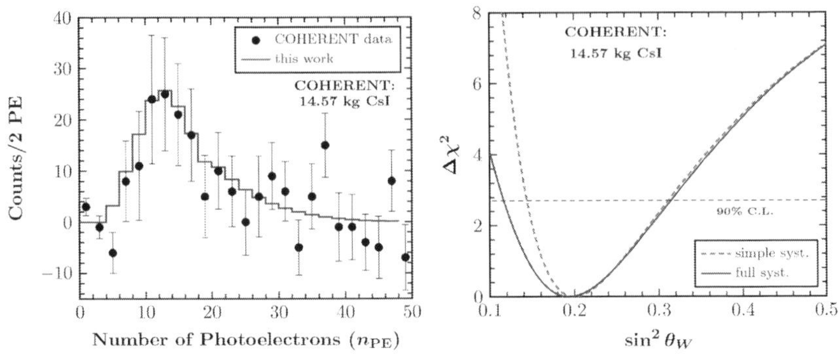

Figure 7. Comparison between the simulated number of CEνNS events and the experimental data by COHERENT **(Left)** and sensitivity of COHERENT on the weak mixing angle **(Right)**. Figure adapted from Papoulias and Kosmas [21] under the terms of the Creative Commons Attribution 4.0 International license.

The discussion made in the previous section emphasized how the CEνNS cross section depends on the nuclear physics effects which are incorporated through the momentum variation of the relevant nuclear form factor. The authors of Aristizabal Sierra et al. [27] demonstrated how the intrinsic nuclear structure uncertainties may have a significant impact to searches beyond the SM such those regarding NSIs, sterile neutrinos and neutrino generalized interactions (GNIs). Starting from the form factor of Equation (18) and expanding in terms of even moments of the charge density distribution one arrives to a model independent expression [104]

$$F_{p,n}(Q^2) \approx 1 - \frac{Q^2}{3!}\langle R_{p,n}^2\rangle + \frac{Q^4}{5!}\langle R_{p,n}^4\rangle - \frac{Q^6}{7!}\langle R_{p,n}^6\rangle + \cdots, \tag{32}$$

with the k-th radial moment defined as

$$\langle R_{p,n}^k\rangle = \frac{\int \rho_{p,n}(\vec{r})\, r^k\, d^3\vec{r}}{\int \rho_{p,n}(\vec{r})\, d^3\vec{r}}, \tag{33}$$

allowing the study of contributions of higher-order moments to nuclear form factors [33]. A sensitivity analysis of the two first moments with current and future COHERENT data is depicted in Figure 8 where the allowed regions are presented at 1σ, 90% and 99% C.L. The calculation in this case was restricted in the physical region [0,6] fm in order to obey the upper limit on $R_n(^{208}Pb) = 5.75 \pm 0.18$ fm from the PREM experiment [105]. The future scenarios considered assume improved statistical uncertainties and more massive detectors in accord with the next generation COHERENT experiments [93] (see [36] for details), while as demonstrated in Patton et al. [104] multi-ton scale detectors will provide significant improvements.

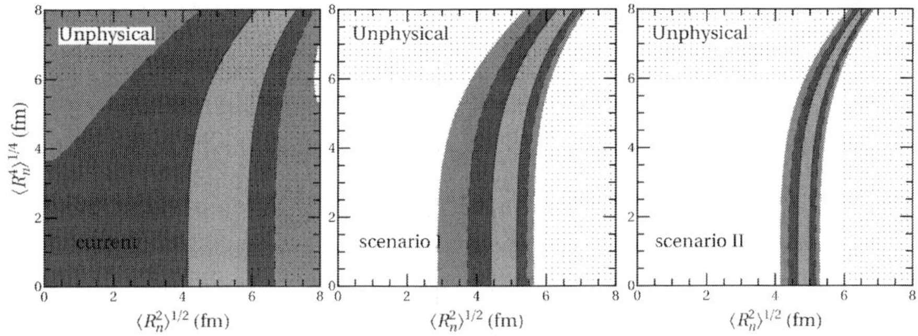

Figure 8. Sensitivity contours in the ⟨R2n⟩1/2⟨Rn2⟩1/2–⟨R4n⟩1/4⟨Rn4⟩1/4 plane from the COHERENT data assuming the current and possible future detector specifications (see the text). The allowed regions are shown at 1σ (turquoise), 90% C.L. (blue) and 99% C.L. (magenta). Taken from Papoulias et al. [36].

The average CsI neutron rms radius has been explored in Cadeddu et al. [32], Huang and Chen [36], and Papoulias et al. [34] using the energy spectrum of the available CEνNS data. The corresponding sensitivity profiles are presented in Figure 9, leading to the best fits at 90% C.L. [36]

$$\begin{aligned}\langle R_n^2\rangle^{1/2} &= 5.64^{+0.99}_{-1.23}\,\text{fm} \quad (\text{current})\,,\\ \langle R_n^2\rangle^{1/2} &= 5.23^{+0.42}_{-0.50}\,\text{fm} \quad (\text{scenario I})\,,\\ \langle R_n^2\rangle^{1/2} &= 5.23^{+0.22}_{-0.22}\,\text{fm} \quad (\text{scenario II})\,,\end{aligned} \tag{34}$$

while the potential of improvement through a more accurate determination of the QF is promising (see e.g., [78]). An independent analysis combining APV and CEνNS data was performed in Cadeddu et al. [106] leading to essentially similar results. Finally, it is worthwhile to mention the reported upper bound on the neutron skin $\Delta R_{np}=\Delta R_n-\Delta R_p=0.7^{+0.9}_{-1.1}$ fm [32].

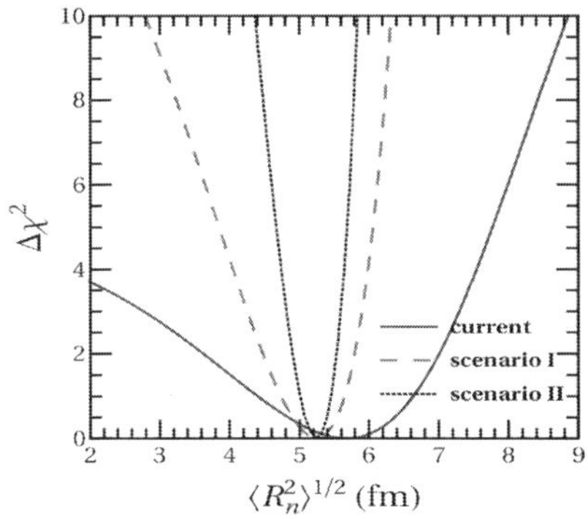

Figure 9. COHERENT sensitivity on the average nuclear rms radius of CsI assuming the current and possible future detector specifications. Taken from Papoulias et al. [36].

Non-standard and Generalized Neutrino Interactions

Non-standard interactions (NSI) [107] appear in several appealing SM extensions [108] involving four-fermion contact interaction, various seesaw realizations [109–111], left-right symmetry [112], gluonic operators [113], etc., constituting an interesting model independent probe of new physics. NSIs may have implications to SN [114], neutrino oscillations [57] and CEνNS [70, 115], while recently NSI terms were explored in the context of GNI [116] and effective field theory (EFT) operators [117, 118]. Finally the RG issue has been partly addressed in the context of NSI in Davidson and Gorbahn [119].

For sufficiently low energies vector-type NSIs arise from the effective four-fermion operators [56]

$$\mathcal{O}^{qV}_{\alpha\beta} = \left(\bar{\nu}_\alpha \gamma^\mu L \nu_\beta\right)\left(\bar{q}\gamma_\mu P q\right) + \text{H.c.}, \tag{35}$$

leading to new contributions to the CEνNS rate from exotic processes of the form

$$\nu_\alpha(\bar{\nu}_\alpha) + (A,Z) \to \nu_\beta(\bar{\nu}_\beta) + (A,Z), \tag{36}$$

where α, $\beta = \{e, \mu, \tau\}$ ($\alpha \neq \beta$), q denotes a first-generation quark q = {u, d} and P = {L, R} is the chiral projection operator. For the case of CEνNS the new interactions are taken into account through the NSI charge with the substitution $Q^V_W \to Q^V_{NSI}$ in Equation (14). The latter contains flavor-preserving non-universal $(\epsilon_{\alpha\alpha}^{qV})$ and flavor changing $(\epsilon_{\alpha\beta}^{qV})$ terms and is expressed as

$$\mathcal{Q}_{\mathrm{NSI}}^{V} = (2\epsilon_{\alpha\alpha}^{uV} + \epsilon_{\alpha\alpha}^{dV} + g_p^V)Z + (\epsilon_{\alpha\alpha}^{uV} + 2\epsilon_{\alpha\alpha}^{dV} + g_n^V)N + \sum_{\alpha\neq\beta}\left[(2\epsilon_{\alpha\beta}^{uV} + \epsilon_{\alpha\beta}^{dV})Z + (\epsilon_{\alpha\beta}^{uV} + 2\epsilon_{\alpha\beta}^{dV})N\right], \quad (37)$$

implying that the NSI CEvNS cross section becomes flavor dependent. There is a reach literature on NSI investigations with the recent COHERENT data. Assuming one nonvanishing coupling at a time, the authors of Papoulias and Kosmas [21] focused on the non-universal terms and obtained the sensitivity profiles shown in the left panel of Figure 10, while the corresponding allowed regions resulting from a combined analysis of the NSI couplings are illustrated in the upper panel of Figure 11 at 90% C.L. Regarding the future prospects of MINER, Ricochet, NUCLEUS and CONNIE, similar studies were conducted concentrating on the non-universal [120] and flavor-changing [121] terms. Indeed, a multitarget strategy can break degeneracies involved between up and down flavor-diagonal NSI terms that survives analysis of neutrino oscillation experiments [14]. Constraints on the corresponding parameters arising from leptoquarks [120], GNI [116], and EFT [117, 118] have been also reported. NSI constraints from CEvNS place meaningful constraints excluding a large part of the existing CHARM constraints and overlap with results coming out of LHC monojet searches (see [120] for a usefull comparison). Regarding the near-term future, a potential improvement on determination of the QF [103] may yield severe constraints. For example, updated bounds are possible by analyzing the number of events [78], the energy spectrum [122] as well as through a combined analysis of both energy and timing COHERENT data [123].

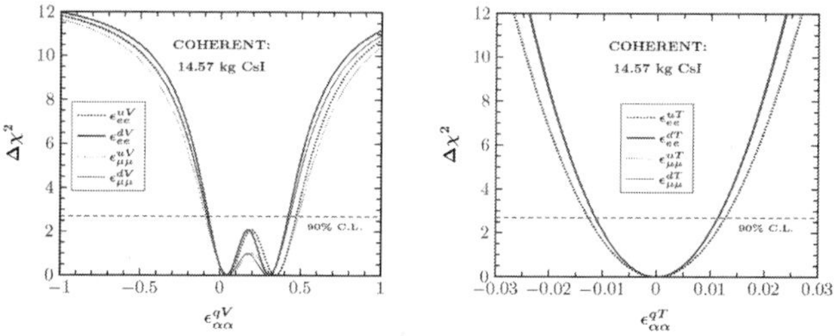

Figure 10. Sensitivity of COHERENT to vector **(Left)** and tensor **(Right)** NSI parameters. Thick (thin) curves correspond to the ν_e (νμ+ν̄ μνμ+ν̄μ) beam.

Figure adapted from Papoulias and Kosmas [21] under the terms of the Creative Commons Attribution 4.0 International license.

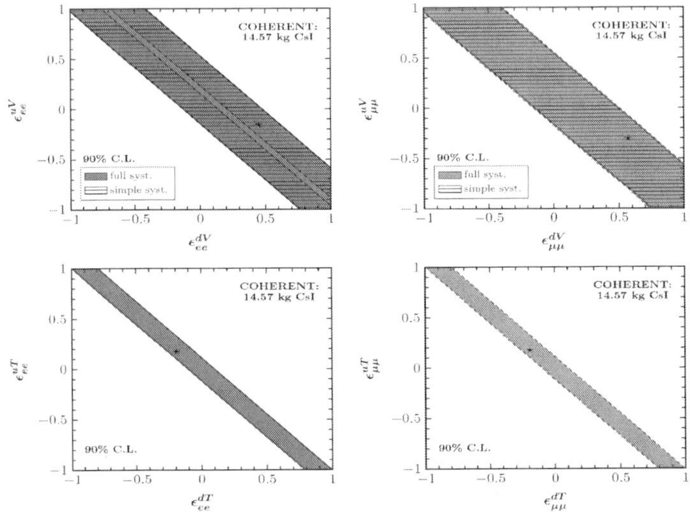

Figure 11. Sensitivity contours in the vector **(Upper panel)** and tensor **(Lower panel)** NSI parameter space. The results are presented at 90% C.L. assuming non-universal couplings only. The left (right) panel corresponds to the ν_e ($\nu\mu+\bar{\nu}$ $\mu\nu\mu+\bar{\nu}\mu$) beam, while the best-fit points are indicated by an asterisk ⋆. Figure adapted from Papoulias and Kosmas [21] under the terms of the Creative Commons Attribution 4.0 International license.

Novel tensor-type interactions are predicted in the general context of NSI [124] and GNI [116] which induce terms of the form

$$\mathcal{O}_{\alpha\beta}^{qT} = \left(\bar{\nu}_\alpha \sigma^{\mu\nu} \nu_\beta\right)\left(\bar{q}\sigma_{\mu\nu} q\right) + \text{H.c.} \quad (38)$$

Such interactions violate the chirality constraint allowing for a wide class of new interactions, e.g., relevant to neutrino EM properties (see [125, 126]). Contrary to the vector NSI case, for tensorial interactions there is absence of interference with the SM interactions. In the approximation of a vector-type translation the corresponding tensor NSI charge has been expressed as [124]

$$\mathcal{Q}_{\text{NSI}}^{T} = (2\epsilon_{\alpha\alpha}^{uT} + \epsilon_{\alpha\alpha}^{dT})Z + (\epsilon_{\alpha\alpha}^{uT} + 2\epsilon_{\alpha\alpha}^{dT})N \quad (39)$$

while a more systematic interpretation has been carried out in Aristizabal Sierra et al. [116].

To account for the new contributions in the presence of tensorial NSI, the CEvNS cross is written [21]

$$\left(\frac{d\sigma}{dT_N}\right)_{\text{SM+NSI}_{\text{tensor}}} = \mathcal{G}^T_{\text{NSI}}(E_\nu, T_N)\left(\frac{d\sigma}{dT_N}\right)_{\text{SM}}, \tag{40}$$

with the tensor NSI factor defined as

$$\mathcal{G}^T_{\text{NSI}} = 1 + 4\left(\frac{\mathcal{Q}^T_{\text{NSI}}}{\mathcal{Q}^V_W}\right)^2 \frac{1 - \frac{MT_N}{4E_\nu^2}}{1 - \frac{MT_N}{2E_\nu^2}}. \tag{41}$$

From the analysis of the COHERENT data, the sensitivity profiles accounting to tensor NSIs, assuming one non-zero coupling at a time, are illustrated in the right panel of Figure 10 (see also [21]). The corresponding allowed regions coming out from a two parameter analysis are presented in the lower panel of Figure 11 at 90% C.L. The result is more stringent as compared to the analysis carried out in the framework of GNI for reasons discussed above. On the other hand, comparing with the vector NSI case the absence of SM-tensor NSI interference causes the allowed regions to appear with more narrow bands.

The Novel NSI Mediators Z′ (Vector) and ϕ (Scalar)

Theories beyond the SM with an additional U(1)′ symmetry have been comprehensively investigated. Regarding CEvNS related studies a novel massive mediator predicted in these concepts is expected to induce a detectable distortion to the nuclear recoil spectrum, provided that its mass is comparable to the momentum transfer. The study of models with new vector or scalar interactions that involve hidden sector particles may be also accessible at CEvNS experiments [127]. Such frameworks are interesting since they may play a central role in explaining anomalies with regards to B-meson decays at the LHCb experiment [128] and at DM searches [129].

We first examine the case of a new massive vector boson Z′. Restricting ourselves to the neutrino sector with only left-handed neutrinos the Lagrangian reads [130]

$$\mathcal{L}_{\text{vec}} = Z'_\mu\left(g^{qV}_{Z'}\bar{q}\gamma^\mu q + g^{\nu V}_{Z'}\bar{\nu}_L\gamma^\mu\nu_L\right) + \frac{1}{2}M^2_{Z'}Z'_\mu Z'^\mu. \tag{42}$$

The arising cross sections imply a re-scaling of the SM one according to the expression

$$\left(\frac{d\sigma}{dT_N}\right)_{\mathrm{SM}+Z'} = \mathcal{G}_{Z'}^2(T_N)\left(\frac{d\sigma}{dT_N}\right)_{\mathrm{SM}}, \tag{43}$$

with the Z′ factor taking the form

$$\mathcal{G}_{Z'} = 1 - \frac{1}{2\sqrt{2}G_F}\frac{\mathcal{Q}_{Z'}}{\mathcal{Q}_W^V}\frac{g_{Z'}^{\nu V}}{2MT_N + M_{Z'}^2}. \tag{44}$$

Here, $g_{Z'}^{\nu V}$ denotes the neutrino-vector coupling, while the respective Z′ charge reads [129]

$$\mathcal{Q}_{Z'} = \left(2g_{Z'}^{uV} + g_{Z'}^{dV}\right)Z + \left(g_{Z'}^{uV} + 2g_{Z'}^{dV}\right)N. \tag{45}$$

However, in the general case the $\nu - Z'$ coupling is flavor dependent $(g_{Z'}^{\nu V})_{\alpha\beta}$. Denton et al. [16] has explored this possibility and concluded that for a sufficiently small momentum transfer with respect to $M_{Z'}$, the new physics contributions can be addressed in the form of NSIs

$$\epsilon_{\alpha\beta}^{qV} = \frac{(g_{Z'}^{\nu V})_{\alpha\beta}\, g^{qV}}{2\sqrt{2}G_F M_{Z'}^2} \tag{46}$$

where the Z′ has been integrated out. Unlike the NSI case that can only modify the energy spectrum by a global factor, the additional momentum dependence expected due to the new light mediators discussed here can be well encoded in the detected signature and subsequently lead to conclusive indications of the new physics nature.

We now turn our attention on new interactions induced by the presence of a CP-even mediator. In particular, we consider a new real scalar boson ϕ with mass M_ϕ, based on the Lagrangian [130]

$$\mathcal{L}_{\mathrm{sc}} = \phi\left(g_\phi^{qS}\bar{q}q + g_\phi^{\nu S}\bar{\nu}_R\nu_L + \mathrm{H.c.}\right) - \frac{1}{2}M_\phi^2\phi^2, \tag{47}$$

with g_ϕ^{qS} and $g_\phi^{\nu S}$ representing the scalar-quark and scalar-neutrino couplings, respectively. In this framework the SM CEνNS cross section acquires an additive contribution due to the boson exchange that can be quantified in terms of the respective cross section

$$\left(\frac{d\sigma}{dT_N}\right)_{\text{scalar}} = \frac{G_F^2 M^2}{4\pi} \frac{\mathcal{G}_\phi^2 M_\phi^4 T_N}{E_\nu^2 \left(2MT_N + M_\phi^2\right)^2} F^2(T_N), \tag{48}$$

with the scalar factor G_ϕ being

$$\mathcal{G}_\phi = \frac{g_\phi^{\nu S} \mathcal{Q}_\phi}{G_F M_\phi^2}. \tag{49}$$

Analogously to the previous case, the corresponding scalar charge is defined as [130]

$$\mathcal{Q}_\phi = \sum_{\mathcal{N},q} g_\phi^{qS} \frac{m_\mathcal{N}}{m_q} f_{T,q}^{(\mathcal{N})}, \tag{50}$$

where the form factors $f_{T,q}^{(\mathcal{N})}$ capture the effective low-energy coupling of ϕ to the nucleon N={p,n} (m_N is the nucleon mass) for the quark q.

As discussed previously, different new physics signatures are expected to leave different imprints on the event and recoil spectrum. Contrary to the Z' scenario discussed above, the Dirac structure of the $\phi\nu\bar{\nu}$ vertex accounting for the scalar mediator is different (chirality-flipping) with respect to the SM one (chirality-conserving). Indeed, there is no interference between vector (or axial-vector) neutrino interactions and (pseudo-)scalar, tensor neutrino interactions [29, 131]. Therefore, the absence of interference between SM-scalar interactions gives rise to an overall modification of the expected CEνNS spectrum (see [129]). Moreover, comparing the vector and scalar cross sections it becomes evident that the corresponding signals are expected to be well distinguishable. The scalar effects are not pronounced at eV-thresholds, while on the contrary they are expected to be stronger at recoil energies of the order of keV. A thorough classification of the new physics signatures with respect to vector and scalar interactions is given in Aristizabal Sierra et al. [132] providing also key information on the possibility of breaking isospin-related degeneracies from combined measurements with different detector material.

Assuming universal couplings, one finds the equalities [130]

$$g_{Z'}^2 = \frac{g_{Z'}^{\nu V} \mathcal{Q}_{Z'}}{3A}, \quad g_\phi^2 = \frac{g_\phi^{\nu S} \mathcal{Q}_\phi}{(14A + 1.1Z)} \tag{51}$$

Using the COHERENT data, bounds have been put on the parameter planes $(g_{Z'}^2, M_{Z'})$ and (g_ϕ^2, M_ϕ) for the vector and scalar mediators, respectively [21]. In the left panel of Figure 12, the limits are shown at 90% C.L., where a degenerate area appears that cannot be excluded by the data is found due to the cancellations involved in Equation (44). However as shown in Liao and Marfatia [13] this degeneracy can be broken in the context of NSI, while for heavy mediator masses, $M_{Z'} \gg \sqrt{2MT_N}$ ~50 MeV, it remains unbroken and depends on the ratio

$$\frac{g_{Z'}^2}{M_{Z'}^2} \approx 2\sqrt{2}G_F \frac{Q_W^V}{3A} \tag{52}$$

For the case of light mediator masses $M_{Z'} \ll \sqrt{2MT_N}$, it holds

$$g_{Z'}^2 \approx 4\sqrt{2}G_F \frac{Q_W^V}{3A} MT_N \tag{53}$$

which implies that the bound is only sensitive to the coupling. The latter could be drastically improved by combining data from different detectors [133]. The case of a scalar field mediating CEvNS is explored in the right panel of Figure 12 where the extracted bounds in the (M_ϕ, g^2_ϕ) plane are depicted at 90% C.L. Significant improvements are possible through powerful analyses that are based on both energy and timing COHERENT data [17] as well as by taking into account improved quenching factors [78]. The future of CEvNS experiments will offer a complementary probe to various existing limits in the low- and high-energy regime. The currently best results for a low-energy light vector mediator of $M_{Z'}$<10MeV have been recently reported by the CONNIE Collaboration [134]. The attainable sensitivity is expected to be competitive with existing bounds from neutrino-electron scattering, dark photon searches at BaBar and LHCb results (see [30, 120]). Before closing our discussion it is important to note that, very recently CP violating effects have been also analyzed with the current and future COHERENT data in the context of light vector mediator scenarios [27]. The latter have been also found to be applicable to reactor or solar/atmospheric neutrino searches with important implications on multi-ton dark matter detectors.

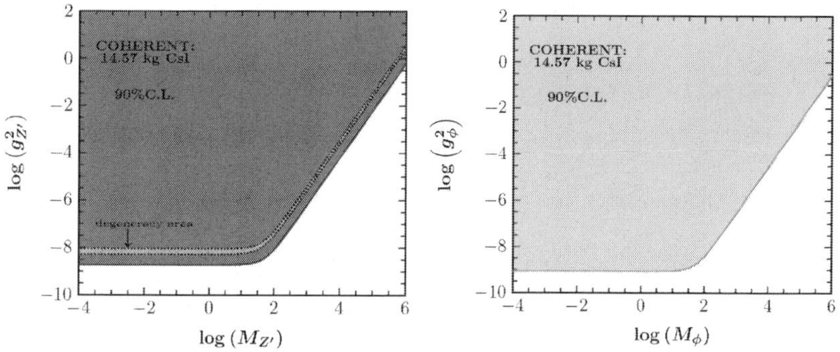

Figure 12. Excluded regions at 90% C.L. by the COHERENT experiment assuming simplified scenarios involving a Z' vector mediator **(Left panel)** and a ϕ scalar mediator **(Right panel)**. The mediator masses are in units of MeV. Figure adapted from Papoulias and Kosmas [21] under the terms of the Creative Commons Attribution 4.0 International license.

Studying Electromagnetic Neutrino Interactions

Non-trivial electromagnetic (EM) properties of massive neutrinos constitute an interesting probe to look for physics beyond the SM and at the same time they are crucial for distinguishing between the Dirac or Majorana nature of neutrinos [135]. The two main phenomenological parameters observable at a neutrino experiment are the effective neutrino magnetic moment and the neutrino charge radius (the possibility of a neutrino millicharge is explored in [23, 106]). Assuming Majorana neutrinos, the EM neutrino vertex is described by the electromagnetic field tensor $F_{\alpha\beta}$ of the effective Hamiltonian [3, 136].

$$H_{\mathrm{EM}}^{\mathrm{M}} = -\frac{1}{4}\nu_L^T C^{-1}\lambda\sigma^{\alpha\beta}\nu_L F_{\alpha\beta} + \mathrm{H.c.}, \tag{54}$$

while for the case of Dirac neutrinos one has

$$H_{\mathrm{EM}}^{\mathrm{D}} = \frac{1}{2}\bar{\nu}_R\lambda\sigma^{\alpha\beta}\nu_L F_{\alpha\beta} + \mathrm{H.c.} \tag{55}$$

It is important to note that for Majorana (Dirac) neutrinos $\lambda = \mu - i\epsilon$ is an antisymmetric complex (general complex) matrix. The two imaginary matrices μ (magnetic moment) and ϵ (electric dipole moment) obey the respective properties $\mu^T = -\mu$ ($\mu = \mu^\dagger$) while $\epsilon^T = -\epsilon$ ($\epsilon = \epsilon^\dagger$). It thus becomes evident that, unlike the Dirac case, for Majorana neutrinos the diagonal moments are vanishing μMii= ϵMii = 0μiiM= ϵiiM = 0.

For a low-energy neutrino scattering experiment the observable neutrino magnetic moment (flavor dependent) is in fact a combination of the neutrino transition magnetic moments (TMMs) discussed above. In the mass basis it reads [137]

$$\left(\mu_{\nu}^{M}\right)^{2} = \tilde{a}_{-}^{\dagger}\tilde{\lambda}^{\dagger}\tilde{\lambda}\tilde{a}_{-} + \tilde{a}_{+}^{\dagger}\tilde{\lambda}\tilde{\lambda}^{\dagger}\tilde{a}_{+} . \tag{56}$$

In Equation (56) the following transformations have been introduced

$$\tilde{a}_{-} = U^{\dagger}a_{-}, \qquad \tilde{a}_{+} = U^{\mathsf{T}}a_{+}, \qquad \tilde{\lambda} = U^{\mathsf{T}}\lambda U , \tag{57}$$

where the 3−vectors a_+ and a_- denote positive and negative helicity states, respectively, while the magnetic moment matrix $\lambda\,(\tilde{\lambda})$ in the flavor (mass) basis is written as [138]

$$\lambda = \begin{pmatrix} 0 & \Lambda_{\tau} & -\Lambda_{\mu} \\ -\Lambda_{\tau} & 0 & \Lambda_{e} \\ \Lambda_{\mu} & -\Lambda_{e} & 0 \end{pmatrix}, \qquad \tilde{\lambda} = \begin{pmatrix} 0 & \Lambda_{3} & -\Lambda_{2} \\ -\Lambda_{3} & 0 & \Lambda_{1} \\ \Lambda_{2} & -\Lambda_{1} & 0 \end{pmatrix} \tag{58}$$

with $\Lambda_{\alpha} = |\Lambda_{\alpha}|\, e^{i\zeta_{\alpha}}$ and $\Lambda_{i} = |\Lambda_{i}|\, e^{i\zeta_{i}}$ being the TMMs in the flavor and mass basis, respectively, where ζ_α and ζ_i denote the corresponding CP-phases.

The potential EM neutrino properties appear in the form of an effective neutrino magnetic moment that is conveniently expressed in the mass basis according to Equation (56) in terms of fundamental parameters (TMMs, CP-violating phases and neutrino mixing angles). The latter induce an additive contribution to the SM cross section [139]

$$\left(\frac{d\sigma}{dT_N}\right)_{\text{SM+EM}} = \mathcal{G}_{\text{EM}}(E_{\nu}, T_N)\left(\frac{d\sigma}{dT_N}\right)_{\text{SM}} , \tag{59}$$

where the EM factor reads [21]

$$\mathcal{G}_{\text{EM}} = 1 + \frac{1}{G_F^2 M}\left(\frac{\mathcal{Q}_{\text{EM}}}{\mathcal{Q}_W^V}\right)^2 \frac{\frac{1-T_N/E_{\nu}}{T_N}}{1-\frac{MT_N}{2E_{\nu}^2}} . \tag{60}$$

Here, the E_M charge Q_{EM} is written in terms of the fine structure constant a_{EM} and the effective neutrino magnetic moment, as [115]

$$\mathcal{Q}_{\text{EM}} = \frac{\pi a_{\text{EM}}\mu_{\nu_{\alpha}}}{m_e} Z . \tag{61}$$

Moreover, the effect of a neutrino charge radius can be taken into consideration through a shift in the definition of the weak mixing angle [140]

$$\sin^2\theta_W \rightarrow \sin^2\overline{\theta_W} + \frac{\sqrt{2}\pi\, a_{\rm EM}}{3G_F}\langle r^2_{\nu_\alpha}\rangle \qquad (62)$$

where by $\sin^2\overline{\theta_W}$ it is denoted the low energy value of the weak mixing angle, e.g., $\hat{s}^2_Z = 0.2382$.

The presence of a neutrino magnetic moment is expected to yield a distortion in the recoil spectrum during the CEνNS process, i.e., a detectable excess of events especially for low recoil energies. The left panel of Figure 13 shows the χ^2 profile of the effective neutrino magnetic moment extracted by the first light of COHERENT data. A similar analysis has been performed in order to quantify the sensitivity of COHERENT on the neutrino charge radius as shown in the right panel of the same figure. Note that, an essential improvement due to a more accurate treatment of the QF is possible (see [78] for more details).

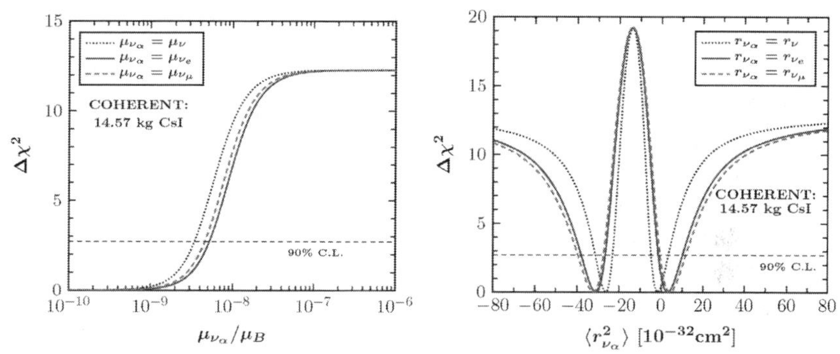

Figure 13. Constraints on electromagnetic neutrino properties by the COHERENT experiment. **(Left)** Sensitivity to the effective neutrino magnetic moment $\mu_{\nu\alpha}$. **(Right)** Sensitivity to the neutrino charge-radius ⟨r2να⟩⟨rνα2⟩. Figure adapted from Papoulias and Kosmas [21] under the terms of the Creative Commons Attribution 4.0 International license.

The authors of Miranda et al. [22] performed a comprehensive analysis on the sensitivity of various existing and future CEνNS experiments and extracted constraints on the different components Λ_i of the neutrino TMM matrix. In particular, their study focused on existing and next generation experimental setups of COHERENT as well as on the expected data from the future reactor-based experiments: CONUS, CONNIE, TEXONO, MINER, and RED100. In a similar manner, Billard et al. [120] extracted constraints focusing on the NUCLEUS and Ricochet detectors at the Chooz NPP, however assuming the effective neutrino magnetic moment. Miranda et al. [22] performed a systematic combined analysis with regards to the TMMs exploring also the effects of the CP violating phases of the complex matrix given in Equation (58). As a concrete example, Figure 14 shows the contours in the ($|\Lambda_i|-|\Lambda_j|$) parameter plane for the case of current and next generation reactor-based CEvNS experiments. It is worth mentioning that these bounds are comparable to existing ones from low energy solar neutrino data at Borexino phase-II [141]. Figure 15 shows the sensitivity contours in the $(\langle r^2_{\nu_\alpha}\rangle, \langle r^2_{\nu_\beta}\rangle)$ plane that resulted from the COHERENT data. A similar analysis is performed in Papoulias [78] and Khan and Rodejohann [122], while for a comprehensive fit including energy and timing data the reader is referred to Cadeddu et al. [106].

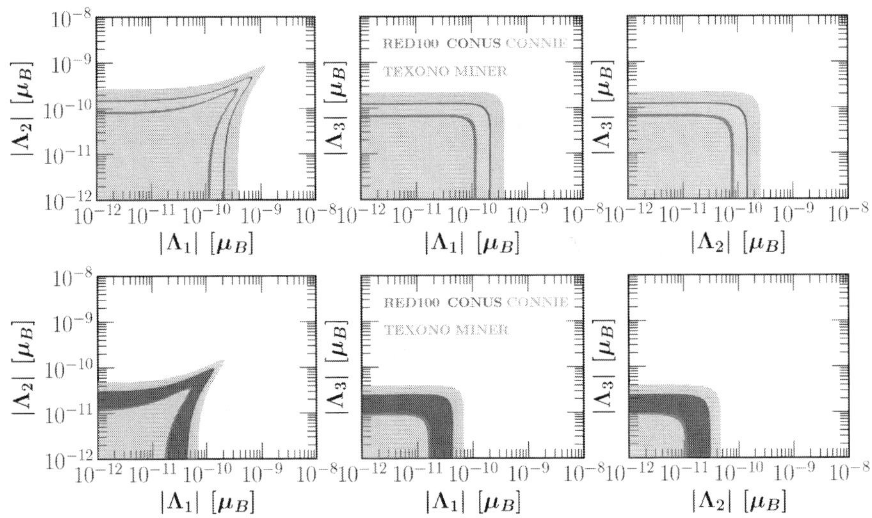

Figure 14. Sensitivity contours at 90% C.L. in the $|\Lambda_i| - |\Lambda_j|$ parameter space that follow from the expected results of the current and future reactor based CEvNS experiments (see the text). The calculation assumes vanishing $|\Lambda_k|$ and vanishing CP phases ζ_i. Figure adapted from Miranda et al. [22] under the terms of the Creative Commons Attribution 4.0 International license.

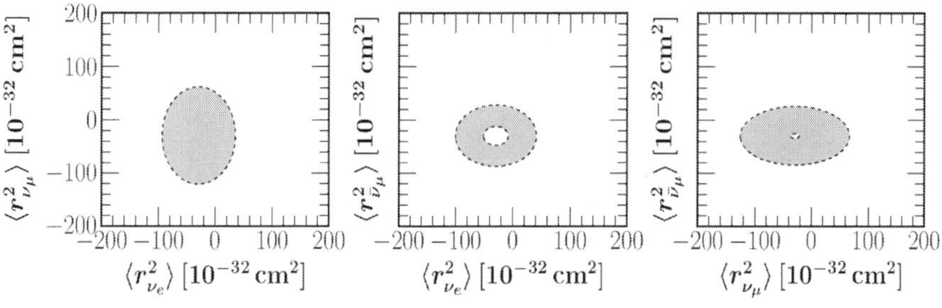

Figure 15. Sensitivity in the (⟨r2να⟩,⟨r2νβ⟩)(⟨rνα2⟩,⟨rνβ2⟩) parameter space at 90% C.L. from the analysis of the COHERENT data. Taken from Papoulias [78].

The Existence of the Sterile Neutrinos

The three-neutrino paradigm has been put in rather solid grounds from the interpretation of solar and atmospheric oscillation data. On the other hand, controversial anomalies such as those coming from recent reactor data as well as existing anomalies implied by the LSND and MiniBooNE experiments inspired a reach phenomenology beyond the three-neutrino oscillation picture, based on the existence of a fourth sterile neutrino state with eV-scale mass ($m_{1,2,3} \ll m_4$) and tiny mixing angles. To accommodate sterile neutrinos the lepton mixing matrix is minimally extended so that the flavor eigenstates ν_α, (α = e, μ, τ, s, $\cdots$) are related to the mass eigenstates ν_i (i = 1, 2, 3, 4, $\cdots$) by the unitary transformation $\nu_\alpha = \sum_i U_{\alpha i} \nu_i$. Then, for the short-baseline CEνNS experiments the survival probability of an active neutrino at a distance L is written [26]

$$P_{\alpha \to e,\mu,\tau} = 1 - 4\left|U_{\alpha 4}\right|^2 \left(1 - \sum_{\beta = e,\mu,\tau} \left|U_{\beta 4}\right|^2\right) \sin^2(\Delta) , \tag{63}$$

with the abbreviation $\Delta \equiv \Delta m^2 L/4E_\nu$ and the mass splittings under the approximation $\Delta m^2_{41} \approx \Delta m^2_{42} \approx \Delta m^2_{43} \equiv$≡$\Delta m^2$. At this point it should be stressed that neutrino-nucleus scattering experiments are favorable facilities to probe sterile neutrinos being complementary to dedicated experiments such as MINOS/MINOS+ [142], MiniBooNE [143], Daya-Bay [144], Juno [145], and NEOS [146]. Indeed, due to the purely neutral-current character of the CEνNS process it is not necessary to disentangle between active-sterile neutrino mixing [147].The possibility of investigating sterile neutrinos in the simplest (3+1) scheme through the CEνNS process was examined for the first time in Anderson et al. [148], relying on an SNS source. A combined sterile neutrino analysis was performed in Kosmas et al. [24] highlighting the complementarity between accelerator and reactor neutrino sources, by focusing on COHERENT and TEXONO experiments, respectively (see Figure 16). Moreover, a detailed study of various reactor-based CEvNS proposals has been carried out in Cañas et al. [25], showing how such future measurements can be exploited to solve the reactor antineutrino anomaly. After the first observation of CEvNS by the COHERENT experiment, Papoulias and Kosmas [21] reported the first constraints under the assumption of a universal new mixing angle, extracting the conclusion that the current sensitivity is rather poor. By exploiting timing data the potential of a future measurement at the next generation of COHERENT with a 100 kg CsI detector has been demonstrated in Blanco et al. [26], concluding that the prospects of probing the exclusion regions in the (Δm241Δm412–sin22θeμsin22θeμ) plane from the latest MiniBooNE [143] and LSND [149] are promising. Finally, focusing at CONUS in Berryman [150] it was shown that the complementarity between terrestrial-cosmological experiments may resolve the tension raised by astrophysical observations regarding the existence of sterile neutrinos.

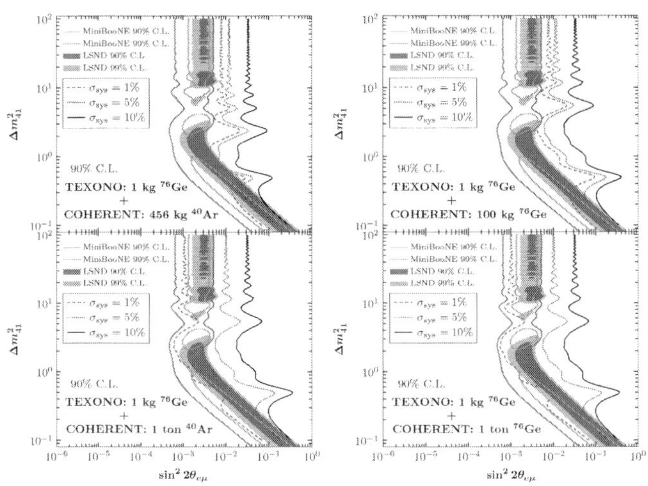

Figure 16. Exclusion curves at 90% C.L. in the (Δm241Δm412–sin22θeμsin22θeμ) parameter plane from a combined analysis of COHERENT and TEXONO experiments. The results are compared to existing constraints from MiniBooNE and LSND. Figure reproduced from Kosmas et al. [24] with the permission of the American Physical Society and with updated results from MiniBooNE [143].

Summary of Constraints

Emphasis has been put on the physics beyond the SM by devoting a great part to the past and current research efforts and by concentrating on the various channels contributing to CEvNS processes and their interpretation. Through a χ^2 sensitivity analysis, based on the recoil or timing spectra of the COHERENT data, the current limits are listed at 90% C.L. in Table 2. For a given parameter set SS, the best fit is found through the minimum value χ2min(S)χmin2(S). The limits involve electroweak (weak-mixing angle), nuclear (nuclear radius), and physics beyond the SM (NSIs and EM neutrino properties). Significant improvements are expected through a more accurate determination of the QF and from a better control of the systematic uncertainties. The reported constraints on electroweak and NSIs have been extracted with various analysis methods, i.e., by combining existing APV measurements or global oscillation constraints with the recent COHERENT data, emphasizing the complementarity of CEvNS data in the low energy regime.

Table 2. Constraints on electroweak, nuclear and new physics parameters at 90% C.L. after the first CEvNS measurement by the COHERENT experiment.

Parameter	Dataset	References	Limit (90% C.L.)
$\sin^2\theta_W$[a]	COHERENT + APV	[101]	$0.239^{+0.006}_{-0.007}$
R_n			$5.42^{+0.50}_{-0.50}$
ϵ_{ee}^{uV}			0.028 – 0.60
ϵ_{ee}^{dV}	COHERENT + oscillation	[151]	0.030 – 0.55
$\epsilon_{\mu\mu}^{uV}$			−0.088 – 0.37
$\epsilon_{\mu\mu}^{dV}$			−0.075 – 0.33
ϵ_{ee}^{uT}			−0.013 – 0.013
ϵ_{ee}^{dT}	COHERENT (recoil)	[21]	−0.011 – 0.011
$\epsilon_{\mu\mu}^{uT}$			−0.013 – 0.013
$\epsilon_{\mu\mu}^{dT}$			−0.011 – 0.011
μ_{ν}			< 43
μ_{ν_e}	COHERENT (recoil)	[21]	< 52
μ_{ν_μ}			< 46
$\langle r^2_{\nu_e}\rangle$			−63 – 12
$\langle r^2_{\nu_\mu}\rangle$			−7 – 9
$\langle r^2_{\nu_{e\mu}}\rangle$	COHERENT (timing and recoil)	[152]	< 22
$\langle r^2_{\nu_{e\tau}}\rangle$			< 37
$\langle r^2_{\nu_{\mu\tau}}\rangle$			< 26

[a] *The limit is shown at 1σ.*

The limits are presented in units of. fm for the nuclear rms radius, 10^{-10} μ_B for the neutrino magnetic moment and 10^{-32} cm^2 for the neutrino charge radius.

CONNECTION OF CENNS WITH DARK MATTER, CLFV PROCESSES AND ASTROPHYSICS

Neutrino-nucleus scattering is one of the dominant processes taking place in SN environment and thus the emitted neutrinos can be an extremely useful tool for deep sky investigations. Moreover, SN constitute an ideal source for flavor physics applications since all flavors are involved. Going beyond the SM, potential FCNC under stellar conditions can modify the percentage of the neutrino flavors in the interior of massive stars [58]. The latter, may drastically affect a plethora of other processes governing the explosive-stellar nucleosynthesis [153], causing significant alteration of the evolution phenomena [114, 154]. If large enough, the modified neutrino energy-densities arriving at the terrestrial SN-neutrino detectors can be tested at CEνNS experiments [155]. It should be stressed that a SN neutrino burst can be well detected by the current technology DM detectors.

Direct Dark Matter detection experiments are expected to be sensitive to astrophysical neutrinos from the Sun, the Atmosphere and from core-collapse SN (e.g., diffuse supernova background, DSNB) [156, 157]. The neutrino-floor [158], being an irreducible background determines the criteria for using the appropriate detector material, threshold, mass, etc. Figure 17 illustrates the differential and integrated event rate of CEνNS expected at a ton-scale DM detector, calculated in the framewrok of the DSM assuming only SM interactions. Future precision measurements at such rare-event facilities may become sensitive to nuclear structure effects which in principle can be explored by experiments looking for CEνNS. Therefore precise information on the nuclear form factors becomes very relevant for DM detectors especially for those involving multi-ton mass scale [35]. For example, alterations are expected at high recoil energies of neutrino-induced interactions at direct DM detection searches [39] which on the other hand may be limited by the current uncertainties of the Atmospheric and DSNB neutrinos. Models involving light mediators are well testable at CEνNS searches and may offer a key solution to LMA-Dark [151, 159] as well as implications to DM searches [28, 133] and to the neutrino floor [40]. In the same spirit, combined analyses of oscillation and CEνNS data [15, 18] concluded that the LMA-D solution is excluded at 3.1σ (3.6σ) for NSI with up (down) quarks. Finally, it has been recently pointed out that potential DM-induced signatures from dark photon decay could be also detectable at CEνNS experiments, explaining an excess in the timing distribution of the COHERENT signal [44].

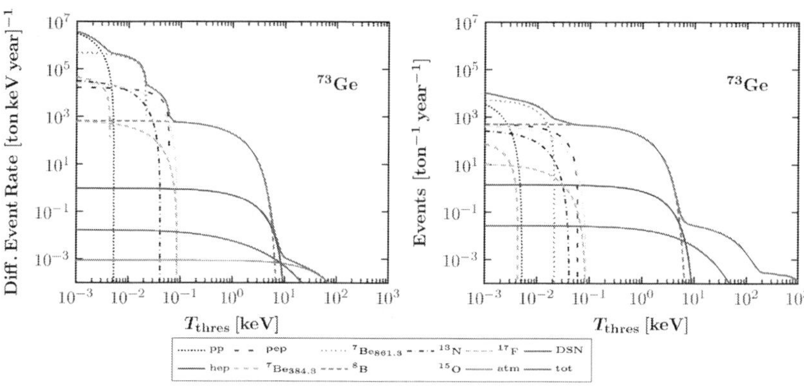

Figure 17. Differential event rate **(Left)** and total number of events above threshold **(Right)** expected due to CEvNS from solar, Atmospheric and DSNB neutrinos at a Germanium detector. Figure adapted from Papoulias et al. [39] under the terms of the Creative Commons Attribution 4.0 International license.

In the case of cLFV processes of μ → e transitions, especially coherent $\mu^- \to e^-$ conversion in the field of nuclei has attracted much interest in the context of new physics mechanisms discussed in this article [160]. For example, μ → e conversion has been studied in the context of the inverse seesaw [109] and new Z′ mediators [161]. It is given by

$$\mu^- + (A, Z) \to e^- + (A, Z), \qquad (64)$$

which might have close relations to the process given in Equation (36) in the neutral sector. When the final nuclear state coincides the ground state this process could be a coherent channel, which in fact dominates by its enhancement by a factor of the square of the number of nucleons in nuclei. The cLFV processes are known to be highly suppressed in the SM even with lepton mixing due to the small neutrino masses, down to O(10−54) O(10-54) [162]. However, many theoretical models involving NSI predict sizable rates which the future experiments could reach [163]. The future experiments aiming to search for $\mu^- \to e^-$ conversion are under preparation at J-PARC, Japan (COMET) [164] and Fermilab, in the USA (Mu2e) [165]. They expect to measure a characteristic peak of outgoing electrons (at energy $E_e \approx m_\mu$) emitted from muonic atoms in a target. These experiments are aiming at sensitivities of the order of O(10−17)O(10-17) to O(10−18) O(10-18), which is a factor of 10,000 or more improvement over the current experimental limits. Therefore they have excellent potential to establish or rule out the presence of new physics in the near future.

It is important to notice that theoretically the $\mu^- \rightarrow e^-$ branching ratio depends on the nuclear form factor which can be probed from CEνNS measurements as discussed in Sect 3. For the relevant nuclei such as ^{27}Al and ^{48}Ti the nuclear form factors at q≈mμ=0.53 fm−1 have values 0.63 and 0.53, respectively, i.e., well far from the approximation of point like nucleus (see [81] for a detailed discussion). The incoherent channels of $\mu \rightarrow e$ conversion can be studied with the matrix elements described in section 2 (see for example [82, 166]). Once $\mu^- \rightarrow e^-$ conversion is observed, it holds significant potential for constraining the parameters of the NSI Lagrangian of the lepton-nucleus interactions [167]. It may shed light on FCNC processes in the leptonic sector [168–170], and particularly on the existence of the charged-lepton mixing which is analogous to neutrino oscillations at short baseline experiments.

SUMMARY AND CONCLUSIONS

In this review article, we made an attempt to summarize the main research efforts devoted to the conventional and exotic neutrino-nucleus interactions, in the recent years. The standard process of neutral-current neutrino-nucleus scattering, mediated by the neutral Z-boson boson presents two channels the elastic and inelastic scattering of neutrinos (and anti-neutrinos) off a nuclear isotope (A, Z), with A nucleons and Z protons. In the elastic process, the initial and final states of the target nucleus are the same and the detectable signal is an energy recoil, whereas in the case of the inelastic channel, the final nucleus is an excited state with the signal being a de-excitation product (gammas). We have mainly concentrated on beyond the SM neutrino-nucleus interactions, and especially on the prospects of extracting new physics from the operating prominent rare-event detectors looking for the coherent elastic neutrino-nucleus scattering. Such channels may involve lepton LFV in neutral-currents. This is motivated by the recent measurements of CEνNS events at the COHERENT experiment, the analysis and interpretation of which may imply the necessity of including non-standard neutrino-nucleus interactions. Toward this end, we discussed the impact of non-standard interactions and novel Z' or ϕ mediators to the CEνNS event rates providing an estimation of the attainable sensitivities at current and future experiments. With regards to neutrino oscillations constraints on NSIs from neutral current interactions at CEνNS experiments are complementary since the former are only to sensitive to differences between the diagonal terms. It is furthermore expected that the next generation of the currently operating experiments like the COHERENT, TEXONO, MINER, CONUS, RED100, νGEN, Ricochet, NUCLEUS etc., will be of benefit to unravel open issues of the leptonic sector. The studies covered in this review article have evident connection with neutrino astronomy, SN physics, direct DM detection and cLFV processes. To understand these new interactions the proton and neutron weak nuclear form factors play key roles. This opens up the necessity of measuring the neutron nuclear form factors by appropriately designed and appreciably sensitive experiments such as those looking for CEνNS processes.

Author Contributions

DP: sensitivity analysis. TK: incoherent formalism. YK: experimental physics.

Conflict of Interest

The authors declare that the research was conducted in the absence of any commercial or financial relationships that could be construed as a potential conflict of interest.

ACKNOWLEDGMENTS

The authors are grateful to Valentina De Romeri and Jorge Terol Calvo for fruitful discussions as well as to M. Cadeddu and F. Dordei for useful correspondence. DP is supported by the Spanish grants SEV-2014-0398 and FPA2017-85216-P (AEI/FEDER, UE), PROMETEO/2018/165 (Generalitat Valenciana) and the Spanish Red Consolider MultiDark FPA2017-90566-REDC. YK acknowledges support by the JSPS KAKENHI Grant No. 18H04231.

Footnotes

1. ^Typically 25-150 MeV for most nuclei.
2. ^BASKET [90] is a synergy of Ricochet and NUCLEUS that is developing a Li_2WO_4[Mo] Scintillating bolometer.
3. ^See e.g., talk by Ran Han: 10.5281/zenodo.3464505

Appendix

A. Multipole Operators

The Donnelly-Walecka multipole decomposition method yields a set of eight linearly independent irreducible tensor operators which are typically expressed in terms of the Spherical Bessel functions, j_l, and combined with the Spherical Harmonics, $Y^L{}_M$, or the vector Spherical Harmonics, Y^L_M [171]:

$$M^J_M(\kappa\mathbf{r}) = \delta_{LJ} j_L(\kappa r) Y^L_M(\hat{r}), \tag{A1}$$

$$\mathbf{M}^{(L1)J}_M(\kappa\mathbf{r}) = j_L(\kappa r)\mathbf{Y}^{(L1)J}_M(\hat{r}), \tag{A2}$$

with

$$\mathbf{Y}^{(L,1)J}_M(\hat{r}) = \sum_{M_L,\lambda} \langle LM_L 1\lambda | JM\rangle Y^L_{M_L}(\hat{r})\mathbf{e}_\lambda \tag{A3}$$

As a consequence of the V-A structure of electroweak interactions

$$\hat{\mathcal{J}}_\mu = \hat{J}_\mu - \hat{J}^5_\mu = (\hat{\rho}, \hat{\mathbf{J}}) - (\hat{\rho}^5, \hat{\mathbf{J}}^5), \tag{A4}$$

four operators are associated to the vector component $\hat{J}_\lambda = (\hat{\rho}, \hat{\mathbf{J}})$ and four to the axial-vector component $\hat{J}^5_\lambda = (\hat{\rho}^5, \hat{\mathbf{J}}^5)$ of the hadronic current

$$\hat{\mathcal{M}}_{JM}(\kappa) = \hat{M}^{coul}_{JM} - \hat{M}^{coul5}_{JM} = \int d\mathbf{r} M^J_M(\kappa\mathbf{r})\hat{\mathcal{J}}_0(\mathbf{r}), \tag{A5}$$

$$\hat{\mathcal{L}}_{JM}(\kappa) = \hat{L}_{JM} - \hat{L}^5_{JM} = i\int d\mathbf{r}\left(\frac{1}{\kappa}\nabla M^J_M(\kappa\mathbf{r})\right)\cdot\hat{\mathcal{J}}(\mathbf{r}), \tag{A6}$$

$$\hat{\mathcal{T}}^{el}_{JM}(\kappa) = \hat{T}^{el}_{JM} - \hat{T}^{el5}_{JM} = \int d\mathbf{r}\left(\frac{1}{q}\nabla\times\mathbf{M}^{JJ}_M(\kappa\mathbf{r})\right)\cdot\hat{\mathcal{J}}(\mathbf{r}), \tag{A7}$$

$$\hat{\mathcal{T}}^{mag}_{JM}(\kappa) = \hat{T}^{mag}_{JM} - \hat{T}^{mag5}_{JM} = \int d\mathbf{r}\mathbf{M}^{JJ}_M(\kappa\mathbf{r})\cdot\hat{\mathcal{J}}(\mathbf{r}) \tag{A8}$$

where $\kappa = |\mathbf{q}|$ denotes the 3-momentum transfer. Note that, the vector component yields the Coulomb M^{coul}_{JM}, longitudinal L_{JM}, transverse electric T^{el}_{JM} (normal parity $\pi = (-)^J$) and transverse magnetic T^{mag}_{JM} (abnormal parity $\pi = (-)^{J+1}$), while regarding the axial-vector component M^{coul5}_{JM}, L^5_{JM}, T^{el5}_{JM} have abnormal parity and T^{mag5}_{JM} has normal parity. The matrix elements of the above operators involve momentum dependence of the nucleon form factors $F_X(Q^2), X = 1, A, P$ and $\mu^V(Q^2)$

$$\hat{M}^{coul}_{JM}(\kappa\mathbf{r}) = F^V_1(Q^2)M^J_M(\kappa\mathbf{r}), \tag{A9}$$

$$\hat{L}_{JM}(\kappa\mathbf{r}) = \frac{q_0}{\kappa}\hat{M}^{coul}_{JM}(\kappa\mathbf{r}), \tag{A10}$$

$$\hat{T}^{el}_{JM}(\kappa\mathbf{r}) = \frac{\kappa}{m_N}\left[F_1^V(Q^2)\Delta'^J_M(\kappa\mathbf{r}) + \frac{1}{2}\mu^V(Q^2)\Sigma^J_M(\kappa\mathbf{r})\right], \quad \text{(A11)}$$

$$i\hat{T}^{mag}_{JM}(\kappa\mathbf{r}) = \frac{\kappa}{m_N}\left[F_1^V(Q^2)\Delta^J_M(\kappa\mathbf{r}) - \frac{1}{2}\mu^V(Q^2)\Sigma'^J_M(\kappa\mathbf{r})\right], \quad \text{(A12)}$$

$$i\hat{M}^5_{JM}(\kappa\mathbf{r}) = \frac{\kappa}{m_N}\left[F_A(Q^2)\Omega^J_M(\kappa\mathbf{r}) + \frac{1}{2}\left(F_A(Q^2) + q_0F_P(Q^2)\right)\Sigma''^J_M(\kappa\mathbf{r})\right], \quad \text{(A13)}$$

$$-i\hat{L}^5_{JM}(\kappa\mathbf{r}) = \left[F_A(Q^2) - \frac{\kappa^2}{2m_N}F_P(Q^2)\right]\Sigma''^J_M(\kappa\mathbf{r}), \quad \text{(A14)}$$

$$-i\hat{T}^{el5}_{JM}(\kappa\mathbf{r}) = F_A(Q^2)\Sigma'^J_M(\kappa\mathbf{r}), \quad \text{(A15)}$$

$$\hat{T}^{mag5}_{JM}(\kappa\mathbf{r}) = F_A(Q^2)\Sigma^J_M(\kappa\mathbf{r}). \quad \text{(A16)}$$

It becomes evident that only seven are linearly independent

$$T_1^{JM} \equiv M^J_M(\kappa\mathbf{r}) = \delta_{LJ}\, j_L(\kappa\mathbf{r})Y^L_M(\hat{r}), \quad \text{(A17)}$$

$$T_2^{JM} \equiv \Sigma^J_M(\kappa\mathbf{r}) = \mathbf{M}^{JJ}_M \cdot \boldsymbol{\sigma}, \quad \text{(A18)}$$

$$T_3^{JM} \equiv \Sigma'^J_M(\kappa\mathbf{r}) = -i\left[\frac{1}{\kappa}\nabla \times \mathbf{M}^{JJ}_M(\kappa\mathbf{r})\right]\cdot\boldsymbol{\sigma}, \quad \text{(A19)}$$

$$T_4^{JM} \equiv \Sigma''^J_M(\kappa\mathbf{r}) = \left[\frac{1}{\kappa}\nabla M^J_M(\kappa\mathbf{r})\right]\cdot\boldsymbol{\sigma}, \quad \text{(A20)}$$

$$T_5^{JM} \equiv \Delta^J_M(\kappa\mathbf{r}) = \mathbf{M}^{JJ}_M(\kappa\mathbf{r})\cdot\frac{1}{\kappa}\nabla, \quad \text{(A21)}$$

$$T_6^{JM} \equiv \Delta'^J_M(\kappa\mathbf{r}) = -i\left[\frac{1}{\kappa}\nabla \times \mathbf{M}^{JJ}_M(\kappa\mathbf{r})\right]\cdot\frac{1}{\kappa}\nabla, \quad \text{(A22)}$$

$$T_7^{JM} \equiv \Omega^J_M(\kappa\mathbf{r}) = M^J_M(\kappa\mathbf{r})\boldsymbol{\sigma}\cdot\frac{1}{\kappa}\nabla. \quad \text{(A23)}$$

In the proton-neutron representation, $T_i^{JM}(\kappa\mathbf{r}),\ i = 1, 2, \cdots, 7$ can be written in closed form [65]

$$\langle j_1 || T_i^J || j_2 \rangle = e^{-y} y^{\beta/2} \sum_{\mu=0}^{n_{max}} \mathcal{P}^{i,J}_{\mu} y^{\mu}, \qquad i = 1, \cdots, 7. \quad \text{(A24)}$$

B. Coefficients for Calculating the Charge Density Distribution and Form Factors in the Context of FOP

The coefficients θ_λ of the polynomial $\Phi(|\mathbf{q}|b, Z)$ are evaluated, as

$$\theta_\lambda = \frac{\sqrt{\pi}}{4^\lambda} \sum_{\substack{(n,l)j \\ (2n+l>\lambda)}}^{N_{max}} \sum_{m=s}^{2n} \frac{(2j+1)n!C^m_{nl}\Lambda_\lambda(m+l,0)(l+m)!}{2\Gamma(n+l+\frac{3}{2})} \quad \text{(A25)}$$

In the latter expression, $\Gamma(x)$ denotes the Gamma function while the definition of the index s is

$$s = \begin{cases} 0, & \text{if } \lambda - l \leq 0 \\ \lambda - l & \text{if } \lambda - l > 0 \end{cases} \tag{A26}$$

and

$$\Lambda_k(n,l) = \frac{(-)^k}{k!}\binom{n+l+1/2}{n-k}, \quad C_{nl}^{m} = \sum_{k=0}^{m} \Lambda_{m-k}(n,l)\Lambda_k(n,l) \tag{A27}$$

The corresponding coefficients f_λ are written as

$$f_\lambda = \sum_{(n,l)j} \frac{\pi^{1/2}(2j+1)n!\,C_{nl}^{\lambda-l}}{2\Gamma\left(n+l+\frac{3}{2}\right)}. \tag{A28}$$

As a concrete example the coefficients θ_λ and f_λ for even-even nuclei up to ^{50}Sn are listed in Table A1.

Table A1. Calculated coefficients f_λ (θ_λ) for the determination of the proton/neutron density distributions (nuclear form factors).

Z (N)	(nl)j	λ = 0		λ = 1		λ = 2		λ = 3		λ = 4	
2	$0s_{1/2}$	2	(2)								
6	$0p_{3/2}$	2	(6)	$\frac{8}{3}$	$(-\frac{2}{3})$						
8	$0p_{1/2}$	2	(8)	4	(−1)						
14	$0d_{5/2}$	2	(14)	4	(−3)	$\frac{8}{5}$	$(\frac{1}{10})$				
18	$0d_{3/2}$	2	(18)	4	$(-\frac{13}{3})$	$\frac{8}{3}$	$(\frac{1}{6})$				
20	$1s_{1/2}$	5	(20)	0	(−5)	4	$(\frac{1}{4})$				
22	$1p_{1/2}$	5	(22)	$\frac{10}{3}$	(−6)	$\frac{4}{3}$	$(\frac{13}{3})$	$\frac{8}{15}$	$(-\frac{1}{120})$		
30	$0f_{7/2}$	5	(30)	$\frac{10}{3}$	(−10)	$\frac{4}{3}$	$(\frac{5}{6})$	$\frac{8}{7}$	$(-\frac{1}{56})$		
34	$1p_{3/2}$	5	(34)	10	(−12)	−4	$(\frac{6}{5})$	$\frac{232}{105}$	$(-\frac{29}{840})$		
40	$0f_{5/2}$	5	(40)	10	(−15)	−4	$(\frac{3}{2})$	$\frac{8}{3}$	$(-\frac{1}{24})$		
50	$0h_{9/2}$	5	(50)	10	$(-\frac{65}{3})$	−4	$(\frac{5}{2})$	$\frac{8}{3}$	$(-\frac{5}{56})$	$\frac{32}{189}$	$(\frac{1}{1512})$

Table adapted from Papoulias and Kosmas [[illegible]] under the terms of the Creative Commons Attribution 4.0 International license.

REFERENCES

1. Ejiri H, Suhonen J, Zuber K. Neutrino-nuclear responses for astro-neutrinos, single beta decays and double beta decays. Phys Rept. (2019) **797**:1–102. doi: 10.1016/j.physrep.2018.12.001
2. Schechter J, Valle JWF. Neutrino masses in SU(2) x U(1) theories. Phys Rev D. (1980) **22**:2227. doi: 10.1103/PhysRevD.22.2227
3. Schechter J, Valle JWF. Neutrino decay and spontaneous violation of lepton number. Phys Rev D. (1982) **25**:774. doi: 10.1103/PhysRevD.25.774
4. Kosmas TS, Oset E. Charged current neutrino nucleus reaction cross-sections at intermediate-energies. Phys Rev C. (1996) **53**:1409–15. doi: 10.1103/PhysRevC.53.1409
5. Ejiri H. Nuclear spin isospin responses for low-energy neutrinos. Phys Rept. (2000) **338**:265–351. doi: 10.1016/S0370-1573(00)00044-2
6. Balasi KG, Langanke K, Martínez-Pinedo G. Neutrino–nucleus reactions and their role for supernova dynamics and nucleosynthesis. Prog Part Nucl Phys. (2015) **85**:33–81. doi: 10.1016/j.ppnp.2015.08.001
7. Freedman DZ. Coherent neutrino nucleus scattering as a probe of the weak neutral current. Phys Rev D. (1974) **9**:1389–92. doi: 10.1103/PhysRevD.9.1389
8. Tubbs DL, Schramm DN. Neutrino opacities at high temperatures and densities. Astrophys J. (1975) **201**:467–88. doi: 10.1086/153909
9. Drukier A, Stodolsky L. Principles and applications of a neutral current detector for neutrino physics and astronomy. Phys Rev D. (1984) **30**:2295. doi: 10.1103/PhysRevD.30.2295
10. Akimov D, Albert JB, An P, Awe C, Barbeau PS, Becker B, et al. Observation of coherent elastic neutrino-nucleus scattering. Science. (2017) **357**:1123–6. doi: 10.1126/science.aao0990
11. Akimov D, Albert JB, An P, Awe C, Barbeau PS, Becker B, et al. COHERENT collaboration data release from the first observation of coherent elastic neutrino-nucleus scattering. arXiv:1804.09459.
12. Aristizabal Sierra D, Balantekin AB, Caratelli D, Cogswell B, Collar JI, Dahl CE, et al. Proceedings of the magnificent CEvNS workshop 2018. arXiv:1910.07450.
13. Liao J, Marfatia D. COHERENT constraints on nonstandard neutrino interactions. Phys Lett B. (2017) **775**:54–7. doi: 10.1016/j.

physletb.2017.10.046

14. Dent JB, Dutta B, Liao S, Newstead JL, Strigari LE, Walker JW. Accelerator and reactor complementarity in coherent neutrino-nucleus scattering. Phys Rev D. (2018) **97**:035009. doi: 10.1103/PhysRevD.97.035009
15. Aristizabal Sierra D, Rojas N, Tytgat MHG. Neutrino non-standard interactions and dark matter searches with multi-ton scale detectors. JHEP. (2018) **1803**:197. doi: 10.1007/JHEP03(2018)197
16. Denton PB, Farzan Y, Shoemaker IM. Testing large non-standard neutrino interactions with arbitrary mediator mass after COHERENT data. JHEP. (2018) **1807**:037. doi: 10.1007/JHEP07(2018)037
17. Dutta B, Liao S, Sinha S, Strigari LE. Searching for beyond the standard model physics with COHERENT energy and timing data. Phys Rev Lett. (2019) **123**:061801. doi: 10.1103/PhysRevLett.123.061801
18. Coloma P, Gonzalez-Garcia MC, Maltoni M, Schwetz T. COHERENT enlightenment of the neutrino dark side. Phys Rev D. (2017) **96**:115007. doi: 10.1103/PhysRevD.96.115007
19. Gonzalez-Garcia MC, Maltoni M, Perez-Gonzalez YF, Zukanovich Funchal R. Neutrino discovery limit of dark matter direct detection experiments in the presence of non-standard interactions. JHEP. (2018) **1807**:019. doi: 10.1007/JHEP07(2018)019
20. Kosmas TS, Miranda OG, Papoulias DK, Tortola M, Valle JWF. Probing neutrino magnetic moments at the Spallation Neutron Source facility. Phys Rev D. (2015) **92**:013011. doi: 10.1103/PhysRevD.92.013011
21. Papoulias DK, Kosmas TS. COHERENT constraints to conventional and exotic neutrino physics. Phys Rev D. (2018) **97**:033003. doi: 10.1103/PhysRevD.97.033003
22. Miranda OG, Papoulias DK, Tórtola M, Valle JWF. Probing neutrino transition magnetic moments with coherent elastic neutrino-nucleus scattering. JHEP. (2019) **1907**:103. doi: 10.1007/JHEP07(2019)103
23. Parada A. New constraints on neutrino electric millicharge from elastic neutrino-electron scattering and coherent elastic neutrino-nucleus scattering. arXiv:1907.04942.
24. Kosmas TS, Papoulias DK, Tortola M, Valle JWF. Probing light sterile neutrino signatures at reactor and Spallation Neutron Source neutrino experiments. Phys Rev D. (2017) **96**:063013. doi: 10.1103/

PhysRevD.96.063013

25. Cañas BC, Garcés EA, Miranda OG, Parada A. The reactor antineutrino anomaly and low energy threshold neutrino experiments. Phys Lett B. (2018) **776**:451–6. doi: 10.1016/j.physletb.2017.11.074
26. Blanco C, Hooper D, Machado P. Constraining sterile neutrino interpretations of the LSND and MiniBooNE anomalies with coherent neutrino scattering experiments. arXiv:1901.08094.
27. Aristizabal Sierra D, De Romeri V, Rojas N. CP violating effects in coherent elastic neutrino-nucleus scattering processes. JHEP. (2019) **09**:069. doi: 10.1007/JHEP09(2019)069
28. Dent JB, Dutta B, Liao S, Newstead JL, Strigari LE, Walker JW. Probing light mediators at ultralow threshold energies with coherent elastic neutrino-nucleus scattering. Phys Rev D. (2017) **96**:095007. doi: 10.1103/PhysRevD.96.095007
29. Farzan Y, Lindner M, Rodejohann W, Xu XJ. Probing neutrino coupling to a light scalar with coherent neutrino scattering. JHEP. (2018) **1805**:066. doi: 10.1007/JHEP05(2018)066
30. Abdullah M, Dent JB, Dutta B, Kane GL, Liao S, Strigari LE. Coherent elastic neutrino nucleus scattering as a probe of a Z' through kinetic and mass mixing effects. Phys Rev D. (2018) **98**:015005. doi: 10.1103/PhysRevD.98.015005
31. Brdar V, Rodejohann W, Xu XJ. Producing a new fermion in coherent elastic neutrino-nucleus scattering: from neutrino mass to dark matter. JHEP. (2018) **1812**:024. doi: 10.1007/JHEP12(2018)024
32. Cadeddu M, Giunti C, Li YF, Zhang YY. Average CsI neutron density distribution from COHERENT data. Phys Rev Lett. (2018) **120**:072501. doi: 10.1103/PhysRevLett.120.072501
33. Ciuffoli E, Evslin J, Fu Q, Tang J. Extracting nuclear form factors with coherent neutrino scattering. Phys Rev D. (2018) **97**:113003. doi: 10.1103/PhysRevD.97.113003
34. Huang XR, Chen LW. Neutron skin in CsI and low-energy effective weak mixing angle from COHERENT data. arXiv:1902.07625. (2019). doi: 10.1103/PhysRevD.100.071301
35. Aristizabal Sierra D, Liao J, Marfatia D. Impact of form factor uncertainties on interpretations of coherent elastic neutrino-nucleus scattering data. JHEP. (2019) **1906**:141. doi: 10.1007/JHEP06(2019)141
36. Papoulias D, Kosmas T, Sahu R, Kota V, Hota M. Constraining

nuclear physics parameters with current and future COHERENT data. arXiv:1903.03722.

37. Arcadi G, Lindner M, Martins J, Queiroz FS. New physics probes: atomic parity violation, polarized electron scattering and neutrino-nucleus coherent scattering. arXiv:1906.04755.
38. Cadeddu M, Dordei F, Giunti C, Kouzakov K, Picciau E, Studenikin A. Potentialities of a low-energy detector based on ^{4}He evaporation to observe atomic effects in coherent neutrino scattering and physics perspectives. arXiv:1907.03302. (2019). doi: 10.1103/PhysRevD.100.073014
39. Papoulias DK, Sahu R, Kosmas TS, Kota VKB, Nayak B. Novel neutrino-floor and dark matter searches with deformed shell model calculations. Adv High Energy Phys. (2018) **2018**:6031362. doi: 10.1155/2018/6031362
40. Bœhm C, Cerdeño DG, Machado PAN, Olivares-Del Campo A, Reid E. How high is the neutrino floor? JCAP. (2019) **1901**:043. doi: 10.1088/1475-7516/2019/01/043
41. Link JM, Xu XJ. Searching for BSM neutrino interactions in dark matter detectors. JHEP. (2019) **08**:004. doi: 10.1007/JHEP08(2019)004
42. Ge SF, Shoemaker IM. Constraining photon portal dark matter with texono and coherent data. JHEP. (2018) **1811**:066. doi: 10.1007/JHEP11(2018)066
43. Ng KCY, Beacom JF, Peter AHG, Rott C. Solar atmospheric neutrinos: a new neutrino floor for dark matter searches. Phys Rev D. (2017) **96**:103006. doi: 10.1103/PhysRevD.96.103006
44. Dutta B, Kim D, Liao S, Park JC, Shin S, Strigari LE. Dark matter signals from timing spectra at neutrino experiments. arXiv:1906.10745.
45. Wong HT. Neutrino-nucleus coherent scattering and dark matter searches with sub-keV germanium detector. Nucl Phys A. (2010) **844**:229C–33C. doi: 10.1016/j.nuclphysa.2010.05.040
46. Aguilar-Arevalo A, Bertou X, Bonifazi C, Butner M, Cancelo G, Castaneda Vazquez A, et al. Results of the engineering run of the coherent neutrino nucleus interaction experiment (CONNIE). JINST. (2016) 11:P07024. doi: 10.1088/1748-0221/11/07/P07024
47. Agnolet G, Baker W, Barker D, Beck R, Carroll TJ, Cesar J, et al. Background studies for the MINER coherent neutrino scattering reactor experiment. Nucl Instrum Meth A. (2017) **853**:53–60. doi: 10.1016/j.

nima.2017.02.024

48. Belov V, Brudanin V, Egorov V, Filosofov D, Fomina M, Gurov Y, et al. The νGeN experiment at the Kalinin nuclear power plant. JINST. (2015) 10:P12011. doi: 10.1088/1748-0221/10/12/P12011
49. Private communication with CONUS collaboration.
50. Billard J, Carr R, Dawson J, Figueroa-Feliciano E, Formaggio JA, Gascon J, et al. Coherent neutrino scattering with low temperature bolometers at chooz reactor complex. J Phys G. (2017) **44**:105101. doi: 10.1088/1361-6471/aa83d0
51. Strauss R, Rothe J, Angloher G, Bento A, Gütlein A, Hauff d, et al. The ν-cleus experiment: a gram-scale fiducial-volume cryogenic detector for the first detection of coherent neutrino-nucleus scattering. Eur Phys J C. (2017) **77**:506. doi: 10.1140/epjc/s10052-017-5068-2
52. Barranco J, Miranda OG, Moura CA, Valle JWF. Constraining non-standard interactions in nu(e) e or anti-nu(e) e scattering. Phys Rev D. (2006) **73**:113001. doi: 10.1103/PhysRevD.73.113001
53. Friedland A, Lunardini C, Maltoni M. Atmospheric neutrinos as probes of neutrino-matter interactions. Phys Rev D. (2004) **70**:111301. doi: 10.1103/PhysRevD.70.111301
54. Friedland A, Lunardini C, Pena-Garay C. Solar neutrinos as probes of neutrino matter interactions. Phys Lett B. (2004) **594**:347. doi: 10.1016/j.physletb.2004.05.047
55. Friedland A, Lunardini C. A Test of tau neutrino interactions with atmospheric neutrinos and K2K. Phys Rev D. (2005) **72**:053009. doi: 10.1103/PhysRevD.72.053009
56. Miranda OG, Nunokawa H. Non standard neutrino interactions: current status and future prospects. New J Phys. (2015) **17**:095002. doi: 10.1088/1367-2630/17/9/095002
57. Farzan Y, Tortola M. Neutrino oscillations and non-standard interactions. Front Phys. (2018) **6**:10. doi: 10.3389/fphy.2018.00010
58. Amanik PS, Fuller GM, Grinstein B. Flavor changing supersymmetry interactions in a supernova. Astropart Phys. (2005) **24**:160–82. doi: 10.1016/j.astropartphys.2005.06.004
59. Esteban-Pretel A, Tomas R, Valle JWF. Interplay between collective effects and non-standard neutrino interactions of supernova neutrinos. Phys Rev D. (2010) **81**:063003. doi: 10.1103/PhysRevD.81.063003

60. Akimov DYu, Belov VA, Bolozdynya A, Efremenko YuV, Konovalov AM, Kumpan AV, et al. Coherent elastic neutrino scattering on atomic nucleus: recently discovered type of low-energy neutrino interaction. Phys Usp. (2019) **62**:166–78.
61. Papoulias DK, Kosmas TS. Standard and nonstandard neutrino-nucleus reactions cross sections and event rates to neutrino detection experiments. Adv High Energy Phys. (2015) **2015**:763648. doi: 10.1155/2015/763648
62. Bednyakov VA, Naumov DV. Coherency and incoherency in neutrino-nucleus elastic and inelastic scattering. Phys Rev D. (2018) **98**:053004. doi: 10.1103/PhysRevD.98.053004
63. Bednyakov VA, Naumov DV. On coherent neutrino and antineutrino scattering off nuclei. arXiv:1904.03119.
64. Almosly W, Carlsson BG, Suhonen J, Ydrefors E. Neutral-current supernova-neutrino cross sections for $^{204,\,206,\,208}$Pb calculated by Skyrme quasiparticle random-phase approximation. Phys Rev C. (2019) **99**:055801. doi: 10.1103/PhysRevC.99.055801
65. Chasioti VC, Kosmas TS. A unified formalism for the basic nuclear matrix elements in semi-leptonic processes. Nucl Phys A. (2009) **829**:234–52. doi: 10.1016/j.nuclphysa.2009.08.009
66. De Vries H, De Jager CW, De Vries C. Nuclear charge and magnetization density distribution parameters from elastic electron scattering. Atom Data Nucl Data Tabl. (1987) **36**:495–536. doi: 10.1016/0092-640X(87)90013-1
67. Donnelly TW, Walecka JD. Semileptonic weak and electromagnetic interactions with nuclei: isoelastic processes. Nucl Phys A. (1976) **274**:368–412. doi: 10.1016/0375-9474(76)90209-8
68. Donnelly TW, Peccei RD. Neutral current effects in nuclei. Phys Rept. (1979) **50**:1. doi: 10.1016/0370-1573(79)90010-3
69. Lindner M, Rodejohann W, Xu XJ. Coherent neutrino-nucleus scattering and new neutrino interactions. JHEP. (2017) **1703**:097. doi: 10.1007/JHEP03(2017)097
70. Barranco J, Miranda OG, Rashba TI. Probing new physics with coherent neutrino scattering off nuclei. JHEP. (2005) **512**:21. doi: 10.1088/1126-6708/2005/12/021
71. Tanabashi M, Hagiwara K, Hikasa K, Nakamura K, Sumino Y, Takahashi F, et al. Review of particle physics. Phys Rev D. (2018) **98**:030001.

72. Papoulias DK. Exotic lepton flavour violating processes in the field of nucleus. (Ph.D. thesis). Ioannina University (2016). Available online at: http://dx.doi.org/http://hdl.handle.net/10442/hedi/43131

73. Angeli I, Marinova KP. Table of experimental nuclear ground state charge radii: an update. Atom Data Nucl Data Tabl. (2013) **99**:69–95. doi: 10.1016/j.adt.2011.12.006

74. Kortelainen M, Suhonen J, Toivanen J, Kosmas TS. Event rates for CDM detectors from large-scale shell-model calculations. Phys Lett B. (2006) **632**:226–32. doi: 10.1016/j.physletb.2005.10.057

75. Toivanen P, Kortelainen M, Suhonen J, Toivanen J. Large-scale shell-model calculations of elastic and inelastic scattering rates of lightest supersymmetric particles (LSP) on I-127, Xe-129, Xe-131, and Cs-133 nuclei. Phys Rev C. (2009) **79**:044302. doi: 10.1103/PhysRevC.79.044302

76. Papoulias DK, Kosmas TS. Nuclear aspects of neutral current non-standard ν-nucleus reactions and the role of the exotic $\mu^-\rightarrow e^-$ transitions experimental limits. Phys Lett B. (2014) **728**:482–8. doi: 10.1016/j.physletb.2013.12.028

77. Pirinen P, Suhonen J, Ydrefors E. Neutral-current neutrino-nucleus scattering off Xe isotopes. Adv High Energy Phys. (2018) **2018**:9163586. doi: 10.1155/2018/9163586

78. Papoulias DK. COHERENT constraints after the Chicago-3 quenching factor measurement. arXiv:1907.11644.

79. Kosmas TS, Vergados JD. Nuclear densities with fractional occupation probabilities of the states. Nucl Phys A. (1992) **536**:72–86. doi: 10.1016/0375-9474(92)90246-G

80. Amanik PS, McLaughlin GC. Nuclear neutron form factor from neutrino nucleus coherent elastic scattering. J Phys G. (2009) **36**:015105. doi: 10.1088/0954-3899/36/1/015105

81. Kosmas TS, Vergados JD. Nuclear matrix elements for the coherent μ–e conversion process. Phys Lett B. (1988) **215**:460–4. doi: 10.1016/0370-2693(88)91341-X

82. Kosmas TS, Vergados JD. Study of the flavor violating (μ^-, e^-) conversion in nuclei. Nucl Phys A. (1990) **510**:641–70. doi: 10.1016/0375-9474(90)90353-N

83. Helm RH. Inelastic and elastic scattering of 187-Mev electrons from selected even-even nuclei. Phys Rev. (1956) **104**:1466–75. doi:

10.1103/PhysRev.104.1466

84. Piekarewicz J, Linero AR, Giuliani P, Chicken E. Power of two: assessing the impact of a second measurement of the weak-charge form factor of ^{208}Pb. Phys Rev C. (2016) **94**:034316. doi: 10.1103/PhysRevC.94.034316
85. Fricke G, Bernhardt C, Heilig K, Schaller LA, Schellenberg L, Shera EB, et al. Nuclear ground state charge radii from electromagnetic interactions. Atom Data Nucl Data Tabl. (1995) **60**:177–285. doi: 10.1006/adnd.1995.1007
86. Lewin JD, Smith PF. Review of mathematics, numerical factors, and corrections for dark matter experiments based on elastic nuclear recoil. Astropart Phys. (1996) **6**:87–112. doi: 10.1016/S0927-6505(96)00047-3
87. Sprung DWL, Martorell J. The symmetrized Fermi function and its transforms. J Phys A. (1997) **30**:6525–34.
88. Klein S, Nystrand J. Exclusive vector meson production in relativistic heavy ion collisions. Phys Rev C. (1999) **60**:014903. doi: 10.1103/PhysRevC.60.014903
89. Akimov D, Albert JB, An P, Awe C, Barbeau PS, Becker B, et al. First constraint on coherent elastic neutrino-nucleus scattering in argon. arXiv:1909.05913.
90. AlianeA, Avetissovb ICh, Barinovab OP, de la Broisec X, Danevichd FA, Dumoulin L, et al. First test of a Li2WO4(Mo) bolometric detector for the measurement of coherent neutrino-nucleus scattering. Nucl Instrum Methods Phys Res A. **949**:162784. doi: 10.1016/j.nima.2019.162784
91. Bellenghi C, Chiesa D, Di Noto L, Pallavicini M, Previtali E, Vignati M. Coherent elastic nuclear scattering of^{51} Cr neutrinos. Eur Phys J C. (2019) **79**:727. doi: 10.1140/epjc/s10052-019-7240-3
92. Baxter D, Collar JI, Coloma P, Dahl CE, Esteban I, Ferrario P, et al. Coherent Elastic Neutrino-Nucleus Scattering at the European Spallation Source. arXiv:1911.00762 [physics.ins-det].
93. Akimov D, Albert JB, An P, Awe C, Barbeau PS, Becker B, et al. COHERENT 2018 at the spallation neutron source. arXiv:1803.09183.
94. Thornton R. Searching for Muon Neutrino Disappearance at LSND Neutrino Energies with CCM. (2019). Available online at: https://indico.cern.ch/event/782953/contributions/3444547/
95. Hakenmuller J, Buck C, Fülber K, Heusser G, Klages T, Lindner M, et

al. Neutron-induced background in the CONUS experiment. Eur Phys J C. (2019) **79**:699. doi: 10.1140/epjc/s10052-019-7160-2

96. Aguilar-Arevalo A, Bertou X, Bonifazi C, Cancelo G, Castañeda A, Cervantes Vergara B, et al. Exploring low-energy neutrino physics with the coherent neutrino nucleus interaction experiment (CONNIE). arXiv:1906.02200.
97. Angloher G, Ardellier-Desages F, Bento A, Canonica L, Erhart A, Ferreiro N, et al. Exploring CEvNS with NUCLEUS at the Chooz nuclear power plant. arXiv:1905.10258.
98. Akimov DY, Berdnikova AK, Belov VA, Bolozdynya AI, Burenkov AA, Dolgolenko AG, et al. Status of the RED-100 experiment. JINST. (2017) 12:C06018. doi: 10.1088/1748-0221/12/06/C06018
99. Wong HT, Li HB, Li J, Yue Q, Zhou ZY. Research program towards observation of neutrino-nucleus coherent scattering. J Phys. (2006) **39**:266–8. doi: 10.1088/1742-6596/39/1/064
100. Rich GC. Measurement of Low-Energy Nuclear-Recoil Quenching Factors in CsI[Na] and Statistical Analysis of the First Observation of Coherent, Elastic Neutrino-Nucleus Scattering. PhD thesis, North Carolina U., 2017.
101. Cadeddu M, Dordei F. Reinterpreting the weak mixing angle from atomic parity violation in view of the Cs neutron rms radius measurement from COHERENT. Phys Rev D. (2019) **99**:033010. doi: 10.1103/PhysRevD.99.033010
102. Cañas BC, Garcés EA, Miranda OG, Parada A. Future perspectives for a weak mixing angle measurement in coherent elastic neutrino nucleus scattering experiments. Phys Lett B. (2018) **784**:159–62. doi: 10.1016/j.physletb.2018.07.049
103. Collar J, Kavner A, Lewis C. Response of CsI[Na] to nuclear recoils: impact on coherent elastic neutrino-nucleus scattering (CEvNS). arXiv:1907.04828.
104. Patton K, Engel J, McLaughlin GC, Schunck N. Neutrino-nucleus coherent scattering as a probe of neutron density distributions. Phys Rev C. (2012) **86**:024612. doi: 10.1103/PhysRevC.86.024612
105. Horowitz CJ, Ahmed Z, Jen CM, Rakhman A, Souder PA, Dalton MM, et al. Weak charge form factor and radius of 208Pb through parity violation in electron scattering. Phys Rev C. (2012) **85**:032501. doi: 10.1103/PhysRevC.85.032501

106. Cadeddu M, Dordei F, Giunti C, Li YF, Zhang YY. Neutrino, electroweak and nuclear physics from COHERENT elastic neutrino-nucleus scattering with refined quenching factor. arXiv:1908.06045.
107. Bhupal Dev PS, Babu KS, Denton PB, Machado PAN, Argüelles CA, Barrow JL, et al. Neutrino Non-Standard Interactions: A Status Report. (2019). arXiv:1907.00991 [hep-ph]. http://lss.fnal.gov/archive/2019/conf/fermilab-conf-19-299-t.pdf.
108. Babu KS, Dev PSB, Jana S, Thapa A. Non-Standard Interactions in Radiative Neutrino Mass Models. (2019). arXiv:1907.09498 [hep-ph].
109. Deppisch F, Kosmas TS, Valle JWF. Enhanced μ^- - e^- conversion in nuclei in the inverse seesaw model. Nucl Phys B. (2006) **752**:80–92. doi: 10.1016/j.nuclphysb.2006.06.032
110. Malinsky M, Ohlsson T, Zhang H. Non-standard neutrino interactions from a triplet seesaw model. Phys Rev D. (2009) **79**:011301. doi: 10.1103/PhysRevD.79.011301
111. Forero DV, Morisi S, Tortola M, Valle JWF. Lepton flavor violation and non-unitary lepton mixing in low-scale type-I seesaw. JHEP. (2011) **1109**:142. doi: 10.1007/JHEP09(2011)142
112. Das SP, Deppisch FF, Kittel O, Valle JWF. Heavy neutrinos and lepton flavour violation in left-right symmetric models at the LHC. Phys Rev D. (2012) **86**:055006. doi: 10.1103/PhysRevD.86.055006
113. Petrov AA, Zhuridov DV. Lepton flavor-violating transitions in effective field theory and gluonic operators. Phys Rev D. (2014) **89**:033005. doi: 10.1103/PhysRevD.89.033005
114. Amanik PS, Fuller GM. Stellar collapse dynamics with neutrino flavor changing neutral currents. Phys Rev D. (2007) **75**:083008. doi: 10.1103/PhysRevD.75.083008
115. Scholberg K. Prospects for measuring coherent neutrino-nucleus elastic scattering at a stopped-pion neutrino source. Phys Rev D. (2006) **73**:033005. doi: 10.1103/PhysRevD.73.033005
116. Aristizabal Sierra D, De Romeri V, Rojas N. COHERENT analysis of neutrino generalized interactions. Phys Rev D. (2018) **98**:075018. doi: 10.1103/PhysRevD.98.075018
117. Altmannshofer W, Tammaro M, Zupan J. Non-standard neutrino interactions and low energy experiments. arXiv:1812.02778.
118. Bischer I, Rodejohann W. General neutrino interactions from an effective field theory perspective. Nucl Phys B. (2019) **947**:114746.

doi: 10.1016/j.nuclphysb.2019.114746

119. Davidson S, Gorbahn M. Charged lepton flavour change and Non-Standard neutrino Interactions. arXiv:1909.07406.
120. Billard J, Johnston J, Kavanagh BJ. Prospects for exploring new physics in coherent elastic neutrino-nucleus scattering. JCAP. (2018) **1811**:016. doi: 10.1088/1475-7516/2018/11/016
121. Miranda OG, Sanchez Garcia G, Sanders O. Coherent elastic neutrino-nucleus scattering as a precision test for the Standard Model and beyond: the COHERENT proposal case. Adv High Energy Phys. (2019) **2019**:3902819. doi: 10.1155/2019/3902819
122. Khan AN, Rodejohann W. New physics from COHERENT data with improved Quenching Factors. arXiv:1907.12444.
123. Giunti C. General COHERENT constraints on neutrino non-standard interactions. arXiv:1909.00466.
124. Barranco J, Bolanos A, Garces EA, Miranda OG, Rashba TI. Tensorial NSI and unparticle physics in neutrino scattering. Int J Mod Phys A. (2012) **27**:1250147. doi: 10.1142/S0217751X12501473
125. Healey KJ, Petrov AA, Zhuridov D. Nonstandard neutrino interactions and transition magnetic moments. Phys Rev D. (2013) **87**:117301. doi: 10.1103/PhysRevD.87.117301
126. Papoulias DK, Kosmas TS. Neutrino transition magnetic moments within the non-standard neutrino–nucleus interactions. Phys Lett B. (2015) **747**:454–9. doi: 10.1016/j.physletb.2015.06.039
127. Datta A, Dutta B, Liao S, Marfatia D, Strigari LE. Neutrino scattering and B anomalies from hidden sector portals. JHEP. (2019) **1**:91. doi: 10.1007/JHEP01(2019)091
128. Abdullah M, Dalchenko M, Dutta B, Eusebi R, Huang P, Kamon T, et al. Bottom-quark fusion processes at the LHC for probing Z' models and B -meson decay anomalies. Phys Rev D. (2018) **97**:075035. doi: 10.1103/PhysRevD.97.075035
129. Bertuzzo E, Deppisch FF, Kulkarni S, Perez Gonzalez YF, Zukanovich Funchal R. Dark matter and exotic neutrino interactions in direct detection searches. JHEP. (2017) **1704**:073. doi: 10.1007/JHEP04(2017)073
130. Cerdeño DG, Fairbairn M, Jubb T, Machado PAN, Vincent AC, Bœhm C. Physics from solar neutrinos in dark matter direct detection experiments. JHEP. (2016) **1605**:118. doi: 10.1007/JHEP05(2016)118

131. Rodejohann W, Xu XJ, Yaguna CE. Distinguishing between Dirac and Majorana neutrinos in the presence of general interactions. JHEP. (2017) **5**:24. doi: 10.1007/JHEP05(2017)024

132. Aristizabal Sierra D, Dutta B, Liao S, Strigari LE. Coherent elastic neutrino-nucleus scattering in multi-ton scale dark matter experiments: classification of vector and scalar interactions new physics signals. arXiv:1910.12437.

133. Shoemaker IM. COHERENT search strategy for beyond standard model neutrino interactions. Phys Rev D. (2017) **95**:115028. doi: 10.1103/PhysRevD.95.115028

134. Aguilar-Arevalo A, Bertou X, Bonifazi C, Cancelo G, Cervantes-Vergara BA, Chavez C, et al. Light vector mediator search in the low-energy data of the CONNIE reactor neutrino experiment. arXiv:1910.04951.

135. Giunti C, Studenikin A. Neutrino electromagnetic interactions: a window to new physics. Rev Mod Phys. (2015) **87**:531. doi: 10.1103/RevModPhys.87.531

136. Schechter J, Valle JWF. Majorana neutrinos and magnetic fields. Phys Rev D. (1981) **24**:1883–9. doi: 10.1103/PhysRevD.24.1883

137. Grimus W, Schwetz T. Elastic neutrino electron scattering of solar neutrinos and potential effects of magnetic and electric dipole moments. Nucl Phys B. (2000) **587**:45–66. doi: 10.1016/S0550-3213(00)00451-X

138. Tortola M. Constraining neutrino magnetic moment with solar and reactor neutrino data. arXiv:hep-ph/0401135.

139. Vogel P, Engel J. Neutrino electromagnetic form-factors. Phys Rev D. (1989) **39**:3378. doi: 10.1103/PhysRevD.39.3378

140. Hirsch M, Nardi E, Restrepo D. Bounds on the tau and muon neutrino vector and axial vector charge radius. Phys Rev D. (2003) **67**:033005. doi: 10.1103/PhysRevD.67.033005

141. Agostini M, Altenmüller K, Appel S, Atroshchenko V, Bagdasarian Z, Basilico D, et al. Limiting neutrino magnetic moments with Borexino Phase-II solar neutrino data. Phys Rev D. (2017) **96**:091103.

142. Adamson P, Anghel I, Aurisano A, Barr G, Bishai M, Blake A, et al. Search for sterile neutrinos in MINOS and MINOS+ using a two-detector fit. Phys Rev Lett. (2019) **122**:091803. doi: 10.1103/PhysRevLett.122.091803

143. Aguilar-Arevalo AA, Brown BC, Bugel L, Cheng G, Conrad JM,

Cooper RL, et al. Significant excess of electronLike events in the MiniBooNE short-baseline neutrino experiment. Phys Rev Lett. (2018) **121**:221801. doi: 10.1103/PhysRevLett.121.221801

144. Adey D, An FP, Balantekin AB, Band HR, Bishai M, Blyth S, et al. Measurement of the electron antineutrino oscillation with 1958 days of operation at daya bay. Phys Rev Lett. (2018) **121**:241805. doi: 10.1103/PhysRevLett.121.241805

145. An F, An G, An Q, Antonelli V, Baussan E, Beacom J, et al. Neutrino physics with JUNO. J Phys G. (2016) **43**:030401. doi: 10.1088/0954-3899/43/3/030401

146. Ko YJ, Kim BR, Kim JY, Han BY, Jang CH, Jeon EJ, et al. Sterile neutrino search at the NEOS experiment. Phys Rev Lett. (2017) **118**:121802. doi: 10.1103/PhysRevLett.118.121802

147. Formaggio JA, Figueroa-Feliciano E, Anderson AJ. Sterile neutrinos, coherent scattering and oscillometry measurements with low-temperature bolometers. Phys Rev D. (2012) **85**:013009. doi: 10.1103/PhysRevD.85.013009

148. Anderson AJ, Conrad JM, Figueroa-Feliciano E, Ignarra C, Karagiorgi G, Scholberg K, et al. Measuring active-to-sterile neutrino oscillations with neutral current coherent neutrino-nucleus scattering. Phys Rev D. (2012) **86**:013004. doi: 10.1103/PhysRevD.86.013004

149. Aguilar-Arevalo A, Auerbach B, Burman RL, Caldwell DO, Church ED, Cochran AK, et al. Evidence for neutrino oscillations from the observation of anti-neutrino(electron) appearance in a anti-neutrino(muon) beam. Phys Rev D. (2001) **64**:112007.

150. Berryman JM. Constraining sterile neutrino cosmology with terrestrial oscillation experiments. Phys Rev D. (2019) **100**:023540. doi: 10.1103/PhysRevD.100.023540

151. Coloma P, Denton PB, Gonzalez-Garcia MC, Maltoni M, Schwetz T. Curtailing the dark side in non-standard neutrino interactions. JHEP. (2017) **1704**:116. doi: 10.1007/JHEP04(2017)116

152. Cadeddu M, Giunti C, Kouzakov KA, Li YF, Studenikin AI, Zhang YY. Neutrino charge radii from COHERENT elastic neutrino-nucleus scattering. Phys Rev D. (2018) **98**:113010. doi: 10.1103/PhysRevD.98.113010

153. Giannaka PG, Kosmas TS. Electron capture cross sections for stellar nucleosynthesis. Adv High Energy Phys. (2015) **2015**:398796. doi:

10.1155/2015/398796

154. Stapleford CJ, Väänänen DJ, Kneller JP, McLaughlin GC, Shapiro BT. Nonstandard neutrino interactions in supernovae. Phys Rev D. (2016) **94**:093007. doi: 10.1103/PhysRevD.94.093007
155. Biassoni M, Martinez C. Study of supernova ν -nucleus coherent scattering interactions. Astropart Phys. (2012) **36**:151–5. doi: 10.1016/j.astropartphys.2012.05.009
156. Monroe J, Fisher P. Neutrino backgrounds to dark matter searches. Phys Rev D. (2007) **76**:033007. doi: 10.1103/PhysRevD.76.033007
157. O'Hare CAJ. Dark matter astrophysical uncertainties and the neutrino floor. Phys Rev D. (2016) **94**:063527. doi: 10.1103/PhysRevD.94.063527
158. Billard J, Strigari L, Figueroa-Feliciano E. Implication of neutrino backgrounds on the reach of next generation dark matter direct detection experiments. Phys Rev D. (2014) **89**:023524. doi: 10.1103/PhysRevD.89.023524
159. Farzan Y. A model for large non-standard interactions of neutrinos leading to the LMA-Dark solution. Phys Lett B. (2015) **748**:311–5. doi: 10.1016/j.physletb.2015.07.015
160. Kosmas TS, Vergados JD. (μ^-, e^-) conversion: a symbiosis of particle and nuclear physics. Phys Rept. (1996) **264**:251–66. doi: 10.1016/0370-1573(95)00041-0
161. Farzan Y, Heeck J. Neutrinophilic nonstandard interactions. Phys Rev D. (2016) **94**:053010. doi: 10.1103/PhysRevD.94.053010
162. Petcov ST. The processes μ –> e Gamma, μ –> e e anti-e, Neutrino' –> Neutrino gamma in the Weinberg-Salam Model with Neutrino Mixing. Sov J Nucl Phys. (1977) **25**:340. [Erratum: Yad. Fiz. 25:1336 (1977)].
163. Kuno Y, Okada Y. Muon decay and physics beyond the standard model. Rev Mod Phys. (2001) **73**:151–202. doi: 10.1103/RevModPhys.73.151
164. Lee M. COMET muon conversion experiment in J-PARC. Front Phys. (2018) **6**:133. doi: 10.3389/fphy.2018.00133
165. Bernstein RH. The Mu2e experiment. Front Phys. (2019) **7**:1. doi: 10.3389/fphy.2019.00001
166. Chiang HC, Oset E, Kosmas TS, Faessler A, Vergados JD. Coherent and incoherent (μ^-, e^-) conversion in nuclei. Nucl Phys A. (1993) **559**:526–

42. doi: 10.1016/0375-9474(93)90259-Z

167. Davidson S, Pena-Garay C, Rius N, Santamaria A. Present and future bounds on nonstandard neutrino interactions. JHEP. (2003) **303**:11. doi: 10.1088/1126-6708/2003/03/011
168. Kitano R, Koike M, Okada Y. Detailed calculation of lepton flavor violating muon electron conversion rate for various nuclei. Phys Rev D. (2002) **66**:096002. doi: 10.1103/PhysRevD.66.096002
169. Cirigliano V, Kitano R, Okada Y, Tuzon P. On the model discriminating power of μ —> e conversion in nuclei. Phys Rev D. (2009) **80**:013002. doi: 10.1103/PhysRevD.80.013002
170. Davidson S, Kuno Y, Yamanaka M. Selecting $\mu \rightarrow e$ conversion targets to distinguish lepton flavour-changing operators. Phys Lett B. (2019) **790**:380–8. doi: 10.1016/j.physletb.2019.01.042
171. Edmonds AR. Angular Momentum in Quantum Mechanics. Reissue ed. Princeton, NJ: Princeton University Press (1996).

SECTION 2
SEESAW MECHANISM ONTO THE STANDARD MODEL AND NEUTRINOS MASSES

CHAPTER

4

Examining the Neutrino Option

Ilaria Brivio and Michael Trott

Niels Bohr International Academy and Discovery Centre,

Niels Bohr Institute, University of Copenhagen,

Blegdamsvej 17, DK-2100 Copenhagen, Denmark

ABSTRACT

The neutrino option is a scenario where the electroweak scale, and thereby the Higgs mass, is generated simultaneously with neutrino masses in the seesaw model. This occurs via the leading one loop and tree level diagrams matching the seesaw model onto the Standard Model Effective Field Theory. We advance the study of this scenario by determining one loop corrections to the leading order matching results systematically, performing a detailed

Citation: Brivio, I., Trott, M. "Examining the neutrino option". J. High Energ. Phys. 2019, 107 (2019). https://doi.org/10.1007/JHEP02(2019)107

numerical analysis of the consistency of this approach with Neutrino data and the Standard Model particle masses, and by examining the embedding of this scenario into a more ultraviolet complete model. We find that the neutrino option remains a viable and intriguing scenario to explain the origin of observed particle masses.

Keywords: Effective Field Theories, Neutrino Physics, Higgs Physics

INTRODUCTION

The origin of the Higgs potential and the observed neutrino masses are two outstanding questions unanswered by the Standard Model (SM). The smallness of neutrino masses compared to the masses of the quarks and Electroweak gauge bosons can be naturally accommodated in the Seesaw model [2–6]. In this approach, the neutrino-quark mass hierarchy follows from a separation of the scale (denoted M) associated with a lepton number violating sector leading to neutrino masses, compared to the Electroweak scale ($\bar{v}_T \equiv \sqrt{2\langle H^\dagger H\rangle}$) associated with the remaining SM field content.

If the Seesaw model generates neutrino masses, the sensitivity of the Higgs mass term to M can lead to a curious tuning of Lagrangian parameters. Such parameter tuning can be avoided — if the Majorana mass scale and couplings are of a particular form [7, 8]. For the same region of parameter space, the Seesaw model can be a phenomenologically viable boundary condition for the Higgs potential, while supplying an origin for the Electroweak scale as a descendent from the scale M [8]. What leads to the Higgs phase of the SM in this case, is Fermi statistics in the leading one loop matching contribution from the seesaw model to the SM potential terms. This matching necessarily occurs while Neutrino masses are generated by the leading tree level matching, as the Majorana states are integrated out. This is the "neutrino option" for generating neutrino masses and the Higgs potential.

The purpose of this paper is to examine this scenario in more numerical and theoretical detail than ref. [8] and to illustrate how this scenario can be embedded in more concrete Ultraviolet (UV) models.1 We consider the minimal case that can lead to a successful lower energy neutrino phenomenology with two heavy Majorana states leading to two massive light neutrinos. The outline of this paper is as follows. In section 2 we define our notation and conventions for the Seesaw model and the leading tree level matching onto the Standard Model Effective Field Theory (SMEFT). In section 3 we develop the theoretical framework to study the neutrino option at next to leading order (NLO) accuracy, determining the relevant one loop matching into the seesaw model, and discuss the one loop running of the neutrino parameters. In section 4 we report numerical results fixing the low scale neutrino parameters and Higgs mass as inputs, and determine the quartic coupling λ required at the matching scale for the scenario to be self consistent. In section 5 we examine and extend the embedding of the neutrino option into a classically (i.e. massless particle spectrum at leading order) conformal UV completion, as recently proposed by Brdar et al. [1]. We discuss the cut off scale in this particular UV embedding and the potential to avoid fine tuning. We briefly comment on the possibility that this UV framework can simultaneously provide a Dark Matter candidate, and in section 6 we conclude.

THEORETICAL FRAMEWORK

The physics of the SM is such that global lepton number conservation can provide an accidental symmetry protection mechanism for the Higgs potential terms. The operator dimension in the SM, or the SMEFT operator expansion leads to the association of an operators dimension being even (odd) if (ΔB−ΔL)/2 is even (odd) [9, 10] due to the nature of the SM field content. Here ΔB and ΔL are the baryon and lepton number violation of the operator considered. ($H^\dagger H$) is of even dimension with ΔB = ΔL = 0. General lepton number carrying fields associated with the mass scale M, either couple in pairs to $H^\dagger H$, or with a dimensionful coupling expected to be proximate to M, if parameter tuning is avoided. An example of the former case is a fermion field, that then participates in ΔL = 0 interactions with H, while the latter scenario is realized e.g. for a scalar field and leads to L violating couplings. In the scalar case, tree level exchanges contribute to $(H^\dagger H)^2$, with a cancellation of the introduced M dependence (up to ratios of couplings). The coupling of lepton number violating Majorana field takes also place through the portal interaction $H^\dagger l$ (+ h.c.). A heavy scale associated with this lepton number violating field leads to an inverse dependence on th scale M at tree level, starting with the (ΔL = 2) dimension five Weinberg operator, due to these interactions.2 The minimal scenario of this form is the Seesaw model of neutrino mass generation. An expectation is that small neutrino masses result, made only smaller by any small couplings (ω) to heavy particles. Simultaneously, additive contributions to the Higgs mass parameter appear at one loop $\propto \omega^2 M^2/16\pi^2$. These loop corrections are not forbidden, as lepton number is an accidental symmetry.

$$m_\nu \sim \frac{\omega^2\,\bar{v}_T^2}{M}, \qquad m_h \sim \frac{\omega M}{4\pi}, \qquad \bar{v}_T \sim \frac{\omega M}{4\,\sqrt{2}\,\pi\,\sqrt{\lambda}}, \tag{2.1}$$

and the Seesaw model parameter space M ~ 10^7 GeV and $|\omega|$ ~ 10^{-4} , which leads to mv ~ 10^{-11} GeV and mh ~ 10^2 GeV, is particularly interesting.3

This is the pattern of masses expected when the neutrino option is used to generate the Higgs potential. The idea is to use a UV boundary condition to generate the Higgs potential and an effective Electroweak scale. The smallness of the Higgs mass parameter is linked to the small neutrino masses and a set of approximate symmetries: global lepton number and scale invariance. For the latter, an expansion around the approximately scale invariant limit of the SM,4 incorporating the soft breaking of approximate scale invariance feeding into the Higgs potential due to the scale M is done. (In the limit $M \to 0$, $\langle H^\dagger H \rangle \leadsto \Lambda_{QCD}^2$.) We stress however, that the excitation of the Higgs field is not the dilaton of spontaneously broken scale invariance in this approach.

The motivation to consider this possible origin of the Electroweak scale is largely supplied by current experimental results. Neutrino's are known to be massive states, requiring an extension of the SM. It is natural to consider the effects of extending the SM to generate Neutrino masses on the Higgs. Generating the Higgs potential around the scales probed by LHC, with partner states associated with the multiplets of an approximate stabilizing symmetry (such as SUSY), or through lowering the effective Planck scale, is now subject to increasingly severe experimental bounds.5 Conversely, the neutrino option uses the running of the Higgs potential parameters in conjunction with threshold matchings required for Neutrinos to have mass when generating the Higgs potential.

This occurs at scales far above the observed Higgs mass and generates an effective Electroweak scale with scant experimental evidence of any stabilizing symmetry.6 The advantage of this approach is that a simple spectra of new physics states, motivated out of the experimental fact that neutrino's are massive can lead to the observed "mexican-hat" potential at lower scales, once the Higgs potential is run down using the Renormalization Group Equations (RGE) of the SM. A bare value of λ_0 ~ 0.01 is required, in addition to the threshold matching contribution to λ from the seesaw model. This is the main result of this paper's numerical study. As the Higgs is not a pseudo-Goldstone boson of any symmetry — a bare λ parameter is also not forbidden.

The SMEFT

We study the neutrino option in the SMEFT [23, 24], where the SM is extended with higher dimensional operators to capture the low energy limit of the seesaw model

$$\mathcal{L}_{\rm SMEFT} = \mathcal{L}_{\rm SM} + \mathcal{L}^{(5)} + \mathcal{L}^{(6)} + \mathcal{L}^{(7)} + \dots, \qquad \mathcal{L}^{(d)} = \sum_i \frac{C_i^{(d)}}{M^{d-4}} \mathcal{Q}_i^{(d)}, \text{ for } d > 4. \tag{2.2}$$

Here $Q_i^{(d)}$ are suppressed by d−4 powers of the Majorana scale M, that acts as the cut-off, and the $C_i^{(d)}$ are the Wilson coefficients. Our SM notation is consistent with the SMEFT review [25] except for the modified notation for the Higgs potential terms

$$\begin{aligned} V(H^\dagger H) &= -\frac{m^2}{2}(H^\dagger H) + \lambda (H^\dagger H)^2 + \cdots, \\ &= -\frac{m_0^2 + \Delta m^2}{2}(H^\dagger H) + (\lambda_0 + \Delta\lambda)(H^\dagger H)^2 + \cdots. \end{aligned} \tag{2.3}$$

$\tilde{H}_j = \epsilon_{jk}(H^k)^\star$ and the star superscript is generally reserved for complex conjugation on bosonic quantities. Here the bare parameters are m_0, λ_0. In a classical conformal limit for the mass spectrum of the SM (with non conformal renormalized couplings) $m_0^2 \simeq 0$ while λ0 is unconstrained. For the Weinberg operator [26, 27] we use the notation

$$\mathcal{Q}_{\alpha\beta}^{(5)} = \left(\overline{\ell_L^{c,\alpha}}\, \tilde{H}^\star\right)\left(\tilde{H}^\dagger\, \ell_L^\beta\right) \tag{2.4}$$

Our spinor conventions are that the c superscript corresponds to a charge conjugated Dirac four-component spinor $\psi^c = C\overline{\psi}^T$ with $C = -i\gamma_2\gamma_0$ in the chiral basis for the γ_i we use. Chiral projection and c do not commute so we fix notation that ℓ_L^c denotes a doublet lepton field chirally projected and subsequently charge conjugated.

Seesaw Model

We use the notation and conventions of refs. [28, 29] for the Seesaw model. In the Seesaw model, the SM Lagrangian field content is extended with right handed singlet fields $N_{R,p}$ with vanishing SU(3) × $SU(2)_L$ × $U(1)_Y$ charges. As these are singlet fermion fields they have Majorana mass terms [30] of the form

$$\overline{N^c_{R,p}}\, M_{pr}\, N_{R,r} + \overline{N_{R,p}}\, M^\star_{pr}\, N^c_{R,r} \tag{2.5}$$

where the charge conjugate of N_R is N^c_R. We define a field satisfying the Majorana condition as $N_p = N^c_p$ in its mass eigenstate basis as [28, 31]

$$N_p = e^{i\theta_p/2}\, N_{R,p} + e^{-i\theta_p/2}\, (N_{R,p})^c \tag{2.6}$$

With this choice, all Majorana phases θ_p shifted into the effective couplings and the relevant terms in the UV Lagrangian are

$$\mathcal{L}_N = \frac{1}{2}\bar{N}_p(i\partial\!\!\!/ - M_p)N_p - \frac{1}{2}\left[\overline{\ell^\beta_L}\tilde{H}\omega^{p,\dagger}_\beta N_p + \overline{\ell^{c\beta}_L}\tilde{H}^*\omega^{p,T}_\beta N_p + \overline{N_p}\omega^{p,*}_\beta \tilde{H}^T \ell^{c\beta}_L + \overline{N_p}\omega^p_\beta \tilde{H}^\dagger \ell^\beta_L\right] \tag{2.7}$$

Here p = {1, 2} runs over the heavy N_p Majorana states (Mp ~ M), while β = {1, 2, 3} runs over the SM lepton flavors. This formulation of the Seesaw model is mathematically equivalent to the formulation where

$$\mathcal{L}'_N = \frac{1}{2}\bar{N}_p(i\partial\!\!\!/ - M_p)N_p - \left[\overline{\ell^\beta_L}\tilde{H}\omega^{p,\dagger}_\beta N_p + \overline{N_p}\omega^p_\beta \tilde{H}^\dagger \ell^\beta_L\right] \tag{2.8}$$

In this case, the Lagrangian is reduced using the charge conjugation identities and the Majorana condition for the field N_p. Comparing calculations in these two formulations beyond tree level uncovers an interesting subtlety in using the Wick expansion, which is discussed in the appendix.

$L^{(5)}$ Matching

ω^p_β is a $\mathbb{C}_{2\times3}$matrix, related to the physical light neutrino masses and mixings via matching onto the Weinberg operator

$$\mathcal{L}^{(5)} = \frac{c^{(5)}_{\alpha\beta}}{2}\mathcal{Q}^{(5)}_{\alpha\beta} + \text{h.c.}, \qquad c^{(5)}_{\alpha\beta} = \frac{\omega^p_\alpha\,\omega^p_\beta}{M_p}. \tag{2.9}$$

Expanding the Higgs field around its classical background field gives

$$\mathcal{L}^{(5)} \supset -\frac{m_{\nu,k}}{2}\,\overline{\nu'^{c,k}_L}\,\nu'^k_L + \text{h.c.},$$

$$\text{where} \qquad m_{\nu,k} = -\frac{v^2}{2}(U^T)_{k\alpha}\, c_{5,\alpha\beta}\, U_{\beta k},$$

and ν'^k_L are the mass eigenstates of the light neutrinos $\nu^\alpha_L = U_{\alpha k}(\nu, L)\nu'^k_L$. The matrix U(ν, L) rotates the neutrinos from their weak eigenstates to their mass eigenstates. Similarly, the matrix U(e, L) rotates between the charged lepton weak and mass eigenstates. These rotations are not the same in general, leading to physical effects due to an overlap matrix being present in the lepton SU(2)L doublet field `. This is the Pontecorvo-MakiNakagawa-

Sakata (PMNS) matrix [32, 33] defined as

$$\mathcal{U}_{PMNS} = \mathcal{U}^{\dagger}(e, L)\mathcal{U}(\nu, L). \tag{2.10}$$

The three eigenvectors in U(e, L) form a basis for the field C 3 , as they diagonalize $\mathcal{M}_e^{\dagger}\mathcal{M}_e$, a Hermitian positive matrix defined over C 3 . As the physical effects of the U_{PMNS} matrix come about due to the relative orientation of the eigenvectors defining U(e, L) and U(ν, L), we can choose the eigenvectors of U(e, L) such that U(e, L) = diag(1, 1, 1) as a basis for this space, so long as no physical conclusions depend on this choice. We use the parameterization

$$U_{\rm PMNS} \equiv V \cdot \begin{pmatrix} e^{-i\phi/2} & & \\ & e^{-i\phi'/2} & \\ & & 1 \end{pmatrix}, \quad V = \begin{pmatrix} c_2c_3 & s_3c_2 & s_2e^{-i\delta} \\ -c_1s_3 - s_1s_2c_3e^{i\delta} & c_1c_3 - s_1s_2s_3e^{i\delta} & s_1c_2 \\ s_1s_3 - c_1s_2c_3e^{i\delta} & -s_1c_3 - c_1s_2s_3e^{i\delta} & c_1c_2 \end{pmatrix} \tag{2.11}$$

where $s_i = \sin\theta_i$, $c_i = \cos\theta_i$.

The matrix $m_{\nu,k}$ has two non-zero eigenvalues in the case we consider. The lightest neutrino is massless, which is consistent with experimental results, and the mass eigenstates are labeled in descending order of mixing with the νe flavor eigenstate. With a Normal Hierarchy (NH) in masses one has then $0 = m\nu_1 < m\nu_2 < m\nu_3$, while for an Inverted Hierarchy (IH) one has $0 = m\nu_3 < m\nu_1 < m\nu_2$. The two remaining masses are related to the squared mass differences defined in refs. [34, 35] as7

$$\begin{cases} m_{\nu 1} = 0 \\ m_{\nu 2} = \sqrt{\Delta m^2_{21}} \quad \text{(NH)} \\ m_{\nu 3} = \sqrt{\Delta m^2_{3\ell}} \end{cases} \qquad \begin{cases} m_{\nu 3} = 0 \\ m_{\nu 1} = \sqrt{-\Delta m^2_{21} - \Delta m^2_{3\ell}} \quad \text{(IH)} \\ m_{\nu 2} = \sqrt{-\Delta m^2_{3\ell}}\,. \end{cases} \tag{2.12}$$

The notation is such that $\Delta m^2_{3\ell}$is the largest mass splitting eigenvalue in the case of either mass ordering. $\Delta m^2_{3\ell} = m_3^2 - m_1^2 > 0$ for a NH and $\Delta m^2_{3\ell} = m_3^2 - m_2^2 < 0$for an IH.

Casas-Ibarra parameterization for ω

The parameter ω can be written as a function of the light neutrino masses, the PMNS angles and phases, and the heavy Majorana masses M_p, using the Casas-Ibarra parameterization [36]. This is a general result whose derivation makes use of the relations in the previous section. In our case, it reads8

$$\omega_{NH} = \frac{i\sqrt{2}}{v}\,{\rm diag}(\sqrt{M_1}, \sqrt{M_2}) \cdot R \cdot {\rm diag}(0, \sqrt{m_{\nu 2}}, \sqrt{m_{\nu 3}}) \cdot U^{\dagger}_{\rm PMNS}, \tag{2.13}$$

$$\omega_{IH} = \frac{i\sqrt{2}}{v}\,{\rm diag}(\sqrt{M_1}, \sqrt{M_2}) \cdot R \cdot {\rm diag}(0, \sqrt{m_{\nu 1}}, \sqrt{m_{\nu 2}}) \cdot U^{\dagger}_{\rm PMNS}. \tag{2.14}$$

for the normal and inverted hierarchy respectively. The matrix R is orthogonal and can be parameterized as

$$R = \begin{pmatrix} 0 & \sqrt{1-r^2} & r \\ 0 & -r & \sqrt{1-r^2} \end{pmatrix}, \qquad r \in \mathbb{C}, \tag{2.15}$$

so that the rightmost 2 × 2 sub-block is orthogonal. The values of r dictate the relation between measured low energy neutrino parameters and the Lagrangian parameters. Large values of |r| correspond to parameter space where the ω and M are related as

$$\left(\frac{(\omega_{2\beta})}{(\omega_{1\beta})}\right)_{|r|\gg 1} \approx \pm i\sqrt{\frac{M_2}{M_1}}. \tag{2.16}$$

The eigenvalues of $c^{(5)}_{\alpha\beta}$ result from a significant degree of cancellation between Lagrangian parameters as a result of this condition being enforced. Interestingly, the latter emerges naturally if the Majorana states N1, N2 are assumed to form a pseudo-Dirac pair, thereby imposing an approximate lepton number conservation [37]. In the absence of such a symmetry, however, these relations are not invariant under the RGE of the theory, and therefore represent tuned solutions. To keep the discussion general, in the following we restrict to values |r| ≤ 1.

PHENOMENOLOGY OF THE NEUTRINO OPTION AT NLO

Integrating out the Np states at one loop gives a threshold matching to the SMEFT Lagrangian parameters proportional to $|\omega|^2/16\pi^2$ and $|\omega|^4/16\pi^2$. These threshold matchings are used to generate the Higgs potential in the neutrino option, so a consistent treatment of the corrections at one loop order is of interest. A more complete treatment of this matching than ref. [8] at next to leading order includes all corrections which result from figure 1 and the inclusion of the effects of the running of the Weinberg operator. It is also necessary to utilize an alternative numerical strategy than pursued in ref. [8] to increase the numerical stability of the results. We first develop a consistent NLO framework for studying the neutrino option in this section.

One Loop Matching

The one loop matching of the seesaw model onto the SMEFT is given by equating

$$\langle i|\mathcal{L}_{SM} + \mathcal{L}_N|j\rangle = \langle i|\mathcal{L}_{SMEFT}|j\rangle \tag{3.1}$$

for fixed initial (i) and final (f) states at the scale $\mu \simeq M$, and solving for the resulting SMEFT parameters. This determines the Wilson coefficients of the higher dimensional operators and defines contributions to the $L^{(d\leq 4)}$ SM couplings due to matching boundary conditions.

We calculate in dimension regularization with d = 4−2∊, and use MS subtraction. The counterterms renormalize the theories separately on each side of the matching equations, but finite one loop matching results remain. When matching the propagators used are canonically normalized. The first diagram in figure 1a leads to a non-canonical N_p field. A canonical normalization condition can be satisfied by performing a finite renormalization field redefinition of the form

$$N^{(0)}_{R,m} \to \sqrt{R^N_{mn}}\, N^{(r)}_{R,n},$$

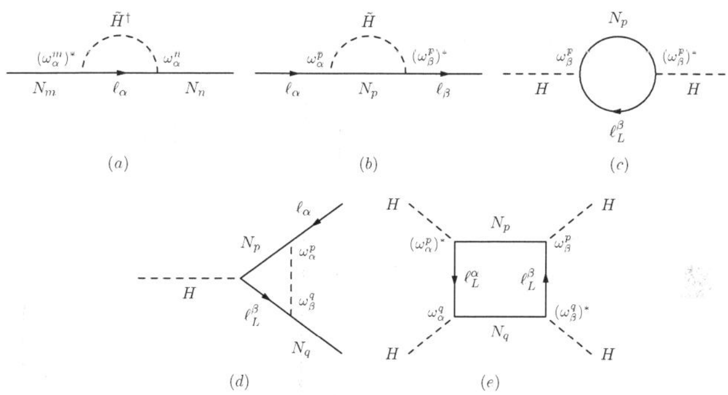

Figure 1. One loop matching diagrams for the seesaw model.

for the S matrix element in eq. (3.1). Here the (0) and (r) superscripts correspond to the un-renormalized and renormalized fields respectively in the on-shell matrix elements in the seesaw model, and

$$R^N_{mn} = \delta_{mn} - \frac{\omega_{\alpha,m}(\omega^\alpha_n)^\dagger}{4\pi^2}\left(1 + \frac{1}{2}\log\left[\frac{-\mu^2}{M_n^2}\right]\right) \tag{3.3}$$

We have expanded in $m^2/p^2 < 1$ where m2 is the value of the Higgs mass parameter before threshold matching and have fixed $p^2 = M^2_n$ due to the N state which will be taken on-shell in the threshold matching. Similarly figure (1)b leads to a non-canonical kinetic term for ` which is restored with the finite renormalization field redefinition

$$(\ell_L^{\alpha})^{(0)} \to \sqrt{R^{\ell}_{\alpha\beta}}\,(\ell_L^{\beta})^{(r)} \tag{3.4}$$

where

$$R_{\ell}^{\alpha\beta} = \delta^{\alpha\beta} - \frac{\omega^{\dagger}_{\alpha,p}\omega^{p}_{\beta}}{64\pi^2}\left(3 + 2\log\left[\frac{\mu^2}{M_p^2}\right]\right) \tag{3.5}$$

Similarly the canonical kinetic term for H is restored with the finite renormalization field redefinition $H^{(0)} \to \sqrt{R^H}\,H^{(r)}$where

$$R^{H} = 1 - \frac{\omega^{\dagger}_{\alpha,p}\,\omega^{p}_{\alpha}}{32\pi^2}\left(1 + 2\log\left[\frac{\mu^2}{M_p^2}\right]\right) \tag{3.6}$$

The three point interactions coupling N to the SM are corrected at one loop. For example, one of the interaction terms has the one loop correction

$$-\frac{1}{2}\overline{N_p}\omega_{\beta}^{p,*}\tilde{H}^T\ell_L^{c\,\beta} \to -\frac{1}{2}\overline{N_p}\omega_{\beta}^{q,*}\tilde{H}^T\ell_L^{c\,\alpha}\left[\sqrt{R_{qp}^{N,\star}}\sqrt{R_{\alpha\beta}^{\ell,\star}}\sqrt{R^{H,\star}} + \frac{\delta_{\alpha\beta}(\omega_q^{\star}\cdot\omega_p)}{16\pi^2}F[\rho_{pq}]\right. \tag{3.7}$$

where

$$F[\rho_{pq}] = \int_0^1 dx \int_0^{1-x} dy \frac{2\sqrt{\rho_{pq}}(y-1)}{x + y\,(x+y-1)\rho_{pq}}, \quad \rho_{pq} = M_p^2/M_q^2. \tag{3.8}$$

Note that this diagram violates lepton number. The corrections for the remaining three point interactions are similar and these (numerically small) effects are required for a complete one loop treatment of the matching of the seesaw model into the SMEFT. The threshold corrections to the SM Higgs potential are

$$\Delta\lambda = \frac{1}{16\pi^2}(\omega_q\cdot\omega^{p*})(\omega_p\cdot\omega^{q*})\left(1 - \frac{M_pM_q\log\frac{M_p^2}{M_q^2} + M_q^2\log\frac{\mu^2}{M_q^2} - M_p^2\log\frac{\mu^2}{M_p^2}}{M_p^2 - M_q^2}\right) \tag{3.9}$$

$$\Delta m^2 = -\frac{|\omega_p|^2 M_p^2}{4\pi^2}\left(1 + \log\frac{\mu^2}{M_p^2}\right) \tag{3.10}$$

where ω_p is the p-th row of ω. Restricting to the case of 2 massive right-handed neutrinos: in the degenerate limit $M_1 = M_2 = M$ and evaluating at $\mu = Me^{-3/4}$ the thresholds to be consistent with the extraction of the threshold correction from the effective potential in MS,10 the result takes the form

$$\Delta\lambda = -\frac{5}{32\pi^2}\left(|\omega_1|^4 + |\omega_2|^4 + 6\,|\omega_1\omega_2^*|^2 + 4\mathrm{Re}(\omega_1\omega_2)^2\right) \tag{3.11}$$

$$\Delta m^2 = \frac{M^2}{8\pi^2}(|\omega_1|^2 + |\omega_2|^2) \tag{3.12}$$

In the more general $M_2 \gtrsim M_1$ case, we sum the contributions for p = q = 1 evaluated at $\mu = M_1$, that for p = q = 2 evaluated at $\mu = M_2$ to be consistent and the mixed term in $\Delta\lambda$ evaluated at $\mu = M_2$ (the only scale where both fields are still dynamical), obtaining:

$$\begin{aligned}\Delta\lambda = &-\frac{5}{32\pi^2}\left[|\omega_1|^4 + |\omega_2|^4 + |\omega_1\omega_2^*|^2\left(1 + \frac{2\,M_1}{M_1 - M_2}\log\frac{M_2^2}{M_1^2}\right)\right],\\ &+\frac{5}{16\pi^2}\left[\mathrm{Re}(\omega_1\omega_2)^2\frac{M_1\,M_2}{M_1^2 - M_2^2}\log\frac{M_1^2}{M_2^2}\right],\end{aligned} \tag{3.13}$$

$$\Delta m^2 = \frac{1}{8\pi^2}\left[M_1^2|\omega_1|^2 + M_2^2|\omega_2|^2\right] \tag{3.14}$$

Figure 2 shows how the thresholds change assuming $M_2 = xM_1$ relative to the degenerate case, as a function of x. For definiteness we have fixed and $\omega_{p\beta} \equiv 1$, but the variation has little dependence on this choice.

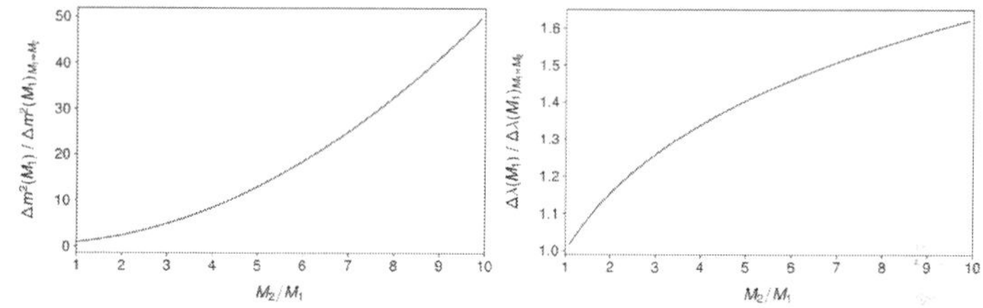

Figure 2. Relative variation of $\sqrt{\Delta m^2}$ (left) and $\Delta\lambda$ (right) assuming M2 = xM1 compared to the degenerate case, as a function of x.

The net effect of a consistent one loop matching to the seesaw scenario $(\propto \omega^2/16\pi^2)$ gives the SMEFT at the matching scale with the Wilson coefficients

$$\begin{aligned}&\frac{(\omega_\alpha^p)^T\omega_\beta^p}{M_p}\left[1-\frac{|\omega_p|^2}{4\pi^2}\left(1+\frac{1}{2}\log\left[\frac{-\mu^2}{M_p^2}\right]\right)\right]-\frac{(\omega_\alpha^p)^T\omega_\beta^p}{32\pi^2 M_p}\left|\sum_{s<n}|\omega_r|^2\left(1+2\right.\right.\\ &-\sum_{s\leq p}\frac{(\omega_\alpha^s)^T\omega_\beta^p+(\omega_\alpha^p)^T\omega_\beta^s}{8\pi^2 M_p}(\omega^p\cdot\omega_s^\star)\left(1+\frac{1}{2}\log\left[\frac{-\mu^2}{M_s^2}\right]-\frac{F[\rho_{sp}]}{4}\right),\\ &-\frac{(\omega_\alpha^s)^T\omega_\beta^p+(\omega_\alpha^p)^T\omega_\beta^s}{8\pi^2 M_p}\frac{\epsilon_{sp}(M_1+M_2)}{M_1-M_2}\mathrm{Re}(\omega_1\cdot\omega_2^\star)\left[1+\frac{1}{2}\log\left[\frac{-\mu^2}{M_2^2}\right]\right],\\ &-\frac{(\omega_\rho^p)^T\omega_\beta^p}{16\pi^2 M_p}\sum_{s\leq p}\frac{\omega_\alpha^s(\omega_\rho^s)^\dagger}{8}\left(3+2\log\left[\frac{\mu^2}{M_s^2}\right]\right)-\frac{(\omega_\alpha^p)^T\omega_\sigma^p}{16\pi^2 M_p}\sum_{s\leq p}\frac{\omega_\beta^s(\omega_\sigma^s)^\dagger}{8}\left(3+2\log\left[\frac{\mu^2}{M_s^2}\right]\right),\end{aligned} \tag{3.15}$$

with s summed over the Ns active states, when each Np mass eigenstate Majorana field is integrated out. $\epsilon_{12} = -\epsilon_{21} = 1$ and $M_{1,2}$ are the tree level masses of the Np states. The potential after the one loop matching is given by

$$V(H^\dagger H) = -\frac{m_0^2 R_H + \Delta m^2}{2}(H^\dagger H) + (\lambda_0 R_H^2 + \Delta\lambda)(H^\dagger H)^2 + \cdots . \quad (3.16)$$

We do not assume that hierarchies among the ω^p_β, or significant effects due to their orientation in flavour space in order to enhance the importance of these one loop corrections. Further, we assume m0 is negligible as we are considering a classically scaleless mass spectrum. In the numerical results presented we absorb the correction to λ0, Mp into the leading order parameters as a result and neglect the correction to $c^{(5)}_{\alpha\beta}$. We adopt this approach as these corrections are smaller than the remaining numerical uncertainties in the RGE evolution of the SMEFT parameters.

Running

The SM potential parameters are compared to measured values run to the scale $\mu = \hat{m}_t$. For the RGE of the SM we use results from appendix B of ref. [44] and evaluate the running of $\{g_1, g_2, g_2, y_t, y_b, y_\tau, \lambda, m^2\}$ as a coupled system, using the RGE computed at n_{RGE} loops. The results in ref. [44] allow numerical studies of the order $n_{RGE} = 3$ for $\{g_1, g_2, g_2, y_t, \lambda, m^2\}$ (without y_b, y_τ dependence) and up to $n_{RGE} = 2$ for $\{y_b, y_\tau\}$.

Neutrino Parameter Running

The seesaw model matched onto $c^{(5)}_{\alpha\beta}(\mu^2 = M^2)$is compared to the values of masses and mixing angles extracted at $\mu = \hat{m}_Z$after running the seesaw parameters with one loop RGE evolution of the Weinberg operator.13 The RGEs of the c5 coefficient are extracted from refs. [45, 46]. Below the scale ˆmZ the neutrino parameters do not run significantly, as we have explicitly verified.

$$16\pi^2\mu\frac{dc_5}{d\mu} = -\frac{3}{2}\left[c_5(Y_e^\dagger Y_e)+(Y_e^\dagger Y_e)^T c_5\right] - \left[3g_2^2 - 4\lambda - 2\left(3Y_u^\dagger Y_u + 3Y_d^\dagger Y_d + Y_e^\dagger Y_e\right)\right] c_5$$
$$\simeq -\frac{3}{2}\left[c_5 \cdot \mathrm{diag}(0,0,y_\tau^2) + \mathrm{diag}(0,0,y_\tau^2)\cdot c_5\right] - \left[3g_2^2 - 4\lambda - 6y_t^2 - 6y_b^2 - 2y_\tau^2\right] c_5 .$$

(3.17)

The Yukawa coupling normalization is $\mathcal{L}_Y = -\left[H^{\dagger j}\overline{d}\, Y_d\, q_j + \widetilde{H}^{\dagger j}\overline{u}\, Y_u\, q_j + H^{\dagger j}\overline{e}\, Y_e\, \ell_j + \text{h.c.}\right]$. In this limit, there is no mixing between the different entries of c5 and the individual elements run according to

$$16\pi^2\mu\frac{dc_{5,\alpha\beta}}{d\mu}=-\kappa_{\alpha\beta}\,c_{5,\alpha\beta},\quad \kappa_{\alpha\beta}=\begin{cases}3g_2^2-4\lambda-6y_t^2-6y_b^2-2y_\tau^2, & \alpha,\beta\neq 3\\ 3g_2^2-4\lambda-6y_t^2-6y_b^2-y_\tau^2/2, & \beta(\alpha)=3,\ \alpha(\beta)\neq 3\\ 3g_2^2-4\lambda-6y_t^2-6y_b^2+y_\tau^2, & \alpha=\beta=3.\end{cases} \tag{3.18}$$

Comparing the extracted eigensystem of c_5 at any given scale μ to experimental results in parameter scans is numerically unstable. Using the β-functions of the measurable neutrino parameters themselves reduces this numerical uncertainty. We use the results in ref. [47] for a generic parameterization14 of the measurable parameters extracted from c_5.

For a normal hierarchy15 with two nonzero masses, the results in ref. [47] reduce to

$$16\pi^2\mu\frac{dm_{\nu,k}}{d\mu}=-m_{\nu,k}\left[3y_\tau^2|V_{3i}|^2+3g_2^2-4\lambda-6y_t^2-6y_b^2-2y_\tau^2\right] \tag{3.19}$$

$$16\pi^2\mu\frac{d\theta_1}{d\mu}=\frac{3y_\tau^2}{2c_2}\,\mathrm{Re}\left[s_3V_{31}V_{33}^*+c_3\,T_{32}\right] \tag{3.20}$$

$$16\pi^2\mu\frac{d\theta_2}{d\mu}=\frac{3y_\tau^2}{2}\,\mathrm{Re}\left[e^{-i\delta}\left(-c_3V_{31}V_{33}^*+s_3\,T_{32}\right)\right] \tag{3.21}$$

$$16\pi^2\mu\frac{d\theta_3}{d\mu}=-\frac{3y_\tau^2}{2}\frac{s_2}{c_2}\,\mathrm{Re}\left[e^{-i\delta}\left(\frac{c_2}{s_2}e^{i\delta}V_{31}V_{32}^*+s_3V_{31}V_{33}^*-c_3\,T_{32}\right)\right] \tag{3.22}$$

Here we have used the notation

$$T_{32}=\frac{2m_{\nu2}m_{\nu3}}{m_{\nu2}^2-m_{\nu3}^2}e^{i\phi'}V_{32}^*V_{33}+\frac{m_{\nu2}^2+m_{\nu3}^2}{m_{\nu2}^2-m_{\nu3}^2}V_{32}V_{33}^*. \tag{3.23}$$

The remaining RGEs are

$$\begin{aligned}16\pi^2\mu\frac{d\delta}{d\mu}=\frac{3y_\tau^2}{2}\,\mathrm{Im}\Bigg[&\frac{V_{31}V_{32}^*}{c_3s_3}-\frac{s_3}{s_1c_2c_3}V_{31}V_{22}^*V_{33}^*-\frac{e^{-i\delta}}{s_2c_1c_2}V_{31}V_{22}V_{33}^*\\&-T_{32}\left(\frac{e^{-i\delta}V_{21}}{c_1c_2s_2}-\frac{c_3V_{21}^*}{s_1s_3c_2}\right)\Bigg]\end{aligned} \tag{3.24}$$

$$16\pi^2\mu\frac{d\phi}{d\mu}=3y_\tau^2\,\mathrm{Im}\left[\frac{c_3V_{31}V_{32}^*}{s_3}+\frac{V_{21}^*V_{33}^*V_{31}}{s_1c_2}+\frac{|V_{31}|^2}{c_1c_2}V_{33}^*+T_{32}\left(\frac{c_3V_{21}^*}{s_1c_2s_3}-\frac{V_{32}^*}{c_1c_2}\right)\right] \tag{3.25}$$

$$16\pi^2\mu\frac{d\phi'}{d\mu}=3y_\tau^2\,\mathrm{Im}\left[\frac{s_3V_{31}V_{32}^*}{c_3}+\frac{|V_{31}|^2V_{33}^*}{c_1c_2}-\frac{s_3}{s_1c_2c_3}V_{31}V_{22}^*V_{33}^*-T_{32}\left(\frac{V_{22}^*}{s_1c_2}+\frac{V_{32}^*}{c_1c_2}\right)\right] \tag{3.26}$$

where V_{ij} denotes the corresponding entry of the V matrix defined in eq. (2.11).

Ref. [47] defined the running of three unphysical phases, that are added to the definition of the U rotation such that

$$U=\mathrm{diag}\left(e^{i\alpha_e},e^{i\alpha_\mu},e^{i\alpha_\tau}\right)\cdot U_{\mathrm{PMNS}} \tag{3.27}$$

This approach is convenient as the field redefinitions that reduce the phases of the PMNS matrix to the minimal set must be re-imposed at each scale μ. The β functions of the unphysical phases are

$$16\pi^2\mu\frac{d\alpha_e}{d\mu} = \frac{3y_\tau^2}{2}\mathrm{Im}\left[\frac{V_{31}V_{32}^*}{s_3c_3} + \frac{|V_{31}|^2V_{33}^*}{c_1c_2} - \frac{s_3}{s_1c_2c_3}V_{31}V_{22}^*V_{33}^* - T_{32}\left(\frac{V_{32}^*}{c_1c_2} + \frac{V_{22}^*c_3}{s_1c_2s_3}\right)\right], \tag{3.28}$$

$$16\pi^2\mu\frac{d\alpha_\mu}{d\mu} = \frac{3y_\tau^2}{2c_2s_1}\mathrm{Im}\left[V_{31}V_{21}^*V_{33}^* - T_{32}V_{22}^*\right], \tag{3.29}$$

$$16\pi^2\mu\frac{d\alpha_\tau}{d\mu} = \frac{3y_\tau^2}{2c_2c_1}\mathrm{Im}\left[|V_{31}|^2V_{33}^* - T_{32}V_{32}^*\right]. \tag{3.30}$$

We have verified analytically that

$$\left(16\pi^2\mu\frac{dc_5}{d\mu}\right) = \sum_k \frac{\partial c_5}{\partial x_k}\left(16\pi^2\mu\frac{dx_k}{d\mu}\right), \quad x_k = \{m_{\nu 1}, m_{\nu 2}, m_{\nu 3}, \theta_1, \theta_2, \theta_3, \delta, \phi, \phi', \alpha_e, \alpha_\mu, \alpha_\tau\}. \tag{3.31}$$

The running of neutrino parameters depends on $\{g_2, \lambda, y_t, y_b, y_\tau\}$. In the numerical results, these are fixed to their value at $\mu = m_t$ neglecting yet higher order running effects. Although the running of the neutrino parameters was studied for completeness in the numerical results shown, we confirm past results that find the neutrino parameters do not run in a numerically significant manner. We have verified that the dependence on the top quark mass in the numerical running of the neutrino parameters is a negligible effect when scanning parameter space.

NUMERICAL STRATEGY AND RESULTS

Consistency of the neutrino option explaining the Higgs potential and Neutrino masses is dictated by choosing a subset of inputs among

$$\{\hat{m}_\nu^i(\mu^2 \ll \langle H^\dagger H\rangle), \hat{\theta}_i, \hat{m}_h(\langle H^\dagger H\rangle), \hat{\lambda}(\langle H^\dagger H\rangle)\}, \tag{4.1}$$

and predicting the remaining quantity(ies). In ref. [8] a consistency test was formulated where $\{\hat{\lambda}(M), \hat{m}_h(M)\}$was fixed, and it was shown that parameters can be chosen such that $\hat{m}_\nu^i(\mu^2 \ll \langle H^\dagger H\rangle)$can be approximately reproduced. It was observed that significant numerical instability is present in this approach. It is necessary to avoid an asymptotic value of $\lambda(M) \to 0$ being used as an input for numerical precision. In a consistency test, any mismatch between a predicted value of a parameter, and an observed value can be accommodated by an extended scenario where the Neutrino option is embedded in a UV theory. Only the parameter λ can receive further classical tree level contributions without breaking a symmetry or adding field content to the scenario. For these reasons, in this paper we formulate a consistency test where $\{\hat{m}_\nu^i(\mu^2 \ll \langle H^\dagger H\rangle), \hat{\theta}_i, \hat{m}_h(\langle H^\dagger H\rangle)\}$ are fixed as inputs, and the required $\lambda(M)$ is then compared to $\lambda_0(M) + \Delta\lambda(M)$.

Numerical Inputs

We enforce that the low energy neutrino parameters taken from the global fit in ref. [35] and given in table. 1 are reproduced. The SM parameters are extracted from ref. [44] solving the reported integral equations for the input parameters $\hat{m}_W, \hat{m}_Z, \hat{m}_h$ at the indicated loop order. The light bottom and τ Yukawa's, $\hat{y}_b, \hat{y}_\tau$, are matched at tree level as higher order corrections are negligible. The value of $g3(\mu = m_t)$ is extracted from eq. 60 of ref. [44] which includes higher order QCD corrections. To these SM results we add the effects of the seesaw model as matched onto the SMEFT. The results in ref. [29] characterize the tree level matching of the seesaw model onto the SMEFT up to sub-leading order ($L^{(7)}$ corrections) but we restrict our attention to the matching onto $L^{(5)}$ in this work.

Table 1. Best fit values of neutrino parameters taken from the global fit in ref. [35]

	Normal Hierarchy		Inverted Hierarchy	
	best fit	3σ range	best fit	3σ range
s_1^2	0.441	0.385–0.635	0.587	0.393–0.640
s_2^2	0.02166	0.01934–0.02392	0.02179	0.01953–0.02408
s_3^2	0.306	0.271–0.345	0.306	0.271–0.345
$\delta(°)$	261	0–360	277	145–391
Δm_{21}^2 (10^{-5} eV^2)	7.50	7.03–8.09	7.50	7.03–8.09
Δm_{3l}^2 (10^{-3} eV^2)	2.524	2.407–2.643	-2.514	(-2.635)–(-2.399)

Table 2. Left table: best fit values of the quantities used as inputs in the numerical analysis, while m_t is varied in the range specified. Right table: matching values for the SM parameters at $\mu = m_t$ obtained from the expressions in appendix A in ref. [44] with the inputs on the left when $m_t = 173.2$ GeV.

	best fit	range
$\hat{G}_F$ [GeV^{-2}]	$1.1663787 \cdot 10^{-5}$	
$\hat{\alpha}_s(m_Z)$	0.1185	
$\hat{m}_Z$ [GeV]	91.1875	
$\hat{m}_W$ [GeV]	80.387	
$\hat{m}_h$ [GeV]	125.09	
$\hat{m}_t$ [GeV]	173.2	171–175
$\hat{m}_b$ [GeV]	4.18	
$\hat{m}_\tau$ [GeV]	1.776	

	tree	1-loop	2-loop
$\hat{\lambda}$	0.1291	0.1276	0.1258
$\hat{m}$ [GeV]	125.09	132.288	131.431
$\hat{g}_1$	0.451	0.463	0.461
$\hat{g}_2$	0.653	0.6435	0.644
$\hat{g}_3$	—	1.22029	—
$\hat{y}_t$	0.995	0.946	0.933
$\hat{y}_b$	0.024	—	—
$\hat{y}_\tau$	0.0102		

The $\omega_{p\beta}$ are required to reproduce the observed neutrino masses and mixings. The range of values for $\Delta m^2(M_1)$, $\Delta\lambda(M_1)$ compatible with this condition is determined scanning the low energy parameter space with a sample of 1000 points randomly selected within the 3σ allowed ranges for $\{\hat{m}_{\nu i}, \hat{\theta}_i, \hat{\delta}, \hat{\phi}, \hat{\phi}'\}$ given in table 1.

Each point determined represents a boundary condition at $\mu = \hat{m}_Z$ for the neutrino parameters' RGE (eqs. (3.19)–(3.30)). For each of them it is then possible to determine the running quantities $\omega_{p\beta}(\mu)$ via the Casas-Ibarra parameterization (eqs. (2.13), (2.14)) and consequently the threshold corrections $\Delta m^2(\mu = M_1)$, $\Delta\lambda(\mu - M_1)$ as a function of the lightest Majorana mass M1 (eqs. (3.14), (3.13)). The parameter r of the Casas-Ibarra parameterization is varied at every point and chosen as a random complex with $|r| \leq 1$. Four independent scans are performed; assuming either normal or inverted neutrino mass hierarchy and either degenerate ($M_1 = M_2$) or nearly-degenerate $(M_1 \lesssim M_2 \lesssim 10\, M_1)$ N_p states.16

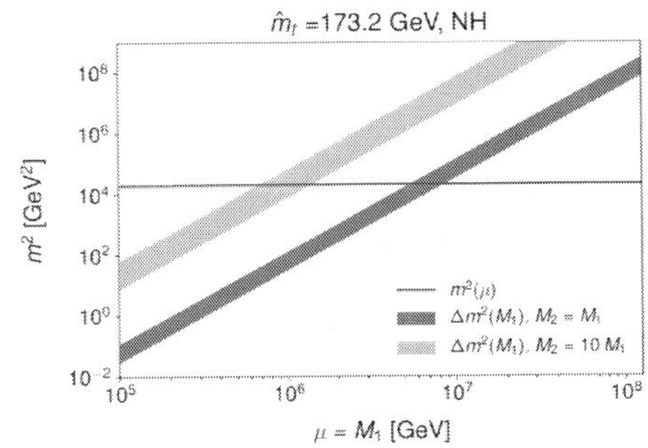

Figure 3. Numerical comparison between the values of the threshold correction $\Delta m^2(M_1)$ compatible with neutrino physics constraints in the degenerate $M_1 = M_2$ case (blue band) and when $M_2 = 10\, M_1$ (orange band) with the running Higgs mass $m^2(\mu)$ determined by the SM RGE and the measured SM parameters (red line). The running effect is not appreciable in the scale shown because $m^2(\mu)$ varies vary little compared to $\Delta m^2(M_1)$. This plot assumes normal ordering of neutrino masses (NH) and $\hat{m}_t = 173.2\,\mathrm{GeV}$.

The values of $\{m^2(\mu), \lambda(\mu)\}$ that are compatible with the measured $\{\hat{m}_h, \hat{\lambda} = \hat{G}_F \hat{m}_h^2/2\}$ are then determined. These are the solutions to the SM RGE system [44] with the matching conditions in table 2 (right) fixed at $\mu = \hat{m}_t$. We consider RGEs with $n_{RGE} = \{1, 2, 3\}$ with order ($n_{RGE} - 1$) matching and three benchmark values for $\hat{m}_t = \{171,\ 173.2,\ 175\}\,\mathrm{GeV}$.

We then compare the results obtained in these steps. Unlike in ref. [8], for the sake of generality we allow here for a term $\lambda_{0(}H^\dagger H)^2$ in the scalar potential. The neutrino option is then realized for values of (M_1, λ_0) that satisfy simultaneously

$$m^2(M_1) \simeq \Delta m^2(M_1), \tag{4.2a}$$

$$\lambda(M_1) \simeq \lambda_0 + \Delta\lambda(M_1) \tag{4.2b}$$

Case m_t = 173.2 GeV and normal neutrino mass hierarchy

The results of the analysis are shown in figures 3, 4 for the case of normal neutrino mass hierarchy and $\hat{m}_t = 173.2\,\mathrm{GeV}$. Figure 3 shows $m^2(\mu)$ (red line) vs. $\Delta m^2(M_1)$ for degenerate Np states (blue band) and for $M_2 = 10\ M_1$ (orange band). Eq. (4.2a) is satisfied in the region where the bands overlap with the RGE curve: for the degenerate case we find17

$$\begin{aligned} 5\cdot 10^6\,\mathrm{GeV} \lesssim M_1 \lesssim 8.3\cdot 10^6\,\mathrm{GeV} \quad (\mathrm{NH}),\\ 4.2\cdot 10^6\,\mathrm{GeV} \lesssim M_1 \lesssim 7\cdot 10^6\,\mathrm{GeV} \quad (\mathrm{IH}), \end{aligned} \tag{4.3}$$

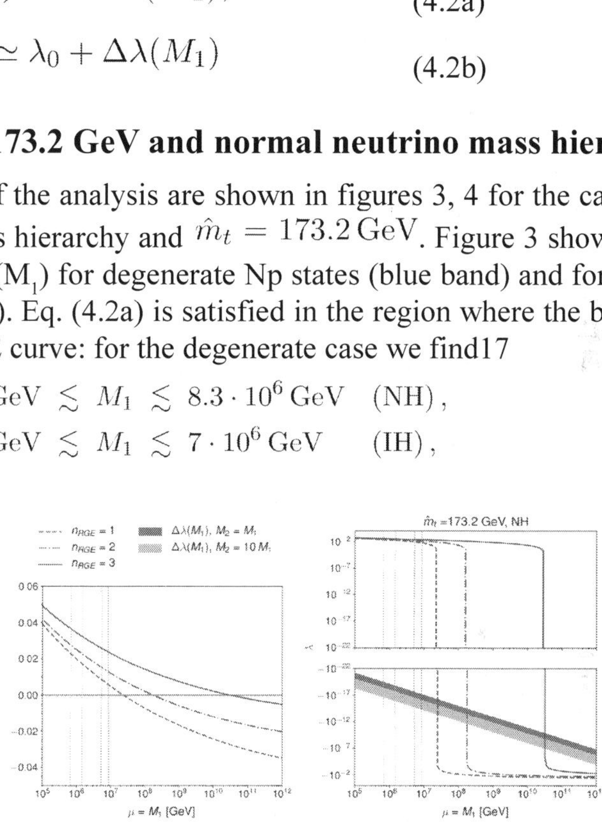

Figure 4. Numerical comparison between the values of the threshold correction $\Delta\lambda(M_1)$ compatible with neutrino physics constraints in the degenerate $M_1 = M_2$ case (blue band) and when $M_2 = 10\ M_1$ (orange band) with the running quartic coupling $\lambda(\mu)$ determined by the SM RGE and the measured SM parameters (red lines). The blue and orange bands overlap completely in the left plot. The left and right plots differ uniquely in the choice of the y-axis scale (linear vs. logarithmic) and the right figure has been split in two symmetric panels for $\lambda > 0$ and $\lambda < 0$. The dotted vertical lines mark the mass regions where the matching for m2 is fulfilled (cf. figure 3) for both the degenerate (blue) and non-degenerate (orange) cases. This figure assumes normal ordering of neutrino masses (NH) and $\hat{m}_t = 173.2\,\mathrm{GeV}$.

while for $M_2/M_1 = x > 1$ lower values of M_1 are allowed. For the benchmark x = 10 the viable mass region is

$$6.4 \cdot 10^5\,\mathrm{GeV} \lesssim M_1 \lesssim 1.4 \cdot 10^6\,\mathrm{GeV} \quad (\mathrm{NH}),$$
$$5.5 \cdot 10^5\,\mathrm{GeV} \lesssim M_1 \lesssim 8.9 \cdot 10^5\,\mathrm{GeV} \quad (\mathrm{IH}), \tag{4.4}$$

and intermediate values are possible for $1 < x < 10$. Notably, $m^2(\mu)$ has negligible dependence on both n_{RGE} and $\hat{m}_t$. Therefore the range

$$5 \cdot 10^5\,\mathrm{GeV} \lesssim M_1 \lesssim 10^7\,\mathrm{GeV} \tag{4.5}$$

is a general prediction of the neutrino option. Specific assumptions about the neutrino mass ordering and M2/M1 refine this range as detailed in figure 6.

Figure 4 shows $\lambda(\mu)$ vs $\Delta\lambda(M_1)$, both in linear (left) and log scale (right). For reference, the regions in eqs. (4.3), (4.4) are marked with blue and orange dotted vertical lines. Within these energy windows the threshold correction $\Delta\lambda(M_1)$ is always negative and very small, while the SM running curve $\lambda(\mu)$ is positive and $\sim O(10^{-2})$. For the neutrino option to be realized it is then necessary that

$$\lambda(M_1) \simeq \lambda_0 \gg \Delta\lambda(M_1), \quad \Rightarrow \quad \lambda_0 \sim \mathcal{O}(10^{-2}) \tag{4.6}$$

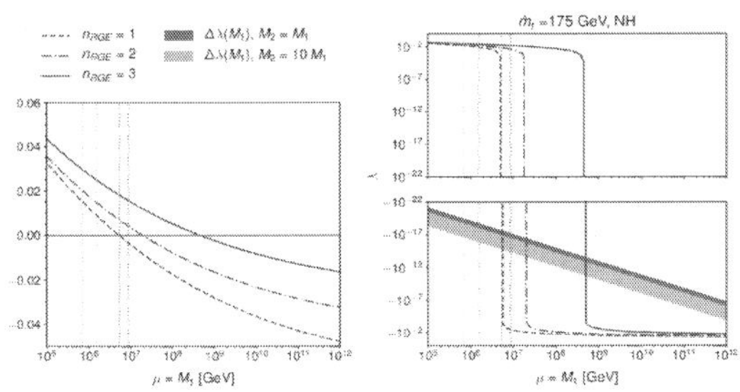

Figure 5. Same as figure 4, for a top mass $\hat{m}_t = 175\,\mathrm{GeV}$.

Varying m_t and other Benchmark Assumptions

Assuming an inverted neutrino mass hierarchy and taking different values of the top quark mass leads to qualitatively similar figures and has a modest impact on the numerical results. The main conclusions, i.e. the identification of the mass range in eq. (4.5) and the necessity of a bare term $\lambda_0 \sim O(10^{-2})$ in the Lagrangian, are general and emerge in all the benchmarks considered.

Note that the 3-loop RGE accuracy is crucial for establishing the condition $\lambda_0 \sim O(10^{-2})$ as, for instance, an analysis restricted to $n_{RGE} = 1$ does admit solutions of eqs. (4.2) with $\lambda_0 \simeq 0$ [8]. This is shown explicitly in figure 5, that reports the matching results for the λ parameter obtained with $\hat{m}_t = 175\,\text{GeV}$. In this case, the $n_{RGE} = 1$ dashed curve matches directly the threshold corrections band in the M1 region where eq. (4.2a) is satisfied. This is consistent with what was observed in ref. [8].

The values of (M_1, λ_0) where the neutrino option can be realized for each of the setups considered are summarized in figure 6. Varying either n_{RGE} or ˆmt mainly impacts $\lambda(\mu)$, with larger n_{RGE} and smaller $\hat{m}_t$ giving a smoother running curve and consequently requiring larger values of λ_0 in the matching. Choosing the inverse neutrino mass hierarchy rigidly results in slightly lower M_1 and slightly larger λ_0. This is easily understood as follows: in the IH case the neutrino masses are larger, which leads to larger $\omega_{p\beta}$ (eq. (2.14)). The m^2 matching relation in eq. (4.2a) selects then a lower M_1 region compared to the NH case. Because $\lambda(\mu)$ is larger there, this has the indirect consequence of requiring a larger λ_0. Finally, the threshold correction $\Delta m^2(M_1)$ is very sensitive to the relative size of the Majorana masses $x = M_2/M_1$ (see figure 2). Larger x lead to lower M1 being selected and, by the argument above, this indirectly requires larger λ_0. This explains the relative position of the blue and orange crosses in figure 6. The correction $\Delta\lambda(M_1)$ does not play any significant role in the numerical analysis as it is always $\ll \lambda(\mu)$ for all the setups considered.

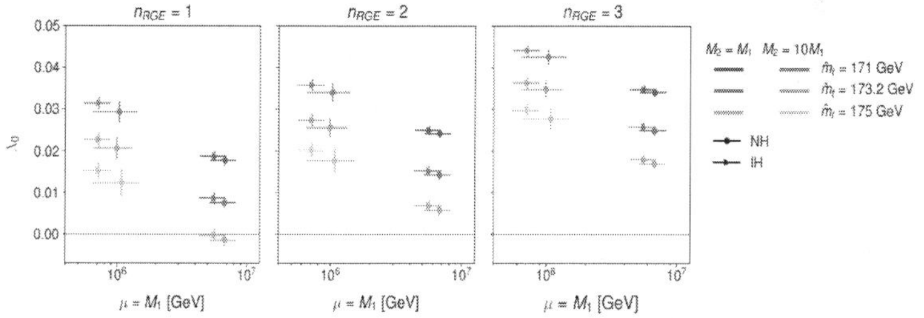

Figure 6. Regions of the (M_1, λ_0) parameter space in which the neutrino option can be realized, for different choices of: the SM RGE order (panels from left to right), the top mass $\hat{m}_t$ (colors from lighter to darker), the neutrino mass ordering (dots vs triangles) and $x = M_2/M_1$ (blue for x = 1, orange for x = 10).

UV EMBEDDINGS

The numerical results of section 4 (see also refs. [1, 8]) show that a threshold matching defining a boundary condition for $H^{\dagger}H$ in the seesaw model can be consistent with the lower scale Higgs mass measured experimentally. This conclusion is robust against using $n_{RGE} = \{1, 2, 3\}$ in the results presented. As the coefficient of $H^{\dagger}H$ is dimensionful, this modifies the usual concerns of the Electroweak scale Hierarchy problem into an alternate framework. For this framework to be embedded into a theoretically successful UV completion requires a UV scenario that can generate the Majorana scale used. This needs to occur in a manner that does not lead to other, larger, threshold matching conditions. Further, the threshold matching to the λ parameter, parametrically $\Delta\lambda \propto \omega^4/16\pi^2$ can be vanishingly small due to integrating out the Majorana Neutrino, as $\omega \ll 1$ in order to separate the Majorana mass scale from the effective observed Electroweak scale. As a direct result, a small matching effect for Δλ can be subdominant to other UV boundary effects [8], or even a bare λ0 parameter, which is not forbidden by a symmetry. Explaining the origin of the required λ(M) in a UV scenario would also advance the embedding of this theoretical framework in a more complete UV scenario.

The Conformal UV Embedding of the Neutrino Option of ref. [1]

Recently a UV framework for the Majorana scale generation was put forth in ref. [1] that addresses most of these theoretical challenges. The idea is to extend the SM with a set of scalar fields to generate the Majorana scale spontaneously by satisfying a GildenerWeinberg [50] condition. The conformal UV completion of the neutrino option (hereafter the LCNO) of ref. [1] is defined as

$$\begin{aligned}
\mathcal{L}_{CNO} &= \frac{1}{2}\partial^{\mu} S \partial_{\mu} S + \frac{1}{2}\partial^{\mu} R \partial_{\mu} R + \overline{N}^p_R i \partial\!\!\!/ N^p_R - V(H, S, R) + \mathcal{L}_{\rm int}, \\
\mathcal{L}_{\rm int} &= -\left(\frac{y_M^{pr}}{2} S\, \overline{N}^p_R N^{r,c}_R + \overline{\ell^{\beta}_L} \tilde{H} \omega^{p,\dagger}_{\beta} N_p + h.c.\right), \\
V(H, S, R) &= \lambda_S S^4 + \lambda_R R^4 + \lambda_{HS} S^2 (H^{\dagger} H) + \lambda_{HR} R^2 (H^{\dagger} H) + \lambda_{SR} S^2 R^2.
\end{aligned} \tag{5.1}$$

Here S,R are real $SU(3) \times SU(2)_L \times U(1)_Y$ singlet scalar fields, and R has an odd charge under a Z_2 symmetry [1].18 The running of the parameters leads to the condition

$$\frac{\partial^4 V(H, S, R)}{\partial^4 S} = \lambda_S(\Lambda_{GW}) = 0, \tag{5.2}$$

being satisfied, which leads to the spontaneous breaking of scale invariance once perturbative corrections are included in the Coleman-Weinberg (CW) potential [52]. S is identified as the pseudo-Goldstone boson of broken scale invariance, the dilaton [53, 54]. S has a large coupling y_M^{pr} with O(1) entries to the only other fields that are pure singlets, i.e. N_R^p. The spontaneous breaking of scale invariance gives

$$\langle H^\dagger H \rangle = \langle R \rangle = 0, \qquad \langle S \rangle \equiv v_S \neq 0, \tag{5.3}$$

and the mass spectrum is [1, 50]

$$M_N^{pr} = y_M^{pr} v_S, \qquad M_R = \sqrt{2\lambda_{SR}} v_S, \qquad M_S = 2\sqrt{2B} v_S. \tag{5.4}$$

The interaction terms of the seesaw model can be written as in eq. (2.7) with all CP violating phases shifted to effective couplings and real diagonal entries in a mass matrix. This is conveniently done while by introducing Np = UN$^{p'}$ states, defining a rotated diagonal coupling matrix $\tilde{y}_M^{ss} = U^\dagger y_M^{pr} U$. (The mass matrix in eq. (2.7) is in the diagonal mass basis with the prime superscripts dropped.) Using this notation the result in eq. (5.3) explicitly depends on the potential parameters through

$$B = \frac{2\lambda_{HS}^2 + 2\lambda_{SR}^2 - \sum_s (\tilde{y}_M^{ss})^4}{32\pi^2}, \tag{5.5}$$

and it is required that B > 0 for physical solutions. The remaining contribution to the effective potential is given by [1, 50]

$$A = \frac{1}{32\pi^2}\left[2\lambda_{HS}^2\left(\log[\lambda_{HS}] - \frac{3}{2}\right) + 2\lambda_{SR}^2\left(\log[2\lambda_{SR}] - \frac{3}{2}\right) - \sum_s y_M^4\left(\log\left[(y_M^s)^2\right] - \frac{3}{2}\right)\right] \tag{5.6}$$

and the relation between v_S and Λ_{GW} is

$$\log\left[\frac{v_S}{\Lambda_{GW}}\right] = -\frac{1}{4} - \frac{A}{2B} \tag{5.7}$$

An additional threshold contribution to H†H of the form

$$\Delta m^2 = \frac{1}{16\pi^2}\left[|\omega_p|^2 M_p^2 - \lambda_{HR} M_R^2\left(1 + 2\log\frac{M_R^2}{M_S^2}\right)\right] \tag{5.8}$$

is present, and the first term in this expression must dominate for the correct sign to be obtained for Δm^2. The following set of consistency conditions are required to hold [1]

$$\begin{aligned} \frac{2\lambda_{HS}^2 + 2\lambda_{SR}^2 - \sum_s (\tilde{y}_M^{ss})^4}{32\pi^2} &> 0, \\ |\omega_p|^2 M_p^2 &> \lambda_{HR} M_R^2, \\ |\lambda_{HS}| &< \frac{1}{16\pi^2}|\omega_p|^2 M_p^2. \end{aligned} \tag{5.9}$$

The parameter space examined in ref. [1] is consistent with these conditions and such that

$$|\omega_p| \sim 10^{-7} - 10^{-3}, \qquad y_M \sim \mathcal{O}(0.1), \qquad \Lambda_{GW}[\text{GeV}] \sim 10^{6} - 10^{9},$$
$$\lambda_{HS} \sim 10^{-16} - 10^{-19} \qquad \lambda_{SR} \sim \mathcal{O}(0.1) \qquad \lambda_{R} \sim \mathcal{O}(0.1). \tag{5.10}$$

Extending the Conformal Neutrino Option

Define a conformal transformation to be a smooth transformation of the metric

$$\tilde{g}_{\mu\nu} = \Omega^2(x) g_{\mu\nu}, \tag{5.11}$$

that preserves the causal structure of the theory. Further define the conformal weight of the scalar fields of the theory, collectively denoted φ, to be $\tilde{\phi} = \phi/\Omega(x)$. A classically conformal UV embedding of the neutrino option requires some further extension due to the existence of gravity. First, this is because the scalar fields Klein-Gordon equations do not satisfy conformal invariance, until gravitational interactions are included. Second, the existence of the Planck scale itself is an explicit scale in the complete Lagrangian, calling into question the conformal starting point, and possibly leading to fine tuning.

$$S_{CNO} = \int d^4x\sqrt{-g}\left[-\frac{m_{pl}^2}{2}\mathcal{R} - \frac{1}{6}H^\dagger H\,\mathcal{R} - \frac{1}{12}S^2\mathcal{R} - \frac{1}{12}R^2\mathcal{R} + \mathcal{L}_{SM} + \mathcal{L}_{CNO} + \cdots\right]. \tag{5.12}$$

Here R is the Ricci scalar, $m_{pl} = 2.44 \times 10^{18}$ GeV is the reduced Planck mass, and $g = \det(g_{\mu\nu})$. Note the scalar kinetic terms flip sign due to the adopted metric convention. At this stage the theory is necessarily non-renormalizable as the interaction terms $\varphi^2 R$ represent an infinite tower of higher dimensional operators characterizing the interactions of the graviton with the dynamical field content. Eq. (5.12) can be studied directly by expanding the metric around flat space in terms of the dynamical graviton field hμν as $g_{\mu\nu} = \eta_{\mu\nu} + h_{\mu\nu}/m_{pl} + \cdots$ which makes it clear that the Lagrangian term represents a the tower of higher dimensional operators. This set of interactions are dependent on the background field values $\langle\varphi^2\rangle$ and at large field values a large mixing of the scalar degrees of freedom with scalar modes of (non-canonically normalized) gravity results. The identification of the scale μ in the CW potential defined in field space, with particle masses leading to the threshold corrections of the neutrino option, ties together running in field space and the running in energy of the theory, that are formally distinct.

The effects of the non-minimal interactions with gravity are more easily studied by performing a transformation from the Jordan frame in eq. (5.12) to the Einstein frame of the theory. Taking $\tilde{g}_{\mu\nu} = f(\phi^2) g_{\mu\nu}$ and

$$f(\phi^2) = \left(1 + \frac{H^\dagger H}{3\,m_{pl}^2} + \frac{R^2}{6\,m_{pl}^2} + \frac{S^2}{6\,m_{pl}^2}\right)^{-1} \tag{5.13}$$

as a further conformal transformation takes the theory to canonical form. The relevant results to study the higher dimensional operators generated already exist in the literature in refs. [55, 56]. One finds the expression

$$\frac{\mathcal{L}_{CNO}}{\sqrt{-g}} = -\frac{m_{pl}^2}{2} R - g^{\mu\nu} f^2 \left((D_\nu H)^\dagger (D_\mu H) + \frac{1}{2}\partial_\nu S \partial_\mu S + \frac{1}{2}\partial_\nu R \partial_\mu R\right) + f^2 V(H,S,R), \\ + f^2 \overline{N}_R^p i \partial\!\!\!/ N_R^p + f^2 \mathcal{L}_{\text{int}} + \cdots . \tag{5.14}$$

We are interested in the possible relation between the values of the parameters in the scalar potential in the CNO and the effects of the higher dimensional operators in eq. (5.14). Although expanding in small field values leads to a series of power corrections $\langle S^2\rangle / m_{pl}^2$ to potential terms, setting a lower bound on allowed λ coupling parameter space (when parameter tuning is avoided), we find the values of parameters examined in ref. [1] are still viable.

Possible Fine Tuning

Addressing the consistency of the Planck scale with a conformal embedding of the neutrino option is less straightforward. On the one hand, the required boundary values of the SM couplings at the Planck scale, including λ, are expected to be generated in a consistent UV theory which includes quantum gravity.19 The demand for such an embedding is reinforced by the fact that the SM interactions have Landau poles above the Planck scale. It is natural to speculate that a UV fixed point can appear in such a theory [57]. If such a conformal field theory in the UV has an interacting fixed point, then the arguments of ref. [58] imply that this scenario can be subject to fine tuning.

Here we review the relevant results of ref. [58] to make the potential issue clear. The idea is that a contribution to the scalar two point function will be generated by the transition in the running behavior of the coupling constants of the theory. Consider the contribution to the scalar two point function determined by an approximately conformal field theory in position space following ref. [58]

$$\langle 0|T\mathcal{O}^{\dagger}(x)\mathcal{O}(0)|0\rangle = \left(\frac{1}{-x^2}\right)^{d-1} f(-x^2\mathcal{N}^2) \tag{5.15}$$

The scale N is a non perturbative scale characterizing the changing in the running of the coupling constants from the IR free to UV fixed point behavior, that is assumed. This scale is not associated with a particle mass. Even so, the key point of ref. [58] is that by assuming non-analytic dependence on the scale N in the function f, threshold matchings proportional to N 2 are generated for $H^{\dagger}H$. We agree that the results of ref. [58] follow from this assumed non-analytic dependence on N. As the function $f(-x^2N^2)$ is fundamentally non-perturbative it is not possible to draw strong conclusions about its analytic, or nonanalytic form, without an explicit UV theory and non-perturbative study.

In the case at hand, we can examine if sensitivity to the scale m_{pl} for $H^{\dagger}H$ might already be present using perturbative methods. The presence of higher dimensional operators in eq. (5.14) indicates this theory is UV incomplete and the presence of the scale m_{pl} in Eqn (5.12) calls into question the starting assumption of conformal invariance. This theory does have a cut off scale associated with mpl. The cut off scale is background field dependent [56] but given by $\Lambda \simeq 6\, m_{pl}$ [59–61]. The cut off scale comes from unitarity violation generated by the scattering diagram of the SU(2)L scalar multiplet coupled to gravity, while the singlet scalar fields do not introduce this cut off dependence [62]. Fundamentally this cut off scale stems from the interaction term $H^{\dagger}HR$. The conformal transformation in eq. (5.13) used to eliminate $H^{\dagger}HR$ taking the theory to the Einstein frame results in a tower of higher dimensional operators effecting the CW potential as non-renormalizable classical potential terms.20 No m_{pl}^2 contribution to $H^{\dagger}H$ is generated by such operators when expanding the effective potential through the particle thresholds leading to the SM. An inverse dependence on the scale $1/m_{pl}$ is present breaking conformal symmetry explicitly.

In effective field theory, non-analytic behavior is usually associated with propagating long distance states of the theory, associated with the poles dictating the properties of the S-matrix. Threshold matchings come about due to fixing that the IR limit of Smatrix elements are reproduced when transitioning through a particle threshold. Nonperturbative matchings can be present, such as the effects of a multi-pole expansion due to underlying structure, which EFT can also be used to represent. Such effects introduce an inverse dependence on a heavy scale when integrated out, as the multi-pole expansion is a perturbative expansion in ratios of Compton wavelengths.

The usual rules of EFT also dictate that a dimension two operator such as H†H should be considered to have a dimensionful parameter $\propto \Lambda^2 = (6\, m_{pl})^2$. If this is the case, then this approach to the Hierarchy problem has severe fine tuning associated with it. Associating this cut off scale with physical particle masses, one does obtain large contributions to the $H^\dagger H$ operator $\propto\Lambda^2$. When not associating this cut off scale with physical particle masses, if a threshold contribution to $H^\dagger H$ is still generated, this would be consistent with the arguments in ref. [58]. However, in this case, the cut-off scale is a sign that the effective field theory is smoothly transitioning from a linear SMEFT description to a non-linear EFT description [64] due to the Higgs field mixing with the scalar component of gravity. This mixing introduces non-linearities into the EFT description that require a different description of asymptotic states above and below the cut off scale. No particles are integrated out at this scale and the resulting background field dependent matching contributions across this threshold do not lead to a large shift in the Higgs mass at low field values. Extrapolating through this scale to large field values is subject to uncertainties due to introduced UV dependence, as the power counting of the EFT breaks down, but these effects do not necessarily generate a severe fine tuning of the Higgs mass parameter.

Comment on Exceptional Parameter Space and Dark Matter

The set of consistency conditions given in eq. (5.9) can be satisfied if alternate parameter space is adopted than considered in ref. [1]. The field content and Lagrangian involving the R scalar field coupled to the SM is the minimal singlet scalar field Dark Matter model [65– 67]. It is interesting to lower the mass of the R scalar to the ~ GeV scale for this reason using such alternate parameter space. The model parameter space with ~ GeV scale exotic scalar states can potentially provide a successful Dark Matter candidate. Viable parameter space for this model's Dark Matter candidate has recently been highly constrained, see refs. [68–70]. The remaining viable parameter space can be consistent with eq. (5.9) at the cost of introducing small, technically natural, scalar couplings. In particular, the allowed resonance region of parameter space where $2m_R<m_h$ can be chosen consistent with the constraints in refs. [68–70] while a "FIMP" scenario [71] is present leading to a successful Dark matter relic density satisfying [72]

$$\frac{\Omega_R\, h^2}{0.12} = 5.3 \times 10^{21}\ \lambda_{HR}^2 \left(\frac{m_R}{\text{GeV}}\right) \tag{5.16}$$

In addition, as the scalar S is a pseudo-Goldstone boson, MS<<MR; a scattering channel $R^2 \longrightarrow S^2$ is always kinematically open that depletes the relic abundance of R. A straightforward calculation yields

$$\sigma_{\rm ann} v_{\rm rel} \sim \frac{\lambda_{SR}^2}{4\pi M_R^2} \sim \frac{1}{16\pi v_S^2} \qquad (5.17)$$

For the exceptional parameter space to be viable, it is also necessary that additional Gildener-Weinberg [50] conditions discussed in ref. [1] must not be satisfied, to avoid spontaneously breaking scale invariance at scales other than ΛGW. Finally, due to the presence of the scalar S, and the self interactions of the R field, the viability of the model is dependent on a nontrivial thermal history and also the balance of freeze-in and freeze out effects. We leave a detailed investigation of this possibility to a future publication.

CONCLUSIONS

In this paper, we have examined the neutrino option, where the Electroweak scale is generated simultaneously with neutrino masses. We have examined the numerical consistency of this scenario using one, two and three loop RGE equations for the SM, and one loop running of the Weinberg operator. We have developed a consistent NLO framework for such studies by determining the full set of ($\propto\omega^2$) one loop corrections to the leading tree level matching. We have identified the requirement of a λ parameter ~O(10^{-2}) at the matching scale and confirmed previous results that indicate that this scenario predicts Majorana states around the scales $5 \cdot 10^5\,\text{GeV} \lesssim M_1 \lesssim 10^7\,\text{GeV}$. We have also extended the conformal neutrino option scenario of Brdar et al. [1] to include the leading couplings to gravity. The conformal neutrino option necessarily includes a cut off scale due to scalargraviton mixing. Nevertheless, we have also argued that this cut off scale does not result in a large fine tuning unless it is an indication of a UV-completion with Planck scale states, and in addition this cut off scale can be directly interpretted as a sign of a transition to a non-linear EFT set up around the Planck scale with no new states. It is far from obvious that calculable Planck scale threshold corrections to the Higgs mass can be demonstrated in this case. Overall, our results strongly support further investigations of the neutrino option and its conformal UV embedding being pursued. Extensions to the Standard Model motivated out of the experimental requirement of neutrino mass generation remarkably remain an option to address the Hierarchy problem.

ACKNOWLEDGMENTS

MT and IB acknowledge generous support from the Villum Fonden and partial support by the Danish National Research Foundation (DNRF91) through the Discovery centre. We thank E. Bjerrum-Bohr, J. Bourjaily, A. Denner, W. Goldberger, H. Haber, A. Manohar, H. Murayama, S. Patil, P. Perez, S. Petcov and W. Skiba for useful discussions and comments.

APPENDIX: EVALUATING CORRELATION FUNCTIONS IN ALTERNATE SEESAW MODEL FORMULATIONS

Evaluating correlation functions when examining the neutrino option can be subtle due to the presence of a Majorana field. Feynman rules for Majorana particles exist in refs. [73– 75], but subtleties can still remain, as we illustrate in this appendix. Consider evaluating a vacuum expectation value of a correlation function using Wick's theorem [76] and comparing the results using the Lagrangians L_N, L'_N. The interaction terms of L_N, L'_N are classically identical,

$$\frac{1}{2}\left[\overline{\ell_L^\beta}\tilde{H}\omega_\beta^{p,\dagger}N_p + \overline{\ell_L^{c\beta}}\tilde{H}^*\omega_\beta^{p,T}N_p + \overline{N_p}\omega_\beta^{p,*}\tilde{H}^T\ell_L^{c\beta} + \overline{N_p}\omega_\beta^p\tilde{H}^\dagger\ell_L^\beta\right] = \left[\overline{\ell_L^\beta}\tilde{H}\omega_\beta^{p,\dagger}N_p + \overline{N_p}\omega_\beta^p\tilde{H}^\dagger\ell_L^\beta\right]$$

Now consider evaluating Π_{HH} using L'_N. Wick's theorem gives

$$\begin{aligned} i\Pi_{HH^\dagger} &= \frac{1}{2!}\langle 0|T\{H^\dagger(x)H(y)\int dz^4\mathcal{L}'_N(z)\int dw^4\mathcal{L}'_N(w)\}|0\rangle, \\ &= \langle 0|TH^\dagger H\overline{\ell_L^\beta}\tilde{H}N_p\overline{N_p}\tilde{H}^\dagger\ell_L^\beta|0\rangle|\omega_p|^2, \\ &= -\frac{i\,|\omega_p|^2M_p^2}{8\pi^2}\left(1+\log\frac{\mu^2}{M_p^2}\right). \end{aligned} \tag{A.1}$$

Due to the presence of a Majorana field, one can use the transformation

$$\overline{\ell_L^\beta}\tilde{H}\omega_\beta^{p,\dagger}N_p \to \overline{N_p}\omega_\beta^{p,\star}\tilde{H}^T\ell_L^{c,\beta}, \qquad \overline{N_p}\omega_\beta^p\tilde{H}^\dagger\ell_L^\beta \to \overline{\ell_L^{c,\beta}}(\omega_\beta^p)^T\tilde{H}^\star N_p, \tag{A.2}$$

before the contractions in Wick's theorem are evaluated when considering the "self square" interaction terms $[\overline{\ell_L^\beta}\tilde{H}\omega_\beta^{p,\dagger}N_p]^2$, $[\overline{N_p}\omega_\beta^p\tilde{H}^\dagger\ell_L^\beta]^2$, also present in the time-ordered exponential.

These transformations follow from transposing the interaction Lagrangian, and the Fermistatistics of the fermionic field operators. Transposition of the Lagrangian is justified as the individual Lagrangian terms are invariant under the (compact Euclideanized SO(4) $\simeq$ $SU(2)_L$ × $SU(2)_R$) Lorentz group, and the global SU(3) × $SU(2)_L$ × $U(1)_Y$. Although this transformation still allows the Majorana field N to be contracted defining a two point function, as the correlation functions

$$\langle 0|\ell^c\,\bar{\ell}|0\rangle = 0, \quad \langle 0|\overline{\ell^c}\,\ell|0\rangle = 0, \quad \langle 0|\ell\,\bar{\ell}^c|0\rangle = 0, \quad \langle 0|\bar{\ell}\,\ell^c|0\rangle = 0,$$

no additional contribution to the sum of the time ordered product in Wick's theorem results. These allowed transformations do lead to further contractions when considering other matrix elements, such as the three point and four point functions evaluated in this work, when using L'_N. A careful use of the Feynman rules of refs. [73, 74] still evaluates the correlation function in each case only using non-zero correlation functions $\ell\bar{\ell}$, $\ell^c\bar{\ell}^c$.

On the other hand, using L_N and neglecting such "self square" interaction terms, directly gives

$$i\Pi_{HH^\dagger} = -\frac{i\,|\omega_p|^2 M_p^2}{16\pi^2}\left(1 + \log\frac{\mu^2}{M_p^2}\right) \tag{A.3}$$

In this case, the "self square" of the manifestly charge symmetric interaction Lagrangian terms include

$$\left[\overline{\ell_L^\beta}\,\tilde{H}\omega_\beta^{p,\dagger}N_p + \overline{\ell_L^{c\beta}}\,\tilde{H}^*\omega_\beta^{p,T}N_p\right]^2. \tag{A.4}$$

Using the charge conjugation identities, transposing the total Lagrangian, and Fermistatistics gives

$$\left[\overline{\ell_L^\beta}\tilde{H}\omega_\beta^{p,\dagger}N_p + \overline{\ell_L^{c\beta}}\tilde{H}^*\omega_\beta^{p,T}N_p\right]^2 = \left[\overline{\ell_L^\beta}\tilde{H}\omega_\beta^{p,\dagger}N_p + \overline{\ell_L^{c\beta}}\tilde{H}^*\omega_\beta^{p,T}N_p\right]\left[\overline{N_p}\omega_\beta^{p,*}\tilde{H}^T\ell_L^{c\beta} + \overline{N_p}\omega_\beta^{p}\tilde{H}^\dagger\ell_L^\beta\right]$$

Using a manifestly charge symmetric interaction Lagrangian results in the additional contributions of this form, due to these allowed manipulations when evaluating the Wick expansion. An additional overall factor of two results evaluating iΠHH, which brings the results obtained using LN , L 0 N into agreement. A sign of the need to perform transformations of this form in evaluating the Wick expansion is the presence of non-zero contributions

$$\langle 0|\ell^c(x)\,\ell(y)|0\rangle, \quad \langle 0|\ell(y)\,\ell^c(x)|0\rangle \neq 0. \tag{A.5}$$

These terms do not have a simple expression in terms of a charge conjugation matrix and gamma matrices, but indicate a transposition is required to evaluate the spin sum in some terms in the Wick expansion. On the other hand, when considering the three point function, the presence of the l and l^c fields leads to a direct evaluation of the Wick expansion in the L_N formulation, while a transformation of the interaction terms present in the Taylor expansion is required to evaluate Wick expansion in the case of L'_N.

Judiciously utilizing the transposition transformations can simplify various calculations when using L_N or L'_N.

REFERENCES

1. V. Brdar, Y. Emonds, A.J. Helmboldt and M. Lindner, The conformal UV completion of the neutrino option, arXiv:1807.11490 [INSPIRE].
2. P. Minkowski, $\mu \rightarrow e\gamma$ at a rate of one out of 109 muon decays?, Phys. Lett. B 67 (1977) 421 [INSPIRE].
3. M. Gell-Mann, P. Ramond and R. Slansky, Complex spinors and unified theories, Conf. Proc. C 790927 (1979) 315 [arXiv:1306.4669] [INSPIRE].
4. R.N. Mohapatra and G. Senjanovi´c, Neutrino mass and spontaneous parity nonconservation, Phys. Rev. Lett. 44 (1980) 912 [INSPIRE].
5. T. Yanagida, Horizontal symmetry and masses of neutrinos, Prog. Theor. Phys. 64 (1980) 1103 [INSPIRE].
6. J. Schechter and J.W.F. Valle, Neutrino masses in $SU(2) \times U(1)$ theories, Phys. Rev. D 22 (1980) 2227 [INSPIRE].
7. F. Vissani, Do experiments suggest a hierarchy problem?, Phys. Rev. D 57 (1998) 7027 [hep-ph/9709409] [INSPIRE].
8. I. Brivio and M. Trott, Radiatively generating the Higgs potential and electroweak scale via the seesaw mechanism, Phys. Rev. Lett. 119 (2017) 141801 [arXiv:1703.10924] [INSPIRE].
9. A. de Gouv^ea, J. Herrero-Garcia and A. Kobach, Neutrino masses, grand unification and baryon number violation, Phys. Rev. D 90 (2014) 016011 [arXiv:1404.4057] [INSPIRE].
10. A. Kobach, Baryon number, lepton number and operator dimension in the Standard Model, Phys. Lett. B 758 (2016) 455 [arXiv:1604.05726] [INSPIRE].
11. R.N. Mohapatra, Mechanism for understanding small neutrino mass in superstring theories, Phys. Rev. Lett. 56 (1986) 561 [INSPIRE].
12. R.N. Mohapatra and J.W.F. Valle, Neutrino mass and baryon number nonconservation in superstring models, Phys. Rev. D 34 (1986) 1642 [INSPIRE].
13. J. Bernabeu, A. Santamaria, J. Vidal, A. Mendez and J.W.F. Valle, Lepton flavor nonconservation at high-energies in a superstring inspired Standard Model, Phys. Lett. B 187 (1987) 303 [INSPIRE].
14. H. Davoudiasl and I.M. Lewis, Right-handed neutrinos as the origin of the electroweak scale, Phys. Rev. D 90 (2014) 033003 [arXiv:1404.6260] [INSPIRE].

15. J.A. Casas, V. Di Clemente and M. Quir´os, The effective potential in the presence of several mass scales, Nucl. Phys. B 553 (1999) 511 [hep-ph/9809275] [INSPIRE].

16. J.A. Casas, V. Di Clemente, A. Ibarra and M. Quir´os, Massive neutrinos and the Higgs mass window, Phys. Rev. D 62 (2000) 053005 [hep-ph/9904295] [INSPIRE].

17. G. Bambhaniya, P. Bhupal Dev, S. Goswami, S. Khan and W. Rodejohann, Naturalness, vacuum stability and leptogenesis in the minimal seesaw model, Phys. Rev. D 95 (2017) 095016 [arXiv:1611.03827] [INSPIRE].

18. C.G. Callan Jr., S.R. Coleman and R. Jackiw, A new improved energy-momentum tensor, Annals Phys. 59 (1970) 42 [INSPIRE].

19. S.R. Coleman and R. Jackiw, Why dilatation generators do not generate dilatations?, Annals Phys. 67 (1971) 552 [INSPIRE].

20. W.A. Bardeen, On naturalness in the Standard Model, in Ontake Summer Institute on Particle Physics, Ontake Mountain, Japan, 27 August–2 September 1995 [INSPIRE].

21. D.M. Capper and M.J. Duff, Trace anomalies in dimensional regularization, Nuovo Cim. A 23 (1974) 173 [INSPIRE].

22. B. Bellazzini, C. Cs´aki and J. Serra, Composite Higgses, Eur. Phys. J. C 74 (2014) 2766 [arXiv:1401.2457] [INSPIRE].

23. W. Buchm¨uller and D. Wyler, Effective Lagrangian analysis of new interactions and flavor conservation, Nucl. Phys. B 268 (1986) 621 [INSPIRE].

24. B. Grzadkowski, M. Iskrzynski, M. Misiak and J. Rosiek, Dimension-six terms in the Standard Model Lagrangian, JHEP 10 (2010) 085 [arXiv:1008.4884] [INSPIRE].

25. I. Brivio and M. Trott, The Standard Model as an effective field theory, arXiv:1706.08945 [INSPIRE].

26. S. Weinberg, Baryon and lepton nonconserving processes, Phys. Rev. Lett. 43 (1979) 1566 [INSPIRE].

27. F. Wilczek and A. Zee, Operator analysis of nucleon decay, Phys. Rev. Lett. 43 (1979) 1571 [INSPIRE].

28. A. Broncano, M.B. Gavela and E.E. Jenkins, The effective Lagrangian for the seesaw model of neutrino mass and leptogenesis, Phys. Lett. B 552 (2003) 177 [Erratum ibid. B 636 (2006) 332] [hep-ph/0210271] [INSPIRE].

29. G. Elgaard-Clausen and M. Trott, On expansions in neutrino effective field theory, JHEP 11 (2017) 088 [arXiv:1703.04415] [INSPIRE].

30. E. Majorana, Teoria simmetrica dell'elettrone e del positrone (in Italian), Nuovo Cim. 14 (1937) 171 [INSPIRE].

31. S.M. Bilenky, J. Hosek and S.T. Petcov, On oscillations of neutrinos with Dirac and Majorana masses, Phys. Lett. B 94 (1980) 495 [INSPIRE].

32. B. Pontecorvo, Mesonium and anti-mesonium, Sov. Phys. JETP 6 (1957) 429 [Zh. Eksp. Teor. Fiz. 33 (1957) 549] [INSPIRE].

33. Z. Maki, M. Nakagawa and S. Sakata, Remarks on the unified model of elementary particles, Prog. Theor. Phys. 28 (1962) 870 [INSPIRE].

34. M.C. Gonzalez-Garcia, M. Maltoni and T. Schwetz, Updated fit to three neutrino mixing: status of leptonic CP-violation, JHEP 11 (2014) 052 [arXiv:1409.5439] [INSPIRE].

35. I. Esteban, M.C. Gonzalez-Garcia, M. Maltoni, I. Martinez-Soler and T. Schwetz, Updated fit to three neutrino mixing: exploring the accelerator-reactor complementarity, JHEP 01 (2017) 087 [arXiv:1611.01514] [INSPIRE].

36. J.A. Casas and A. Ibarra, Oscillating neutrinos and $\mu \rightarrow e, \gamma$, Nucl. Phys. B 618 (2001) 171 [hep-ph/0103065] [INSPIRE].

37. A. Ibarra, E. Molinaro and S.T. Petcov, Low energy signatures of the TeV scale see-saw mechanism, Phys. Rev. D 84 (2011) 013005 [arXiv:1103.6217] [INSPIRE].

38. B. Grzadkowski and M. Lindner, Nonlinear evolution of Yukawa couplings, Phys. Lett. B 193 (1987) 71 [INSPIRE].

39. Yu. F. Pirogov and O.V. Zenin, Two loop renormalization group restrictions on the Standard Model and the fourth chiral family, Eur. Phys. J. C 10 (1999) 629 [hep-ph/9808396] [INSPIRE].

40. W. Grimus and H. Neufeld, Radiative neutrino masses in an $SU(2) \times U(1)$ model, Nucl. Phys. B 325 (1989) 18 [INSPIRE].

41. W. Grimus and L. Lavoura, One-loop corrections to the seesaw mechanism in the multi-Higgs-doublet Standard Model, Phys. Lett. B 546 (2002) 86 [hep-ph/0207229] [INSPIRE].

42. P.S.B. Dev and A. Pilaftsis, Minimal radiative neutrino mass mechanism for inverse seesaw models, Phys. Rev. D 86 (2012) 113001 [arXiv:1209.4051] [INSPIRE].

43. E. Fernandez-Martinez, J. Hernandez-Garcia, J. Lopez-Pavon and M.

Lucente, Loop level constraints on seesaw neutrino mixing, JHEP 10 (2015) 130 [arXiv:1508.03051] [INSPIRE].

44. D. Buttazzo et al., Investigating the near-criticality of the Higgs boson, JHEP 12 (2013) 089 [arXiv:1307.3536] [INSPIRE].

45. K.S. Babu, C.N. Leung and J.T. Pantaleone, Renormalization of the neutrino mass operator, Phys. Lett. B 319 (1993) 191 [hep-ph/9309223] [INSPIRE].

46. S. Antusch, M. Drees, J. Kersten, M. Lindner and M. Ratz, Neutrino mass operator renormalization revisited, Phys. Lett. B 519 (2001) 238 [hep-ph/0108005] [INSPIRE].

47. J.A. Casas, J.R. Espinosa, A. Ibarra and I. Navarro, General RG equations for physical neutrino parameters and their phenomenological implications, Nucl. Phys. B 573 (2000) 652 [hep-ph/9910420] [INSPIRE].

48. A. Pilaftsis and T.E.J. Underwood, Resonant leptogenesis, Nucl. Phys. B 692 (2004) 303 [hep-ph/0309342] [INSPIRE].

49. J.A. Casas, J.M. Moreno, N. Rius, R. Ruiz de Austri and B. Zaldivar, Fair scans of the seesaw. Consequences for predictions on LFV processes, JHEP 03 (2011) 034 [arXiv:1010.5751] [INSPIRE].

50. E. Gildener and S. Weinberg, Symmetry breaking and scalar bosons, Phys. Rev. D 13 (1976) 3333 [INSPIRE].

51. H. Davoudiasl, R. Kitano, T. Li and H. Murayama, The new minimal Standard Model, Phys. Lett. B 609 (2005) 117 [hep-ph/0405097] [INSPIRE].

52. S.R. Coleman and E.J. Weinberg, Radiative corrections as the origin of spontaneous symmetry breaking, Phys. Rev. D 7 (1973) 1888 [INSPIRE].

53. A. Salam and J.A. Strathdee, Nonlinear realizations. 1: the role of Goldstone bosons, Phys. Rev. 184 (1969) 1750 [INSPIRE].

54. A. Salam and J.A. Strathdee, Nonlinear realizations. 2: conformal symmetry, Phys. Rev. 184 (1969) 1760 [INSPIRE].

55. C.P. Burgess, H.M. Lee and M. Trott, Comment on Higgs inflation and naturalness, JHEP 07 (2010) 007 [arXiv:1002.2730] [INSPIRE].

56. F. Bezrukov, A. Magnin, M. Shaposhnikov and S. Sibiryakov, Higgs inflation: consistency and generalisations, JHEP 01 (2011) 016 [arXiv:1008.5157] [INSPIRE].

57. S. Weinberg, Critical phenomena for field theorists, Springer, Boston, MA, U.S.A. (1978), pg. 1 [INSPIRE].
58. G. Marques Tavares, M. Schmaltz and W. Skiba, Higgs mass naturalness and scale invariance in the UV, Phys. Rev. D 89 (2014) 015009 [arXiv:1308.0025] [INSPIRE].
59. T. Han and S. Willenbrock, Scale of quantum gravity, Phys. Lett. B 616 (2005) 215 [hep-ph/0404182] [INSPIRE].
60. C.P. Burgess, H.M. Lee and M. Trott, Power-counting and the validity of the classical approximation during inflation, JHEP 09 (2009) 103 [arXiv:0902.4465] [INSPIRE].
61. J.L.F. Barbon and J.R. Espinosa, On the naturalness of Higgs inflation, Phys. Rev. D 79 (2009) 081302 [arXiv:0903.0355] [INSPIRE].
62. S.R. Huggins and D.J. Toms, One graviton exchange interaction of nonminimally coupled scalar fields, Class. Quant. Grav. 4 (1987) 1509 [INSPIRE].
63. A. Andreassen, W. Frost and M.D. Schwartz, Consistent use of effective potentials, Phys. Rev. D 91 (2015) 016009 [arXiv:1408.0287] [INSPIRE].
64. C.P. Burgess, S.P. Patil and M. Trott, On the predictiveness of single-field inflationary models, JHEP 06 (2014) 010 [arXiv:1402.1476] [INSPIRE].
65. V. Silveira and A. Zee, Scalar phantoms, Phys. Lett. B 161 (1985) 136 [INSPIRE].
66. J. McDonald, Gauge singlet scalars as cold dark matter, Phys. Rev. D 50 (1994) 3637 [hep-ph/0702143] [INSPIRE].
67. C.P. Burgess, M. Pospelov and T. ter Veldhuis, The minimal model of nonbaryonic dark matter: a singlet scalar, Nucl. Phys. B 619 (2001) 709 [hep-ph/0011335] [INSPIRE].
68. J.M. Cline, K. Kainulainen, P. Scott and C. Weniger, Update on scalar singlet dark matter, Phys. Rev. D 88 (2013) 055025 [Erratum ibid. D 92 (2015) 039906] [arXiv:1306.4710] [INSPIRE].
69. GAMBIT collaboration, Status of the scalar singlet dark matter model, Eur. Phys. J. C 77 (2017) 568 [arXiv:1705.07931] [INSPIRE].
70. 70.P. Athron, J.M. Cornell, F. Kahlhoefer, J. McKay, P. Scott and S. Wild, Impact of vacuum stability, perturbativity and XENON1T on global fits of Z2 and Z3 scalar singlet dark matter, Eur. Phys. J. C 78 (2018) 830 [arXiv:1806.11281] [INSPIRE].

71. L.J. Hall, K. Jedamzik, J. March-Russell and S.M. West, Freeze-in production of FIMP dark matter, JHEP 03 (2010) 080 [arXiv:0911.1120] [INSPIRE].
72. N. Bernal, M. Heikinheimo, T. Tenkanen, K. Tuominen and V. Vaskonen, The dawn of FIMP dark matter: a review of models and constraints, Int. J. Mod. Phys. A 32 (2017) 1730023 [arXiv:1706.07442] [INSPIRE].
73. A. Denner, H. Eck, O. Hahn and J. Kublbeck, Feynman rules for fermion number violating interactions, Nucl. Phys. B 387 (1992) 467 [INSPIRE].
74. A. Denner, H. Eck, O. Hahn and J. Kublbeck, Compact Feynman rules for Majorana fermions, Phys. Lett. B 291 (1992) 278 [INSPIRE].
75. H.K. Dreiner, H.E. Haber and S.P. Martin, Two-component spinor techniques and Feynman rules for quantum field theory and supersymmetry, Phys. Rept. 494 (2010) 1 [arXiv:0812.1594] [INSPIRE].
76. G.C. Wick, The evaluation of the collision matrix, Phys. Rev. 80 (1950) 268 [INSPIRE]

CHAPTER 5

Neutrino Masses from a Dark Neutrino Sector Below the Electroweak Scale

Peter Ballett, Matheus Hostert and Silvia Pascoli

Institute for Particle Physics Phenomenology, Department of Physics, Durham University, South Road, Durham DH1 3LE, United Kingdom

ABSTRACT

We consider a minimal extension of the Standard Model which advocates a dark neutrino sector charged under a hidden U(1)′. We show that neutrino masses can arise radiatively in this model. The observed values are compatible with a light dark sector below the electroweak scale and would imply new heavy fermions which may be testable in the next generation of beam dump searches at DUNE, NA62 and SHIP.

Citation: Peter Ballett, Matheus Hostert, and Silvia Pascoli. "Neutrino masses from a dark neutrino sector below the electroweak scale" Phys. Rev. D 99, 091701(R) (2019). https://doi.org/10.1103/PhysRevD.99.091701

INTRODUCTION

Neutrino oscillations have been established by several experiments [1], implying small but nonvanishing neutrino masses. In the Standard Model (SM), neutrinos are strictly massless due to the absence of right-handed neutrino fields, urging for extensions of the theory. The Type-I seesaw mechanism [2], arguably the most popular mechanism to explain the lightness of neutrino masses, relies on the addition of at least two heavy right-handed neutrinos NR. The large scales of NR and/or the smallness of the Yukawa couplings makes the minimal realization of this model difficult to test.

Therefore, searching for variations of the Type-I seesaw where novel and testable phenomena are present is an essential part of solving the neutrino mass puzzle [3]. A few notable examples of such alternatives are the inverse seesaw (ISS) [4] and the linear seesaw (LSS) [5], where the lightness of neutrino masses is explained by an approximate conservation of lepton number, and the extended seesaw (ESS) [6], where new heavy neutral fermions generally appear at small scales. This class of models assumes additional SM gauge neutral fermions that mix with light neutrinos, usually referred to as sterile neutrinos. These, however, need not be completely sterile and might have new gauge interactions shared with the SM fermions [7–15] or not [16–20]. In the latter case, we refer to these new heavy fermions as dark neutrinos. The interest in such particles arises from their novel interactions which may "leak" into the SM sector via neutrino mixing, where they offer a variety of phenomenological and cosmological consequences.

In this article, we consider the new minimal model introduced in Ref. [21]. It introduces two types of new neutral fermions, namely dark neutrinos νD and additional sterile neutrinos N. We impose a hidden U(1)′ gauge symmetry with the associated hidden gauge boson Xμ, which mediates the dark neutrino interactions. The symmetry is subsequently broken by the vacuum expectation value (vev) of a complex dark scalar Φ. As discussed in Ref. [21], the model can exhibit a significantly different phenomenology than the case of neutrino mixing only. Beyond evading many current bounds, such dark neutrinos could explain the MiniBooNE anomaly as discussed in [22] (see also [23]) and lead to novel neutrino scattering signatures [24]. Bounds on dark photons might also be severely weakened. If kinematically allowed, they would mainly decay into heavy neutrinos, which may be invisible or lead to multilepton plus missing energy signatures.

In this article, we discuss the generation of neutrino masses in our dark neutrino model. Crucially, the new gauge symmetry forbids Majorana mass terms for the νD states and, after symmetry breaking, leads to a mass matrix similar to the one in the so-called minimal ISS [25]. As such, this symmetry-enhanced seesaw predicts vanishing light neutrino masses at tree level. Here, we show that it induces their radiative generation via one-loop diagrams involving the new scalar and vector particles [17,25,26]. After identifying the range of heavy neutrino parameters required to explain the observed light neutrino masses, we point out interesting phenomenological consequences.

MODEL SETUP

Following [21], we add two types of heavy neutral fermions to the SM, namely a dark neutrino $\nu_{D,L} \equiv \nu_D$ and a sterile state $N_L \equiv N$. For simplicity, we restrict the discussion to one generation in order to focus on the main features of the model.

We impose a new Abelian gauge symmetry U(1)′ with associated mediator X_μ and introduce a neutral complex scalar Φ. No SM fields are charged under U(1)′. The scalar Φ and the fermion νD carry the same U(1)′ charge, while N remains completely neutral. The gauge-invariant Lagrangian is given by

$$\begin{aligned}\mathcal{L} = {} & \mathcal{L}_{\rm SM} + (D_\mu\Phi)^\dagger(D^\mu\Phi) - V(\Phi,H) \\ & -\frac{1}{4}X^{\mu\nu}X_{\mu\nu} + \bar{N}i\partial\!\!\!/ N + \overline{\nu_D}iD\!\!\!\!/\,\nu_D \\ & -\left[y_\nu^\alpha(\overline{L_\alpha}\cdot\tilde{H})N^c + \frac{\mu'}{2}\bar{N}N^c + y_N\bar{N}\nu_D^c\Phi + \text{H.c.}\right], \end{aligned} \tag{1}$$

where $X^{\mu\nu} \equiv \partial^\mu X^\nu - \partial^\nu X^\nu$, $D_\mu \equiv (\partial_\mu - ig'X_\mu)$, $L_\alpha \equiv (\nu_\alpha^T, \ell_\alpha^T)^T$ is the SM leptonic doublet of flavor α=e, μ, τ and $\tilde{H} \equiv i\sigma_2 H^*$ is the charge conjugate of the SM Higgs doublet. In the neutral fermion sector, we have Yukawa couplings y_ν^α and y_N responsible for L_α−N and ν_D−N interactions, respectively, and a Majorana mass μ′ for N. The latter term violates by two units any lepton number assignment which leaves the Yukawa term L_α−N invariant. As such, it plays a crucial role in the generation of light neutrino masses, as we discuss.

We are interested in the case in which both the neutral component of the fields H and Φ acquire nonvanishing vev's, v_H and v_φ. They induce mixing between active and heavy fermions, and give a mass to the gauge boson X_μ and to the real component of the scalar field φ. We are interested in proposing a model for neutrino masses which is testable in current and future noncollider experiments, and as such we focus on a new physics scale which is below the electroweak one, $v_\varphi < v_H$. In addition to the neutrino portal, this model can accommodate a vector portal arising from vector kinetic mixing term and a scalar portal coming from the cross-coupling term $H^\dagger H \Phi^\dagger \Phi$ in the potential [21]. Kinetic mixing can be reabsorbed in a redefinition of vector fields, leading to a new gauge boson which has vector couplings to the SM fermions proportional to their electric charge. For our neutrino mass generation mechanism, the vector and scalar mixing do not play a relevant role and we set them to zero from here onward, unless otherwise specified. Regarding the vector boson, we refer to it as a Z′, independently of kinetic mixing.

NEUTRINO MASSES

After symmetry breaking, two Dirac mass terms are induced with $m_D \equiv y_\nu^\alpha v_H/\sqrt{2}$ and $\Lambda \equiv y_N v_\varphi/\sqrt{2}$. For one active neutrino ν_α, α=e, μ, τ, the mass matrix is given by

$$\mathcal{L}_{\text{mass}} \supset \frac{1}{2}\begin{pmatrix}\bar{\nu}_\alpha & \bar{N} & \bar{\nu_D}\end{pmatrix}\begin{pmatrix}0 & m_D & 0\\ m_D & \mu' & \Lambda \\ 0 & \Lambda & 0\end{pmatrix}\begin{pmatrix}\nu_\alpha^c \\ N^c \\ \nu_D^c\end{pmatrix} + \text{H.c.} \tag{2}$$

Let us emphasize the fact that in our model the zeros in the ν_D–ν_D and ν_α–ν_D entries are enforced by the U(1)′ symmetry, differently from LSS and ISS models, in which these are generically assumed to be nonzero and small due to the quasipreservation of lepton number. Here, lepton number violation (LNV) may be large, as the μ′ term breaks it by two units. Alternatively, it can be small and technically natural, leading to quasidegenerate heavy neutrinos, see below.

After diagonalization of the mass matrix, the two heavy neutrinos, ν_h (h=4, 5), acquire masses

$$m_{4,5} = \frac{\mu' \mp \sqrt{\mu'^2 + 4(\Lambda^2 + m_D^2)}}{2}.$$

Assuming that $m_D \ll \Lambda$, we focus on two interesting limiting cases.

The ISS-like scenario is defined by $\Lambda \gg \mu'$: the two heavy neutrinos are nearly degenerate with a mass Λ and mass splitting μ′. The relevant mixing parameters are $U_{\alpha 4,5} \sim m_D/\sqrt{2}\Lambda$ and $U_{D4,5} \sim 1/\sqrt{2}$. The ESS-like case has $\Lambda \ll \mu'$: one neutral lepton remains very heavy, $m5 \sim \mu'$, and mainly in the completely neutral direction N, and the other acquires a small mass via the seesaw mechanism in the hidden sector with $m_4 \sim -\Lambda^2/\mu'$ and $U_{D5} \sim \Lambda/\mu'$. The mixing with active neutrinos is given by $U_{\alpha 5} \sim m_D/\mu' \ll U_{\alpha 4} \sim m_D/\Lambda$.

The specific form of the mass matrix in Eq. (2) implies vanishing light neutrino masses at tree level, as its determinant is zero [25,27]. This feature holds to all orders in the seesaw expansion [27–29]. The light neutrino masses, however, are not protected by any symmetry and arise from radiative corrections (for a review of radiative neutrino mass models see, e.g., Ref. [30]).

A. Radiative Corrections

We now show that our model generically leads to the generation of light neutrino masses at one loop. The calculation of the radiative mass term follows Refs. [31,32] with the addition of the loops with the new boson and scalar particles shown in Fig. 1. The self-energy of the Majorana neutrino fields is given by

$$\Sigma_{ij}(\not{q}) = \not{q} P_L \Sigma^L_{ij}(\not{q}) + \not{q} P_R \Sigma^{L*}_{ij}(\not{q}) + P_L \Sigma^M_{ij}(q^2) + P_R \Sigma^{M*}_{ij}(q^2).$$

Using the on-shell renormalization scheme, the renormalized mass matrix for the light neutrinos, massless at tree level, emerges at one loop and is given by [32]

$$m^{\text{one-loop}}_{ij} = \text{Re}[\Sigma^M_{ij}(0)], \qquad i, j < 4.$$

The self-energy can be decomposed as

$$\Sigma^M_{ij}(0) = \Sigma^Z_{ij}(0) + \Sigma^h_{ij}(0) + \Sigma^{G_h}_{ij}(0) + \Sigma^{Z'}_{ij}(0) + \Sigma^{\varphi'}_{ij}(0) + \Sigma^{G_\varphi}_{ij}(0). \tag{3}$$

where Σ^{Z,h,G_h} come from the SM particles, Z0, the Higgs and the associated Goldstone boson, respectively, and $\Sigma^{Z',\varphi',G_\varphi}$ are the new terms present in our model, mediated by the new gauge boson and new scalar components. From it, we write the 3×3 light neutrino mass matrix

$$m_{ij} = \frac{1}{4\pi^2} \sum_{k=4}^{5} \left[C_{ik} C_{jk} \frac{m_k^3}{m_Z^2} F(m_k^2, m_Z^2, m_h^2) + D_{ik} D_{jk} \frac{m_k^3}{m_{Z'}^2} F(m_k^2, m_{Z'}^2, m_{\varphi'}^2) \right], \tag{4}$$

where we defined coupling matrices corresponding to the SM and new physics interaction terms assuming $\chi = \lambda_{\Phi H} = 0$:

$$C_{ik} \equiv \frac{g}{4c_W}\sum_{\alpha=e}^{\tau} U^*_{\alpha i}U_{\alpha k} \quad \text{and} \quad D_{ik} \equiv \frac{g'}{2}U^*_{Di}U_{Dk}. \tag{5}$$

Equivalent expressions can be found for nonvanishing portal couplings, but considering experimental constraints we find that these do not play a role in the neutrino mass generation. It is possible to show that in general $\sum_k m_k C_{ik}C_{jk} = 0$ and $\sum_k m_k D_{ik}D_{jk} = 0$ for any i, j. By virtue of the latter property, the loop function can be written as

$$F(a,b,c) \equiv \frac{3\ln(a/b)}{a/b-1} + \frac{\ln(a/c)}{a/c-1} \tag{6}$$

Turning off the g′ gauge coupling, we recover the expression for the Type-I seesaw case [31]:

$$m_{ij} = \frac{\alpha_W}{16\pi}\sum_{\alpha,\beta=e}^{\tau} U^*_{\alpha i}U^*_{\beta j}U_{\alpha 5}U_{\beta 5}\frac{m_5}{m_W^2} \times \left(m_5^2 F(m_5^2, m_Z^2, m_h^2) - m_4^2 F(m_4^2, m_Z^2, m_h^2)\right) \tag{7}$$

These SM corrections to neutrino masses also arise in the minimal ISS model [25,27]. In the latter, however, no explanation is provided as to why they dominate neutrino masses. Moreover, if we restrict the discussion to scales well below the electroweak one, m5≪10 GeV, bounds on the mixing angles severely constrain the parameter space viable to generate the observed values of the masses.

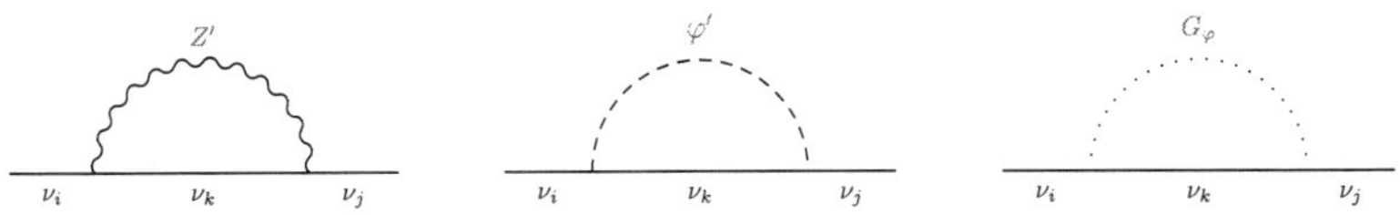

Figure 1. The three contributions to the neutrino self-energy arising from novel bosons in the theory.

For a light Z′, the second term in Eq. (4) dominates

$$m_{ij} \simeq \frac{g'^2}{16\pi^2}U^*_{Di}U^*_{Dj}U^2_{D5}\frac{m_5}{m_{Z'}^2} \times \left(m_5^2 F(m_5^2, m_{Z'}^2, m_{\varphi'}^2) - m_4^2 F(m_4^2, m_{Z'}^2, m_{\varphi'}^2)\right) \tag{8}$$

We notice that the resulting mass matrix has only one nonzero eigenvalue. This suggests that a typical prediction of our model is a normal ordering mass spectrum, in which m3 is given by this radiative mechanism and m2 has another origin, for example, the loops mediated by the SM gauge bosons or by additional particle content. Our simplifying assumption of one generation of hidden fermions is by no means necessary and more generations of new fermions are possible, leading to a much richer structure for the light neutrino mass matrix. The additional μ' terms would not be constrained and could be at different scales, while the Λ terms arise from the U(1)′ breaking and are therefore constrained to be at/below v_{φ}. Therefore, the full model could present a combination of relatively light Majorana νh, mainly in dark direction, some very heavy nearly neutral neutrinos and pseudo-Dirac pairs at intermediate scales. A discussion of this extension is beyond our scope, but we note that it has interesting consequences for both the heavy and light neutrino mass spectra and mixing structure.

Working in a single family case, we derive expressions for Eq. (8) in the seesaw limit for both the ISS- and ESS-like scenarios. In the ISS-like regime and assuming $m_{Z'}$, $m_{\varphi'} \ll \Lambda$, Eq. (8) simplifies to

$$m_3 \simeq \frac{g'^2}{8\pi^2}\frac{m_D^2}{m_{Z'}^2}\mu'\left(3\ln\frac{m_{Z'}^2}{\Lambda^2}+\ln\frac{m_{\varphi'}^2}{\Lambda^2}-4\right) \qquad (9)$$

while for $m_{Z'}$, $m_{\varphi'} \gg \Lambda$ it reduces to

$$m_3 \simeq \frac{g'^2}{16\pi^2}\frac{m_D^2}{\Lambda^2}\mu'\left(3+\frac{m_{\varphi'}^2}{m_{Z'}^2}\right) \qquad (10)$$

As can be expected, neutrino masses are controlled by the LNV parameter μ' and are enhanced with respect to the SM contribution by a factor of $(m_Z/m_{Z'})^2$ in the former, or $(m_Z/\Lambda)^2$ in the latter case.

For the ESS-like regime, taking $m_{Z'}$, $m_{\varphi'} \ll \mu'$, the light neutrino mass is approximately

$$m_3 \simeq \frac{g'^2}{16\pi^2}\frac{m_D^2}{\Lambda^2+m_D^2}\frac{\Lambda^2}{m_{Z'}^2}\mu'\left(3\ln\frac{m_{Z'}^2}{\mu'^2}+\ln\frac{m_{\varphi'}^2}{\mu'^2}\right) \qquad (11)$$

while for $m_{Z'}$, $m_{\varphi'} \gg \mu'$, it is

$$m_3 \simeq \frac{g'^2}{8\pi^2}\frac{m_D^2}{\Lambda^2+m_D^2}\frac{\Lambda^2}{\mu'}\left(3\ln\frac{m_{Z'}^2}{\Lambda^2}+\ln\frac{m_{\varphi'}^2}{\Lambda^2}-4\right) \qquad (12)$$

In this case, the light neutrino masses are controlled mainly by ν5, and the intermediate state ν4 can be much lighter.

SEARCHING FOR THE ORIGIN OF NEUTRINO MASSES

In what follows, we discuss the experimental reach to the heavy neutrinos responsible for neutrino mass generation in our model. Since the vector and scalar portals do not contribute significantly to neutrino masses, we first restrict the study to the case $\chi=\lambda_{\Phi H}=0$. For the sake of simplicity and concreteness, we work with a single generation of light neutrinos and focus on the mixing with the muon neutrino. We emphasize that our model predicts

$$\frac{m_4}{m_5} = -\frac{U_{\alpha 5}^2}{U_{\alpha 4}^2} \qquad (13)$$

implying that both heavy neutrinos should be searched for. For a real mixing matrix one can write $\sum_i^3 U_{Di}^2 \sim U_{\mu 4}^2$ and $U_{D5}^2 \sim 1$ for small $U_{\mu 4}$. Using these relations and Eq. (4), we plot the region of interest for neutrino mass generation in Fig. 2. We require $m_3 = \sqrt{\Delta m^2_{\rm atm}} \sim 0.05$ eV and vary m4/m5 from 1% (ESS-like) to 99% (ISS-like). For the hidden sector parameters, we fix $m_{Z'}=1$ GeV, $m_{\varphi'}=2$ GeV and $g'=1$. By decreasing (increasing) the mass of the Z′, it is possible to shift the band to smaller (larger) values of the mixing angles, although for values smaller than a few hundred MeV, the neutrino masses have a very mild dependence on $m_{Z'}$ [Eqs. (10) and (12)]. Increasing m_4/m_5 to values closer to 100% (i.e., decreasing μ′ below m5/100) shifts the top of the band to larger values of mixing angle and asymptotically recovers lepton number as a symmetry. Although this possibility appears excluded for mZ′=1 GeV, it can be achieved by lowering the mass of the mediator particles. For instance, for $m_{Z'}=m_{\varphi'}/2=100$ MeV and m5<100 MeV, we find that values as small as $\mu' \gtrsim 10^{-3} m^5$ are not covered by the gray region in Fig. 2. Values of $m4/m_5<1\%$ have no effect in the parameter space of ν_5, since in that limit the ν_5 state (mostly in the N direction) dominates the loop contribution.

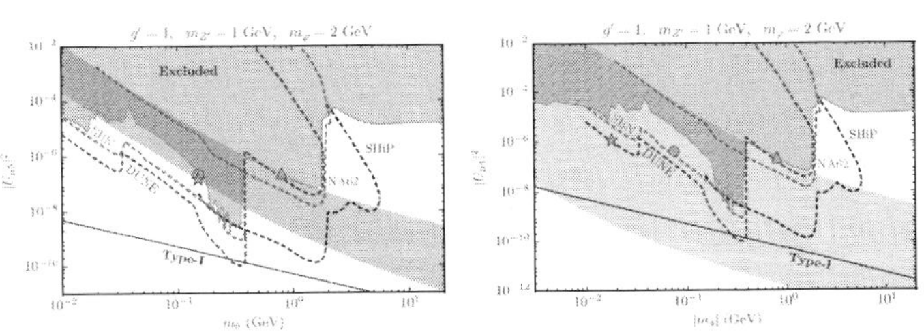

Figure 2. The region of interest for neutrino mass generation in our model in the parameter space of the (left panel) ν_5 and (right panel) ν_4 mass states. We require $m_3 = \sqrt{\Delta m^2_{\rm atm}}$ and vary $1\%<m_4/m_5<99\%$. Our BPs are (Δ) $m_5=800$ MeV, $m_4/m_5=99\%$, (∘) $m_5=150$ MeV, $m_4/m_5=50\%$ and (⋆) $m_5=150$ MeV, $m_4/m_5=12\%$. All bounds and projections displayed assume $\chi=\lambda\Phi H=0$. The dashed black line shows the equivalent Type-I seesaw contribution to the light neutrino mass.

The region labeled as excluded in Fig. 2 is composed of bounds from peak searches [33–35], beam dump [36–41] and collider experiments [42–44]. Current and future neutrino experiments can also cover a large region of parameter space with mh≲2 GeV. For instance, we show the sensitivity of the Short-Baseline Neutrino Program (SBN) [45] and of the Deep Underground Neutrino Experiment (DUNE) near detector [46,47] to heavy neutrinos in decay-in-flight searches. We also show the reach of the NA62 kaon factory operating in beam dump mode [48], and the dedicated beam dump experiment Search for Hidden Particles (SHiP) [49,50], which will cover a much larger region of parameter space from 400 MeV to ≲6 GeV. All bounds and sensitivities shown do not take into account the new invisible decays of the heavy neutrinos. Searches that rely on the visible decay products of the heavy neutrinos need to be revisited if the ν_h can decay invisibly or if new channels mediated by the vector (and/or scalar) portal dominate. In particular, faster decays of ν_h can shift decay-in-flight bounds to lower values of mixing angles, as discussed in detail in Ref. [45]. Peak searches apply as shown provided ν_h does not decay immediately via neutral-current channels with visible charged particles.

Let us first consider the case of subdominant vector and scalar portals. Compared to the "standard" sterile neutrino case, in which ν_h have only SM interactions suppressed by neutrino mixing, the new neutral-current interaction can enhance the ν_h decays into light and heavy neutrinos. A comprehensive analysis is beyond the scope of this article and we focus on three benchmark points (BPs) shown in Fig. 2 to exemplify the most characteristic properties. The BP represented as a triangle (Δ) corresponds to m_5=800 MeV and m_4/m_5=99%. In this case, the two heavy states are very degenerate in mass and decay like a standard sterile neutrino via $|U_{\mu 4}|^2$-suppressed SM charge- and neutral-current interactions. The channel $\nu_5 \to \nu_4 \nu_\alpha \bar{\nu}_\alpha$ via the Z' is phase space suppressed and becomes relevant only for larger mass splittings. The invisible ν4 decay mediated by the Z' is subdominant as it scales as $|U_{\mu 4}|^6$ and becomes important only for larger values of the mixing angles.

For the next BPs we fix m5=150 MeV. If we take m_4/m_5=50%, as we do for the BP represented by the circle (∘), ν5 will predominantly decay to $\nu_4 \nu_\alpha \bar{\nu}_\alpha$ due to the Z' contribution (provided $|U_{\mu 5}|^2 \gtrsim (m_{Z'}/m_Z)^4$). Consequently, the best candidate for detection is the ν_4 via the SM weak decays $\nu_4 \to \nu_\alpha e^+ e^-$. The values of the mixing angles for this BP, $|U_{\mu 4}|^2 \sim 3\times10^{-7}$ and $|U_{\mu 5}|^2 \sim 10^{-7}$, are within reach of the SBN and DUNE experiments. For a larger mass hierarchy, e.g., m_4/m_5=12%, see star BP (⋆), the Z' mediated decay

$\nu_5 \rightarrow \nu_4\overline{\nu_4}\nu_4$ dominates, inducing a large ν_4 population in addition to the states already produced in the beam. The intermediate state ν_4 can further decay as in the previous case into $\nu_4 \rightarrow \nu_\alpha e^+ e^-$. For the mixing angles we are considering, $|U_{\mu 4}|^2 \sim 10^{-6}$ and $|U_{\mu 5}|^2 \sim 10^{-7}$, DUNE will be able to test this BP. Similar considerations apply to the case where $m_5 > m_4 + m_{Z'}$, where now the Z' can be produced on shell in the ν_5 decay. The behavior of ν_4 is as discussed above. If $m_{Z'} < m_4$, then both heavy neutrinos predominantly decay into neutrinos and the Z', which presents a challenge for detection as it produces mainly light neutrinos.

Experimental detection of the Z' and φ' particles in the absence of kinetic and scalar mixing is also daunting. Nevertheless, they can be searched for in the kinematics of charged particles from meson decays [51,52]. Another strategy is to search for the neutrino byproducts of the decay of a Z' produced at accelerator neutrino facilities [53].

If the vector (and scalar) portals are non-negligible, the phenomenology could be significantly richer, as discussed in [21]. In particular, Z'-mediated decays into $\nu_\alpha e{+}e{-}$, and $\nu_\alpha \mu^+\mu^-$ if kinematically allowed, could dominate even for tiny values of χ^2. For instance, for the circle BP, χ^2 as low as 10^{-8} would make the above decays the main channels. Pseudoscalar final states are suppressed due to the vector nature of the Z'. The scalar portal is expected to give subdominant contributions due to the small Higgs-electron Yukawa coupling, although decay chains with intermediate ν4 states may become relevant. Finally, cosmological bounds on heavy neutrino in the 10 MeV–GeV scale may be weakened as they would decay well before big bang nucleosynthesis [54] (see also the discussion in Ref. [55]).

We have focused on the mixing with muon neutrinos as these provide one of the most sensitive avenue to test the model. In the electron sector, direct bounds on the active-heavy mixing are similar, with peak searches from $\pi\pm$ decay being most relevant below $\approx$100 MeV. For cases with large LNV, heavy neutrinos can dominate neutrinoless double beta decay [27], and this sets the strongest constraints in the parameter space. The tau sector is relatively poorly constrained, so greater freedom exists if such entries are relevant for neutrino mass generation.

CONCLUSIONS

We have considered a recently proposed model which invokes the existence of a dark neutrino sector and a new hidden gauge symmetry, focusing on the generation of neutrino masses in this context. The presence of a hidden broken gauge symmetry protects the neutrino mass matrix, leading to a minimal ISS-like structure, and allows for one-loop diagrams involving the new vector and scalar content to generate the correct neutrino masses. Searches for the neutrino mass generation in our model are possible via conventional heavy neutral lepton searches, as well as through exotic signatures arising from the interplay of portal couplings.

ACKNOWLEDGMENTS

The authors would like to thank Kris Moffatt for insights into the neutrino mass generation mechanism. This work was partially supported by Conselho Nacional de Ciência e Tecnologia (CNPq). This project also received support from the European Union's Horizon 2020 research and innovation program under Marie Skłodowska-Curie Grant Agreements No. 690575 (RISE InvisiblesPlus) and No. 674896 (ITN Elusives). S.P. and P.B. are supported by the European Research Council under ERC Grant NuMass (FP7-IDEAS-ERC ERC-CG 617143). S.P. acknowledges partial support from the Wolfson Foundation and the Royal Society.

REFERENCES

1. Y. Fukuda et al. (Super-Kamiokande Collaboration), Phys. Rev. Lett. 82, 2644 (1999); Q.R. Ahmad et al. (SNO Collaboration), 89, 011301 (2002); K. Eguchi et al. (KamLAND Collaboration), 90, 021802 (2003).
2. P. Minkowski, Phys. Lett. 67B, 421 (1977); R.N. Mohapatra and G. Senjanovic, Phys. Rev. Lett. 44, 912 (1980); M. Gell-Mann, P. Ramond, and R. Slansky, in Proceedings of the Supergravity Workshop, Stony Brook, New York, 1979, edited by P. van Nieuwenhuizen and D.Z. Freedman, (North Holland Publ. Co., 1979); ; Conf. Proc. C790927, 315 (1979); T. Yanagida, in Proceedings of the Workshop on the Uni Ed Theories and the Baryon Number in the Universe, Tsukuba, Japan, 1979, (Natl.Lab.High Energy Phys., Tsukuba, 1979); ; Conf. Proc. C7902131, 95 (1979); G. Lazarides, Q. Shafi, and C. Wetterich, Nucl. Phys. B181, 287 (1981); R.N. Mohapatra and G. Senjanovic, Phys. Rev. D 23, 165 (1981); J. Schechter and J.W.F. Valle, 22, 2227 (1980); T.P. Cheng and L.-F. Li, 22, 2860 (1980); R. Foot, H. Lew, X.G. He, and G.C. Joshi, Z. Phys. C 44, 441 (1989).
3. S.M. Boucenna, S. Morisi, and J.W.F. Valle, Adv. High Energy Phys. (London, U.K.) 2014, 831598 (2014).
4. R.N. Mohapatra and J.W.F. Valle, Phys. Rev. D 34, 1642 (1986); M.C. Gonzalez-Garcia and J.W.F. Valle, Phys. Lett. B 216, 360 (1989).
5. D. Wyler and L. Wolfenstein, Nucl. Phys. B218, 205 (1983); E.K. Akhmedov, M. Lindner, E. Schnapka, and J.W.F. Valle, Phys. Lett. B 368, 270 (1996); Phys. Rev. D 53, 2752 (1996).
6. J. Barry, W. Rodejohann, and H. Zhang, J. High Energy Phys. 2011, 091; H. Zhang, Phys. Lett. B 714, 262 (2012).
7. W. Buchmuller, C. Greub, and P. Minkowski, Phys. Lett. B 267, 395 (1991).
8. S. Khalil, J. Phys. G 35, 055001 (2008).
9. P. Fileviez Perez, T. Han, and T. Li, Phys. Rev. D 80, 073015 (2009).
10. S. Khalil, Phys. Rev. D 82, 077702 (2010).
11. C.O. Dib, G.R. Moreno, and N.A. Neill, Phys. Rev. D 90, 113003 (2014).
12. S. Baek, H. Okada, and K. Yagyu, J. High Energy Phys. 2015, 049.
13. V. De Romeri, E. Fernandez-Martinez, J. Gehrlein, P.A.N. Machado, and V. Niro, J. High Energy Phys. 2017, 169.

14. T. Nomura and H. Okada, Lett. High Energy Phys. 1, 10 (2018).
15. V. Brdar and A. Y. Smirnov, J. High Energy Phys. 2019, 045.
16. H. Okada and K. Yagyu, Phys. Rev. D 90, 035019 (2014).
17. C.E. Diaz, S.F. Mantilla, and R. Martinez, arXiv:1712.05433.
18. E. Bertuzzo, P.A.N. Machado, Z. Tabrizi, and R. Zukanovich Funchal, J. High Energy Phys. 2017, 004.
19. T. Nomura and H. Okada, Phys. Rev. D 99, 055033 (2019).
20. E. Bertuzzo, S. Jana, P.A.N. Machado, and R. Zukanovich Funchal, Phys. Lett. B 791, 210 (2019).
21. P. Ballett, M. Hostert, and S. Pascoli, arXiv:1903.07589.
22. P. Ballett, S. Pascoli, and M. Ross-Lonergan, Phys. Rev. D 99, 071701 (2019).
23. E. Bertuzzo, S. Jana, P.A.N. Machado, and R. Zukanovich Funchal, Phys. Rev. Lett. 121, 241801 (2018).
24. C.A. Argüelles, M. Hostert, and Y.-D. Tsai, arXiv:1812.08768.
25. P. S. B. Dev and A. Pilaftsis, Phys. Rev. D 86, 113001 (2012).
26. Y. Zhang, X. Ji, and R.N. Mohapatra, J. High Energy Phys. 2013, 104.
27. J. Lopez-Pavon, S. Pascoli, and C.-f. Wong, Phys. Rev. D 87, 093007 (2013).
28. W. Grimus and L. Lavoura, J. High Energy Phys. 2000, 042.
29. R. Adhikari and A. Raychaudhuri, Phys. Rev. D 84, 033002 (2011).
30. Y. Cai, J. Herrero-García, M.A. Schmidt, A. Vicente, and R.R. Volkas, Front. Phys. 5, 63 (2017).
31. A. Pilaftsis, Z. Phys. C 55, 275 (1992).
32. B.A. Kniehl and A. Pilaftsis, Nucl. Phys. B474, 286 (1996).
33. T. Yamazaki et al., in Proceedings of the 22nd International Conference on High Energy Physics: Vol. 1, Leipzig, Germany, 1984, Zeuthen, Germany: Akad. Wiss., (1984).
34. A.V. Artamonov et al. (E949 Collaboration), Phys. Rev. D 91, 052001 (2015); et al., 91, 059903(E) (2015).
35. J.R. Batley et al. (NA48/2 Collaboration), Phys. Lett. B 769, 67 (2017).
36. G. Bernardi et al., Phys. Lett. 166B, 479 (1986).
37. F. Bergsma et al. (CHARM Collaboration), Phys. Lett. 128B, 361 (1983).

38. J. Badier et al. (NA3 Collaboration), Z. Phys. C 31, 21 (1986).
39. A. Vaitaitis et al. (NuTeV and E815 Collaborations), Phys. Rev. Lett. 83, 4943 (1999).
40. A.M. Cooper-Sarkar et al. (WA66 Collaboration), Phys. Lett. 160B, 207 (1985).
41. P. Astier et al. (NOMAD Collaboration), Phys. Lett. B 506, 27 (2001).
42. P. Abreu et al. (DELPHI Collaboration), Z. Phys. C 74, 57 (1997); et al., 75, 580(E) (1997).
43. M.Z. Akrawy et al. (OPAL Collaboration), Phys. Lett. B 247, 448 (1990).
44. A.M. Sirunyan et al. (CMS Collaboration), Phys. Rev. Lett. 120, 221801 (2018).
45. P. Ballett, S. Pascoli, and M. Ross-Lonergan, J. High Energy Phys. 2017, 102.
46. P. Ballett, T. Boschi, and S. Pascoli, in Proceedings of the Prospects in Neutrino Physics (NuPhys2017), London, 2017, (2018), pp. 124–128.
47. P. Ballett, T. Boschi, and S. Pascoli (to be published).
48. M. Drewes, J. Hajer, J. Klaric, and G. Lanfranchi, in Proceedings of the 53rd Rencontres De Moriond on Electroweak Interactions and Unified Theories (Moriond EW 2018), La Thuile, Italy, 2018, (2018).
49. W. Bonivento et al., arXiv:1310.1762.
50. S. Alekhin et al., Rep. Prog. Phys. 79, 124201 (2016).
51. R. Laha, B. Dasgupta, and J.F. Beacom, Phys. Rev. D 89, 093025 (2014).
52. P. Bakhti and Y. Farzan, Phys. Rev. D 95, 095008 (2017).
53. P. Bakhti, Y. Farzan, and M. Rajaee, Phys. Rev. D 99, 055019 (2019).
54. A.D. Dolgov, S.H. Hansen, G. Raffelt, and D.V. Semikoz, Nucl. Phys. B590, 562 (2000).
55. S. Hannestad, R.S. Hansen, and T. Tram, Phys. Rev. Lett. 112, 031802 (2014); B. Dasgupta and J. Kopp, 112, 031803 (2014); A. Mirizzi, G. Mangano, O. Pisanti, and N. Saviano, Phys. Rev. D 91, 025019 (2015); X. Chu, B. Dasgupta, and J. Kopp, J. Cosmol. Astropart. Phys. 2015, 011; J.F. Cherry, A. Friedland, and I.M. Shoemaker, ; arXiv:1605.06506; X. Chu, B. Dasgupta, M. Dentler, J. Kopp, and N. Saviano, J. Cosmol. Astropart. Phys. 2018, 049; N. Song, M.C. Gonzalez-Garcia, and J. Salvado, 2018, 055.

CHAPTER

6

Improved Upper Limit on the Neutrino Mass from a Direct Kinematic Method by KATRIN

M. Aker,[1,2] K. Altenmüller,[10,3,4] M. Arenz,[5] M. Babutzka,[6] J. Barrett,[7] S. Bauer,[8] M. Beck,[8,12] A. Beglarian,[9] J. Behrens,[6,1,8] T. Bergmann,[10,3,9] U. Besserer,[1,2] K. Blaum,[11] F. Block,[6] S. Bobien,[2] K. Bokeloh,[8]§ J. Bonn,[12,] B. Bornschein,[1,2] L. Bornschein,[1] H. Bouquet,[9] T. Brunst,[10,3] T. S. Caldwell,[13,14] L. La Cascio,[6] S. Chilingaryan,[9] W. Choi,[6] T. J. Corona,[13,14,7] K. Debowski,[15,6] M. Deffert,[6] M. Descher,[6] P. J. Doe,[16] O. Dragoun,[17] G. Drexlin,[6,1]† J. A. Dunmore,[16] S. Dyba,[8] F. Edzards,[10,3] L. Eisenblätter,[9] K. Eite[l,1] E. Ellinger,[15] R. Engel,[1,6] S. Enomoto,[16] M. Erhard,[6] D. Eversheim,[5] M. Fedkevych,[8] A. Felden,[1] S. Fischer,[1,2] B. Flatt,[12] J. A. Formaggio,[7] F. M. Fränkle,[1,13,14] G. B. Franklin,[18] H. Frankrone,[9] F. Friedel,[6] D. Fuchs,[10,3] A. Fulst,[8] D. Furse,[7] K. Gauda,[8] H. Gemmeke,[9] W. Gil,[1] F. Glück,[1] S. Görhardt,[1] S. Groh,[6] S. Grohmann,[2] R. Grössle,[1,2] R. Gumbsheimer,[1] M. Ha Minh,[10,3] M. Hackenjos,[1,2,6] V. Hannen,[8] F. Harms,[6] J. Hartmann,[9] N. Haußmann,[15] F. Heizmann,[6] K. Helbing,[15] S. Hickford,[1,15] D. Hilk,[6] B. Hillen,[8] D. Hillesheimer,[1,2] D. Hinz,[1] T. Höhn,[1] B. Holzapfel,[2] S. Holzmann,[2] T. Houdy,[10,3] M. A. Howe,[13,14] A. Huber,[6] T.

Citation: M. Aker et al. (KATRIN Collaboration). "Improved Upper Limit on the Neutrino Mass from a Direct Kinematic Method by KATRIN" Phys. Rev. Lett. 123, 221802 (2019). https://doi.org/10.1103/PhysRevLett.123.221802

M. James,[2] A. Jansen,[1] A. Kaboth,[7] C. Karl,[10,3] O. Kazachenko,[21] J. Kellerer,[6] N. Kernert,[1] L. Kippenbrock,[16] M. Kleesiek,[6] M. Klein,[1,6] C. Köhler,[10,3] L. Köllenberger,[1] A. Kopmann,[9] M. Korzeczek,[6] A. Kosmider,[1] A. Kovalík,[17] B. Krasch,[1,2] M. Kraus,[6] H. Krause,[1] L. Kuckert,[1] B. Kuffner,[1] N. Kunka,[9] T. Lasserre,[4,3,10] T. L. Le,[1,2] O. Lebeda,[17] M. Leber,[16] B. Lehnert,[19] J. Letnev,[20] F. Leven,[6] S. Lichter,[1] V. M. Lobashev,[21], A. Lokhov,[8,21] M. Machatschek,[6] E. Malcherek,[1] K. Müller,[1] M. Mark,[1] A. Marsteller,[1,2] E. L. Martin,[13,14,16] C. Melzer,[1,2] A. Menshikov,[9] S. Mertens,[10,3,19,1] L. I. Minter,[16] S. Mirz,[1,2] B. Monreal,[22] P. I. Morales Guzmán,[10,]3 K. Müller,[1] U. Naumann,[15] W. Ndeke,[25] H. Neumann,[2] S. Niemes,[1,2] M. Noe,[2] N. S. Oblath,[7] H.-W. Ortjohann,[8] A. Osipowicz,[20] B. Ostrick,[8] E. Otten,[12],* D. S. Parno,[18,16] D. G. Phillips II,[13,14] P. Plischke,[1] A. Pollithy,[10,3] A. W. P. Poon,[19] J. Pouryamout,[15] M. Prall,[8] F. Priester,[1,2] M. Röllig,[1,2] C. Röttele,[1,6,2] P. C.-O. Ranitzsch,[8] O. Rest,[8] R. Rinderspacher,[1] R. G. H. Robertson,[16] C. Rodenbeck,[8] P. Rohr,[9] Ch. Roll,[25] S. Rupp,[1,2,6] M. Ryšavý,[17] R. Sack,[8] A. Saenz,[25] P. Schäfer,[1,2] L. Schimpf,[6] K. Schlösser,[1] M. Schlösser,[1,2] L. Schlüter,[10,3] H. Schön,[2] K. Schönung,[11,1,2] M. Schrank,[1] B. Schulz,[25] J. Schwarz,[1] H. Seitz-Moskaliuk,[6] W. Seller,[20] V. Sibille,[7] D. Siegmann,[10,3] A. Skasyrskaya,[21] M. Slezák,[10,17] A. Špalek,[17] F. Spanier,[1] M. Steidl,[1] N. Steinbrink,[8] M. Sturm,[1,2] M. Suesser,[2] M. Sun,[16] D. Tcherniakhovski,[9] H. H. Telle,[23] T. Thümmler,[1,8] L. A. Thorne,[18] N. Titov,[21] I. Tkachev,[21] N. Trost,[1] K. Urban,[10,3] D. V´enos, [17] K. Valerius,[1,8] B. A. VanDevender,[16] R. Vianden,[5] A. P. Vizcaya Hernández,[18] B. L. Wall,[16] S. Wüstling,[9] M. Weber,[9] C. Weinheimer ,[8] C. Weiss,[24] S. Welte,[1,2] J. Wendel,[1,2] K. J. Wierman,[13,14] J. F. Wilkerson,[13,14] J. Wolf,[6] W. Xu,[7] Y.-R. Yen,[18] M. Zacher,[8] S. Zadorozhny,[21] M. Zboril,[8,17] and G. Zeller[1,2]

[1]Institute for Nuclear Physics (IKP), Karlsruhe Institute of Technology (KIT), Hermann-von-Helmholtz-Platz 1, 76344 Eggenstein-Leopoldshafen, Germany

[2]Institute for Technical Physics (ITEP), Karlsruhe Institute of Technology (KIT), Hermann-von-Helmholtz-Platz 1, 76344 Eggenstein-Leopoldshafen, Germany

[3]Technische Universität München, James-Franck-Straße 1, 85748 Garching, Germany

[4]IRFU (DPhP & APC), CEA, Universit´e Paris-Saclay, 91191 Gif-sur-Yvette, France

[5]Helmholtz-Institut für Strahlen- und Kernphysik, Rheinische Friedrich-Wilhelms-Universität Bonn, Nussallee 14-16, 53115 Bonn, Germany

[6]Institute of Experimental Particle Physics (ETP), Karlsruhe Institute of Technology (KIT), Wolfgang-Gaede-Straße 1, 76131 Karlsruhe, Germany

[7]Laboratory for Nuclear Science, Massachusetts Institute of Technology, 77

Massachusetts Avenue, Cambridge, Massachusetts 02139, USA

[8]Institut für Kernphysik, Westfälische Wilhelms-Universität Münster, Wilhelm-Klemm-Straße 9, 48149 Münster, Germany

[9]Institute for Data Processing and Electronics (IPE), Karlsruhe Institute of Technology (KIT), Hermann-von-Helmholtz-Platz 1, 76344 Eggenstein-Leopoldshafen, Germany

[10]Max-Planck-Institut für Physik, Föhringer Ring 6, 80805 München, Germany

[11]Max-Planck-Institut für Kernphysik, Saupfercheckweg 1, 69117 Heidelberg, Germany

[12]Institut für Physik, Johannes-Gutenberg-Universität Mainz, 55099 Mainz, Germany

[13]Department of Physics and Astronomy, University of North Carolina, Chapel Hill, North Carolina 27599, USA

[14]Triangle Universities Nuclear Laboratory, Durham, North Carolina 27708, USA

[15]Department of Physics, Faculty of Mathematics and Natural Sciences, University of Wuppertal, Gaußstraße 20, 42119 Wuppertal, Germany

[16]Center for Experimental Nuclear Physics and Astrophysics, and Department of Physics, University of Washington, Seattle, Washington 98195, USA

[17]Nuclear Physics Institute of the CAS, v. v. i., CZ-250 68 Řež, Czech Republic
18Department of Physics, Carnegie Mellon University, Pittsburgh, Pennsylvania 15213, USA

[19]Institute for Nuclear and Particle Astrophysics and Nuclear Science Division, Lawrence Berkeley National Laboratory, Berkeley, California 94720, USA

[20]University of Applied Sciences (HFD) Fulda, Leipziger Straße 123, 36037 Fulda, Germany

[21]Institute for Nuclear Research of Russian Academy of Sciences, 60th October Anniversary Prospect 7a, 117312 Moscow, Russia

[22]Department of Physics, Case Western Reserve University, Cleveland, Ohio 44106, USA

[23]Departamento de Química Física Aplicada, Universidad Autonoma de Madrid, Campus de Cantoblanco, 28049 Madrid, Spain

[24]Project, Process, and Quality Management (PPQ), Karlsruhe Institute of Technology (KIT), Hermann-von-Helmholtz-Platz 1, 76344 Eggenstein-Leopoldshafen, Germany

[25]Institut für Physik, Humboldt-Universität zu Berlin, Newtonstraße 15, 12489 Berlin, Germany

ABSTRACT

We report on the neutrino mass measurement result from the first four-week science run of the Karlsruhe Tritium Neutrino experiment KATRIN in spring 2019. Beta-decay electrons from a high-purity gaseous molecular tritium source are energy analyzed by a high-resolution MAC-E filter. A fit of the integrated electron spectrum over a narrow interval around the kinematic end point at 18.57 keV gives an effective neutrino mass square value of $(-1.0^{+0.9}_{-1.1})$ eV^2. From this, we derive an upper limit of 1.1 eV (90% confidence level) on the absolute mass scale of neutrinos. This value coincides with the KATRIN sensitivity. It improves upon previous mass limits from kinematic measurements by almost a factor of 2 and provides model-independent input to cosmological studies of structure formation.

INTRODUCTION.

The observation of flavor oscillations of atmospheric and solar neutrinos [1,2] as well as oscillation studies at reactors and accelerators unequivocally prove neutrinos to possess nonzero rest masses (e.g., [3]), contradicting the standard model (SM) expectation of them being massless. The absolute values mi of the neutrino mass states ν_i (i=1, 2, 3), which cannot be probed by oscillations, are of fundamental importance in cosmological studies [4–6] and for particle physics models beyond the SM [7].

Because of the unique role of primordial neutrinos in the formation of large-scale structures in the universe, observations of matter clustering in different epochs of the universe allow one to probe the neutrino mass sum $\Sigma_i m_i$. The current upper limits depend on the selection of data sets included in the analyses and are valid only within the ΛCDM concordance model [6,8]. Another model-dependent method is provided by the search for neutrinoless double beta decay 0νββ, a process forbidden in the SM due to lepton number violation. It gives access to the effective Majorana neutrino mass (e.g., [9,10]).

A model-independent, direct method for probing the neutrino mass scale in the laboratory is provided by kinematic studies of weak-interaction processes such as β decay of tritium (^{3}H) and electron capture on holmium (^{163}Ho) [11–15]. These investigations yield an incoherent sum of spectra, containing the squares of the neutrino eigenmasses m^2_i as parameters. Each spectral component is weighted by the absolute square of the corresponding electron-flavor matrix element $|U_{ei}|^2$. In the quasidegenerate regime m_i>0.2 eV, the eigenmasses are the same to better than 3%. The mass measured in β decay or electron capture, often called " m(νe)," is the neutrino mass $m_ν≈m_i$ in this regime. Because of its low end point energy (E0=18.57 keV) and favorable half-life (t1/2=12.32 yr), the decay of tritium $^3H→^3He^+ + e^- + νe$ has been investigated by a large number of experiments looking for the small, characteristic shape distortion of the β spectrum close to E_0 due to mν [11,12]. Experimental advances over many decades have steadily increased the sensitivity to the present upper limit of $m_ν<2$ eV (95% confidence level, C.L.) [16]. In this Letter, we report on the first neutrino mass result from the Karlsruhe Tritium Neutrino experiment KATRIN [17–20], which is targeted to advance the sensitivity on $m_ν$ by 1 order of magnitude down to 0.2 eV (90% C.L.) after five years.

EXPERIMENTAL SETUP

KATRIN combines a windowless gaseous molecular tritium source (WGTS), pioneered by the Los Alamos experiment [21], with a spectrometer based on the principle of magnetic adiabatic collimation with electrostatic filtering (MAC-E-filter) [22,23], developed at Mainz and Troitsk [24,25]. These techniques allow the investigation of the end point region of tritium β decay with very high energy resolution, large statistics, and small systematics. KATRIN has been designed and built to refine this direct kinematic method to its ultimate precision level. To improve the sensitivity on mν by 1 order of magnitude calls for an increase in statistics and a reduction of systematic uncertainties by 2 orders of magnitude, as the observable in kinematic studies is the neutrino mass square, $m^2_ν$.

Figure 1 gives an overview of the 70 m long experimental setup located at the Karlsruhe Institute of Technology. The source-related components in contact with tritium, the rear section (RS) (a), the WGTS cryostat (b), as well as the differential (DPS) and cryogenic (CPS) pumping sections (c) are integrated into the extensive infrastructure of Tritium Laboratory Karlsruhe to enable a closed cycle of tritium [26]. High-purity tritium gas from a pressure-controlled buffer vessel is continuously injected at 30 K into the WGTS at the midpoint of its 90 mm diameter, 10 m long stainless steel beam tube. The gas then diffuses to both ends where it is pumped out by a series of turbomolecular pumps (TMPs) in the DPS, yielding the nominal column density ρdnom=$5×10^{17}$ molecules cm^{-2}. In combination with the CPS, housing a large-capacity cryotrap operated at around 3 K, the flow rate of tritium into the following spectrometer and detector section [Figs. 1(d)–1(f)] downstream is negligible, well below the 14 orders of magnitude of flow reduction required to eliminate source-related background by neutral tritium gas [17].

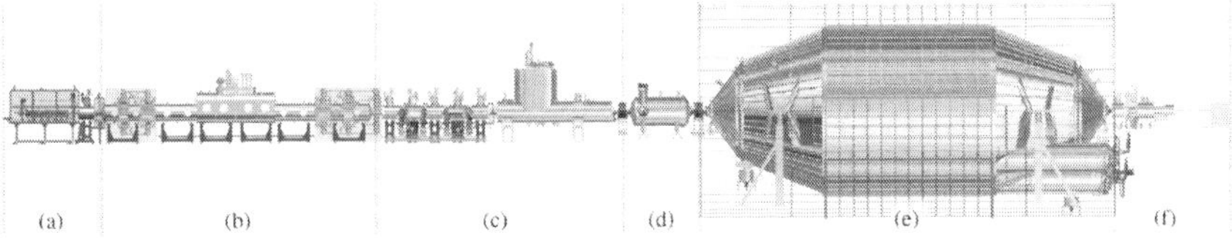

Figure 1. The major components of the KATRIN beam line consist of (a) the rear section for diagnostics, (b) the windowless gaseous tritium source WGTS, c) the pumping section with the DPS and CPS cryostats, and a tandem setup of two MAC-E-filters: (d) the smaller prespectrometer and (e) the larger main spectrometer with its surrounding air coil system. This system transmits only the highest-energy β decay electrons onto (f) the solid-state detector where they are counted.

The source magnetic field (section b) in Fig. 1, B_{WGTS}=2.52 T) as well as other superconducting solenoids [section (c) in Fig. 1] [27] adiabatically guide primary β-decay electrons, secondary electrons, and ions to the spectrometers. A series of blocking and dipole electrodes eliminates ions by an $\vec{E} \times \vec{B}$ drift to the beam tube, so that they cannot generate background in the spectrometer section [19].

High-precision electron spectroscopy is achieved by the MAC-E-filter technique, where electrons of charge q are guided by the magnetic field, collimated by its gradient and filtered by an electrostatic barrier, the retarding potential energy qU. The resulting high-pass filter transmits only electrons with enough energy to overcome the barrier qU and allows the scanning of the tritium β-decay spectrum in an integral mode.

The tandem configuration of MAC-E-filters performs a two-step filter process: first, the smaller prespectrometer is operated at fixed high voltage (HV) of −10.4 kV in this work to act as a prefilter to reject electrons that carry no information on m_ν. In a second step, a variable qU is applied to the main spectrometer for precision filtering of β-decay electrons close to E_0. Its huge size guarantees fully adiabatic motion to the central "analyzing plane," where the minimum magnetic field B_{min} and the maximum retarding energy qU coincide for the filtering process to occur. Elevating the two spectrometers to a negative HV forms a strong Penning trap which can give rise to background [28,29]. This is avoided by operating both at an ultrahigh vacuum (UHV) regime of 10^{-11} mbar using nonevaporable getter (NEG) pumps and TMPs [30]. A defining property of a MAC-E-filter is $\Delta E/E$, the filter width at energy E, which is given by the ratio B_{min}/B_{max} of the minimum to maximum magnetic field in nonrelativistic approximation. The present ratio (0.63 mT/4.24 T) is equivalent to ΔE=2.8 eV at E_0. This value constrains the size V_{ft} of the flux tube around B_{min} and, consequently, the overall background rate, which is proportional to V^{ft} to the first order. A large air coil system of 12.6 m diameter [31] is used to adjust B_{min} and V_{ft}. After the potential of the spectrometer vessel is elevated, an offset of up to −200V can be applied to the wire electrode system mounted on the inner surface of the vessel to define qU.

Electrons transmitted through the spectrometers are finally counted in a radially and azimuthally segmented monolithic silicon detector array with 148 pixels [32] as a function of qU. To optimize the signal-to-background ratio, transmitted electrons are postaccelerated by a potential of +10 kV before they impinge on the detector.

COMMISSIONING MEASUREMENTS

Over the past years, we have commissioned the entire setup by a series of dedicated long-term measurements [19,26,27,33] which have demonstrated that all specifications [18] are met, or even surpassed by up to 1 order of magnitude, except for the background rate R_{bg}.

A major benchmark is to operate the source at ρd_{nom} at a stability level of 10^{-3}/h so that variations of the column density ρd can be neglected. This calls for a stable gas injection rate via capillaries [26] and a constant beam-tube temperature. For the latter, a stability level of better than 10^{-3}/h has been achieved by a two-phase beam-tube cooling system at 30 (100) K using neon (argon) as a cooling fluid [34]. In mid-2018, measurements at 1% DT concentration within a 99% D_2 carrier gas at ρd_{nom} have verified the required level of source stability [35]. This “first tritium” campaign has allowed us to collect the first integral electron spectra which agree well with the model expectation.

In this spectral comparison, the response function f(E−qU) [20] plays a fundamental role [see Eq. (1)], as it converts the theoretical β spectrum into the experimental spectrum. It describes the probability of transmission of an electron with initial energy E as a function of its surplus energy E−qU. For an ensemble, it depends on the angular spread of electrons and the amount of neutral gas they pass in the source, where they can undergo inelastic scattering processes with a total cross section σ (3.64×10^{-18}cm^2 at 18.57 keV, adopted from [36]). We measure f(E−qU) using monoenergetic electrons with a small angular spread produced in a dedicated photoelectron source (e gun) [37] located at the RS. These electrons span a 50 eV wide range of surplus energies E−qU and pass through the integral column density ρd of the source. This allows us to measure the characteristics of single (s=1) and multiple (s=2,3,…) inelastic scattering in T2. In Fig. 2 (top), we display the measured response f(E−qU) for a narrow-angle photoelectron source as well as the calculated response fcalc(E−qU) for isotropically emitted β-decay electrons for the normal integrating MAC-E mode for $\rho d \approx \rho d_{nom}$. The sharp rise with the filter width ΔE to a plateau extending up to 11 eV results from "no loss" (energy loss δE=0) e-gun electrons, which leave the source without scattering (s=0) with a probability exp(−ρd·σ). At larger E−qU, s-fold scattering (s=1, 2, 3) is visible. In Fig. 2 (bottom), the differential data from the MAC-E-TOF mode [38] are shown, where the electron time of flight (TOF) is recorded. This allows us to even better assess the s-fold inelastic scattering and to obtain the energy-loss function of electrons ϵ(δE) by a deconvolution with the no loss peak at δE=E−qU=0.

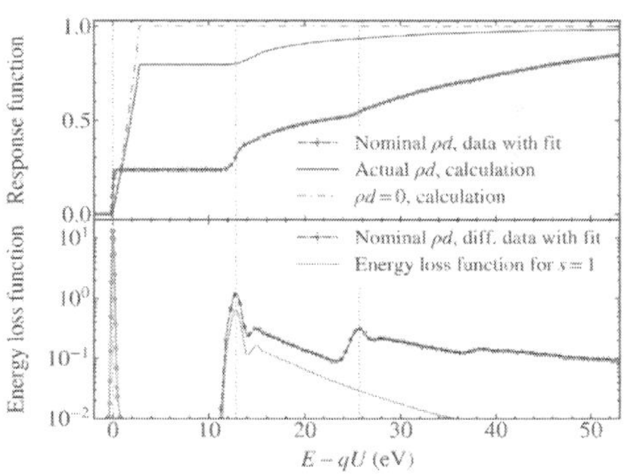

Figure 2. (top) Measured and calculated response functions f(E−qU) for electron surplus energies E−qU at different ρd values of T2. Measured f(E−qU) for a narrow-angle photoelectron source close to ρdnom and fit (cyan line); and calculated fcalc(E−qU) for isotropically emitted β decay electrons up to θmax at ρdexp (1.11×10^{17}cm^{-2}), the set point of our scans (red line), and in the limit of vanishing ρd=0 (grey, dashed-dotted line). (bottom) Differential distributions of energy losses δE from the MAC-E-TOF mode after a selection 35 μs≤TOF≤ 50 μs at ρd≈ρdnom and fit (cyan line). The "no loss" peak at δE=E−qU=0 is followed by peaks with s=2 (s=3) scattering at twice (triple) the δE value of s=1. The energy loss function ϵ(δE) for s=1 is obtained by deconvolution (orange line).

As the background rate R_{bg} exceeds its design goal of 0.01 counts per second (cps), we have studied the nature and origin of background processes so as to implement mitigation measures. Up to now, source-related backgrounds have not been observed, so that spectrometer-related processes [39] dominate R_{bg} apart from a small detector-related contribution [32]. Electrons generated at the spectrometer surface by cosmic muons and environmental gamma rays are inhibited from entering the inner flux tube by magnetic and electric barriers [40,41]. Thus, R_{bg} originates from excited or unstable neutral atoms which can propagate freely in the UHV environment. Accordingly, R_{bg} is observed to have an almost constant rate per unit volume in the flux tube.

Detailed investigations of the background behavior [39,42] revealed that a significant part of R_{bg} is due to Rydberg atoms sputtered off the inner spectrometer surfaces by 206Pb-recoil ions following α decays of 210Po. These processes follow the decay chain of the long-lived 222Rn progeny $210P_b$, which was surface-implanted from ambient air (activity ≈1 Bq/m^2) during the construction phase. A small fraction of these Rydberg states is ionized by black-body radiation when propagating over the magnetic flux tube. The resulting sub-eV scale electrons are accelerated to qU by the MAC-E-filter and form a Poisson component to R_{bg}.

The other part of background events stems from α decays of single 219Rn atoms (t_1/2=3.96s) emanating from the NEG pumps located in two of the three pump ports of the main spectrometer which release a large number of electrons up to the keV scale in the flux tube, where they are stored due to its magnetic bottle characteristics. By scattering off residual gas, the stored electrons subsequently produce secondary background electrons. This process continues until the stored electron is cooled off to an energy of a few eV when it can escape and also contribute itself to R_{bg} at qU. Owing to its origin from a small number of $219R_n$ decays, this background results in extended time periods where the background level is enhanced yielding a small non-Poissonian component [43]. Liquid-nitrogen cooled copper baffles at the inlet of the NEG pumps act as a countermeasure [44]. The coverage of the inner surface of the main spectrometer with a monolayer of H_2O, originating from an imperfect bake out of the prespectrometer, has led to the formation of a thin layer of H_2O covering the baffle surface. Therefore, the retention of $219R_n$ in this work is hampered such that R_{bg} retains a small non-Poissonian component.

MEASUREMENTS OF THE TRITIUM Β SPECTRUM

In the following, we report on our first high-purity tritium campaign from April 10 to May 13, 2019 which demonstrates the functionality of all system components and of the extensive tritium infrastructure at large source activity (2.45×10^{10}Bq) and tritium throughput (4.9 g/day). As a result of radiochemical reactions of T2 with the previously unexposed inner metal surface of the injection capillary, we observe drifts in the source column density. To limit these drifts to a level of $\pm2\times10^{-2}$ over our campaign, we keep the column density at an average value of $\rho d_{exp}=1.11\times10^{17}$molecules cm^{-2}, which is about a factor of 5 smaller than ρd_{nom}.

At this setting, the smaller value of $\rho d_{exp}\cdot\sigma$ (0.404) reduces the amount of inelastic scattering of electrons off neutral gas, see Fig. 2. The relative fractions of the six hydrogen isotopologues injected into the source are continuously monitored by laser-Raman spectroscopy with 10^{-3} precision [45]. The average isotopic tritium purity ϵT (0.976) of our analyzed data sample is derived from the composition of the tritiated species T2 (0.953), HT (0.035), and DT (0.011), with inactive species (D2, HD, and H2) being present only in trace amounts.

Because of the large number of β decay s and ionization processes, a cold magnetized plasma of electrons (meV to keV scale) and ions (meV scale) is formed which interacts with the neutral gas. The strong solenoidal field BWGTS and the resulting large longitudinal conductance of the plasma allow the coupling of its potential to the surface of the rear wall (RW) located at the RS and, thus, to control the starting energies of β-decay electrons over the volume [46].

Biasing the gold-plated RW disk with small areal variation of the work function to −0.15 V relative to the grounded beam tube gives a very good radial homogeneity of the source potential.

This is verified during initial tritium scans with fits of E0 over detector pixel rings, which do not show a significant radial variation.

Additional information on plasma effects is provided by comparing the line shape and position of quasimonoenergetic conversion electrons (L3-32) from 83 mKr runs in T2 to 83 mKr runs without the carrier gas at 100 K [47].

We do not identify sizeable shifts (<0.04 eV) or broadening (<0.08 eV) of lines so that the contribution of plasma effects at ρdexp to the systematic error budget in Table I can be neglected.

Table 1. 1σ systematic uncertainties (σ_{syst}) for $m^2\nu$ in eV^2, averaged over positive and negative errors, using the method of MC propagation.

Effect	**Relative uncertainty**	σ(m2ν) **in** eV2
Source properties		
ρd·σ	0.85%	0.05
Energy loss ϵ(δE)	O(1%)	Negligible
Beamline		0.05
BWGTS	2.5%	
Bmin	1%	
Bmax	0.2%	
Final state distribution	O(1%)	0.02
Fluctuations in scan k		0.05
HV stacking	2 ppm	
ρd variation	0.8%	
Isotopologue fractions	0.2%	
Background		
Background slope	1.7%/keV	0.07
Non-Poisson background	6.4%	0.30
Total syst. uncertainty		0.32

The integral tritium β-decay spectrum is scanned repeatedly in a range from [E_0−90 eV, E_0+50eV] by applying a set of nonequidistant HV settings to the inner electrode system. Each scan over this range takes a net time of about 2 h and is performed in alternating upward and downward directions to compensate for any time-dependent drift of the system to first order. At each HV set point, the transmitted electrons are counted over time intervals varying from 17 to 576 s with typical values of ~300s for points close to E_0. When setting a new HV value, we make use of a custom-made postregulation system for voltage stabilization and elimination of high-frequency noise. At the same time, a custom-made HV divider [48] continuously monitors the retarding voltage with ppm precision.

To limit the influence of systematic uncertainties for this Letter, we analyze a scan range covering the region of 40 eV below E_0 (22 HV set points) and 50 eV above (5 HV set points). The nonuniform measuring time distribution in this interval is shown in Fig. 3(c). It maximizes the sensitivity for $m^2\nu$ by focusing on the narrow region below E_0, where the imprint of the neutrino mass on the spectrum is most pronounced [20]. Shorter time intervals with a set point 200 V below E_0 are interspersed to monitor the source activity, in addition to other measures [49].

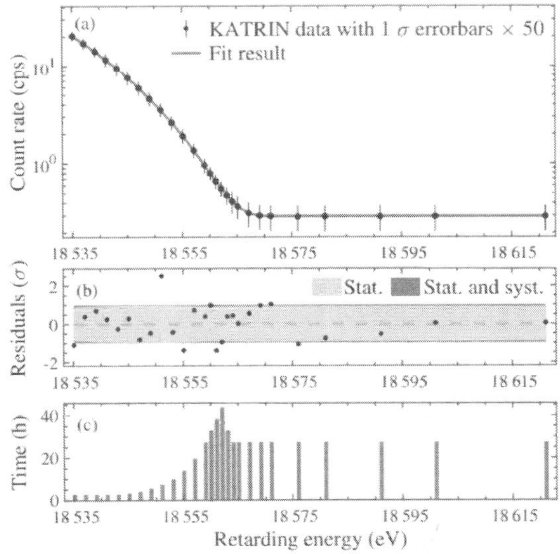

Figure 3. (a) Spectrum of electrons R(⟨qU⟩) over a 90 eV-wide interval from all 274 tritium scans and best-fit model Rcalc(⟨qU⟩) (line). The integral β decay spectrum extends up to E_0 on top of a flat background R_{bg}. Experimental data are stacked at the average value ⟨qU⟩l of each HV set point and are displayed with 1σ statistical uncertainties enlarged by a factor of 50. (b) Residuals of R(⟨qU⟩) relative to the 1σ uncertainty band of the best fit model. (c) Integral measurement time distribution of all 27 HV set points.

DATA ANALYSIS

For each tritium scan with its 27 HV set points, we apply quality cuts to relevant slow-control parameters to select a data set with stable run conditions. This results in 274 scans with an overall scanning time of 521.7 h. We also define a list of 117 detector pixels (out of 148), which excludes those pixels that are noisy or shadowed by beam line instrumentation in the β-electron path along the magnetic flux tube. For the digitized, calibrated and pile-up-corrected detector spectra, a broad region of interest (ROI) between 14 and 32 keV is defined. The ROI takes into account the detector energy resolution and its elevated potential (+10 kV) and allows us to include a large fraction of electrons backscattered at the detector in the narrow scan region close to E0 resulting in a negligible contribution to the systematic uncertainty budget [32].

The long-term stability of the scanning process is verified by fits to single scans to extract their effective β-decay end points. The 274 fit values show no time-dependent behavior and follow a Gaussian distribution (σ=0.25 eV) around a mean value of E0=18 573.7 eV. In view of this and the very good overall stability of the slow-control parameters for our data set, we merge the data of all 274 scans over all 117 pixels into one single 90-eV-wide spectrum, which is displayed in Fig. 3(a) in units of cps.

The underlying process corresponds to the "stacking" of events at the mean HV set points ⟨qU⟩l (l=1–27). The small Gaussian spread (rms=34 mV) of the actual HV value $qU_{l,k}$ during a scan k relative to ⟨qU⟩ l, the average of all scans, is a minor systematic effect which is accounted for in the analysis. The resulting stacked integral spectrum, R(⟨qU⟩), comprises 2.03×10^6 events, with 1.48×10^6 β-decay electrons below E_0 and a flat background ensemble of 0.55×10^6 events in the 90 eV scan interval. This high-statistics data set allows us to show 1σ error bars enlarged by a factor of 50 in Fig. 3.

The experimental spectrum is well described by our detailed model of the KATRIN response to β-decay electrons and background. It contains four free parameters: the signal amplitude As, the effective β-decay end point E0, the background rate Rbg, and the neutrino mass square m_2ν. We leave E_0 and As unconstrained, which is equivalent to a "shape-only" fit. The goodness-of-fit is illustrated in Fig. 3(b) from the scatter of residuals around the error band of the model.

The four-parameter fit procedure over the averaged HV set points ⟨qU⟩ l compares the experimental spectrum R(⟨qU⟩) to the model R_{calc}(⟨qU⟩). The latter is the convolution of the differential β-electron spectrum R_β(E) with the calculated response function fcalc(E−⟨qU⟩), with an added energy-independent background rate R_{bg}

$$R_{\text{calc}}(\langle qU\rangle) = A_s N_T \int R_\beta(E) f_{\text{calc}}(E - \langle qU\rangle) dE + R_{\text{bg}}. \tag{1}$$

Here, N_T denotes the calculated number of tritium atoms in the source multipliedwiththeacceptedsolidangleofthesetup $\Delta\Omega/4\pi = (1-\cos\theta_{\text{max}})/2$ and the detector efficiency $[\theta_{\text{max}} = \arcsin\sqrt{(B_{\text{WGTS}}/B_{\text{max}})} = 50.4°]$, its uncertainty is absorbed by the fit parameter As.

The electron spectrum $R_\beta(E)$ from the super allowed β decay of molecular tritium is calculated using Fermi's Golden Rule

$$R_\beta(E) = \frac{G_F^2 \cdot \cos^2\Theta_C}{2\pi^3} |M_{\text{nucl}}^2| F(E, Z') \times (E + m_e)\sqrt{(E + m_e)^2 - m_e^2} \times \sum_j \zeta_j \varepsilon_j \sqrt{\varepsilon_j^2 - m_\nu^2}\,\Theta(\varepsilon_j - m_\nu), \quad (2)$$

with the square of the energy-independent nuclear matrix element $|M^2_{nucl}|$, the Fermi constant GF, the Cabibbo angle Θ_C, the electron mass me, the Fermi function F(E,Z′=2), and the neutrino energy $\epsilon_j = E_0 - E - V_j$. In addition, our calculations incorporate radiative corrections (for details, see [12,20]), and we account for thermal Doppler broadening at 30 K.

When calculating $R_\beta(E)$, we sum over a final-state distribution (FSD) which is given by the probabilities ζ_j with which the daughter ion $^3HeT^+$ is left in a molecular (i.e., a rotational, vibrational, and electronic) state with excitation energy Vj. For this analysis, we first confirm the most recent theoretical FSD calculations [50,51] using new codes for solving the electronic and rovibrational problems within the Born-Oppenheimer approximation. We then refine the FSD by adopting a more efficient treatment of the rovibrational part and an update of other kinematics-related quantities, such as molecular masses, as well as recoil parameters (momenta and kinetic energy shifts). Most importantly, we treat all isotopologues (T_2, HT, and DT) in a consistent way with initial angular momenta distributions Jκ (κ=0,…,3) at 30 K for the electronic bound states n=1,…,6. The FSD includes higher excitation energies up to the continuum based on [50], but their contribution to our analysis interval [E0−40 eV] is at an overall level of 10^{-4} only. Accordingly, the FSD uncertainties in our narrow analysis interval of 40 eV below E_0 only contribute at the level of 0.02 eV2 to the total systematics budget on m2ν (see Table I).

The response function $f_{calc}(E-qU)$ used in the analysis is shown as the red curve in Fig. 2 (top). It corresponds to β-decay electrons born with energies close to E_0 and emitted isotropically up to θ_{max} in the source gas. Compared to the e-gun beam, they possess a different distribution of energy losses due to their broader range of pitch angles θ and the varying amount of source gas (ρ_d) they traverse. These processes are studied on the basis of gas dynamical simulations [52] which yield an approximately triangular-shaped longitudinal source profile.

After modeling the energy loss of β-decay electrons through the source by making use of ρd·σ and ϵ(δE), their subsequent propagation is tracked by the kassiopeia simulation software [53]. It incorporates a detailed beam line model which takes account of the small radial inhomogeneities of B_{min} and qU at the analyzing plane. The full model provides the detailed shape of ΔE and the distribution of electron pitch angles up to θmax from the parameters of the magnetic field triplet (B_{WGTS}, B_{min}, B_{max}).

The energy-independent part of R_{calc}(⟨qU⟩), R_{bg}, comes from a fit of the spectrum R(⟨qU⟩) over our 90 eV scan range. The fit value R_{bg}=(0.293±0.001) cps is largely constrained by the 5 HV set points above E_0 and agrees with data from independent background runs taken with an empty source before the tritium measurements.

The resulting model, R_{calc}(⟨qU⟩), is then fitted to R(⟨qU⟩). To ensure that this proceeds without bias, we employ a twofold "blinding" scheme. The first blinding step leaves the data untouched, but a modification is applied during the building of the model R_{calc}(⟨qU⟩). The FSD part describing rovibrational excitations of the electronic ground state is replaced with a Gaussian distribution with parameters not accessible to the analysis at first. As a result, fits with the blinded FSD do not reveal the unbiased value of m^2v. The "true" FSD is revealed only at the last step ("unblinding") after having fixed all model inputs and systematic uncertainties.

The second measure to mitigate biasing is to perform the full analysis, including parameter fitting, using Monte Carlo–based (MC) data sets first, before turning to the experimental data. For each experimental scan k, we generate a "MC twin," R_{calc}(⟨qU⟩)k, from its averaged slow-control parameters to procure Rβ(E)k, fcalc(E−⟨qU⟩)k, and Rbg,k. Analysis of MC twins allows us to verify the accuracy of our parameter inference by recovering the correct input MC values for m^2_v. This approach is also used to assess statistical (σ_{stat}) and systematic (σ_{syst}) uncertainties and to compute our expected sensitivity.

In the following, we report on the results of two independent analyses with different strategies to propagate systematic uncertainties: the "Covariance Matrix" and the "MC propagation" approaches.

In the covariance method, we fit the experimental spectrum R(⟨qU⟩) with the model R_{calc}(⟨qU⟩) by minimizing the standard χ^2 estimator. To propagate the systematic uncertainties, a covariance matrix is computed after performing O(104) simulations of R_{calc}(⟨qU⟩), while varying the relevant parameters for each calculation according to the likelihood given by their uncertainties [35,54,55]. The resulting systematic uncertainties agree with the values shown in Table I, which is based on the second approach. The sum of all matrices encodes the total uncertainties of $R_{calc}(\langle qU \rangle)$ and their HV set point dependent correlations. The χ^2 estimator is then minimized to determine the four best-fit parameters, and the shape of χ^2 function is used to infer the uncertainties. The results of this fit are displayed in Fig. 3. We obtain a goodness-of-fit of χ^2=21.4 for 23 d.o.f., corresponding to a p value of 0.56.

The MC-propagation approach is a hybrid Bayesian-frequentist method, adapted from Refs. [56–58]. We fit the experimental spectrum $R(\langle qU \rangle)$ with the model $Rcalc(\langle qU \rangle)$ by minimizing the negative Poisson-likelihood function. The goodness-of-fit of −2lnL=23.3 for 23 d.o.f. corresponds to a p value of 0.44. To propagate the systematic uncertainties, we repeat the fit 105 times, while varying the relevant parameters in each fit according to their uncertainties given in column 2 of Table I.

We report the 1σ width of the fit parameters as their systematic uncertainty in the third column of Table I. In order to simultaneously treat statistical and all systematic uncertainties, each of the 10^5 fits is performed on a statistically fluctuated MC copy of the true data set, leading to the distributions of m^2_v and E^0 shown in Fig. 4. The strong correlation (0.97) between the two parameters is an expected feature in kinematic studies of β decay [11,12]. The final-best fit is given by the mode of the fit-parameter distributions, and the 1σ total error is determined by integrating the distributions up to 16% from either side.

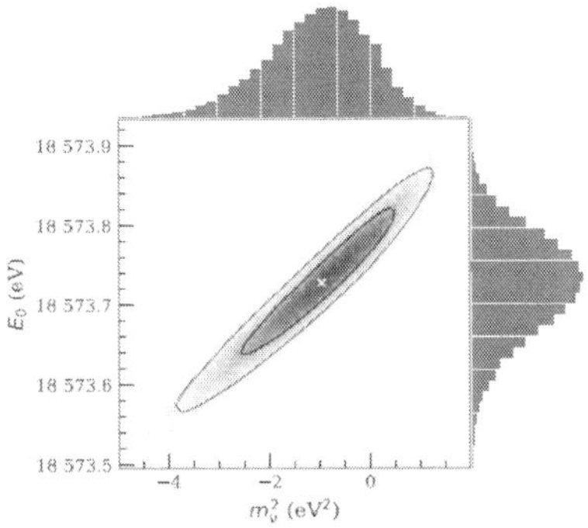

Figure 4. Scatter plot of fit values for the mass square m^2_v and the effective β-decay end point E_0 together with 1σ (black) and 2σ (blue) error contours around the best fit point (cross). It follows from a large set of pseudo experiments emulating our experimental data set and its statistical and systematical uncertainties.

RESULTS

The two independent methods agree to within a few percent of the total uncertainty. As best fit value for the neutrino mass, we find $m^2_v=(-1.0+0.9-1.1)\ eV^2$. This best fit result corresponds to a 1σ statistical fluctuation to negative values of m2v possessing a p value of 0.16.

The total uncertainty budget of m^2v is largely dominated by σ_{stat} ($0.97 eV^2$) as compared to σ_{syst} ($0.32 eV^2$). As displayed in Table I, the dominant contributions to σ_{syst} are found to be the non-Poissonian background from radon and the uncertainty on the background slope, which is constrained from the wide-energy integral scans of the earlier "first tritium" data [35]. Uncertainties of the column density, energy-loss function, final-state distribution, and magnetic fields play a minor role in the budget of σsyst. Likewise, the uncertainties induced via fluctuations of ϵT and HV parameters during a scan are negligibly small compared to σ_{stat}. The statistical (systematic) uncertainty of our first result on m^2v is smaller by a factor of 2 (6) compared to the final results of Troitsk and Mainz [24,25].

The methods of Lokhov and Tkachov (LT) [59] and of Feldman and Cousins (FC) [60] are then used to calculate the upper limit on mv. Both procedures avoid empty confidence intervals for nonphysical negative best-fit estimates of m^2v. For this first result, we follow the LT method. For a statistical fluctuation into the nonphysical region the method returns a confidence belt that coincides with the experimental sensitivity and avoids a shrinking upper limit for more negative values of m^2v. Using the LT construction, we derive an upper limit of mv<1.1 eV (90% C.L.) as the central result of this Letter. By construction, it is identical to the expected sensitivity. For completeness, we also note the FC upper limits mv<0.8(0.9) eV at 90% (95%) C.L.

For the effective end point, our two analysis methods both obtain the best-fit value E_0=(18 573.7±0.1) eV (see Fig. 4). At this level of precision, a consistency check on the energy scale of KATRIN can be performed by comparing our experimental Q value for molecular tritium with that based on measurements of the 3He-3H atomic mass difference [61]. Our result for the Q value of (18575.2±0.5) eV is obtained from our best-fit value for E0 by adding the center-of-mass molecular recoil of T2 (1.72 eV) [11], as well as the relative offset (−0.2±0.5 eV) of the source potential to the work function of the inner electrode. The calculated Q value from the 3He-3H atomic mass difference is (18575.72±0.07) eV when accounting for the different binding energies and kinematic variables of atomic and molecular tritium [11]. The consistency of both Q values underlines the robustness of the energy scale in our scanning process of molecular tritium.

CONCLUSION AND OUTLOOK

The reported upper limit m_v<1.1 eV (90% C.L.) improves upon previous results [24,25] by almost a factor of 2 after a measuring period of only four weeks while operating at reduced column density. It is based on a purely kinematic method. As such, it has implications for both particle physics and cosmology. For the former, it narrows down the allowed range of quasidegenerate neutrino mass models by a direct method. For the latter, this model-independent limit can be used as laboratory-based input for studies of structure evolution in ΛCDM and other cosmological models.

Our result shows the potential of KATRIN to probe mv by a direct kinematic method. After 1000 days of data taking at nominal column density and further reductions of systematics and R_{bg}, we will reach a sensitivity of 0.2 eV (90% C.L.) on mv, augmented by searches for physics beyond the SM, such as for sterile neutrino admixtures with masses from the eV to the keV scale.

ACKNOWLEDGEMENTS

We acknowledge the support of the Helmholtz Association, Ministry for Education and Research BMBF (Grants No. 5A17PDA, No. 05A17PM3, No. 05A17PX3, No. 05A17VK2, and No. 05A17WO3), Helmholtz Alliance for Astroparticle Physics (HAP), Helmholtz Young Investigator Group (Grant No. VH-NG-1055), Max Planck Research Group (MaxPlanck@TUM), and Deutsche Forschungsgemeinschaft DFG (Research Training Groups Grants No. GRK 1694 and No. GRK 2149, and Graduate School Grant No. GSC 1085—KSETA) in Germany; Ministry of Education, Youth and Sport (Grants No. CANAM-LM2011019, and No. LTT19005) in the Czech Republic; and the United States Department of Energy through Grants No. DE-FG02-97ER41020, No. DE-FG02-94ER40818, No. DE-SC0004036, No. DE-FG02-97ER41033, No. DE-FG02-97ER41041, No. DE-AC02-05CH11231, No. DE-SC0011091, and No. DE-SC0019304, and the National Energy Research Scientific Computing Center.

REFERENCES

1. Y. Fukuda et al. (Super-Kamiokande Collaboration), Evidence for Oscillation of Atmospheric Neutrinos, Phys. Rev. Lett. 81, 1562 (1998).
2. Q.R. Ahmad et al. (SNO Collaboration), Direct Evidence for Neutrino Flavor Transformation from Neutral-Current Interactions in the Sudbury Neutrino Observatory, Phys. Rev. Lett. 89, 011301 (2002).
3. I. Esteban, M.C. Gonzalez-Garcia, A. Hernandez-Cabezudo, M. Maltoni, and T. Schwetz, Global analysis of three-flavour neutrino oscillations: Synergies and tensions in the determination of θ23, δCP, and the mass ordering, J. High Energy Phys. 2019, 106.
4. S. Hannestad, Primordial neutrinos, Annu. Rev. Nucl. Part. Sci. 56, 137 (2006).
5. J. Lesgourgues and S. Pastor, Neutrino cosmology and Planck, New J. Phys. 16, 065002 (2014).
6. A. Loureiro et al., On The Upper Bound of Neutrino Masses from Combined Cosmological Observations and Particle Physics Experiments, Phys. Rev. Lett. 123, 081301 (2019).
7. S.F. King, A. Merle, S. Morisi, Y. Shimizu, and M. Tanimoto, Neutrino mass and mixing: From theory to experiment, New J. Phys. 16, 045018 (2014).
8. N. Aghanim et al. (Planck Collaboration), Planck 2018 Results. VI. Cosmological Parameters, Planck 2018 results. VI. Cosmological parameters arXiv:1807.06209v2.
9. A.S. Barabash, Main features of detectors and isotopes to investigate double beta decay with increased sensitivity, Int. J. Mod. Phys. A 33, 1843001 (2018).
10. M.J. Dolinski, A.W. Poon, and W. Rodejohann, Neutrinoless double-beta decay: Status and prospects, Annu. Rev. Nucl. Part. Sci. 69, 219 (2019).
11. E.W. Otten and C. Weinheimer, Neutrino mass limit from tritium β *decay*, Rep. Prog. Phys. 71, 086201 (2008).
12. G. Drexlin, V. Hannen, S. Mertens, and C. Weinheimer, Current direct neutrino mass experiments, Adv. High Energy Phys. 2013, 293986 (2013).
13. L. Gastaldo et al., The electron capture in 163Ho *experiment* ECHo, Eur. Phys. J. Spec. Top. 226, 1623 (2017).

14. A. Nucciotti, The use of low temperature detectors for direct measurements of the mass of the electron neutrino, Adv. High Energy Phys. 2016, 9153024 (2016).
15. A. Ashtari Esfahani et al. (Project 8 Collaboration), Determining the neutrino mass with cyclotron radiation emission spectroscopy—Project 8, J. Phys. G 44, 054004 (2017).
16. M. Tanabashi et al. (Particle Data Group), Review of particle physics, Phys. Rev. D 98, 030001 (2018).
17. A. Osipowicz et al. (KATRIN Collaboration), KATRIN: A Next Generation Tritium Beta Decay Experiment with Sub-EV Sensitivity for the Electron Neutrino Mass. Letter of Intent, KATRIN: A Next generation tritium beta decay experiment with sub-eV sensitivity for the electron neutrino mass. Letter of intent arXiv:hep-ex/0109033.
18. J. Angrik et al. (KATRIN Collaboration), FZKA Scientific Report No. 7090, 2005.
19. M. Arenz et al. (KATRIN Collaboration), First transmission of electrons and ions through the KATRIN beamline, J. Instrum. 13, P04020 (2018).
20. M. Kleesiek et al., β-decay spectrum, response function and statistical model for neutrino mass measurements with the KATRIN experiment, Eur. Phys. J. C 79, 204 (2019).
21. R.G.H. Robertson, T.J. Bowles, G.J. Stephenson, D.L. Wark, J.F. Wilkerson, and D.A. Knapp, Limit on ¯ve *Mass from Observation of the* β *Decay of Molecular Tritium*, Phys. Rev. Lett. 67, 957 (1991).
22. V.M. Lobashev and P.E. Spivak, A method for measuring the anti-electron-neutrino rest mass, Nucl. Instrum. Methods Phys. Res., Sect. A 240, 305 (1985).
23. A. Picard, H. Backe, H. Barth, J. Bonn, B. Degen, T. Edling, R. Haid, A. Hermanni, P. Leiderer, T. Loeken, A. Molz, R.B. Moore, A. Osipowicz, E.W. Otten, M. Przyrembel, M. Schrader, M. Steininger, and C. Weinheimer, A solenoid retarding spectrometer with high resolution and transmission for keV electrons, Nucl. Instrum. Methods Phys. Res., Sect. B 63, 345 (1992).
24. C. Kraus, B. Bornschein, L. Bornschein, J. Bonn, B. Flatt, A. Kovalik, B. Ostrick, E.W. Otten, J.P. Schall, T. Thümmler, and C. Weinheimer, Final results from phase II of the Mainz neutrino mass search in tritium β *decay*, Eur. Phys. J. C 40, 447 (2005).
25. V.N. Aseev, A.I. Belesev, A.I. Berlev, E.V. Geraskin, A.A. Golubev,

N. A. Likhovid, V. M. Lobashev, A. A. Nozik, V. S. Pantuev, V. I. Parfenov, A. K. Skasyrskaya, F. V. Tkachov, and S. V. Zadorozhny, Upper limit on the electron antineutrino mass from the Troitsk experiment, Phys. Rev. D 84, 112003 (2011).

26. F. Priester, M. Sturm, and B. Bornschein, Commissioning and detailed results of KATRIN inner loop tritium processing system at Tritium Laboratory Karlsruhe, Vacuum 116, 42 (2015).
27. M. Arenz et al. (KATRIN Collaboration), The KATRIN superconducting magnets: Overview and first performance results, J. Instrum. 13, T08005 (2018).
28. F.M. Fränkle et al., Penning discharge in the KATRIN pre-spectrometer, J. Instrum. 9, P07028 (2014).
29. M. Beck et al., Effect of a sweeping conductive wire on electrons stored in the Penning trap between the KATRIN spectrometers, Eur. Phys. J. A 44, 499 (2010).
30. M. Arenz et al. (KATRIN Collaboration), Commissioning of the vacuum system of the KATRIN Main Spectrometer, J. Instrum. 11, P04011 (2016).
31. M. Erhard et al., Technical design and commissioning of the KATRIN large-volume air coil system, J. Instrum. 13, P02003 (2018).
32. J.F. Amsbaugh et al., Focal-plane detector system for the KATRIN experiment, Nucl. Instrum. Methods Phys. Res., Sect. A 778, 40 (2015).
33. M. Arenz et al., Calibration of high voltages at the ppm level by the difference of 83 mKr *conversion electron lines at the KATRIN experiment*, Eur. Phys. J. C 78, 368 (2018).
34. S. Grohmann, T. Bode, M. Hötzel, H. Schön, M. Süsser, and T. Wahl, The thermal behaviour of the tritium source in KATRIN, Cryogenics 55–56, 5 (2013).
35. M. Aker et al. (KATRIN Collaboration), First Operation of the KATRIN Experiment with Tritium, First operation of the KATRIN experiment with tritium arXiv:1909.06069v1.
36. J.W. Liu, Total cross sections for high-energy electron scattering by h2 *(*1Σ+g), n2 *(*1Σ+g), and o2 *(*3Σ−g), Phys. Rev. A 35, 591 (1987).
37. J. Behrens et al., A pulsed, mono-energetic and angular-selective UV photo-electron source for the commissioning of the KATRIN experiment, Eur. Phys. J. C 77, 410 (2017).
38. J. Bonn, L. Bornschein, B. Degen, E. Otten, and C. Weinheimer, A

high resolution electrostatic time-of-flight spectrometer with adiabatic magnetic collimation, Nucl. Instrum. Methods Phys. Res., Sect. A 421, 256 (1999).

39. F.M. Fraenkle (KATRIN collaboration), Background processes in the KATRIN main spectrometer, J. Phys. 888, 012070 (2017).
40. K. Altenmüller et al. (KATRIN Collaboration), Muon-induced background in the KATRIN main spectrometer, Astropart. Phys. 108, 40 (2019).
41. K. Altenmüller et al. (KATRIN Collaboration), Gamma-induced background in the KATRIN main spectrometer, Eur. Phys. J. C 79, 807 (2019).
42. N. Trost, Modeling and Measurement of Rydberg-State Mediated Background at the KATRIN Main Spectrometer, Karlsruhe Institute of Technology, 2018.
43. F.M. Fränkle, L. Bornschein, G. Drexlin, F. Glück, S. Görhardt, W. Käfer, S. Mertens, N. Wandkowsky, and J. Wolf, Radon induced background processes in the KATRIN pre-spectrometer, Astropart. Phys. 35, 128 (2011).
44. S. Görhardt et al., Impact of a cryogenic baffle system on the suppression of radon-induced background in the KATRIN Pre-Spectrometer, J. Instrum. 13, T10004 (2018).
45. M. Schlösser, H. Seitz, S. Rupp, P. Herwig, C.G. Alecu, M. Sturm, and B. Bornschein, In-line calibration of raman systems for analysis of gas mixtures of hydrogen isotopologues with sub-percent accuracy, Anal. Chem. 85, 2739 (2013).
46. A.F. Nastoyashchii, N.A. Titov, I.N. Morozov, F. Glück, and E.W. Otten, Effects of plasma phenomena on neutrino mass measurements process using a gaseous tritium β-source, Fusion Sci. Technol. 48, 743 (2005).
47. D. Vénos, J. Sentkerestiová, O. Dragoun, M. Slezák, M. Ryšavý, and A. špalek, Properties of 83 mKr *conversion electrons and their use in the KATRIN experiment*, J. Instrum. 13, T02012 (2018).
48. T. Thummler, R. Marx, and C. Weinheimer, Precision high voltage divider for the KATRIN experiment, New J. Phys. 11, 103007 (2009).
49. M. Babutzka et al., Monitoring of the operating parameters of the KATRIN Windowless Gaseous Tritium Source, New J. Phys. 14, 103046 (2012).

50. A. Saenz, S. Jonsell, and P. Froelich, Improved Molecular Final-State Distribution of HeT+ *for the* β-Decay Process of T2 , Phys. Rev. Lett. 84, 242 (2000).
51. N. Doss, J. Tennyson, A. Saenz, and S. Jonsell, Molecular effects in investigations of tritium molecule β *decay endpoint experiments*, Phys. Rev. C 73, 025502 (2006).
52. L. Kuckert, F. Heizmann, G. Drexlin, F. Glück, M. Hötzel, M. Kleesiek, F. Sharipov, and K. Valerius, Modelling of gas dynamical properties of the KATRIN tritium source and implications for the neutrino mass measurement, Vacuum 158, 195 (2018).
53. D. Furse et al., Kassiopeia: A modern, extensible C++ *particle tracking package*, New J. Phys. 19, 053012 (2017).
54. R. J. Barlow, Statistics: A Guide to the Use of Statistical Methods in the Physical Sciences (Wiley, Chichester, 1989).
55. G. D'Agostini, On the use of the covariance matrix to fit correlated data, Nucl. Instrum. Methods Phys. Res., Sect. A 346, 306 (1994).
56. G. Cowan, K. Cranmer, E. Gross, and O. Vitells, Asymptotic formulae for likelihood-based tests of new physics, Eur. Phys. J. C 71, 1554 (2011); Erratum, 73, 2501(E) (2013).
57. R. D. Cousins and V. L. Highland, Incorporating systematic uncertainties into an upper limit, Nucl. Instrum. Methods Phys. Res., Sect. A 320, 331 (1992).
58. P. Harris and M. G. Cox, On a Monte Carlo method for measurement uncertainty evaluation and its implementation, Metrologia 51, S176 (2014).
59. A. V. Lokhov and F. V. Tkachov, Confidence intervals with a priori parameter bounds, Phys. Part. Nucl. 46, 347 (2015).
60. G. J. Feldman and R. D. Cousins, A Unified approach to the classical statistical analysis of small signals, Phys. Rev. D 57, 3873 (1998).
61. E. G. Myers, A. Wagner, H. Kracke, and B. A. Wesson, Atomic Masses of Tritium and Helium-3, Phys. Rev. Lett. 114, 013003 (2015).

CHAPTER 7

Distinguishing Dirac and Majorana Neutrinos by their Decays Via Nambu-Goldstone bosons in the Gravitational-Anomaly model of Neutrino Masses

Lena Funcke[1], Georg Raffelt[2] and Edoardo Vitagliano[2]

[1]Perimeter Institute for Theoretical Physics, 31 Caroline Street North, Waterloo, Ontario N2L 2Y5, Canada

[2]Max-Planck-Institut für Physik (Werner-Heisenberg-Institut), Föhringer Ring 6, 80805 München, Germany

INTRODUCTION

A completely new approach to explain small Dirac or Majorana neutrino masses [1] relies on new physics at the low-energy frontier of particle physics

Citation: Lena Funcke, Georg Raffelt, and Edoardo Vitagliano. "Distinguishing Dirac and Majorana neutrinos by their decays via Nambu-Goldstone bosons in the gravitational-anomaly model of neutrino masses" Phys. Rev. D 101, 015025 (2020). https://doi.org/10.1103/PhysRevD.101.015025

instead of high-energy extensions of the Standard Model. The key idea is that a hypothetical topological vacuum susceptibility of gravity induces fermion condensation [2,3], which can give rise to effective fermion masses. If this effect is realized in nature, it is most important for the lightest fermions and could be the unique origin of neutrino masses.

Phenomenologically, neutrino condensation would be accompanied by the appearance of pseudoscalar Nambu-Goldstone bosons ϕ, similar to light mesons in QCD, that can be interpreted as neutrino-antineutrino bound states [1,3] and that we call gravi-Majorons.[1] An important difference to conventional Nambu-Goldstone bosons is that the ννϕ vertex, due to its low-energy origin, "melts" for high-energy off-shell situations so that constraints based on scattering processes, e.g., Majoron production in supernovae, typically do not apply [1,5]. On the other hand, decays of the type $\nu_i \to \overset{(-)}{\nu}_j + \phi$ proceed in the usual way.

While this scenario is exotic, it may become empirically motivated in the near future. One predicts a neutrinoless Universe after photon decoupling, or at least all neutrinos in the lowest mass state, due to decays and annihilations into gravi-Majorons [1]. If near-future cosmological observations [6] fail to detect a hot dark matter component on the minimal level expected from oscillation experiments, we may be forced to contemplate the absence of the usual cosmic neutrino background. Similar questions arise if the KATRIN experiment [7,8] detects a neutrino mass in conflict with cosmological limits.

The gravitational mass model works for both Dirac and Majorana neutrinos. Therefore, one important question remains how we can experimentally distinguish between these two possibilities. In high-energy models, neutrinoless double-beta (0νββ) decay [9] is the most promising approach and one that remains viable in our scenario. Moreover, in the current paper we predict that the low-energy gravitational mass model offers an additional opportunity through fast $\nu_i \to \overset{(-)}{\nu}_j + \phi$ decays.

It is not new that details, e.g., of radiative decays νi→ $\overset{(-)}{\nu}_j + \gamma$ depend on the Dirac vs Majorana nature [10,11]. The γ spectrum in relativistic decays depends on this property, inherited from the angular γ distribution relative to the spin of the mother neutrino in its rest frame. However, radiative decays of light neutrinos are usually too slow to be of any practical interest.

We argue that the $\nu_i \to \overset{(-)}{\nu}_j + \phi$ decays in the gravitational mass model are in principle fast enough to distinguish between the Majorana and Dirac cases by using the flux and spectrum of the daughter neutrinos. A Dirac neutrino νi decays into a neutrino, either active ν_j or sterile N_j, whereas a Majorana neutrino always decays into an active state, which however in a detector appears as either a neutrino νj or an antineutrino $\bar{\nu}_j$. This is possible because a Majorana neutrino does not have a defined lepton number, and what we call antineutrino is simply a state with right-handed helicity.

A detector that can distinguish neutrinos from antineutrinos could identify the Dirac or Majorana nature by looking at neutrino vs antineutrino appearance, assuming an asymmetry at the source. Moreover, given the source spectrum, the energy spectrum depends on the nature of the mass term independently of an asymmetry at the source. If the mass spectrum is degenerate, Majorana and Dirac particles could be distinguished because spin-flip is not suppressed compared to spin conservation. As a result, Majorana neutrinos would decay to antineutrinos, whereas Dirac neutrinos would decay to sterile states. While these methods are not yet experimentally feasible, with the current data we can put strong bounds on the Majorana case using null results from experiments searching for antineutrino appearance from the Sun.

The paper is structured as follows. In Sec. II we explain the key phenomenological differences between conventional Majoron-like models and the gravi-Majoron scenario. In Sec. III we compute the neutrino decay rate and in Sec. IV we comment on their impact on solar, IceCube, and supernova neutrinos. We summarize our findings in Sec. V.

DIFFERENCES TO TRADITIONAL MAJORON-LIKE MODELS

A. Dirac or Majorana Neutrino Nature

In contrast to the original Majoron model [4], the gravitational neutrino mass mechanism works for both Dirac and Majorana neutrinos. Therefore we can in principle distinguish three possible scenarios:

(1) *Pure left-handed Majorana case:* If neutrinos are identical to their antiparticles and no right-handed states exist, neutrinos are massless in the Standard Model. The neutrino condensate then generates small left-handed Majorana masses.

(2) *Pure Dirac case:* If neutrinos are distinct from their antiparticles, right-handed states exist and the condensate generates small effective Dirac masses.

(3) *Mixed case:* If neutrinos are identical to their antiparticles and right-

handed states exist, the condensate generates Dirac masses as well as left- and right-handed Majorana masses. In this case, the active and sterile Dirac states have masses mν~meV−eV and are substantially mixed.

For simplicity, our computations in Sec. III will focus on the minimal cases (1) and (2). In the former case, the effective left-handed Majorana mass terms violate isospin by one unit but are allowed after electroweak symmetry breaking. Here, we note that only the condensate but not gravity is assumed to violate isospin. In the latter case, the Yukawa couplings to the SM Higgs doublet are assumed to vanish based on chiral symmetry protection, so that the gravitational mechanism is the only origin of the observed neutrino masses. Note, however, that even in the presence of nonzero Yukawa couplings or other hard neutrino mass origins (e.g., the various seesaw mechanisms [12]), neutrino condensation would still take place as long as $m_{\nu,hard}<\Lambda_G$. Here, ΛG is the scale of the topological vacuum susceptibility of gravity, which is constrained by model-independent phenomenological requirements to lie within the range of Λ_G~meV−eV [1,5,13]. Since the susceptibility induces fermion condensation, Λ_G also sets the neutrino condensation scale, $\langle\bar{\nu}_i\nu_j\rangle \sim \Lambda_G^3$, and thus the order of the absolute neutrino mass scale (see Ref. [1] for a discussion of the hierarchy).

Concerning the gravi-Majorons, the pure Dirac case (2) gives rise to 14+1 (pseudo) Nambu-Goldstone bosons ϕ≡{ϕk,ην} from spontaneously breaking the neutrino flavor symmetry $U(3)_V\times U(3)_A\rightarrow U(1)^3_V$. Here, the ην boson comes from the anomalous U(1)A part [3] and the 14 ϕ_k bosons come from the $SU(3)_V\times SU(3)_A$ part [1]. The ην gets a mass through the chiral gravitational anomaly, $m_{\eta\nu}\sim\Lambda_G\sim T_G$, while the ϕ_k are massless if neutrinos have only effective (gravitational) masses and no hard masses.[2] The analogous left-handed Majorana case (1) would yield fewer bosons due to breaking a smaller initial flavor symmetry.

The off-diagonal couplings between gravi-Majorons and neutrinos are responsible for neutrino decay. In typical Majoron models such couplings are heavily suppressed by (mν,Dirac/mν,Majorana)4≪1 because the Majorons only couple to the heavy right-handed neutrino states [14]. In our case, there is no such suppression and we can treat the off-diagonal couplings as free parameters, which are proportional to the unknown entries of the neutrino-mass diagonalization matrix.

To briefly comment on the mixed case (3), this active-sterile neutrino scenario could be motivated by short-baseline anomalies [15]. While light sterile neutrinos generally conflict with cosmological constraints on neutrino masses and the effective number of neutrino species [16], these tensions disappear in the gravitational mass model because (i) the mass bounds are weakened [1,13] and (ii) the active relic neutrino background is massless before photon decoupling and therefore uncoupled to the sterile states (see Appendix A for more details).

We note that light sterile neutrinos would strongly distort the parameter space for 0νββ decay [17] and could even make the decay vanish completely [18]. In contrast, gravi-Majorons do not affect the 0νββ decay because their emission is strongly suppressed (see Sec. II C). Finally, there was a debate[3] whether the 0νββ decay rate might be altered because the gravitational neutrino masses are not hard masses generated at high-energy scales but effective ones generated at the low-energy condensation scale ΛG. However, since there is no momentum flow through the masses generated by the condensate, we expect them to be indistinguishable from hard neutrino masses in the 0νββ process.

B. Late Neutrinoless Universe

Interactions between neutrinos and Nambu-Goldstone bosons are strongly constrained by cosmological data. For example, almost the entire parameter space of the "neutrinoless Universe" model [19], which evades the cosmological bounds $\sum_i m_{\nu i} \lesssim 0.2$ eV [20], was ruled out by early-Universe neutrino free-streaming constraints [21–25] and precision measurements of the primordial radiation density [26]. In contrast, gravi-Majorons are not ruled out because they only arise in the late Universe after photon decoupling. We will discuss the details of this high-temperature suppression of the new gravitational effects in Appendix A.

When the Universe cools down to the low phase-transition temperature $T_G \lesssim T_{decoupling} \sim 0.3$ eV, the neutrino condensate forms and the emerging gravi-Majorons $\phi \equiv \{\phi k, \eta\nu\}$ [1,3] start to interact with neutrinos through the same nonperturbative gravitational vertex that is responsible for the neutrino masses,

$$\mathcal{L}_{\text{int}} = \sum_{k=1}^{8} \partial_\mu \phi_k \sum_{i,j=1}^{3} \frac{g_{ij,k}}{m_i + m_j} \bar{\nu}_i \gamma^\mu \gamma_5 \nu_j$$
$$+ \sum_{k=9}^{14} \partial_\mu \phi_k \sum_{i \neq j=1}^{3} \frac{g_{ij,k}}{m_i - m_j} \bar{\nu}_i \gamma^\mu \nu_j$$
$$+ i\eta_\nu \sum_{i,j=1}^{3} g_{ij,15} \bar{\nu}_i \gamma_5 \nu_j + \text{H.c.} \quad (1)$$

Here, the dimensionless couplings $g_{ij,k}$ are normalized in such a way that we recover the pseudoscalar-coupling Lagrangian in Eq. (2). The couplings correspond to the generators of the broken flavor symmetry, i.e., the Gell-Mann matrices plus the identity. The U(3)A generators yield diagonal and off-diagonal axial vector-like couplings proportional to m_i+m_j (affecting 8+1 of the 14+1 bosons), while the U(3)V generators yield off-diagonal vector-like couplings proportional to mi−mj (affecting the other 6 bosons), similar to familon models (see, e.g., Ref. [27]). Note that these couplings do not yield long-range forces in macroscopic systems, since the pseudoscalar coupling is spin-dependent and the scalar coupling is flavor-changing. In the following, we neglect the sum over k and conservatively assume that the couplings are to a single Nambu Goldstone boson.

After the cosmological phase transition, the previously massless neutrinos become massive, quickly decay into the lowest mass eigenstate, and annihilate into gravi-Majorons through the process $\nu + \bar{\nu} \rightarrow \phi + \phi$ [1]. Thus, the early-Universe massless neutrino "radiation" converts into massless gravi-Majoron radiation after photon decoupling, since all massive ϕ bosons decay into massless ones. This almost complete annihilation could only be evaded in the hypothetical presence of neutrino asymmetries in the Dirac case, weakening the cosmological neutrino mass bounds to $\sum m_\nu$<4.8 eV at 95% CL [13].

To conclude, unless there are large primordial neutrino asymmetries, the gravitational neutrino mass model [1] predicts a late neutrinoless Universe and could be falsified by a cosmological neutrino mass detection, e.g., by the upcoming DESI or Euclid surveys [6]. The KATRIN beta-decay experiment [7,8] could provide a hint toward our predicted neutrinoless Universe if it detects an unexpectedly large neutrino mass scale. We stress that KATRIN's measurement of the electron energy spectrum would be unaffected by the gravi-Majorons (unlike the bosons considered in Ref. [28]) because their interactions take place on much longer timescales than beta decay. The neutrinoless Universe scenario would also make it impossible for the recently proposed PTOLEMY experiment [29] to detect the relic neutrino background. Only in the case of substantial neutrino asymmetries, PTOLEMY could detect the relic neutrinos with an enhanced (suppressed) detection rate for normal (inverted) neutrino mass ordering, since the lightest mass eigenstate contains a large (small) fraction of the electron-neutrino flavor eigenstate.

C. Stellar and Laboratory Bounds

One might naively expect that gravi-Majorons are ruled out by astrophysical observations because their couplings are severely constrained by stellar processes. As explained in Ref. [5], the least suppressed two-photon coupling of the gravi-Majorons is $g_{\phi\gamma} \sim (\Lambda_G/me)^3/\Lambda_G \sim 10^{-10}GeV^{-1}$ for $\Lambda_G \sim 0.1$ eV, which at first sight is already ruled out by constraints from solar axion experiments, $g_{\phi\gamma} \lesssim 0.88\times10^{-10}$ GeV^{-1} at 95% CL for $m_\phi \lesssim 0.02$ eV [30]. However, the gravi-Majoron production in astrophysical environments with $E \gg m_\nu$ is additionally suppressed by $(\Lambda_G/E)n$ due to the high-energy softening of the gravitational vertex, where n is an unknown power-law exponent that has to be fixed by phenomenological requirements (as discussed in [1,5] and Appendix A). Thus, the common relation between axionlike predictions for laboratory and solar axion experiments does not apply [5]. By similar arguments one also evades other astrophysical bounds on the gravitational $\nu\bar{\nu}\gamma, \nu\bar{\nu}\phi$, $\nu\bar{\nu}\nu\bar{\nu}$, $e\bar{e}\phi$, and $e\bar{e}\nu\bar{\nu}$ couplings, as we comment on in Appendix A.

Regarding laboratory experiments, the most important bounds on the emission of conventional M ajoron-like bosons comes from $0\nu\beta\beta$ decay [31] and leptonic decays of mesons [32]. Following the argumentation above, these bounds do not apply to gravi-Majorons because they and their couplings to the virtual intermediate neutrinos dissolve in these high-energy processes. However, short-distance fifth-force experiments can put important bounds on our new gravitational interactions, as discussed in [1,5] and Appendix A.

We finally stress that the gravitational vertex suppression is due to the large four-momentum transfer in the processes under consideration, which is why the suppression does not apply to neutrino oscillations or to the neutrino decays considered in Sec. III.

NEUTRINO DECAY RATE

As predicted by the gravitational neutrino mass model [1], the presence of gravi-Majorons implies neutrino decay through the vertex of Eq. (1). The crucial new point of this paper is that the Dirac and Majorana neutrino cases yield different decay channels, so that the composition of the daughter neutrinos depends on the type of masses generated through the gravitational mechanism.

A. Pseudoscalar Decay Channels

For the two neutrino decay channels, $\nu_i \rightarrow \nu_j + \phi$ and $\nu_i \rightarrow \overset{(-)}{\nu}_j + \phi$ with $m_i > m_j$, the axial derivative coupling in Eq. (1) is equivalent to the pseudoscalar coupling[4] [36]

$$\mathcal{L}_{\text{int}} = i\phi \sum_{ij} g_{ij} \bar{\nu}_i \gamma_5 \nu_j + \text{H.c.} \tag{2}$$

as we have checked explicitly in Appendix B.

The differential rate Γ for the decay $\nu_i \rightarrow \bar{\nu}_j + \phi$ in a generic reference frame is

$$d\Gamma = \frac{1}{2E_i}(2\pi)^4 \delta^{(4)}(p_i - p_j - k)|\mathcal{M}|^2 \frac{d^3\mathbf{p}_j}{(2\pi)^3 2E_j} \frac{d^3\mathbf{k}}{(2\pi)^3 2\omega}. \tag{3}$$

where pi, p_j and k are the four-momenta of ν_i, ν_j and ϕ, respectively. The squared amplitudes for both the helicity-preserving and the helicity-changing processes are given in Appendix B. After integration over the gravi-Majoron momenta and the direction of the neutrino, the differential rate becomes

$$\frac{d\Gamma}{dE_j} = \frac{1}{16\pi E_i |\mathbf{p}_i|}|\mathcal{M}|^2. \tag{4}$$

In the laboratory frame, where Ei≫mi, we find for Majorana neutrinos

$$\frac{d\Gamma^{\text{M}}_{\nu_i \rightarrow \nu_j + \phi}}{dE_j} = \frac{g_{ij}^2}{4\pi E_i |\mathbf{p}_i|} m_i m_j (A - 2) \tag{5a}$$

$$\frac{d\Gamma^{\text{M}}_{\nu_i \rightarrow \bar{\nu}_j + \phi}}{dE_j} = \frac{g_{ij}^2}{4\pi E_i |\mathbf{p}_i|} m_i m_j \left(\frac{m_i^2 + m_j^2}{m_i m_j} - A\right) \tag{5b}$$

where

$$A = \left(\frac{m_i E_j}{m_j E_i} + \frac{m_j E_i}{m_i E_j}\right) \tag{6}$$

Dirac neutrinos decay either into active (left-handed) neutrinos or into sterile (right-handed) neutrinos N_j. The differential decay rates equal 1/4 times the rates for Majorana neutrinos,

$$\frac{d\Gamma^{D}_{\nu_i \to \nu_j + \phi}}{dE_j} = \frac{1}{4}\frac{d\Gamma^{M}_{\nu_i \to \nu_j + \phi}}{dE_j} \quad (7a)$$

$$\frac{d\Gamma^{D}_{\nu_i \to N_j + \phi}}{dE_j} = \frac{1}{4}\frac{d\Gamma^{M}_{\nu_i \to \bar{\nu}_j + \phi}}{dE_j} \quad (7b)$$

The differential rates must be integrated over the allowed energy range for the daughter neutrino νj,

$$\frac{E_i}{2}\left(1+\frac{1}{x_{ij}^2}\right) - \frac{|\mathbf{p}_i|}{2}\left(1-\frac{1}{x_{ij}^2}\right) \le E_j \le \frac{E_i}{2}\left(1+\frac{1}{x_{ij}^2}\right) + \frac{|\mathbf{p}_i|}{2}\left(1-\frac{1}{x_{ij}^2}\right) \quad (8)$$

where $x_{ij} \equiv m_i/m_j > 1$. In the lab frame with Ei≫mi, Eq. (8) reduces to

$$\frac{E_i}{x_{ij}^2} \le E_j \le E_i. \quad (9)$$

The decay rates for Majorana neutrinos are then

$$\Gamma^{M}_{\nu_i \to \nu_j + \phi} = \frac{g_{ij}^2 m_i m_j}{4\pi E_i}\left(\frac{x_{ij}}{2} - 2 + \frac{2}{x_{ij}}\log x_{ij} + \frac{2}{x_{ij}^2} - \frac{1}{2x_{ij}^3}\right) \quad (10a)$$

$$\Gamma^{M}_{\nu_i \to \bar{\nu}_j + \phi} = \frac{g_{ij}^2 m_i m_j}{4\pi E_i}\left(\frac{x_{ij}}{2} - \frac{2}{x_{ij}}\log x_{ij} - \frac{1}{2x_{ij}^3}\right) \quad (10b)$$

Our results agree with the literature [33–35] if one accounts for the different normalization of the couplings.

There are two important limits for these expressions: the case of a large mass hierarchy $m_i \gg m_j$ and the degenerate case $m_i \simeq m_j$. The latter case remains allowed in the gravitational mass model because the cosmological bounds on neutrino masses are substantially weakened (Sec. II B). In the limit of a large mass hierarchy, one finds

$$\Gamma^{M}_{\nu_i \to \nu_j + \phi} = \Gamma^{M}_{\nu_i \to \bar{\nu}_j + \phi} = \frac{\Gamma^{M}_{\text{tot}}}{2} \quad (11)$$

where the total decay rate is

$$\Gamma^{M}_{\text{tot}} = \frac{g_{ij}^2 m_i}{4\pi}\frac{m_i}{E_i} \quad (12)$$

Here, the first factor is the total decay rate in the rest frame of ν_i and m_i/E_i is the Lorentz factor.

The total decay rate in Eq. (12) yields the neutrino rest-frame lifetimes $\tau_i=1/\Gamma_i$ of

$$\frac{\tau_3}{m_3} \simeq \frac{4\times10^{-11}}{g_{32}^2+g_{31}^2}\frac{\text{s}}{\text{eV}} \quad (13a)$$

$$\frac{\tau_2}{m_2} \simeq \frac{1\times10^{-9}}{g_{21}^2}\frac{\text{s}}{\text{eV}} \quad (13b)$$

For example, the smallest possible, normal-ordered neutrino masses of m_1=0 meV, m_2=9 meV, and m_3=50 meV [37] give

$$\tau_3 \simeq \frac{2\times10^{-12}\ \text{s}}{g_{32}^2+g_{31}^2} \quad (14a)$$

$$\tau_2 \simeq \frac{9\times10^{-12}\ \text{s}}{g_{21}^2} \quad (14b)$$

The degenerate limit, $m_i \simeq m_j$, gives

$$\Gamma^{M}_{\nu_i\to\bar{\nu}_j+\phi} = 2\Gamma^{M}_{\nu_i\to\nu_j+\phi} = \frac{g_{ij}^2(m_i-m_j)^3}{3\pi m_i^2}\frac{m_i}{E_i} \quad (15a)$$

B. Scalar Decay Channels

The scalar decay channels arising from the derivative vectorlike couplings in Eq. (1) are studied in analogous fashion. In this case, the derivative vectorlike coupling can be computed using a scalar coupling,

$$\mathcal{L}_{\text{int}} = i\phi\sum_{i\neq j} g_{ij}\bar{\nu}_i\nu_j + \text{H.c.}, \quad (16)$$

analogous to the pseudoscalar coupling in Eq. (2).

The decay rates for Majorana neutrinos are

$$\Gamma^{M}_{\nu_i\to\nu_j+\phi} = \frac{g_{ij}^2 m_i m_j}{4\pi E_i}\left(\frac{x_{ij}}{2}+2+\frac{2}{x_{ij}}\log x_{ij}-\frac{2}{x_{ij}^2}-\frac{1}{2x_{ij}^3}\right) \quad (17a)$$

$$\Gamma^{M}_{\nu_i\to\bar{\nu}_j+\phi} = \frac{g_{ij}^2 m_i m_j}{4\pi E_i}\left(\frac{x_{ij}}{2}-\frac{2}{x_{ij}}\log x_{ij}-\frac{1}{2x_{ij}^3}\right) \quad (17b)$$

Similar to the pseudoscalar case, the differential decay rates for Dirac neutrinos equal 1/4 times the rates for Majorana neutrinos. In the limit of a large mass hierarchy, one finds again

$$\Gamma^{M}_{\nu_i\to\nu_j+\phi} = \Gamma^{M}_{\nu_i\to\bar{\nu}_j+\phi} = \frac{\Gamma^{M}_{\text{tot}}}{2} \quad (18)$$

where the total decay rate is

$$\Gamma^{\rm M}_{\rm tot} = \frac{g^2_{ij} m_i}{4\pi} \frac{m_i}{E_i} \quad (19)$$

this is identical to the decay rate found for the pseudoscalar decay channels. The degenerate limit, mi≃mj, gives instead

$$\Gamma^{\rm M(D)}(\nu_i \to \nu_j + \phi) = \frac{g^2}{\zeta^{\rm M(D)}\pi} \frac{(m_i^2 - m_j^2)}{E_i} \quad (20)$$

where $\zeta^{M}=1$ and $\zeta^{D}=4$, while the spin-flip process is strongly suppressed by a factor of $[(m_i^2 - m_j^2)/m_i^2]^3$ [38]. We observe that the scalar decay rate of Majorana (M) and Dirac (D) neutrinos in Eq. (2.6) of Ref. [38] needs to be corrected. Our result corrects for an inconsistency between the coupling definition in their Lagrangian and their rate (if neutrinos are Majorana particles), as well as an incorrect expansion of their Eq. (2.2).[5]

In the degenerate mass scenario, the main difference between the scalar and pseudoscalar decay channels is that helicity-flipping processes are strongly suppressed in the former case but not in the latter—see Eq. (15). Since spin-flip processes are unsuppressed in the pseudoscalar case and their rate is different from the spin-conserving processes, we can identify several different observables to distinguish Majorana and Dirac neutrinos, as we will discuss in Sec. IV. Given that the vectorlike coupling is either equivalent to the axial vectorlike coupling (in the hierarchical scenario) or cannot distinguish Majorana from Dirac neutrinos (in the degenerate scenario), we will focus for the rest of the paper on the axial vectorlike couplings, i.e., the pseudoscalar decay channels.

To show how the cases for Majorana and Dirac neutrinos are different, in Fig. 1 we plot the energy distributions $F(E_j/E_i)$ of the daughter neutrinos produced in the pseudoscalar decay of the parent neutrinos ν_i, which is proportional to the differential rate and normalized to the total decay rate. To describe the figure, let us begin with the decay of Majorana neutrinos with energy Ei. In the upper panel of Fig. 1, we show the hierarchical case, $m_i \gg m_j$. The spectrum of neutrinos and antineutrinos is a box given by the sum of two triangles, one corresponding to outgoing νj and one to outgoing $\bar{\nu}_j$. In the degenerate case $m_i \simeq m_j$ shown in the lower panel of Fig. 1, the two triangles are distorted, but the sum of the ν_j and $\bar{\nu}_j$ distributions is still a box, as the sum of Eqs. (5a) and (5b) is independent of energy. In the Dirac case, the spectrum of active daughter neutrinos is only the orange triangle because the complement to the box, the blue triangle, corresponds to undetectable sterile daughter neutrinos.

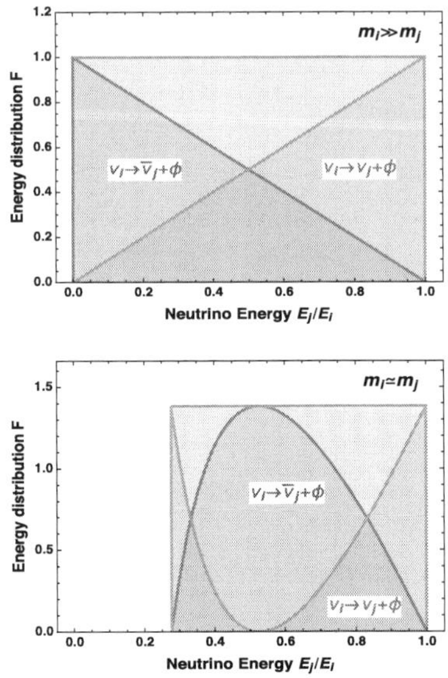

Figure 1. Normalized energy distributions $F(E_j/E_i)$ of the daughters ν_j and $\bar{\nu}_j$ produced in the pseudoscalar decay of Majorana neutrinos $\nu_i \to \nu_j + \phi$ (orange), $\nu_i \to \bar{\nu}_j + \phi$ (blue), and their sum (green), for hierarchical (upper panel) and degenerate (lower panel) neutrino masses. For the Dirac case, the blue curve instead refers to the decay to undetectable sterile states N_j.

CONSTRAINTS AND DETECTION OPPORTUNITIES

One of the possibilities to distinguish Majorana from Dirac neutrinos is given by the appearance or disappearance of neutrinos or antineutrinos from a source whose flux is asymmetric between ν and $\overline{\nu}$. For example, the Sun emits only neutrinos [39], besides a tiny contribution of antineutrinos coming from heavy-element decay, which is much smaller than the geoneutrino background [40–42]. The detection of antineutrinos could point toward the decay of Majorana neutrinos. Notice, however, that other explanations for such a detection would be possible, e.g., a spin precession due to a neutrino magnetic dipole moment [43]. New liquid scintillator detectors will significantly increase the current best experimental limits [44].

A second possibility relies on the observation of the differential energy spectrum from a source whose flux is known. In this case, there is no need for an asymmetry between the initial neutrino and antineutrino fluxes because the differential energy spectrum is differently affected by the decay depending on the nature of the neutrino mass.Finally, the observation of a change in the flavor pattern in the neutrino flux from a known source could in principle distinguish Majorana and Dirac neutrinos if it is possible to distinguish neutrinos from antineutrinos in the detector. Otherwise, it is only possible to observe the decay, as the latter changes the neutrino composition in terms of mass eigenstates of the flux.

In the following, we will quantitatively examine the first possibility, which yields bounds on the Majorana case of our neutrino decay scenario. Moreover, we will qualitatively discuss the other two proposed methods, which turn out to be not experimentally feasible yet.

A. Solar, Atmospheric, and Long-Baseline Neutrinos

The current constraints on neutrino decay for a normal nondegenerate mass ordering are

$$\frac{\tau_3}{m_3} > 2 \times 10^{-10}\ \frac{\text{s}}{\text{eV}} \tag{21a}$$

$$\frac{\tau_2}{m_2} > 1.0 \times 10^{-3}\ \frac{\text{s}}{\text{eV}} \tag{21b}$$

The bound on τ_3 at 90% CL is obtained through an analysis of long-baseline neutrinos [45,46] (comparable constraints can be put on invisible decays into sterile neutrinos, using atmospheric and long-baseline neutrino data [47]). The bound on τ_2 at 99% CL has been recently obtained with updated solar neutrino measurements [48], which have improved the previous estimate [49]. It is important to notice that these bounds only apply to invisible neutrino decays, i.e., the decay products are assumed not to cause significant signals in the detectors [47,49]. Moreover, we stress again that these limits on the lifetimes assume a hierarchical ordering [38,50]. The constraints in Eqs. (21a) and (21b) require the off-diagonal couplings to obey

$$\sqrt{g_{32}^2 + g_{31}^2} < 9 \times 10^{-2} \quad \text{and} \quad g_{21} < 3 \times 10^{-4}. \tag{22}$$

These bounds on the couplings are valid both for Majorana neutrinos and, once multiplied by a factor of 2, also for Dirac neutrinos. In the Majorana case, additional strong bounds can be obtained because solar neutrinos would decay to antineutrinos showing up at experiments like SNO [51] and KamLAND [52]. Also Borexino [53] can put competitive bounds on the lifetime of solar neutrinos. However, the analysis in [53] only yields bounds on the conversion of ν to $\overline{\nu}$ assuming either equal shapes for the solar ν and the converted $\overline{\nu}$ or a unknown shape for the $\overline{\nu}$ spectrum, while no dedicated analysis is carried out for neutrino decay.

The analysis of KamLAND data gives bounds on the lifetime of ν_2 Majorana neutrinos of [52]

$$\frac{\tau_2}{m_2} > 6.7 \times 10^{-2}\ \frac{\mathrm{s}}{\mathrm{eV}} \quad \text{(degenerate)} \tag{23a}$$

$$\frac{\tau_2}{m_2} > 1.1 \times 10^{-3}\ \frac{\mathrm{s}}{\mathrm{eV}} \quad \text{(hierarchical)} \tag{23b}$$

at 90% CL, which translate to bounds on the coupling

$$g_{21} < 4 \times 10^{-5} \quad \text{(degenerate)} \tag{24a}$$

$$g_{21} < 3 \times 10^{-4} \quad \text{(hierarchical).} \tag{24b}$$

Concerning ν_3 decay, there is as yet no dedicated analysis in the context of Majoron models [37], presumably because it was not known until recently that there is a small component of ν_3 in electron neutrinos. To estimate the flux of v3 from the Sun, we observe that this mass eigenstate is not affected by matter effects, so its component is simply $|U_{e3}|^2=0.02$ [54]. In conclusion, approximately 2% of the neutrinos coming from the Sun are ν_3. The bounds on the lifetime of ν_3 Majorana neutrinos are then

$$\frac{\tau_3}{m_3} > 1.3 \times 10^{-4}\ \frac{\mathrm{s}}{\mathrm{eV}} \quad \text{(degenerate)} \tag{25a}$$

$$\frac{\tau_3}{m_3} > 2.2 \times 10^{-5}\ \frac{\mathrm{s}}{\mathrm{eV}} \quad \text{(hierarchical)} \tag{25b}$$

at 90% CL, and the bounds on the coupling are finally

$$\sqrt{g_{32}^2 + g_{31}^2} < 3 \times 10^{-4} \quad \text{(degenerate)} \tag{26}$$

$$\sqrt{g_{32}^2 + g_{31}^2} < 2 \times 10^{-3} \quad \text{(hierarchical)} \tag{27}$$

These bounds apply to any model where neutrinos can decay to a light pseudoscalar.

B. Icecube and Supernova Neutrinos

Neutrino decays imply a distinct flavor composition of long-traveling astrophysical neutrinos because all neutrinos arrive in the lightest mass state. Observable decay effects require $\Gamma_i(m_i/E) \gtrsim D^{-1}$ and therefore [36]

$$g_{ij} \gtrsim 5 \times 10^{-8} \left(\frac{50\ \mathrm{meV}}{m_i}\right) \left(\frac{E}{10\ \mathrm{TeV}}\right)^{1/2} \left(\frac{100\ \mathrm{Mpc}}{D}\right)^{1/2}. \tag{28}$$

As mentioned in Ref. [1], the relatively weak constraints in Eq. (22) from atmospheric, long-baseline, and solar neutrinos therefore imply that a deviation from an equal neutrino flavor ratio $(\nu_e:\nu_\mu:\nu_\tau)=(1:1:1)$ could be measured at experiments such as IceCube. While normal mass ordering would imply a dominance of νe due to $\nu_1=(0.68:0.11:0.21)$, inverted ordering would yield a dominance of ν_μ and ν_τ due to ν3=(0.02:0.54:0.44) [54]. An equal flavor ratio would not be allowed because only the intermediate eigenstate ν_2 has an almost equal flavor content.

An additional question is whether one can distinguish Majorana from Dirac neutrinos at IceCube. At low energies, the detector cannot distinguish neutrinos from antineutrinos. Nevertheless, for neutrinos with energies around 6.3 PeV in the laboratory frame, the Glashow resonance makes IceCube more efficient for $\bar{\nu}_e$ detection [55]. Such a resonance is the s-channel of the process $\bar{\nu}_e + e^- \rightarrow W^- \rightarrow \bar{\nu}_\alpha + l_\alpha^-$ and allows the detector to distinguish neutrinos from antineutrinos. However, the main problem here is the unknown asymmetry of the neutrino and antineutrino fluxes. For analyzing IceCube data, it is customary to assume equal fluxes for ν_α and $\bar{\nu}_\alpha$, both in energy and flavor [56]. In this case, it would be impossible to distinguish Majorana from Dirac neutrinos. However, in general, these fluxes are expected to be different. Neutrinos are produced by the decay of charged pions via the decay chain [57]

$$\begin{array}{l} \pi^+ \rightarrow \mu^+ + \nu_\mu \\ \qquad \downarrow \\ \qquad e^+ + \nu_e + \bar{\nu}_\mu \end{array} \tag{29}$$

and the corresponding charge-conjugate process. If there is an asymmetry in charged-pion production, there will be an asymmetry in neutrino fluxes. For example, the production mechanism $p + \gamma \rightarrow \Delta^+ \rightarrow n + \pi^+$ involving proton collisions on photons of the environment (p_γ sources) produces no π− and thus no ¯νe [55]. Note that the ¯νe from neutron beta decay can be neglected here due to the large neutron lifetime compared to its propagation timescale and the short π^+ and μ^+ lifetimes.

For normal mass ordering, complete neutrino decay is currently disfavored by IceCube data at the 2σ level [58]. Moreover, a single event detected in the energy range of the Glashow resonance could rule out complete decay for inverted mass ordering [59]. Even though such bounds are stronger than the ones from solar neutrinos obtained in this paper, we stress that solar neutrinos already allow one to distinguish Majorana from

Dirac neutrinos, making the results previously discussed complementary to IceCube results. In the far future, a better understanding of neutrino production in astrophysical sources would give us tools to further distinguish the Majorana from the Dirac scenario.The same might be true for supernova neutrinos, which are expected to decay into the lightest mass eigenstate while traveling to Earth. While this decay scenario requires modified analyses of original supernova neutrino spectra [1], it could also be probed through the future detection of the supernova relic neutrino flux, i.e., the redshifted neutrino background from all past supernovae. In Ref. [60] it was argued that a complete decay scenario can potentially enhance the supernova relic neutrino background density up to the current experimental detection bound, so that its measurement might be feasible with near-future experiments.

CONCLUSIONS

In this paper, we have explored the possibility of distinguishing Majorana from Dirac neutrinos by their gravi-Majoron decays, assuming a gravitational origin of neutrino masses as proposed in Ref. [1]. Interactions between neutrinos and conventional Nambu-Goldston bosons are strongly constrained by cosmology, astrophysics, and laboratory experiments. In contrast, the gravi-Majorons we consider are not ruled out because they only arise in the very late Universe and effectively decouple from high-energy processes. The detection of an unexpectedly large absolute neutrino mass scale in beta-decay experiments could provide a hint toward the gravitational mass mechanism because it evades all cosmological mass bounds. Such a large absolute neutrino mass scale would imply a degenerate mass hierarchy, allowing us to distinguish Majorana from Dirac neutrinos through their decay. In the Majorana case, we put strong quantitative bounds on the decay of the heaviest neutrino, $\tau_3/m_3 > 2.2\times10^{-5}$s/eV at 90% CL, using data from experiments searching for antineutrino appearance from the Sun. We also qualitatively explored the possibility of distinguishing Dirac from Majorana neutrinos with future IceCube data; however, such a possibility requires a better understanding of neutrino production in astrophysical sources. The future will bring more data in neutrino astronomy, and perhaps even a supernova, to further study such an exotic, yet fascinating, scenario.

NOTES

[1]This suggestive terminology is a bit of a misnomer because Majorons [4] were originally invoked to explain Majorana masses, whereas it is a key point of

our discussion that ϕ bosons appear for both the Dirac and Majorana option.

[2]This discussion neglects tiny corrections for some ϕk masses due to weak SU(2)W effects, $m_{\phi k} \sim 10^{-33} eV(\Lambda_G/0.1eV)^3$.

[3]We thank Nima Arkani-Hamed for raising this question and Gia Dvali and Misha Shifman for further discussions. We also note that this question has not yet been unambiguously answered.

[4]Notice the Lagrangian in [33–35] includes a factor of 1/2.

[5]We thank Nicole Bell for communications on this point.

[6]Note that the analogous gravitational four-fermion vertices for charged fermions are additionally suppressed by the masses of these fermions, which renders them cosmologically irrelevant [1].

[7]Notice there is a special case in which the relative CP phase of the decaying neutrino and the daughter neutrino is either −1 or +1. If this scenario is realized, Majorana neutrino can only have either pseudoscalar or pseudovector coupling, not both.

ACKNOWLEDGMENTS

We thank Gia Dvali and Aaron Vincent for discussions and acknowledge partial support by the Deutsche Forschungsgemeinschaft through Grant No. SFB 1258 (Collaborative Research Center "Neutrinos, Dark Matter, Messengers"), as well as the European Union through Grant No. H2020-MSCA-ITN-2015/674896 (Innovative Training Network "Elusives"). Research at Perimeter Institute is supported by the Government of Canada through the Department of Innovation, Science and Economic Development and by the Province of Ontario through the Ministry of Research, Innovation and Science.H2020 Marie Sklodowska-Curie Actions

APPENDICES

Appendix A: High-Energy Softening of the Gravitational Vertex

In this appendix, we explain and constrain the high-energy softening of the nonperturbative gravitational vertex that gives rise to the coupling in Eq. (1). Some of the constraints have already been discussed in Refs. [1,5].

1. Cosmological Constraints

In the early Universe before photon decoupling, i.e., for $T \gtrsim 0.3$ eV, neutrino self-interactions $\lambda_{ij}=\{g_{ij},y_{ij}\}$ are strongly constrained by data from the cosmic microwave background, $\lambda_{ij} \lesssim 10^{-7}$ [36,61]. Since we only compare relative temperatures here, we can neglect the ratio of (4/11)1/3 between the neutrino and photon temperatures. Moreover, in the even earlier Universe before neutrino freeze-out, i.e., for $T \gtrsim 1$ MeV, the existence of new degrees of freedom (e.g., sterile neutrinos) is excluded if they had been in equilibrium with photons or electrons [26]. Thus, the interaction rate $\Gamma=\langle\sigma_v v_v\rangle n_v \sim \lambda^4_{ij} T$ needs to be smaller than the Hubble expansion rate $H(T)\sim T^2/MP$, which yields the constraint $\lambda_{ij}<(T/M_P)^{1/4}\sim 10^{-8}$ for T=meV. Note that new neutrino physics at lower temperatures becomes irrelevant for this constraint, since energy is conserved in the relic neutrino sector after neutrinos freeze out.

The mentioned cosmological constraints do not rule out the 't-Hooft-like gravitational four-neutrino vertex [1,5] that gives rise to the interactions λij in Eq. (1) after the phase-transition, because this effective low-energy vertex is strongly temperature-suppressed before the neutrino phase-transition, $\lambda_{ij}(T)\propto(\Lambda_G/T)n$.[6] Here, n is an unknown power-law exponent that has to be fixed by phenomenological requirements.

For comparison, the nonperturbatively generated QCD axion mass is suppressed by $m_a\propto(\Lambda_{QCD}/T)11$ for $T \gg m_q$, where ΛQCD is the QCD scale and mq are the quark masses [62]. However, we note that the analogy between the QCD and gravitational scenarios does not go too far because the perturbative QCD coupling becomes strong at $T\sim\Lambda_{QCD}$, while the perturbative gravitational coupling is expected to be weak and cosmologically decoupled at the phase-transition temperature $T_G\sim\Lambda_G$ [1,5]. While quarks and pions couple via the slowly running perturbative QCD coupling, our gravitational vertex is purely nonperturbative and thus expected to decrease much faster than in the perturbative QCD case. Thus, our analogy is only based on possible similarities between the nonperturbative and anomaly structures of these two theories [1].

The mentioned cosmological constraints require $\lambda_{ij}(T_0)\lesssim 10^{-4}$ for a minimal suppression of n=1 and a phase-transition temperature of $T_G \gtrsim T_0$, where $T_0\sim 0.2$ meV is the cosmic temperature today. Already for $n\geq 2$, the zero-temperature couplings can be as large as $\lambda_{ij}(T0)\sim 1$. We note in this context that a simplified cosmological analysis of the neutrino mass model demonstrates that the phase-transition is favored to happen at very low redshifts, $z\sim 0$ [13].

2. Stellar and Laboratory Constraints

The gravi-Majoron production in high-energy astrophysical environments with E≫mv is suppressed by (Λ_G/E)n, where n=1 is already sufficient to satisfy the astrophysical constraints on $g_{\phi\gamma\gamma}$ mentioned in Sec. II C. The gravitational $\nu\bar{\nu}\gamma$, $\nu\bar{\nu}\phi$, $\nu\bar{\nu}\nu\bar{\nu}$, $e\bar{e}\phi$, and $e\bar{e}\nu\bar{\nu}$ couplings are also strongly suppressed in stellar processes. However, the constraints on direct couplings to electrons are stronger than in the two-photon case, due to the absence of the electron-loop suppression. The most stringent bounds on Majoron-like models usually come from Compton or bremsstrahlung processes in red giant cores, $g_{e\bar{e}\phi} \lesssim 10^{-9}\ \mathrm{GeV}^{-1}$ [63], which equally apply to $g_{e\bar{e}\nu\bar{\nu}}$. In our case, these astrophysical bounds cannot be trivially translated into bounds on n, since astrophysical processes typically involve different momentum transfers. In particular, these processes becomes less n-suppressed but more phase-space suppressed at low momentum transfer [64]. We defer a detailed analysis of such constraints to later studies and only observe here that the constraints should be satisfied for a relatively low suppression factor.

The most stringent laboratory bounds on fifth forces mediated by the scalar analogs of gravi-Majorons [5] are $|\alpha| \lesssim 10^6\ (10^{-2})$ for $\Lambda_G \sim 0.1$ eV (1 meV) [65], where |α| is the strength of the Yukawa-like correction to the Newtonian gravitational constant (i.e., normalized to gravity). These bounds on fifth forces between hadrons translate into constraints on the power-law exponent of the large-mass decoupling, n≳1.6 (1.7), due to $|\alpha| \sim (\Lambda_G/mp)^{2n}(M_P/m_p)^2$ at distances $r \sim \Lambda_G^{-1}$ [5]. Here, we used $(\Lambda_G/m_p) \sim 10^{-10}(10^{-12})$ and denoted MP and mp as the Planck and proton masses, respectively. The high-energy suppression of the predicted fifth force between electrons is much weaker due to $(\Lambda_G/m_e) \sim 2\times10^{-7}$ (2×10^{-9}), but the experimental bounds on such a force are weaker as well, $|\alpha| \lesssim 10^{22}$ (1018) [66], translating to n≳1.2 (1.2). Thus, all of the laboratory constraints are compatible with the model predictions for n≳2 [5].

Appendix B: Matrix Element for Neutrino Decay

In this appendix, we calculate the matrix element for the decay process $\nu_i \to \overset{(-)}{\nu}_j + \phi$ of a Dirac or a Majorana neutrino with fixed initial spin and derivative coupling to a gravi-Majoron[7]

$$\mathcal{L}_{\rm int} = \partial_\mu \phi \sum_{ij} \frac{g_{ij}}{m_i + m_j} \bar{\nu}_i \gamma^\mu \gamma_5 \nu_j + \mathrm{H.c.} \qquad \text{(B1)}$$

For the Dirac case, fixing the initial spin is equivalent to choosing an active *or* a sterile initial state, whereas for a Majorana neutrino it simply means choosing if it is a neutrino (left-handed) or an antineutrino (right-handed). Using the Feynman rules of Ref. [67], we find

$$\mathcal{M}^{\mathrm{D}} = g_{\phi,ij}\bar{u}_{\nu_j}\gamma^\mu\gamma^5 u_{\nu_i} J^\phi_\mu \quad \text{(Dirac)} \tag{B2a}$$

$$\mathcal{M}^{\mathrm{M}} = g_{\phi,ij}\bar{u}_{\nu_j}\gamma^\mu\gamma^5 u_{\nu_i} J^\phi_\mu \times 2 \quad \text{(Majorana)} \tag{B2b}$$

where we have neglected global phases and $J^\phi_\mu = k_\mu$ is the gravi-Majoron current. Note that the matrix element for the Majorana case is twice as large as for the Dirac case because the Hermitian conjugate in the Lagrangian also contributes to the amplitude, so that the rate will be four times larger for a decaying Majorana neutrino. This can be compared, e.g., to radiative decays induced by standard model interactions, for which the one of Majorana neutrinos is two times larger than the decay width of Dirac neutrinos [68]. The global constant can be reabsorbed in the coupling definition.

Let us consider the case of a decaying Majorana neutrino. Squaring the amplitude one finds

$$|\mathcal{M}^{\mathrm{M}}|^2 = \frac{g_{ij}^2}{(m_i + m_j)^2}\mathrm{Tr}[(\not{p}_i + m_i)(1 + h_i\gamma^5\not{S}_i)\gamma^\mu\gamma^5 \times(\not{p}_j + m_j)(1 + h_j\gamma^5\not{S}_j)\gamma^\nu\gamma^5]k_\mu k_\nu, \tag{B3}$$

where p_i, p_j, and k are the four-momenta of ν_i, ν_j, and ϕ respectively and we used [69]

$$u_\nu\bar{u}_\nu = \frac{1}{2}(\not{p} + m_\nu)(1 + h\gamma^5\not{S}) \tag{B4}$$

with the spin vector

$$S = \left(\frac{|\mathbf{p}|}{m_\nu}, \frac{E_\nu}{m_\nu}\hat{\mathbf{p}}\right) \tag{B5}$$

The squared amplitude in Eq. (B3) has 16 terms, 8 of which contain an odd number of γμ and 4 of which have a γ5 so that they do not contribute for symmetry reasons. The remaining terms give

$$|\mathcal{M}^{\mathrm{M}}|^2 = 4g_{ij}^2[(p_i \cdot p_j - m_i m_j)(1 + h_i h_j S_i \cdot S_j) - h_i h_j (p_i \cdot S_j)(p_j \cdot S_i)]. \tag{B6}$$

Analogously, one can compute the decay through a pseudoscalar coupling

$$\mathcal{L}_{\mathrm{int}} = i\phi \sum_{ij} g_{ij} \bar{\nu}_i \gamma_5 \nu_j + \mathrm{H.c.} \tag{B7}$$

and find explicitly that it has the same squared amplitude as in Ref. [33]. Notice that there are two factors to be taken into account. The coupling in our interaction Lagrangian is twice as large as in Ref. [33]; furthermore, we use a different spinor normalization, so an additional factor 1/(4mimj) has to be included. Our results agree with Ref. [33] but differ from the ones reported in Ref. [38], where the couplings are defined as in our Lagrangian but the rate is quoted directly from Ref. [33]; the results in Ref. [38] would be correct if they considered exclusively Dirac neutrinos. In the following we will consider only the pseudoscalar coupling, as the derivative coupling process is equivalent to the latter.

After substituting Eq. (B5) in (B8), one can evaluate the squared amplitude in the ν_i rest frame (where the spin vector S=(0,S), so that S·p=0), showing that the emission of a certain helicity is not isotropic,

$$|\mathcal{M}^{\mathrm{M}}|^2 = 4g_{ij}^2 \left[(m_i E_j - m_i m_j) \left(1 - h_i h_j E_j \frac{\mathbf{p}_j}{|\mathbf{p}_j|} \cdot \mathbf{S}_i \right) - h_i h_j \left(m_i \frac{|\mathbf{p}_j|}{m_j} \right) (\mathbf{p}_j \cdot \mathbf{S}_i) \right]. \tag{B8}$$

The different angular distributions in the rest frame translate into different energy spectra in the laboratory frame for different helicities [11]. We can evaluate the squared amplitudes in the laboratory frame, in which $E_i \gg m_i$, finding for the helicity conserving decay

$$|\mathcal{M}^{\mathrm{M}}_{\nu_i \to \nu_j + \phi}|^2 = 4g_{ij}^2 m_i m_j (A - 2) \tag{B9}$$

and for the helicity flipping decay

$$|\mathcal{M}^{\mathrm{M}}_{\nu_i \to \bar{\nu}_j + \phi}|^2 = 4g_{ij}^2 m_i m_j \left(\frac{m_i^2 + m_j^2}{m_i m_j} - A \right), \tag{B10}$$

where

$$A \equiv \left(\frac{m_i E_j}{m_j E_i} + \frac{m_j E_i}{m_i E_j} \right). \tag{B11}$$

The decay rate for Dirac neutrinos to an active or a sterile neutrino neutrino are found by dividing these expression by 4 and interpreting $\bar{\nu}_j$ as a sterile state N_j. These are the expressions that are relevant for the decay rate discussed in the main text.

The vectorlike coupling case is obtained following the same steps. Starting from the Lagrangian

$$\mathcal{L}_{\text{int}} = \partial_\mu \phi \sum_{i \neq j} \frac{g_{ij}}{m_i - m_j} \bar{\nu}_i \gamma^\mu \nu_j + \text{H.c.} \tag{B12}$$

one finds for the helicity conserving decay

$$|\mathcal{M}^{\text{M}}_{\nu_i \to \nu_j + \phi}|^2 = 4g_{ij}^2 m_i m_j (A + 2) \tag{B13}$$

and for the helicity flipping decay

$$|\mathcal{M}^{\text{M}}_{\nu_i \to \bar{\nu}_j + \phi}|^2 = 4g_{ij}^2 m_i m_j \left(\frac{m_i^2 + m_j^2}{m_i m_j} - A \right) \tag{B14}$$

The decay rate for Dirac neutrinos are again found by dividing these expression by 4 and interpreting $\bar{\nu}_j$ as a sterile state N_j.

REFERENCES

1. G. Dvali and L. Funcke, Small neutrino masses from gravitational θ-term, Phys. Rev. D 93, 113002 (2016).
2. G. Dvali, Three-Form Gauging of Axion Symmetries and Gravity, Three-form gauging of axion symmetries and gravity arXiv:hep-th/0507215.
3. G. Dvali, S. Folkerts, and A. Franca, How neutrino protects the axion, Phys. Rev. D 89, 105025 (2014).
4. Y. Chikashige, R. Mohapatra, and R. Peccei, Are there real Goldstone bosons associated with broken lepton number?, Phys. Lett. B 98, 265 (1981).
5. G. Dvali and L. Funcke, Domestic Axion, Domestic axion arXiv:1608.08969.
6. T. Brinckmann, D. C. Hooper, M. Archidiacono, J. Lesgourgues, and T. Sprenger, The promising future of a robust cosmological neutrino mass measurement, J. Cosmol. Astropart. Phys. 2019, 059.
7. G. Drexlin, V. Hannen, S. Mertens, and C. Weinheimer, Current direct neutrino mass experiments, Adv. High Energy Phys. 2013, 1 (2013).
8. M. Aker et al. (KATRIN Collaboration), An Improved Upper Limit on the Neutrino Mass from a Direct Kinematic Method by KATRIN, Phys. Rev. Lett. 123, 221802 (2019).
9. M. J. Dolinski, A. W. P. Poon, and W. Rodejohann, Neutrinoless double-beta decay: Status and prospects, Annu. Rev. Nucl. Part. Phys. 69, 219 (2019).
10. A. B. Balantekin and B. Kayser, On the properties of neutrinos, Annu. Rev. Nucl. Part. Sci. 68, 313 (2018).
11. A.B. Balantekin, A. de Gouvêa, and B. Kayser, Addressing the Majorana vs Dirac question with neutrino decays, Phys. Lett. B 789, 488 (2019).
12. S. F. King, Neutrino mass models, Rep. Prog. Phys. 67, 107 (2004).
13. C. S. Lorenz, L. Funcke, E. Calabrese, and S. Hannestad, Time-varying neutrino mass from a supercooled phase transition: Current cosmological constraints and impact on the Ωm−σ8 *plane*, Phys. Rev. D 99, 023501 (2019).
14. J. Schechter and J. W. F. Valle, Neutrino decay and spontaneous violation of lepton number, Phys. Rev. D 25, 774 (1982).

15. S. Gariazzo, C. Giunti, M. Laveder, Y.F. Li, and E.M. Zavanin, Light sterile neutrinos, J. Phys. G 43, 033001 (2016).
16. J. Schechter and J.W.F. Valle, Neutrino decay and spontaneous violation of lepton number, Phys. Rev. D 25, 774 (1982).
17. I. Girardi, A. Meroni, and S.T. Petcov, Neutrinoless double beta decay in the presence of light sterile neutrinos, J. High Energy Phys. 2013, 146.
18. A. de Gouvêa, See-saw energy scale and the LSND anomaly, Phys. Rev. D 72, 033005 (2005).
19. J.F. Beacom, N.F. Bell, and S. Dodelson, Neutrinoless Universe, Phys. Rev. Lett. 93, 121302 (2004).
20. A.J. Cuesta, V. Niro, and L. Verde, Neutrino mass limits: Robust information from the power spectrum of galaxy surveys, Phys. Dark Universe 13, 77 (2016).
21. S. Hannestad, Structure formation with strongly interacting neutrinos—Implications for the cosmological neutrino mass bound, J. Cosmol. Astropart. Phys. 2005, 011.
22. B. Follin, L. Knox, M. Millea, and Z. Pan, First Detection of the Acoustic Oscillation Phase Shift Expected from the Cosmic Neutrino Background, Phys. Rev. Lett. 115, 091301 (2015).
23. L. Lancaster, F.-Y. Cyr-Racine, L. Knox, and Z. Pan, A tale of two modes: Neutrino free-streaming in the early universe, J. Cosmol. Astropart. Phys. 2017, 033.
24. I.M. Oldengott, T. Tram, C. Rampf, and Y.Y.Y. Wong, Interacting neutrinos in cosmology: Exact description and constraints, J. Cosmol. Astropart. Phys. 2017, 027.
25. C.D. Kreisch, F.-Y. Cyr-Racine, and O. Doré, The Neutrino Puzzle: Anomalies, Interactions, and Cosmological Tensions, The neutrino puzzle: Anomalies, interactions, and cosmological tensions arXiv:1902.00534.
26. N. Aghanim et al. (Planck Collaboration), Planck 2018 Results. VI. Cosmological Parameters, Planck 2018 results. VI. Cosmological parameters arXiv:1807.06209.
27. J.L. Feng, T. Moroi, H. Murayama, and E. Schnapka, Third generation familons, b *factories, and neutrino cosmology*, Phys. Rev. D 57, 5875 (1998).
28. G. Arcadi, J. Heeck, F. Heizmann, S. Mertens, F.S. Queiroz, W. Rodejohann, M. Slezák, and K. Valerius, Tritium beta decay with

additional emission of new light bosons, J. High Energy Phys. 2019, 206.

29. S. Betts et al., in Proceedings, 2013 Community Summer Study on the Future of U.S. Particle Physics: Snowmass on the Mississippi (CSS2013): Minneapolis, MN, USA, (2013).
30. V. Anastassopoulos et al. (CAST Collaboration), New CAST limit on the axion-photon interaction, Nat. Phys. 13, 584 (2017).
31. R. Arnold et al. (NEMO-3 Collaboration), Measurement of the 2νββ *decay half-life of* 150Nd *and a search for* 0νββ *decay processes with the full exposure from the NEMO-3 detector*, Phys. Rev. D 94, 072003 (2016).
32. P. S. Pasquini and O. L. G. Peres, Bounds on neutrino-scalar Yukawa coupling, Phys. Rev. D 93, 053007 (2016); Erratum, 93, 079902 (2016).
33. C. W. Kim and W. P. Lam, Some remarks on neutrino decay via a Nambu-Goldstone boson, Mod. Phys. Lett. A 05, 297 (1990).
34. M. Lindner, T. Ohlsson, and W. Winter, A combined treatment of neutrino decay and neutrino oscillations, Nucl. Phys. B607, 326 (2001).
35. Z. Moss, M. H. Moulai, C. A. Argüelles, and J. M. Conrad, Exploring a nonminimal sterile neutrino model involving decay at IceCube, Phys. Rev. D 97, 055017 (2018).
36. S. Hannestad and G. G. Raffelt, Constraining invisible neutrino decays with the cosmic microwave background, Phys. Rev. D 72, 103514 (2005).
37. M. Tanabashi et al. (Particle Data Group), Review of particle physics, Phys. Rev. D 98, 030001 (2018).
38. J.F. Beacom and N.F. Bell, Do solar neutrinos decay?, Phys. Rev. D 65, 113009 (2002).
39. C. Giunti and C. W. Kim, Fundamentals of Neutrino Physics and Astrophysics (Oxford University Press, Oxford, 2007).
40. E. Vitagliano, I. Tamborra, and G. Raffelt, Grand Unified Neutrino Spectrum at Earth, Grand unified neutrino spectrum at Earth arXiv:1910.11878.
41. R.A. Malaney, B. S. Meyer, and M.N. Butler, Solar antineutrinos, Astrophys. J. 352, 767 (1990).
42. F. Mantovani, L. Carmignani, G. Fiorentini, and M. Lissia, Antineutrinos from the Earth: The reference model and its uncertainties, Phys. Rev. D 69, 013001 (2004).

43. E.K. Akhmedov and J. Pulido, Solar neutrino oscillations and bounds on neutrino magnetic moment and solar magnetic field, Phys. Lett. B 553, 7 (2003).
44. S.J. Li, J.J. Ling, N. Raper, and M.V. Smirnov, Sensitivity to neutrino-antineutrino transitions for boron neutrinos, Nucl. Phys. B944, 114661 (2019).
45. A.M. Gago, R.A. Gomes, A.L.G. Gomes, J. Jones-Perez, and O.L.G. Peres, Visible neutrino decay in the light of appearance and disappearance long baseline experiments, J. High Energy Phys. 2017, 022.
46. M.V. Ascencio-Sosa, A.M. Calatayud-Cadenillas, A.M. Gago, and J. Jones-Pérez, Matter effects in neutrino visible decay at future long-baseline experiments, Eur. Phys. J. C 78, 809 (2018).
47. M.C. Gonzalez-Garcia and M. Maltoni, Status of oscillation plus decay of atmospheric and long-baseline neutrinos, Phys. Lett. B 663, 405 (2008).
48. B. Aharmim et al. (SNO Collaboration), Constraints on neutrino lifetime from the Sudbury neutrino observatory, Phys. Rev. D 99, 032013 (2019).
49. R. Picoreti, M.M. Guzzo, P.C. de Holanda, and O.L.G. Peres, Neutrino decay and solar neutrino seasonal effect, Phys. Lett. B 761, 70 (2016).
50. J.M. Berryman, A. de Gouvêa, and D. Hernández, Solar neutrinos and the decaying neutrino hypothesis, Phys. Rev. D 92, 073003 (2015).
51. B. Aharmim et al. (SNO Collaboration), Electron antineutrino search at the Sudbury neutrino observatory, Phys. Rev. D 70, 093014 (2004).
52. K. Eguchi et al. (KamLAND Collaboration), High Sensitivity Search for ¯ve's from the Sun and Other Sources at KamLAND, Phys. Rev. Lett. 92, 071301 (2004).
53. G. Bellini et al. (Borexino Collaboration), Study of solar and other unknown anti-neutrino fluxes with Borexino at LNGS, Phys. Lett. B 696, 191 (2011).
54. F. Capozzi, E. Lisi, A. Marrone, and A. Palazzo, Current unknowns in the three neutrino framework, Prog. Part. Nucl. Phys. 102, 48 (2018).
55. H. Nunokawa, B. Panes, and R. Zukanovich Funchal, How unequal fluxes of high energy astrophysical neutrinos and antineutrinos can fake new physics, J. Cosmol. Astropart. Phys. 2016, 036.
56. M.G. Aartsen et al. (IceCube Collaboration), Observation of High-

Energy Astrophysical Neutrinos in Three Years of IceCube Data, Phys. Rev. Lett. 113, 101101 (2014).

57. T.K. Gaisser, R. Engel, and E. Resconi, Cosmic Rays and Particle Physics (Cambridge University Press, Cambridge, England, 2016).
58. M. Bustamante, J.F. Beacom, and W. Winter, Theoretically Palatable Flavor Combinations of Astrophysical Neutrinos, Phys. Rev. Lett. 115, 161302 (2015).
59. M. Bustamante, J.F. Beacom, and K. Murase, Testing decay of astrophysical neutrinos with incomplete information, Phys. Rev. D 95, 063013 (2017).
60. G.L. Fogli, E. Lisi, A. Mirizzi, and D. Montanino, Three generation flavor transitions and decays of supernova relic neutrinos, Phys. Rev. D 70, 013001 (2004).
61. M. Archidiacono and S. Hannestad, Updated constraints on non-standard neutrino interactions from Planck, J. Cosmol. Astropart. Phys. 2014, 046.
62. P. Sikivie, Axion cosmology, Lect. Notes Phys. 741, 19 (2008).
63. G. Raffelt and A. Weiss, Red giant bound on the axion-electron coupling revisited, Phys. Rev. D 51, 1495 (1995).
64. G.G. Raffelt, Stars as Laboratories for Fundamental Physics (Chicago University Press, Chicago, 1996).
65. Y.J. Chen, W.K. Tham, D.E. Krause, D. Lopez, E. Fischbach, and R.S. Decca, Stronger Limits on Hypothetical Yukawa Interactions in the 30–8000 nm Range, Phys. Rev. Lett. 116, 221102 (2016).
66. M.B. Wise and Y. Zhang, Lepton flavorful fifth force and depth-dependent neutrino matter interactions, J. High Energy Phys. 2018, 053.
67. J. Gluza and M. Zralek, Feynman rules for Majorana neutrino interactions, Phys. Rev. D 45, 1693 (1992).
68. D. Gorbunov and M. Shaposhnikov, How to find neutral leptons of the νMSM?, J. High Energy Phys. 2007, 015; Erratum and Addendum, 2013, 101.
69. A.J. Long, C. Lunardini, and E. Sabancilar, Detecting non-relativistic cosmic neutrinos by capture on tritium: Phenomenology and physics potential, J. Cosmol. Astropart. Phys. 2014, 038.

SECTION 3
DIRECT NEUTRINO MASS SEARCHES, DOUBLE β-DECAY, AND NEUTRINO OSCILLATIONS

CHAPTER

8

Search for Neutrinoless Double-β Decay with the Complete EXO-200 Dataset

G. Anton[1], I. Badhrees[2,a], P.S. Barbeau[3], D. Beck[4], V. Belov[5], T. Bhatta[6], M. Breidenbach[7], T. Brunner[8,9], G.F. Cao[10], W.R. Cen[10], C. Chambers[11,b], B. Cleveland[12,c], M. Coon[4], A. Craycraft[11], T. Daniels[13], M. Danilov[5,d], L. Darroch[8], S.J. Daugherty[14], J. Davis[7], S. Delaquis[7,*], A. Der Mesrobian-Kabakian[12], R. DeVoe[15], J. Dilling[9], A. Dolgolenko[5], M.J. Dolinski[16], J. Echevers[4], W. Fairbank, Jr.[11], D. Fairbank[11], J. Farine[12], S. Feyzbakhsh[17], P. Fierlinger[18], D. Fudenberg[15], P. Gautam[16], R. Gornea[2,9], G. Gratta[15], C. Hall[19], E.V. Hansen[16], J. Hoessl[1], P. Hufschmidt[1], M. Hughes[20], A. Iverson[11], A. Jamil[21], C. Jessiman[2], M.J. Jewell[15], A. Johnson[7], A. Karelin[5], L.J. Kaufman[7,e], T. Koffas[2], R. Krücken[9], A. Kuchenkov[5], K.S. Kumar[22,f], Y. Lan[9], A. Larson[6], B.G. Lenardo[15], D.S. Leonard[23], G.S. Li[15,†], S. Li[4], Z. Li[21], C. Licciardi[12], Y.H. Lin[16], R. MacLellan[6], T. McElroy[8], T. Michel[1], B. Mong[7], D.C. Moore[21], K. Murray[8], O. Njoya[22], O. Nusair[20], A. Odian[7], I. Ostrovskiy[20], A. Piepke[20], A. Pocar[17], F. Retière[9], A.L. Robinson[12], P.C. Rowson[7], D. Ruddell[13], J. Runge[3], S. Schmidt[1], D. Sinclair[2,9], A.K. Soma[20], V. Stekhanov[5], M. Tarka[17], J. Todd[11], T. Tolba[10], T.I. Totev[8], B. Veenstra[2], V. Veeraraghavan[20], P. Vogel[24], J.-L. Vuilleumier[25], M. Wagenpfeil[1], J. Watkins[2], M. Weber[15], L.J. Wen[10], U. Wichoski[12], G. Wrede[1], S.X. Wu[15], Q. Xia[21], D.R. Yahne[11], L. Yang[4], Y.-R. Yen[16], O.Ya. Zeldovich[5], and T. Ziegler[1] (EXO-200 Collaboration)

Citation: G. Anton et al. (EXO-200 Collaboration). "Search for Neutrinoless Double-β Decay with the Complete EXO-200 Dataset" Phys. Rev. Lett. 123, 161802 (2019). https://doi.org/10.1103/PhysRevLett.123.161802

[1]Erlangen Centre for Astroparticle Physics, Friedrich-Alexander-University Erlangen-Nürnberg, Erlangen 91058, Germany

[2]Physics Department, Carleton University, Ottawa, Ontario K1S 5B6, Canada

[3]Department of Physics, Duke University, and Triangle Universities Nuclear Laboratory, Durham, North Carolina 27708, USA

[4]Physics Department, University of Illinois, Urbana-Champaign, Illinois 61801, USA

[5]Institute for Theoretical and Experimental Physics named by A.I. Alikhanov of National Research Center "Kurchatov Institute," 117218 Moscow, Russia

[6]Department of Physics, University of South Dakota, Vermillion, South Dakota 57069, USA

[7]SLAC National Accelerator Laboratory, Menlo Park, California 94025, USA

[8]Physics Department, McGill University, Montreal H3A 2T8, Quebec, Canada

[9]TRIUMF, Vancouver, British Columbia V6T 2A3, Canada

[10]Institute of High Energy Physics, Beijing 100049, China

[11]Physics Department, Colorado State University, Fort Collins, Colorado 80523, USA

[12]Department of Physics, Laurentian University, Sudbury, Ontario P3E 2C6, Canada

[13]Department of Physics and Physical Oceanography, University of North Carolina at Wilmington, Wilmington, North Carolina 28403, USA

[14]Physics Department and CEEM, Indiana University, Bloomington, Indiana 47405, USA

[15]Physics Department, Stanford University, Stanford, California 94305, USA

[16]Department of Physics, Drexel University, Philadelphia, Pennsylvania 19104, USA

[17]Amherst Center for Fundamental Interactions and Physics Department, University of Massachusetts, Amherst, Massachusetts 01003, USA

[18]Physik Department and Excellence Cluster Universe, Technische Universität München, Garching 80805, Germany

[19]Physics Department, University of Maryland, College Park, Maryland 20742, USA

[20]Department of Physics and Astronomy, University of Alabama, Tuscaloosa, Alabama 35487, USA

[21]Wright Laboratory, Department of Physics, Yale University, New Haven, Connecticut 06511, USA

[22]Department of Physics and Astronomy, Stony Brook University, SUNY, Stony Brook, New York 11794, USA

[23]IBS Center for Underground Physics, Daejeon 34126, Korea

[24]Kellogg Lab, Caltech, Pasadena, California 91125, USA

[25]LHEP, Albert Einstein Center, University of Bern, Bern CH-3012, Switzerland

[*]Deceased.

[†]Corresponding author. ligs@stanford.edu

[a]Permanent address: King Abdulaziz City for Science and Technology, Riyadh, Saudi Arabia.

[b]Present address: Physics Department, McGill University, Montreal, Quebec, Canada.

[c]Also at SNOLAB, Sudbury, Ontario, Canada.

[d]Present address: P. N. Lebedev Physical Institute of the Russian Academy of Sciences, Moscow, Russia.

[e]Also at Physics Department and CEEM, Indiana University, Bloomington, Indiana, USA.

[f]Present address: Physics Department, University of Massachusetts, Amherst, Massachusetts, USA.

ABSTRACT

A search for neutrinoless double-β decay (0νββ) in 136Xe is performed with the full EXO-200 dataset using a deep neural network to discriminate between 0νββ and background events. Relative to previous analyses, the signal detection efficiency has been raised from 80.8% to 96.4±3.0%, and the energy resolution of the detector at the Q value of 136Xe 0νββ has been improved from σ/E=1.23% to 1.15±0.02% with the upgraded detector. Accounting for the new data, the median 90% confidence level 0νββ half-life sensitivity for this analysis is 5.0×1025 yr with a total 136Xe exposure of 234.1 kg yr. No statistically significant evidence for 0νββ is observed, leading to a lower limit on the 0νββ half-life of 3.5×10^{25}yr at the 90% confidence level.

ARTICLE TEXT

Double- β decay is a second-order weak transition in which two neutrons simultaneously decay into two protons. Although the mode with emission of two electrons and two antineutrinos (2νββ) has been observed in several nuclides in which single- β decay was suppressed [1], the hypothetical neutrinoless mode (0νββ) [2] is yet to be discovered. The search for 0νββ is recognized as the most sensitive probe for the Majorana nature of neutrinos [3–6]. Its observation would provide direct evidence for a beyond-the-Standard-Model process that violates lepton number conservation as well as helps constrain the absolute mass scale of neutrinos [7].

Recent experiments probing a number of nuclides [8–12] have set lower limits on the 0νββ half-life with sensitivities in the range of 10^{25}–10^{26}yr at a 90% confidence level (CL). Exploiting the advantages of a liquid xenon (LXe) cylindrical time projection chamber (TPC) filled with LXe enriched to 80.6% in 136Xe [13], the EXO-200 [14] experiment achieved a sensitivity of 3.7×1025 yr with the most recent 0νββ search [8], whereas the most sensitive search to date for the same isotope reached 5.6×1025 yr [9]. Here, we report on a search with similar sensitivity to the previous best search.

In December 2018, the EXO-200 experiment completed data taking with the upgraded detector ("Phase II," May 2016 to December 2018) after collecting an exposure similar to that of its first run ("Phase I," September 2011 to February 2014). This Letter reports a search for 0νββ using the full EXO-200 dataset that, after data quality cuts [13], totals 1181.3 d of live time. This represents an approximately 25% increase in exposure relative to the previous search [8] that already included nearly half of the Phase II dataset. In addition to the new data acquired between June 2017 and October 2018, this search introduces several analysis developments to optimize the detector sensitivity to 0νββ, including the incorporation of a deep neural network (DNN) to discriminate between background and signal events.

In the EXO-200 detector, a common cathode splits the LXe TPC into two drift regions, each with a radius of ∼18 cm and a drift length of ∼20 cm. The TPC is enclosed by a radiopure thin-walled copper vessel. The electric field in the drift regions was raised from 380 V/cm in Phase I to 567 V/cm in Phase II to improve the energy resolution because it was found that, after the detector was restarted, higher voltage values on the cathode were stable. The ionization produced from interactions in the LXe is read out after being drifted to crossed-wire planes at each anode, whereas the scintillation light produced at the interaction time is collected by arrays of large area avalanche photodiodes (LAAPDs) [15] located behind the wire planes.

The underground location of the experiment, which is the Waste Isolation Pilot Plant (WIPP) near Carlsbad, New Mexico, provides an overburden of 1624^{+22}_{-21} m of water equivalent [16]. In addition to several layers of passive shielding (including ~50 cm of HFE-7000 cryofluid [17], 5.4 cm of copper, and ~25 cm of lead in all directions [14]), an active muon veto system with scintillator panels on four sides allows prompt identification of >94% of the cosmic ray muons passing through the TPC. This system is also used in this analysis to reject background events arising from cosmogenically produced 137Xe, which primarily decays via β emission with a half-life of 3.82 min [8,18].

Each TPC event is reconstructed by grouping charge and light signals into individual energy deposits. Ionization signals measured by two wire planes, an induction plane (V wires) and a collection plane (U wires), provide information about coordinates x and y perpendicular to the drift field. The z position, along the drift direction, is obtained from the time delay between the prompt light and the delayed charge signals together with the measured electron drift velocity [19]. Events reconstructed with single and multiple energy deposit(s) are referred to as "single site" (SS) and "multisite" (MS). 0νββ events are predominantly SS, whereas backgrounds are mostly MS. Although the main procedures for spatial reconstruction are the same as in previous searches [8,13,20], the 0νββ detection efficiency has been raised to 97.8±3.0% (96.4±3.0%) in Phase I (Phase II) from 82.4±3.0% (80.8±2.9%) [8] by relaxing two selection criteria. First, the time required for events to be separated from all other reconstructed events has been reduced from >1 s to >0.1 s. This time cut is still at least two orders of magnitude longer than expected from typical time-correlated backgrounds seen in the detector [16,21], whereas the 0νββ efficiency loss due to accidental coincidence is reduced from 7% to 0.5%. Second, the search presented here includes events containing deposits without a detected V-wire signal if these deposits contribute <40% of the total event energy, which were removed in the previous analyses. Because of the higher energy threshold for signal detection on the V wires (~200 keV) versus the U wires (~90 keV), a significant number of events with small energy deposits are well reconstructed by the U wires but incompletely on the V wires, resulting in events with full z reconstruction but incomplete xy reconstruction for smaller energy deposits. Relaxing this 3D cut criterion only recovers MS events and retrieves almost all potential 0νββ events with incomplete xy reconstruction due to small, separated energy deposits from bremsstrahlung. Although 0νββ primarily induces SS events, the smaller fraction of MS 0νββ events can be distinguished from the dominant γ backgrounds using a discriminator for MS events (described below), resulting in an enhancement in the 0νββ half-life sensitivity.

Events within the fiducial volume (FV) are required to lie within a hexagon in the xy plane with an apothem of 162 mm. They are further required to be more than 10 mm away from the cylindrical polytetrafluoroethylene (PTFE) reflector, as well as the cathode and the V-wire planes. This FV contains 3.31×10^{26} atoms of ^{136}Xe, with an equivalent mass of 74.7 kg. Although the incomplete xy-matched energy deposits may fall outside the FV, this effect is determined by detector simulations to have a negligible effect on the estimated detection efficiency due to the energy requirements imposed on these events. The 136Xe exposure of the entire dataset after data quality cuts and accounting for live time loss due to vetoing events coincident with the muon veto is 234.1 kg yr, or 1727.5 mol yr, with 117.4 (116.7) kg yr in Phase I (Phase II).

The detector response to 0νββ decays and background interactions is modeled by a detailed Monte Carlo (MC) simulation based on geant4 [22]. This MC simulation models the energy deposits produced by interactions in the LXe, and then it propagates the ionization through the detector to produce waveforms associated to readout channels. These simulated waveforms are input to the same reconstruction and analysis framework used for data waveforms. Calibration data with external γ sources located 9 (11) cm away from the FV at set positions around the cathode (behind the anodes) [13] were regularly taken to validate the analysis.

After the previous EXO-200 0νββ search [8], a small fraction of the observed candidate events presented light-to-charge ratios that were not fully consistent with their expected distributions. Using calibration and 2νββ data, the distribution of the light-to-charge ratio is measured and found to be approximately Gaussian around the mean ratio. While keeping the maximal search sensitivity, a cut is imposed requiring that events are within 2.5σ of the mean of the distribution. This improves the previous cut [13], primarily aimed at removing α decays, by also removing poorly reconstructed β and γ events with an anomalous light-to-charge ratio. All systematic errors associated with the signal detection efficiency are summarized in Table 1.

Table 1. Summary of systematic error contributions.

Source	**Phase I (%)**	**Phase II (%)**
Background errors		
Spectral shape agreement	5.8	4.4
Background model	4.0	4.6
Other [8]	1.5	1.2
Total error	7.1	6.5
Signal detection efficiency		
Fiducial volume	2.8	2.6
Partial 3D cut	<0.4	<0.4
Light-to-charge ratio	0.9	0.9
Denoising misreconstructed	⊠	1.0
Other [13]	<1.0	<1.0
Total error	3.1	3.1

The reconstructed energy is determined by combining the anticorrelated charge and light signals [23] to optimize the resolution at the 0νββ decay energy of $Q_{\beta\beta}$=2457.83±0.37 keV [24]. An offline denoising algorithm [25], previously introduced to account for excess APD readout noise observed in Phase I, has been further optimized with measurements of the light response of the detector and adapted for Phase II data. In addition, a proper modeling of mixed signals from the induced and collected charge in wires is introduced to the signal finder in the event reconstruction process. The resulting energy measurement shows good spectral agreement between the data and the simulation for SS and MS events using ^{228}Th, ^{226}Ra, and ^{60}Co calibration sources, as shown in Fig. 1. The electronics upgrade carried out before Phase II data taking resulted in substantially improved resolution and stability, as illustrated in Fig. 2. The average detector resolution is determined by uniformly weighting all calibration data from several positions and accounting for the detector live time. The averages for Phase I and Phase II are $\sigma/E(Q_{\beta\beta})$=1.35±0.09% and 1.15±0.02%, respectively.

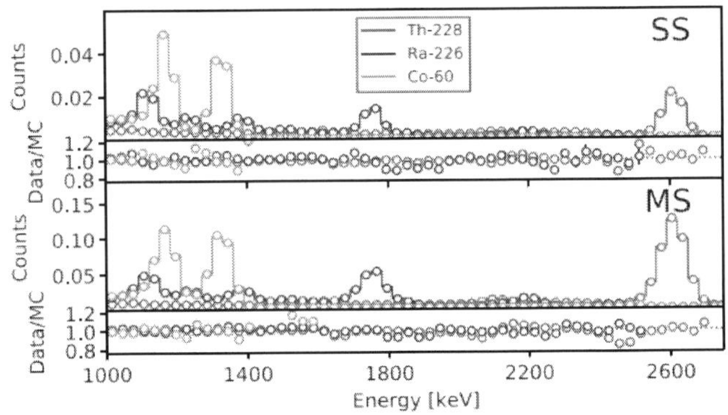

Figure 1. Comparison of energy distributions in data (circles) and MC (lines) for SS (top half) and MS (bottom half) events from calibration sources positioned near the cathode.

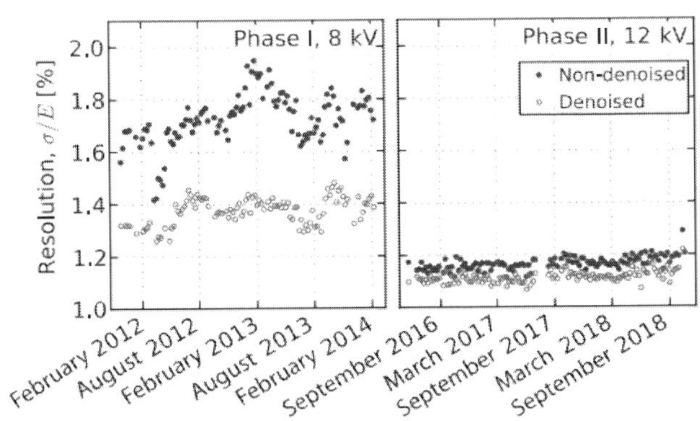

Figure 2. Energy resolution of SS events measured using a 228Th calibration source deployed to a position near the cathode. The effect of the denoising algorithm and weekly variation of the resolution at the 2615 keV 208Tl γ line are shown for both Phase I and Phase II. The resolution worsened slightly after a xenon recovery in July 2017 due to a power outage. A degraded resolution due to an increase of excess noise is visible in the last weeks of Phase II.

All data, including those previously reported, were blinded to hide all candidate 0νββ SS events having energy within $Q_{\beta\beta}\pm2\sigma$. No information about such events is used in the development of the techniques for this analysis. New background discriminators are studied to optimize the sensitivity of this search while minimizing the systematic errors. The search for 0νββ is performed with a simultaneous maximum-likelihood (ML) fit to the SS and MS energy spectra, with the discriminators added as additional fit dimensions. Although Phase I and Phase II are fit independently and then combined by summing their individual profile likelihoods for various signal hypotheses, both use the same background model developed in [8] composed of decays originating in the detector and surrounding materials. Systematic errors are included in the ML fit as nuisance parameters constrained by normal distributions. The median 90% CL sensitivity is estimated using toy datasets (simulated trial experiments) generated from the MC probability density functions (PDFs) of the background model.

The primary topological discrimination of backgrounds is the SS and MS event classification. Figure 3 shows the agreement between the source calibration data and MC for the "SS fraction": SS/(SS+MS). Because the relaxed 3D cut recovers MS events, the SS fraction near $Q_{\beta\beta}$ is lowered from 24% (23%) to 12% (14%) for the ^{228}Th (^{226}Ra) source as compared to previous searches. Systematic errors related to the SS fractions are determined by comparisons between the data and MC. Taking into account different calibration sources at various positions, these systematics are evaluated to be 5.8% (4.6%) for Phase I (Phase II).

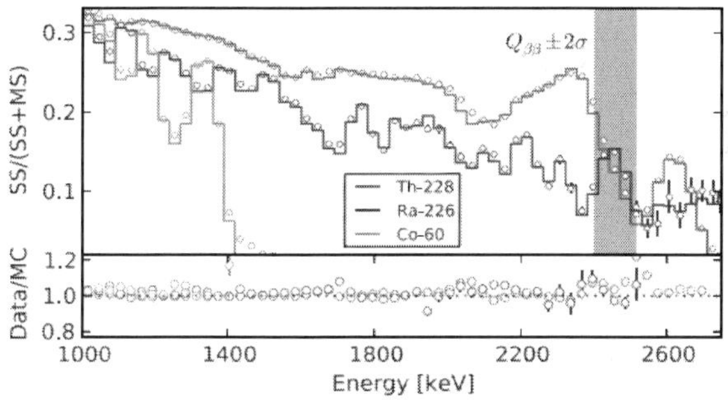

Figure 3. SS fractions for MC (lines) and data (circles) in Phase II using calibration sources positioned near the cathode.

Motivated by the results in [26], this analysis introduces a new discriminator for SS and MS events using a DNN that relies on the waveforms of U-wire signals and is found to outperform the searches in [8,20]. The training inputs for the DNN are gray scale images built by arranging neighboring channels next to each other and encoding the amplitudes of U-wire waveforms as pixel values. The training data are produced in a MC simulation for two classes of events in equal weights: background-like events composed of γ events with uniform energy distribution between 1000 and

3000 keV, and signal-like 0νββ events with a random decay energy restricted to the same energy limits. The locations of the simulated interactions of both types are drawn uniformly from the detector volume to focus discrimination only on the topological event characteristics. This dataset is divided into 90% for training and 10% for validation. The DNN architecture is inspired by the Inception architecture proposed by Google [27] and implemented with the Keras library [28] using the Tensorflow back end [29].

The agreement for the DNN discriminator between the data and MC is improved when signals from U-wire waveforms are first identified by the signal finder in the EXO-200 reconstruction framework and then used to regenerate the images. Because there is no spatial dependence in training the DNN for signal- and background-like events, the standoff distance (minimum distance between the event position and the closest material surface excluding the cathode) is incorporated in the search as a third fit dimension for both SS and MS events. Figure 4 shows a comparison of these two discriminators between simulated and observed data distributions for the ^{226}Ra calibration source, as well as for the measured background-subtracted 2νββ distribution. While keeping as much discriminating power as possible, the binning used for each variable is selected to minimize systematic errors arising from imperfections in the MC simulation.

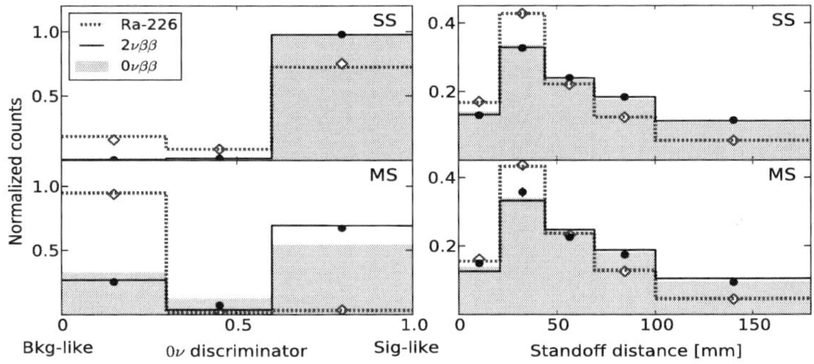

Figure 4. Comparison between data (dots) and MC (solid and dashed lines) for the DNN 0νββ discriminator (left) and standoff distance (right). Shown are the distributions from the ^{226}Ra calibration source (blue) and the background-subtracted 2νββ spectrum from low background data (black). The simulated distributions for 0νββ events are indicated by the red filled regions. The difference in DNN distributions between 0νββ and 2νββ events in MS is due to the higher rate of bremsstrahlung at higher electron energy.

Because the fit cannot resolve the detailed location of backgrounds arising from materials far from the LXe vessel, the ^{238}U, ^{232}Th, and ^{60}Co contributions from such components are assigned to fewer representative locations. For example, all far ^{238}U are represented by the decays in the air gap between the cryostats and the lead shielding in the background model. To account for the errors introduced by this approximation, ^{238}U simulated in the cryostats is used to represent all 238U from remote locations. This is taken to represent an extreme deviation from the more realistic case used in the analysis. The resulting change in the expected number of events near $Q_{\beta\beta}$ is taken as the systematic error of the background model. This is evaluated to be 4.0% (4.6%) in Phase I (Phase II) by adding contributions from ^{238}U, ^{232}Th, and ^{60}Co in quadrature. In addition, toy studies were used to find the average bias in the expected number of events near $Q_{\beta\beta}$ arising from the measured spectral differences between the data and MC for the energy, the DNN 0νββ discriminator, and the standoff distance. The differences between the data and MC for their distributions obtained with the γ calibration sources are used to correct the predicted PDFs, whereas differences in the background-subtracted 2νββ distribution are used for β-like components. The relative differences between the results with toy datasets generated from the corrected PDFs but fit without this correction are added in quadrature for all contributors and sum to 5.8% (4.4%) in Phase I (Phase II). Table I summarizes the contributions to background errors, including other sources unchanged from previous searches.

The measured rate of radon decays in the LXe is used to constrain the appropriate background components arising from these atoms, as described in [13]. The relative rate of cosmogenically produced backgrounds is also constrained [16]. In addition, a possible difference between the energy scale from γ calibration sources (Eγ) and from single- or double- β decays (Eβ) is accounted for by a factor (B) that scales the energy of the β-like components in the ML fit: Eβ=BEγ. B is allowed to freely float and is found to be consistent with unity to the subpercent level in both phases.

The 90% CL median sensitivity for this 0νββ search with the DNN 0νββ discriminator is evaluated to be 5.0×10^{25}yr. The coverage is validated with toy MC studies and found to agree with Wilks's theorem [30,31]. A secondary analysis is performed using a boosted decision tree (BDT) discriminator for MS events, and the BDT discriminator designed in [8] for SS events as the second fit dimension. The BDT for MS is built on variables containing information on the energy fraction of the most energetic deposit, the spatial spread among deposits, and the number of deposits. The

BDT analysis provides comparable but slightly worse (~3%) sensitivity, suggesting that the discrimination power of the DNN discriminator can be mostly accounted for by careful construction of BDT variables. The DNN analysis was selected as the primary analysis prior to unblinding because it had the best sensitivity.

After unblinding the dataset, the SS candidate events within $Q_{\beta\beta}\pm2\sigma$ were examined, which led us to find that one event (originally with energy in this region) was misreconstructed by the denoising algorithm. Its impact on the 0νββ detection efficiency was investigated, and it was found to only affect Phase II with an efficiency loss of <1.0%. A conservative 1.0% error was also added to the signal detection efficiency.

ML fits are performed to Phase I and Phase II separately, and the best-fit results are shown in Fig. 5. No statistically significant evidence for 0νββ is observed. The best-fit background contributions in counts to $Q_{\beta\beta}\pm2\sigma$ are shown in Table 2. The rates normalized over the total fiducial Xe mass, including all isotopes, are $(1.7\pm0.2)\times10^{-3}$ and $(1.9\pm0.2)\times10^{-3}$ $kg^{-1}yr^{-1}keV^{-1}$ for Phase I and Phase II, respectively. The lower limit on thc ^{136}Xe 0νββ half-life is derived by profiling over all nuisance parameters, and it results in T1/2>1.7×10^{25} yr (T1/2>4.3×10^{25}yr) at 90% CL in Phase I (Phase II), whereas the combined limit is T1/2>3.5×10^{25} yr. This corresponds to an upper limit on the Majorana neutrino mass of $\langle m_{\beta\beta}\rangle$<(93−286) meV [5] using the nuclear matrix elements of [32–36] and the phase space factor from [37].

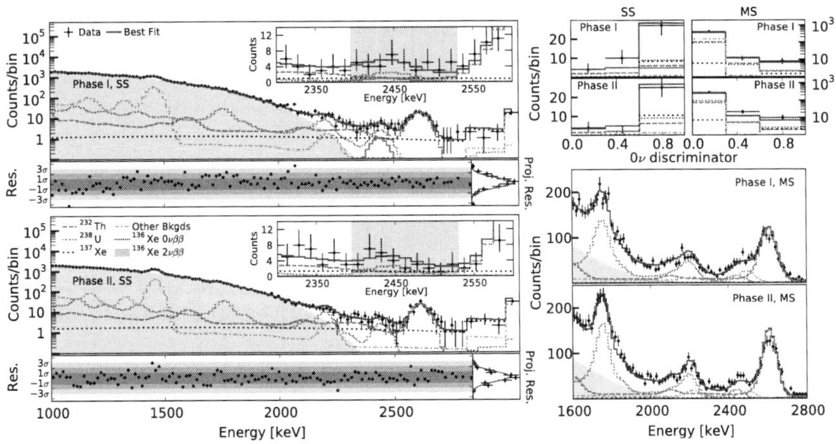

Figure 5. Best fits to the low background (bkgd) data SS energy spectrum for Phase I (top left) and Phase II (bottom left). The energy bins are 15 and 30 keV below and above 2800 keV, respectively. The inset shows an enlarged view

around the best-fit value for $Q_{\beta\beta}$. (Top right) Projection of events in the range of 2395 to 2530 keV on the DNN fit dimension for SS and MS events. (Bottom right) MS energy spectra. Best-fit residuals typically follow normal distributions, with small deviations taken into account in the spectral shape systematic errors.

Table 2. Best-fit background contributions to $Q_{\beta\beta}\pm2\sigma$ versus observed number of events in data.

	238U	232Th	137Xe	**Total**	**Data**
Phase I	12.6	10.0	8.7	32.3±2.3	39
Phase II	12.0	8.2	9.3	30.9±2.4	26

EXO-200 has concluded its operations, reaching a sensitivity to the Majorana neutrino mass of 78–239 meV, which is similar to the most sensitive searches for 0νββ to date [9–12]. The analysis presented here utilizes a DNN, which maximally makes use of detailed event topology information for background rejection, leading to an ~25% improvement relative to the sensitivity using only event energy and simple SS and MS discriminators. This performance results from the unique capabilities of a monolithic LXe TPC, which includes good energy resolution, near maximal signal detection efficiency, and strong topological discrimination of backgrounds. This combination holds promise for nEXO [38,39], which is the planned tonne-scale successor to EXO-200, designed to achieve a sensitivity to the 0νββ half-life of $\sim10^{28}$ yr in ^{136}Xe.

ACKNOWLEDGEMENTS

EXO-200 is supported by the DOE and NSF in the U.S., the NSERC in Canada, the SNF in Switzerland, the IBS in Korea, the RFBR (18-02-00550) in Russia, the DFG in Germany, and CAS and ISTCP in China. The EXO-200 data analysis and simulation use resources of the National Energy Research Scientific Computing Center (NERSC). We gratefully acknowledge the KARMEN collaboration for supplying the cosmic-ray veto detectors, as well as the WIPP for their hospitality.

REFERENCES

1. M. Tanabashi et al. (Particle Data Group), Phys. Rev. D 98, 030001 (2018).
2. W.H. Furry, Phys. Rev. 56, 1184 (1939).
3. W. Rodejohann, Int. J. Mod. Phys. E 20, 1833 (2011).
4. J.D. Vergados, H. Ejiri, and F. Simkovic, Rep. Prog. Phys. 75, 106301 (2012).
5. S. Dell'Oro, S. Marcocci, M. Viel, and F. Vissani, Adv. High Energy Phys. 2016, 2162659 (2016).
6. M.J. Dolinski, A.W.P. Poon, and W. Rodejohann, [Annu. Rev. Nucl. Part. Phys. (to be published)] arXiv:1902.04097.
7. J. Engel and J. Menéndez, Rep. Prog. Phys. 80, 046301 (2017).
8. J.B. Albert et al. (EXO-200 Collaboration), Phys. Rev. Lett. 120, 072701 (2018).
9. A. Gando et al. (KamLAND-Zen Collaboration), Phys. Rev. Lett. 117, 082503 (2016).
10. C.E. Aalseth et al. (MAJORANA Collaboration), Phys. Rev. Lett. 120, 132502 (2018).
11. M. Agostini et al. (GERDA Collaboration), Phys. Rev. Lett. 120, 132503 (2018).
12. C. Alduino et al. (CUORE Collaboration), Phys. Rev. Lett. 120, 132501 (2018).
13. J.B. Albert et al. (EXO-200 Collaboration), Phys. Rev. C 89, 015502 (2014).
14. M. Auger et al., J. Instrum. 7, P05010 (2012).
15. R. Neilson et al., Nucl. Instrum. Methods Phys. Res., Sect. A 608, 68 (2009).
16. J.B. Albert et al. (EXO-200 Collaboration), J. Cosmol. Astropart. Phys. 2016, 029.
17. 3M HFE-7000, https://www.3m.com/3M/en_US/company-us/all-3m-products/~/3M-Novec-7000-Engineered-Fluid/?N=5002385+329066 7267=rud.
18. G.C. Carlson, W.C. Schick, W.L. Talbert, and F.K. Wohn, Nucl. Phys. A125, 267 (1969).
19. J.B. Albert et al. (EXO-200 Collaboration), Phys. Rev. C 95, 025502

(2017).
20. J.B. Albert et al. (EXO-200 Collaboration), Nature (London) 510, 229 (2014).
21. J.B. Albert et al. (EXO-200 Collaboration), Phys. Rev. C 92, 015503 (2015).
22. J. Allison et al., IEEE Trans. Nucl. Sci. 53, 270 (2006).
23. E. Conti et al., Phys. Rev. B 68, 054201 (2003).
24. M. Redshaw, E. Wingfield, J. McDaniel, and E.G. Myers, Phys. Rev. Lett. 98, 053003 (2007).
25. C.G. Davis et al. (EXO-200 Collaboration), J. Instrum. 11, P07015 (2016).
26. S. Delaquis et al., J. Instrum. 13, P08023 (2018).
27. C. Szegedy et al., arXiv:1409.4842.
28. F. Chollet et al., https://keras.io, 2015.
29. M. Abadi et al., https://www.tensorflow.org, 2015.
30. S.S. Wilks, Ann. Math. Stat. 9, 60 (1938).
31. G. Cowan, Statistical Data Analysis (Oxford Science Publications/ Clarendon, Oxford, England, 1998).
32. J. Barea, J. Kotila, and F. Iachello, Phys. Rev. C 91, 034304 (2015).
33. N.L. Vaquero, T.R. Rodriguez, and J.L. Egido, Phys. Rev. Lett. 111, 142501 (2013).
34. J. Engel, F. Simkovic, and P. Vogel, Phys. Rev. C 89, 064308 (2014).
35. J. Menéndez, A. Poves, E. Caurier, and F. Nowacki, Nucl. Phys. A818, 139 (2009).
36. M.T. Mustonen and J. Engel, Phys. Rev. C 87, 064302 (2013).
37. J. Kotila and F. Iachello, Phys. Rev. C 85, 034316 (2012).
38. S. Al Kharusi et al. (nEXO Collaboration), arXiv:1805.11142.
39. J.B. Albert et al. (nEXO Collaboration), Phys. Rev. C 97, 065503 (2018).

CHAPTER

9

Detailed Studies of ^{100}Mo Two-Neutrino Double Beta Decay in NEMO-3

R. Arnold[1], C. Augier[2], A. S. Barabash[3], A. Basharina-Freshville[4], S. Blondel[2], S. Blot[5], M. Bongrand[2], D. Boursette[2], V. Brudanin[6,7], J. Busto[8], A. J. Caffrey[9], S. Calvez[2], M. Cascella[4], C. Cerna[10], J. P. Cesar[11], A. Chapon[12], E. Chauveau[10], A. Chopra[4], L. Dawson[4], D. Duchesneau[13], D. Durand[12], R. Dvornický[6,19], V. Egorov[6], G. Eurin[2,4], J. J. Evans[5], L. Fajt[14], D. Filosofov[6], R. Flack[4], X. Garrido[2], C. Girard-Carillo[2], H. Gómez[2], B. Guillon1[2], P. Guzowski[5], R. Hodák1[4], A. Huber[10], P. Hubert[10], C. Hugon[10], S. Jullian[2], O. Kochetov[6], S. I. Konovalov[3], V. Kovalenko[6], D. Lalanne[2], K. Lang[11], Y. Lemière[12], T. Le Noblet[13], Z. Liptak[11], X. R. Liu[4], P. Loaiza[2], G. Lutter[10], M. Macko[10,14], C. Macolino[2], F. Mamedov[14], C. Marquet[10], F. Mauger[12], A. Minotti[13], B. Morgan[15], J. Mott[4,27], I. Nemchenok[6], M. Nomachi[16], F. Nova[11], F. Nowacki[1], H. Ohsumi[17], G. Oliviéro[12], R. B. Pahlka[11], C. Patrick[4], F. Perrot[10], A. Pin[10], F. Piquemal[10,18], P. Povinec[19], P. Pˇridal[14], Y. A. Ramachers[15], A. Remoto[13], J. L. Reyss[20], C. L. Riddle[9], E. Rukhadze[14], R. Saakyan[4], A. Salamatin6,

Citation: Arnold, R., Augier, C., Barabash, A.S. et al. "Detailed studies of two-neutrino double beta decay in NEMO-3". Eur. Phys. J. C 79, 440 (2019). https://doi.org/10.1140/epjc/s10052-019-6948-4

R. Salazar[11], X. Sarazin[2], J. Sedgbeer2[1], Yu. Shitov[6], L. Simard[2,22], F. Šimkovic[6,19], A. Smetana[14], K. Smolek[14], A. Smolnikov[6], S. Söldner-Rembold[5], B. Soulé[10], I. Štekl[14], J. Suhonen[23], C. S. Sutton[24], G. Szklarz[2], H. Tedjditi[8], J. Thomas[4], V. Timkin[6], S. Torre[4], Vl. I. Tretyak[25], V. I. Tretyak[6,a], V. I. Umatov[3], I. Vanushin[3], C. Vilela[4], V. Vorobel[26], D. Waters4, F. Xie[4], A. Žukauskas[26]

[1] IPHC, ULP, CNRS/IN2P3, 67037 Strasbourg, France

[2] LAL, Université Paris-Sud, CNRS/IN2P3, Université Paris-Saclay, 91405 Orsay, France

[3] NRC "Kurchatov Institute", ITEP, 117218 Moscow, Russia

[4] UCL, London WC1E 6BT, UK

[5] University of Manchester, Manchester M13 9PL, UK

[6] JINR, 141980 Dubna, Russia

[7] National Research Nuclear University MEPhI, 115409 Moscow, Russia

[8] Aix Marseille Université, CNRS, CPPM, 13288 Marseille, France

[9] Idaho National Laboratory, Idaho Falls, ID 83415, USA

[10] CENBG, Université de Bordeaux, CNRS/IN2P3, 33175 Gradignan, France

[11] University of Texas at Austin, Austin, TX 78712, USA

[12] LPC Caen, ENSICAEN, Université de Caen, CNRS/IN2P3, 14050 Caen, France

[13] LAPP, Université de Savoie, CNRS/IN2P3, 74941 Annecy-le-Vieux, France

[14] Institute of Experimental and Applied Physics, Czech Technical University in Prague, 12800 Prague, Czech Republic

[15] University of Warwick, Coventry CV4 7AL, UK

[16] Osaka University, 1-1 Machikaneyama Toyonaka, Osaka 560-0043, Japan

[17] Saga University, Saga 840-8502, Japan

[18] Laboratoire Souterrain de Modane, 73500 Modane, France

[19] FMFI, Comenius University, 842 48 Bratislava, Slovakia

[20] LSCE, CNRS, 91190 Gif-sur-Yvette, France

[21] Imperial College London, London SW7 2AZ, UK

[22] Institut Universitaire de France, 75005 Paris, France

[23] Jyväskylä University, 40351 Jyvaskyla, Finland

[24] MHC, South Hadley, MA 01075, USA

[25] Institute for Nuclear Research, Kiev 03028, Ukraine

[26] Faculty of Mathematics and Physics, Charles University in Prague, 12116 Prague, Czech Republic

[27] Present Address: Boston University, Boston, MA 02215, USA

ABSTRACT

The full data set of the NEMO-3 experiment has been used to measure the half-life of the two-neutrino double beta decay of ^{100}Mo to the ground state of $T_{1/2} = \left[6.81 \pm 0.01\ (\text{stat})^{+0.38}_{-0.40}\ (\text{syst})\right] \times 10^{18}$ year. The two-electron energy sum, single electron energy spectra and distribution of the angle between the electrons are presented with an unprecedented statistics of 5×1055×105 events and a signal-to-background ratio of ~~ 80. Clear evidence for the Single State Dominance model is found for this nuclear transition. Limits on Majoron emitting neutrinoless double beta decay modes with spectral indices of n=2,3,7n=2,3,7, as well as constraints on Lorentz invariance violation and on the bosonic neutrino contribution to the two-neutrino double beta decay mode are obtained.

INTRODUCTION

Spontaneous nuclear double beta decay is a second order weak interaction process that was theoretically considered for the first time by Goeppert-Mayer [1]. It can occur in some even-even nuclei when two bound neutrons simultaneously undergo beta decay and are transformed into two bound protons emitting two electrons and two (anti)neutrinos. Two-neutrino double beta decay, 2νββ, is one of the rarest directly observed radioactive processes with half-lives ranging from 7×10^{18} to 2×10^{21} years [2, 3].

The decay rate of 2νββ decay can be expressed as

$$1/T_{1/2}^{2\nu} = g_A^4 G^{2\nu} |M^{2\nu}|^2, \quad (1)$$

where g_A is the axial-vector coupling constant, G^{2v} is a phase space factor, and M^{2v} is a nuclear matrix element (NME). Measurement of the 2νββ half-life gives direct access to the value of the NME for this process and therefore provides experimental input into nuclear models that are used to evaluate NMEs. Moreover, 2νββ may provide answers to the question of g_A quenching in nuclear matter that is currently being actively discussed

[4,5,6,7,8]. Detailed studies of $2\nu\beta\beta$ may therefore be useful to improve NME calculations for the neutrinoless mode of double beta decay, $0\nu\beta\beta$, the process which violates total lepton number and is one of the most sensitive probes of physics beyond the Standard Model. A recent review of the $0\nu\beta\beta$ NME calculation methods, challenges and prospects can be found in [9].

Previous measurements have shown that the ^{100}Mo $2\nu\beta\beta$ half-life is shorter compared to other $\beta\beta$ isotopes [10,11,12,13,14,15,16,17], and it is therefore a promising nucleus for precise studies of the process. Here we present the most accurate to date study of ^{100}Mo $2\nu\beta\beta$ decay including single electron energy and angular distributions of the electrons emitted in the decay with an unprecedented statistics of 5×10^5 events. The impact of the single electron energy spectra on nuclear models that are used to calculate the NME is also presented.

Searches for most commonly discussed $0\nu\beta\beta$ mechanisms (exchange of a light Majorana neutrino, right-handed currents, super-symmetry) with NEMO-3 have been reported earlier in [18, 19]. In this paper we present results obtained for ^{100}Mo $0\nu\beta\beta$ decay accompanied by the emission of Majoron bosons with spectral indices $2n\geq2$, as well as constraints on contributions from bosonic neutrinos and from Lorentz invariance violation to $2\nu\beta\beta$ spectra of ^{100}Mo.

THE NEMO-3 DETECTOR

The NEMO-3 detector, its calibration and performance are described in detail in [20] and more recently in [19]. A combination of tracking and calorimetric approaches allows for a full reconstruction of $\beta\beta$ event topology. A tracking chamber is used to reconstruct electron tracks, their origin and end points. The electron energies and arrival times are measured with a plastic scintillator calorimeter. The cylindrical detector measuring 3 m in height and 5 m in diameter is made up of 20 wedge-shaped sectors of identical size. Each sector hosts 7 thin foil strips containing a $\beta\beta$ isotope. The source foils are positioned in the middle of the tracking detector at a radius of 1 m and have a height of 2.48 m.

The tracking detector is based on a wire chamber made of 6180 open drift cells operating in Geiger mode with helium as the main working gas with the addition of ethanol (4%), argon (1%) and water vapour (0.15%). The wire cells are strung vertically parallel to the source foils and have average transverse and longitudinal resolutions of 0.5 mm and 0.8 cm (σ)

respectively. The tracking volume is surrounded by a segmented calorimeter composed of 1940 optical modules made of 10 cm thick polystyrene scintillator blocks coupled to low radioactivity photomultiplier tubes (PMT). The energy resolution of optical modules for 1 MeV electrons ranges from 5.8 to 7.2% and the time resolution is 250 ps (σ). The detector was calibrated by deploying ^{207}Bi, 9090Sr and ^{232}U sources during the course of data collection. The stability of the PMT gains was monitored by a dedicated light injection system that was run every 12 hours.

The NEMO-3 detector is supplied with a solenoid which generates a 25 G magnetic field parallel to the tracking detector wires and provides charge identification by track curvature. The detector is surrounded by passive shielding consisting of a 19 cm thick iron plates to suppress the external gamma ray flux, and of borated water, paraffin and wood to moderate and absorb environmental neutrons.

One of the unique advantages of the NEMO-3 technology is the ability to unambiguously identify electrons, positrons, gamma- and delayed alpha-particles. This approach leads to a strong suppression of backgrounds by eliminating events that do not exhibit a ββ topology. In addition, it allows for an efficient background evaluation by selecting event topologies corresponding to specific background channels. An electron is identified by a reconstructed prompt track in the drift chamber matching to a calorimeter deposit. Extrapolating the track to the foil plane defines the event vertex in the source.

The track extrapolation to the calorimeter identifies the impact point of the electron track with the corresponding optical module and is used to correct the reconstructed energy of the electron deposited in the scintillator. The track curvature in the magnetic field is used to distinguish electrons from positrons.

A γ-ray is identified as an energy deposit in the calorimeter without an associated track in the drift chamber. An α-particle is identified by a short straight track delayed with respect to the prompt electron in order to tag ^{214}Bi →→ ^{214}Po delayed coincidences.

The NEMO-3 detector took data at the Modane Underground Laboratory (LSM) in the Frejus tunnel at a depth of 4800 m w.e. enabling the cosmic muon flux suppression by a factor of $>10^6$. The detector hosted source foils of 7 different ββ isotopes. The two isotopes with the largest mass were ^{100Mo}(6.914 kg) [19] and 82Se (0.932 kg) [21] with smaller amounts of ^{48}Ca, ^{96}Zr, ^{116}Cd, ^{130}Te and ^{150}Nd [22,23,24,25,26].

Two types of purified molybdenum foils were installed in NEMO-3, metallic and composite. Both foil types were enriched in ^{100}Mo with the isotopic enrichment factor ranging from 95.14±0.05 to 98.95±0.05%. The average enrichment factor was 97.7% for metallic foils and 96.5% for composite foils.

The metallic foils contained 2479±52479±5 g of 100100Mo. The mean metallic foil density is 58 mg/cm^2 with a total foil surface of 43,924 cm^2. The composite foils contained 4435±13g of ^{100}Mo.

They were produced by mixing a fine molybdenum powder with polyvinyl alcohol (PVA) glue and deposited between Mylar foils of 19 μm thickness. The average surface density of the composite foils is 66 mg/cm2cm2 and the total foil surface area is 84,410 cm^2.

Monte Carlo (MC) simulations are performed with a GEANT-3 based [27] program using the DECAY0 [28] event generator. The time-dependent status and performance of the detector are taken into account in modelling the detector response.

The data presented here were collected between February 2003 and October 2010 with a live time of 4.96 years and a total exposure of 34.3 kg year of ^{100}Mo. This is the same exposure as that used for 0νββ results published earlier [19].

BACKGROUND MODEL

Trace quantities of naturally-occurring radioactive isotopes can occasionally produce two-electron events and thus can mimic ββ-decay events. The largest contributions come from isotopes that are progenies of ^{238}U (^{234}mPa, ^{214}Pb, ^{214}Bi, ^{210}Bi) and of ^{232}Th (^{228}Ac, ^{212}Bi, ^{208}Tl), as well as ^{40}K.

The background is categorized as internal if it originates from radioactive decays inside the ββ source foils, see Fig. 1a. Two electrons can be produced via ββ-decay followed by a Møller scattering, ββ-decay to an excited state with the subsequent internal conversion or due to Compton scattering of the de-excitation photon.

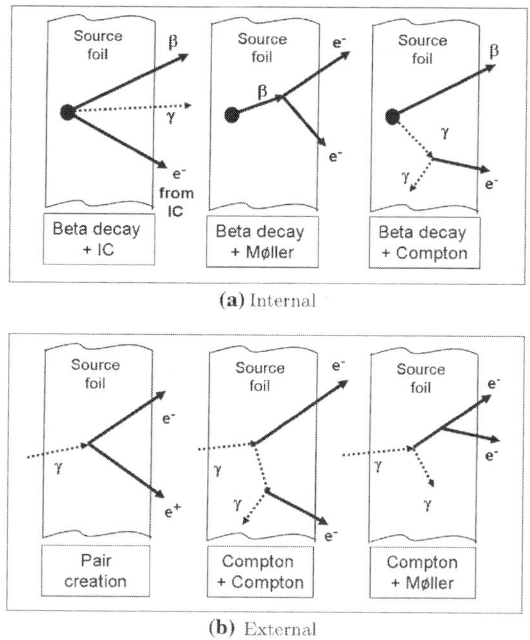

Figure 1. Mechanisms of internal (**a**) and external (**b**) background production in the source foil

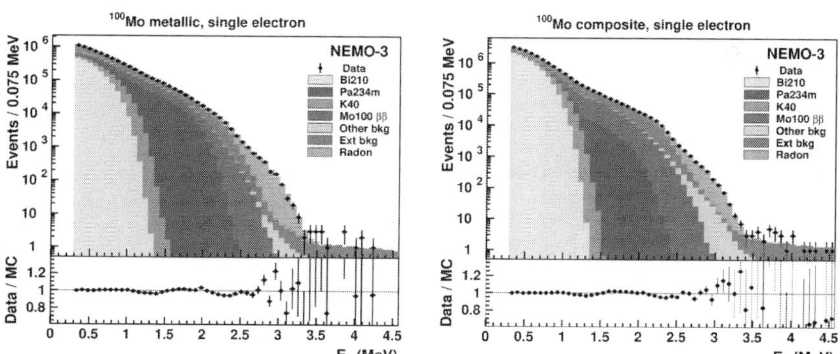

Figure 2. Single electron events energy spectra for metallic and composite molybdenum. The error bars correspond to statistical uncertainty only

Decays inside the tracking detector volume form a separate background category. The main source of this background is radon, ^{222}Rn. The decay of radon progenies near the source foil can produce signal-like events in an analogous manner to internal background decays.

The last background category is due to the external γ-ray flux produced by decay of radioactive isotopes in detector components, the surrounding area and due to neutron interactions in the shield and material of the detector. The PMT glass is the main source of these γ-rays.

They can produce two-electron events due to e^+e^- pair creation in the source foil and subsequent charge misidentification, double Compton scattering or Compton scattering followed by Møller scattering, see Fig. 1b.

A detailed discussion of the NEMO-3 background model is presented in [29] and results of screening measurements can be found in [19, 20, 29]. Here we follow the same background model as that presented for the ^{100}Mo 0νββ analysis [19].

However, radioactive isotopes contributing to the low energy region of the ^{100}Mo 2νββ spectrum were not relevant for the 0νββ analysis in [19] and are therefore discussed in more detail below. The background in question comcs from traces of ββ-decaying isotopes ^{210}Bi, 40K and 234mPa in ^{100}Mo foils.

In addition, ^{100}Mo 2νββ decay to the 0+101+ excited state of 100100Ru is also taken into account as a source of internal background. The experimental half-life value of T1/2=6.7−0.4+0.5×10^{20} years [3] is used to evaluate this contribution.

The activities of β-emitters in ^{100}Mo foils are determined from the fit to the electron energy distribution for a single electron event sample, which is shown in Fig. 2 separately for metallic and composite foils.

To disentangle the ^{210}Bi contribution from the source foils and the surface of the tracker wires the activity measured in [29] is used for the latter. Figure 2 shows the sum of both contributions. Secular equilibrium is assumed between ^{214}Pb and ^{214}Bi. The same is done between ^{228}Ac, ^{212}Bi and ^{208}Tl, where the branching ratio of 35.94% is taken into account. There is sufficiently good agreement between data and MC for the single electron energy spectrum.

The observed deviations of MC from data are within 6% and are not significant when the systematic uncertainty on the external background is taken into account.

The results of the internal ^{100}Mo foil contamination measurements carried out with the NEMO-3 detector are shown in Table 1.

Table 1. 100100Mo source foil contamination activities measured with the NEMO-3 detector. Activities of 214214Bi and 208208Tl are from [19]

Source	^{100}Mo metallic	^{100}Mo composite
^{214}Bi internal, mBq/kg	0.060 ± 0.019	0.305 ± 0.038
^{214}Bi mylar, mBq/kg	—	1.05 ± 0.06
^{208}Tl, mBq/kg	0.087 ± 0.004	0.128 ± 0.003
^{234m}Pa, mBq/kg	11.40 ± 0.06	2.10 ± 0.03
^{40}K , mBq/kg	8.67 ± 0.05	13.57 ± 0.04
^{210}Bi, mBq/m^2	5.51 ± 0.03	19.42 ± 0.03

TWO-NEUTRINO DOUBLE BETA DECAY OF ^{100}MO

Candidate ββ events are selected by requiring two reconstructed electron tracks, each associated with an energy deposited in an individual optical module. The energy deposited by the electron in a single optical module should be greater than 300 keV. Each PMT must be flagged as stable according to the light injection survey [19]. The tracks must both originate from the 100100Mo source foil, and their points of intersection with the plane of the source foil must be within 4 cm transverse to and 8 cm along the direction of the tracker wires, in order to ensure that the two tracks are associated to a common event vertex. The track curvatures must be consistent with electrons moving outwards from the source foil. The timing and the path length of the electrons must be consistent with the hypothesis of simultaneous emission of two electrons from a common vertex in the ^{100}Mo source foil [19]. There should be no γγ-ray hits and αα-particle tracks in the event.

After the above event selection there are 501,534 ^{100}Mo two-electron candidate events, with 193,699 coming from the metallic foils and 307,835 from the composite foils. Table 2 shows the number of expected background and candidate signal events in ^{100}Mo foils. The number of 2νββ events is obtained from a binned log-likelihood fit to the two-electron energy sum distribution under the single state dominance (SSD) nuclear model, as detailed below. The average signal-to-background ratio is S/B=79, with S/B=63 for the metallic foils and S/B=94 for the composite foils. The detector acceptance and selection efficiency for 2νββ ^{100}Mo events calculated using MC simulations is ϵ=(2.356±0.002)%, with ϵ_{met}=(2.472±0.003)% and ϵ_{com}=(2.292±0.002)% for the metallic and composite molybdenum foils respectively. Using the above values gives the ^{100}Mo 2νββ-decay half-life of $T_{1/2}=(6.65\pm0.02)\times10^{18}$ year for the metallic foils and $T_{1/2}=(6.91\pm0.01)\times10^{18}$ year for the composite foils. The difference between the two sample measurements may be explained by inaccuracy of the thin foil modelling and is taken into account in estimation of the systematic uncertainty in Sect. 4.2. We consider the mean value over the two data samples as the more reliable half-life estimation

$$T_{1/2} = (6.81 \pm 0.01) \times 10^{18} \text{ year.} \quad (2)$$

Table 2. Expected number of background events in the two-electron channel and the number of ^{100}Mo 2νββ **candidate events in molybdenum foils**.

Source	Metallic	Composite	Total ^{100}Mo
^{228}Ac, ^{212}Bi, ^{208}Tl	49.5 ± 0.5	142.3 ± 1.3	191.8 ± 1.4
^{214}Pb, ^{214}Bi	14.2 ± 0.1	177.2 ± 0.7	191.3 ± 0.7
^{40}K	101.4 ± 2.5	296.0 ± 7.3	397.5 ± 7.7
^{234m}Pa	1783.8 ± 11.8	656.7 ± 4.3	2440.5 ± 12.5
^{210}Bi	25.6 ± 1.4	90.3 ± 2.8	115.9 ± 3.1
Radon	434.3 ± 6.2	590.3 ± 5.2	1024.6 ± 8.1
Ext Bkg	562.7 ± 9.7	1238.6 ± 14.7	1801.3 ± 17.6
$\beta\beta\ 0_1^+$	48.6 ± 0.8	71.1 ± 1.0	119.7 ± 1.3
Tot bkg	3020 ± 17	3263 ± 18	6283 ± 25
$\beta\beta$ g.s.	190683 ± 117	304571 ± 144	495254 ± 186
Data	193699	307835	501534

The two-electron energy sum spectra and the distributions of cosine of the angle between two electrons emitted from 100100Mo foil are shown in Fig. 3, separately for the metallic and composite foils as well as for the total 100100Mo sample.

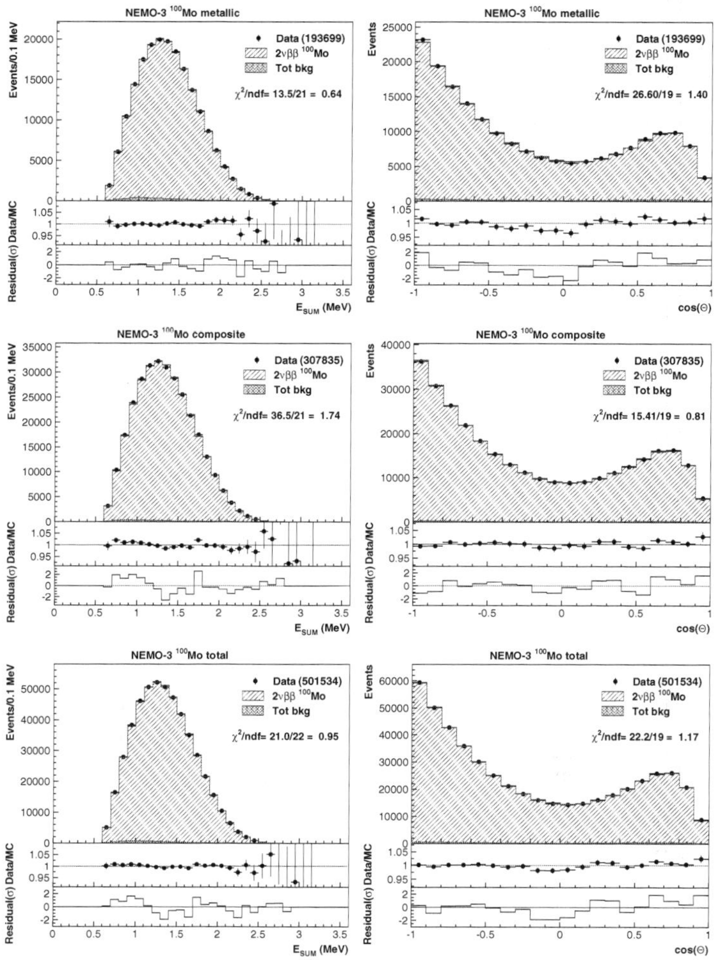

Figure 3. Distributions of two-electron summed kinetic energy and the opening angle between two electron tracks in 100100Mo foils after an exposure of 34.3 kg year. Data are compared to the MC prediction of the SSD model (see text), where the resulting event numbers are taken from a binned log-likelihood fit

The electron energy measured in the calorimeter is smaller than the energy at the point of origin due to energy losses in the foil and in the drift chamber. For instance in the case of ^{100}Mo 2νββ decay the mean electron track length from the source foil to the calorimeter is 75 cm and the mean energy loss of electrons in the drift chamber is 43 keV. The single and summed electron energy distributions are presented for the measured values of the electron kinetic energy Ee and sum of the measured electron kinetic energies E_{SUM}, respectively, i.e., without correction for the energy loss.

The angular distribution is corrected with the well-measured distribution of the opening angle between two electrons emitted in ^{207}Bi decay. The MC distribution of the cosine of the angle between two electron tracks has been reweighted based on data collected in the regular energy calibration runs performed with ^{207}Bi sources. The correction is biggest for small opening angles, and is at the level of 4% on average.

Role of intermediate nuclear states in ^{100}Mo 2νββ transition

The nuclear ββ decay (A,Z) →→ (A,Z+2) is realized via two subsequent virtual ββ transitions through the complete set of states of intermediate nucleus (A,Z+1). In the case of ^{100}Mo 2νββ transition between the ground states of the parent (^{100}Mo) and daughter (^{100}Ru) nuclei with spin-parity 0^+ the process is governed by two Gamow-Teller transitions through 1^+ states of ^{100}Tc. Nuclear theory does not predict a priori whether there is a dominance of transition through the 1^+ ground state (SSD hypothesis [30,31,32]) or through higher lying excited states, namely from the region of the Gamow-Teller resonance (HSD hypothesis). The SSD versus HSD analysis is feasible as the ground state of ^{100}Tc has spin-parity JP=1^+ and is lying close to the ground state of ^{100}Mo.

The evidence in favour of SSD in ^{100}Mo 2νββ decay was already observed at the beginning of NEMO-3 data analysis [33]. Further hints for the SSD model in the ^{100}Mo 2νββ decay were obtained in charge-exchange experiments by observing a strong Gamow-Teller transition to the 1+1+ ground state of ^{100}Tc in the ^{100}Mo(^{33}He,t)^{100}Tc reaction [34]. It was estimated that this transition could contribute as much as 80% to the total value of the ^{100}Mo 2νββ matrix element.

It was shown in [31, 32] that SSD and HSD models can be directly distinguished by making high precision kinematics measurements of 2νββ decay products. The distribution of the individual electron energies was shown to have the most discriminating power, especially in the low energy part of the spectrum. Figure 4 shows the individual electron energy spectra for three nuclear models, with SSD-3 being a modification of the SSD model where a finer structure of intermediate states is accounted for [35, 36].

Figure 5 shows the energy sum and angular distribution of the final state electrons where the data are fitted with the HSD model. The tension between the data and the model is evident already from these distributions with χ^2/ndf=4.57 (pvalue=5.3×10^{-12}) and χ^2/ndf=1.98 (p_{value}=0.007) for the energy sum and angular distributions respectively. However, the strongest evidence comes from the single electron energy distributions shown in Fig. 6 for the three models, HSD, SSD and SSD-3, fitted to the data. It is clear from the distributions and χ^2 values that the HSD model can be ruled out with high confidence while SSD and SSD-3 provide a fairly good description of the data.

The difference between SSD and SSD-3 in describing the data is maximised with a cut on the electron energy sum of E_{SUM}>1.4 MeV as shown in Fig. 7, which also increases the signal-to-background ratio. There is a slight preference of the SSD-3 model over SSD in this case, contrary to the results obtained without this cut demonstrated at Fig. 6. Due to systematic effects connected to the energy reconstruction and electron energy loss simulations discussed below these two models cannot be discriminated against each other. The SSD is chosen as the baseline model and is used to estimate the ^{100}Mo 2νββ half-life (see Sect. 4 and Fig. 3). We note that differences in the low energy part of the single electron spectra (Fig. 4) affect the selection efficiency of ^{100}Mo 2νββ events. Consequently, the measured half-life for the SSD model is 14% shorter than the analogous result for the HSD model. The SSD-3 model would give a 1.8% shorter half-life than that of the SSD model.

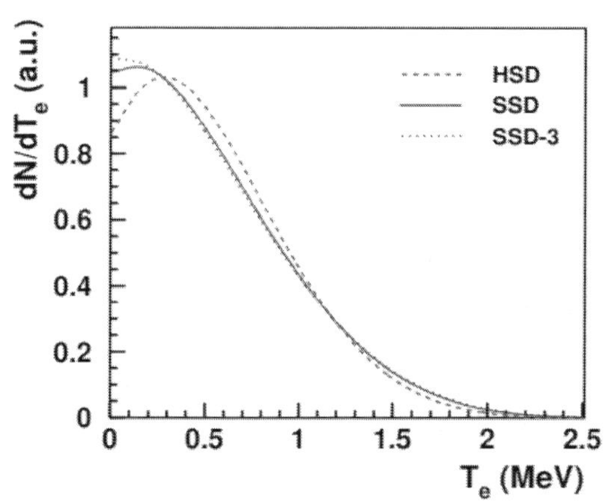

Figure 4. Theoretical distributions of the individual electron kinetic energy for three models of ^{100}Mo 2νββ decay: HSD, SSD and SSD-3

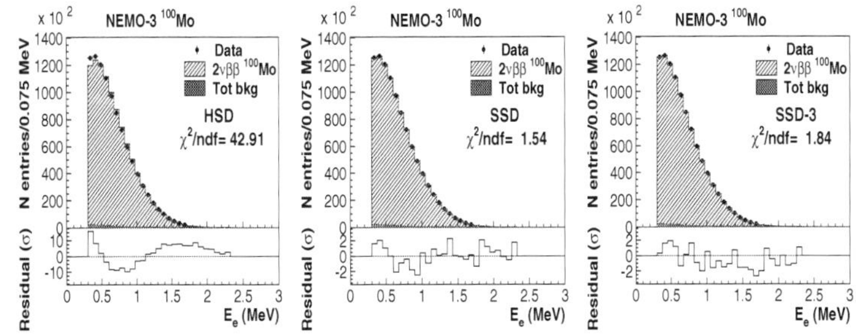

Figure 5. Two-electron events. Energy sum and cosine of the angle between the two electrons for HSD model

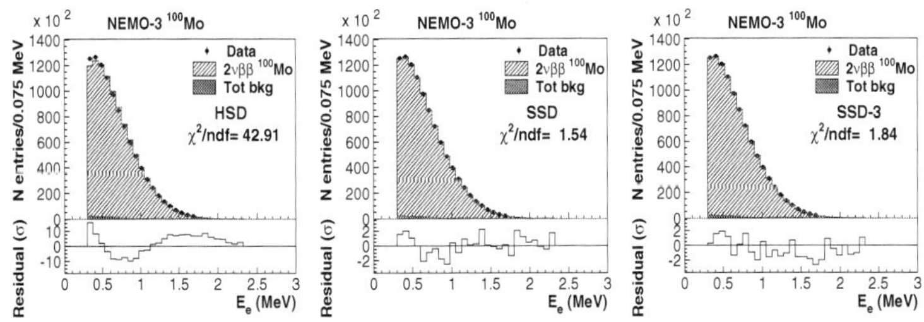

Figure 6. Distribution of individual electron kinetic energy in the ββ channel from ^{100}Mo foils compared with MC spectra under the HSD, SSD and SSD-3 nuclear models. The HSD hypothesis is excluded (χ^2/ndf=1159/27) while the data are consistent with the SSD and SSD-3 models (χ^2/ndf=41.5/27 and χ^2/ndf=49.7/27 respectively)

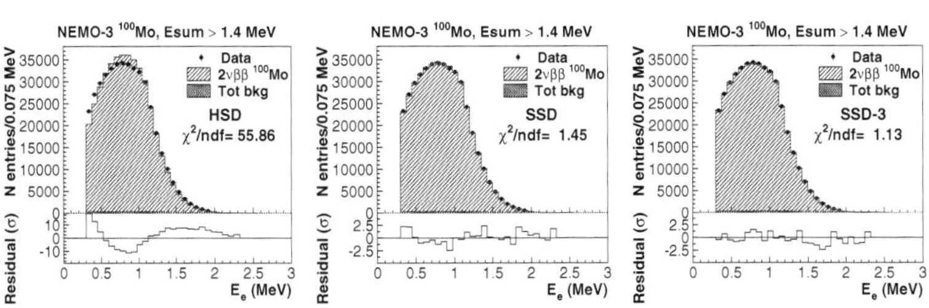

Figure 7. Distribution of individual electron kinetic energy in the ββ channel from ^{100}Mo foils with the cut on the summed electron energy E_{SUM}>1.4 MeV to maximise the signal-to-background ratio. The data are compared with MC spectra under the HSD, SSD and SSD-3 nuclear models. The HSD hypothesis is excluded (χ^2/ndf=1508/27) while the data are consistent with the SSD and SSD-3 models (χ^2/ndf=39/27 and χ^2/ndf=30.6/27respectively)

Systematic uncertainties on ^{100}Mo 2νββ half-life

Apart from the statistical uncertainties on the fitted number of signal events, the measurement of the 2νββ decay half-life is subject to a number of systematic uncertainties.

The uncertainty on the reconstruction and selection efficiency including the detector acceptance effects is evaluated by carrying out dedicated calibrations with ^{207}Bi sources whose activities were known with a 5% uncertainty. Consequently, the systematic error on the signal efficiency is taken to be 5%.

Limited precision of MC simulation program in modelling of multiple scattering processes and electron energy losses in molybdenum ββ source foils also contribute to the total systematic error. Corresponding uncertainty is evaluated as the difference between the mean half-life value and the values obtained with metallic (−2.3%) and composite (+1.5+1.5%) foils.

The 1.8% half-life value difference between the SSD and SSD-3 nuclear models is taken as a systematic error due to the ^{100}Mo 2νββ decay model.

The uncertainty on the energy scale translates into an error on the half-life measurement of 0.6%.

The ^{100}Mo mass uncertainty gives directly the corresponding uncertainty of the half-life value and is estimated to be 0.2%.

The error on the activities of external backgrounds, radon and the foil contamination with ^{214}Bi and ^{208}Tl is 10% as shown in [19]. The uncertainty on the backgrounds from 40K in the source foils as well as from ^{210}Bi is estimated to be 4%. The observed discrepancy in the 234m234mPa decay scheme reported in [37] and [38] lead to a 30% normalisation uncertainty on the activity from this isotope. The 7.5% error on the rate of the ^{100}Mo 2νββ decay to the excited states [3] is also taken into account. Overall, due to a high signal-to-background ratio the uncertainty on all background contributions produces only a 0.2% systematic uncertainty on the ^{100}Mo 2νββ half-life determination.

The systematic uncertainties on the measured 2νββ ^{100}Mo half-life are summarised in Table 3. The individual sources of the systematic error are assumed to be uncorrelated and the total uncertainty is obtained to be [+5.6,−5.8+5.6,−5.8]%.

The final value of the half-life for the 2νββ decay of ^{100}Mo under the SSD model is:

$$T_{1/2} = \left[6.81 \pm 0.01 \text{ (stat)}^{+0.38}_{-0.40} \text{ (syst)}\right] \times 10^{18} \text{ year.} \quad (3)$$

This value is in good agreement with the world average value of $(7.1\pm0.4)\times10^{18}$year [3] and with a recent result obtained using low-temperature scintillating bolometers ($Li_2^{100}MoO_4$), $[6.90\pm0.15(\text{stat})\pm0.37(\text{syst})]\times10^{18}$ year [17].

SEARCH FOR NEW PHYSICS WITH CONTINUOUS ^{100}MO BB ENERGY SPECTRA

Deviations in the shape of the 2νββ energy spectra can provide hints of new physics. Below we report on results of searches for physics beyond the Standard Model that can modify the two-electron energy sum distribution of the ^{100}Mo 2νββ decay due to emission of Majoron bosons, the existence of a bosonic component in the neutrino states and possible Lorentz invariance violation.

The shape of the two-electron energy sum distribution in various types of decays is characterized by the spectral index n [39], being determined by the phase space $G\sim(Q_{\beta\beta}-T)^n$, where $Q_{\beta\beta}$ is the full energy released in the decay minus two electron masses and T is the sum of kinetic energies of two emitted electrons. The ordinary 2νββ decay has a spectral index of n=5. Any modification from this functional form can be an indication of new physics.

A number of grand unification theories predict the existence of a massless or light boson which couples to the neutrino. Neutrinoless ββ decay can proceed with the emission of one or two Majoron bosons resulting in a continuous energy sum spectrum with spectral index n≠5. The decay accompanied by a single Majoron emission has n=1,2n=1,2 and 3, while models with two Majoron emissions predict n=3 and 7 (see [40] and references therein). The results for the neutrinoless ββ decay with the emission of a Majoron corresponding to the spectral index n=1n=1 have already been published in [18, 19]. The Majoron-accompanied 0νββ decay modes with spectral indices n=2,3 and 7 are considered here.

It was noted in [41] that violation of the Pauli exclusion principle resulting in a bosonic component in the neutrino states can be tested by looking at the shape of the energy and angular distributions of the electrons emitted in ββ decay. For the two-electron energy sum distribution the corresponding index would be n=6.

Lorentz invariance is a fundamental symmetry. However, new physics at very high energies close to the Planck scale can manifest itself in small effects at low energies, including Lorentz invariance violation. Consequently, searches for non-Lorentz invariant effects have attracted active theoretical and experimental effort [42,43,44,45]. The possibility to test Lorentz invariance with ββ decay was discussed in [46, 47]. In case of 2νββ decay the Lorentz invariance violation may be manifested as a modification of the conventional electron sum spectrum due to an additional contribution of the Lorentz-violating perturbation with a spectral shape of n=4.

Table 3. Summary of systematic uncertainties on the mcasured 2νββ ^{100}Mo half-life

Source of uncertainty	Effect on $T_{1/2}^{2\nu}$ (%)
Absolute normalization of ϵ_{2e}	±5
Thin source foil modelling	[+1.5, −2.3]
^{100}Mo decay model	±1.8
Energy calibration	±0.6
^{100}Mo mass	±0.2
Background uncertainty	±0.2
Total	[+5.6, −5.8]

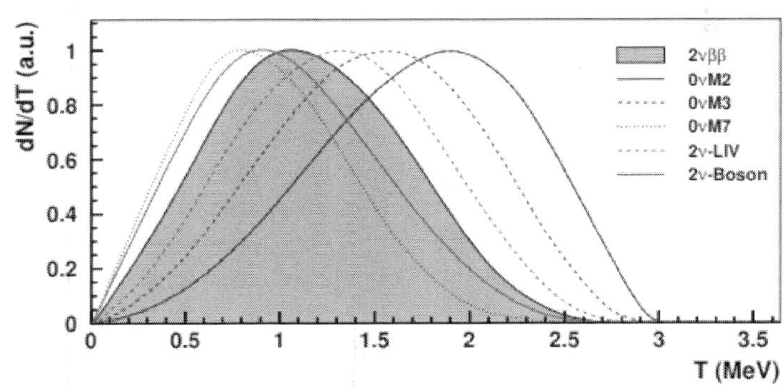

Figure 8. Spectrum of the of kinetic energy sum of two electrons for the standard ^{100}Mo 2νββ decay (spectral index n=5) compared to the spectra for neutrinoless ββ decay with the emission of one or two Majorons 0νMn (n=2,3,7); shape of the perturbation to the standard 2νββ decay due to Lorentz invariance violation 2ν-LIV (n=4) and spectrum for 2νββ decay with bosonic neutrino 2ν-Boson (n=6n=6)

The theoretical distributions of the two-electron energy sum for different modes of ^{100}Mo ββ decay discussed above are shown in Fig. 8. The difference in the shape of the distributions due to different spectral indices n is used to evaluate possible contributions from physics beyond the Standard Model. No significant deviations from the expected ^{100}Mo 2νββ spectral shape (n=5) have been observed and therefore limits on new physics parameters have been set using the full energy sum spectrum of the full ^{100}Mo data set. The contributions of the ββ decay modes with spectral indices n=2,3,6,7 are constrained with a modified frequentist CLs method [48, 49] using a profile likelihood fitting technique (COLLIE software package [50]). A profile likelihood scan is used for the distribution with the spectral index n=4 in order to explore possibility of negative as well as positive Lorentz-violating perturbation.

The systematic uncertainties on background contributions discussed in Sect. 4.2, the 5% uncertainty on the detector acceptance and selection efficiency for signal, a possible distortion in the shape of the two-electron energy sum spectrum due to the energy calibration accuracy, as well as a 5% error on the modelling of the energy loss of electrons are taken into account in limit setting without imposing a constraint on the normalization of standard 2νββ contribution.

The limits on the half-lives for different 0νββ modes with Majoron(s) emission, and for the bosonic neutrino admixture obtained with the CLs method are given in Table 4.

Table 4. Lower bounds on half-lives (1021 year) at 90% C.L. from 0νββ searches with Majoron emission (spectral indices n=2,3,7), and searches for the bosonic neutrino admixture. The ranges in the expected half-life limits are from the ±1σ range of the systematic uncertainties on the background model, signal efficiency and distortions in the shape of the energy spectrum

Decay mode		Expected		Observed
	-1σ	Median	$+1\sigma$	
Majoron n=2	13	9.2	6.2	9.9
Majoron n=3	6.1	4.3	2.9	4.4
Majoron n=7	1.8	1.3	0.88	1.2
$2\nu\beta\beta$ Bosonic ν	1.7	1.2	0.83	1.2

The half-life limits on the Majoron 0νββ modes are translated into the upper limits on the lepton number violating parameter ⟨gee⟩, which is proportional to the coupling between the neutrino and the Majoron boson, using the relation,

$$1/T_{1/2} = |\langle g_{ee}\rangle|^m G|M|^2, \tag{4}$$

where G is the phase space (which includes the axial-vector coupling constant gA), M is the nuclear matrix element, and m=2(4) is the mode with the emission of one (two) Majoron particle(s). The M and G values are taken from [51]. For the single Majoron emission and n=3, M and G are taken from [52]. There are no NME and phase space calculations available for n=2.

The upper limits on the Majoron-neutrino coupling constant ⟨gee⟩ are shown in Table 5. One can see that the NEMO-3 results presented here are the current best limits for n=3n=3 and the single Majoron emission mode and are comparable with the world's best results from the EXO-200 [53] and GERDA [54] experiments for the other two modes.

Table 5. Upper limits on the Majoron-neutrino coupling constant ⟨gee⟩ ⟨gee⟩ from NEMO-3 (100100Mo, this work) and EXO-200 (136136Xe) [53] and GERDA (7676Ge) [54] experiments. All limits are at 90% C.L. The ranges are due to uncertainties in NME calculations

n	Mode	^{100}Mo	^{136}Xe [49]	^{76}Ge [50]
n=3	χ^0	0.013 – 0.035	0.06	0.047
n=3	$\chi^0\chi^0$	0.59 – 5.9	0.6 – 5.5	0.7 – 6.6
n=7	$\chi^0\chi^0$	0.48 – 4.8	0.4 – 4.7	0.8 – 7.1

The contribution of bosonic neutrinos to the 2νββ-decay rate can be parametrised as [41]:

$$W_{tot} = \cos^4\chi W_f + \sin^4\chi W_b, \tag{5}$$

where W_f and W_b are the weights in the neutrino wave-function expression corresponding to the two fermionic and two bosonic antineutrino emission respectively. The purely fermionic, $T_{1/2}^f$, and purely bosonic, $T^{1/2}{}_{b}$, half-lives are calculated under the SSD model to be [41] :

$$T^f_{1/2}(0^+ g.s.) = 6.8 \times 10^{18} \text{ year,}$$
$$T^b_{1/2}(0^+ g.s.) = 8.9 \times 10^{19} \text{ year.} \tag{6}$$

Using the NEMO-3 half-life limit of $T^b_{1/2}(0^+ g.s.) > 1.2 \times$ ×1021 year (Table 4) an upper limit on the bosonic neutrino contribution to the ^{100}Mo 2νββ decay to the ground state can be evaluated as:

$$\sin^2 \chi < 0.27 \, (90\% \, \text{C.L.}). \qquad (7)$$

Although this limit is stronger than the bound obtained earlier in [41], the 2νββ transition of ^{100}Mo to the ground state is not very sensitive to bosonic neutrino searches due to a small value of the expected bosonic-to-fermionic decay branching ratio r0(0$^+$g.s.)=0.076. The ^{100}Mo 2νββ decay to the first excited 2^+_1 state has a branching ratio of $r_0(2^+_1)$=7.1 [41] and is therefore potentially more promising despite a lower overall decay rate. The current best experimental limit for this process is $T_{1/2}(2+1)>2.5\times10^{21}$ year [55]. This bound is still an order of magnitude lower than the theoretically expected half-life value of $T^b_{1/2}(2^+_1) = 2.4 \times 10^{22}$ year for purely bosonic neutrino, and two orders of magnitude lower than the corresponding expected value for purely fermionic neutrino, $T^f_{1/2}(2^+_1) = 1.7 \times 10^{23}$ year [41].

The Standard Model Extension (SME) provides a general framework for Lorentz invariance violation (LIV) [42]. In this model, the size of the Lorentz symmetry breakdown is controlled by SME coefficients that describe the coupling between standard model particles and background fields. Experimental limits have been set on hundreds of these SME coefficients from constraints in the matter, photon, neutrino and gravity sectors [42]. The first search for LIV in 2νββ decay was carried out in [56]. The two-electron energy sum spectrum of ^{136}Xe was used to set a limit on the parameter $\mathring{a}^{(3)}_{of}$, which is related to a time-like component of this LIV operator. The value of this parameter was constrained to be $-2.65 \times 10^{-5}\,\text{GeV} < \mathring{a}^{(3)}_{of} < 7.6 \times 10^{-6}\,\text{GeV}$ by looking at deviations from the predicted energy spectrum of ^{136}Xe 2νββ decay [56].

In this work we adopt the same method, using the phase space calculations from [57], and perform a profile likelihood scan over positive and negative contributions of LIV to two-electron events by altering the ^{100}Mo 2νββ energy sum spectrum with positive and negative values of $\mathring{a}^{(3)}_{of}$. The result of this scan is shown in Fig. 9.

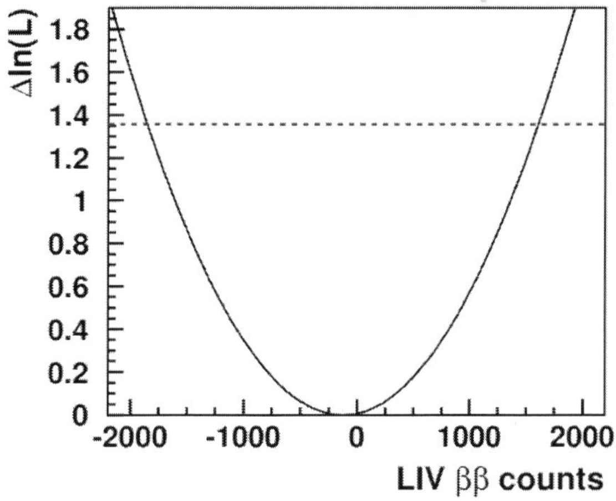

Figure 9. Profile likelihood scan over observed two-electron LIV counts in ^{100}Mo 2νββ energy sum spectrum. The 90% CL exclusion limit is shown with the dashed line

The minimum of the profile log-likelihood function corresponds to −135 counts and is not statistically significant even at 1σ level. The 90% CL exclusion limit is shown in Fig. 9 with the dashed line and gives −1798 and 1527 events for negative and positive contributions to the deviation from the ^{100}Mo 2νββ energy sum spectrum respectively. The corresponding constraint on $\mathring{a}^{(3)}_{of}$ is calculated using equations (2)–(6) in [56]. The result for ^{100}Mo obtained with a full set of NEMO-3 data is

$$-4.2 \times 10^{-7}\ \text{GeV} < \mathring{a}^{(3)}_{of} < 3.5 \times 10^{-7}\ \text{GeV (90\% C.L.)}. \quad (8)$$

A summary of the best available constraints on LIV and CPT violation parameters can be found in compilation [42].

SUMMARY

The results of the 2νββ decay of ^{100}Mo with the full data set of the NEMO-3 experiment corresponding to a 34.3 kg××year exposure are presented. The summed energy of two electrons, the single electron energy and the angular distributions between the two electrons have been studied with an unprecedented statistical precision (5×10^5 events). The single electron energy distribution has been used to discriminate between different nuclear models providing direct experimental input into NME calculations. The HSD model is excluded with high confidence, while the SSD model is consistent with the NEMO-3 data. The corresponding half-life for the 2νββ decay of ^{100}Mo is found to be

$$T_{1/2} = \left[6.81 \pm 0.01\ (\text{stat})^{+0.38}_{-0.40}\ (\text{syst})\right] \times 10^{18}\ \text{y} \tag{9}$$

Deviations from the expected shape of the ^{100}Mo 2νββ energy sum spectrum have been studied to obtain constraints on parameters for physics beyond the Standard Model. The most stringent upper limit to date has been obtained for the Majoron-neutrino coupling parameter ⟨gee⟩ for the decay mode with a single Majoron particle emission and the spectral index n=3. For other 0νββ modes with two Majoron bosons emission a comparable sensitivity with the world's best limits has been achieved. The most stringent constraints on the bosonic neutrino admixture and Lorentz invariance violation in 2νββ decay have been set.

DATA AVAILABILITY STATEMENT

This manuscript has no associated data or the data will not be deposited. [Authors' comment: The data are available from the corresponding author upon reasonable request.]

ACKNOWLEDGEMENTS

We thank the staff of the Modane Underground Laboratory for their technical assistance in running the experiment. We are grateful to S.V. Semenov for providing the spectra of SSD-3 model. We acknowledge support by the grants agencies of the Czech Republic (Grant Number EF16_013/0001733), CNRS/IN2P3 in France, RFBR in Russia (Project No.19-52-16002 NCNIL a), APVV in Slovakia (Project No. 15-0576), STFC in the UK and NSF in the USA.

REFERENCES

1. M. Goeppert-Mayer, Phys. Rev. 48, 512 (1935)
2. M. Tanabashi et al. (Particle Data Group), Phys. Rev. D 98, 030001 (2018)
3. A. Barabash, Nucl. Phys. A 935, 52 (2015)
4. J. Barea, J. Kotila, F. Iachello, Phys. Rev. C 91, 034304 (2015)
5. P. Pirinen, J. Suhonen, Phys. Rev. C 91, 054309 (2015)
6. J. Kostensalo, M. Haaranen, J. Suhonen, Phys. Rev. C 95, 044313 (2017)
7. J. Suhonen, Phys. Rev. C 96, 055501 (2017)
8. J. Suhonen, Front. Phys. 5, 55 (2017)
9. J. Engel, J. Menéndez, Rep. Prog. Phys. 80, 046301 (2017)
10. H. Ejiri et al., Phys. Lett. B 258, 17 (1991)
11. D. Dassie et al., Phys. Rev. D 51, 2090 (1995)
12. A. De Silva, M.K. Moe, M.A. Nelson, M.A. Vient, Phys. Rev. C 56, 2451 (1997)
13. M. Alston-Garnjost et al., Phys. Rev. C 55, 474 (1997)
14. V.D. Ashitkov et al., JETP Lett. 74, 529 (2001)
15. R. Arnold et al., Phys. Rev. Lett. 95, 182302 (2005)
16. L. Cardani et al., J. Phys. G 41, 075204 (2014)
17. E. Armengaud et al., Eur. Phys. J. C 77, 785 (2017)
18. R. Arnold et al., Phys. Rev. D 89, 111101(R) (2014)
19. R. Arnold et al., Phys. Rev. D 92, 072011 (2015)
20. R. Arnold et al., Nucl. Instrum. Methods A 536, 79 (2005)
21. R. Arnold et al., Eur. Phys. J. C 78, 821 (2018)
22. R. Arnold et al., Phys. Rev. D 93, 112008 (2016)
23. J. Argyriades et al., Nucl. Phys. A 847, 168 (2010)
24. R. Arnold et al., Phys. Rev. D 95, 012007 (2017)
25. R. Arnold et al., Phys. Rev. Lett. 107, 062504 (2011)
26. R. Arnold et al., Phys. Rev. D 94, 072003 (2016)
27. GEANT Detector Description and Simulation Tool, CERN Program Library Long Writeup W5013 (1995)
28. O.A. Ponkratenko, V.I. Tretyak, Yu.G. Zdesenko, Phys. Atom. Nucl.

63, 1282 (2000)

29. R. Argyriades et al., Nucl. Instrum. Methods A 606, 449 (2009)
30. J. Abad et al., Ann. Fis. A 80, 9 (1984)
31. F. Šimkovic et al., J. Phys. G 27, 2233 (2001)
32. P. Domin, S. Kovalenko, F. Šimkovic, S.V. Semenov, Nucl. Phys. A 753, 337 (2005)
33. Yu.A. Shitov, Phys. At. Nucl. 69, 2090 (2006)
34. J.H. Thies et al., Phys. Rev. C 86, 044309 (2012)
35. S.V. Semenov, Phys. Part. Nucl. 48, 1018 (2017)
36. S.V. Semenov, Phys. Part. Nucl. 49, 698 (2018)
37. E. Browne, J.K. Tuli, Nucl. Data Sheets 108, 681 (2007)
38. Y.A. Akovali, Nucl. Data Sheets 71, 181 (1994)
39. P. Barnet, C.P. Burgess, R.N. Mohapatra, Nucl. Phys. B 449, 25 (1995)
40. R. Arnold et al., Nucl. Phys. A 765, 483 (2006)
41. A.S. Barabash et al., Nucl. Phys. B 783, 90 (2007)
42. V.A. Kostelecky, N. Russell, Rev. Mod. Phys. 83, 11 (2011)
43. S. Liberati, Class. Quantum Gravity 30, 133001 (2013)
44. J. Ellis, N.E. Mavromatos, Astropart. Phys. 43, 50 (2013)
45. J.D. Tasson, Rep. Prog. Phys. 77, 062901 (2014)
46. J.S. Diaz, V.A. Kostelecky, R. Lehnert, Phys. Rev. D 88, 071902 (2013)
47. J.S. Diaz, Phys. Rev. D 89, 036002 (2014)
48. T. Junk, Nucl. Instrum. Methods A 434, 435 (1999)
49. A.L. Read, J. Phys. G 28, 2693 (2002)
50. W. Fisher, FERMILAB-TM-5595 (2010)
51. M. Hirsch et al., Phys. Lett. B 372, 8 (1996)
52. C. Barbero et al., Phys. Lett. B 392, 419 (1997)
53. J.B. Albert et al., Phys. Rev. D 90, 092004 (2014)
54. M. Agostini et al., Eur. Phys. J. C 75, 416 (2015)
55. R. Arnold et al., Nucl. Phys. A 925, 25 (2014)
56. J.B. Albert et al., Phys. Rev. D 93, 072001 (2016)
57. S. Stoica, AIP Conf. Proc. 1894, 020023 (2017)

CHAPTER 10

Quasi-Dirac Neutrino Oscillations

Gaetana Anamiati, Renato M. Fonseca, and Martin Hirsch

AHEP Group, Instituto de Física Corpuscular, C.S.I.C./Universitat de València Edifício de Institutos de Paterna, Apartado 22085, E-46071 València, Spain

ABSTRACT

Dirac neutrino masses require two distinct neutral Weyl spinors per generation, with a special arrangement of masses and interactions with charged leptons. Once this arrangement is perturbed, lepton number is no longer conserved and neutrinos become Majorana particles. If these lepton number violating perturbations are small compared to the Dirac mass terms, neutrinos are quasi-Dirac particles. Alternatively, this scenario can be characterized by the existence of pairs of neutrinos with almost degenerate masses, and a lepton mixing matrix which has 12 angles and 12 phases. In this work we discuss the phenomenology of quasi-Dirac neutrino oscillations and derive limits on the relevant parameter space from various experiments. In one parameter

Citation: Gaetana Anamiati, Renato M. Fonseca, and Martin Hirsch. "Quasi-Dirac neutrino oscillations" Phys. Rev. D 97, 095008 (2018). https://doi.org/10.1103/PhysRevD.97.095008

perturbations of the Dirac limit, very stringent bounds can be derived on the mass splittings between the almost degenerate pairs of neutrinos. However, we also demonstrate that with suitable changes to the lepton mixing matrix, limits on such mass splittings are much weaker, or even completely absent. Finally, we consider the possibility that the mass splittings are too small to be measured and discuss bounds on the new, nonstandard lepton mixing angles from current experiments for this case.

INTRODUCTION

Neutrino oscillation experiments cannot distinguish Dirac from Majorana neutrinos, hence it is still unknown whether or not lepton number is conserved. Other processes, such as neutrinoless double beta decay [1,2], need to be probed in order to answer this question. However, while the nature of neutrinos is often seen as a dichotomy, presenting two sharply distinct scenarios, the Dirac neutrino case can be seen as a limit of the more general Majorana case in which lepton number violating mass terms are zero, and this limit can be approached smoothly.

In practice, one can start with a model of 2n Majorana neutrinos and get a phenomenology arbitrarily close to the one of a model of n Dirac neutrinos. This can already be seen with only one generation of active (ν) and sterile neutrinos (N^c). In the basis $(\nu,N^c)^T$ the most general mass matrix reads:

$$m_\nu = \begin{pmatrix} m_L & m_D \\ m_D & m_R \end{pmatrix} \qquad (1)$$

If $m_L=m_R=0$, lepton number is preserved and neutrinos are Dirac particles. This limit can alternatively be characterized by two exactly degenerate mass eigenstates composed in equal parts of ν and N^c: $\nu_1=1/\sqrt{2}(\nu+N^c)$ and $\nu_2=i/\sqrt{2}(-\nu+N^c)$.[1] Small deviations from the limit $m_L=m_R=0$ lead to a quasi-Dirac scenario where lepton number is no longer exactly preserved.

Let us rewrite Eq. (1) using:

$$\varepsilon = \frac{(m_L + m_R)}{2m_D}, \qquad (2)$$

$$\theta = \frac{(m_L - m_R)}{4m_D} \qquad (3)$$

As long as ϵ and θ are much smaller than one, we obtain:

$$m_{1,2} \simeq m_D(1 \pm \varepsilon) \quad (4)$$

$$\nu_1 \simeq 1/\sqrt{2}[(1+\theta)\nu + (1-\theta)N^c], \quad (5)$$

$$\nu_2 \simeq i/\sqrt{2}[(-1+\theta)\nu + (1+\theta)N^c] \quad (6)$$

Departures from the Dirac case therefore can manifest themselves as either new mass splittings or new mixing angles (or, in general, both). Moreover, as this simple example shows, mass splittings and mixing angles are completely independent of each other. Note that for small values of ϵ and θ, lepton number violation is naturally suppressed, as expected. This can be most easily seen in our one generation scenario for the double beta decay observable $\langle m\nu \rangle$: for θ=0 (ϵ=0) it is straightforwardly calculated to be $\langle m_\nu \rangle \simeq \epsilon m_D$ ($\langle m_\nu \rangle \simeq 2\theta m_D$).

We have therefore the following situation. Oscillation experiments cannot distinguish a model with n Majorana neutrinos (containing n Weyl spinors) from one with n Dirac neutrinos (containing 2n Weyl spinors) with matching masses and mixing angles. Nevertheless, once we add to a model with Dirac neutrinos small sources of lepton number violation, oscillation probabilities will change. Some illustrative examples are shown in Fig. 1. We plot there the electron neutrino survival probability for low-energy (reactor) neutrinos at distances up to (and slightly larger than) the typical distances of the KamLAND experiment [3]. In all plots the black lines show the expectation for the current global best fit point [4] for the ordinary neutrino parameters in the standard three generation case, to which we have added either a nonzero mass splitting to a Dirac state (top row) or one particular new quasi-Dirac angle (bottom row).

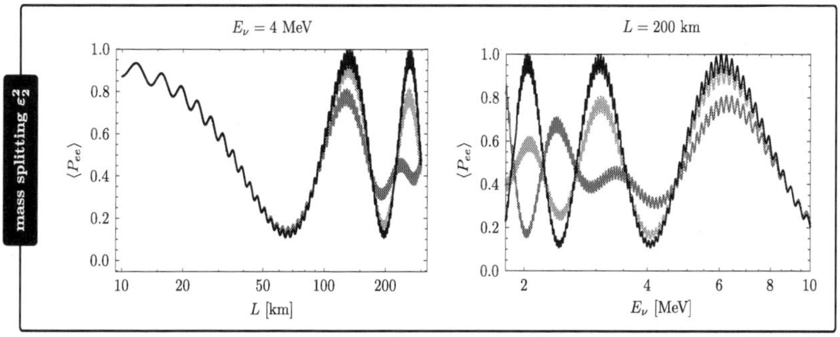

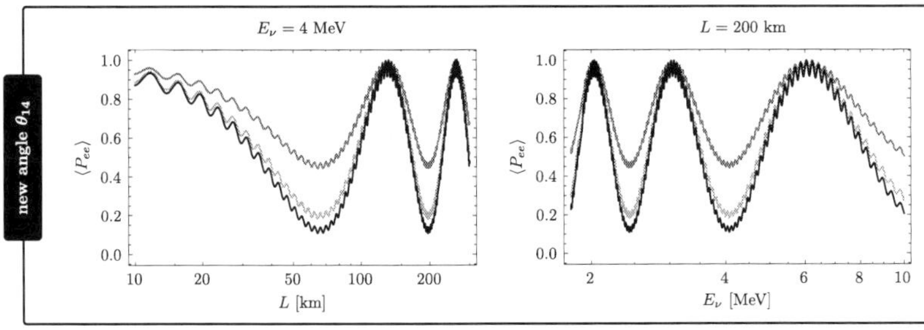

Figure 1. Electron neutrino survival probability for quasi-Dirac neutrinos with a fixed energy Eν=4 MeV as a function of distance (left), and for fixed distance L=200 km as function of Eν (right). The standard 3-generation neutrino oscillation parameters have been fixed at their best fit point values [4], to which a small perturbation has been added. In the top row, we show the effect of mass splittings: $\epsilon_2^2 = 0$ (black), $\epsilon_2^2 = 10^{-5}$ eV^2 (orange) and $\epsilon_2^2 = 2 \times 10^{-5}$ eV^2 (red). In the bottom row, it is possible to see the effect of introducing a nonstandard angle: θ_{14}=0 (black), θ_{14}=π/8 (orange) and θ_{14}=π/4 (red). The exact definition of ϵ_2 and θ_{14} will be given latter in Sec. II.

In Sec. II we will discuss the general parametrization of masses and mixing angles for scenarios with three generations of quasi-Dirac neutrinos. However, from the examples shown in Fig. 1 one can read off already some basic facts about oscillations of quasi-Dirac neutrinos, which we will work out in greater detail in Sec. III. First, small nonzero values of ϵ's are equivalent to introducing new, large oscillation lengths. Thus, the best constraints on ϵ will come from oscillation experiments with the largest possible baselines. And second, even if mass splittings are negligibly small, the new, nonstandard angles which appear in this setup (called θ above) may affect oscillation probabilities in a way similar to standard angles, hence creating parameter degeneracies. For example, as Fig. 1 shows, from Pee alone one cannot provide limits on a single angle. (In this example variations of θ14 can be compensated by varying θ12.) Even by combining more than one oscillation probability, constraints can only be derived for certain combinations of angles and phases of the mixing matrix. We will discuss this in detail in Sec. III B. Constraints on mass splittings are discussed in Sec. III A.

A word on nomenclature. The terminologies *quasi-Dirac* and *pseudo-Dirac* neutrinos appear nearly interchangeably in the literature. We prefer to define quasi-Dirac (QD) neutrinos as being a mixture of active and sterile states, in contrast with pseudo-Dirac (PD) neutrinos[2] which are composed of active states only. In both cases, the structure of mass and mixing matrices must be such that the lepton sector is close to preserving one or more U(1) symmetries.

With this definition, quasi-Dirac and pseudo-Dirac neutrinos are then very different objects, both theoretically and phenomenologically. Let us briefly mention that various aspects of pseudo-Dirac neutrinos have been considered in the literature: Magnetic moments and double beta decay [7], possible mass textures [8–10], and oscillatory behavior [11–14]. We note in passing that models of pseudo-Dirac neutrinos require neutrino mass matrices which no longer fit the solar and atmospheric neutrino oscillation data [15–17].[3]

Many more papers discussed the phenomenology of quasi-Dirac neutrinos. For example, double beta decay was first discussed in this context in [6], while [19] and [20,21] consider quasi-Dirac neutrinos as a possible explanation of the atmospheric and solar neutrino problems, respectively. More ambitiously, explaining atmospheric, solar and LSND neutrino oscillations simultaneously was discussed in [22–24]. However, all these proposals are by now ruled out experimentally, since they predict too much oscillations into sterile neutrinos. Limits on quasi-Dirac neutrino parameters, on the other hand, have been derived from solar neutrino data [25] as well as from solar, atmospheric neutrino data and cosmology [26]. Furthermore, in [27–30] QD neutrinos have been discussed in the context of neutrino telescopes, such as IceCube.

Quasi-Dirac neutrino oscillations were also discussed in [31], where it was claimed that to leading order in $m_{R,L}/m_D$ the flavor composition of the mass eigenstates does not change (only mass splittings appear), hence oscillations for n=3 pairs of quasi-Dirac neutrinos are described by the standard mixing matrix. This assertion was taken to be true by others [27–29,32], yet we want to stress that this claim is not correct, as can be seen from the Eqs. (4)–(6). Already for one generation, these expressions show that the mass splitting and the departure from maximal mixing are both linearly dependent on $m_{R,L}$ and, more importantly, they are controlled by orthogonal combinations of these two parameters. As such, it is even possible to have no mass splittings at all and at the same time have arbitrary mixing angles.

There are also a number of more theoretical papers discussing how quasi-Dirac neutrinos could arise. One possibility is the so-called "singular" seesaw where the mass matrix for the singlet neutrinos (Nc) has a determinant equal or close to zero [33]. Quasi-Dirac neutrinos from such a singular seesaw with additional type-II seesaw contributions have been discussed in [34]. Another possibility [32] involves introducing additional singlets (S), as it is done for the inverse seesaw mechanism [35]. A double seesaw is then responsible for producing very light S states which, together with the active states, form quasi-Dirac neutrinos [32]. The authors of [36] use a Dirac seesaw to explain the necessary smallness of the Dirac neutrino mass terms first, and then generate quasi-Dirac states by the addition of a very small seesaw type-II term. The "mirror world" model of [37] is another way to obtain these particles.

In models with extended gauge groups quasi-Dirac neutrinos can also appear. An example is the E6 inspired 331 model of [38]. Here, several electroweak triplets of the gauge group SU(3)L are needed to accomodate the Standard Model leptons, and the observed active light neutrinos are automatically quasi-Dirac states [39]. A very different idea, based on supergravity has been discussed in [40]. There it was pointed out that if neutrino Dirac terms are generated from the Kähler potential (instead of the superpotential), neutrinos would be quasi-Dirac, since Majorana terms come from higher order Kähler potential terms and thus are expected to be suppressed. This idea [40] is particularly attractive, since it could, at least in principle, explain the observed smallness of the Dirac neutrino mass terms.

In addition to n active neutrinos, models of Dirac neutrinos require the introduction of n Weyl spinors transforming trivially under the electroweak gauge group. For this reason, the study of quasi-Dirac neutrinos necessarily has some overlap with the physics of sterile neutrinos. Many experiments have searched for sterile neutrinos. Most famously, the SNO neutral current measurement rules out dominant contributions of sterile neutrinos to the solar neutrino oscillations [41]. Super-Kamiokande searched for steriles in atmospheric neutrinos [42]. OPERA [43], MINOS and DayaBay [44], IceCube [45] and NOvA [46] published searches for sterile neutrinos. For a more complete list of references see the recent reviews [47,48]. Note, however, that constraints on sterile are usually derived assuming best fit point values for the standard oscillation parameters, to which two new parameters (one angle and one mass splitting) are added in the fit. This approach does not cover the general quasi-Dirac neutrino parameter space. In particular, keeping the standard neutrino parameters fixed can lead to misleading conclusions about limits for the new/extra parameters.

There are also some hints for the existence of sterile neutrinos. However, all these hints point to a new and *much larger* mass scale in oscillations, i.e. $\Delta m^2 \simeq O(1)\ eV^2$. Since these indications imply masses and mixings very different from those of the standard oscillations, they can not be explained by quasi-Dirac neutrinos. We thus do not discuss these hints any further and refer only to the recent review [48].

The rest of this paper is organized as follows. In Sec. II we discuss the basics of quasi-Dirac oscillations, constructing general expressions for the mixing matrix for the three generation case. In Sec. III we discuss constraints on the new, nonstandard parameters from various neutrino experiments. Constraints on quasi-Dirac mass splittings are discussed in Sec. III A, while in Sec. III B we discuss the constraints on angles, for the case in which mass splittings are negligible. We then close with a short summary and discussion.

DEFINITIONS FOR QUASI-DIRAC NEUTRINO OSCILLATIONS

Dirac neutrinos can be described either in the weak or in the mass basis. The two pictures are equivalent. We will choose the latter one. Consider then a lepton-number preserving model with three active and three sterile neutrinos (ν and N^c).[4] In the basis where the charged lepton mass matrix is diagonal, the relevant part of the Lagrangian reads

$$\mathcal{L} = \frac{g}{\sqrt{2}} \overline{\ell_L} \mathbb{1} \gamma^\mu \nu W_\mu^- + \nu^T m_\nu N^c + \text{H.c.} \quad [\text{flavor basis}]. \tag{7}$$

In order to diagonalize the matrix mv, both active and sterile neutrinos must be rotated, $\nu \to V\nu$ and $N^c \to V_N N^c$, such that $m_\nu^{(\text{diag})} = V^T m_\nu V_N$:

$$\mathcal{L} = \frac{g}{\sqrt{2}} \overline{\ell_L} V \gamma^\mu \nu W_\mu^- + \nu^T m_\nu^{(\text{diag})} N^c + \text{H.c.} \quad [\text{mass basis 1}]. \tag{8}$$

Strictly speaking, the neutrino mass matrix is not yet diagonal since it is still mixing different states (active and sterile neutrinos). This can be solved by rewriting ν_i and N_i^c $(i = 1,\ 2,\ 3)$ as $\psi_i \equiv 1/\sqrt{2}(\nu_i + N_i^c)$, and $\psi_{i+3} \equiv i/\sqrt{2}(-\nu_i + N_i^c)$:

$$\mathcal{L} = \frac{g}{\sqrt{2}} \overline{\ell_L} \Omega \gamma^\mu \psi W_\mu^- + \sum_{j=1}^{6} m_j^\psi \psi_j \psi_j + \text{H.c.} \quad [\text{mass basis 2}]. \tag{9}$$

where the masses and the 3×6 mixing matrix Ω have a special form (V is a 3×3 square matrix):

$$m_j^\psi = (m_1, m_2, m_3, m_1, m_2, m_3) \quad [\text{Dirac limit}]. \tag{10}$$

$$\Omega = \frac{1}{\sqrt{2}} \begin{bmatrix} \ddots & & \ddots & \ddots & & \ddots \\ & V & & \cdot & iV & \\ \ddots & & \ddots & \ddots & & \ddots \end{bmatrix} \quad \text{[Dirac limit]}. \tag{11}$$

If the pattern of masses and mixing in Eqs. (10) and (11) is perturbed, neutrinos are no longer Dirac particles and lepton number is violated. Note that this is equivalent to switching on the lepton number violating masses m_L and m_R in Eq. (1). We shall now look into the possible departures from the Dirac limit as seen from the mass basis.

In the case of masses, it is possible to split the three pairs of (m_i^ψ, m_{i+3}^ψ), hence we may introduce three ϵ_i such that

$$(m_i^\psi)^2, (m_{i+3}^\psi)^2 \to m_i^2 - \frac{\varepsilon_i^2}{2}, m_i^2 + \frac{\varepsilon_i^2}{2}, \tag{12}$$

with the understanding that, for quasi-Dirac neutrinos, the ϵi are small in comparison to the atmospheric and solar mass scales. In total there arc now five mass parameters relevant for oscillation experiments: the usual Δm^2_{Atm} and $\Delta m^2_{\odot}$, plus three new ϵ_i mass splittings. (As usual, the overall mass scale of neutrinos does not enter the oscillation probabilities.)

Let us now turn our attention to a generic mixing matrix Ω with dimensions n×m. Such a matrix can be described by 2nm real numbers, yet orthonormality of rows ($\Omega\Omega^\dagger=1$) imposes n^2 conditions on them, and furthermore it is possible to absorb n phases into the charged lepton fields, hence there is a total of n(2m−n−1) real physical degrees of freedom in Ω. For a 3×6 matrix, this corresponds to 12 angles and 12 phases, but note that 5 of these phases cannot be observed in neutrino oscillation experiments (they correspond to column phases). The matrix Ω can be explicitly parametrized as follows [49] (called below the SV parametrization). First, consider an elementary rotation in the (i,j) entries given by the complex number $\tilde{\theta}_{ij} \equiv \theta_{ij} \exp i\phi_{ij}$ such that, in the (1,2) case, it has the form

$$R(\tilde{\theta}_{12}) = \begin{pmatrix} \cos\theta_{12} & -e^{i\phi_{12}}\sin\theta_{12} & 0 & \cdots \\ e^{-i\phi_{12}}\sin\theta_{12} & \cos\theta_{12} & 0 & \cdots \\ 0 & 0 & 1 & \cdots \\ \vdots & \vdots & \vdots & \ddots \end{pmatrix}. \tag{13}$$

In the SV parametrization, the ith row of Ω ($\equiv\Omega_i$) is then given by the expression

$$\Omega^T_{SV,i} = \prod_{a=1}^{i} \prod_{b=a+1}^{6} R(\tilde{\theta}_{ab}) e^{(i)},$$

(14)

where $e^{(i)}$ is a column vector with entries $e_j^{(i)} = \delta_{ij}$. We do not give here Ω_{SV} in full because the expression is very lengthy.

For a particular arrangement of the 24 angles and phases in Eq. (14), Ω takes the special form (11) which is associated with the Dirac limit. Note that, as usual, one can write the 3×3 square matrix V with three angles and one phase:

$$V = \begin{pmatrix} 1 & 0 & 0 \\ 0 & \cos\theta_{23} & \sin\theta_{23} \\ 0 & -\sin\theta_{23} & \cos\theta_{23} \end{pmatrix} \begin{pmatrix} \cos\theta_{13} & 0 & \sin\theta_{13} e^{-i\delta} \\ 0 & 1 & 0 \\ -\sin\theta_{13} e^{i\delta} & 0 & \cos\theta_{13} \end{pmatrix} \times \begin{pmatrix} \cos\theta_{12} & \sin\theta_{12} & 0 \\ -\sin\theta_{12} & \cos\theta_{12} & 0 \\ 0 & 0 & 1 \end{pmatrix}. \quad (15)$$

(15)

Unfortunately, it is very complicated to describe the Dirac limit in the SV parametrization. Hence we make a small modification by introducing the following 6×6 rotation matrix:

$$\Omega(\theta_{ij}, \phi_{ij}) = \Omega_{SV}(\theta_{ij}, \phi_{ij}) U, \qquad U = \frac{1}{\sqrt{2}} \begin{pmatrix} 1 & i1 \\ 1 & -i1 \end{pmatrix}.$$

(16)

With this definition, the mixing matrix in Eq. (11), with V parametrized as in Eq. (15), corresponds to $\Omega(\theta_{ij},\phi_{ij})$, as in (16), with $\theta_{i4}=\theta_{i5}=\theta_{i6}=0$ (i=1, 2, 3), $\phi_{12}=\phi_{23}=0$ and $\phi_{13}=\delta$. In other words, with this definition the Dirac limit for Ω simply corresponds to keeping only the standard three generation neutrino mixing angles nonzero.

We can then write the probability of neutrino oscillation from a flavor α to a flavor β for an energy E and after a length L as[5]:

$$P(\nu_\alpha \to \nu_\beta) = \left| \sum_{j=1}^{6} \Omega_{\beta j} \Omega^*_{\alpha j} \exp\left(-\frac{i m_j^2 L}{2E}\right) \right|^2$$

(17)

Note that this expression is insensitive to column rephasings $\Omega \to \Omega \mathrm{diag}(e^{i\kappa 1}, e^{i\kappa 2}, e^{i\kappa 3}, e^{i\kappa 4}, e^{i\kappa 5}, e^{i\kappa 6})$. It is easy to show that Eq. (17) reduces to the standard oscillation formula in the Dirac limit.

CURRENT EXPERIMENTAL LIMITS AND FUTURE PROSPECTS

As discussed in the previous section, the full parameter space for a system of 3 pairs of QD neutrinos has 30 free parameters: Two independent Δm^2_{ij} plus one overall mass scale, three ϵ^2_i, twelve angles and twelve phases. Even discounting the five Majorana phases and the overall mass scale, which can not be probed in oscillation experiments, the remaining number of parameters is much too large to fit simultaneously.

Nearly all experimental data, on the other hand, is consistent with the standard picture of only three active neutrino species participating in oscillations [4], i.e. two mass squared differences (Δm^2_{Atm} and $\Delta m^2_{\odot}$), three mixing angles (θ_{23}, θ_{12} and θ_{13}) plus one phase (δ) are sufficient to describe the data. As mentioned in the introduction, there are also some hints for sterile neutrinos with a mass scale of the order of $\Delta m^2 \sim O(eV)$ [47,48]. However, all these hints are at most of the order of (2–3)σ, we will thus not take them into account in the following. Instead, since the standard three generation picture seems to describe the data well, we will consider "small" perturbations and derive limits on particular combinations of nonstandard parameters.

In order to deal effectively with the large number of parameters controlling quasi-Dirac neutrino oscillations, we will consider two simplified scenarios:

- (1) First, we take one nonzero ϵ_i at a time. In these one-parameter extensions, very stringent limits on ϵ_i are found, in agreement with earlier analysis, see for example [25,26]. We then extend this analysis to two new parameters: One mass splitting plus one new angle. This second step allows us to identify "blind spots" in the oscillation experiments, i.e. degenerate minima in particular directions in parameter space, where limits on mass splittings are much worse than in the one parameter fits. We then discuss a particular parametrization of these degenerate directions in parameter space, where the effects of ϵi can be decoupled from oscillation experiments nearly completely.
- (2) In the second setup, we discuss the limit where mass splittings are too small to be measured in oscillation experiments, hence there are just angles and phases of Ω to deal with. In this situation, it can be shown that from the 24 parameters in Ω only 13 combinations enter the oscillation probabilities of active neutrinos. Moreover, since there is only very limited information

on oscillations involving ν_τ, we can in practice restrict ourselves to experiments involving νe's and ν_μ's. There are then only 7 combinations of the 24 angles θij and phases ϕij which appear in the oscillation probabilities. We discuss the construction of these 7 quantities, the current constraints and possible tests for quasi-Dirac neutrinos in this limit.

In our analysis we do not take into account the data from every existing oscillation experiment. Given the scarcity of data on τ neutrinos, we ignored it altogether, concentrating instead on the available charged current data for e and μ neutrinos and anti-neutrinos. Also, we focus on those experiments, which should provide the most important constraints for quasi-Dirac neutrinos. First, we consider KamLAND [3] since it fixes most accurately the so-called solar mass splitting $\Delta m^2_\odot$. From the solar neutrino experiments we fit Super-K elastic scattering data [50] and from Borexino the measured pp [51] and 7Be fluxes [52]. This choice is motivated by the fact that Super-K [50] has the most accurate data in the high energy range, while [51,52] fix the low energy part of the solar neutrino spectrum.

We take the atmospheric neutrino data from [53]. We concentrate in our fit on the sub-GeV sample, since the lowest energetic neutrinos will be most sensitive to small values of ϵ^2_i, as demonstrated in Fig. 2. In addition, we use data from the MINOS Collaboration (muon and anti-muon neutrino survival) [54] and T2K (muon neutrino survival and muon to electron neutrino transition) [55], since these two experiments determine Δm2Atm better than atmospheric data. And, finally, we take into account data from DayaBay [56], since from the three current reactor neutrino experiments DayaBay determines θ_{13} with the smallest error.

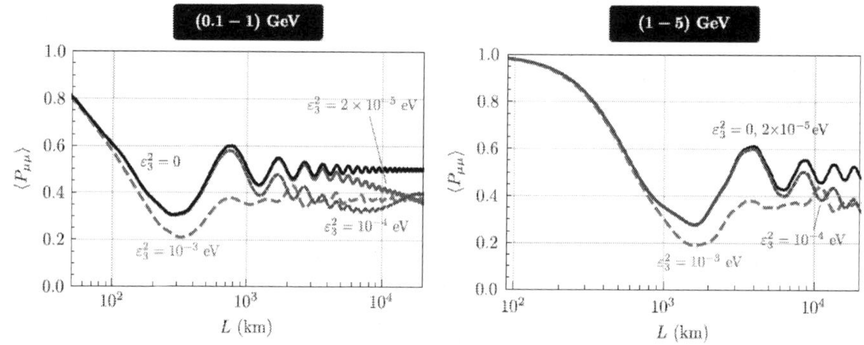

Figure 2. Averaged atmospheric muon neutrino survival probability for neutrinos with energies E_ν=(0.1–1) GeV (left) and E_ν=(1–5) GeV (right), as a

function of distance (L), for different choices of ϵ_{23}. Lower neutrino energies are more sensitive to small ϵ_3 values. This plot is calculated with the simplifying assumptions of $\sin^2\theta_{23}=1/2$, $\theta_{13}=0$ and $\Delta m^2_{\odot}=0$.

In our fits, we use a simple $\chi 2$ method to determine the allowed ranges of model parameters. We take statistical and systematic errors from the experimental publications, to which we added a further (small) systematic error for the uncertainties in our theoretical calculations. This latter systematic error was chosen such that our simulations reproduce the allowed parameter ranges for the standard oscillation parameters, determined by the respective experiment, within typically (1–1.5)σ C.L. ranges. Note that we do not attempt to do a precision global fit for standard neutrino oscillation parameters. Rather, we consider reproducing the experimental results for the standard case as a test for the reliability of our derived limits.

A. Limits On Mass Splittings ϵi

1. One Parameter Limits

We will first discuss limits derived on ϵi assuming one ϵi≠0 at a time and taking all nonstandard angles to be zero. Table I shows limits on ϵ2i and the corresponding experimental data sets used to derive the limits.

Table 1. 95% upper limits on ϵ2i derived from different experimental data sets. Two numbers are given for each case; the first one is the limit obtained marginalizing over two standard oscillation parameters (see text), the second (in brackets) is the limit obtained for the best fit point value of the standard oscillation parameters. For a discussion see text.

Experiment	ϵ21 [eV2]	ϵ22 [eV2]	ϵ23 [eV2]
KamLAND	7.7(3.4)×10−6	1.7(1.0)×10−5	...
Solar+KamLAND	1.7(1.3)×10−11	1.7(1.5)×10−11	...
DayaBay+MINOS+T2K	...	1.5(0.9)×10−4	1.3(0.074)×10−3
Super-K+DayaBay+MINOS+T2K	...	1.9(1.8)×10−5	1.2(1.1)×10−5
JUNO	1.7(0.07)×10−5	2.3(0.09)×10−5	6.0(2.2)×10−5

For each case listed in Table 1, we have calculated the upper limits on the ϵ^2_i twice: (a) marginalizing over two of the standard neutrino oscillation parameters and (b) for the best fit point value of the standard parameters. Marginalization over standard oscillation parameters leads to less stringent limits. However, the importance of this marginalization procedure differs widely for different experiments. For example, in the case of KamLAND, bounds on ϵ21 of the order of roughly 10^{-5} are derived marginalizing over the allowed ranges of $\Delta m^2_{\odot}$ and $\sin^2\theta_{12}$, while for the best fit values of these last two parameters, the limits are more stringent by "only" roughly a factor 2.

As the table shows, the strongest constraints on ϵ^2_1 and ϵ^2_2 come from solar neutrino data. This is easily understood from Fig. 3, which shows the electron neutrino survival probability as a function of neutrino energy for different values of ϵ_2^2 and for the best fit values of the standard solar oscillation parameters. For low values of neutrino energies, vacuum oscillations dominate and so very small ϵ_2^2 can be probed up to a scale essentially determined by the Earth-Sun distance ($\sim 10^{-12}$,−11 eV2). Note that a nonzero ϵ^2_2 reduces P_{ee}, but this reduction could be hidden in the relatively large error bar of the low-energy measurements.[6] Nevertheless, at higher neutrino energies, a similar reduction of Pee is produced due to matter effects in the sun. Since Super-K data provides a very accurate measurement of Pee at these higher energies, one can rule out values of ϵ^2_2 which can not be excluded by the Borexino measurements alone. The situation is very similar for ϵ^2_1: limits on ϵ^2_1 and ϵ^2_2 are then of the order of 10^{-11}eV^2 if we take the best fit values of the standard solar oscillation parameters; slightly less stringent numbers are obtained when marginalizing over the standard parameter uncertainties.

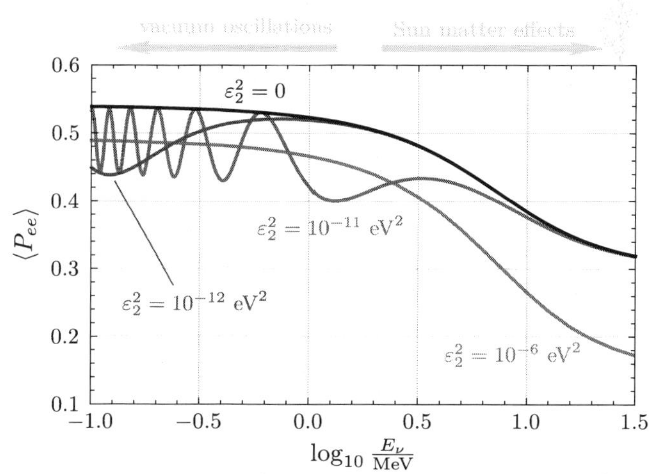

Figure 3. Solar neutrino survival probability as a function of neutrino energy, for different choices of ϵ22. Solar angle and mass splitting have been fixed at their best fit values in this plot [4].

Solar neutrino experiments have essentially no sensitivity to ϵ^2_3. This is simply due to the smallness of θ_{13} ($\sin^2\theta_{13} \simeq 0.0215$ [4]). Thus, we have to rely on experiments testing the atmospheric scale to derive limits on ϵ^2_3. Table 1 quotes numbers for two cases.

In the first scenario, we have combined data from DayaBay [56], T2K [55] and MINOS [54]: DayaBay fixes most accurately θ_{13}, while both MINOS and T2K measure Δm^2_{Atm} with rather small errors. Here, the limit on ϵ_2^2 is (not surprisingly) less stringent than the one derived from KamLAND (or solar). Depending on whether or not Δm^2_{Atm} and $\sin^2\theta_{23}$ are used at their best fit value or marginalized over, we get very different limits on ϵ23. This is due to the fact that when scanning over the standard oscillation parameters, the χ^2 function has two almost degenerate minima: one for small values of ϵ^2_3 and another for ϵ^2_3 of the order 10^{-3}eV2. However, as the table also shows (second case), this nonstandard solution is excluded, once we add Super-K atmospheric neutrino sub-GeV data to the fit. With the combination of these four experiments limits on ϵ^2_2 and ϵ^2_3 are again of order 10^{-5}eV2.

In the last line of Table I we give our forecast of the sensitivity of the planned experiment JUNO [57]. JUNO will measure $\Delta m^2_\odot$ and Δm^2_{Atm} very precisely and thus it will also be able to derive limits on any ϵ^2_i. However, our results indicate that, despite being a very precise experiment, JUNO will not lead to a major improvement over existing limits on ϵ2i. Here, it is important to stress that limits using the best fit point and limits marginalizing over standard parameter uncertainties are very different. This can be traced back again to a near-degeneracy in the χ^2-function: For ε^2_i of the order of Δm^2_{ij} one has two only slightly different oscillation lengths contributing in the fit, which can give a better description than a single oscillation length.

Insummary,stronglimitsonmasssplittingscanbederivedfromatmospheric and solar neutrino data $(\varepsilon_3^2 \sim \mathcal{O}(10^{-5})\ \mathrm{eV}^2$ and $\varepsilon_{1,2}^2 \sim \mathcal{O}(10^{-11})\ \mathrm{eV}^2$, respectively) in the case where no other extra parameter is added to the standard neutrino oscillation picture.

2. Two Parameter Case

While the discussion in the previous subsection seems to show that constraints on QD mass splittings are very stringent, we will now see that this conclusion is valid only under the assumption that no other nonstandard parameter is different from zero.

As a simple example, consider the electron neutrino survival probability at distances short enough that the effects of $\Delta m^2_\odot$ can be neglected.[7] We shall consider the particular example of a nonzero ϵ^2_3 and a nonzero θ_{16} angle, defined in Sec. II. One finds that

$$P_{ee} = 1 - 2c_{13}^2 c_{16}^2 (\Delta m_{ee}^- (s_{16} + c_{16} s_{13})^2 + \Delta m_{ee}^+ (s_{16} - c_{16} s_{13})^2), \quad (18)$$

where c_{ij} and s_{ij} are short-hands for $\cos\theta_{ij}$ and $\sin\theta_{ij}$ and

$$\Delta m_{ee}^{\pm} \equiv c_{12}^2 \sin^2\left[\frac{L}{4E}(\Delta m_{31}^2 \pm \varepsilon_3^2/2)\right] + s_{12}^2 \sin^2\left[\frac{L}{4E}(\Delta m_{32}^2 \pm \varepsilon_3^2/2)\right]. \quad (19)$$

It is straightforward to see that the above expression for the neutrino survival probability remains (nearly) unchanged if we swap θ_{13} by θ_{16}. This is true up to very small terms proportional to $\Delta(P_{ee}) \propto (\Delta m^-_{ee} - \Delta m^+_{ee})(c_{13} - c_{16}) s_{13} s_{16}$. More specifically, in the limit where $\Delta m^{\pm}_{ee}$ have the same value, P_{ee} is only a function of the combination $\sin^2\theta_{13} + \sin^2\theta_{16}$. Thus, there will be a near-degeneracy of the relevant χ^2 function involving these two angles, and so values (or limits) derived for one of these parameters, without varying the other, will be misleading.

There is, however, another more interesting degeneracy associated to Eq. (18). In calculating this expression we have used a certain parametrization for the mass splitting, which we may call the symmetric parametrization: $m_i, m_{i+3} \to \sqrt{m_i^2 - \varepsilon_i^2/2}, \sqrt{m_i^2 + \varepsilon_i^2/2}$. Choosing $\sin\theta_{13} = \tan\theta_{16}$, the second term inside the bracket in Eq. (18) vanishes (this choice corresponds to $\Omega_{e6} = 0$). So, by adjusting Δm_{31}^2 and ε_3^2 we can keep $\Delta m_{31}^2 - \varepsilon_3^2/2$ constant and equal to the best fit point value of Δm^2_{Atm}, in which case there will be no upper limit on ε_3^2 itself coming from the electron neutrino survival experiments.

Note that we could have defined $m_i, m_{i+3} \to m_i, \sqrt{m_i^2 + \varepsilon_i^2}$.[8] We call this the asymmetric parametrization. Rewriting Eq. (18) with this parametrization, the first term inside the bracket would not depend on ε_3^2 at all, so it becomes obvious that for the choice of $\sin\theta_{13} = \tan\theta_{16}$ all dependence of Pee on ε_3^2 disappears. Figure 4 shows these parameter degeneracies in the space ($\varepsilon_3^2, \sin^2\theta_{13}, \sin^2\theta_{16}$), using only the DayaBay data (on the left column). The underlying scan was done in the asymmetric parametrization, which is numerically simpler to implement. The plot in the upper and middle panel show clearly that there is no upper limit on ε_3^2 in this scan. The lower plot shows the degeneracy in parameter space under the exchange of $\theta_{13} \leftrightarrow \theta_{16}$.

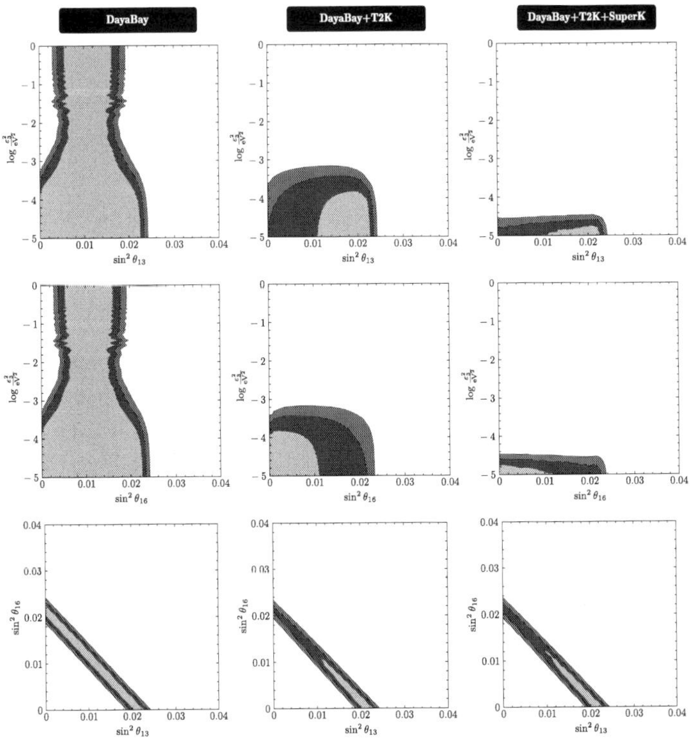

Figure. 4. Allowed parameter ranges for ε_3^2, $\sin^2\theta_{13}$ and $\sin^2\theta_{16}$ for different experiments. The parameter planes always marginalize over the parameter not shown and all calculations used the best fit point value for Δm^2_{Atm}. In the plots on the left only DayaBay data is taken into account; the middle panel combines DayaBay with T2K and the panel to the right shows the combination of DayaBay, T2K with Super-K atmospheric neutrino data. The different coloured regions present the 1, 2 and 3σ C.L. allowed regions (cyan, blue and red). For discussion see text.

We can break this particular degeneracy in parameter space, by adding more experiments. T2K measures two probabilities: (a) The muon neutrino survival probability, $P_{\mu\mu}$, and (b) the electron neutrino appearance probability, $P_{\mu e}$, both at values of L/E which give access to the atmospheric neutrino mass scale, Δm^2_{Atm}. If the only nonstandard angle different from zero is θ_{16}, then $P_{\mu\mu}$ will not depend on θ16 at all, while $P_{\mu e}$ will have θ_{16}-dependence which is different from the one of Pee. Thus, adding T2K data to the scan is enough to break the degeneracy in $\theta_{13} \leftrightarrow \theta_{16}$ hence an upper limit on ε_3^2reappears. The middle column of Fig. 4 illustrates this point; it shows the results of a combined scan over ε_3^2, $\sin^2\theta_{13}$ and $\sin^2\theta_{16}$ for DayaBay plus

T2K data. By comparison of the right with the middle column of Fig. 4, one can clearly see that the addition of Super-K data generates a strong upper limit on ϵ_3^2, for this particular choice of parameter subspace.

Given these results, one might wonder if there are particular directions in parameter space for which oscillation experiments become completely blind to QD neutrino mass splittings. Recall that the blind (or: degenerate) direction discussed above for P_{ee} corresponds to the particular choice of $\Omega_{e6}=0$. In a similar way, for example, $P_{\mu\mu}$ would loose any sensitivity to ϵ_3^2 if $\Omega_{\mu 6}=0$. Thus, with some special choice of θ_{16} and θ_{26} such that both Ω_{e6} and $\Omega_{\mu 6}$ are zero at the same time, one can indeed make DayaBay and T2K blind to variations of ϵ_3^2.

While it is possible, in principle, to calculate the combination of angles θij and phases ϕ_{ij} (defined in section II) associated to these blind directions, in the following we will consider a simpler alternative. Consider a unitary rotation of the columns i and i+3 of the mixing matrix Ω. Since we are not interested in column phases, such a rotation is governed by just two parameters (φ_i and β_i):

$$\begin{pmatrix} \vdots & \vdots \\ \text{col.}\, i\, \text{of}\, \Omega & \text{col.}\, i+3\, \text{of}\, \Omega \\ \vdots & \vdots \end{pmatrix} \rightarrow \begin{pmatrix} \vdots & \vdots \\ \text{col.}\, i\, \text{of}\, \Omega & \text{col.}\, i+3\, \text{of}\, \Omega \\ \vdots & \vdots \end{pmatrix} \times \begin{pmatrix} \cos\varphi_i & e^{i\beta_i}\sin\varphi_i \\ -e^{-i\beta_i}\sin\varphi_i & \cos\varphi_i \end{pmatrix} \quad (20)$$

(20)

Now, recall that in the Dirac limit [see Eq. (11)] the columns i and i+3 of the mixing matrix Ω are proportional to each other, $\Omega_{\alpha,i}=-i\Omega_{\alpha,i+3}=V_{\alpha,i}/\sqrt{2}$. This means that applying the (φ_i,β_i) transformation to the Dirac neutrino mixing matrix and, without loss of generality setting βi=0, we obtain:

$$\begin{pmatrix} \vdots & \vdots \\ \Omega_{\alpha,i} & \Omega_{\alpha,i+3} \\ \vdots & \vdots \end{pmatrix} = \begin{pmatrix} \vdots & \vdots \\ \frac{\cos\varphi_i+\sin\varphi_i}{\sqrt{2}} V_{\alpha,i} & i\frac{\cos\varphi_i-\sin\varphi_i}{\sqrt{2}} V_{\alpha,i} \\ \vdots & \vdots \end{pmatrix}.$$

(21)

From the last of these equations it can be seen that the ith column [the (i+3)th column] of Ω vanishes, if one chooses φi=3π/4 ($\varphi_i=\pi/4$).

Figure 5 shows a scan over the allowed parameter space in the plane (ϵ^2_3,φ3) using DayaBay, T2K, MINOS and Super-K atmospheric neutrino data. In agreement with the above discussion, there is a blind spot where no limit on ϵ^2_3 exist. This blind direction corresponds to the choice of $\varphi_3=\pi/4$.[9] Figure 5 also shows that the addition of JUNO data can lead only to a marginally improved limit.

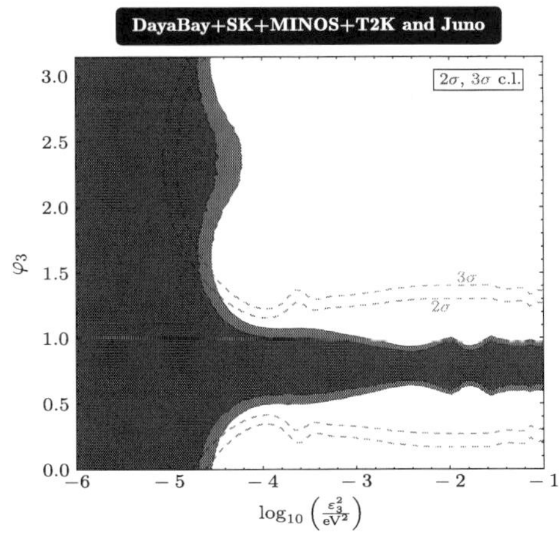

Figure 5. Allowed parameter space in the plane (ε_3^2,φ3) using DayaBay, T2K, MINOS and Super-K atmosheric neutrino data. The coloured plane shows the 2 and 3σ C.L. allowed regions (blue and red). The dashed lines show the expected limits for JUNO. The asymmetric parametrization of the mass splitting $(m_3, \sqrt{m_3^2 + \varepsilon_3^2})$ was used, so for $\varphi_3 = \pi/4$ there is no sensitivity to ε_3^2.

We now turn to a discussion of ε_1^2 and ε_2^2. For these two parameters, again solar neutrino physics provides the most important constraints. As above for ϵ23, we can define a rotation angle φ_1 (φ_2) between the columns 1 and 4 (2 and 5) of the mixing angle which will mitigate the effects of a nonzero ε_1^2 (ε_2^2). Figure (6) shows the P_{ee} probability for solar neutrinos as a function of neutrino energy for two different values of ϵ^2_2 and various values of φ_2.[10] The results for ϵ^2_1 and φ1 are completely analogous. As the figure shows, for $\varphi_2 = \pi/4$ again the effects of ϵ^2_2 completely decouple from the oscillation probability.

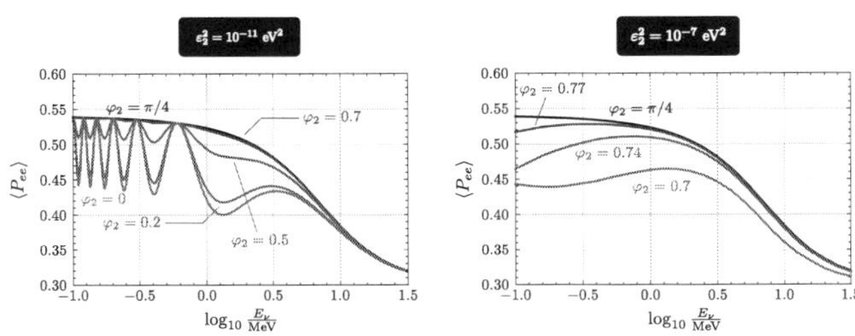

Figure 6. Average solar neutrino survival probability as a function of neutrino energy, for different choices of φ_2 and two different values of ϵ^2_2 and φ_2.

Figure 7 shows the allowed parameter space in the two planes (φ_1,ϵ^2_1) and (φ_2,ϵ^2_2) using solar data and combining solar data with KamLAND. These plots have been calculated using the best fit point values for $\Delta m^2_{\odot}$ and $\sin^2\theta_{12}$ from the global fit [4]. The plots show in all cases that there exists a slight preference, between (2–2.5)σ in all cases, for nonzero values of ϵ^2_i. Note that the preferred solution of the solar data in the region of $\varphi_1 \sim 3/4\pi$ and $\epsilon^2_1 \sim (10^{-4.5}–10^{-4})$ is ruled out by KamLAND. However, even combining solar and KamLAND data some preference for nonzero ϵ^2_i of the order of very roughly $10^{-10.5}eV^2$ remains.

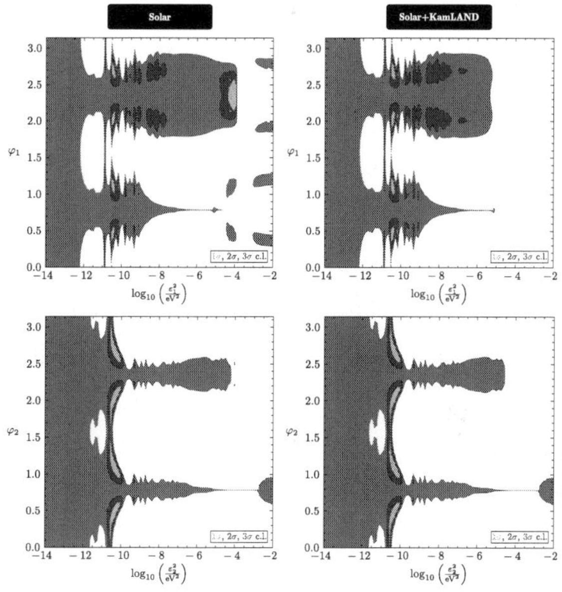

Figure 7. Allowed parameter range in the space φ_1,ϵ^2_1 (top) and φ_2,ϵ^2_2 (bottom). To the left: Solar data, to the right solar data+KamLAND. This plot uses the best fit point values for $\Delta m^2\odot$ and $\sin^2\theta_{12}$ from the global fit. This combination of data shows a slight preference for a nonzero value of the mass splitting, for a discussion see text.

We have traced back this preference for nonzero mass splittings in solar data to the well-known difference in the best fit points from $\Delta m^2_{\odot}$ in solar and KamLAND data. As can be seen also in the latest global fits [4], solar data prefers a $\Delta m^2_{\odot}$ around $(4–5)\times10^{-5}eV^2$, while KamLAND prefers $\Delta m^2_{\odot} \simeq 7.6\times10^{-5}eV^2$. This tension between the two data sets is roughly of the order of 2σ, with the error bar dominated by the larger error on $\Delta m^2_{\odot}$ in the solar data set. We have therefore recalculated the constraints on (φ_1,ϵ^2_1) from solar data for a value of $\Delta m^2_{\odot}$=4×10^{-5} eV^2. Figure 8 shows the results of such a scan. As can be seen, in this calculation there is no longer any preference for a nonzero value of ϵ^2_1.

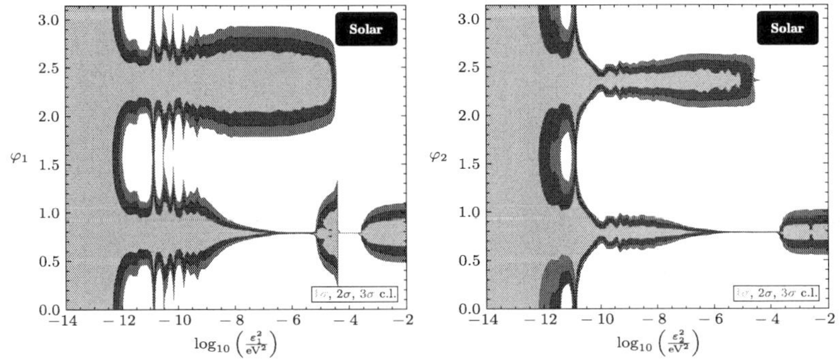

Figure 8. Allowed parameter range in the space φ^1,ϵ^2_1 (left) and φ_2,ϵ^2_2 (right) from solar data, using $\Delta m^2_{\odot}$=4×10^{-5}eV^2.

Note that such a low value of Δm2⊙ is ruled out by many σ from the KamLAND data. Thus, a small nonzero mass splitting could provide, in principle, a solution for the observed tension between solar and KamLAND data.

In summary, by introducing one mass splitting at a time, we extracted bounds for the pairs of parameters (ϵ_i,φ_i), i=1, 2, 3. In the limit where φ_i is (2±1)/4π, one column of the mixing matrix vanishes and therefore the mass splitting ϵi becomes unobservable. For this reason, one expects that for reasonably large values of ϵ_i there must be tight limits on |φ_i−(2±1)/4π|, meaning that φ_i has to be quite far from the Dirac limit (φ_i=0). On the other hand, for a small enough value of the mass splitting ϵ_i, the associated oscillation length eventually become larger than the baseline of the relevant experiments, and in that case φi becomes unconstrained.

B. Quasi-Dirac Neutrinos in the Limit ϵ_i→0

If masses are degenerate as in Eq. (10), then the oscillation probability formula will not change under unitary rotations of the columns i and i+3 of the mixing matrix, see Eq. (20). In other words,

$$\begin{pmatrix} \vdots & \vdots \\ \text{col.}i\text{ of }\Omega & \text{col.}i+3\text{ of }\Omega \\ \vdots & \vdots \end{pmatrix} \to \begin{pmatrix} \vdots & \vdots \\ \text{col.}i\text{ of }\Omega & \text{col.}i+3\text{ of }\Omega \\ \vdots & \vdots \end{pmatrix} U(i) \quad (22)$$

for unitary matrices U(i) (i=1, 2, 3) leaves $P(\nu_\alpha \to \nu_\beta)$ unchanged. Hence, there is a U(2)3 redundancy in our description of Ω, and in turn this means that out of the 24 parameters describing the mixing matrix, oscillation experiments are only sensitive to 13.[11] This number is further reduced to 7 if we ignore tau neutrinos. In this latter case the oscillation probabilities can be written as:

$$P(\nu_e \to \nu_e) = 1 + (1 - X_1 - X_2)X_2\mathcal{A}_{12} + (1 - X_1 - X_2)X_1\mathcal{A}_{13} + X_1X_2\mathcal{A}_{23}. \quad (23)$$

$$P(\nu_\mu \to \nu_\mu) = 1 + (1 - X_3 - X_4)X_4\mathcal{A}_{12} + (1 - X_3 - X_4)X_3\mathcal{A}_{13} + X_3X_4\mathcal{A}_{23}. \quad (24)$$

$$P(\nu_e \to \nu_\mu) = -(X_6 + \text{Re}X_7)\mathcal{A}_{12} - (X_5 + \text{Re}X_7)\mathcal{A}_{13} + \text{Re}X_7\mathcal{A}_{23} + \text{Im}X_7(\mathcal{B}_{12} - \mathcal{B}_{13} + \mathcal{B}_{23}), \quad (25)$$

with the oscillating factors $\mathcal{A}_{ij} \equiv -4\sin^2[(m_i^2 - m_j^2)L/(4E)]$ and $\mathcal{B}_{ij} \equiv 2\sin[(m_i^2 - m_j^2)L/(2E)]$ the 7 parameters X_i defined as follows:

$$X_1 \equiv |\Omega_{e3}|^2 + |\Omega_{e6}|^2, \qquad X_2 \equiv |\Omega_{e2}|^2 + |\Omega_{e5}|^2, \quad (26)$$

$$X_3 \equiv |\Omega_{\mu3}|^2 + |\Omega_{\mu6}|^2, \qquad X_4 \equiv |\Omega_{\mu2}|^2 + |\Omega_{\mu5}|^2, \quad (27)$$

$$X_5 \equiv |\Omega_{e3}\Omega_{\mu3}^* + \Omega_{e6}\Omega_{\mu6}^*|^2, \qquad X_6 \equiv |\Omega_{e2}\Omega_{\mu2}^* + \Omega_{e5}\Omega_{\mu5}^*|^2, \quad (28)$$

$$X_7 \equiv (\Omega_{e3}\Omega_{\mu3}^* + \Omega_{e6}\Omega_{\mu6}^*)(\Omega_{e2}^*\Omega_{\mu2} + \Omega_{e5}^*\Omega_{\mu5}). \quad (29)$$

As a side remark, we would like to point out here that a similar approach could, in principle, be used in the presence of one ϵ_i: in this case, instead of 7, there would be 9 combinations of angles and phases to take into account.[12]

Note that the Xi defined above can take any value in our framework, provided that the following constraints are obeyed:

(1) the first six X_i are non-negative real numbers;

(2) neither X_1+X_2 nor X_3+X_4 can be larger than 1;

(3) $X_5 \le X_1X_3$ and $X_6 \le X_2X_4$;

(4) the norm of X_7 is fixed by X_5 and X_6 ($|X_7|^2 = X_5X_6$), so even though X_7 is a complex parameter, only $\arg(X_7)$ is an independent degree of freedom;

(5) $X_5 + X_6 + 2\cos[\arg(X_7)]\sqrt{X_5 X_6}$ cannot be bigger than $(1-X_1-X_2)(1-X_3-X_4)$.

$$\hat{X}_1 \equiv |\Omega_{e3}|^2 + |\Omega_{e6}|^2, \quad \hat{X}_2 \equiv |\Omega_{e2}|^2 + |\Omega_{e5}|^2, \quad \hat{X}_3 \equiv |\Omega_{e1}|^2. \tag{30}$$

$$\hat{X}_4 \equiv |\Omega_{\mu 3}|^2 + |\Omega_{\mu 6}|^2, \quad \hat{X}_5 \equiv |\Omega_{\mu 2}|^2 + |\Omega_{\mu 5}|^2, \quad \hat{X}_6 \equiv |\Omega_{\mu 1}|^2, \tag{31}$$

$$\hat{X}_7 \equiv |\Omega_{e3}\Omega^*_{\mu 3} + \Omega_{e6}\Omega^*_{\mu 6}|^2, \quad \arg(\hat{X}_8) \equiv \arg(\Omega_{e1}\Omega^*_{\mu 1}\Omega^*_{e4}\Omega_{\mu 4}), \tag{32}$$

$$\arg(\hat{X}_9) \equiv \arg[\Omega_{e1}\Omega^*_{\mu 1}(\Omega^*_{e2}\Omega_{\mu 2} + \Omega^*_{e5}\Omega_{\mu 5})]. \tag{33}$$

These conditions are a consequence of the definitions of the Xi and the fact that the rows of the mixing matrix Ω are orthonormal ($\Omega\Omega^{\dagger}=1$). Taking them into account, we are able to pick out all valid points in the Xi parameter space, without ever referencing back to specific entries of the mixing matrix.

For reference, the values of these X_i parameters in the Dirac limit as a function of the standard θ_{12}, θ_{13}, θ_{23} and δ parameters [see Eq. (15)] are the following:

$$X_1 = \sin^2\theta_{13}, \quad X_2 = \sin^2\theta_{12}\cos^2\theta_{13}, \quad X_3 = \cos^2\theta_{13}\sin^2\theta_{23}, \tag{34}$$

$$X_4 = \sin^2\theta_{12}\sin^2\theta_{13}\sin^2\theta_{23} + \cos^2\theta_{12}\cos^2\theta_{23} - \frac{1}{2}\sin 2\theta_{12}\sin 2\theta_{23}\sin\theta_{13}\cos\delta, \tag{35}$$

$$X_5 = X_1 X_3, \quad X_6 = X_2 X_4, \tag{36}$$

$$X_7 = \sin\theta_{12}\sin\theta_{13}\cos^2\theta_{13}\sin\theta_{23}(-\sin\theta_{12}\sin\theta_{13}\sin\theta_{23} + e^{-i\delta}\cos\theta_{12}\cos\theta_{23}). \tag{37}$$

Using $\Delta m^2_{Atm}=2.55\times10^{-3}$eV and $\Delta m^2_{\odot}=7.56\times10^{-5}$, we performed a 7-dimensional scan over all X_i. Electron neutrino survival data from at KamLAND, DayaBay, SuperK and Borexino was used, together with muon neutrino survival data at MINOS and T2K and $\nu_\mu \to \nu_e$ T2K data. The allowed values in the planes (X_1,X_2), (X_3,X_4) and (X_5,X_6) are shown in Fig. 9. The marginalized χ^2 function for $X_1,\ldots,6$ is shown in Fig. (10); the marginalized $\chi^2[\arg(X7)]$ function is not shown as it is essentially flat from 0 to 2π.

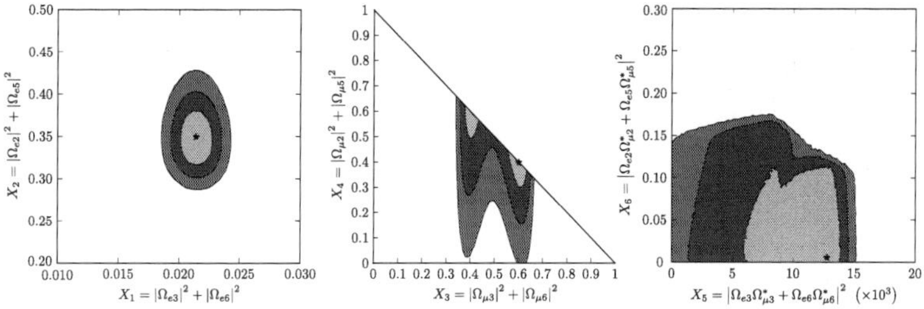

Figure 9. One, two and three σ regions in the (X1,X2), (X3,X4) and (X5,X6) w hich are allowed by electron neutrino survival data from at KamLAND, Daya-Bay, SuperK and Borexino, muon neutrino survival data at MINOS and T2K and muon to electron transition data from T2K. This plots were obtained from a 7-dimensional scan of the Xi defined in Eqs. (26)–(29) and marginalizing over 5 variables.

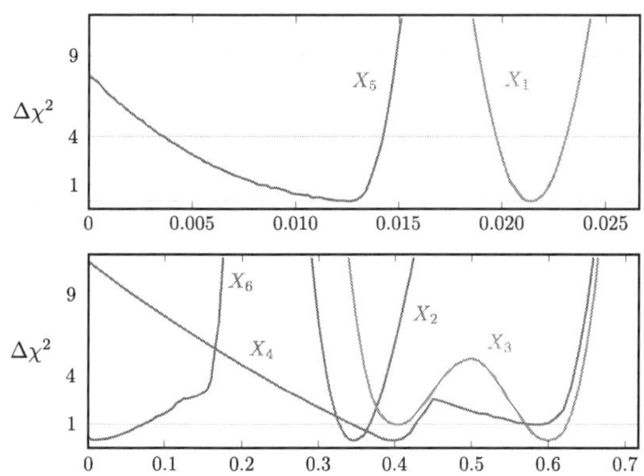

Figure 10. Marginalized $\Delta\chi^2$ values for the variables X_1,…,6 defined in Eqs. (26)–(29). The χ^2 function for X_7 is essentially flat. These parameters are a function of the entries of the mixing matrix only; no mass splittings ϵ_i were considered.

Overall, the bounds on these 7 parameters are broadly consistent with the standard three neutrino oscillation picture. In other words, by substituting in expressions (34)–(37) the numbers obtained for θ_{12}, θ_{13}, θ_{23} and δ from global fits [4], we get values for the X_i roughly in agreement with Figs. 9 and 10. In order to see clearly that current data is consistent with the Dirac limit, note that in this latter case there are only 4 independent parameters. Thus, it follows that the standard three neutrino oscillation picture must correspond to three relations among the 7 X_i. These are

$$X_5 = X_1X_3, \qquad X_6 = X_2X_4 \quad \text{and}$$
$$\text{Re}(X_7) = \frac{1}{2}(1 - X_1 - X_2 - X_3 - X_4 + X_1X_4 + X_2X_3), \tag{38}$$

and from Fig. 11 one can see that oscillation data is compatible with each of these equalities within ~1σ. The three together are disfavored only at min(χ^2_{Dirac})−min(χ^2)=1.9 so, assuming no mass splittings ϵ^i, there is currently no significant indication for quasi-Dirac neutrinos.

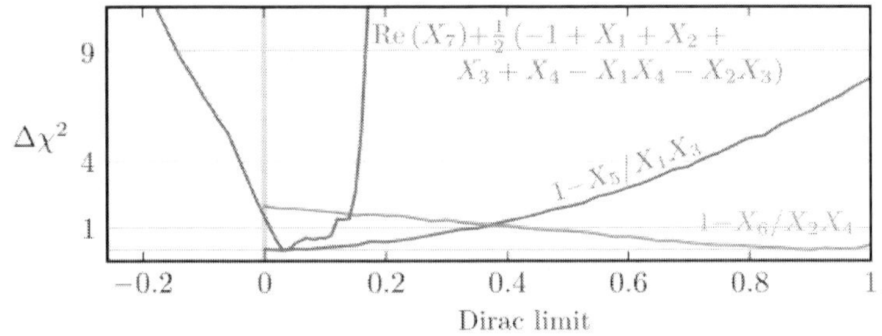

Figure 11. $\Delta\chi^2$ functions for the three combinations of parameters which, when equal to 0 simultaneously, signal the Dirac limit [see Eq. (38)].

We would like to point out that even in the absence of new mass scales, quasi-Dirac neutrinos can, in principle, be distinguished from some other scenarios through oscillation experiments. In particular, consider 3 active neutrinos and a nonunitary 3×3 mixing matrix V. The oscillation probabilities are then given by the expressions

$$P(\nu_e \to \nu_e) = F(J^{12}_{ee}, J^{13}_{ee}, J^{23}_{ee}) + J^{12}_{ee}A_{12} + J^{13}_{ee}A_{13} + J^{23}_{ee}A_{23}, \tag{39}$$

$$P(\nu_\mu \to \nu_\mu) = F(J^{12}_{\mu\mu}, J^{13}_{\mu\mu}, J^{23}_{\mu\mu}) + J^{12}_{\mu\mu}A_{12} + J^{13}_{\mu\mu}A_{13} + J^{23}_{\mu\mu}A_{23}, \tag{40}$$

$$\begin{aligned} P(\nu_e \to \nu_\mu) &= F'(J^{12}_{ee}, J^{13}_{ee}, J^{23}_{ee}, J^{12}_{\mu\mu}, J^{13}_{\mu\mu}, J^{23}_{\mu\mu}, J^{12}_{e\mu}, J^{13}_{e\mu}, J^{23}_{e\mu}) \\ &+ \text{Re}(J^{12}_{e\mu})A_{12} + \text{Re}(J^{13}_{e\mu})A_{13} \\ &+ \text{Re}(J^{23}_{e\mu})A_{23} + \text{Im}(J^{12}_{e\mu})B_{12} \\ &+ \text{Im}(J^{13}_{e\mu})B_{13} + \text{Im}(J^{23}_{e\mu})B_{23}, \end{aligned} \tag{41}$$

where $J^{ij}_{\alpha\beta} = V^*_{\alpha i}V^*_{\beta j}V_{\beta i}V_{\alpha j}$. The exact form of the functions F and F′ which control the 0-distance neutrino behavior is not important for the present discussion. The more important point is that with a nonunitary V one can have an oscillatory behavior which is impossible to reproduce with quasi-Dirac neutrinos, and vice-versa.

For example, the B_{ij} coefficients in $P(\nu_e \to \nu_\mu)$ do not need to be related for a nonunitary V, while for quasi-Dirac neutrinos they must be the same [up to a minus sign—see Eq. (25)]. On the other hand, note that from the J^{ij}_{ee} and $J^{ij}_{\mu\mu}$ one can extract the modulus of the absolute value of all Jijeμ, hence by measuring $P(\nu_e \rightarrow \nu_e)$ and $P(\nu_\mu \rightarrow \nu_\mu)$, as well as the coefficients B_{ij} in $P(\nu_e \rightarrow \nu_\mu)$, the coefficients of the oscillatory factors A_{ij} in $P(\nu_e \rightarrow \nu_\mu)$ are fixed for a nonunitary V (up to ±signs). However, for quasi-Dirac neutrinos no such constraint exists. So, with this short theoretical argument, one can conclude that in principle these two nonstandard neutrino scenarios can be distinguished through oscillation experiments.

SUMMARY

In general, neutrinos can have lepton number violating (Majorana) and lepton number conserving (Dirac) mass terms. If the lepton number violating mass terms are smaller than the lepton number preserving ones, neutrinos are *quasi-Dirac* particles. Phenomenologically, this corresponds to the existence of three pairs of neutrinos with slightly different masses, hence oscillation experiments are sensitive not only to the usual solar and atmospheric mass scales, but also to three small mass splittings ϵi. Furthermore, for quasi-Dirac neutrinos there are more than 3+1 angles and phases to be considered. In this work, we have analyzed the constraints on these quasi-Dirac neutrino parameters imposed by current neutrino oscillation data and also briefly discussed the potential of the future JUNO experiment to improve upon existing constraints.

In Sec. II we have discussed a fully general parametrization of the lepton sector for three generations of quasi-Dirac neutrinos. In addition to the charged lepton masses, there is a total of 6 masses, 12 angles and 12 phases. Oscillation experiments are not sensitive to the overall neutrino mass scale nor to 5 of the phases (which are of the Majorana type). Hence we are left with a 24-dimensional model space, compared to the six-dimensional space for an ordinary three generation case ($\Delta m^2_{\odot}$, Δm^2_{Atm}, θ_{12}, θ_{13}, θ_{23} and δ).

It is numerically too costly to handle such a large number of parameters at the same time, hence we analyzed several different special cases. First, we took a single mass splitting $\epsilon^2_i \neq 0$. If we split two neutrinos with mass mi into a quasi-degenerate pair of particles with masses $\sqrt{m_i^2 - \varepsilon_i^2/2}$ and $\sqrt{m_i^2 + \varepsilon_i^2/2}$ then a new oscillation length L∝1/ϵ2i appears which is associated to the conversion of active to sterile neutrinos. Very stringent limits on ϵ2i in such one parameter extensions can be derived, of the order of 10^{-11}eV^2 for $\varepsilon^2_{1.2}$ (from solar neutrino data) and 10^{-5}eV^2 for ϵ^2_3 (dominated by Super-K atmospheric neutrino data).

Next, we considered the case when one mass splitting and one of the nonstandard angles are allowed to take nonzero values at the same time. As we have shown, in this situation degeneracies of the χ^2 function can occur, implying that from a single experiment in many cases it will no longer be possible to derive meaningful limits on individual parameters. These degeneracies can be resolved by considering data from more than one experiment, accessing different $P(\nu_\alpha \rightarrow \nu_\beta)$.

We then considered the possibility of nullifying the effects of the ϵi completely by changing some particular combinations of the angles θ_{ij} of our parametrization. Instead of pursuing the exact form of these rather complex parameter combinations, we discussed a simpler definition, describing 3 angles φ_i associated to rotations between the columns i and i+3 of the quasi-Dirac mixing matrix, such that in the limit where these angles are equal to $\pi/4$ ($3\pi/4$) the i+3(i) column of the mixing matrix vanishes and hence the associated neutrino mass disappears from the oscillation probability formula. We stress that for these particular parameter combinations no limits on the ϵ_i can be derived from oscillation experiments. The regions in the planes (ϵ_i,φ_i) which are allowed by various experiments are shown in Figs. 5 and 7. In this context, it is interesting to note that the tension between the value of the solar mass scale preferred by global fits ($\sim 7.6\times10^{-3}$eV^2) and the lower one preferred by solar data ($\sim 4\times10^{-3}$ eV2) might be resolved by a nonzero value for either ϵ_1 or ϵ_2.

Lastly, we considered the possibility that the mass splittings ϵ_i are too small to be measured in oscillation experiments. Even in this scenario, one can have departures from the lepton-number-conserving Dirac scenario due to the new angles θ_{ij} (and phases ϕ_{ij}). As mentioned above, there is a large number of such parameters. However, it can be shown that with 3 pairs of neutrinos with the same mass, oscillations will only depend on a total of 13 combinations of angles and phases. Additionally, if we focus just on electron and muon neutrinos, this number is further reduced to 7, corresponding to 6 angles and 1 phase. In the text we called these parameter combinations X1⋯7 and stressed that they cannot be identified with θ_{12}, θ_{13}, θ_{23} nor δ, as these quantities by themselves are not physical. Instead, the 7 Xi correspond to combinations of these and additional θ_{ij} angles and ϕ_{ij} phases.

In Sec. III B we made a 7-dimensional scan of these Xi parameters in the absence of mass splittings. Their exact definitions, as well as the limits imposed on them by current data can be found there. Crucially, for Dirac neutrinos there are only 4 parameters. Hence the Dirac limit corresponds to 3 relations among the 7 Xi. By testing these relations, we find that min(χ^2_{Dirac})−min(χ^2)=1.9, i.e. current data is compatible with the Dirac scenario. Progress on tests for quasi-Diracness can be made in the future with a more precise measurement of P($\nu_e \rightarrow \nu_\mu$) and P($\nu_\mu \rightarrow \nu_\mu$). Thus, more statistics taken in T2K, MINOS+ or NOνA and, in particular, the future precise measurements possible at DUNE should provide more sensitive probes for this particular setup of quasi-Dirac neutrinos without new mass scales.

NOTES

[1] Note the factor i in ν^2. One could equally well choose the two mass eigenstates to be $\pm m_D$ instead.

[2] This distinction is based on two early papers on the subject [5,6]. Wolfenstein [5] discussed pairs of active neutrinos, which almost preserve lepton number (due to a relative CP-sign) if the mixing angle between them is close to maximal and the mass splitting is small. He called such particles "pseudo-Dirac" neutrinos. Near the end of the paper, Wolfenstein then extended the terminology to mass matrices which contain both, active and sterile states. In [6], on the other hand, Valle proposed to use the terminology "quasi-Dirac" neutrinos for active-sterile pairs, to differentiate them from "pseudo-Dirac" (active-active pairs).

[3] PD neutrinos have mass matrices with entries close to zero on the diagonal. This is similar to the case discussed in [15–17] for the Zee model [18].

[4] Fields with no flavor indices should be seen as vectors.

[5] This is true as long as the rows of Ω are orthonormal, i.e. $\Omega\Omega^\dagger$=1.

[6] Due to the annual variation of the Earth-Sun distance, for values of ϵ22 larger than ~10−10,−9 eV2, the oscillations are averaged over, so only an overall reduction of survival probability is seen.

[7] To a good approximation, this is the situation in the DayaBay experiment.

[8] Numerically this leads to the same limits on ϵ2i, as long as the mass splitting is much smaller than the relevant Δm2ij (i.e., the solar or atmospheric scale).

[9] Shifting ϵ23 one could alternatively define (√m23+ϵ23,m3). In that case, the blind spot occurs at φ3=3π/4 instead.

[10] This probability is averaged over the variations of the Earth-Sun distance, and neutrino production point inside the Sun.

[11] The counting goes as follows: each U(2) describes 4 redundancies in the parameters, hence there is a total of 12 redundancies in U(2)3. However, one of them corresponds to the irrelevance of multiplying Ω by an overall phase; that was already taken care of when row phases were removed from the mixing matrix. Hence we are left with 24−12+1 real parameters which affect the neutrino oscillation probabilities if no ϵi's are introduced.

[12]Consideranonzeroϵ_1(for$\epsilon_{i=2,3}\neq 0$,thechangestothefollowingexpressions are trivial). Then the $P(\nu_e \to \nu_e)$, $P(\nu_e \to \nu_\mu)$ and $P(\nu_\mu \to \nu_\mu)$ probabilities depend only on the following quantities refer eq. (30), (31), (32) and (33).

ACKNOWLEDGMENTS

We would like to thank Mariam Tórtola for patiently explaining to us, on numerous occasions, several details concerning the various neutrino oscillation experiments. This work was funded by the Spanish state through the Projects No. FPA2014-58183-P and No. SEV-2014-0398 (from the Ministerio de Economía, Industria y Competitividad), as well as PROMETEOII/2014/084 (from the Generalitat Valenciana). R.F. was also financially supported through the Grant Juan de la Cierva-formación FJCI-2014-21651.

REFERENCES

1. I. Avignone, T. Frank, S.R. Elliott, and J. Engel, Double beta decay, Majorana neutrinos, and neutrino mass, Rev. Mod. Phys. 80, 481 (2008).
2. F.F. Deppisch, M. Hirsch, and H. Päs, Neutrinoless double-beta decay and physics beyond the standard model, J. Phys. G 39, 124007 (2012).
3. S. Abe et al. (KamLAND Collaboration), Precision Measurement of Neutrino Oscillation Parameters with KamLAND, Phys. Rev. Lett. 100, 221803 (2008).
4. P.F. de Salas et al., Status of Neutrino Oscillations 2017, , Status of neutrino oscillations 2017, arXiv:1708.01186.
5. L. Wolfenstein, Different varieties of massive Dirac neutrinos, Nucl. Phys. B186, 147 (1981).
6. J.W.F. Valle, Neutrinoless double-β *decay with quasi-Dirac neutrinos*, Phys. Rev. D 27, 1672 (1983).
7. S.T. Petcov, On pseudo-Dirac neutrinos, neutrino oscillations and neutrinoless double β-decay, Phys. Lett. 110B, 245 (1982).
8. M. Doi, M. Kenmoku, T. Kotani, H. Nishiura, and E. Takasugi, Pseudo Dirac neutrino, Prog. Theor. Phys. 70, 1331 (1983).
9. G. Dutta and A.S. Joshipura, Pseudo Dirac neutrinos in the seesaw model, Phys. Rev. D 51, 3838 (1995).
10. A.S. Joshipura and S.D. Rindani, Phenomenology of pseudo Dirac neutrinos, Phys. Lett. B 494, 114 (2000).
11. S.M. Bilenky and B. Pontecorvo, Neutrino oscillations with large oscillation length in spite of large (Majorana) neutrino masses?, Lett. Nuovo Cimento 37, 467 (1983); [38] Sov. J. Nucl. Phys. 38, 248 (1983)].
12. S.M. Bilenky and S.T. Petcov, Massive neutrinos and neutrino oscillations, Rev. Mod. Phys. 59, 671 (1987); Erratum, 60, 575 (1988).
13. C. Giunti, C.W. Kim, and U.W. Lee, Oscillations of pseudo Dirac neutrinos and the solar-neutrino problem, Phys. Rev. D 46, 3034 (1992).
14. Y. Nir, Pseudo-Dirac solar neutrinos, J. High Energy Phys. 2000, 039.
15. P.H. Frampton, M.C. Oh, and T. Yoshikawa, Zee model confronts SNO data, Phys. Rev. D 65, 073014 (2002).
16. X.-G. He, Is the Zee model neutrino mass matrix ruled out?, Eur. Phys.

J. C 34, 371 (2004).

17. B. Brahmachari and S. Choubey, Viability of bimaximal solution of the Zee mass matrix, Phys. Lett. B 531, 99 (2002).
18. A. Zee, A theory of lepton number violation and neutrino Majorana masses, Phys. Lett. 93B, 389 (1980); Erratum, 95B, 461 (1980).
19. A. Geiser, Pseudo-Dirac neutrinos as a potential complete solution to the neutrino oscillation puzzle, Phys. Lett. B 444, 358 (1998).
20. W. Krolikowski, Option of three pseudo-Dirac neutrinos, Acta Phys. Pol. B 31, 663 (2000).
21. K.R.S. Balaji, A. Kalliomaki, and J. Maalampi, Revisiting pseudo-Dirac neutrinos, Phys. Lett. B 524, 153 (2002).
22. E. Ma and P. Roy, Model of four light neutrinos in the light of all present data, Phys. Rev. D 52, R4780 (1995).
23. S. Goswami and A.S. Joshipura, Neutrino anomalies and quasi Dirac neutrinos, Phys. Rev. D 65, 073025 (2002).
24. K.S. Babu and R.N. Mohapatra, $L_e+L_\mu-L_\tau-L_s$ *symmetry and a mixed* 2+2 *scenario for neutrino oscillations*, Phys. Lett. B 522, 287 (2001).
25. A. de Gouvea, W.-C. Huang, and J. Jenkins, Pseudo-Dirac neutrinos in the new standard model, Phys. Rev. D 80, 073007 (2009).
26. M. Cirelli, G. Marandella, A. Strumia, and F. Vissani, Probing oscillations into sterile neutrinos with cosmology, astrophysics and experiments, Nucl. Phys. B708, 215 (2005).
27. J.F. Beacom, N.F. Bell, D. Hooper, J.G. Learned, S. Pakvasa, and T.J. Weiler, Pseudo-Dirac Neutrinos: A Challenge for Neutrino Telescopes, Phys. Rev. Lett. 92, 011101 (2004).
28. A. Esmaili, Pseudo-Dirac neutrino scenario: Cosmic neutrinos at neutrino telescopes, Phys. Rev. D 81, 013006 (2010).
29. A. Esmaili and Y. Farzan, Implications of the pseudo-Dirac scenario for ultra high energy neutrinos from GRBs, J. Cosmol. Astropart. Phys. 2012, 014.
30. A.S. Joshipura, S. Mohanty, and S. Pakvasa, Pseudo-Dirac neutrinos via a mirror world and depletion of ultrahigh energy neutrinos, Phys. Rev. D 89, 033003 (2014).
31. M. Kobayashi and C.S. Lim, Pseudo Dirac scenario for neutrino oscillations, Phys. Rev. D 64, 013003 (2001).

32. Y.H. Ahn, S.K. Kang, and C.S. Kim, A model for pseudo-Dirac neutrinos: leptogenesis and ultra-high energy neutrinos, J. High Energy Phys. 2016, 092.
33. G.J. Stephenson, Jr., J.T. Goldman, B.H.J. McKellar, and M. Garbutt, Large mixing from small: Pseudo-Dirac neutrinos and the singular see-saw, Int. J. Mod. Phys. A 20, 6373 (2005).
34. K.L. McDonald and B.H.J. McKellar, The type-II singular see-saw mechanism, Int. J. Mod. Phys. A 22, 2211 (2007).
35. R. Mohapatra and J. Valle, Neutrino mass and baryon-number nonconservation in superstring models, Phys. Rev. D 34, 1642 (1986).
36. D. Chang and O.C.W. Kong, Pseudo-Dirac neutrinos, Phys. Lett. B 477, 416 (2000).
37. V. Berezinsky, M. Narayan, and F. Vissani, Mirror model for sterile neutrinos, Nucl. Phys. B658, 254 (2003).
38. L.A. Sanchez, W.A. Ponce, and R. Martinez, SU(3)c×SU(3)L×U(1) X *as an* E6 *subgroup*, Phys. Rev. D 64, 075013 (2001).
39. R.M. Fonseca and M. Hirsch, Lepton number violation in 331 models, Phys. Rev. D 94, 115003 (2016).
40. S. Abel, A. Dedes, and K. Tamvakis, Naturally small Dirac neutrino masses in supergravity, Phys. Rev. D 71, 033003 (2005).
41. Q.R. Ahmad et al. (SNO Collaboration), Direct Evidence for Neutrino Flavor Transformation from Neutral-Current Interactions in the Sudbury Neutrino Observatory, Phys. Rev. Lett. 89, 011301 (2002).
42. K. Abe et al. (Super-Kamiokande Collaboration), Limits on sterile neutrino mixing using atmospheric neutrinos in Super-Kamiokande, Phys. Rev. D 91, 052019 (2015).
43. N. Agafonova et al. (OPERA Collaboration), Limits on muon-neutrino to tau-neutrino oscillations induced by a sterile neutrino state obtained by OPERA at the CNGS beam, J. High Energy Phys. 2015, 069.
44. P. Adamson et al. (MINOS, Daya Bay Collaborations), Limits on Active to Sterile neutrino oscillations from disappearance searches in the MINOS, Daya Bay, and Bugey-3 experiments, Phys. Rev. Lett. 117, 151801 (2016); et al., Publisher's Note, 117, 209901 (2016).
45. M.G. Aartsen et al. (IceCube Collaboration), Search for sterile neutrino mixing using three years of IceCube DeepCore data, Phys. Rev. D 95, 112002 (2017).

46. P. Adamson et al. (NOvA Collaboration), Search for active-sterile neutrino mixing using neutral-current interactions in NOvA, Phys. Rev. D 96, 072006 (2017).
47. K.N. Abazajian et al., Light Sterile Neutrinos: A White Paper, , Light sterile neutrinos: A white paper, arXiv:1204.5379.
48. S. Gariazzo, C. Giunti, M. Laveder, Y.F. Li, and E.M. Zavanin, Light sterile neutrinos, J. Phys. G 43, 033001 (2016).
49. J. Schechter and J.W.F. Valle, Neutrino masses in SU(2)×U(1) *theories*, Phys. Rev. D 22, 2227 (1980).
50. K. Abe et al. (Super-Kamiokande Collaboration), Solar neutrino measurements in Super-Kamiokande-IV, Phys. Rev. D 94, 052010 (2016).
51. G. Bellini et al. (BOREXINO Collaboration), Neutrinos from the primary proton-proton fusion process in the Sun, Nature (London) 512, 383 (2014).
52. G. Bellini et al., Precision Measurement of the 7Be *Solar Neutrino Interaction Rate in* Borexino, Phys. Rev. Lett. 107, 141302 (2011).
53. T. Kajita, The measurement of neutrino properties with atmospheric neutrinos, Annu. Rev. Nucl. Part. Sci. 64, 343 (2014).
54. P. Adamson et al. (MINOS Collaboration), Measurement of Neutrino and Antineutrino Oscillations Using Beam and Atmospheric Data in MINOS, Phys. Rev. Lett. 110, 251801 (2013).
55. K. Abe et al. (T2K Collaboration), Measurements of neutrino oscillation in appearance and disappearance channels by the T2K experiment with 6.6×1020 *protons on target*, Phys. Rev. D 91, 072010 (2015).
56. F.P. An et al. (Daya Bay Collaboration), Measurement of electron antineutrino oscillation based on 1230 days of operation of the Daya Bay experiment, Phys. Rev. D 95, 072006 (2017).
57. F. An et al. (JUNO Collaboration), Neutrino physics with JUNO, J. Phys. G 43, 030401 (2016).

CHAPTER

11

Three-Flavor Oscillations with Accelerator Neutrino Beams

Mauro Mezzetto [1] and Francesco Terranova [2,3]

[1]INFN sez. di Padova, Via F. Marzolo 8, I-35131 Padova, Italy

[2]Department of Physics, Universe of Milano-Bicocca, Piazza della Scienza 3, I-20126 Milano, Italy

[3]INFN sez. di Milano-Bicocca, Piazza della Scienza 3, I-20126 Milano, Italy

ABSTRACT

The three-flavor neutrino oscillation paradigm is well established in particle physics thanks to the crucial contribution of accelerator neutrino beam experiments. In this paper, we review the most important contributions of these experiments to the physics of massive neutrinos after the discovery of θ13 and future perspectives in such a lively field of research. Special emphasis is given to the technical challenges of high power beams and the oscillation results of T2K, OPERA, ICARUS, and NO ν A. We discuss in detail the role of accelerator neutrino experiments in the precision era of

Citation: Mezzetto, M.; Terranova, F. "Three-Flavor Oscillations with Accelerator Neutrino Beams". Universe 2020, 6, 32. https://doi.org/10.3390/universe6020032

neutrino physics in view of DUNE and Hyper-Kamiokande, the program of systematic uncertainty reduction and the development of new beam facilities.

Keywords: neutrino oscillations; accelerator neutrino beams; neutrino detectors

INTRODUCTION

High intensity neutrino beams at energies above 100 MeV are produced at proton accelerators since 1962 and played a key role in the foundation of the electroweak theory and the study of nuclear structure. Unlike other neutrino sources, they provide an unprecedented level of control of the flavor at source, flux and energy, which has been prominent in confirming the very existence of neutrino oscillations and measuring with high precision the oscillation parameters. Accelerator neutrino beams [**1**] are mostly sources of ν_μ beams because the vast majority of neutrinos produced at accelerators originate by the decay in flight of multi-GeV pions. As a consequence, they play a role in neutrino oscillation physics only if the oscillation length Lo is within the earth diameter for the neutrino energy range E accessible to accelerators (0.1–100 GeV). The existence of at least one oscillation length smaller than the earth diameter and involving νμ in the GeV-energy range has been established by Super-Kamiokande in 1998 [**2,3**]. In the modern formalism that describes neutrino oscillation (see **Section 2**), it corresponds to saying that both the mixing angle between the second and third neutrino mass eigenstates (θ_{23}) and their squared mass difference $(\Delta m^2_{23} \equiv m^2_3 - m^3_2)$ are different from zero. Such statement opened up the modern field of neutrino oscillations at accelerators. Since $\Delta m^2_{23} \simeq 2.3 \times 10^{-3}$ eV^2 and the oscillation length is:

$$L_o(\mathrm{km}) \simeq 2.48E(\mathrm{GeV})/\Delta m^2_{23}(\mathrm{eV}^2)$$

oscillations are visible for a GeV neutrino if the source-to-detector distance (“baseline”, L) is $L \simeq L_0 \simeq O(1000\mathrm{km})$. It is fortunate that oscillations are visible at propagation lengths smaller than the earth diameter because both the source and the detector can be artificial, i.e., designed by the experimentalist to maximise sensitivity and precision. On the other hand, it is unfortunate that the baselines of interest are hundreds of km long (“long-baseline experiments”): since neutrinos cannot be directly focused, the divergence of the ν beam is large and the transverse size of the flux at the detector is much larger than the size of the detector itself. As a consequence, the number of neutrinos crossing the detector decreases as L−2 and such a drop must be compensated by brute force, increasing the number of neutrinos at source. All beams discussed in this paper are high power beams driven by proton accelerators operating at an average power >100 kW.

Even if the inception of the study of massive neutrinos at accelerators can be dated back to 1998, the golden age of accelerator neutrino physics has just started, triggered by the discovery of θ13. In 2012 reactor (Daya-Bay [**4**], RENO [**5**]) and accelerator (T2K [**6**]), experiments demonstrated that even the mixing angle between the first and third neutrino mass eigenstate is non-zero. Indeed, θ13 is quite large (θ13≃8°) so that the entire neutrino mixing matrix (PMNS—see **Section 2**) has a very different structure compared with the Cabibbo–Kobayashi–Maskawa matrix. This result corroborates once more the fact that the Yukawa sector of the Standard Model is far from being understood both in the quark and lepton sector. From the experimental point of view, a large θ_{13} opens up the possibility of observing simultaneously $\nu_\mu \rightarrow \nu_\tau$ and $\nu_\mu \rightarrow \nu_e$ oscillations in long baseline experiments and, in general, three family interference effects. This is the reason why neutrino physicists refer to the current and future generation of beam experiments as the experiments for the "precision era" of neutrino physics. These modern experiments must disentangle the missing parameters of the Standard Model (SM) Lagrangian in the neutrino sector (the CP violating phase, the ordering of the mass eigenstates) from small perturbations in the νμ oscillation probability at the GeV scale. Since the baselines of these experiments range from 295 km (T2K) to 1300 km (DUNE), beam intensity and statistics still play an important role. However, systematics will soon become the limiting factor for precision neutrino physics, especially in the DUNE and Hyper-Kamiokande era.

In this paper, we will review the most important results appeared after the discovery of θ_{13} in long baseline beams of increasing power. We will start in Section 3 from the CNGS, a 500 kW power beam operated up to 2012 to study $\nu_\mu \rightarrow \nu_\tau$ oscillations. Subleading $\nu_\mu \rightarrow \nu_e$ oscillations are currently investigated by two very high power beams at J-PARC and Fermilab serving T2K (Section 4) and NOνA (Section 5), respectively. These acceleration complexes will be upgraded to the status of "Super-beams" (>1 MW power) to carry on the ambitious physics program of DUNE and Hyper-Kamiokande (Section 6). The ancillary systematic reduction program using current and novel acceleration techniques will be discussed in Section 7.

THREE FAMILY NEUTRINO OSCILLATIONS

The Standard Model (SM) of the electroweak interactions has been developed when no evidence of massive neutrinos was available. In its original formulation, therefore, the SM neutrinos were neutral massless fermions with left-handed (LH) chirality or, equivalently, LH helicity. The antineutrinos were their right-handed (RH) partners. The discovery of neutrino oscillations falsified the SM in its original formulation although the theory can be minimally extended to account for massive neutral leptons. In the minimally extended SM [7], massive neutral leptons are treated as (massive) quarks: their flavor eigenstates are linear combination of mass eigenstates and the linear operator that mixes the flavor and mass eigenfunctions is a 3×3 complex matrix. In the quark sector, this matrix is called the Cabibbo–Kobayashi–Maskawa (CKM) matrix. The corresponding matrix in the lepton sector is the Pontecorvo–Maki–Nakagawa–Sakata (PMNS) matrix $U_{\alpha,i}$. The α index runs over the flavor eigenstates (α=e,μ,τ) and the i index runs over the mass eigenstates i=1,2,3. The flavor fields $\nu_{\alpha}L(x)\equiv\nu\alpha$ are thus:

$$\nu_\alpha = \sum_i U_{\alpha i}\nu_i. \quad (1)$$

Here, we dropped the dependence of the field on space-time and the subscript L. Note that, in the (minimally extended) SM, only fields with LH chirality (νL) appear in the charged currents (CC) that describe the coupling of fermions with the W± bosons. In the original SM, νL describe neutrinos with LH chirality that can be observed as particles with LH helicity. In the minimally extended SM, neutrinos are massive and therefore neutrinos with LH chirality may be observed with both spin antiparallel (LH helicity) and parallel (RH helicity) to the direction of motion. Direct mass measurements and oscillation data, however, suggest mi≪me by at least five order of magnitudes. As a consequence, for any practical purpose, the helicity of a LH chirality field can be considered LH, as well. The neutrinos fields in the SM are Dirac fields, as for the case of quarks. The anti-neutrino fields describe particles that are different from the neutrinos ("anti-neutrinos") and only RH anti-neutrinos couple with the W bosons in CC currents. Unlike quarks, where particles are electrically charged and must be different from antiparticles of opposite electric charge, neutrinos are neutral particles and the anti-particle fields may be the same as for the particles. In this case, the neutrino fields are Majorana fields and not Dirac fields. This subtlety is immaterial for massless neutrinos because the RH neutrino field can be identified with the RH anti-neutrino field without changing the value of any observable. In the extended SM, there are observables that are sensitive to whether the neutrino are standard Dirac particles or "Majorana particles" (i.e., elementary fermions that are described by Majorana fields), although the correspondent measurements are extremely challenging due to the smallness of mi. The minimally extended SM Lagrangian is built applying the quark formalism to neutrinos and, hence, in this model neutrinos are Dirac particles. In this case, as for the CKM, the PMNS matrix is unitary ($U^{\dagger}U=\mathbb{1}$) and can be parameterized [**3**] by three angles and one complex phase. The parameterization that has been adopted for the PMNS put the complex phase in the 1–3 sector, i.e., in the rotation matrix between the first and third mass eigenstate:

$$\begin{aligned} U &= \begin{pmatrix} 1 & 0 & 0 \\ 0 & c_{23} & s_{23} \\ 0 & -s_{23} & c_{23} \end{pmatrix} \begin{pmatrix} c_{13} & 0 & s_{13}e^{-i\delta_{CP}} \\ 0 & 1 & 0 \\ -s_{13}e^{i\delta_{CP}} & 0 & c_{13} \end{pmatrix} \begin{pmatrix} c_{12} & s_{12} & 0 \\ -s_{12} & c_{12} & 0 \\ 0 & 0 & 1 \end{pmatrix} \\ &= \begin{pmatrix} c_{12}c_{13} & s_{12}c_{13} & s_{13}e^{-i\delta_{CP}} \\ -s_{12}c_{23} - c_{12}s_{13}s_{23}e^{i\delta_{CP}} & c_{12}c_{23} - s_{12}s_{13}s_{23}e^{i\delta_{CP}} & c_{13}s_{23} \\ s_{12}s_{23} - c_{12}s_{13}c_{23}e^{i\delta_{CP}} & -c_{12}s_{23} - s_{12}s_{13}c_{23}e^{i\delta_{CP}} & c_{13}c_{23} \end{pmatrix}. \end{aligned} \tag{2}$$

In Equation (**2**), the three rotation angle are labeled $\theta_{13},\theta_{23},\theta_{12}$ and $c_{ij}\equiv\cos\theta_{ij}$, $s_{ij}\equiv\sin\theta_{ij}$. In this parameterization, which is in use since more than 20 years, $\theta_{ij}\in[0,\pi/2]$ and $\delta CP\equiv\delta\in[0,2\pi]$. Clearly, physical observables are parameterization-independent and therefore the range of the angles and the choice of the complex phase in the 1–3 sector is done without loss of generalization. All parameters of the minimally extended SM can be completely determined by neutrino oscillation experiments except for an overall neutrino mass scale (see **Table 1**). The neutrino oscillation probability between two flavor eigenstates $P(\nu_\alpha\rightarrow\nu_\beta)$ is the probability to observe a flavor β in a neutrino detector located at a distance L from the source. The source produces neutrinos with flavor α and energy E. The oscillation probability is given by:

$$\begin{aligned} P(\nu_\alpha \rightarrow \nu_\beta) &= \delta_{\alpha\beta} - 4\sum_{i<j} \mathrm{Re}\left[U_{\alpha i}U^*_{\beta i}U^*_{\alpha j}U_{\beta j}\right] \sin^2\left(\frac{\Delta m^2_{ji}L}{4E}\right) \\ &\quad + 2\sum_{i<j} \mathrm{Im}\left[U_{\alpha i}U^*_{\beta i}U^*_{\alpha j}U_{\beta j}\right] \sin\left(\frac{\Delta m^2_{ji}L}{2E}\right), \end{aligned} \tag{3}$$

where $\Delta m^2_{ji} \equiv m^2_j - m^2_i$ and $L\simeq ct$ is the distance travelled by the neutrino. For antineutrino oscillations, U must be replaced by U^* in Equation (**3**), which corresponds to changing the sign of the third term. As a consequence, oscillation experiments can reconstruct all rotation angles and the CP phase. They also can measure the squared mass differences among eigenstates although the absolute mass scale must be measured in an independent manner. Note in particular that CP violation in the leptonic sector can be established just measuring the difference between $P(\nu_\alpha\rightarrow\nu_\beta)$ and $P(\bar{\nu}_\alpha \rightarrow \bar{\nu}_\beta)$ if $\alpha \neq \beta$ if $\alpha\neq\beta$. Since the leading oscillation terms, i.e., the second term of Equation (**3**), depend on a squared sign, determining the signs of Δm_{ij} is also very challenging. In fact, the sign of Δm^2_{12} has been determined at the beginning of the century by solar neutrino experiments. The oscillation probabilities here are strongly perturbed by matter effects in the sun, which are sensitive to the sign of Δm^2_{12}. As discussed in **Section 5**, matter effects on the earth are employed by long-baseline experiments to determine the sign of Δm^2_{23}3, which is not fully established yet.

Table 1. Current value of the mass and mixing parameters as extracted from neutrino oscillation data under the hypothesis of normal hierarchy (current best fit) [**14**].

Parameter	Best Estimate	Units
θ_{12}	$33.82^{+0.78}_{-0.76}$	degree
θ_{23}	$48.3^{+1.1}_{-1.9}$	degree
θ_{13}	8.61 ± 0.13	degree
δ	222^{+38}_{-28}	degree
Δm^2_{12}	$7.39^{+0.21}_{-0.20} \times 10^{-5}$	eV^2
Δm^2_{23}	$2.52 \pm 0.03 \times 10^{-3}$	eV^2

The choice of Dirac fields for the description of neutrinos in the minimally extended SM is the simplest from the technical point of view, but it is very disputable in the light of the gauge principle [**8**]. The most general Lagrangian compatible with the SU(2)⊗U(1) gauge symmetry and containing all SM particles (including massive neutrinos) allows for "Majorana mass terms" as $-\frac{1}{2} m_i \nu^T_{iL} C \nu_{iL} + \text{h.c.}$, where C is the charge conjugation matrix. It is possible to show that these terms imply neutrinos to be Majorana particles and must be removed by hand from the minimally extended SM based on Dirac fields. A vast class of SM extensions is motivated by the existence of Majorana particles. Several of these models are also able to explain the smallness of neutrino masses through the "see-saw mechanism" [**9,10,11,12,13**]. Neutrino oscillations, however, cannot address these kind of issues. If neutrinos are Majorana particles, the PMNS matrix is only approximately unitary and can be parameterized with three rotation angles and three phases as:

$$U = U_{\text{Dirac}} P_{\text{Majorana}} \quad (4)$$

where:

$$P_{\text{Majorana}} = \begin{pmatrix} e^{i\alpha_1} & 0 & 0 \\ 0 & e^{i\alpha_2} & 0 \\ 0 & 0 & 1 \end{pmatrix} \quad (5)$$

and U_{Dirac} is the PMNS in the minimally extended SM (Equation (**2**)). On the other hand, the oscillation formula of Equation (**3**) does not depend on these additional phases. The oscillation measurements are therefore robust against the Dirac vs. Majorana nature of neutrinos, but, on the other hand, do not provide information to determine such nature.

THE CNGS NEUTRINO BEAM

In 1998, disappearance measurements of atmospheric muon neutrinos—i.e., measurements of $\nu_\mu \rightarrow \nu_\mu$ with neutrinos produced by interactions of cosmic rays with the earth atmosphere—provided evidence for non-zero value of both Δm^2_{23} and θ_{23}. At that time, θ_{13} was not known and only upper limits from the Chooz reactor experiment [**15**] were available (θ_{13}<12°). In those years, there was no evidence of oscillations in appearance mode, i.e., no measurements testifying non-zero values of $P(\nu\alpha \rightarrow \nu_\beta)$ for any $\alpha \neq \beta$. Such evidence was still lacking in 2002, when solar and reactor experiments employing νe and $\bar{\nu}_e$ whose energy was well below the kinematic threshold for the production of muons, demonstrated that $\theta_{12} \simeq 35°$ and Δm^2_{23} $\simeq 7\times10^{-5}$ eV2. In order to observe in a neutrino detector the appearance of new flavors ($\alpha \neq \beta$), the energy of the neutrino must be well above the kinematic threshold for the production of the lepton of flavor β. For solar and reactor neutrinos, this is not possible since the energies of their νe and $\bar{\nu}_e$ are much smaller than m_μ. Atmospheric neutrinos, on the other hand, may be the right natural source since their energy spectrum extends well above the μ and even the τ kinematic threshold. At that time, there was no evidence of $\nu_\mu \rightarrow \nu_e$ oscillations at the atmospheric energy scale and, therefore, the disappearance of ν_μ was ascribed to $\nu_\mu \rightarrow \nu_\tau$ oscillations. The CERN-to-Gran Sasso (CNGS) [**16,17**] neutrino beam was designed to provide evidence of $\nu_\mu \rightarrow \nu_\tau$ oscillations in appearance mode, i.e., observing explicitly the appearance of ν_τ by detecting τ leptons produced by CC events in the detector. The CNGS beam exploited the Gran Sasso laboratories (LNGS), an existing underground laboratory located 730 km from CERN. This constraint fixed the baseline of the CNGS experiments: OPERA and

ICARUS. The CNGS has been the only long-baseline beam whose energy ($E \simeq 17$ GeV) was well above the kinematic threshold for the production of the tau lepton (3.5 GeV for CC scattering in nuclei). The energy was tuned maximizing the number of observable ν_τ CC events:

$$\int dE\Phi(E)\epsilon(E)\sigma(E)P(\nu_\mu \to \nu_\tau) \simeq \sin^2 2\theta_{23} \int dE\Phi(E)\epsilon(E)\sigma(E) \sin^2\left(1.27 \frac{\Delta m^2_{23} \cdot 730 \text{ km}}{E(\text{GeV})}\right) \quad (6)$$

where $\Phi(E)$ is the CNGS ν_μ flux, $\epsilon(E)$ is the detector efficiency, and $\sigma(E)$ is the ν_τ CC cross-section. In this formula, we expressed the oscillation phase $\Delta m^2_{23}L/4E$ in practical units: km for L and GeV for E. In particular, a high energy beam is beneficial to produce tau leptons with a large cross-section ($\sigma(E) \sim E$ for E well above 3.5 GeV) and boosted in the laboratory frame. Such boost is needed to observe the decay in flight of the taus in the OPERA nuclear emulsion sheets (see below) and increases substantially $\epsilon(E)$. On the other hand, an energy much larger than $1.27 \cdot \Delta m^2_{23} 730\text{km}$ reduces the oscillation probability as $\sim E-2$ due to the fixed baseline of LNGS. The actual CNGS beam energy has thus been optimized accounting for these requirements and 17 GeV is the value that maximizes Equation (**6**). As a matter of fact, the CNGS observes oscillations off the peak of the oscillation maximum since:

$$\sin^2\left(1.27 \frac{\Delta m^2_{23} \cdot 730 \text{ km}}{17 \text{ GeV}}\right) \simeq 0.016. \quad (7)$$

In the framework of the minimally extended SM, the CNGS thus provides a direct test of the oscillation phenomenon in appearance mode ($\nu_\mu \rightarrow \nu_\tau$) and a test of the oscillation pattern in a parameter phase far from the oscillation maximum.

The CNGS neutrino beam (see **Figure 1**) was produced by 400 GeV/c protons extracted from the SPS accelerator at CERN and transported along a 840 m long beam-line to a carbon target producing kaons and pions [**18**]. The only potential source of $\nu\tau$ was the decay of Ds particles in the proximity of the target, which contributed in a negligible manner ($<10^{-4}$) to the neutrino flux. The positively charged π/K was energy-selected and guided with two focusing lenses ("horn" and "reflector"), in the direction of the Gran Sasso Laboratories in Italy. These particles decay into μ and ν_μ in a 1 km long evacuated decay pipe. All the hadrons, i.e., protons that have not interacted in the target, pions, and kaons that have not decayed in flight, are absorbed in a hadron stopper. Only neutrinos and muons cross the 18 m long block of graphite and iron. The muons, which are ultimately absorbed downstream in around 500 m of rock, were monitored by muon detector stations.

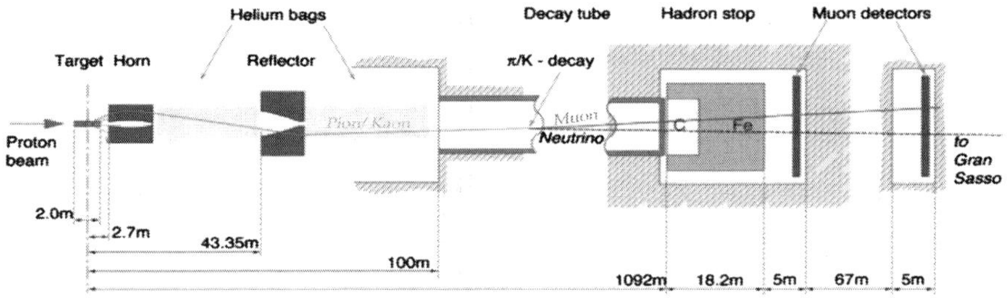

Figure 1. Layout of the CNGS neutrino beam.

Unlike all other long baseline facilities, the CNGS was not equipped with a near detector due to the cost and complexity of building a large underground experimental area located after the CNGS decay pipe. In addition, since the CNGS was designed to study the appearance of tau neutrinos in a beam with a negligible contamination of ν_τ, the need for a near detector was less compelling than other long-baseline beams. The CNGS beam ran from 2008 to 2012, providing to OPERA an integrated exposure of 17.97×10^{19} protons on target (pot). OPERA exploited this large flux using real-time ("electronic") detector interleaved with emulsion cloud chambers (ECC).

The OPERA [**18**] target had a total mass of about 1.25 kt composed of two identical sections. Each section consisted of 31 walls of Emulsion Cloud Chamber (ECC) bricks, interleaved by planes of horizontal and vertical scintillator bars (Target Tracker) used to select ECC bricks in which a neutrino interaction had occurred. Each ECC brick consisted of 57 emulsion films interleaved with lead plates (1 mm thickness), with a (12.7×10.2) cm^2 cross-section and a total thickness corresponding to about 10 radiation lengths. Magnetic muon spectrometers were located downstream of each target section, and were instrumented by resistive plate chambers and drift tubes.

Over the years, a total of 19,505 neutrino interaction events in the target were recorded by the electronic detectors, of which 5603 were fully reconstructed in the OPERA emulsion films [**19**].

The construction and data analysis of OPERA was a tremendous challenge due to the need of identifying tau leptons on an-event-by-event basis. Even with current technologies, the only mean to achieve this goal by detecting the decay kink with very high purity is by using nuclear emulsions. This technique has been exploited in the 1990s by the CHORUS experiment to search for short baseline $\nu\mu \rightarrow \nu_\tau$ oscillations [**20**] and extended to a much larger scale by OPERA. Nuclear emulsions consist of AgBr crystals scattered in a gelatine binder. After the passage of a charged particle, the crystal produces electron–hole pairs proportional to the deposited energy. During a chemical-physical process known as development, the reducer in the developer gives electrons to the crystal through the latent image center and creates silver metal filaments scavenging silver atoms from the crystal. This process increases the number of metal silver atoms by many orders of magnitude (10^8–10^{10}). The grains of silver atoms reach a diameter about 0.6 mm and, therefore, become visible with an optical microscope.

The development procedure is irreversible, but it must be performed in a finite time scale (a few years) to prevent aging of the emulsions. As a consequence, the electronic detectors must locate the ECC brick in real time and a semi-automatic system must remove the brick while OPERA is still taking data and bring this component to the development facility. The scanning of the emulsions can be done at any time since the particle tracks are permanently fixed on the bulk of the emulsions, but a fast scanning time is a major asset to provide feedback on the detector performance and optimize the analysis chain.

From the analysis point of view, OPERA is a rare event search experiment since just a handful of fully reconstructed tau events are expected from Equation (**6**). The tau events in OPERA are reconstructed as events with a primary track that shows a kink with respect to the initial direction. The kink testifies the decay in flight of the tau lepton in the lead of the ECC. In particular, OPERA has been seeking for tau leptons in most of its high branching ratio decay modes: $\tau \rightarrow \mu$ (i.e., $\tau \rightarrow \mu\nu\mu\nu\tau$ with a visible muon from the primary vertex), $\tau \rightarrow e$, $\tau \rightarrow h$ ("one prong", i.e., hadronic decay with only one visible charged hadron) and "three prongs" (i.e., hadronic decay with three visible charged hadrons). Rare event searches are plagued by potential biases due to the tuning of the selection procedure in the course of data taking. In order to minimize bias, the OPERA Collaboration followed two approaches. The most conservative one was implemented during data taking and the initial stage of data analysis, bringing to a 5σ discovery of tau appearance in 2015. Here, tau leptons were selected using topological and kinematic cuts defined a priori in the design phase of the experiment. The efficiency of these cuts were evaluated by a full simulation of OPERA performed with GENIE [**21**] (neutrino interactions), NEGN [**22**] (kinematic properties of final states) and GEANT4 [**23**] (detector response). The simulation was validated with a control sample based on ν_μ CC events producing charmed particles in the final state. The 2015 analysis [**24**] is based on 5 ν_τ CC candidates whose background is summarized in **Table 2**.

Table 2. Summary of expected events and background contributions in the 2015 OPERA analysis.

Decay Mode	Exp. Bkg	Exp. Signal	Observed
$\tau \to h$	0.04 ± 0.01	0.52 ± 0.10	3
$\tau \to 3h$	0.17 ± 0.03	0.75 ± 0.14	1
$\tau \to \mu$	0.004 ± 0.001	0.61 ± 0.12	1
$\tau \to e$	0.03 ± 0.01	0.78 ± 0.16	0
total	0.25 ± 0.05	2.64 ± 0.53	5

The events observed in OPERA for the standard 2015 analysis are consistent with the "three flavor neutrino oscillation paradigm", i.e., the minimally extended SM accounting for neutrino oscillations. In particular, given the superior background rejection capability of the ECC based analysis, ν_τ appearance was established with a statistical significance of 5.1 σ with just five candidates (see **Figure 2**). For that particular analysis, the probability of detecting five events or more is 6.4%.

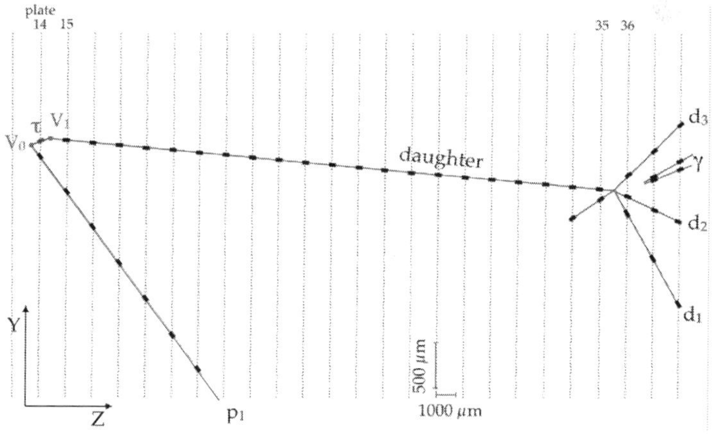

Figure 2. The 5th τ candidate (τ→h) observed in one of the OPERA ECC in 2015.

In 2018, OPERA developed a more sophisticated analysis reaping on the improvement in the knowledge of the detector response and background subtraction techniques accumulated during data taking. The final OPERA analysis is based on looser kinematic cuts applied to four variables: the distance between the decay vertex and the downstream face of the lead plate containing the primary vertex (z_{dec}), the momentum (p) and transverse momentum (p_T) of the secondary particle and the 3D angle θk_{ink} between the parent particle and its daughters. The cuts are independent of the tau decay mode (z_{dec}<2.6 mm, θk_{ink}>0.02 rad, p>1 GeV) except for the transverse momentum (pT>0.15 GeV for $\tau \rightarrow h$ and p_T>0.1 GeV for $\tau \rightarrow e$ or μ). In addition, p<15 GeV is requested for the tau decay into muon to suppress the high energy tail of ν_μ CC events of the CNGS. After these cuts, a multivariate statistical approach is employed to separate signal from background and to produce a single-variable discriminant for the oscillation hypothesis test and parameters estimation. In this analysis, 10 observed events were selected. The expectation in the minimally extended SM was 6.8±0.75 events plus a background of 2.0±0.4 events. This analysis, which represents the final result of OPERA on search for $\nu_\mu \rightarrow \nu_\tau$ appearance, establishes oscillations at 6.1 σ confidence level [**25**].

Beyond tau appearance, OPERA performed analyses in ν_μ disappearance mode and in $\nu_\mu \rightarrow \nu_e$ appearance mode. Even if these searches are not competitive with dedicated experiments equipped with a near detector and electron-sensitive far detectors with much larger mass than OPERA (see **Section 4** and **Section 5**), the three neutrino analysis of OPERA [**19**] provides a consistency test of the minimally extended SM in an energy range between the atmospheric oscillation scale (few GeV neutrinos) and the energy scale of neutrino telescopes like IceCube/DeepCore. The OPERA and ICARUS (see below) experiments, in addition, provided significant constraints to the existence of light sterile neutrinos and, in general, to extension of the three family oscillation mechanism.

The CNGS served for several years also the ICARUS experiment [**26**] that took data at LNGS from 2010 to 2012. ICARUS was a 600 ton liquid Argon TPC (see **Figure 3**) composed of a large cryostat divided into two identical, adjacent half-modules, each one with internal dimensions of 3.6 m (width) × 3.9 m (height) × 19.6 m (length). The inner detector structure of each submodule (300 ton) consisted of two TPCs separated by a common cathode. Each TPC is made of three parallel wire planes, 3 mm apart and the relative angles between the two wire planes is 60°. The distance between the wires is 3 mm (wire pitch). These wire planes constitute the anode of the TPC. The TPC volume is filled with high purity liquid Argon and the electrons produced by the charged particles drift toward the anode wires.

A uniform electric field perpendicular to the wires drives the electrons in the LAr volume of each half-module by means of a HV system. The cathode plane, parallel to the wire planes, is placed in the center of the LAr volume of each half-module at a distance of about 1.5 m from the wires of each side. This distance defines the maximum drift path. Between the cathode and the anode, field-shaping electrodes are installed to guarantee the uniformity of the field along the drift direction. At the nominal voltage of 75 kV, corresponding to an electric field of 500 V/cm, the maximum drift time in LAr is about 1 ms. As a consequence, the purity of the liquid Argon must be such that the electron lifetime is comparable with the drift time, i.e., that electrons can drift in the active volume before recombining with electronegative impurities. Such a long drift length requires Argon purity <0.01 part-per-billion (ppb). The space resolution that can be achieved in ICARUS is much worse than the ECC (1 mm versus a few microns), but this detector technology can be readout as a standard TPC and the observation and selection of CC events can be performed in real time. In addition, this technology is scalable to masses that are unpractical for the ECC ($\gg$1 kton).

Figure 3. The ICARUS detector, operated underground at LNGS during the run of CNGS, was moved to Fermilab in 2018 and installed at a shallow depth to search for sterile neutrinos in the framework of the Fermilab Short Baseline Programme (SBN) [**27**].

In fact, the mass of the TPC operated at LNGS—the largest LArTPC ever built—was still smaller than what was needed for the observation of ν_τ CC events on a kinematical basis. Still, ICARUS achieved important results in the study of $\nu_\mu \rightarrow \nu_e$ oscillations beyond the minimally extended

SM (limits to sterile neutrino oscillations) [**28**] and in the clarification of the neutrino velocity anomaly reported in 2011 by OPERA [**29**]. Even more, the successful operation of ICARUS in a deep underground laboratory at the 600 t scale, opened up the possibility of building multi-kton detectors based on the LArTPC technology, paving the way to the design and construction of ProtoDUNE-SP and DUNE (see **Section 6**). At the end of the ICARUS data taking at LNGS, the Argon recirculation and purification systems reached a free electron lifetime exceeding 15 ms, corresponding to about 20 parts per trillion of O2-equivalent contamination [**30**].

T2K AND ITS UPGRADES

T2K

The T2K experiment is the successor of the former K2K experiment, [**31**], the first long-baseline neutrino experiment that confirmed the Super-Kamiokande discovery of neutrino oscillations [**2**] by detecting a reduction of the νμ flux together with a distortion of the energy spectrum in an accelerator neutrino beam generated at KEK [**32**].

With respect to K2K, T2K had a number of substantial improvements, the most important being (a) a new neutrino beam created by the Japan Proton Accelerator Research Complex (J-PARC), designed to be about 50 times more intense than the K2K neutrino beam; (b) a beam line configuration where the far detector (again SK) was at an off-axis angle of 2.5° and a baseline of 295 km; (c) a new close detector system, allowing a much more effective measurement of neutrino beam components, backgrounds, and neutrino cross-sections.

The T2K neutrino beam is generated at J-PARC where a 30 GeV proton beam impinges onto a graphite target [**33**]. The primary proton beam recently achieved an average power of 492 kW. The secondary hadrons, mostly pions, are charge-selected and focused by a system of three magnetic horns [**34**] and decay in a 96 m long decay volume. The axis of the beam is displaced by 2.5o from the SK detector.

Thanks to the kinematics of the pion two-body decay, the resulting neutrino beam is narrower than an on-axis beam [**35**], better optimized to the expected neutrino oscillations (as already measured by SK and K2K), and with a smaller contamination of intrinsic νe (mostly generated by kaon decays, whose kinematics is different from pion decays, producing less neutrinos at the off-axis angle of 2.5o).

Two detectors are located 280 m downstream of the beam target, to measure neutrino interactions before they start oscillating. The INGRID detector [**36**] is located on-axis, and its main purpose is to monitor the direction and the flux stability of the neutrino beam.

The ND280 detector system is located as the same off-axis angle of SK and its main purpose is to measure all the neutrino beam components. It consists of three time projection chambers (TPCs) [**37**] interleaved with two fine-grained tracking detectors [**38**], a π^{0}-optimized detector (P0D) [**39**], and an electromagnetic calorimeter [**40**]. ND280 is mounted inside the magnet of the former UA1 and Nomad experiments, donated by CERN to the T2K Collaboration; it allows for independently measuring both neutrino and antineutrino interaction rates. A side muon range detector instruments the magnet yoke [**41**].

An important set of external data for T2K comes from the NA61 experiment that measured hadroproduction by 30 GeV protons on a graphite target with a thickness of 4% of a nuclear interaction length [**42**]. These measurements drastically reduce systematic errors on the evaluation of the neutrino beam flux, resulting in a better determination of neutrino interaction rates by ND280 [**43**].

The neutrino beam is peaked around 0.6 GeV, at the energy where the oscillation probability is maximal. At those energies, the dominant process are Charged Current Quasi-Elastic scatterings (CCQE), where a charged lepton of the same flavor of the incoming neutrino is produced. In this class of interactions, the measurement of the momentum and direction of the outgoing lepton is sufficient to reconstruct the energy of the incoming neutrino. It has to be noted that the 295 km baseline is too short to allow good sensitivity to the measurement of neutrino mass ordering (see **Section 5**) and that the $\nu\mu$ neutrino energy is well below the τ production threshold.

The far detector is the Super-Kamiokande (SK) detector [**44**], a deep underground 50 kton water Cherenkov detector, 39 m in diameter and 42 m tall, equipped with 11,129 inward facing 20-inch photomultiplier tubes (PMTs) and 1885 outward-facing 8-inch PMTs mainly used as a veto. The Cherenkov light emitted in water by charged particles above a momentum threshold allows for measure with great precision the momentum, the direction and the vertex of the leptons produced by neutrino interactions. The different topologies of the muon and electron tracks produce different patterns of hit phototubes; in this way, electron and muons are identified with great precision by SK. The quasi-elastic topology, where a single lepton is produced, is particularly favorable for the characteristics of the SK detector.

The first important result by T2K was published in 2011 [**45**] by observing an indication of $\nu_\mu \rightarrow \nu_e$ appearance in data accumulated with 1.43×10^{20} proton on target collected from January 2010 to March 2011. Six events passed all the selection criteria in SK, with an expected background of 1.5±0.3 events, corresponding to a significance of about 2.5σ. It represented the first indication of a non-zero value of θ13:0.03<sin22θ13<0.28 at 90% CL for δCP = 0 and normal neutrino mass hierarchy, a fundamental milestone for any subsequent development in neutrino oscillation physics.

T2K had a long stop of more than one year soon after this result due to the catastrophic earthquake–tsunami that occurred in Japan in March 2011.

In this period, three reactor experiments published evidence of non-zero values of θ13 by measuring the flux of reactor antineutrinos with a clear indication of $\overline{\nu}_e$ disappearance. Eventually, the θ_{13} determination by reactor experiments achieved a much better precision than T2K and today the global best-fit to θ_{13} is [**3**] $\sin^2\theta_{13}$=0.0215 with a 3σ allowed range of 0.0190–0.0240. The precise, unambiguous determination of θ13 by reactors greatly improved the sensitivity of long baseline experiments to δCP and neutrino mass ordering. This is nicely illustrated in **Figure 4**, where the confidence regions measured by T2K without any input from reactor experiments are compared with the confidence region from reactors [**46**], indicating how better δCP is constrained after the inclusion of reactor data.

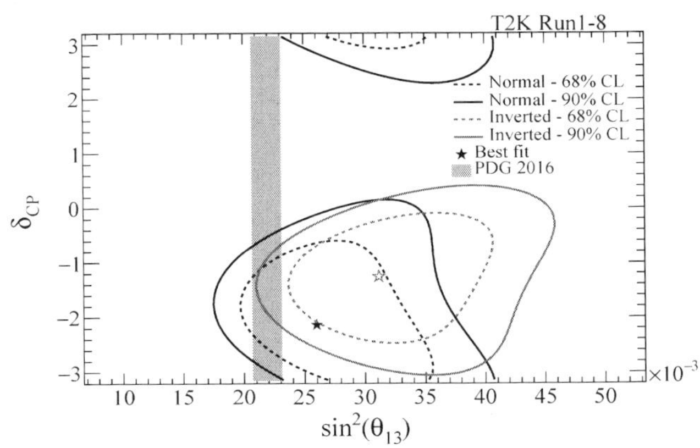

Figure 4. Confidence regions (68% and 90%) in the sin2θ13−δCP plane as measured by T2K without any prior information on θ13 compared with the 68% confidence regions from reactor experiments, shown by a yellow vertical band.

The most recent T2K published results are based on SK data collected between 2009 and 2017 in (anti)neutrino mode with a neutrino beam exposure of 14.7×1020 (7.6×1020) protons on target [**46**].

Neutrino fluxes, interaction cross sections and detector responses are simulated, and a fit to the near detector data constrains them. The results of this fit are propagated to the far detector as a Gaussian covariance on the systematic parameters. The fit of the far detector data then constrains the oscillation parameters.

Primary proton interactions on the T2K target system are simulated with FLUKA2011 [**47**] and GEANT3 [**48**] together with the decay of hadrons and muons that produce neutrinos. The modeling is constrained by thin target hadron production data, including recent charged pion and kaon measurements from the NA61/SHINE experiment at CERN [**42**,**49**,**50**]. The systematic errors of beam prediction at this stage are of 8% and 7.3% for the neutrino and antineutrino beams, respectively. A significant reduction of these errors is expected with the inclusion of recent hadroproduction measurement of the T2K replica target by NA61/SHINE [**51**]. It is extremely important to have flux systematics as low as possible at this stage for any physics analysis at the close detector and to attenuate the strong correlation between neutrino fluxes and neutrino cross-sections at the close detector.

The neutrino interaction model is based on NEUT [**52**]. Simulated data are also generated with alternative models and fitted with NEUT in order to evaluate the systematic effects of wrong modelling to the oscillation results.

Events selected as charged current (CC) interactions in ND280 are used to constrain the flux and neutrino interaction uncertainties.

A total of 14 event samples are defined: six neutrino samples depending of the number of hadrons (0, 1, or more) and whether the interaction happens in FGD1 or FGD2, this latter being 40% enriched in water; eight samples are defined in the antineutrino mode, depending on the number of tracks (0 or 1), the charge of the outgoing muon and again on the position of the interaction vertex.

It is very important to have the possibility of defining such a large number of event samples; they are sensitive to different combinations of beam fluxes and neutrino interactions, helping to cover all the analysis parameters and to break parameter degeneracies and ambiguities.

A global likelihood is computed for the fit of all the event samples. The global fits greatly reduce systematic errors and increases by 10%–15% the interaction rate predictions at the far detector.

SK data are subdivided in five independent samples: events with a single muon-like or a single electron-like rings in neutrino and antineutrino beam modes plus a sample with an electron-like ring together with a pion track candidate in neutrino mode. Events are reconstructed with a new reconstruction algorithm, firstly introduced in [**53**] to suppress neutral current πo backgrounds, where a maximum likelihood algorithm exploiting the timing and charge information of all the photosensors simultaneously. This new alghorithm improves efficiency and purity of the selected samples and allows a fiducial volume expansion of 25% (14%) for ν_e (ν_μ) events.

The observed number of CCQE events with an electron candidate at SK are 74 in the neutrino run and seven in the anti-neutrino run, to be compared with the expectations for δCP=0(−π/2), normal hierarchy, of 61.4 (73.5) and 9.0 (7.9) respectively. The of the two samples are 8.8% and 7.1%, respectively. The selected ν_μ candidates in the neutrino systematic errors and antineutrino modes are 240 and 68, respectively, with 5.1% and 4.3% systematic errors. Finally, 15 events are selected in the single-electron single-pion sample, with a prediction of 6.9 events and a systematic error of 18.4%. It should be noted that systematic errors already at this stage (29% of the design statistics) are sizable and are approaching the values of statistical errors.

Oscillation fits to the five samples of data show a preference for the normal mass ordering with a posterior probability of 87%. The oscillation angle θ_{23} is measured as $\sin^2(\theta_{23}) = 0.526^{+0.032}_{-0.036}$ for normal ordering, central value for inverted ordering is $\sin^2(\theta_{23})$=0.530. Assuming the normal (inverted) mass ordering the best fit to the atmospheric mass splitting is $\Delta m^2_{32}=(2.463\pm0.07)\times10^{-3}$ ($\Delta m^2_{13}=(2.432\pm0.07)\times10^{-3}$) eV²/c4. For δCP, the best-fit value assuming the normal (inverted) mass ordering is −1.87(−1.43), the 2σ confidence regions are (−2.99,−0.59) for normal ordering and (−1.81,−1.01) for inverted ordering. Both intervals rule out the CP conserving points, δCP=0 and δCP=π.

A recent preprint of the T2K collaboration, which includes 2018 data and improves by a factor 2.2 the pot collected in antineutrino mode, reports for the first time a closed 99.73% (3σ) interval on the CP-violating phase δCP [**54**].

Other notable results of the T2K Collaboration are several measurements of neutrino cross-sections [**55**], searches for light sterile neutrinos [**56**], and searches for heavy neutrinos [**57**].

T2K II

The exciting results achieved by the T2K experiment so far convinced the collaboration to propose an upgrade of the experimental setup, T2K II [**58**], in order to reach a 3σ sensitivity for a significant fraction of the possible δCP values by the year 2026, when the next generation Hyper-Kamiokande experiment is expected to start.

The proposed upgrades, all re-usable by the Hyper-Kamiokande experiment, will be

- Upgrade the J-PARC main ring to the power of 1.3 MW. This would allow for collecting 20×10^{21} pot by 2026, three times more the original beam request of 7.7×10^{21} pot and more than six times the statistics collected so far in neutrino+antineutrino modes.
- Push the magnetic horns system to its ultimate performances, bringing its operation current from 250 to 320 kA. This upgrade would provide a 10% increment of neutrino fluxes reducing the wrong-sign contamination by 5%–10%
- Upgrade the ND280 close detector as proposed in [**59**]. A new tracker, consisting of a 2-ton horizontal plastic scintillator target sandwiched between two new horizontal TPCs, will substitute the present PØD. The plastic scintillator, called Super-FGD [**60**], consists of a matrix of 1×1×1 cm3 cubes made of extruded plastic scintillator, where each cube is crossed by three wavelength shifting fibers along the three directions, allowing a 3D reconstruction of the events.

The TPCs will be similar in design to the existing ones, with two major improvements: the Micromegas detector will be constructed with the "resistive bulk" technique, thereby allowing a lower density of readout pads and eliminating the discharges (sparks). The field cage will be realized with a layer of solid insulator laminated on a composite material. This will minimize the dead space and maximize the tracking volume.

This tracker would be surrounded by Time-of-Flight counters (made of plastic scintillator) to measure the direction of the tracks and reject out of fiducial volume events. The total mass of active target for neutrino interactions increases from 2.2 to 4.3 tons, allowing the doubling of the expected statistics for a given exposure. Furthermore, the new configuration will allow a better coverage of the neutrino quasi-elastic interactions, improving the acceptance for high angle tracks.

NOνA IN THE NuMI BEAM

The Neutrinos at the Main Injector (NuMI) neutrino beam [**61**] is currently the most powerful neutrino beam in the world and is running with a peak hourly-averaged power of 742 kW. It has been the workhorse of accelerator neutrino physics in the US since 2005 and served several experiments both at Fermilab (MINERνA, ArgoNeuT) and at far locations (MINOS, MINOS+ in the Soudan mine and NOνA on surface at Ash River, MN). In this review, we will mostly focus on its performance and results as neutrino beam for NOνA, which commenced data taking on February 2014 [**62**].

The NUMI beam is produced by the 120 GeV protons extracted from the Fermilab Main Injector. Like the CNGS, it thus leverages the acceleration chain of proton colliders steering the protons toward a solid state target.

The protons from the Main Injector hit the graphite target and the produced hadrons are focused by two magnetic horns and then enter a 675 m long decay volume. The maximum proton energy that can be achieved at the Main Injector also offers a greater flexibility in the choice of the mean neutrino energy. This feature was successfully exploited during the MINOS data taking [**63**,**64**], when the uncertainty on Δm223 was still quite large and the location of the first oscillation maximum was unclear. After 2014, however, NUMI was operated mostly in "medium energy" mode, producing neutrinos with a mean energy of about 2 GeV. This operation mode maximizes the number of oscillated neutrinos in NOνA, which is located off-axis (14 mrad) 810 km far from Fermilab.

Compared with the NUMI run at the time of MINOS, the current facility exploits some modification of the acceleration complex that originates from the shutdown of Tevatron. The Fermilab Recycler is now used extensively for neutrino physics and stores the Booster cycles during acceleration in the Main Injector. In this way, the two most time-consuming operations of NUMI (the transfer from the Booster to the next accelerator, and the acceleration in the Main Injector) can be carried out in parallel rather than in series. As a consequence, the overall cycle time is reduced from 2 to 1.3 s. From February 2014 to June 2019, NOνA accumulated 8.85×1020 pot in neutrino mode and 12.3×1020 pot in anti-neutrino mode, reversing the currents in the focusing horns. The purity of the νμ (ν¯μ) beam is 96% (83%), respectively.

The NOvA experiment measures the energy spectra of neutrino interactions in two detectors. The Near Detector is placed in the proximity of NUMI at a distance of 1 km from the production target and is located

100 m underground. It is a 290 t liquid scintillator detector that measures 3.8 m × 3.8 m × 12.8 m and is followed by a muon range stack. The 14 kton far detector (see **Figure 5**) is the largest neutrino detector ever operated on surface and measures 15 m × 15 m × 60 m. Both detectors use liquid scintillator contained in PVC cells that are 6.6 cm × 3.9 cm (0.15 radiation lengths per cell) in cross-section and span the height and width of the detectors in planes of alternating vertical and horizontal orientation. The light is read out by wavelength-shifting fibers connected to APDs. Operation at surface, with practically no overburden (three meters of water equivalent, 130 kHz of cosmic-ray activity) is possible thanks to the short duration of the proton burst that allows for exploiting the time correlation between the events observed in the detector and the occurrence of the proton extraction to the target in NUMI. The analysis [**65**] is built upon the set of all APD signals above threshold around the 10 μs beam spill. The cosmic ray background is directly measured by a dedicated dataset recorded in the 420 μs surrounding the beam spill. This control sample is employed for calibration, study of stability response, and for the training of the event reconstruction algorithms.

Figure 5. The NOνA detector at Ash River.

NOνA was designed to search for non-null values of θ13 and measure the leading oscillation parameters at the atmospheric scale $(\Delta m^2_{23}, \theta_{23})$. Even if the discovery of θ13 has been achieved by Daya Bay, Double-Chooz, RENO and T2K well before the start of the NOVA data taking, the physics reach of this experiment has been significantly boosted by the large value of this angle. NOVA is the only long baseline experiment currently exploiting matter effects to determine the neutrino mass pattern. As mentioned in **Section 2**, the presence of a dense medium can perturb the oscillation probability of neutrinos because matter is rich in electrons and deprived of muons and tau leptons. As a consequence, the evolution of the flavor wavefunction is perturbed by the fact that ν_μ and ν_τ can interact only by neutral currents (NC) with the electrons of matter while both NC and CC elastic scattering can occur for νe. Elastic weak scattering can occur also in the nuclei, but, in this case, the νe have the same interaction probabilities as for ν_μ or ν_τ. The matter perturbations thus are only due to the density of the electrons in the medium and are parameterized by an effective potential $V = \sqrt{2}G_F n_e(x)$. Here, GF is the Fermi constant and ne(x) is the electron matter density at the point x. Unlike propagation in the sun, where ne(x) is strongly varying and can cause highly nonlinear phenomena (MSW resonant flavor conversions [**66**]), the electron density in the earth crust and mantle is nearly constant. The oscillation formula corrected by matter effects in conditions interesting for long baseline experiment can be written as [**67,68**]:

$$\begin{aligned} P(\nu_\mu \to \nu_e) \simeq\ & \sin^2\theta_{23}\frac{\sin^2 2\theta_{13}}{(A-1)^2}\sin^2[(A-1)\Delta_{31}] + \alpha^2\cos^2\theta_{23}\frac{\sin^2 2\theta_{12}}{A^2}\sin^2(A\Delta_{31}) \\ & + \alpha\frac{\cos\theta_{13}\sin 2\theta_{12}\sin 2\theta_{13}\sin 2\theta_{23}\cos\delta_{CP}}{A(1-A)}\cos\Delta_{31}\sin(A\Delta_{31})\sin[(1-A)\Delta_{31}] \\ & - \alpha\frac{\cos\theta_{13}\sin 2\theta_{12}\sin 2\theta_{13}\sin 2\theta_{23}\sin\delta_{CP}}{A(1-A)}\sin\Delta_{31}\sin(A\Delta_{31})\sin[(1-A)\Delta_{31}], \end{aligned} \quad (8)$$

where $\alpha \equiv \Delta m^2_{21}/\Delta m^2_{31}$, $\Delta_{31} \equiv \Delta m^2_{31}L/4E$ and the electron density is embedded in $A \equiv 2VE/\Delta m^2_{31} = 2\sqrt{2}G_F n_e E/\Delta m^2_{31}$. The probability for $\bar{\nu}_\mu \to \bar{\nu}_e$ oscillations is the same as for neutrinos except for a change of the sign of the CP-violating phase δ_{CP} and of the matter parameter A. Equation (**8**) is an approximate formula that results from expanding to second order in the small quantities α and $\sin\theta_{13}$ under the hypothesis of a constant electron density ne. Exact formulas can be obtained integrating the propagation in steps of x and are commonly embedded in software as GLoBES [**69**]. Still, Equation (**8**) is an excellent approximation for all current long-baseline experiments. The change of sign of A is hence a fake CP violation effect due to the presence of electrons in matter and should be decoupled by genuine CP effects. On the other hand, fake CP effects depend on the sign of Δm^2_{23} and, therefore, provide an elegant tool to determine whether $m_1<m_2<m_3$ ("normal ordering", NO) or $m_3<m_1<m_3$ ("inverse ordering", IO). All other ordering options are already excluded by solar neutrino data that imply $\Delta m^2_{12} > 0$. Distinguishing Δm223<0 (IO) from Δm223>0 (NO) is extremely interesting not only for building theory models of neutrino masses and mixing. IO would imply that the eigenstate that is maximally mixed with νe is indeed the largest one. IO thus eases remarkably the measurement of the absolute neutrino masses and of neutrinoless double beta decay, which are entirely based on νe. NOVA is the only running experiment sensitive to matter effects because A=0.046 in T2K and A=0.12 in NOνA. Similarly, the CNGS is not sensitive to matter effect because it is off the peak of the oscillation maximum (Δ31≪1) [70] and hence:

$$\frac{\sin^2(1-A)\Delta_{31}}{(A-1)^2} \simeq \Delta_{31}$$

In the design phase of NOνA, it was not clear whether it would had been possible (even with a combination with T2K) to disentangle mass ordering effects from genuine CP effects because accidental cancellations [71,72,73] arise for some values of δ and $\mathrm{sign}(\Delta m^2_{23})$ ("$\delta - sign(\Delta m^2_{23})$ ambiguity"). Matter effects enhance the electron neutrino appearance probability in the case of normal mass hierarchy (NO) and suppress it for inverted mass hierarchy (IO). CP-conserving oscillations occur if δ=0 or π, while the rate of νe CC is enhanced around δ=3π/2, and suppressed around δ=π/2. At NOνA, the impact of these factors on the νe appearance probability are of similar magnitudes, which can lead to degeneracies between them, especially in neutrino runs alone. For antineutrinos, the mass hierarchy and CP phase have the opposite effect on the oscillation probability. Increasing values of $\sin^2\theta_{23}$ increase the appearance probabilities for νe and $\bar{\nu}_e$ alike. Still, the current best fit of δ are in the most favorable region of the parameter space and, if the central values are confirmed, NOνA will be able to gain a strong evidence of NO in the next few years combining neutrino and antineutrino runs [74].

NOνA is a detector with a very low density so that electromagnetic showers develop for several PVC cells easing the identification of νe CC events. In addition, NOνA is equipped with a near detector, which opens up the possibility of measuring neutrinos at the far detector also in disappearance mode ($\nu_\mu \rightarrow \nu_\mu$). The energy of the beam is well below the kinematic threshold for τ production and therefore the contamination from ντ CC events is negligible.

As for T2K, the simulation of NOνA employs a detailed description of the beamline based on GEANT4. Again, the computed neutrino flux is corrected according to constraints on the hadron spectrum from thin-target hadroproduction data using the PPFX tools developed for the NuMI beam by the MINERνA collaboration [75]. Neutrino interactions in the detector are simulated with GENIE [21] while the detector response is simulated with GEANT4. Unlike T2K, NOνA employs in a systematic manner analyses techniques inherited from the field of computer vision. The detector hits are formed into clusters by grouping hits in time and space to isolate individual interactions. A Convolutional Visual Network classifier employs the hits from these clusters, without any further reconstruction, as input and applies a series of trained operations to extract the classifying features from the image [**76**]. The classifier feeds an artificial neutral network, i.e., an input layer of perceptrons linked to additional hidden layers and an output node. The network is then trained to recognize different neutrino topologies: ν_e CC, ν_μ CC, and NC. The ν_μ CC and ν_e CC samples correspond to events whose score exceed a given threshold. Muon identification in ν_μ CC events is performed employing the dE/dX and multiple scattering of the track, the length of the track and the fraction of plane where the track is mip-like [**65**]. The NOνA analysis selects ν_μ CC events with an efficiency of 32.2%. In terms of purity, the final selected sample is 92.7% ν_μ CC. The energy of the muon is estimated from the range and the hadronic energy from the number of hits. The energy resolution for the whole sample of ν_μ CC is 9.1%. The quality of the simulation and event selection procedures are validated using the ν_μ CC reconstructed spectrum at the near detector, which is not distorted by oscillations. Possible discrepancies (in the 4%–13% range for ν_μ CC) are corrected to reach complete agreement with the near detector (ND) data. The Δm223 and sin22θ23 parameters are extracted from the spectrum of the ν_μ CC events at the far detector compared with the corresponding spectrum at the near detector. The selection of the νe CC sample is particularly difficult due to the presence of electromagnetic components in NC events (e.g., π0 production) or short muon in ν_μ CC events. The near detector ν_e CC selected sample consists indeed of 42% ν_e CC, 30% NC and 28% ν_μ CC. The selection efficiency—i.e., the probability to select a true ν_e CC event—amounts to 67.4%. The estimator of the energy of the electron candidate has a resolution of 11%. Since the NC and cosmogenic background is mostly in the low energy range of the electron spectrum and the ν_e CC from the νe contamination of the beam dominates at large energies, the analysis is performed only for events whose energy is between 1 and 4 GeV. Again, the quality of the simulation is validated by ND data. In particular, νμ selected candidates are used to cross-check the ν_e selection efficiency and correct the expected νe signal at the far detector (FD). ν_e at ND are artificially created in the simulation replacing the muon with an electron of the same energy and direction. The ν_e CC selection algorithm is applied to this sample and the results match expectations within 2%. The intrinsic ν_e component of the beam (0.7% of all neutrinos) represents a nearly irreducible background. ND data are used to constrain such ν_e component beyond the estimate performed with GEANT4. ND data are also used to constrain the observed ν_μ CC and NC events (see **Figure 5** of Reference [**65**]). For the 2018 NOνA analysis, corresponding to 8.85×1020 pot, the overall error budget for $\sin^2\theta_{23}$, Δm^2_{23} and δ are 4×10^{-2}, 9×10^{-5} eV2 and 0.67π, respectively. At present, the systematic budget is smaller than the statistical uncertainty and amounts to 1×10^{-2}, 4×10^{-5} eV2 and 0.12π for $\sin^2\theta_{23}$, Δm^2_{23} and δ, respectively. However, it is likely that NOνA will be systematic limited in the forthcoming years, well before the start of DUNE.

At the time of writing [**77**], the T2K and NOνA results [**65,78,79,80**] are compatible for all oscillation parameters (see **Figure 6**). NOνA data indicate a slight preference (1.8 σ) for normal hierarchy (Δm^2_{23}>0), which corroborates the hint from atmospheric neutrino data [**14**].

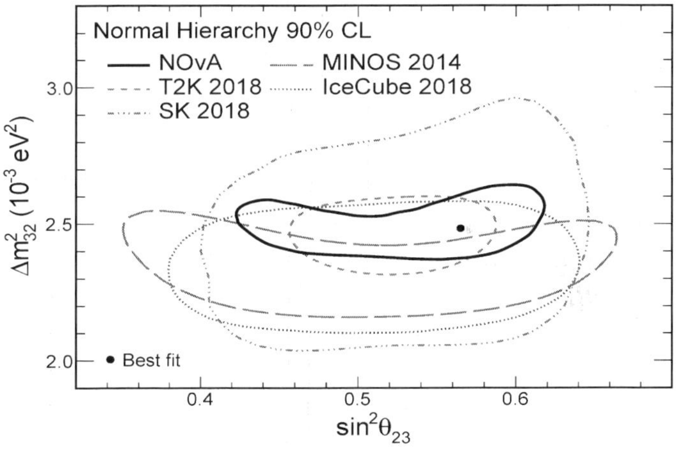

Figure 6. Δm^2_{23} versus $\sin^2\theta_{23}$ (90% CL) measured by NOνA, T2K, MINOS+, and atmospheric neutrino experiments.

DUNE AND HYPER-KAMIOKANDE

The optimal strategy to perform precision measurements and three family interference effects in the neutrino sector has been debated for decades. Accelerator neutrino beams became the technology of choice already in the late 1990s. At that time, experimental data start pointing toward two different mass scales (the atmospheric mass scale Δm^2_{23} and the solar mass scale Δm^2_{12}), but the ratio of the scales ($\alpha\simeq0.03$ in Equation (**8**)) was not too small to prevent the observation of perturbations driven by Δm^2_{12} in the leading $\nu_\mu \rightarrow \nu_e$ transitions at the atmospheric scale. As a consequence, an intense and controlled ν_μ source of artificial neutrinos at the O(GeV) energy scale would have been the optimal facility to search, for instance, for CP violation in the leptonic sector. Unfortunately, at that time, θ_{13} was not known (the Chooz limits indicated $\theta_{13}<12°$), and the overall size of the leading $\nu_\mu \rightarrow \nu_e$ oscillation could have been arbitrarily small. For $\theta_{13} \ll 12°$, the intensity of conventional beam would have been prohibitive at baselines O(1000 km) and the intrinsic contamination would have represented a major limitation to seek for $\nu\mu \rightarrow \nu e$ oscillations at the per-mill level or below [**81**,**82**]. This consideration triggered an intense R&D toward new acceleration concepts, namely the Neutrino Factory (see **Section 8**) and the Beta Beam [**83**,**84**] that aimed at providing pure and intense sources of ν_e and ν_μ. The discovery of θ_{13} at 8° set the scale of the leading $\nu_\mu \rightarrow \nu_e$ oscillation at the level of 7% ($\sin^2 2\theta_{13} \simeq 0.07$), which can be studied further increasing the power of current accelerator beams. Since 2012, therefore, R&D efforts have focused toward the design of "Superbeams": conventional neutrino beams with beam power exceeding 1 MW.

Two Superbeams will drive the field of accelerator neutrino physics in the next twenty years: the Long Baseline Neutrino Facility (LBNF) at Fermilab serving the DUNE experiment in South Dakota and the J-PARC Neutrino Beamline upgrade serving Hyper-Kamiokande in Japan. Both facilities are able to address CP violation in a significant region of the δ parameter space, perform precision measurements of oscillation parameters (e.g., the octant of θ_{23}, which is still unknown), determine the mass hierarchy using matter effect in beam (DUNE) or atmospheric (Hyper-Kamiokande) neutrinos and, in general, establish stringent tests for the minimally extended SM and its competitors. Furthermore, both DUNE and Hyper-Kamiokande are sensitive to natural neutrino sources and can perform a rich astroparticle physics program.

DUNE

LBNF [**85**] is a new neutrino facility built upon the Fermilab acceleration complex, as it was in the past for NUMI. In the next few years, this complex will undergo a significant upgrade (Proton-Improvement-Plan PIP-II) that includes the construction and operation of a new 800 MeV superconducting linear accelerator. As for NUMI, the primary proton beam will be extracted from the Main Injector in the energy range of 60–120 GeV. With the Main

Injector upgrades already implemented for NOνA as well as with the expected implementation of PIP-II, each extraction will deliver 7.5×10^{13} protons in one machine cycle (0.7 s/60 GeV–1.2 s/120 GeV) to the LBNF target in 10 μs. The complex delivers to DUNE an average power of 1.2 MW for 120 GeV protons. Neutrinos are produced after the protons hit a solid target and produce mesons which are subsequently focused by three magnetic horns into a 194 m long helium-filled decay pipe where they decay into muons and neutrinos. The facility is equipped with a near detector located 300 m after the absorber. Unlike its predecessor, LBNF has been designed to provide neutrinos over a wide energy range in order to cover both the first and the second oscillation maximum in DUNE. The beamline is designed to cope with a 2.4 MW power beam without retrofitting since an additional upgrade of the proton acceleration chain is planned (PIP-III) to be operational by 2030.

The 40 kt DUNE Far Detector [**86**] consists of four LArTPC detector modules, each with fiducial mass of about 10 kt, installed approximately 1.5 km underground and on-axis with respect to the LBNF beam. Each of the LArTPCs fits inside a cryostat of internal dimensions 18.9 m (width) × 17.8 m (height) × 65.8 m (length) that contains a total Argon mass of about 17.5 kt. The four identically sized modules provide flexibility for staging construction and for evolution of LArTPC technology. The baseline design has been recently validated at CERN by means of a 770 ton prototype (ProtoDUNE-SP [**87**], with a fiducial mass of about 400 ton) using charged particle beams and cosmic rays. This "single phase" (SP) designs inherits the most important techniques developed by ICARUS (see **Section 3**) for the charge readout, HV, and the Argon purification system. Unlike ICARUS, ProtoDUNE-SP (see **Figure 7**) is engineered to be scalable up to a DUNE module. It means that the components of the TPC are modular and can be assembled on-site in the underground laboratory and the cryogenic system has been greatly simplified. ProtoDUNE-SP demonstrated that it is possible to achieve an electron lifetime comparable with ICARUS using a cryostat based on a very cheap technology developed for industrial applications. A "membrane cryostat" is made of a corrugated membrane that contains the liquid and gaseous Argon, a passive insulation that reduces the heat leak, and a reinforced concrete structure to which the pressure is transferred. A secondary barrier system embedded in the insulation protects it from potential spills of liquid Argon, and a vapor barrier over the concrete protects the insulation from the moisture of the concrete. This system is used for a long time for the transportation of liquified natural gases and has been tailored for use with high purity Argon during the R&D for DUNE. Unlike ICARUS, the charge readout electronic is located inside the cold volume to reduce noise, and the light detection system is based on Silicon Photomultipliers (SiPMs) embedded in the anode wire planes instead of large area PMTs.

Figure 7. The ProtoDUNE-SP and ProtoDUNE-DP detectors at CERN.

For the second module, DUNE is considering an innovative Double Phase LArTPC where the electrons are multiplied in the gas phase. The Double Phase (DP) readout allows for a lower energy threshold and a reduction of complexity and cost due to the longer drift length: in DP, the distance between the cathode and the anode is 6 m and the drift takes place in the vertical direction while the distance of the horizontal planes in a SP module is 3.5 m. This technique is currently under validation at CERN (ProtoDUNE-DP [**88**]).

Both DUNE and Hyper-Kamiokande have been designed firstly to provide a superior sensitivity to CP violation and mass hierarchy compared with the previous generation of long-baseline experiments. DUNE, in particular, is able to establish mass ordering at 5σ in 2.5 years of data taking whatever is the value of δ [**89**]. Similarly, it can establish CP violation in the leptonic sector at 5 (3) σ level in 10 (13) years for 50% (75%) of all possible values of δ. If the value of δ is large, as suggested by current hints, CP violation can be established in about six years of data taking: statistics and control of the systematics will thus be crucial to improve the precision in the measurement of δ (about 20° for δ=−π/2). **Table 3** summarizes the sensitivity for a selection of physics observables as a function of time. The number of years are computed assuming the standard DUNE staging scenario: the start of data taking of DUNE will commence after the commissioning of two DUNE modules (20 kton) with an average beam power of 1.2 MW (T0). In the next year (T0 + 1y), the third module will be in data taking (30 kton). DUNE will reach its full mass (40 kton) at T0 + 3y. The beam will be upgraded to 2.4 MW three years later, i.e., at T0 + 6y.

Table 3. Time to achieve the most relevant physics milestones in DUNE for the current detector staging scenario (see text).

Milestone	Exposure (years)
5σ mass hierarchy at $\delta = -\pi/2$	1
5σ mass hierarchy at any δ	2.5
3σ CP violation at $\delta = -\pi/2$	3
3σ CP violation at 50% δ coverage	5
5σ CP violation at $\delta = -\pi/2$	7
5σ CP violation at 50% δ coverage	10
3σ CP violation at 75% δ coverage	13
δ resolution of 10° for $\delta = 0$	8

As noted above, this class of experiments is able to test in a single setup the consistency of the minimally extended SM employing the rate of ν_μ and ν_e in neutrino and antineutrino runs. It thus has a rich search program of physics beyond the Standard Model. Natural sources are exploited in DUNE for the study of Supernovae Neutrino Bursts (SNB), atmospheric neutrinos, and the search for proton decay. In particular, DUNE is the only large mass experiment based on Argon. It has a particular sensitivity to the ν_e component of the SNB through the $\nu_e\,{}^{40}\mathrm{Ar} \rightarrow e^-\,{}^{40}\mathrm{K}^*$ reaction and to proton decay in the $p \rightarrow K^+\bar{\nu}$ channel where the kaon is below the kinematic threshold in Cherenkov detectors.

Hyper-Kamiokande

The possibility of a 5σ sensitivity on the measurement of leptonic CP violation, the very reach program of astrophysical measurements, low energy neutrino multimessenger physics and proton decay, and the strength of the double Noble prize (Super)Kamiokande experiment, motivated a large international collaboration to propose the Hyper-Kamiokande (Hyper-K) project [**90**].

The design of Hyper-Kamiokande is a cylindrical tank with a diameter of 74 m and height of 60 m. The total (fiducial) mass of the detector is 258 (187) kt, for a fiducial mass eight times larger than Super-Kamiokande. A schematic view of Hyper-Kamiokande is shown in **Figure 8**.

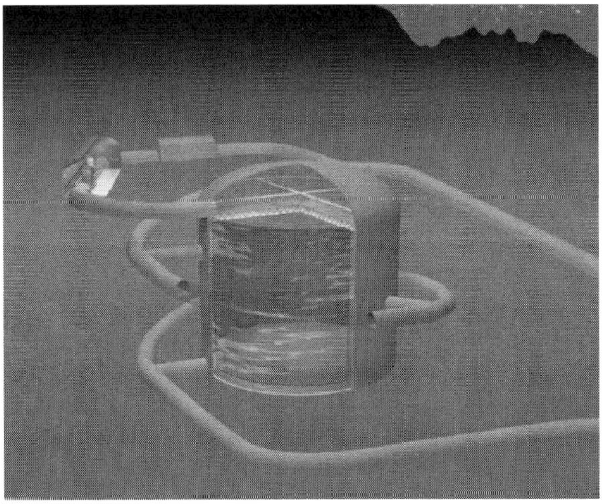

Figure 8. Layout of the Hyper-Kamiokande detector.

The Hyper-Kamiokande detector will be located in Tochibora mine, 8 km south of Super-Kamiokande, but at the same distance from J-PARC (295 km) and at the same off-axis angle (2.5∘). With an overburden of 1750 meters-water-equivalent, the muon flux is about five times higher than in Super-Kamiokande. Hyper-Kamiokande will be instrumented with 40,000 inward facing PTMSs, the 50 cm diameter Hamamatsu R12860 PMTs that have about two times higher photon detection efficiency than those of Super-Kamiokande, for a photocoverage fraction of 40%.

Another possibility is to have a photocoverage fraction of 20% complemented by 5000 Multi-PMT derived from the design of KM3NeT [**91**]. Within a 50 cm diameter, they will contain 19 7.7-cm PMTs together with integrated readout and calibration [**92**]. They can allow an increased granularity, enhancing event reconstruction in particular for multi-ring events, and better timing, which could further reduce the dark hit background and event reconstruction. The improved granularity should also guarantee an increase of the fiducial volume. Simulations show that this second setting-up guarantees very similar performances of the original 40% coverage as far as concerns beam oscillation studies.

The outer detector veto will be similar to that of SK, with 20 cm PMTs resulting in a 1% photocathode coverage of the inner wall.

The neutrino beam configuration of Hyper-K would be the same as T2K, with an upgraded power of the main 30 GeV proton ring at 1.3 MW.

The close detector system is designed to be again ND280, in the upgraded version described in **Section 4.1.1**.

A water Cherenkov (WC) near detector can be used to measure the cross-section on H2O directly. It could be placed at intermediate distance, ~2 km, from the interaction target, in order to have a neutrino flux more similar to the flux at SK and an interaction rate not critical for the performance of a WC detector. These additional WC measurements are important to achieve the low systematic errors required by Hyper-K, and complements those of the ND280 magnetized tracking detector.

The studies of the sensitivity to CP violation assume an integrated beam power 1.3MW×10^8 s corresponding to 2.7×10^{22} pot with the 30 GeV J-PARC proton beam. The selection criteria of ν_e and ν_μ candidate events are based on those established in the T2K experiment. The total systematic uncertainties of the number of expected events are assumed to be 3.2% for the νe appearance and 3.9% for the $\overline{\nu}_e$ appearance, assuming a 1:3 ratio of beam power for the $\nu:\overline{\nu}$ modes. These systematic error values roughly correspond to the statistical errors at full statistics. **Figure 9** shows the expected significance to exclude the CP conserving cases at full statistics.

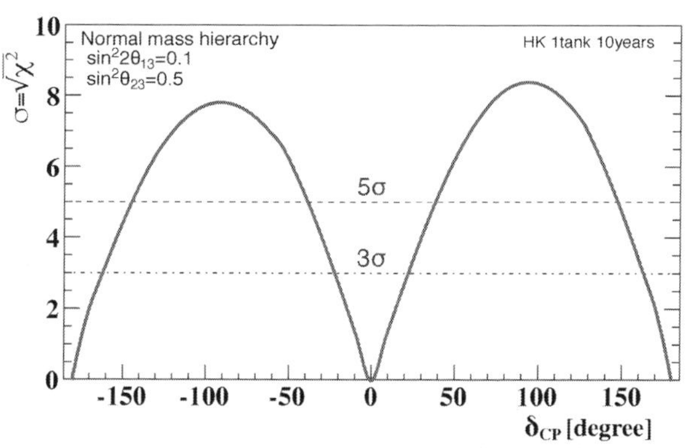

Figure 9. Sensitivity of Hyper-K in measuring δCP [**90**].

CP violation in neutrino oscillations can be observed with >5(3) σ significance for 57(80)% of the possible values of δ_{CP}. Exclusion of δCP=0 can be obtained with a significance of 8σ in the case of maximal CP violation with δCP=−π/2, the favored value of T2K. By combining the analysis of beam neutrinos with the analysis of atmospheric neutrinos,

Hyper-K would have a ~5σ sensitivity in rejecting the false value of neutrino mass ordering.

A powerful possible upgrade of Hyper-K would consist of building a second identical detector in Korea at a distance of ~1100 km and larger overburden (~2700 mwe) from J-PARC [**93**]. This second far detector can be exposed by a 1–3 degree off-axis neutrino beams from J-PARC and would be then mostly sensitive to neutrino oscillations at the second oscillation maximum.

This two-baseline setup could break the degeneracy of oscillation parameters and achieve a better precision in measuring δ_{CP}.

Among the non-oscillation physics goals of Hyper-K, it is worth to noting the 90% sensitivity to proton decays of 7.8×1034 years for the p→e+πo decay channel (existing limit is 1.6×1034 years) and 3.2×1034 years for the p→μ+πo channel (existing limit: 0.7×1034 years). Hyper-K would collect (50−80)×103 events for a 10 kpc supernova explosion and (2−3)×103 events for a supernova at the Large Magellanic Cloud where SN1987a was located. It could achieve a 4.2σ statistical sensitivity in detecting supernova relic neutrinos (SRN), integrating 10 years of data taking. These numbers are computed not assuming gadolinium doping in the detector that would be highly beneficial in all these channels [**94**]. In particlular with 0.1% by mass of gadolinium dissolved in water, the threshold for SRN detection would decrease from 16 to 10 MeV, allowing for exploring the history of supernova bursts back to the epoch of red shift (z)~1. The number of detected SRN events with gadolinium loading would become about 280, compared with 70 events without loading.

THE SYSTEMATIC REDUCTION PROGRAM

Neutrino oscillation experiments mostly rely on the near-far detector cancellation to mitigate the effects of systematic errors. This technique is the cornerstone of long-baseline experiments, and it consists of building an additional detector at a distance from the source where oscillations have not developed yet. The near detectors thus measure the ν_μ and ν_e interaction rate at source and are compared with the rate at the far detector. If the near detector is close to the source (a few hundreds meters) and identical to the far detector, it provides a normalization for the ν_μ and ν_e rates that can be used to extract $\nu_\mu \rightarrow \nu_e$ and $\nu_\mu \rightarrow \nu_\mu$ probabilities accounting for the finite efficiency of the detector for ν_μ and ν_e CC events, provided that the corresponding cross-sections are properly known. If the detectors can measure both ν_μ and ν_e and the beam is monochromatic, the remaining systematics after the near-far cancellation is $\sim\sigma_{\nu e}\epsilon_{\nu e}/\sigma_{\nu\mu}\epsilon_{\nu\mu}$. Unfortunately, other effects make the near-far detector comparison more challenging if systematics must be kept well below 10%. Firstly, the beam spectrum observed at the far location is different from the beam at source as observed in the near detector. This is due to the large difference between the angular acceptance of the two detectors. Such effect is exacerbated if the far detector is located off-axis with respect to the beam. In addition, high power beams have such large rates at close sites that pile-up effects may jeopardize the efficiency of the detector, increase the background of interactions in the surrounding material, and change the performance of PID algorithms. This is particularly disturbing for technologies that are intrinsically slow (liquid Argon TPC) or difficult to be operated at high pile-up rates (Cherenkov detectors). Pile-up effects can be mitigated by using fast neutrino detectors at the price of having the near detector different from the far detector. This is the method of choice of T2K for the ND280 and its upgrades (see **Section 4**). Finally, since no beam is monochromatic, the incoming neutrino energy must be reconstructed from the kinematics of final state particles, which in turn depend on final state interactions and nuclear effects. The simulation of these effects from first principles is beyond the state of electroweak nuclear physics and the models implemented in the neutrino generators must be tuned and validated by experimental data from either the near detector or dedicated cross-section experiments.

Over the years, the most rewarding strategy has been constraining the near detector observables embedding all available information on flux, cross-section, energy reconstruction, and detector effects combining both measurements on-site (near detector) and data recorded in dedicated facilities (hadroproduction experiments, cross-section experiments, detector test-beam). In spite of the complexity of this approach, current long baseline experiments can claim an overall systematic budget at ~5% level for νμ events and 8%–9% for νe events (see **Section 4**).

The next generation of Super-beams (DUNE and Hyper-Kamiokande) has a physics reach that will be mostly dominated by the residual systematic errors if they remain at the same values of T2K and NOνA (statistical errors at full statistics will be below 3%). Systematic errors have to be roughly halved if the investments for the huge exposures of next generation experiments have to be fully exploited. For the sake of illustration, **Figure 10** shows the impact on the νe normalization uncertainty for DUNE as a function of the exposure expressed in (mass) × (beam power) × year, i.e., kton × MW × years [**86**].

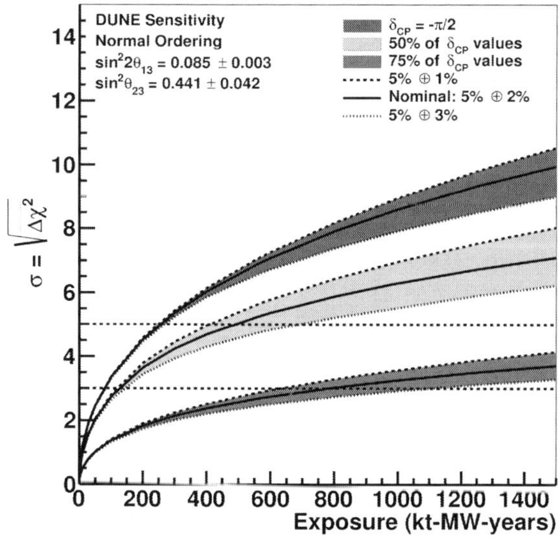

Figure 10. CP violation sensitivity for DUNE as a function of the exposure (see text). The width of the band corresponds to the difference in sensitivity between ν_e signal normalization uncertainty of 1% and 3%. The nominal uncertainty of 5% on the νμ disappearance mode and 2% on the νe appearance mode is shown as the solid line. The three curves show the significance when $\delta_{CP}=-\pi/2$ and the minimum significance for 50% and 75% of true δ_{CP} values (CP coverage).

A vigorous systematic reduction program is therefore extremely cost effective and, in addition, provides robustness against hidden systematic biases in Super-beam experiments. This program is under development and follows three lines of research.

Near Detectors

Both for DUNE and Hyper-Kamiokande, the near detector will comprise a high granularity setup similar to ND280. Even if the technology envisaged in the upgrade of ND280 and in the DUNE high-pressure TPC are different from the far detector, they allow for a superior reconstruction of final state particles in CC events and can be operated at high rate. In DUNE, the high pressure (gas Argon) TPC is complemented by a modular liquid Argon TPC that is readout by pixels after a small (50 cm) drift length [**95**]. Unlike the on-axis detector (3DST embedded in the former KLOE magnet and calorimeter), this setup is movable horizontally in the plane perpendicular to the beam axis. It thus samples different beam angles to retrieve the relation between the incident and reconstructed neutrino energies in the detector (““PRISM concept” [**96**]). The same concept is under consideration in the Hyper-Kamiokande intermediate detector. This detector is located at a larger distance (~2 km) than the N280 to reduce pile-up nuisance. In its nominal position, it samples a beam whose spectrum is the same as for the far detector. The intermediate detector, however, is movable in the vertical axis and thus implements the PRISM concept in the energy range of interest for T2K and Hyper-K [**90**].

Hadroproduction Experiments

Hadroproduction experiments measure the total and differential yields of secondary particles (mostly pions, protons, and kaons) produced by high energy protons impinging on a target. They play a key role in the determination of the initial flux of neutrino beams because the secondary production is dominated by non-perturbative QCD effects, and it is thus the most systematic prone part of the beamline simulation. Dedicated hadroproduction experiments have been routinely performed since the 1990s as ancillary experiments for long and short baseline facilities and are essential for the DUNE/Hyper-K program. Following the experience of the HARP experiment that measured hadroproduction in the K2K [**97**] and MiniBoone [**98**] targets, the NA61 experiment [**51**] at CERN (see **Figure 11**) has been extended beyond the CERN Long Shutdown 2 [**99**] to perform measurement campaigns with thin and replica targets of T2K, NOνA, Hyper-K, and DUNE.

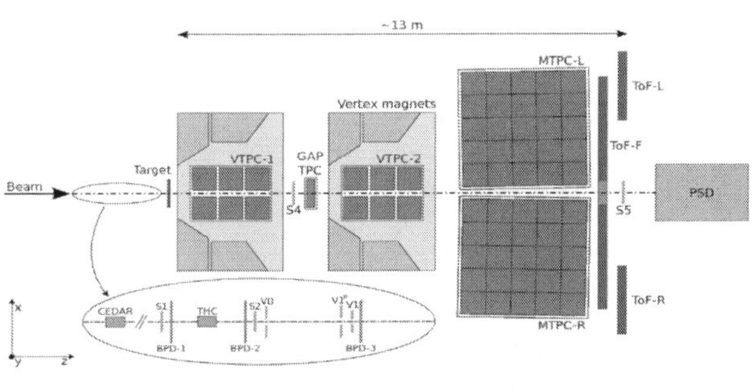

Figure 11. Layout of the NA61/SHINE experiment [**99**].

Cross-Section Experiments

In the last decade, our knowledge of total and differential cross-sections at the GeV scale has grown significantly through dedicated experiments (SciBoone, Minerva, ArgoNeut), experiments operated in short baseline beams to search for sterile neutrinos (MiniBooNE, MicroBooNE), and the near detectors of short baseline experiments [**100,101**]. The results of these experiments showed significant disagreements with theory predictions and with the neutrino interaction generators used by long-baseline experiments. They hence played a prominent role in the systematic reduction program of T2K and NOνA. In the DUNE/Hyper-K era, their role will be even more important [**102,103**]. In order to reduce the systematic budget at the percent level, we cannot rely only on cancellation effects in the ratio:

$$\frac{\int dE\, \Phi_{\nu_e}(E)\sigma_{\nu_e}(E)\epsilon_{\nu_e}(E)}{\int dE\, \Phi_{\nu_\mu}(E)\sigma_{\nu_\mu}(E)\epsilon_{\nu_\mu}(E)} \qquad (9)$$

but we have to master each term of the integrals and, in particular $\sigma_{\nu e}$ and $\sigma_{\nu\mu}$ (currently known at the 25% and 10% level, respectively), and the corresponding differential cross-sections. Even if $\sigma_{\nu e}$ is still statistics dominated and can be improved by employing the near detectors of DUNE and T2K-II, the dominant systematic contribution to the muon and electron neutrino cross-sections will soon become the uncertainty on the flux of the short baseline beam used for the measurement of $\sigma_{e/\mu}$. This scenario advocates for a new generation of cross-section experiments with a highly controlled beam. The most promising candidates are the "monitored neutrino beams" [**104**] where the flux is measured in a direct manner monitoring the rate of leptons in the decay tunnel. In particular, the CERN NP06/ENUBET [**105**] experiment (ERC Consolidator Grant, PI A. Longhin) is aimed at designing a beamline where the positrons produced at large angles by $K^+ \to e^+\pi^0\nu_e$ decays are monitored at single particle level to measure the flux of ν_e. Monitored neutrino beams may achieve a precision of <1% in the flux determination and deliver 10^4 (10^6) ν_e (ν_μ) CC events per year in a 500 ton detector as ProtoDUNE-SP (CERN) or ICARUS (Fermilab). In addition, the ENUBET beam is a narrow band beam where the position of the neutrino interaction vertex in the neutrino detector is strongly correlated to the energy of the neutrino. It can therefore provide a 8% (22%) precision measurement of the energy of the incoming neutrino at 3 (1) GeV without relying on the kinematics of the final state particle. Energy reconstruction in total and differential cross-sections is thus not biased by nuclear effects.

The ENUBET beamline can be exploited not only by LArTPCs but also by dedicated high-granularity detectors specifically designed for cross-section measurements at low and high Z. The impact of the ENUBET beamline on the measurement of the νe cross-section is depicted in **Figure 12**.

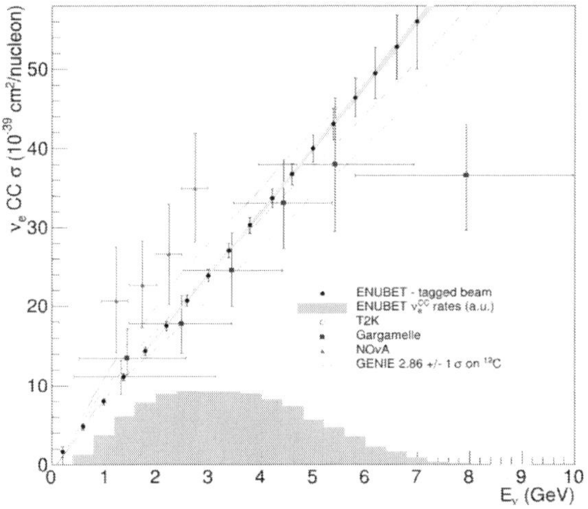

Figure 12. Impact of ENUBET on the knowledge of the ν_e cross-section. The ENUBET measurement and the statistical errors are shown with black dots. The orange bands represent the current systematic uncertainty from GENIE. The yellow band is the ENUBET systematic uncertainty on the flux. Past measurements from Gargamelle, T2K and NOνA are also shown. The gray area shows the ν_e CC energy spectrum of ENUBET.

A much larger statistics of ν_e can be achieved by changing the core technology of accelerator neutrino beams. R&D on muon colliders and Neutrino Factories are ongoing for more than 30 years. In a Neutrino Factory

[**106,107**], it is possible to produce ν_e ($\overline{\nu}_e$) at the same pace as ν_μ ($\overline{\nu}_\mu$) from the three-body decay of muons stored in the accumulator ring. In its simplest design (NUSTORM [**108,109**]), a low energy Neutrino Factory can deliver up to 10^6 ν_e CC events per year to the same detectors employed for ENUBET. As a consequence, NUSTORM would represent both a powerful source for cross-section measurements and a major step forward in the design of facilities for the post-DUNE era (see **Section 8**).

A GLIMPSE TO THE LONG-TERM FUTURE

It is hard to envisage how the field of accelerator neutrino beams will be shaped by the results of DUNE and Hyper-Kamiokande. A confirmation of the persistent sterile neutrino hints [**110**] would change in a substantial manner the three-family oscillation paradigm considered in this paper and would urge a re-definition of the entire experimental strategy. Still, even in the framework of the minimally extended Standard Model, the Superbeams will not say the final word on lepton mixing. The PMNS precision in 2030 will still be very far from the precision achieved in the CKM. In flavor physics, the main tool to explore physics beyond the Standard Model are unitarity tests and inconsistencies between the mixing parameters measured by different channels: both tools require precision. In addition, the striking difference between the PMNS and CKM indicates that the entire Yukawa sector of the SM is not understood and that the origin of such difference may reside in physics at energy scales not accessible at accelerators. A percent level measurement of the PMNS combined with the study of neutrino non-standard interactions (NSI) and the $\nu_e \rightarrow \nu_\tau$ transitions (unitarity tests) are the main route toward these scales and require employing neutrino oscillations at an unprecedented level of precision. Any upgrade of DUNE and Hyper-Kamiokande will be of little impact because, at full statistics, they will be limited by systematic errors and those errors cannot be further significantly reduced, being related to the intrinsic limitations of conventional neutrino beams. New concepts in neutrino beams, as Neutrino Factories and/or Beta Beams will then become necessary for any further development after the third generation of long baseline neutrino experiments. A bridge between the DUNE/Hyper-K era and the Neutrino Factory era will likely occur by upgrading existing facilities. In the European context, a special role will be played by the European Spallation Source, the most powerful proton source devised for material science studies. The infrastructure and technology of ESS can be upgraded to serve the neutrino science community with a low energy Superbeam operating at the second oscillation maximum (~350 MeV) and, thus, with the largest CP sensitivity. The ESS also represents an interesting first stage facility for the implementation of the accelerator complex serving a Neutrino Factory and a high energy Muon Collider.

CONCLUSIONS

In the last ten years, accelerator neutrino beams consolidated their role in the study of neutrino oscillations. In this paper, we reviewed the most important experimental results provided by neutrino beam experiments and interpreted these results in the framework of the minimally extended

Standard Model. Neutrino beams have established in a conclusive manner the appearance of tau neutrinos ($\nu_\mu \rightarrow \nu_\tau$ oscillations) at CNGS and of electron neutrinos ($\nu_\mu \rightarrow \nu_e$ oscillations) at J-PARC and NUMI. T2K and NOνA lead the measurements of the oscillation parameters at the atmospheric scale (Δm^2_{23} and θ_{23}). In particular, the T2K+NOνA combined measurements provide hints both of normal hierarchy and of CP violation. These hints will be checked conclusively in the next generation of Superbeam, namely by DUNE and Hyper-Kamiokande. Our field is now in a stage where a superior control of systematics and a precision knowledge of neutrino interactions in matter is of paramount importance to ground the physics reach of the Superbeams. These needs foster a vigorous systematic reduction program based on movable near detectors, hadroproduction experiments, and a new generation of facilities for the study of cross-sections at the GeV scale.

AUTHOR CONTRIBUTIONS

Investigation, M.M., F.T., Writing—Review and Editing, M.M., F.T. All authors have read and agreed to the published version of the manuscript.

FUNDING

This work is supported by the EU H2020 program under Grant Nos. 674896 and 777419, by the COST action CA15139 and by RISE-GA822070-JENNIFER2 2020.

ACKNOWLEDGMENTS

The authors gratefully acknowledge the members of the OPERA, ICARUS, T2K, NOνA, ESSνSB, DUNE and HyperK Collaborations for many useful discussions and details on the experiments and analyses.

CONFLICTS OF INTEREST

The authors declare no conflict of interest.

REFERENCES

1. Kopp, S.E. Accelerator-based neutrino beams. Phys. Rep. 2007, 439, 101–159.
2. Fukuda, Y.; et al. [Super-Kamiokande Collaboration] Evidence for oscillation of atmospheric neutrinos. Phys. Rev. Lett. 1998, 81, 1562.
3. Tanabashi, M.; et al. [Particle Data Group] Review of Particle Physics. Phys. Rev. D 2018, 98, 030001.
4. An, F.P.; et al. [Daya Bay Collaboration] Observation of Electron-Antineutrino Disappearance at Daya Bay. Phys. Rev. Lett. 2012, 108, 171803.
5. Ahn, J.K.; et al. [RENO Collaboration] Observation of Reactor Electron Antineutrinos Disappearance in the RENO Experiment. Phys. Rev. Lett. 2012, 108, 191802.
6. Abe, K.; et al. [T2K Collaboration] Evidence of electron neutrino appearance in a muon neutrino beam. Phys. Rev. D 2013, 88, 032002.
7. Gonzalez-Garcia, M.C.; Maltoni, M. Phenomenology with Massive Neutrinos. Phys. Rept. 2008, 460, 1–129.
8. Altarelli, G.; Feruglio, F. Models of neutrino masses and mixings. New J. Phys. 2004, 6, 106.
9. Minkowski, P. $\mu \rightarrow e\gamma$ at a Rate of One Out of 10^9 Muon Decays? Phys. Lett. B 1977, 67, 421–428.
10. Gell-Mann, M.; Ramond, P.; Slansky, R. Complex Spinors and Unified Theories. Conf. Proc. C 1979, 790927, 315–321.
11. Yanagida, T. Horizontal gauge symmetry and masses of neutrinos. Prog. Theor. Phys. 1980, 64, 1103–1105.
12. Mohapatra, R.N.; Senjanovic, G. Neutrino Mass and Spontaneous Parity Nonconservation. Phys. Rev. Lett. 1980, 44, 912–915.
13. Schechter, J.; Valle, J.W.F. Neutrino Masses in SU(2) x U(1) Theories. Phys. Rev. D 1980, 22, 2227–2235.
14. Esteban, I.; Gonzalez-Garcia, M.C.; Hernandez-Cabezudo, A.; Maltoni, M.; Schwetz, T. Global analysis of three-flavour neutrino oscillations: Synergies and tensions in the determination of $_{23}$, $_{CP}$, and the mass ordering. J. High Energy Phys. 2019, 1901, 106.
15. Apollonio, M.; et al. [CHOOZ Collaboration] Search for neutrino oscillations on a long baseline at the CHOOZ nuclear power station. Eur. Phys. J. C 2003, 27, 331.

16. Acquistapace, G.; Baldy, J.L.; Ball, A.E.; Bonnal, P.; Buhler-Broglin, M.; Carminati, F.; Cennini, E.; Elsener, K.; Ereditato, A.; Falaleev, V.; et al. The CERN Neutrino Beam to Gran Sasso (NGS). Available online: http://inspirehep.net/record/473325/ (accessed on 8 February 2020).
17. Gschwendtner, E.; Cornelis, K.; Efthymiopoulos, I.; Pardons, A.; Vincke, H.; Wenninger, J.; Krätschmer, I. CNGS, CERN Neutrinos to Gran Sasso, Five Years of Running a 500 Kilowatt Neutrino Beam Facility at CERN. In Proceedings of the IPAC2013, Shanghai, China, 12–17 May 2013.
18. Acquafredda, R.; Adam, T.; Agafonova, N.; Alvarez Sanchez, P.; Ambrosio, M.; Anokhina, A.; Aoki, S.; Ariga, A.; Ariga, T.; Arrabito, L.; et al. The OPERA experiment in the CERN to Gran Sasso neutrino beam. J. Instrum. 2009, 4, P04018.
19. Agafonova, N.; et al. [OPERA Collaboration] Final results on neutrino oscillation parameters from the OPERA experiment in the CNGS beam. Phys. Rev. D 2019, 100, 051301.
20. Eskut, E.; et al. [CHORUS Collaboration] Final results on $\nu_\mu \rightarrow \nu_\tau$ oscillation from the CHORUS experiment. Nucl. Phys. B 2008, 793, 326–343.
21. Andreopoulos, C.; Bell, A.; Bhattacharya, D.; Cavanna, F.; Dobson, J.; Dytman, S.; Gallagher, H.; Guzowski, P.; Hatcher, R.; Kehayias, P.; et al. The GENIE Neutrino Monte Carlo Generator. Nucl. Instrum. Methods. A 2010, 614, 87.
22. Autiero, D. The OPERA event generator and the data tuning of nuclear re-interactions. Nucl. Phys. B Proc. Suppl. 2005, 139, 253.
23. Allison, J.; Amako, K.; Apostolakis, J.; Arce, P.; Asai, M.; Aso, T.; Bagli, E.; Bagulya, A.; Banerjee, S.; Barrand, G.; et al. Recent developments in Geant4. Nucl. Instrum. Methods. A 2016, 835, 186–225.
24. Agafonova, N.; et al. [OPERA Collaboration] Discovery of τ Neutrino Appearance in the CNGS Neutrino Beam with the OPERA Experiment. Phys. Rev. Lett. 2015, 115, 121802.
25. Agafonova, N.; et al. [OPERA Collaboration] Final Results of the OPERA Experiment on ν_τ Appearance in the CNGS Neutrino Beam. Phys. Rev. Lett. 2018, 120, 211801.
26. Amerio, S.; et al. [ICARUS Collaboration] Design, construction and tests of the ICARUS T600 detector. Nucl. Instrum. Meth. A 2004, 527,

329.

27. Antonello, M.; et al. [MicroBooNE, LAr1-ND and ICARUS-WA104 Collaborations] A Proposal for a Three Detector Short-Baseline Neutrino Oscillation Program in the Fermilab Booster Neutrino Beam. Available online: https://arxiv.org/abs/1503.01520 (accessed on 11 February 2020).
28. Antonello, M.; et al. [ICARUS Collaboration] Search for anomalies in the ν_e appearance from a ν_μ beam. Eur. Phys. J. C 2013, 73, 2599.
29. Antonello, M.; et al. [ICARUS Collaboration] Measurement of the neutrino velocity with the ICARUS detector at the CNGS beam. Phys. Lett. B 2012, 713, 17–22.
30. Antonello, M.; Baibussinov, B.; Benetti, P.; Boffelli, F.; Bubak, A.; Calligarich, E.; Centro, S.; Cesana, A.; Cieślik, K.; Cline, D.B.; et al. Experimental observation of an extremely high electron lifetime with the ICARUS-T600 LAr-TPC. J. Instrum. 2014, 9, P12006.
31. Oyama, Y. [K2K Collaboration]. K2K (KEK to Kamioka) Neutrino-Oscillation Experiment at KEK PS. Available online: https://arxiv.org/abs/hep-ex/9803014 (accessed on 10 February 2020).
32. Ahn, M.H.; et al. [K2K Collaboration] Indications of neutrino oscillation in a 250 km long baseline experiment. Phys. Rev. Lett. 2003, 90, 041801.
33. Abe, K.; et al. [T2K Collaboration] The T2K Experiment. Nucl. Instrum. Meth. A 2011, 659, 106.
34. Sekiguchi, T.; et al. [T2K Collaboration] Development and operational experience of magnetic horn system for T2K experiment. Nucl. Instrum. Meth. A 2015, 789, 57.
35. The Off-Axis Neutrino Beam Concept Was Discussed for the First Time in the BNL E889 Proposal. Available online: http://www.hep.princeton.edu/~mcdonald/nufact/e889/ (accessed on 10 February 2020).
36. Abe, K.; et al. [The T2K Collaboration] Measurements of the T2K neutrino beam properties using the INGRID on-axis near detector. Nucl. Instrum. Meth. A 2012, 694, 211.
37. Abgrall, N.; et al. [T2K ND280 TPC Collaboration] Time Projection Chambers for the T2K Near Detectors. Nucl. Instrum. Meth. A 2011, 637, 25.
38. Amaudruz, P.A.; et al. [T2K ND280 FGD Collaboration] The T2K

Fine-Grained Detectors. Nucl. Instrum. Meth. A 2012, 696, 1.

39. Assylbekov, S.; Barr, G.; Berger, B.E.; Berns, H.; Beznosko, D.; Bodek, A.; Bradford, R.; Buchanan, N.; Budd, H.; Caffari, Y.; et al. The T2K ND280 Off-Axis Pi-Zero Detector. Nucl. Instrum. Meth. A 2012, 686, 48.
40. Allan, D.; et al. [T2K UK Collaboration] The Electromagnetic Calorimeter for the T2K Near Detector ND280. J. Instrum. 2013, 8, P10019.
41. Aoki, S.; Barr, G.; Batkiewicz, M.; Błockid, J.; Brinsonf, J.D.; Colemanf, W.; Dąbrowska, A.; Danko, I.; Dziewiecki, M.; Ellison, B.; et al. The T2K Side Muon Range Detector (SMRD). Nucl. Instrum. Meth. A 2013, 698, 135.
42. Abgrall, N.; et al. [NA61/SHINE Collaboration] Measurements of $\varpi^{\pm}$, $K^{\pm}$, K0S, and proton production in proton–carbon interactions at 31 GeV/c with the NA61/SHINE spectrometer at the CERN SPS. Eur. Phys. J. C 2016, 76, 84.
43. Abe, K.; et al. [T2K Collaboration] T2K neutrino flux prediction. Phys. Rev. D 2013, 87, 012001.
44. Fukuda, Y.; et al. [Super-Kamiokande Collaboration] The Super-Kamiokande detector. Nucl. Instrum. Meth. A 2003, 501, 418.
45. Abe, K.; et al. [T2K Collaboration] Indication of Electron Neutrino Appearance from an Accelerator-produced Off-axis Muon Neutrino Beam. Phys. Rev. Lett. 2011, 107, 041801.
46. Abe, K.; et al. [T2K Collaboration] Search for CP Violation in Neutrino and Antineutrino Oscillations by the T2K Experiment with 2.2×10^{21} Protons on Target. Phys. Rev. Lett. 2018, 121, 171802.
47. Ferrari, A.; Sala, P.R.; Fasso, A.; Ranft, J. FLUKA: A Multi-Particle Transport Code (Program Version 2005). Available online: http://inspirehep.net/record/701721 (accessed on 8 February 2020).
48. Brun, R.; Bruyant, F.; Carminati, F.; Giani, S.; Maire, M.; McPherson, A.; Patrick, G.; Urban, L. GEANT Detector Description and Simulation Tool. Available online: https://cds.cern.ch/record/1073159/files/cer-002728534.pdf (accessed on 8 February 2020).
49. Abgrall, N.; et al. [NA61/SHINE] Measurement of Production Properties of Positively Charged Kaons in Proton-Carbon Interactions at 31 GeV/c. Phys. Rev. C 2012, 85, 035210.
50. Abgrall, N.; et al. [NA61/SHINE] Measurements of Cross Sections and

Charged Pion Spectra in Proton-Carbon Interactions at 31 GeV/c. Phys. Rev. C 2011, 84, 034604.

51. Abgrall, N.; et al. [NA61/SHINE Collaboration] Measurements of $\varpi^{\pm}$, $K^{\pm}$ and proton double differential yields from the surface of the T2K replica target for incoming 31 GeV/c protons with the NA61/SHINE spectrometer at the CERN SPS. Eur. Phys. J. C 2019, 79, 100.
52. Hayato, Y. A neutrino interaction simulation program library NEUT. Acta Phys. Polon. B 2009, 40, 2477–2489.
53. Abe, K.; et al. [T2K Collaboration] Measurements of neutrino oscillation in appearance and disappearance channels by the T2K experiment with 6.6×10^{20} protons on target. Phys. Rev. D 2015, 91, 072010.
54. Abe, K.; et al. [T2K Collaboration] Constraint on the Matter-Antimatter Symmetry-Violating Phase in Neutrino Oscillations. arXiv 2019, arXiv:1910.03887.
55. Avanzini, M.B.; et al.; [T2K Collaboration] T2K neutrino-nucleus cross-section results. PoS NOW 2018, 016.
56. Abe, K.; et al. [T2K Collaboration] Search for light sterile neutrinos with the T2K far detector Super-Kamiokande at a baseline of 295 km. Phys. Rev. D 2019, 99, 071103.
57. Abe, K.; et al. [T2K Collaboration] Search for heavy neutrinos with the T2K near detector ND280. Phys. Rev. D 2019, 100, 052006.
58. Abe, K.; et al. [T2K Collaboration] Sensitivity of the T2K accelerator-based neutrino experiment with an Extended run to 20 × 1021 POT. Available online: https://arxiv.org/abs/1607.08004 (accessed on 10 February 2020).
59. Abe, K.; et al. [T2K Collaboration] T2K ND280 Upgrade—Technical Design Report. Available online: https://arxiv.org/abs/1901.03750 (accessed on 10 February 2020).
60. Blondel, A.; Cadoux, F.; Fedotov, S.; Khabibullin, M.; Khotjantsev, A.; Korzenev, A.; Kostin, A.; Kudenko, Y.; Longhin, A.; Mefodiev, A.; et al. A fully active fine grained detector with three readout views. J. Instrum. 2018, 13, P02006.
61. Adamson, P.; Andersonc, K.; Andrewsc, M.; Andrewsc, R.; Anghelog, I.; Augustine, D.; Aurisanob, A.; Avvakumovab, S.; Ayresg, D.S.; Baller, B.; et al. The NuMI Neutrino Beam. Nucl. Instrum. Meth. A 2016, 806, 279–306.

62. Ayres, D.S.; et al. [NOνA Collaboration] The NOνA Technical Design Report. Available online: https://www.osti.gov/biblio/935497 (accessed on 8 February 2020).
63. Michael, D.G.; et al. [MINOS Collaboration] The Magnetized steel and scintillator calorimeters of the MINOS experiment. Nucl. Instrum. Meth. A 2016, 596, 190–228.
64. Adamson, P.; et al. [MINOS Collaboration] Combined analysis of ν_μ disappearance and $\nu_\mu \rightarrow \nu_e$ appearance in MINOS using accelerator and atmospheric neutrinos. Phys. Rev. Lett. 2014, 112, 191801.
65. Acero, M.A.; et al. [NOνA Collaboration] New constraints on oscillation parameters from ν_e appearance and ν_μ disappearance in the NOνA experiment. Phys. Rev. D 2018, 98, 032012.
66. Giunti, C.; Kim, C.W. Fundamentals of Neutrino Physics and Astrophysics; Oxford University Press: Oxford, UK, 2007.
67. Cervera, A.; Donini, A.; Gavela, M.B.; Gomez Cadenas, J.J.; Hernandez, P.; Mena, O.; Rigolin, S. Golden measurements at a neutrino factory. Nucl. Phys. B 2000, 579, 17.
68. Freund, M. Analytic approximations for three neutrino oscillation parameters and probabilities in matter. Phys. Rev. D 2001, 64, 053003.
69. Huber, P.; Lindner, M.; Winter, W. Simulation of long-baseline neutrino oscillation experiments with GLoBES (General Long Baseline Experiment Simulator). Comput. Phys. Commun. 2005, 167, 195.
70. Migliozzi, P.; Terranova, F. Next generation long baseline experiments on the path to leptonic CP violation. Phys. Lett. B 2003, 563, 73.
71. Fogli, G.L.; Lisi, E. Tests of three flavor mixing in long baseline neutrino oscillation experiments. Phys. Rev. D 1996, 54, 3667–3670.
72. Minakata, H.; Nunokawa, H. Exploring neutrino mixing with low-energy superbeams. J. High Energy Phys. 2001, 10, 001.
73. Barger, V.; Marfatia, D.; Whisnant, K. Breaking eight fold degeneracies in neutrino CP violation, mixing, and mass hierarchy. Phys. Rev. D 2002, 65, 073023.
74. Sanchez, M. NOνA Results and Prospects. Available online: https://zenodo.org/record/1286758#.XkHmGDH-02w (accessed on 10 February 2020).
75. Aliaga, L.; et al. [MINERνA Collaboration] Neutrino Flux Predictions for the NuMI Beam. Phys. Rev. D 2017, 94, 092005, Addendum: Phys.

Rev. D 2017 95 039903.

76. Acero, M.A.; et al. [NOvA Collaboration] First Measurement of Neutrino Oscillation Parameters using Neutrinos and Antineutrinos by NOvA. Phys. Rev. Lett. 2019, 123, 151803.
77. Whittington, D. Results from NOνA. Talk at WIN2019. In Proceedings of the 27th International Workshop on Weak interactions and Neutrinos (WIN2019), Bari, Italy, 3–8 June 2019.
78. Adamson, P.; et al. [NOvA Collaboration] First measurement of muon-neutrino disappearance in NOvA. Phys. Rev. D 2016, 93, 051104.
79. Adamson, P.; et al. [NOvA Collaboration] Constraints on Oscillation Parameters from ν_e Appearance and ν_μ Disappearance in NOνA. Phys. Rev. Lett. 2017, 118, 231801.
80. Adamson, P.; et al. [NOvA Collaboration] Measurement of the neutrino mixing angle $_{23}$ in NOνA. Phys. Rev. Lett. 2017, 118, 151802.
81. Mezzetto, M.; Schwetz, T. $_{13}$: Phenomenology, present status and prospect. J. Phys. G 2010, 37, 103001.
82. Battiston, R.; Mezzetto, M.; Migliozzi, P.; Terranova, F. European facilities for accelerator neutrino physics: Perspectives for the decade to come. Riv. Nuovo Cim. 2010, 33, 313.
83. Zucchelli, P. A novel concept for a $\nu^{-}e/\nu_e$ neutrino factory: The beta beam. Phys. Lett. B 2002, 532, 166.
84. Lindroos, M.; Mezzetto, M. Beta Beams: Neutrino Beams; World Scientific: Singapore, 2009.
85. Acciarri, R.; et al. [DUNE Collaboration] Long-Baseline Neutrino Facility (LBNF) and Deep Underground Neutrino Experiment (DUNE): Conceptual Design Report, Volume 1: The LBNF and DUNE Projects. Available online: https://arxiv.org/abs/1601.05471 (accessed on 10 February 2020).
86. Abi, B.; et al. [DUNE Collaboration] The DUNE Far Detector Interim Design Report, Volume 2: Single-Phase Module. Available online: http://eprints.whiterose.ac.uk/136071/ (accessed on 10 February 2020).
87. Abi, B.; et al. [DUNE Collaboration] The Single Phase ProtoDUNE Technical Design Report. Available online: https://arxiv.org/abs/1706.07081 (accessed on 10 February 2020).
88. De Bonis, I.; Del Amo Sanchez, P.; Duchesneau, D.; Pessard, H.; Bordoni, S.; Ieva, M.; Lux, T.; Sanchez, F.; Jipa, A.; Lazanu, I.; et

al. LBNO-DEMO: Large-Scale Neutrino Detector Demonstrators for Phased Performance Assessment in View of A Long-Baseline Oscillation Experiment. Available online: https://arxiv.org/abs/1409.4405 (accessed on 8 February 2020).

89. Abi, B.; et al. [DUNE Collaboration] The DUNE Far Detector Interim Design Report Volume 1: Physics, Technology and Strategies. Available online: https://arxiv.org/abs/1807.10334 (accessed on 10 February 2020).
90. Abe, K.; et al. [Hyper-Kamiokande Collaboration] Hyper-Kamiokande Design Report. Available online: https://arxiv.org/abs/1805.04163 (accessed on 10 February 2020).
91. Vivolo, D. [KM3NeT Collaboration]. The KM3NeT Digital Optical Module. Springer Proc. Phys. 2018, 213, 192.
92. Ruggeri, A.C. [Hyper-Kamiokande proto Collaboration]. R&D Studies of New Photosensors for the Hyper-Kamiokande Detector. Springer Proc. Phys. 2018, 213, 325.
93. Abe, K.; et al.; [Hyper-Kamiokande Collaboration] Physics potentials with the second Hyper-Kamiokande detector in Korea. Progr. Theor. Exp. Phys. 2018, 2018, 63C01.
94. Watanabe, H.; et al. [Super-Kamiokande Collaboration] First Study of Neutron Tagging with a Water Cherenkov Detector. Astropart. Phys. 2009, 31, 320.
95. Bross, A. The DUNE Near Detector. In Proceedings of the 21st International Workshop on Neutrinos from Accelerators (NUFACT2019), Daegu, Korea, 25–31 August 2019.
96. Bhadra, S.; et al. [nuPRISM Collaboration] Letter of Intent to Construct a nuPRISM Detector in the J-PARC Neutrino Beamline. Available online: https://arxiv.org/abs/1412.3086 (accessed on 10 February 2020).
97. Catanesi, M.G.; et al. [HARP Collaboration] Measurement of the production cross-section of positive pions in p-Al collisions at 12.9-GeV/c. Nucl. Phys. B 2006, 732, 1.
98. Catanesi, M.G.; et al. [HARP Collaboration] Measurement of the production cross-section of positive pions in the collision of 8.9-GeV/c protons on beryllium. Eur. Phys. J. C 2007, 52, 29.
99. Turko, L. NA61/SHINE Experiment—Program beyond 2020. Particles 2018, 1, 296–304.

100. Alvarez-Ruso, L.; Sajjad Atharb, M.; Barbaroc, M.B.; Cherdackd, D.; Christye, M.E.; Colomaf, P.; Donnellyg, T.W.; Dytmanh, S.; de Gouvêai, A.; Hill, R.J.; et al. NuSTEC White Paper: Status and challenges of neutrino–nucleus scattering. Prog. Part. Nucl. Phys. 2018, 100, 1.
101. Katori, T.; Martini, M. Neutrino–nucleus cross sections for oscillation experiments. J. Phys. G 2018, 45, 013001.
102. Huber, P.; Mezzetto, M.; Schwetz, T. On the impact of systematical uncertainties for the CP violation measurement in superbeam experiments. J. High Energy Phys. 2008, 0803, 021.
103. Coloma, P.; Huber, P.; Kopp, J.; Winter, W. Systematic uncertainties in long-baseline neutrino oscillations for large $_{13}$. Phys. Rev. D 2013, 87, 033004.
104. Longhin, A.; Ludovici, L.; Terranova, F. A novel technique for the measurement of the electron neutrino cross section. Eur. Phys. J. C 2015, 75, 155.
105. Acerbi, F.; et al. [ENUBET Collaboration] The ENUBET Project. Available online: https://cds.cern.ch/record/2645532# (accessed on 8 February 2020).
106. Geer, S. Neutrino beams from muon storage rings: Characteristics and physics potential. Phys. Rev. D 1998, 57, 6989.
107. Choubey, S.; et al. [IDS-NF Collaboration] International Design Study for the Neutrino Factory, Interim Design Report. Available online: https://arxiv.org/abs/1112.2853 (accessed on 10 February 2020).
108. Adey, D.; et al. [nuSTORM Collaboration] nuSTORM - Neutrinos from STORed Muons: Proposal to the Fermilab PAC. Available online: https://arxiv.org/abs/1308.6822 (accessed on 10 February 2020).
109. Adey, D.; Bayes, R.; Bross, A.; Snopok, P. nuSTORM and a path to a Muon Collider. Ann. Rev. Nucl. Part. Sci. 2015, 65, 145.
110. Giunti, C.; Lasserre, T. eV-scale Sterile Neutrinos. Ann. Rev. Nucl. Part. Sci. 2019, 69, 163–190.

CHAPTER

12

Probing Nonstandard Lepton number Violating Interactions in Neutrino Oscillations

Patrick D. Bolton[1,] and Frank F. Deppisch[1,2]

[1]Department of Physics and Astronomy, University College London, Gower Street, London WC1E 6BT, United Kingdom

[2]Institut für Hochenergiephysik, Österreichische Akademie der Wissenschaften, Nikolsdorfer Gasse 18, 1050 Wien, Austria

ABSTRACT

We discuss lepton number violating processes in the context of long-baseline neutrino oscillations. We summarize and compare neutrino flavor oscillations in quantum mechanics and quantum field theory, both for standard oscillations and for those that violate lepton number. When the active neutrinos are Majorana in nature, the required helicity reversal gives a strong suppression by the neutrino mass over the energy, $(m_\nu/E_\nu)^2$. Instead, the

Citation: Patrick D. Bolton and Frank F. Deppisch. "Probing nonstandard lepton number violating interactions in neutrino oscillations" Phys. Rev. D 99, 115011 (2019). https://doi.org/10.1103/PhysRevD.99.115011

presence of nonstandard lepton number violating interactions incorporating right-handed lepton currents at production or detection alleviate the mass suppression while also factorizing the oscillation probability from the total rate. Such interactions arise from dimension-six operators in the low energy effective field theory of the Standard Model. We derive general and simplified expressions for the lepton number violating oscillation probabilities and use limits from MINOS and KamLAND to place bounds on the interaction strength in interplay with the unknown Majorana phases in neutrino mixing. We compare the bounds with those from neutrinoless double beta decay and other microscopic lepton number violating processes and outline the requirements for future short- and long-baseline neutrino oscillation experiments to improve on the existing bounds.

INTRODUCTION

The seminal confirmation of neutrino flavor oscillations by the Super-Kamiokande and SNO experiments in 1998 and 2001, respectively, initiated a golden era in the experimental and theoretical studies of massive neutrinos in the Standard Model (SM) [1,2]. Five of the six (or eight) independent mixing parameters describing the three Dirac (or Majorana) neutrinos have now been pinned down by solar, atmospheric, accelerator, and reactor neutrino oscillation experiments [3].

There are still a few pieces of the neutrino mass puzzle that remain unknown: First, the value of the Dirac phase δ_{CP} which controls the magnitude of CP violation in the neutrino sector. Second, the sign of the atmospheric squared mass difference Δm^2_{23} which will decipher the normal ordering (NO) or inverted ordering (IO) of the neutrinos. Both quantities will be determined in the coming years by accelerator and reactor oscillation experiments such as NOvA, DUNE, Hyper-Kamiokande, and JUNO [4–7]. Neutrino oscillations are insensitive to the absolute scale of the neutrino masses m_0—beta decay experiments such as KATRIN aim to combine their results with precision measurements of the cosmic microwave background and large-scale galaxy clustering to place a stringent upper bound on this scale [8–11].

Last but by no means the least is the question of the fundamental nature of the light neutrinos. For massive neutrinos—or generally any massive fermion that is a singlet under gauge transformations—it is possible to write two Lorentz invariant mass terms. The first, proposed by Dirac, is the same as that of the charged fermions. The second, proposed by Majorana, violates lepton number by two units [12,13]. On the grounds of naturalness the latter has been favored for some time, with effective models using the high scale of new physics (NP) to generate the tiny neutrino masses—also known as the seesaw mechanism. Common UV-complete versions of this mechanism incorporate heavy Majorana right-handed (RH) neutrinos into the larger gauge structure of a grand unified theory [14–17].

While these theories generically predict lepton number violating (LNV) phenomena, none have yet been observed in nature. If the light neutrinos are indeed Majorana, a standard explanation of this nonobservation is to promote the nonanomalous combination of baryon number (B) minus lepton number (L) to a gauge group $U(1)_{B-L}$. This symmetry is spontaneously broken at low energies and thereby suppresses (B−L)-violating processes by the high symmetry-breaking scale [18]. Other models which predict additional particles such as the Majoron and supersymmetric partners have been excluded at high significance by astronomical and collider observations [19–23]. If the neutrinos are instead purely Dirac, the absence of an LNV signal could be explained if B−L remains a global symmetry after the spontaneous symmetry breaking of a left-right (LR) symmetric model [24]. There is also the intriguing possibility that neutrinos are quasi-Dirac [25–27].

The most promising current effort to detect an LNV process is via searches for neutrinoless double beta (0νββ) decay which is the nuclear decay process $(Z,A)\rightarrow(Z+2,A)+2e^-$. If observed, the absence of outgoing neutrinos implies there to be an internal neutrino propagator, which is possible only if the neutrino is its own antiparticle, i.e., if neutrinos are Majorana fermions. Even if unrelated NP were to trigger 0νββ decay, a positive signal would automatically imply Majorana neutrinos via the Schechter-Valle black-box theorem [28]. Current and future 0νββ experiments such as EXO-200, KamLAND-Zen, CUORE, GERDA-II, and SuperNEMO are aiming to push the lower limit on the 0νββ decay half-life $T^{0\nu\beta\beta}_{1/2}$ up to 1026 years for various nuclear isotopes [29]. This lower limit can be converted to an upper limit on the neutrino mass parameter $m_{\beta\beta}$ if the nuclear matrix element and associated systematic uncertainty are known for the relevant isotope. A large portion of the region in the $m_{\beta\beta}-m_0$ parameter space corresponding to the quasidegenerate arrangement of neutrino masses ($m_1\simeq m_2\simeq m_3$) has been excluded at 90% C.L. by the KamLAND-Zen and GERDA-II experiments. Future experiments will be able to exclude the region corresponding to the IO of neutrino masses ($m_3\ll m_1<m_2$). The NO region ($m_1<m_2\ll m_3$) can be probed unless the Majorana CP-violating (CPV) phases conspire to cancel the different contributions to $m_{\beta\beta}$ [30].

Contributions to 0νββ decay from LNV physics beyond the SM have been studied extensively [31–49]. Much of this past work has used an effective coefficient ϵ to parametrize the strength of the LNV nonstandard interaction (NSI) with respect to the SM dimension-six Fermi interaction [50]. Alternatively one can introduce Wilson coefficients for the relevant dimension-d combinations of SM fields, suppressed by the scale of NP raised to the power (4−d). A general analysis of LNV channels has been conducted in the effective field theory formalism for operators up to dimension eleven [51]. Experimental signatures of alternative LNV processes have been explored in the literature, for instance the decays $K^+ \rightarrow \pi^- \mu^+ \mu^+$, τ^-, and the lepton flavor and number violating (LNFV) muon conversion process $\mu^- + (Z, A) \rightarrow e^+ +$(Z−2,A) [30,52,53]. While these channels are less sensitive to LNV compared to 0νββ decay, searches have been conducted by a variety of experiments [54–56]. The contributions of lepton number conserving (LNC) but potentially lepton flavor violating (LFV) NSI to standard processes such as meson decays, neutrino oscillations, and neutrino scattering have also been considered exhaustively in the literature [50,57–82]. For processes which are insensitive to lepton number—for example, the outgoing neutrino is not detected in charged pion decays $\pi^\pm \rightarrow \ell_\alpha^\pm \overset{(-)}{\nu}_\alpha$—precision probes of SM predictions such as lepton universality and Cabibbo-Kobayashi-Maskawa unitarity can be used to constrain both LNC and LNV NSI [64].

Much of the literature has so far only considered the effect of LNC NSI on neutrino oscillations. This is for good reason—neutrino oscillation experiments are typically only concerned with neutrino flavor at production and detection, inferring the process $\nu_\alpha \rightarrow \nu_\beta$ (or $\bar{\nu}_\alpha \rightarrow \bar{\nu}_\beta$) from the accompanying charged lepton ℓ_α^+ (ℓ_α^-) at production and ℓ_β^- (ℓ_β^+) at detection. In many cases there is no detector at the neutrino source to identify the initial composition of flavors—this and the associated energy distribution must be inferred from separate measurements and Monte Carlo simulations [83–85]. Often there is also no way to determine the charge of the outgoing lepton at the far detector and therefore to discern the incoming lepton as a neutrino or antineutrino. This probe of lepton number is not a priority for most oscillation experiments because "$\nu_\alpha \rightarrow \bar{\nu}_\beta$" is heavily suppressed if the mechanism is the standard Majorana neutrino helicity reversal [86–89]. As any other helicity suppression this introduces a factor of $\sim(m_\nu/E_\nu)^2$ to the rate of the process, where mν is the neutrino mass scale and Eν is the neutrino energy. Hence the small neutrino masses (mν~0.1 eV) and typical large oscillation experiment neutrino energies (Eν~5 MeV–2 GeV) combine to suppress the magnitude of the and $\bar{\nu}\alpha \rightarrow \nu\beta$ processes by $\sim 10^{-21}$–10^{-16}. On the other hand, the highly suppressed amplitude of such a process can be exploited—any positive signal of LNV such as an excess of "wrong"-signed charged leptons at the far detector would then strongly imply NP. Experiments that have been sensitive to the charge of outgoing leptons, such as the long-baseline (LBL) oscillation experiment MINOS and the LBL reactor/solar oscillation experiment KamLAND, can thus be used to constrain LNV NSI.

In this paper we begin Sec. II with a brief discussion and derivation of the $\nu_\alpha \rightarrow \nu_\beta$ oscillation probability in quantum mechanics (QM) and quantum field theory (QFT). Using the latter we will study the $\nu_\alpha \rightarrow \bar{\nu}_\beta$ process for Majorana neutrinos, obtaining the expected $\sim(m_\nu/E_\nu)^2$ suppression of the total rate. We also show that the total rate cannot be factorized into an oscillation probability in a similar way to $\nu_\alpha \rightarrow \nu_\beta$ oscillations. In Sec. III we consider the impact of an LNV NSI—rather than the well-studied LNC NSI—on neutrino oscillations. We will demonstrate that the total rate is no longer suppressed and factorizes when the chirality of the production and detection vertices are opposite. We write down a general expression for the nonstandard oscillation probability and a simplified expression in the two-neutrino (2ν) mixing approximation, specifically for the ν_μ–ν_τ sector. This allows us in Sec. IVA to use a limit from the MINOS experiment on the $\nu_\mu \rightarrow \bar{\nu}_\mu$ process to place bounds on the simplified 2ν parameter space of this effective model. We generalize to the three-neutrino (3ν) mixing scheme and reevaluate constraints from MINOS in Sec. IVB along with those from the KamLAND experiment in Sec. IVC. We compare these constraints to those from microscopic LNV processes such as 0νββ decay, μ^-–e^+ conversion, and radiative neutrino masses in Sec. IVD. To conclude, we summarize our results and briefly outline the potential for future oscillation experiments with similar sensitivity to improve on these bounds.

SUMMARY OF NEUTRINO OSCILLATIONS IN QUANTUM MECHANICS AND QUANTUM FIELD THEORY

Neutrino oscillations are derived straightforwardly in QM. Their origin lies in the mismatch between kinetic and charged-current (CC) interaction terms in the SM Lagrangian,

$$\mathcal{L}_{\text{SM}} \supset \left(\frac{1}{2}\right)\bar{\nu}_i(i\partial\!\!\!/ - m_i)\nu_i - \frac{g}{\sqrt{2}}(\bar{\nu}_\alpha \gamma^\mu P_L \ell_\alpha W_\mu + \text{H.c.}), \quad (1)$$

where P_L is the left-handed (LH) chirality projector and we have employed four-component spinor notation for the fields. The kinetic term is diagonal in the basis of definite neutrino mass, labeled by the index i, while the interaction term is diagonal in the basis in which the neutrino and associated charged lepton have the same flavor, labeled by the index α.

The Lagrangian in Eq. (1) is valid for both Dirac and Majorana neutrinos (up to the indicated factor of 1/2 in the Majorana kinetic term). The former and latter are defined by

$$\nu^D = \nu_L + \nu_R, \qquad \nu^M = \nu_L + (\nu_L)^c. \tag{2}$$

Here ν^D has a new RH component ν^R, whereas ν^M has the charge conjugate of its LH component (up to an arbitrary intrinsic charge parity which is commonly set to unity). For Dirac neutrinos the SM CC interaction term shown in Eq. (1) creates negative helicity neutrinos $|\nu(q,-)\rangle$ and annihilates positive helicity antineutrinos $|\bar{\nu}(q,+)\rangle$. For Majorana neutrinos $|\bar{\nu}(q,+)\rangle$ is equivalent to $|\nu(q,+)\rangle$. The creation and annihilation of the other 2 degrees of freedom in the Dirac case, or the wrong helicity degree of freedom in the Majorana case, are suppressed by $\sim(m_\nu/E_\nu)$ at the amplitude level [89].

The following discussion is valid in both cases because the interaction Lagrangian remains the same. The fields in the flavor basis can be rotated to those in the mass basis using the Pontecorvo-Maki-Nakagawa-Sakata mixing matrix U,

$$\nu_\alpha(x) = \sum_i U_{\alpha i}\nu_i(x). \tag{3}$$

While oscillation experiments produce neutrinos in a particular flavor eigenstate $|\nu\alpha\rangle$, for oscillations to take place this must be a coherent superposition of the mass eigenstates. The time evolution of each massive state is governed by the Schrödinger equation, resulting in

$$\begin{aligned}|\nu_\alpha(t)\rangle &= \sum_i U^*_{\alpha i} e^{-iE_q t}|\nu_i(t_0)\rangle \\ &= \sum_i \sum_\beta U^*_{\alpha i} e^{-iE_q t} U_{\beta i}|\nu_\beta(t_0)\rangle,\end{aligned} \tag{4}$$

where $E_q = \sqrt{|\mathbf{q}|^2 + m_i^2} \approx |\mathbf{q}| + \frac{m_i^2}{2|\mathbf{q}|}$ is the energy of each ultrarelativistic massive neutrino with mass mi and shared three-momentum q. An oscillation probability can be derived by computing the square of the overlap between the time-evolved initial flavor state $|\nu_\alpha(T)\rangle$ and an arbitrary final flavor state $|\nu\beta\rangle$,

$$P_{\nu_\alpha \to \nu_\beta}(L.E_{\mathbf{q}}) = |\langle \nu_\beta | \nu_\alpha(T)\rangle|^2 = \left|\sum_i U^*_{\alpha i} U_{\beta i} e^{iE_{\mathbf{q}}L}\right|^2 = \sum_i |U_{\alpha i}|^2 |U_{\beta i}|^2 + 2\mathrm{Re}\sum_{i<j} U^*_{\alpha i} U_{\beta i} U_{\alpha j} U^*_{\beta j} e^{-i\frac{\Delta m^2_{ij}}{2E_{\mathbf{q}}}L}, \tag{5}$$

where we have taken T≃L and L is the oscillation baseline.

This simple derivation was the first attempt to quantify oscillation phenomena in a quantum mechanical framework [90–92]. It was soon realized, however, that this description relies on numerous unphysical assumptions, namely the use of plane wave states. Assuming the propagating |νi⟩ states to be plane waves of equal momenta q forces the external particles at production to have definite energies and momenta. Energy-momentum conservation at production is then in tension with the creation of three |νi⟩ with different energies $E_{\mathbf{q}} = \sqrt{|\mathbf{q}|^2 + m_i^2}$ [93]. While it is possible to derive Eq. (5) without the equal momentum assumption, an overall uncertainty in the energy and momentum of the neutrino mass eigenstates is still a necessary component [94,95].

Any rigorous treatment of oscillation in QM must therefore describe the |νi⟩ states with wave packets, which is also known as the internal wave packet model [96–98]. As Eq. (5), the oscillation probability is proportional to the overlap of the wave packets. The loss of coherence seen at long distance in oscillation experiments qualitatively arises from dispersion of the wave packets, which propagate at different group velocities $\mathbf{v} = \frac{\partial E_{\mathbf{q}}}{\partial \mathbf{q}}\big|_{\mathbf{q}=\mathbf{Q}}$ with Q the mean momentum [99–102].

Despite these improvements there are still fundamental issues with this picture. First, it is not apparent what shape the neutrino wave packets should take. Second, a finite energy-momentum uncertainty requires the localization in space and time of the production and detection interaction processes, which is not accounted for [101]. It is also difficult to define a meaningful Fock space for the flavor eigenstate neutrinos [103]. Finally, the QM approach does not consider the possible entanglement between the outgoing ν_i and $\ell_\alpha^\pm$ at production [104].

A fully consistent framework in which to study neutrino oscillations is QFT, also known as the external wave packet model [105]. In this formalism the entire production, propagation, and detection process can be described by a macroscopic Feynman diagram, as shown in Fig. 1. The external interacting particles are wave packets centered on the production and detection points xP and xD, while the neutrinos are described by an internal propagator. The neutrino wave packets used in QM can be derived from the external particle wave packets in QFT; in QFT, however, the coherence conditions at each stage in the process are explicit [102,106,107].

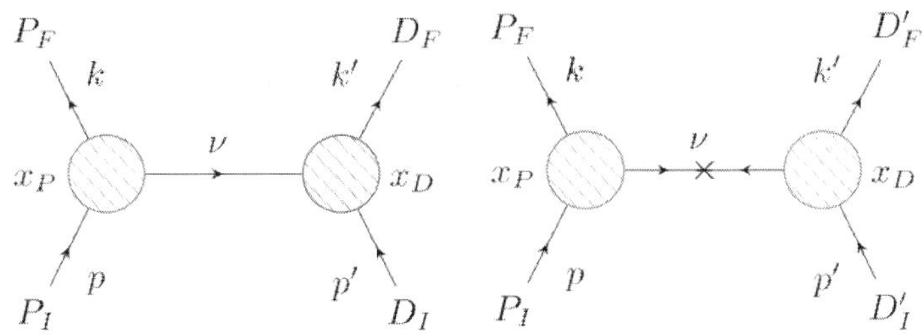

Figure 1. Generic Feynman diagrams depicting the $\nu_\alpha \to \nu_\beta$ process (left) and $\nu_\alpha \to \bar{\nu}_\beta$ process (right), which requires a helicity reversal if the interaction vertices are SM-like.

To calculate an oscillation probability in QFT we must first compute an overall rate. To do this one constructs the S-matrix element at first order in the Fermi coupling $G_F/\sqrt{2}=g^2/8m^2_W$,

$$iA_{\nu_\alpha\to\nu_\beta}(T,\mathbf{L}) = \langle P_F, D_F|\hat{T}\left\{\int d^4x_1 \int d^4x_2\, \mathcal{L}_P(x_1)\mathcal{L}_D(x_2)\right\}|P_I, D_I\rangle, \quad (6)$$

where P_I, D_I, P_F, and D_F are the initial and final state particles at production and detection, respectively, and x_1 and x_2 are spacetime points in the vicinity of x_P and x_D. This tree level process is depicted to the left of Fig. 1. The external asymptotic states $\psi \in \{P_I, D_I, P_F, D_F\}$ are now described by the wave packets

$$|\psi\rangle = \int [dp] f^{(\psi)}_{\mathbf{P}_\psi}(\mathbf{p})|\psi,\mathbf{p}\rangle, \qquad [dp] = \frac{d^3p}{(2\pi)^3\sqrt{2E}}. \quad (7)$$

where $f^{(\psi)}_{\mathbf{P}_\psi}(\mathbf{p})$ is the momentum distribution function of an external particle ψ with mean momentum $P\psi$. To create and annihilate an internal neutrino the production and detection Lagrangian terms must take the general form

$$\mathcal{L}_P(x) = \sum_i U^*_{\alpha i}\bar{\nu}_i(x)\tilde{\mathcal{L}}_P(x),$$
$$\mathcal{L}_D(x) = \sum_i \tilde{\mathcal{L}}_D(x)U_{\beta i}\nu_i(x), \tag{8}$$

where $\tilde{\mathcal{L}}_P(x)$ and $\tilde{\mathcal{L}}_D(x)$ are the "reduced" Lagrangian interaction terms for production and detection with the neutrino fields and Pontecorvo-Maki-Nakagawa-Sakata matrix elements removed.

The total rate is proportional to the spin-averaged S-matrix element squared,

$$\Gamma^{\text{tot}}_{\nu_\alpha \to \nu_\beta} \propto \langle |\mathcal{A}_{\nu_\alpha \to \nu_\beta}(T, \mathbf{L})|^2 \rangle, \tag{9}$$

which can be expanded as

$$\begin{aligned}\langle |\mathcal{A}_{\nu_\alpha \to \nu_\beta}|^2 \rangle &= \text{Tr}\left|\sum_i U^*_{\alpha i}U_{\beta i}\mathcal{A}_i\right|^2 \\ &= \sum_i |U_{\alpha i}|^2|U_{\beta i}|^2\text{Tr}|\mathcal{A}_i|^2 \\ &\quad + 2\text{Re}\sum_{i<j} U^*_{\alpha i}U_{\beta i}U_{\alpha j}U^*_{\beta j}\text{Tr}[\mathcal{A}^\dagger_j\mathcal{A}_i],\end{aligned} \tag{10}$$

where T_r denotes the Dirac trace. The Ai factors are

$$\mathcal{A}_i = \int \frac{d^4q}{(2\pi)^4}\tilde{\Phi}_D(q)\frac{\not{q} + m_i}{q^2 - m_i^2 + i\epsilon}\tilde{\Phi}_P(q)e^{-iq\cdot(x_D - x_P)}, \tag{11}$$

where $\tilde{\Phi}_P$ and $\tilde{\Phi}_D$ are integrals quantifying the overlap of external wave packets at production and detection, respectively, explicitly written as

$$\begin{aligned}\tilde{\Phi}_P(q) &= \int d^4x'_1 e^{iq\cdot x'_1}\int [dp][dk] f^{(P_I)}_{\mathbf{p}}(\mathbf{p}) f^{(P_F)*}_{\mathbf{K}} \\ &\quad \times (\mathbf{k})e^{-i(p-k)\cdot x'_1}\tilde{\mathcal{M}}_P, \\ \tilde{\Phi}_D(q) &= \int d^4x'_2 e^{-iq\cdot x'_2}\int [dp'][dk'] f^{(D_I)}_{\mathbf{p}'}(\mathbf{p}') f^{(D_F)*}_{\mathbf{K}'} \\ &\quad \times (\mathbf{k}')e^{-i(p'-k')\cdot x'_2}\tilde{\mathcal{M}}_D,\end{aligned} \tag{12}$$

where $x'_1 = x_1 - x_P,\ x'_2 = x_2 - x_D$, and $\tilde{\mathcal{M}}_P, \tilde{\mathcal{M}}_D$ are reduced matrix elements defined by

$$\tilde{\mathcal{M}}_P = \langle P_F, \mathbf{k}|\tilde{\mathcal{L}}_P(x_1)|P_I, \mathbf{p}\rangle,$$
$$\tilde{\mathcal{M}}_D = \langle D_F, \mathbf{k}'|\tilde{\mathcal{L}}_D(x_2)|D_I, \mathbf{p}'\rangle. \tag{13}$$

The trace appearing in the interference term of Eq. (10) is now

$$
\begin{aligned}
&\mathrm{Tr}[\mathcal{A}_j^\dagger \mathcal{A}_i] \\
&= \int \frac{d^4q}{(2\pi)^4} \int \frac{d^4q'}{(2\pi)^4} \frac{\mathrm{Tr}[\underline{\tilde{\Phi}}_D \tilde{\Phi}_D (\not{q} + m_i) \tilde{\Phi}_P \underline{\tilde{\Phi}}_P (\not{q}' + m_j)]}{(q^2 - m_i^2 + i\epsilon)(q'^2 - m_j^2 + i\epsilon)} \\
&\quad \times e^{-i(q-q')\cdot(x_D - x_P)}, \qquad (14)
\end{aligned}
$$

where the underline of $\underline{\tilde{\Phi}}_{P,D}$ is shorthand for $\tilde{\Phi}^\dagger_{P,D}\gamma^0$. We now require the well-known limit of the d3q integrals as L→∞, first given in Ref. [108]. If ψ(q) is a twice differentiable function, for large L=|L| and A_i>0 the integral

$$
\begin{aligned}
\int \frac{d^3q}{(2\pi)^3} \frac{\psi(\mathbf{q}) e^{i\mathbf{q}\cdot\mathbf{L}}}{A_i - \mathbf{q}^2 + i\epsilon} &= -\frac{1}{4\pi L} \psi\left(\sqrt{A_i}\frac{\mathbf{L}}{L}\right) e^{i\sqrt{A_i}L} \\
&\quad + \mathcal{O}\left((\sqrt{A_i}L)^{-\frac{3}{2}}\right). \qquad (15)
\end{aligned}
$$

For A_i<0 the integral falls as L^{-2} and can be neglected. Applying Eq. (15) to Eq. (14) with $A_i = E_{\mathbf{q}}^2 - m_i^2$ gives

$$
\begin{aligned}
\mathrm{Tr}[\mathcal{A}_j^\dagger \mathcal{A}_i] = \frac{1}{64\pi^4 L^2} &\int dE_{\mathbf{q}} \int dE'_{\mathbf{q}} \\
&\times \mathrm{Tr}[\underline{\tilde{\Phi}}_{jD} \tilde{\Phi}_{iD} (\not{q}_i + m_i) \tilde{\Phi}_{iP} \underline{\tilde{\Phi}}_{jP} (\not{q}_j + m_j)] \\
&\times e^{i(|\mathbf{q}_i| - |\mathbf{q}_j|)L}. \qquad (16)
\end{aligned}
$$

which has effectively set the neutrinos to be on-shell. As has been stated before in the literature [109,110], if the production and detection processes are of the same chirality, the trace on the right-hand side of Eq. (16) can be factorized in the relativistic limit (m_i≈0) as

$$
\begin{aligned}
&\mathrm{Tr}[\underline{\tilde{\Phi}}_{jD} \tilde{\Phi}_{iD} (\not{q}_i + m_i) \tilde{\Phi}_{iP} \underline{\tilde{\Phi}}_{jP} (\not{q}_j + m_j)] \\
&\quad \approx \mathrm{Tr}[\tilde{\Phi}_{iD} (\not{q}_i + m_i) \underline{\tilde{\Phi}}_{jD}] \mathrm{Tr}[\underline{\tilde{\Phi}}_{jP} (\not{q}_j + m_j) \tilde{\Phi}_{iP}]. \qquad (17)
\end{aligned}
$$

Furthermore, writing $(\not{q} + m)$ as a spinor sum and expanding $|\mathbf{q}| = \sqrt{E_{\mathbf{q}}^2 - m_i^2}$ for m_i≈0, i.e.,

$$
\begin{aligned}
&P_L(\not{q} + m_i)P_R \\
&= u_{iL}(q,-)\bar{u}_{iL}(q,-) + u_{iL}(q,+)\bar{u}_{iL}(q,+) \\
&= (q^0 + |\mathbf{q}|)\begin{pmatrix} 0 & 0 & 0 & 0 \\ 0 & 0 & 0 & 1 \\ 0 & 0 & 0 & 0 \\ 0 & 0 & 0 & 0 \end{pmatrix} + (q^0 - |\mathbf{q}|)\begin{pmatrix} 0 & 0 & 1 & 0 \\ 0 & 0 & 0 & 0 \\ 0 & 0 & 0 & 0 \\ 0 & 0 & 0 & 0 \end{pmatrix} \\
&\approx 2E_{\mathbf{q}}\left\{1 - \left(\frac{m_i}{2E_{\mathbf{q}}}\right)^2\right\}\begin{pmatrix} 0 & 0 & 0 & 0 \\ 0 & 0 & 0 & 1 \\ 0 & 0 & 0 & 0 \\ 0 & 0 & 0 & 0 \end{pmatrix} \\
&\quad + 2E_{\mathbf{q}}\left(\frac{m_i}{2E_{\mathbf{q}}}\right)^2\begin{pmatrix} 0 & 0 & 1 & 0 \\ 0 & 0 & 0 & 0 \\ 0 & 0 & 0 & 0 \\ 0 & 0 & 0 & 0 \end{pmatrix}. \qquad (18)
\end{aligned}
$$

we can see that for $\nu_\alpha \to \nu_\beta$ the propagation of positive helicity neutrinos is doubly suppressed by $(mi/2E_q)^2$ compared to the propagation of negative helicity neutrinos.

Neglecting the positive helicity neutrino contribution to the spinor sum in Eq. (18), the negative helicity spinors can be absorbed into the overlap integrals on either side in the traces of Eq. (17),

$$\Phi_P \equiv \frac{\bar{u}_{jL}^{(-)}(q)}{\sqrt{2E_{\mathbf{q}}}}\tilde{\Phi}_{iP}, \qquad \Phi_P^* \equiv \tilde{\Phi}_{jP}\frac{u_{jL}^{(-)}(q)}{\sqrt{2E_{\mathbf{q}}}}, \tag{19}$$

where in the relativistic limit the mass eigenstate indices can be neglected [102]. The trace appearing in the interference term can now be expressed as

$$\begin{aligned}\mathrm{Tr}[\mathcal{A}_j^\dagger \mathcal{A}_i] = \frac{1}{64\pi^4 L^2}\int dE_{\mathbf{q}}\int dE'_{\mathbf{q}}\langle|\Phi_P|^2\rangle \\ \times \langle|\Phi_D|^2\rangle 4E_{\mathbf{q}}E'_{\mathbf{q}}e^{i(|\mathbf{q}_i|-|\mathbf{q}_j|)L},\end{aligned} \tag{20}$$

and we see that the contributions from production, propagation, and detection have factorized at the squared amplitude level. The total rate for the $\nu_\alpha \to \nu_\beta$ process is now related to the differential production flux, oscillation probability, and detection cross section as

$$\begin{aligned}\Gamma^{\mathrm{tot}}_{\nu_\alpha\to\nu_\beta}(L, E_{\mathbf{q}}) = \frac{1}{4\pi L^2}\int dE_{\mathbf{q}}\frac{d\Gamma^{\mathrm{prod}}_{\nu_\alpha}(E_{\mathbf{q}})}{dE_{\mathbf{q}}} \\ \cdot P_{\nu_\alpha\to\nu_\beta}(L, E_{\mathbf{q}})\cdot\sigma^{\mathrm{det}}_{\nu_\beta}(E_{\mathbf{q}}),\end{aligned} \tag{21}$$

where we have neglected experimental considerations such as the detector efficiency and fiducial volume [111]. The oscillation probability can now be solved for through rearrangement of Eq. (21). As Ref. [102] shows, in the case of continuous fluxes of incoming particles, the differential production flux and detection cross section have the proportionality

$$\begin{aligned}\frac{d\Gamma^{\mathrm{prod}}_{\nu_\alpha}}{dE_{\mathbf{q}}} \propto \sum_i |U_{\alpha i}|^2\langle|\Phi_P|^2\rangle E_{\mathbf{q}}|\mathbf{q}_i|, \\ \sigma^{\mathrm{det}}_{\nu_\beta} \propto \sum_j |U_{\beta j}|^2\langle|\Phi_D|^2\rangle E_{\mathbf{q}}|\mathbf{q}_j|^{-1}.\end{aligned} \tag{22}$$

Using that $\langle|\Phi_P|^2\rangle$ and $\langle|\Phi_D|^2\rangle$ are independent of the mass mi, while also taking $|q_i|\approx|q_j|$ in the relativistic or quasidegenerate mass limit (equivalent to the inequality $||q_i|-|q_j||\ll|q_i|,|q_j|$), any dependence $P_{\nu\alpha\to\nu\beta}$ has on the specific form of the wave packets cancels [102]. In this limit $P_{\nu\alpha\to\nu\beta}$ is given by Eq. (5), confirming the result of the naive QM approach under the above conditions.

For Majorana neutrinos the process $\nu_\alpha \to \bar{\nu}_\beta$ is possible and suppressed by $(m_i/2E_q)^2$. To see this one can construct an amplitude such as Eq. (6) where the production and detection Lagrangian terms are the same

$$\mathcal{L}_P(x) = \mathcal{L}_D(x) = \sum_i U^*_{\alpha i}\bar{\nu}_i(x)\tilde{\mathcal{L}}(x), \tag{23}$$

which is nonzero for Majorana neutrinos because $\bar{\nu}_i(x)$ both creates and annihilates $|\nu_i\rangle$. The Feynman diagram for this process is depicted in Fig. 1 (right). The amplitude can again be squared and expressed as a sum of amplitudes Ai. Using the Majorana fermion Feynman rules of Ref. [112], the amplitudes Ai are identical to Eq. (11) but with Φ_D replaced by Φ^M_D. Φ_D contains the reduced matrix element $\tilde{\mathcal{M}}_D \propto \bar{u}_\beta(p_\beta)\Gamma$, whereas Φ^M_D contains $\tilde{\mathcal{M}}_D \propto \bar{u}_\beta(p_\beta)$, where iΓ is the vertex factor for the detection process and C is the charge conjugation matrix. For left-handed SM CC interactions the trace $\mathrm{Tr}[\bar{\Phi}^M_{iD}(\not{q}_i + m_i)\,\Phi_{iP}\bar{\Phi}_{jP}(\not{q}_j + m_j)\bar{\Phi}^M_{jD}]$ appearing in the squared amplitude now contains the factor

$$\begin{aligned} P_R(\not{q}_i + m_i)P_R &= u_{iR}(q,-)\bar{u}_{iL}(q,-) + u_{iR}(q,+)\bar{u}_{iL}(q,+) \\ &= 2E_q\left(\frac{m_i}{2E_q}\right)\begin{pmatrix} 0&0&0&0\\0&0&0&0\\0&0&0&0\\0&0&0&1\end{pmatrix} \\ &+ 2E_q\left(\frac{m_i}{2E_q}\right)\begin{pmatrix} 0&0&0&0\\0&0&0&0\\0&0&1&0\\0&0&0&0\end{pmatrix} \end{aligned} \tag{24}$$

squared, and so is singly suppressed by $(m_i/2E_q)^2$ compared to the propagation of negative helicity neutrinos in Eq. (18) [113–115].

We also see that in the relativistic limit the trace vanishes instead of factorizing into components corresponding to production, oscillation, and detection. Strictly speaking it is therefore impossible to define an oscillation probability if helicity reversal is the dominant mechanism contributing to $\nu_\alpha \to \bar{\nu}_\beta$, only a probability for the entire process [109].

NONSTANDARD LEPTON NUMBER VIOLATING INTERACTIONS

We will now consider interactions at production and detection which are different from the usual SM CC interaction. CC-like and neutral current (NC)–like NSI which conserve lepton number, but introduce a new source of LFV to the SM have been studied extensively in the literature; see Refs. [50,81,82] for recent reviews. Data from LBL experiments such as MINOS, NOνA, T2K, and KamLAND and short-baseline (SBL) reactor experiments such as Daya Bay, RENO, and Double Chooz have already been exploited to probe the flavor structure of the ϵ coefficients parametrizing the NSI [57,61,116–120]. Future prospects of next-generation experiments such as DUNE, T2HK, and T2HKK have also been explored [121–132]. It is solely the flavor of the charged leptons at detection that enables these constraints to be made, with the charge of the outgoing charged lepton being irrelevant. The Dirac or Majorana nature of the light neutrinos is therefore not probed.

If an experiment was indeed sensitive to the sign of the charged lepton $\ell_\beta^\pm$ produced in the CC-like interaction of the incoming neutrino at detection, it would be able to distinguish between neutrinos and antineutrinos, or in other words negative and positive helicity Majorana neutrinos. This has been possible in the past—a magnetized far detector was used to determine the charge from the curvature of tracks in the steel scintillator near and far detectors of MINOS. The prompt energy deposit from a ℓ^+_β and neutron capture on hydrogen, measurable for KamLAND, was also used as a distinct signal from the ℓ^-_β case [133]. It is therefore worth discussing the LNV equivalents of the CC-like and NC-like NSI discussed previously, as the nonobservation of an excess of wrong-signed charged leptons by these experiments is able to say something about the possible size of an LNV equivalent ϵ coefficient.

The operators to the left of Table I trigger the standard LNC NSI studied in the literature, regardless of neutrinos being Dirac or Majorana. These are given in the notation of Refs. [134,135], which enumerate the effective operators of dimension d≤6 generated below the electroweak scale. The authors determine the matching conditions and anomalous dimensions required to evolve the Wilson coefficients of these low energy effective field theory (LEFT) operators up to the usual SM effective operators satisfying SU(3)c×SU(2)L×U(1)$_Y$. The LNC operators in Table I are matched to ΔL=0 operators in the 59 operator basis of the SM effective field theory (SMEFT). The LNV operators to the right of Table I can also trigger an interaction at detection for an incoming neutrino. It is not possible to match these operators to ΔL=0 dimension-six SMEFT operators above the electroweak scale, but instead to dimension-seven |ΔL|=2 operators with an additional Higgs field H to conserve hypercharge U(1)$_Y$. These dimension-seven operators have been considered previously along with the complete list of LNV odd-dimensional effective operators up to dimension eleven in Ref. [51]. The operators in Table I correspond to terms in SU(2)L expansions of the operators $\mathcal{O}_{3_a}$, $\mathcal{O}_{4_b}$, and $\mathcal{O}_8$. The dimension-seven operators have also been considered more specifically in Ref. [136], but parametrized there in terms of a Wilson coefficient $\tilde{C}^{ud} = C^{ud}/\Lambda^3_{NP}$, where ΛNP is the scale of NP.

Table 1. LNC (left) and LNV (right) dimension-six CC-like interactions in the LEFT notation of Ref. [134]. The indices p, r, s, and t denote flavor.

$\Delta L = 0 + \text{H.c.}$		$\lvert\Delta L\rvert = 2 + \text{H.c.}$	
$\mathcal{O}^{V,LL}_{\nu edu}$	$(\bar{\nu}_{Lp}\gamma^\mu e_{Lr})(\bar{d}_{Ls}\gamma_\mu u_{Lt})$	$\mathcal{O}^{S,LL}_{\nu edu}$	$(\nu^T_{Lp}Ce_{Lr})(\bar{d}_{Rs}u_{Lt})$
$\mathcal{O}^{V,LR}_{\nu edu}$	$(\bar{\nu}_{Lp}\gamma^\mu e_{Lr})(\bar{d}_{Rs}\gamma_\mu u_{Rt})$	$\mathcal{O}^{T,LL}_{\nu edu}$	$(\nu^T_{Lp}C\sigma^{\mu\nu}e_{Lr})(\bar{d}_{Rs}\sigma_{\mu\nu}u_{Lt})$
$\mathcal{O}^{S,RR}_{\nu edu}$	$(\bar{\nu}_{Lp}e_{Rr})(\bar{d}_{Ls}u_{Rt})$	$\mathcal{O}^{S,LR}_{\nu edu}$	$(\nu^T_{Lp}Ce_{Lr})(\bar{d}_{Ls}u_{Rt})$
$\mathcal{O}^{T,RR}_{\nu edu}$	$(\bar{\nu}_{Lp}\sigma^{\mu\nu}e_{Rr})(\bar{d}_{Ls}\sigma_{\mu\nu}u_{Rt})$	$\mathcal{O}^{V,RL}_{\nu edu}$	$(\nu^T_{Lp}C\gamma^\mu e_{Rr})(\bar{d}_{Ls}\gamma_\mu u_{Lt})$
$\mathcal{O}^{S,RL}_{\nu edu}$	$(\bar{\nu}_{Lp}e_{Rr})(\bar{d}_{Rs}u_{Lt})$	$\mathcal{O}^{V,RR}_{\nu edu}$	$(\nu^T_{Lp}C\gamma^\mu e_{Rr})(\bar{d}_{Rs}\gamma_\mu u_{Rt})$

In this paper we will adopt the following notation. A general Lagrangian, normalized to the Fermi interaction, and which can trigger an LNV interaction at production or detection, can be written as

$$\mathcal{L}_{P,D} = -\frac{G_F}{\sqrt{2}}\left\{ j_{V-A}J^\dagger_{V-A} + \sum_{\rho,\sigma}\varepsilon^{(\rho,\sigma)} j_{(\rho)}J^\dagger_{(\sigma)} \right\} + \text{H.c.}, \tag{25}$$

where the second term includes all possible Lorentz contractions of the leptonic current $j_{(\rho)} = (\bar{\nu}\mathcal{O}^{(\rho)}\ell)$ and hadronic current $J_{(\sigma)} = (\bar{u}\mathcal{O}^{(\sigma)}d)$, where (ρ,σ) run over the usual scalar (S), pseudoscalar (P), vector (V), axial vector (A), and tensor (T) Dirac structures. We choose linear combinations of these proportional to the chirality projection operators P_L and P_R.

The $\epsilon^{(\rho,\sigma)}$ coefficients parametrize the strength of the NSI compared to $G_F/\sqrt{2}$. We can define these coefficients as matrices in the flavor basis ($\epsilon^{(\rho,\sigma)}$) or mass basis ($\gamma^{(\rho,\sigma)}$) of the neutrino field, related by the rotation

$$\varepsilon^{(\rho,\sigma)}_{\beta\alpha} \equiv \sum_i U_{\alpha i}\gamma^{(\rho,\sigma)}_{\beta i}, \tag{26}$$

where U is the usual Pontecorvo-Maki-Nakagawa-Sakata matrix. It is now straightforward to see the effect of these interactions on oscillations. If the production and detection leptonic currents are of the same chirality, $\nu_\alpha \rightarrow \nu_\beta$ will be unsuppressed, while for Majorana neutrinos $\nu_\alpha \rightarrow \bar{\nu}_\beta$ will be suppressed by $(m_i/2E_q)^2$. If the production and detection are of opposite chirality, $\nu_\alpha \rightarrow \bar{\nu}_\beta$ will now instead be factorizable and suppressed by $|\epsilon^{(\rho,\sigma)}|^2$. A commonly used example is a V−A (L) leptonic current at production and a V+A (R) leptonic current at detection arising from a LR symmetric model with an additional broken SU(2)R gauge symmetry [137].

An LBL oscillation experiment sensitive to the sign of the outgoing lepton at the far detector, while not detecting the $\ell^{\pm}_{\alpha}$ at the production process (occurring in the beam pipe), could measure the ratio

$$R_{\alpha\beta} \equiv \frac{N_{\ell_\beta^+}}{N_{\ell_\beta^-}} = \frac{\Gamma_{\nu_\alpha \to \bar{\nu}_\beta} + \Gamma_{\bar{\nu}_\alpha \to \bar{\nu}_\beta}}{\Gamma_{\nu_\alpha \to \nu_\beta} + \Gamma_{\bar{\nu}_\alpha \to \nu_\beta}}, \quad (27)$$

where $N_{\ell_\beta^\pm}$ is the number of detected $\ell^{\pm}_{\beta}$ at the far detector. If we assume NP to be the cause of the first term in the numerator of Eq. (27), and that this total rate is factorizable, the terms in $R_{\alpha\beta}$ can be decomposed as

$$R_{\alpha\beta} = \frac{\int dE_{\mathbf{q}} \sum_{\rho,\sigma} \left(\frac{d\Gamma_{\nu_\alpha}}{dE_{\mathbf{q}}} \cdot P^{(\rho,\sigma)}_{\nu_\alpha \to \bar{\nu}_\beta} \cdot \sigma_{\bar{\nu}_\beta} + \frac{d\Gamma_{\bar{\nu}_\alpha}}{dE_{\mathbf{q}}} \cdot P^{(\rho,\sigma)}_{\bar{\nu}_\alpha \to \bar{\nu}_\beta} \cdot \sigma_{\bar{\nu}_\beta} \right)}{\int dE_{\mathbf{q}} \sum_{\rho,\sigma} \left(\frac{d\Gamma_{\nu_\alpha}}{dE_{\mathbf{q}}} \cdot P^{(\rho,\sigma)}_{\nu_\alpha \to \nu_\beta} \cdot \sigma_{\nu_\beta} + \frac{d\Gamma_{\bar{\nu}_\alpha}}{dE_{\mathbf{q}}} \cdot P^{(\rho,\sigma)}_{\bar{\nu}_\alpha \to \nu_\beta} \cdot \sigma_{\nu_\beta} \right)}, \quad (28)$$

where the (ρ,σ) superscript denotes a NP effect. All probabilities are given this superscript because the process $\nu_\alpha \to \bar{\nu}_\beta$ reduces the number of neutrinos that undergo να→νβ, and thus the sum $\sum_\beta P^{(\rho,\sigma)}_{\nu_\alpha \to \nu_\beta}$ does not equal unity. The standard oscillation probability must be scaled by a normalization factor depending on the NP and that ensures $\sum_\beta (P^{(\rho,\sigma)}_{\nu_\alpha \to \nu_\beta} + P^{(\rho,\sigma)}_{\nu_\alpha \to \bar{\nu}_\beta}) = 1$,

$$P^{(\rho,\sigma)}_{\nu_\alpha \to \nu_\beta} = \frac{1}{\mathcal{N}^{(\rho,\sigma)}_\alpha} \left| \sum_i U^*_{\alpha i} U_{\beta i} e^{-i\frac{m_i^2}{2E_{\mathbf{q}}}L} \right|^2,$$

$$P^{(\rho,\sigma)}_{\nu_\alpha \to \bar{\nu}_\beta} = \frac{1}{\mathcal{N}^{(\rho,\sigma)}_\alpha} \left| \sum_i U^*_{\alpha i} \gamma^{(\rho,\sigma)}_{\beta i} e^{-i\frac{m_i^2}{2E_{\mathbf{q}}}L} \right|^2, \quad (29)$$

where the normalization factor is

$$\mathcal{N}^{(\rho,\sigma)}_\alpha = 1 + \sum_\beta \left| \sum_i U^*_{\alpha i} \gamma^{(\rho,\sigma)}_{\beta i} e^{-i\frac{m_i^2}{2E_{\mathbf{q}}}L} \right|^2. \quad (30)$$

Expanding the denominator in Eq. (29) it is straightforward to see that $\sum_\beta P^{(\rho,\sigma)}_{\nu_\alpha \to \nu_\beta} \approx 1 - \mathcal{O}(\epsilon^4)$ and $\sum_\beta P^{(\rho,\sigma)}_{\nu_\alpha \to \bar{\nu}_\beta} \approx \mathcal{O}(\epsilon^2) - \mathcal{O}(\epsilon^6)$, so taking $\epsilon \to 0$ recovers the SM prediction. We assume that the NP is a small effect, $\epsilon^{(\rho,\sigma)} \ll 1$, and thus neglect this modification to the probabilities, i.e., set $\mathcal{N}^{(\rho,\sigma)}_\alpha \approx 1$. The factor $\mathcal{N}^{(\rho,\sigma)}_\alpha$ cancels in the ratio $R_{\alpha\beta}$ regardless. As another simplification we take the NSI coefficients to be real in this work.

Expanding the probability to the right of Eq. (29) gives

$$\begin{aligned} P^{(\rho,\sigma)}_{\nu_\alpha \to \bar{\nu}_\beta} &\approx \left| \sum_i U^*_{\alpha i} \gamma^{(\rho,\sigma)}_{\beta i} e^{-i\frac{m_i^2}{2E_{\mathbf{q}}}L} \right|^2 \\ &= \sum_\lambda F^{(\alpha)}_\lambda (L, E_{\mathbf{q}}, \zeta, \eta) (\varepsilon^{(\rho,\sigma)}_{\beta\lambda})^2 \\ &\quad + \sum_{\lambda<\lambda'} G^{(\alpha)}_{\lambda\lambda'} (L, E_{\mathbf{q}}, \zeta, \eta) \varepsilon^{(\rho,\sigma)}_{\beta\lambda} \varepsilon^{(\rho,\sigma)}_{\beta\lambda'}. \end{aligned} \tag{31}$$

where we have rotated the $\gamma^{(\rho,\sigma)}_{\beta i}$ back into the flavor basis using Eq. (26) and λ, λ′ sum over flavor. The number of flavor and mass indices has been kept general and could include sterile states. The $F^{(\alpha)}_\lambda$ and $G^{(\alpha)}_{\lambda\lambda'}$ are functions of the baseline, neutrino energy, generic mixing parameters ζ, and Majorana phases η. For a general mixing matrix U these functions take the form

$$\begin{aligned} F^{(\alpha)}_\lambda (L, E_{\mathbf{q}}, \zeta) &= \sum_i |U_{\alpha i}|^2 |U_{\lambda i}|^2 \\ &\quad + 2\mathrm{Re}\left\{ \sum_{i<j} U^*_{\alpha i} U^*_{\lambda i} U_{\alpha j} U_{\lambda j} e^{-i\frac{\Delta m^2_{ij}}{2E_{\mathbf{q}}}L} \right\}, \\ G^{(\alpha)}_{\lambda\lambda'} (L, E_{\mathbf{q}}, \zeta) &= 2\mathrm{Re}\left\{ \sum_i |U_{\alpha i}|^2 U^*_{\lambda i} U_{\lambda' i} \right\} \\ &\quad + 2\mathrm{Re}\left\{ \sum_{i<j} (U^*_{\alpha i} U^*_{\lambda i} U_{\alpha j} U_{\lambda' j} \right. \\ &\quad \left. + U^*_{\alpha i} U^*_{\lambda' i} U_{\alpha j} U_{\lambda j}) e^{-i\frac{\Delta m^2_{ij}}{2E_{\mathbf{q}}}L} \right\}. \end{aligned} \tag{32}$$

In the 3ν mixing scheme these are complicated functions of the three mixing angles, three squared splittings, and two Majorana phases. If one were to consider atmospheric or accelerator $\nu_\mu \to \bar{\nu}_{\mu,\tau}$ oscillations, the 2ν mixing approximation is valid because the oscillation Hamiltonian is dominated by the atmospheric mass splitting. The mixing matrix in this case is

$$U = \begin{pmatrix} \cos\vartheta & \sin\vartheta e^{i\eta} \\ -\sin\vartheta e^{-i\eta} & \cos\vartheta \end{pmatrix}, \tag{33}$$

where ϑ is the single mixing angle (approximately corresponding to θ_{23}) and η is a Majorana phase [138], while the squared mass splitting is δm^2 (corresponding to Δm223). $F^{(\mu)}_\lambda$ and $G^{(\mu)}_{\lambda\lambda'}$ take on the simplified forms,

$$\begin{aligned} F^{(\mu)}_\mu (L, E_{\mathbf{q}}, \delta m^2, \vartheta, \eta) &= 1 - \sin^2(2\vartheta)\sin^2\varphi, \\ F^{(\mu)}_\tau (L, E_{\mathbf{q}}, \delta m^2, \vartheta, \eta) &= \sin^2(2\vartheta)\sin^2\varphi, \\ G^{(\mu)}_{\mu\tau} (L, E_{\mathbf{q}}, \delta m^2, \vartheta, \eta) &= 2\sin(2\vartheta)\sin^2\varphi \\ &\quad \times (\sin\eta \cot\varphi - \cos\eta \cos(2\vartheta)), \end{aligned} \tag{34}$$

where $\varphi = \eta - \frac{\delta m^2 L}{4E_{\mathbf{q}}}$. We will make use of these functions in the next section when studying the results from MINOS in the 2ν mixing approximation.

CONSTRAINTS ON LEPTON NUMBER VIOLATING NONSTANDARD INTERACTIONS

We will now use the parametrization discussed above to put constraints on the ϵ coefficient parameter space of these LNV NSI. We will first discuss constraints from the MINOS experiment in the 2ν mixing approximation in Sec. IVA, moving onto analysis of both MINOS and KamLAND in the 3ν mixing scheme in Secs. IVB and IVC. In Sec. IVD we will compare these constraints to the more common limits from microscopic LNV processes such as 0νββ decay, μ^-–e^+ conversion, rare meson decays, and the radiative generation of neutrino masses. We summarize all limits in Table III.

A. Constraints from the Minos Experiment in the Two Neutrino Mixing Approximation

The MINOS experiment initially took data from 2005 to 2012 and used the low energy NuMI beam to detect neutrinos with a near detector at Fermilab and a far detector at a baseline of L=735 km from the source at the Soudan mine [139]. After a break the experiment continued from 2013 to 2016 as MINOS+, using the medium energy NuMI beam [140]. The experiment observed the disappearance of ν_μ produced from $\pi+$ decays (in the focusing beam configuration) and $\bar{\nu}_\mu$ from π^- decays (defocusing), allowing the atmospheric mixing parameters dominating $\nu_\mu \rightarrow \nu_\mu$ disappearance to be probed. The experiment also confirmed ν_e and $\bar{\nu}_e$ appearance, constraining the reactor mixing angle θ_{13}. Most importantly for our discussion, charged lepton sign identification was possible in the near and far detectors through the use of 1.3 T toroidal magnetic fields— $\nu_\mu, \bar{\nu}_\mu, \nu_e$, and $\bar{\nu}_e$ events could therefore be distinguished from the curvature of the outgoing μ^-, μ^+, e^-, and e^+ tracks, respectively.

Before MINOS began taking data, the expected fluxes of ν_μ and $\bar{\nu}_\mu$ in the focusing and defocusing configurations were predicted to high precision from hadron production data and *in situ* measurements. These were more recently updated in Ref. [85]. In the focusing configuration the background of $\bar{\nu}_e$ produced from pion decays upstream of the target and avoiding deflection by the magnetic field is non-negligible and an important systematic error to correct [141]. There are also $\bar{\nu}_e$ produced downstream from secondary interactions in the beam pipe wall [142].

The ratio in Eq. (27) can be split into a signal part $S_{\mu\mu}$ arising from the $\nu_\mu \to \bar{\nu}_\mu$process and a background part $B_{\mu\mu}$ arising from the standard oscillation of the background antineutrinos $\bar{\nu}_\mu \to \bar{\nu}_\mu$. The MINOS analysis of Ref. [142] removes the predicted energy-dependent value of $B_{\mu\mu}$ from the total measured $R_{\mu\mu}$ and derives the constraint $S_{\mu\mu} \lesssim 0.026$. From

$$S_{\mu\mu} \approx \frac{\int dE_q \sum_{\rho,\sigma} \frac{d\Gamma_{\nu_\mu}}{dE_q} \cdot P^{(\rho,\sigma)}_{\nu_\mu \to \bar{\nu}_\mu} \cdot \sigma_{\bar{\nu}_\mu}}{\int dE_q \sum_{\rho,\sigma} \frac{d\Gamma_{\nu_\mu}}{dE_q} \cdot P^{(\rho,\sigma)}_{\nu_\mu \to \nu_\mu} \cdot \sigma_{\nu_\mu}} \lesssim 0.026, \tag{35}$$

we can put corresponding constraints on the range of possible $\varepsilon^{(\rho,\sigma)}_{\mu\lambda}$. The factorized form of Eq. (35) of course assumes the chirality of the production and detection processes to be opposite. For example, we could consider a V−A leptonic current at production (ρ=σ=L) and a V+A leptonic current at detection (ρ=R). The two possible Lorentz contractions with the hadronic current are σ=R and σ=L; in a LR symmetric model the former corresponds to the exchange of a WR boson, the latter to $W_L - W_R$ mixing if the boson mass eigenstates are mismatched, as depicted in Fig. 2 [89]. To simplify this work we will only consider these two cases, setting $\varepsilon^{(\rho,\sigma)}_{\mu\alpha}$ for all other (ρ,σ) to zero and cleaning up the notation with $\varepsilon^{(R,L)}_{\beta\alpha} = \varepsilon^{(R,R)}_{\beta\alpha} \equiv \varepsilon_{\beta\alpha}$. It is worth noting that there is a subtle difference between the LNV NSI being at production and detection because the outgoing lepton $\ell^{\pm}_{\alpha}$ at production is not measured. Hence in theory one must sum over the different initial flavors as $\sum_\alpha S_{\alpha\mu} \lesssim 0.026$., and MINOS is sensitive to $\varepsilon_{e\lambda}$ and $\varepsilon_{\tau\lambda}$. However, for the π± energies in the NuMI beam it is kinematically impossible to produce a τ±, and similar to standard π± decays the production of an electron is chirality suppressed to that of a muon. We therefore neglect this subtlety and assume that an NSI at production is probed in a similar way to that at detection.

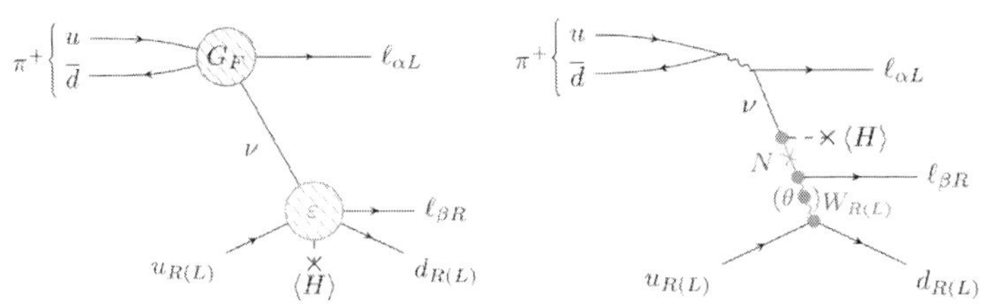

Figure 2. Left: simplified Feynman diagram of the oscillation process under consideration. Shown are the effective interactions at production and detection—a low energy SM CC interaction and LNV NSI, respectively. Right: a possible UV completion in a LR symmetric scenario.

Working with the 2ν mixing expansion of $P_{\nu_\mu \to \bar{\nu}_\mu}$ given in Eqs. (31) and (34), Eq. (35) can be used to put a constraint on the ($\epsilon_{\mu\mu}$,$\epsilon_{\mu\tau}$) parameter space. To do this we integrate the NuMI ν_μ differential fluxes of Ref. [85], normalized probabilities, and cross sections in the numerator and denominator of Eq. (35) over 500 MeV bins in the range 0–20 GeV. The cross sections,

$$\sigma_{\bar{\nu}_\mu p \to \ell_\beta^+ n}(E_{\mathbf{q}}) \simeq \sigma_{\nu_\mu n \to \ell_\beta^- p}(E_{\mathbf{q}}) \simeq \frac{G_F^2 |V_{ud}|^2}{\pi}(g_V^2 + 3g_A^2)E_{\mathbf{q}}^2, \tag{36}$$

are assumed to be equal in the quasielastic scattering limit [143].

In Fig. 3 (left) we first plot the allowed regions in the ($\epsilon_{\mu\mu}$,$\epsilon_{\mu\tau}$) parameter space for fixed L/Eq=735/3 km GeV^{-1}, using best fit values for $\delta m^2 \approx \Delta m_{23}^2$, $\vartheta \approx$, ϑ≈θ23 (in the NO scheme), G_F, V_{ud}, g_V, g_A, and four different values of the Majorana phase η [30]. This is equivalent to assuming the ν_μ flux to be sharply peaked at 3 GeV and evaluating the oscillation probability and cross section at this energy. Appreciable constraints are possible for η=0 and π, and are marginally better in the $\epsilon_{\mu\tau}$ direction. They are of order $|\epsilon_{\mu\mu}| \lesssim 0.2$ and $|\epsilon_{\mu\tau}| \lesssim 0.1$. When η=(n+1/2)π for n∈Z a specific direction in the parameter space alleviates the constraints. The narrow ellipse arises because F(μ)τ≪1 for the best fit parameters and these particular values of η. In Fig. 3 (right) we depict the allowed regions after the full numerical integration of the numerator and denominator of $S_{\mu\mu}$. For η=π/4 and π/2 the narrow bands of allowed values are reduced to ellipses more similar to the ellipses at η=0 and π. The orientations of the ellipses also change marginally. Upper bounds are in the ranges $|\epsilon_{\mu\mu}| \lesssim 0.2$–0.5 and $|\epsilon_{\mu\mu}| \lesssim 0.2$–0.6.

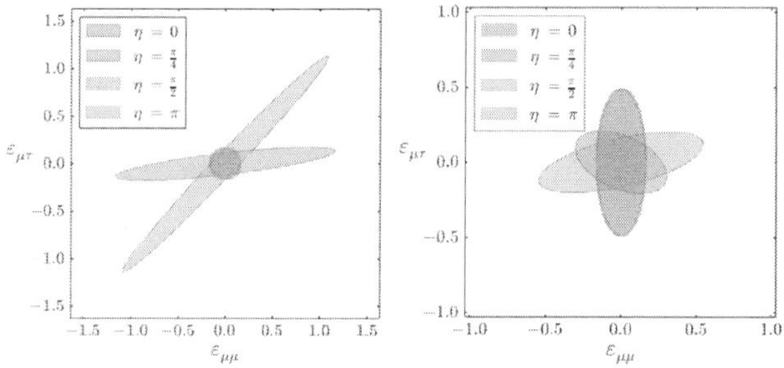

Figure 3. Allowed regions in the ($\epsilon_{\mu\mu}$,$\epsilon_{\mu\tau}$) parameter space for fixed L/Eq=735/3 km GeV−1 and four values of η in the 2ν mixing approximation of the ν_μ−ν_τ sector (left). Allowed region in $\epsilon_{\mu\mu}$−$\epsilon_{\mu\tau}$ for fixed L=735 km and again four values of η, found by integrating over the NuMI beam neutrino energies (right).

While we have so far restricted our analysis to the MINOS experiment, it is interesting to consider what constraints could be made at different baselines for an experiment similar in design to MINOS. To the left of Fig. 4 we set η=0 and examine the allowed regions in the $\epsilon_{\mu\mu}$-$\epsilon_{\mu\tau}$ space for different values of the baseline L, derived for illustrative purposes from the MINOS limit $S_{\mu\mu} \lesssim 0.026$. We see that at zero distance this limit bounds $|\epsilon_{\mu\mu}| \lesssim 0.16$, while $\epsilon_{\mu\tau}$ remains unbounded. This is clear from the expansion of $P_{\nu_\mu \to \bar{\nu}_\mu}$—the functions $F_\tau^{(\mu)}$ and $G_{\mu\tau}^{(\mu)}$ are directly proportional to $\sin(\eta - \frac{\delta m^2 L}{4E_q})$ which vanishes at L=0. The first term in $F_\mu^{(\mu)}$ always contributes to $|\epsilon_{\mu\mu}|^2$ while only oscillation terms contribute to $|\epsilon_{\mu\tau}|^2$ and $\epsilon_{\mu\mu}\epsilon_{\mu\tau}$. At the larger baselines of 200, 600, and 800 km the functions $F_\tau^{(\mu)}$ and $G_{\mu\tau}^{(\mu)}$ are nonzero, and thus the bounded areas again become ellipses. As the L increases the bounded area becomes more and more circular, improving the bound in the $\epsilon_{\mu\tau}$ direction. For η=π/2 shown in Fig. 4 (right), $F_\tau^{(\mu)}$ and $G_{\mu\tau}^{(\mu)}$ are nonzero at L=0, and so the bound is an ellipse at zero distance. Unlike for η=0 the bound in the ϵμμ direction improves as L increases.

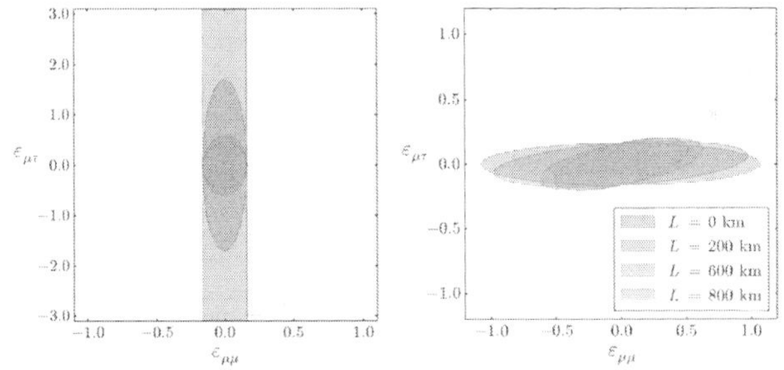

Figure 4. Allowed regions in the (ϵμμ,ϵμτ) parameter space for η=0 (left) and η=π/2 (right) for four different values of the baseline L.

Finally, it is interesting to look at the bounds on $\epsilon_{\mu\mu}$ as a function of L and for $\epsilon_{\mu\tau}$=0. In Fig. 5 (left) we plot allowed values of $\epsilon_{\mu\mu}$ along the x axis as a function of L along the y axis. For η=0 (and η=n_π, n∈Z) the oscillating part of $P_{\nu_\alpha \to \bar{\nu}_\beta}$ exactly cancels the oscillating part of $P_{\nu\alpha} \to \nu_\beta$, and thus the bound on $\epsilon_{\mu\mu}$ does not depend on L. For η=π/2 the constraint at zero distance is far less stringent, but decreases appreciably from 0 km up to 1000 km. For η=π/4 the constraint worsens as L reaches ~800km but improves at larger baselines. For L≳2000 km the constraints for nonzero η slowly oscillate but are roughly the same as for η=0, i.e., $|\epsilon_{\mu\mu}| \lesssim 0.15$. We show in Fig. 5 (right) a similar plot for $\epsilon_{\mu\tau}$, setting $\epsilon_{\mu\mu}$=0 and shading the allowed regions as a function of the baseline. At zero baseline $\epsilon_{\mu\tau}$ is unbounded for $\eta=0$, as discussed previously. For large baseline the upper limits converge to $|\epsilon_{\mu\mu}| \lesssim 0.16$.

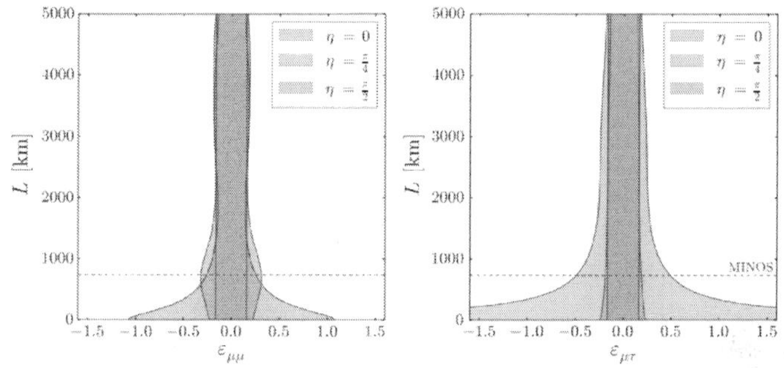

Figure 5. Constraints on $\epsilon_{\mu\mu}$ for $\epsilon_{\mu\tau}=0$ (left) and $\epsilon_{\mu\tau}$ for $\epsilon_{\mu\mu}=0$ (right) and as a function of the baseline L for three values of the Majorana phase η. The baseline of MINOS is indicated by the dashed line.

We summarize the constraints made in the 2ν mixing approximation on the two coefficients $\epsilon_{\mu\mu}$ and $\epsilon_{\mu\tau}$ in Table II. Here we allow one coefficient at a time to be nonzero, compute an upper bound in the fixed energy approximation (middle), and integrate over the energy (right). The lower and upper values are the best and worst upper bounds, respectively, as the phase η is varied. One can see that $\epsilon_{\mu\tau}$ is unbounded for a specific value of η when the energy is fixed. In this analysis we have of course taken the best fit values of the standard mixing parameters to be fixed. A rigorous fit to the data would need to let these parameters vary along with the ϵ coefficients, as taken into account for LNC NSI in Refs. [61,69,81,82].

Table 2. Upper bounds from the MINOS experiment on the LNV NSI coefficients in the 2ν mixing approximation. The range indicates the best and worst upper bound depending on the choice of the Majorana phase. Middle: bounds derived at a fixed neutrino energy of 3 GeV. Right: bounds derived by integrating over the energy-dependent NuMI flux, probability and cross section.

NSI coefficient	Fixed energy upper bound	Integrated upper bound
$\lvert\varepsilon_{\mu\mu}\rvert$	0.11–0.76	0.15–0.55
$\lvert\varepsilon_{\mu\tau}\rvert$	0.12–∞	0.16–0.66

B. Constraints from the Minos Experiment in the Three-Neutrino Mixing Scheme

We now generalize the analysis to the 3ν mixing scheme. We use the standard parametrization of the Pontecorvo-Maki-Nakagawa-Sakata mixing matrix

$$U = \underbrace{\begin{pmatrix} 1 & 0 & 0 \\ 0 & c_{23} & s_{23} \\ 0 & -s_{23} & c_{23} \end{pmatrix}}_{R_{23}} \cdot \underbrace{\begin{pmatrix} c_{13} & 0 & s_{13}e^{-i\delta_{CP}} \\ 0 & 1 & 0 \\ -s_{13}e^{i\delta_{CP}} & 0 & c_{13} \end{pmatrix}}_{W_{13}} \cdot \underbrace{\begin{pmatrix} c_{12} & s_{12} & 0 \\ -s_{12} & c_{12} & 0 \\ 0 & 0 & 1 \end{pmatrix}}_{R_{12}} \cdot \underbrace{\begin{pmatrix} 1 & 0 & 0 \\ 0 & e^{i\frac{\alpha_2}{2}} & 0 \\ 0 & 0 & e^{i\frac{\alpha_3}{2}} \end{pmatrix}}_{D}, \quad (37)$$

where $s_{ij}=\sin\theta_{ij}$, $c_{ij}=\cos\theta_{ij}$, and (α_2, α_3) are Majorana phases. We now look to probe the 3×3 flavor structure of the NSI coefficient matrix $\epsilon_{\beta\alpha}$ in which all elements are taken to be real. We again expand the effective nonstandard oscillation probability as in Eqs. (31) and (32),

$$\begin{aligned} P_{\nu_\alpha \to \bar{\nu}_\beta} &\approx \left| \sum_i^3 U^*_{\alpha i} \gamma_{\beta i} e^{-i\frac{m_i^2}{2E_q}L} \right|^2 \\ &= \sum_{\lambda=e,\mu,\tau} F_\lambda^{(\alpha)}(L, E_q, \zeta, \alpha_2, \alpha_3)\varepsilon^2_{\beta\lambda} \\ &+ \sum_{\lambda<\lambda'=e,\mu,\tau} G_{\lambda\lambda'}^{(\alpha)}(L, E_q, \zeta, \alpha_2, \alpha_3)\varepsilon_{\beta\lambda}\varepsilon_{\beta\lambda'}, \end{aligned} \quad (38)$$

where $F_\lambda^{(\alpha)}$ and $G_{\lambda\lambda'}^{(\alpha)}$ are now complicated functions of the baseline L, neutrino energy Eq, neutrino mixing parameters ζ, and Majorana phases α2 and α_3. Using the best fit values for the mixing parameters θ_{12}, θ_{23}, θ_{13}, Δm^2_{12}, and Δm^2_{23} and setting δ_{CP} to the value hinted at by the most recent data (again in the NO scheme), we numerically rotate to the flavor space coefficients as performed in the second line of Eq. (38) to find $F_\lambda^{(\alpha)}$ and $G_{\lambda\lambda'}^{(\alpha)}$ as functions of L, E_q, α_2, and α_3. It is useful to compare the two Majorana phases used here to the single phase in the 2ν mixing approximation. Comparing the numerical expression of $F^{(\mu)}_\mu$ in the 3ν scheme to that in Eq. (34) for the 2ν scheme and taking the limits $\Delta m^2_{12} \to 0$, $\Delta m^2_{13} \to \Delta m^2_{23}$, we find that ϑ≈θ23, $\delta m^2 \approx \Delta m^2_{23}$ which we had already assumed in Sec. IVA, and that $\eta \approx (\alpha_3 - \alpha_2)/2$.

We now return to the interpretation of the MINOS bound $S_{\mu\mu} \lesssim 0.026$. This ratio is defined in the same manner as Eq. (35)—multiplying Eq. (38) by the differential production flux and detection cross section and integrating over the NuMI beam energy between 0 and 20 GeV. Now in the 3ν mixing scheme, $S_{\mu\mu} < 0.026$ can be converted into an allowed region in the $(\epsilon_{\mu e},\epsilon_{\mu\mu},\epsilon_{\mu\tau})$ parameter space which depends on the value of the Majorana phases α2 and α3. First, and in order to compare with bounds in the 2ν mixing approximation, we set $\epsilon_{\mu e}$=0 and depict in Fig. 6 (left) the allowed regions in the $(\epsilon_{\mu\mu},\epsilon_{\mu\tau})$ parameter space for α_2=0 and three different values of α_3. The ellipses are of similar size to those for the 2ν mixing scheme but have shapes and orientations (which we define as the ellipse eccentricity e and anticlockwise angle Θ from the positive $\epsilon_{\mu\mu}$ axis to the semimajor axis, respectively) with similar dependences on the Majorana phases. We show these dependences as contour plots in Fig. 6 (middle and right). We can see that the dependence of the angle Θ is approximately the same as $\eta\approx(\alpha_3-\alpha_2)/2$, which is evident from the diagonal lines of roughly constant Θ along $\alpha_3=\alpha_2+C$. For $\alpha_2=\alpha_3=0$ for example, we see that the constraint is better in the $\epsilon_{\mu\mu}$ direction—this is similar to η=0 in the 2ν scheme. Likewise, for $\alpha_2=0$, $\alpha_3=\pi$, the constraint is better in the $\epsilon_{\mu\tau}$ direction which is similar to η=π. We see that the largest eccentricity occurs at $\alpha_2\approx\alpha_3\approx\pi$—this coincides with the semimajor axis pointing in the $\epsilon_{\mu\tau}$ direction, and therefore the upper bound on $|\epsilon_{\mu\tau}|$ can be slightly larger than the upper bound on $|\epsilon_{\mu\mu}|$.

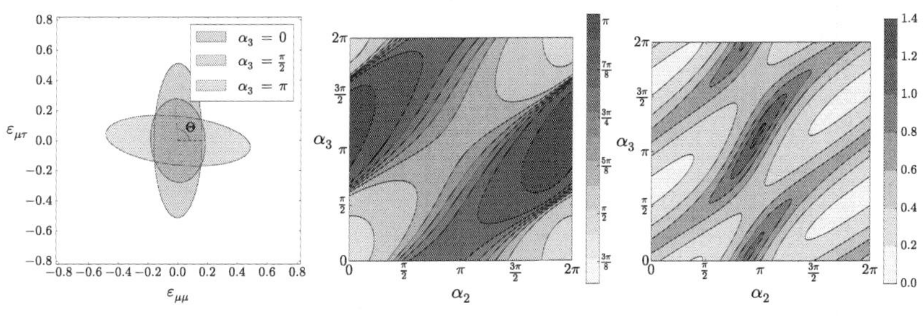

Figure. 6. Left: allowed regions in the $(\epsilon_{\mu\mu},\epsilon_{\mu\tau})$ parameter space for $\epsilon_{\mu e}$=0, L=735 km, α_2=0, and three values of the Majorana phase α_3. Middle: angle Θ from the positive $\epsilon_{\mu\mu}$ axis to the semimajor axis of the constraint ellipse as a function of the phases. Right: eccentricities of the constraint ellipse also as a function of the phases.

In Fig. 7 we plot the upper bounds on each NSI coefficient in the μ sector as a function of the Majorana phases when taking each to be nonzero at a time. We can see the values of α_2 and α_3 where the upper bounds are more or less stringent. For ϵμe (left) we can see that for most of the parameter space the upper bound is of order $|\epsilon_{\mu e}|\lesssim1$, but worsens for particular values of the phases to $|\epsilon_{\mu e}|\lesssim3.4$. Upper bounds for the other two coefficients are in the ranges $|\epsilon_{\mu\mu}|\lesssim0.2$–0.6 and $|\epsilon_{\mu\tau}|\lesssim0.2$–0.7. We summarize these constraints, along with the results from the following sections, to the right of Table III.

The lower and upper values are the best and worst upper bounds depending on the value of the Majorana phases α_2 and α_3.

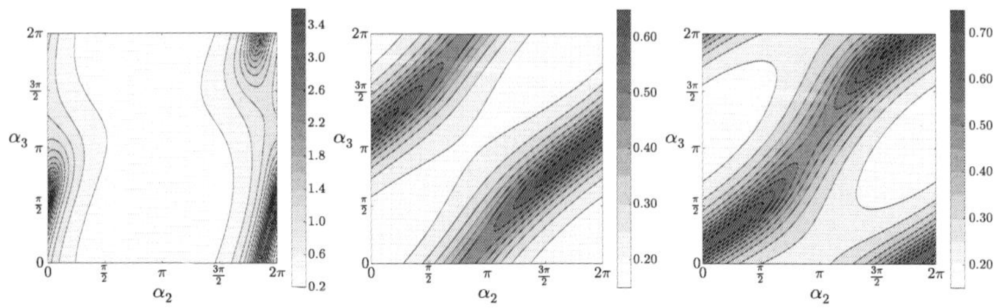

Figure 7. Upper bounds on the NSI coefficients ϵμe (left), ϵμμ (center), and ϵμτ (right), as a function of the Majorana phases α2 and α3, derived from the MINOS limit Sμμ<0.026. Best fit values for the mixing parameters θ12, θ13 and θ23, Δm212 and Δm223, and δCP are taken in the NO.

Table 3. Upper bounds on the LNV NSI flavor coefficients in the e and μ sectors. Left: bounds derived from conventional microscopic LNV processes, with 0νββ decay being the most effective and the e sector and μ^-–e^+ conversion loosely constraining the μ sector. Right: bounds from LBL oscillation experiments MINOS and KamLAND. Two values indicate the variation in the upper bound as (α_2,α_3) are varied.

NSI coefficient	Previous upper bound	Process	LBL upper bound	LBL experiment
$\lvert\varepsilon_{ee}\rvert$	2.1×10^{-9}–6.3×10^{-9}		0.017	
$\lvert\varepsilon_{e\mu}\rvert$	2.9×10^{-9}–∞	$0\nu\beta\beta$ (^{76}Ge)	0.017	KamLAND
$\lvert\varepsilon_{e\tau}\rvert$	2.6×10^{-9}–∞		0.015	
$\lvert\varepsilon_{\mu e}\rvert$	$\sim 4 \times 10^{3}$–1×10^{4}		0.22–3.47	
$\lvert\varepsilon_{\mu\mu}\rvert$	$\sim 6 \times 10^{3}$–∞	$\mu^- - e^+$	0.16–0.63	MINOS
$\lvert\varepsilon_{\mu\tau}\rvert$	$\sim 5 \times 10^{3}$–∞		0.16–0.71	

While we have so far considered only MINOS, OPERA was another LBL oscillation experiment to employ a magnetic field in the far detector [144,145]. Unlike MINOS, OPERA measured neutrinos from the CNGS beam at CERN that were above the threshold for τ± production. The main physics goal of the experiment was to confirm ν_τ appearance—around 10 τ± events were recorded over four years of data taking [146]. Unfortunately the experiment was only able to distinguish the charge of one τ^- event at 5σ significance, while the other charges were undetermined. The statistics are therefore too low to comment on OPERA's sensitivity to lepton number. A future high-statistics LBL experiment above the τ± threshold like OPERA would therefore be able to probe the τ sector NSI coefficients, $\epsilon_{\tau e}$, $\epsilon_{\tau\mu}$, and $\epsilon_{\tau\tau}$.

C. Constraints from the kamLAND Experiment in the Three-Neutrino Mixing Scheme

Operating for a window of 185.5 days between March 4 and December 1, 2002, the LBL reactor neutrino experiment KamLAND conducted a search for solar $\overline{\nu}_e$ with the characteristic flux of 8B νe. The analysis of Ref. [133] hence assumed the $\overline{\nu}_e$ of interest to be descendant from the solar νe, either through spin precession in the Sun's magnetic field due to a nonzero neutrino transition magnetic moment or neutrino decay. Hence the bound derived on the ratio $S_{ee} \lesssim 2.8\times10^{-4}$ at 90% C.L. can be used in a similar fashion to MINOS to place constraints on the NSI coefficients.

The initial νe produced from the beta decay of ^8B and propagate from the solar core to the solar surface, and then through the vacuum to the KamLAND detector. The oscillation probability must therefore take into account the resonant conversion of solar νe to ν_μ and ν_τ through the Mikheyev-Smirnov-Wolfenstein (MSW) effect, a consequence of the slowly decreasing matter potential from the Sun's core to surface. We make the approximation that the conversion is adiabatic and utilize the $\Delta m_{12}^2 \ll \Delta m_{23}^2$ hierarchy to write the standard $\nu_\alpha \rightarrow \nu_\beta$ conversion probability in a similar form to that in Ref. [82],

$$P^{\text{eff}}_{\nu_\alpha\rightarrow\nu_\beta} \approx \left| \sum_{i,j=1,2}^{3} (R^{23}W^{13})^*_{\alpha i}(R^{23}W^{13})_{\beta j}\mathcal{U}_{ij}(x) \right|^2 + |U_{\alpha 3}|^2|U_{\beta 3}|^2, \tag{39}$$

where R^{23} and W13 are Euler rotations making up the standard parametrization of the Pontecorvo-Maki-Nakagawa-Sakata matrix and U is a 2×2 unitary matrix satisfying

$$i\frac{d}{dx}\mathcal{U}(x) = \frac{\hat{M}^2_{2\times2}}{2E_q}\mathcal{U}(x). \tag{40}$$

$\hat{M}^2_{2\times2}$ is the effective 2×2 squared mass matrix

$$\hat{M}^2_{2\times2} = \frac{\Delta m_{12}^2 + c_{13}^2 A_{CC}}{2} + \frac{1}{2}\begin{pmatrix} -\cos 2\theta_{12}\Delta m_{12}^2 + c_{13}^2 A_{CC} & \sin 2\theta_{12}\Delta m_{12}^2 \\ \sin 2\theta_{12}\Delta m_{12}^2 & \cos 2\theta_{12}\Delta m_{12}^2 - c_{13}^2 A_{CC} \end{pmatrix} \tag{41}$$

where $A_{CC}=2\sqrt{2}G_F E_q N_e$ and Ne is the electron density in the Sun. In order to construct See we require the nonstandard oscillation equivalent of Eq. (39). This can be derived from Eq. (31), but an exact formula taking into account the MSW effect, even in the $\Delta m_{12}^2 \ll \Delta m_{23}^2$ limit, is beyond the scope of this work. The possibility that the NSI occurs at production would also complicate the analysis, because $\overline{\nu}_e$ would experience a different matter effect while propagating through the Sun—we therefore concentrate on the NSI being at detection.

We can safely assume that, by the time the solar neutrinos reach Earth, they make up an incoherent mixture of flavor eigenstates. This effectively washes out any dependence on the Majorana phases, and we can approximate the nonstandard oscillation probability as being the effective standard probability $P^{\text{eff}}_{\nu_e\to\nu_\beta}$ (taking into account the MSW effect) multiplied by the NSI coefficient $|\epsilon_{e\beta}|^2$. This takes into account for example the oscillation $\nu_e \to \nu_{\mu,\tau}$ and then the LNV NSI process $\nu_{\mu,\tau} p \to e^+ n$ at detection. The signal ratio See and the limit set on it by KamLAND now becomes

$$S_{ee} \approx \frac{\int dE_{\mathrm{q}} \sum_\beta \frac{d\Gamma_{\nu_e}}{dE_{\mathrm{q}}} \cdot P^{\text{eff}}_{\nu_e\to\nu_\beta} \cdot \varepsilon^2_{e\beta} \cdot \sigma_{\bar{\nu}_\beta}}{\int dE_{\mathrm{q}} \frac{d\Gamma_{\nu_e}}{dE_{\mathrm{q}}} \cdot P^{\text{eff}}_{\nu_e\to\nu_e} \cdot \sigma_{\bar{\nu}_e}} \lesssim 2.8 \times 10^{-4}. \quad (42)$$

We now use Eq. (42) to set limits on NSI coefficients in the e sector, ϵ_{ee}, $\epsilon_{e\mu}$, and $\epsilon_{e\tau}$. Using the ^{8}B flux predicted by the solar model of Ref. [147], we integrate the differential flux, oscillation probability, and low energy cross section (which we assume to scale as E2q) over 0.02 MeV bins in the range 8.3–14.8 MeV, for both the numerator and the denominator.

Taking each LNV NSI coefficient to be nonzero at a time, we show the derived upper bounds in Table III. Because there is no dependence on the Majorana phases, Eq. (42) sets a single possible upper bound on each coefficient. For ϵee for example, the numerator and denominator are identical except for the factor of $|\epsilon_{ee}|^2$ in the numerator—the upper bound on |ϵee| is therefore the square root of 2.8×10^{-4}. The upper bounds on $|\epsilon_{e\mu}|$ and $|\epsilon_{e\tau}|$ are slightly different because the numerators of See contain different oscillation probabilities from the denominator.

D. Comparison with Constraints from other Lepton Number Violating Processes

We will now briefly compare the constraints from the LBL experiments MINOS and KamLAND to those from conventional searches for LNV processes. Neutrinoless double beta ($0\nu\beta\beta$) decay continues to be the most promising method to verify the Majorana nature of the light neutrinos. It is a highly useful probe because the black-box theorem ensures that any positive signal of $0\nu\beta\beta$ decay confirms the Majorana case even when the standard light neutrino exchange process is not the dominant mechanism [28,38,39]. An extensive part of the literature has studied nonstandard mechanisms, including the LNV dimension-six operator considered in this work [34,40,45,136]. Here we simply extend this to the 3×3 flavor structure of the ϵ coefficient in order to compare with the MINOS and KamLAND upper bounds.

Using from Ref. [32] the general expression for the 0νββ decay inverse half-life when a RH interaction is present at one of the interaction vertices (as shown in Fig. 8),

$$[T_{1/2}^{0\nu\beta\beta}]^{-1} = C_{mm}\frac{|\sum U_{ei}^2 m_i|^2}{m_e^2} + C_{\gamma\gamma}\left|\sum_i^3 U_{ei}\gamma_{ei}^*\right|^2 + C_{m\gamma}\mathrm{Re}\left[\sum_{i,j}^3 U_{ei}^2 m_i U_{ej}^* \gamma_{ej}\right], \quad (43)$$

where $C_{mm}=1.12\times10^{-13}$, $C_{\gamma\gamma}=4.44\times10^{-9}$, and $C_{m\gamma}=2.19\times10^{-11}$ are phase space and nuclear matrix element factors given by [32]. The inverse half-life, in a similar manner to the oscillation probability, can be expanded as

$$[T_{1/2}^{0\nu\beta\beta}]^{-1} = X(\zeta,\alpha_2,\alpha_3,m_0) + \sum_\lambda Y_\lambda(\zeta,\alpha_2,\alpha_3,m_0)\varepsilon_{e\lambda} + \sum_\lambda F_\lambda^{(e)}(\zeta,\alpha_2,\alpha_3)\varepsilon_{e\lambda}^2 + \sum_{\lambda<\lambda'} G_{\lambda\lambda'}^{(e)}(\zeta,\alpha_2,\alpha_3)\varepsilon_{e\lambda}\varepsilon_{e\lambda'}, \quad (44)$$

where $F_\lambda^{(e)}(\zeta,\alpha_2,\alpha_3)$ and $G_{\lambda\lambda'}^{(e)}(\zeta,\alpha_2,\alpha_3)$ are the functions in Eq. (38) at zero distance. The contribution from light neutrino exchange is $X(\zeta,\alpha_2,\alpha_3,m_0)$ which is a function of the Majorana phases and lightest neutrino mass. $Y_\lambda(\zeta,\alpha_2,\alpha_3,m_0)$ is the contribution from interference between the light neutrino exchange and the nonstandard mechanism toward terms linear in the ϵ coefficients.

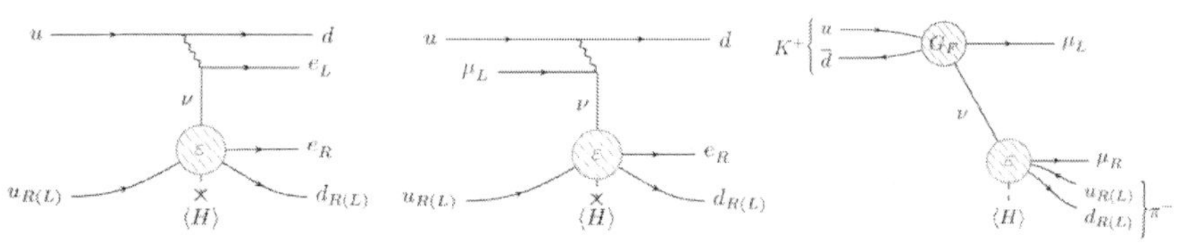

Figure 8. Other processes capable of probing LNV NSI coefficients. Left: 0νββ decay. Middle: μ^-–e^+ conversion. Right: kaon decay $K^+\to\pi^-\mu^+\mu^+$.

We now set $T_{1/2}^{0\nu\beta\beta} > 5.3\times10^{25}$ derived from the 76Ge experiment GERDA-II [148,149]. While KamLAND-Zen set a more stringent limit of $T_{1/2}^{0\nu\beta\beta} > 1.07\times10^{26}$ y with ^{136}Xe [150], the exact number we use is not crucial for the following constraints and discussion. Setting each $\epsilon e\lambda$ to be nonzero at a time, Eq. (44) can be solved to find an upper bound on the coefficient as a function of α_2 and α_3. These are displayed in the contour plots of Fig. 9 for a smallest neutrino mass of m_1=0 eV in the NO scheme. The associated best and worst upper bounds are shown in Table III. It can be seen that regardless of the value of the Majorana phases, the upper bound on $|\epsilon_{ee}|$, as found previously in the literature, is of order 10^{-9}. For very finely tuned values of α_2 and α_3, however, $|\epsilon e\mu|$ and $|\epsilon_{e\tau}|$ are unbounded. Comparing all upper bounds in the e sector, we see that 0νββ decay is unequivocally the best method to probe ϵ_{ee}. For a large portion of the (α_2,α_3) parameter space it is also better at constraining $\epsilon_{e\mu}$ and $\epsilon_{e\tau}$. However, for certain fine-tuned values of the phases these coefficients become unbounded and KamLAND can provide a better upper bound.

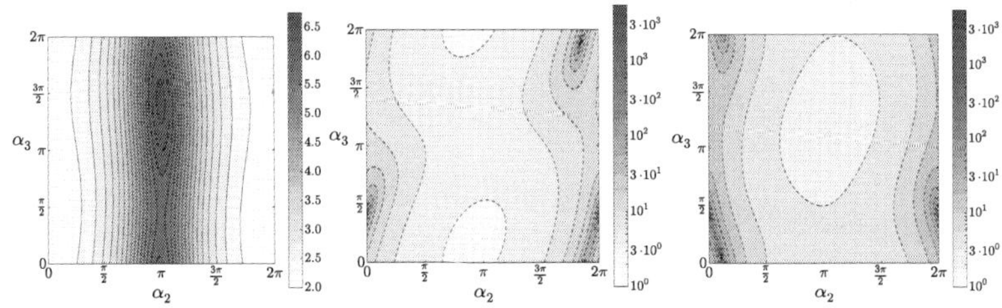

Figure 9. Upper bounds on the NSI coefficients εee (left), εeμ (center), and εeτ (right) all multiplied by 109 as a function of the Majorana phases α2 and α3, found from the 76Ge 0νββ decay limit T0νββ1/2>5.3×1025 y. Best fit values for the mixing angles θ12, θ13, and θ23, squared mass splittings Δm212 and Δm223, Dirac CP phase δCP are taken in the NO, with a lightest neutrino mass of m1=0 eV.

While 0νββ decay is certainly the most sensitive process to test for LNV as outlined above, it can only probe the e sector, whereas other observables may shed light on other flavor coefficients. An interesting process in this regard is the LNFV conversion of captured muons in nuclei, μ^-+(Z,A)→e^++(Z−2,A). Proposed many years ago by Pontecorvo [151], it has gained recent interest due to the upcoming searches for the LNC but LFV muon conversion μ^-+(Z,A)→e^-+(Z,A) by the COMET [152] and Mu2e [153] collaborations, which promise an increased experimental sensitivity by 4 orders of magnitude. While it is doubtful that the current limit $R^{\mathrm{Ti}}_{\mu e} \lesssim 10^{-11}$ [154] on the LNFV mode conversion rate can be improved in a similar fashion due to different background considerations [155], μ^-−e^+ conversion is an important complementary probe to 0νββ decay.

To estimate the sensitivity of the LNFV $\mu^- - e^+$ conversion process on the LNV NSI coefficients considered in this paper, we follow the estimate in [136]. In this approach, and using our notation, the conversion rate is approximated as

$$R_{\mu e} \approx |\xi_{\mu e}|^2 \frac{G_F^2}{2} \frac{Q^6}{q^2} \tag{45}$$

where the effective parameter $\xi_{\mu e}$ is defined by

$$\begin{aligned} |\xi_{\mu e}|^2 \equiv & \left|\sum_i (U_{ei}^* \gamma_{\mu i} + U_{\mu i}^* \gamma_{ei})\right|^2 \\ = & \sum_\lambda (F_\lambda^{(e)}(\zeta, \alpha_2, \alpha_3)\varepsilon_{\mu\lambda}^2 + F_\lambda^{(\mu)}(\zeta, \alpha_2, \alpha_3)\varepsilon_{e\lambda}^2) \\ & + \sum_{\lambda<\lambda'} (G_{\lambda\lambda'}^{(e)}(\zeta, \alpha_2, \alpha_3)\varepsilon_{\mu\lambda}\varepsilon_{\mu\lambda'} \\ & + G_{\lambda\lambda'}^{(\mu)}(\zeta, \alpha_2, \alpha_3)\varepsilon_{e\lambda}\varepsilon_{e\lambda'}). \end{aligned} \tag{46}$$

The two terms in the first line of Eq. (46) take into account that the LNV NSI can be at the interaction vertex of either the incoming μ^- or the outgoing e+ (the latter being shown in Fig. 8). The only difference between the two diagrams is the exchange $U_{ei}^* \gamma_{\mu i} \leftrightarrow U_{\mu i}^* \gamma_{ei}$.

In Eq. (45), q denotes the momentum scale of the intermediate neutrino in the process, $q \approx 100$ MeV, and Q is the energy release of the emitted positron determining the size of the phase space with $Q \approx 15.6$ MeV [156]. Due to the process being incoherent and partially going to excited final nuclear states, Q also approximately convolutes the nuclear matrix element of this transition. Here, it should be emphasized that the relevant nuclear matrix elements have not been calculated in detail and Eq. (45) can only be regarded as a very rough estimate of the order of magnitude of the conversion rate. Nevertheless, using the experimental limit Eq. (45), we estimate a limit on $|\xi_{\mu e}|$ of order $|\xi_{\mu e}| \lesssim 104$. This is barely stringent enough for the EFT assumptions to be consistent, with the limit corresponding to an effective operator scale $\Lambda = (|\xi_{\mu e}| GF)^{-1/2} \approx 3$ GeV>q. Even with the most optimistic improvement of the future sensitivity to $R_{\mu e} \lesssim 10^{-16}$ [156], the coefficient $|\xi_{\mu e}|$ will only be probed at $|\xi_{\mu e}| \sim 30$, corresponding to a scale $\Lambda_{NP} \sim 50$ GeV.

We summarize the constraints on the individual coefficients in the μ sector (when each is taken to be nonzero) in Table III. These are also of order 104. Similar to 0νββ decay, the coefficients with flavor indices not corresponding to the flavor of external charged leptons are unbounded for particular values of the Majorana phases. Because the functions preceding $\epsilon^2_{\mu\lambda}$ in Eq. (46) are $F^{(e)}_{\lambda}(\zeta, \alpha_2, \alpha_3)$, the dependences on the Majorana phases for the μ^-–e^+ upper bounds are identical to those for 0νββ decay in Fig. 9, except being scaled by a factor of ~1012. From Eq. (46) it is clear that μ^-–e^+ can also probe the e sector coefficients, but sets bounds larger than 0νββ decay by a factor of 10^{12}. However, it is of small interest to note that the dependence on the Majorana phases is the same as that for the MINOS upper bounds because the functions preceding $\epsilon^2_{e\lambda}$ in Eq. (46) are $F^{(\mu)}_{\lambda}(\zeta, \alpha_2, \alpha_3)$.

From the rare LNV meson decays such as $K^{\pm}\rightarrow\pi^{\mp}\mu^{\pm}\mu^{\pm}$ and $B^{+}\rightarrow D^{-}\mu^{+}\mu^{+}$ and rare τ decays such as $\tau^{-}\rightarrow\pi^{-}\pi^{-}\mu^{+}$ it is also possible to probe the LNV coefficients considered in this work as well as those in the τ sector [52,109,157]. However, at present these processes have sensitivities at or worse than that of μ^-–e^+ conversion. Comparing the constraints in the μ sector in Table III, we see that the constraints from MINOS (and similar future LBL oscillations experiments) are currently far more stringent than microscopic LNV processes.

To conclude this section we will briefly mention the analyses in Refs. [51,136] on the contribution toward the Majorana neutrino mass Mαβ from the operators in Table I. If the discussion is shifted back to the Wilson coefficients of the operators, i.e., we set

$$\varepsilon_{\beta\alpha} = \frac{v}{G_F}\frac{C_{\beta\alpha}}{\Lambda^3_{\rm NP}} \qquad (47)$$

and assuming that the Wilson coefficients $C_{\beta\alpha}$~O(1), the minimum sum of neutrino masses $\sum m_\nu > 6 \times 10^{-2}$ eV sets a lower bound on ΛNP. For the scalar and tensor Dirac structures the loop mass contains one quark Yukawa coupling, and $\Lambda_{NP}>1\times10^{6}$ TeV. For a vector Dirac structure the loop gets a contribution from the charged lepton Yukawa coupling and both quark Yukawa couplings if they are right-handed, giving $\Lambda_{NP}>6\times10^{3}$ TeV. This can be decreased significantly to $\Lambda_{NP}\sim1$ TeV if the couplings to third generation quarks are suppressed by some flavor symmetry [136].

CONCLUSIONS

In this paper, we have investigated the effect of lepton number violating nonstandard interactions on long-baseline neutrino oscillations. If the light active neutrinos are of Majorana nature the $\nu_\alpha \rightarrow \bar{\nu}_\beta$ process become possible, either through a $(m_\nu/E_\nu)^2$ suppressed light neutrino helicity reversal or an LNV charged current (CC)-like interaction at production or detection. The Majorana neutrino case is favored from a model-building perspective and is currently being probed by a wide range of neutrinoless double beta ($0\nu\beta\beta$) decay experiments.

We first studied the different derivations of neutrino oscillations in quantum mechanics and quantum field theory. The QFT model is a more complete and physically consistent picture, taking into account the coherence of overlapping wave packets at production and detection. In this picture we derived the $(m_\nu/E_\nu)^2$ suppression associated with the Majorana helicity reversal, showing that the total rate in this case cannot be factorized into a production flux, oscillation probability, and detection cross section. In the QFT picture we next studied the LNV interaction at the detection process. It is immediately clear that if the chirality of the production and detection processes are opposite, an "oscillation probability" $P_{\nu_\alpha \to \bar{\nu}_\beta}$ can be factorized from the total rate of the process, justifying the use of a simple nonstandard oscillation formula. The (mv/Ev)2 suppression is replaced by a $|\epsilon|^2$ suppression, where ϵ parametrizes the strength of the LNV NSI compared to the Fermi coupling GF.

Using a bound made by the MINOS experiment on $\nu_\mu \to \bar{\nu}_\mu$ oscillations, we put limits on the ϵ flavor coefficients in the case of a LH SM interaction at production and a RH leptonic current (connected in turn to a LH or RH hadronic current) at detection. The limits are also valid for the LNV NSI being at production. In the two-neutrino (2ν) mixing scheme which is approximately valid for the ν_μ–ν_τ sector, we derived a simplified expression for the nonstandard oscillation probability in terms of the coefficients $\epsilon_{\mu\mu}$ and $\epsilon_{\mu\tau}$. Multiplying the nonstandard oscillation probability by the NuMI beam flux and quasielastic CC cross section, and integrating over the energy range 0–20 GeV, we derived upper bounds on the absolute values of the two coefficients. While the value of the single Majorana CPV phase η alters the upper bounds somewhat, we conservatively obtained $|\epsilon_{\mu\mu}| \lesssim 0.6$ and $|\epsilon_{\mu\tau}| \lesssim 0.7$. If a future experiment such as MINOS were to be at a smaller baseline, we found that the value of η has a larger impact on the upper bound.

We subsequently generalized the constraints to the three-neutrino (3ν) scheme, using the best fit values for the 3ν mixing parameters (including the most recent hinted value of the Dirac phase δ_{CP}), and exploited the MINOS bound to constrain the μ sector parameter space, $(\epsilon_{\mu e}, \epsilon_{\mu\mu}, \epsilon_{\mu\tau})$. Likewise, we used a KamLAND measurement limiting the number of solar $\overline{\nu}_e$ from the source of solar ^8B νe to place constraints on the e sector parameter space, (ϵee,ϵeμ,ϵeτ). For the latter we took into account the MSW effect and decoherence of the propagating neutrinos. We again found upper limits on the absolute values of the flavor coefficients but which do not vary as a function of the Majorana phases α_2 and α_3. We briefly mentioned that a future high statistics experiment above the τ threshold (and with a magnetic field in the far detector such as OPERA) could put constraints on the τ sector parameter space, ($\epsilon_{\tau e}$,$\epsilon_{\tau\mu}$,$\epsilon_{\tau\tau}$).

Of course, as has been considered before in the literature, these coefficients can be constrained by experimental searches of microscopic LNV processes. In order to compare with the KamLAND results we used the most recent 0νββ decay half-life from the 76Ge experiment GERDA-II to put upper bounds on the electron sector. While 0νββ decay still provides the most competitive constraint on |ϵee|, for particular values of the Majorana phases $|\epsilon_{e\mu}|$ and $|\epsilon_{e\tau}|$ become unbounded, where otherwise KamLAND sets a finite upper bound. Similarly, μ^-–e^+ conversion sets very loose bounds on the muon sector coefficients—the upper bounds from MINOS are superior for all values of the Majorana phases. We have focused on the LNV signal observables, but in general the LNV NSI can be constrained along with LNC NSI in processes that are insensitive to lepton number, such as charged pion decay. Both LNC and LNV CC-like NSI can induce deviations from lepton universality and Cabibbo-Kobayashi-Maskawa unitarity.

While the CPV properties of $\nu_\alpha \to \bar{\nu}_\beta$ oscillations were considered in Ref. [113] in the context of the light neutrino helicity-reversal mechanism, future work could study the CPV properties of complex ϵ coefficients. It would also be interesting to consider phenomenology of LNV NC-like matter oscillations, which also generically arise from dimension-six operators in the LEFT of the SM. Constraints on extensions of the 3ν scheme to additional sterile states may also be possible given the improving precision and possible charge-sign sensitivity of future SBL reactor experiments.

ACKNOWLEDGMENTS

P. D. B. and F. F. D. acknowledge support from the Science and Technology Facilities Council (STFC).

REFERENCES

1. Y. Fukuda et al. (Super-Kamiokande Collaboration), Phys. Rev. Lett. 81, 1562 (1998).
2. Q.R. Ahmad et al. (SNO Collaboration), Phys. Rev. Lett. 87, 071301 (2001).
3. P.F. de Salas, D.V. Forero, C.A. Ternes, M. Tortola, and J.W.F. Valle, Phys. Lett. B 782, 633 (2018).
4. M.A. Acero et al. (NOvA Collaboration), Phys. Rev. D 98, 032012 (2018).
5. R. Acciarri et al. (DUNE Collaboration), arXiv:1512.06148.
6. K. Abe et al. (Hyper-Kamiokande Proto-Collaboration), Prog. Theor. Exp. Phys. 2015, 053C02 (2015).
7. F. An et al. (JUNO Collaboration), J. Phys. G 43, 030401 (2016).
8. A. Osipowicz et al. (KATRIN Collaboration), arXiv:hep-ex/0109033.
9. F. Capozzi, E. Di Valentino, E. Lisi, A. Marrone, A. Melchiorri, and A. Palazzo, Phys. Rev. D 95, 096014 (2017).
10. J. Lesgourgues and S. Pastor, Adv. High Energy Phys. 2012, 608515 (2012).
11. S. Vagnozzi, E. Giusarma, O. Mena, K. Freese, M. Gerbino, S. Ho, and M. Lattanzi, Phys. Rev. D 96, 123503 (2017).
12. P.B. Pal, Am. J. Phys. 79, 485 (2011).
13. E. Akhmedov, arXiv:1412.3320.
14. M. Gell-Mann, P. Ramond, and R. Slansky, (World Scientific, Singapore, 1979), p. 315.
15. J. Schechter and J.W.F. Valle, Phys. Rev. D 22, 2227 (1980).
16. T.P. Cheng and L.-F. Li, Phys. Rev. D 22, 2860 (1980).
17. F.T. Avignone, III, S.R. Elliott, and J. Engel, Rev. Mod. Phys. 80, 481 (2008).
18. R.N. Mohapatra and R.E. Marshak, Phys. Rev. Lett. 44, 1316 (1980); 44, 1644(E) (1980).
19. Y. Farzan, Phys. Rev. D 67, 073015 (2003).
20. G.F. Giudice, A. Masiero, M. Pietroni, and A. Riotto, Nucl. Phys. B396, 243 (1993).
21. G. Aad et al. (ATLAS Collaboration), J. High Energy Phys. 2014, 176.

22. G. Aad et al. (ATLAS Collaboration), Phys. Rev. D 93, 052002 (2016).
23. F.F. Deppisch, P.S.B. Dev, and A. Pilaftsis, New J. Phys. 17, 075019 (2015).
24. P.D. Bolton, F.F. Deppisch, C. Hati, S. Patra, and U. Sarkar, arXiv:1902.05802.
25. G. Anamiati, R.M. Fonseca, and M. Hirsch, Phys. Rev. D 97, 095008 (2018).
26. G. Anamiati, M. Hirsch, and E. Nardi, J. High Energy Phys. 2016, 010.
27. A. Das, P.S.B. Dev, and R.N. Mohapatra, Phys. Rev. D 97, 015018 (2018).
28. J. Schechter and J.W.F. Valle, Phys. Rev. D 25, 2951 (1982).
29. S. Dell'Oro, S. Marcocci, M. Viel, and F. Vissani, Adv. High Energy Phys. 2016, 2162659 (2016). 10.1155/2016/2162659.
30. C. Patrignani et al. (Particle Data Group), Chin. Phys. C 40, 100001 (2016).
31. R.N. Mohapatra, Phys. Rev. D 34, 3457 (1986).
32. K. Muto, E. Bender, and H.V. Klapdor, Z. Phys. A 334, 187 (1989).
33. K.S. Babu and R.N. Mohapatra, Phys. Rev. Lett. 75, 2276 (1995).
34. M. Hirsch, H.V. Klapdor-Kleingrothaus, and O. Panella, Phys. Lett. B 374, 7 (1996).
35. H. Pas, M. Hirsch, S.G. Kovalenko, and H.V. Klapdor-Kleingrothaus, Phys. Lett. B 453, 194 (1999).
36. H. Pas, M. Hirsch, H.V. Klapdor-Kleingrothaus, and S.G. Kovalenko, Phys. Lett. B 498, 35 (2001).
37. G. Prezeau, M. Ramsey-Musolf, and P. Vogel, Phys. Rev. D 68, 034016 (2003).
38. M. Hirsch, S. Kovalenko, and I. Schmidt, Phys. Lett. B 642, 106 (2006).
39. M. Duerr, M. Lindner, and A. Merle, J. High Energy Phys. 2011, 091.
40. F.F. Deppisch, M. Hirsch, and H. Pas, J. Phys. G 39, 124007 (2012).
41. J.D. Vergados, H. Ejiri, and F. Simkovic, Rep. Prog. Phys. 75, 106301 (2012).
42. J.C. Helo, M. Hirsch, S.G. Kovalenko, and H. Pas, Phys. Rev. D 88, 011901 (2013).
43. J.C. Helo, M. Hirsch, H. Ps, and S.G. Kovalenko, Phys. Rev. D 88, 073011 (2013).

44. P. S. B. Dev, S. Goswami, M. Mitra, and W. Rodejohann, Phys. Rev. D 88, 091301 (2013).
45. W.-C. Huang and J. Lopez-Pavon, Eur. Phys. J. C 74, 2853 (2014).
46. F. F. Deppisch, L. Graf, J. Harz, and W.-C. Huang, Phys. Rev. D 98, 055029 (2018).
47. F. Šimkovic, M. I. Krivoruchenko, and S. Kovalenko, Phys. Part. Nuclei 48, 1023 (2017).
48. L. Graf, F. F. Deppisch, F. Iachello, and J. Kotila, Phys. Rev. D 98, 095023 (2018).
49. R. Cepedello, F. F. Deppisch, L. González, and M. Hirsch, arXiv:1811.00031.
50. O. G. Miranda and H. Nunokawa, New J. Phys. 17, 095002 (2015).
51. A. de Gouvea and J. Jenkins, Phys. Rev. D 77, 013008 (2008).
52. C. Dib, V. Gribanov, S. Kovalenko, and I. Schmidt, Phys. Lett. B 493, 82 (2000).
53. A. Atre, V. Barger, and T. Han, Phys. Rev. D 71, 113014 (2005).
54. R. Aaij et al. (LHCb Collaboration), Phys. Rev. Lett. 108, 101601 (2012).
55. M. Moulson (NA62 Collaboration), Proc. Sci. 2013, 013.
56. J. R. Batley et al. (NA48/2 Collaboration), Phys. Lett. B 769, 67 (2017).
57. N. Fornengo, M. Maltoni, R. Tomas, and J. W. F. Valle, Phys. Rev. D 65, 013010 (2001).
58. Z. Berezhiani and A. Rossi, Phys. Lett. B 535, 207 (2002).
59. S. Davidson, C. Pena-Garay, N. Rius, and A. Santamaria, J. High Energy Phys. 2003, 011.
60. J. Barranco, O. G. Miranda, and T. I. Rashba, J. High Energy Phys. 2005, 021.
61. M. Blennow, T. Ohlsson, and J. Skrotzki, Phys. Lett. B 660, 522 (2008).
62. J. Kopp, M. Lindner, T. Ota, and J. Sato, Phys. Rev. D 77, 013007 (2008).
63. S. Antusch, J. P. Baumann, and E. Fernandez-Martinez, Nucl. Phys. B810, 369 (2009).
64. C. Biggio, M. Blennow, and E. Fernandez-Martinez, J. High Energy Phys. 2009, 090.
65. C. Biggio, M. Blennow, and E. Fernandez-Martinez, J. High Energy

Phys. 2009, 139.

66. J. Kopp, P.A.N. Machado, and S.J. Parke, Phys. Rev. D 82, 113002 (2010).
67. A. Friedland, M.L. Graesser, I.M. Shoemaker, and L. Vecchi, Phys. Lett. B 714, 267 (2012).
68. T. Ohlsson, Rep. Prog. Phys. 76, 044201 (2013).
69. T. Ohlsson, H. Zhang, and S. Zhou, Phys. Lett. B 728, 148 (2014).
70. A. Esmaili and A. Yu. Smirnov, J. High Energy Phys. 2013, 026.
71. S.K. Agarwalla, P. Bagchi, D.V. Forero, and M. Trtola, J. High Energy Phys. 2015, 060.
72. I. Girardi and D. Meloni, Phys. Rev. D 90, 073011 (2014).
73. Y. Farzan and I.M. Shoemaker, J. High Energy Phys. 2016, 033.
74. D.V. Forero and P. Huber, Phys. Rev. Lett. 117, 031801 (2016).
75. M. Lindner, W. Rodejohann, and X.-J. Xu, J. High Energy Phys. 2017, 097.
76. P.O. Ludl and W. Rodejohann, J. High Energy Phys. 2016, 040.
77. J. Salvado, O. Mena, S. Palomares-Ruiz, and N. Rius, J. High Energy Phys. 2017, 141.
78. B. Dutta, S. Liao, L.E. Strigari, and J.W. Walker, Phys. Lett. B 773, 242 (2017).
79. I.M. Shoemaker, Phys. Rev. D 95, 115028 (2017).
80. W. Rodejohann, X.-J. Xu, and C.E. Yaguna, J. High Energy Phys. 2017, 024.
81. Y. Farzan and M. Tortola, Front. Phys. 6, 10 (2018).
82. I. Esteban, M.C. Gonzalez-Garcia, M. Maltoni, I. Martinez-Soler, and J. Salvado, J. High Energy Phys. 2018, 180.
83. A. Ferrari, A. Guglielmi, and P.R. Sala, Nucl. Phys. B, Proc. Suppl. 168, 169 (2007).
84. J. Cao, Nucl. Phys. B, Proc. Suppl. 229–232, 205 (2012).
85. L. Aliaga et al. (MINERvA Collaboration), Phys. Rev. D 94, 092005 (2016); et al., 95, 039903(A) (2017).
86. J.N. Bahcall and H. Primakoff, Phys. Rev. D 18, 3463 (1978).
87. J. Schechter and J.W.F. Valle, Phys. Rev. D 23, 1666 (1981).
88. L.F. Li and F. Wilczek, Phys. Rev. D 25, 143 (1982).

89. P. Langacker and J. Wang, Phys. Rev. D 58, 093004 (1998).
90. S.M. Bilenky and B. Pontecorvo, Lett. Nuovo Cimento 17, 569 (1976).
91. S. Eliezer and A.R. Swift, Nucl. Phys. B105, 45 (1976).
92. H. Fritzsch and P. Minkowski, Phys. Lett. 62B, 72 (1976).
93. E.K. Akhmedov and A.Yu. Smirnov, Phys. At. Nucl. 72, 1363 (200 ``9).
94. R.G. Winter, Lett. Nuovo Cimento 30, 101 (1981).
95. C. Giunti, C.W. Kim, and U.W. Lee, Phys. Rev. D 44, 3635 (1991).
96. S. Nussinov, Phys. Lett. 63B, 201 (1976).
97. B. Kayser, Phys. Rev. D 24, 110 (1981).
98. K. Kiers, S. Nussinov, and N. Weiss, Phys. Rev. D 53, 537 (1996).
99. C. Giunti and C.W. Kim, Phys. Rev. D 58, 017301 (1998).
100. M. Zralek, Acta Phys. Pol. B 29, 3925 (1998).
101. M. Beuthe, Phys. Rep. 375, 105 (2003).
102. E.K. Akhmedov and J. Kopp, J. High Energy Phys. 2010, 008; 2013, 52.
103. C. Giunti, Eur. Phys. J. C 39, 377 (2005).
104. A.G. Cohen, S.L. Glashow, and Z. Ligeti, Phys. Lett. B 678, 191 (2009).
105. C. Giunti, C.W. Kim, J.A. Lee, and U.W. Lee, Phys. Rev. D 48, 4310 (1993).
106. W. Grimus, P. Stockinger, and S. Mohanty, Phys. Rev. D 59, 013011 (1998).
107. C.Y. Cardall, Phys. Rev. D 61, 073006 (2000).
108. W. Grimus and P. Stockinger, Phys. Rev. D 54, 3414 (1996).
109. D. Delepine, V.G. Macias, S. Khalil, and G.L. Castro, Phys. Lett. B 693, 438 (2010).
110. D. Delepine, V.G. Macias, S. Khalil, and G.L. Castro, Phys. Rev. D 79, 093003 (2009).
111. A.M. Ankowski and C. Mariani, J. Phys. G 44, 054001 (2017).
112. A. Denner, H. Eck, O. Hahn, and J. Kublbeck, Nucl. Phys. B387, 467 (1992).
113. Z.-z. Xing, Phys. Rev. D 87, 053019 (2013).
114. Z.-z. Xing and Y.-L. Zhou, Phys. Rev. D 88, 033002 (2013).

115. A. de Gouvea, B. Kayser, and R.N. Mohapatra, Phys. Rev. D 67, 053004 (2003).
116. A. Friedland and C. Lunardini, Phys. Rev. D 74, 033012 (2006).
117. W. Altmannshofer, M. Tammaro, and J. Zupan, arXiv:1812.02778.
118. M. Aartsen et al. (IceCube Collaboration), Phys. Rev. D 97, 072009 (2018).
119. P. Adamson et al. (MINOS Collaboration), Phys. Rev. D 95, 012005 (2017).
120. J. Liao and D. Marfatia, Phys. Lett. B 775, 54 (2017).
121. H. Oki and O. Yasuda, Phys. Rev. D 82, 073009 (2010).
122. A. de Gouva and K.J. Kelly, Nucl. Phys. B908, 318 (2016).
123. P. Coloma, J. High Energy Phys. 2016, 016.
124. M. Masud, A. Chatterjee, and P. Mehta, J. Phys. G 43, 095005 (2016).
125. M. Blennow, S. Choubey, T. Ohlsson, D. Pramanik, and S.K. Raut, J. High Energy Phys. 2016, 090.
126. K. Huitu, T.J. Krkkinen, J. Maalampi, and S. Vihonen, Phys. Rev. D 93, 053016 (2016).
127. S. Fukasawa, M. Ghosh, and O. Yasuda, Phys. Rev. D 95, 055005 (2017).
128. M. Masud and P. Mehta, Phys. Rev. D 94, 013014 (2016).
129. J. Liao, D. Marfatia, and K. Whisnant, J. High Energy Phys. 2017, 071.
130. S.K. Agarwalla, S.S. Chatterjee, and A. Palazzo, Phys. Lett. B 762, 64 (2016).
131. P. Bakhti, A.N. Khan, and W. Wang, J. Phys. G 44, 125001 (2017).
132. I. Bischer and W. Rodejohann, Phys. Rev. D 99, 036006 (2019).
133. K. Eguchi et al. (KamLAND Collaboration), Phys. Rev. Lett. 92, 071301 (2004).
134. E.E. Jenkins, A.V. Manohar, and P. Stoffer, J. High Energy Phys. 2018, 016.
135. E.E. Jenkins, A.V. Manohar, and P. Stoffer, J. High Energy Phys. 2018, 084.
136. C.-Q. Geng and D. Huang, J. High Energy Phys. 2017, 103.
137. R.N. Mohapatra and J.C. Pati, Phys. Rev. D 11, 566 (1975).
138. N. Cabibbo, Phys. Rev. Lett. 10, 531 (1963).

139. G. Pawloski (MINOS Collaboration), Proc. Sci. 2017, 004.
140. P. Adamson et al. (MINOS Collaboration), Phys. Rev. Lett. 110, 251801 (2013).
141. S.E. Kopp (MINOS Collaboration), AIP Conf. Proc. 981, 142 (2008).
142. I. Danko (MINOS Collaboration), arXiv:0910.3439.
143. J.A. Formaggio and G.P. Zeller, Rev. Mod. Phys. 84, 1307 (2012).
144. R. Acquafredda et al., J. Instrum. 4, P04018 (2009).
145. G. Galati (OPERA Collaboration), Proc. Sci. 2016, 010.
146. N. Agafonova et al. (OPERA Collaboration), Phys. Rev. Lett. 120, 211801 (2018).
147. J.N. Bahcall, E. Lisi, D.E. Alburger, L. De Braeckeleer, S.J. Freedman, and J. Napolitano, Phys. Rev. C 54, 411 (1996).
148. M. Agostini et al., Nature (London) 544, 47 (2017).
149. M. Agostini et al. (GERDA Collaboration), Phys. Rev. Lett. 120, 132503 (2018).
150. A. Gando et al. (KamLAND-Zen Collaboration), Phys. Rev. Lett. 117, 082503 (2016); et al., 117, 109903(E) (2016).
151. B. Pontecorvo, Zh. Eksp. Teor. Fiz. 53, 1717 (1967); [Sov. Phys. JETP 26, 984 (1968)].
152. G. Adamov et al. (COMET Collaboration), arXiv:1812.09018.
153. L. Bartoszek et al. (Mu2e Collaboration), arXiv:1501.05241.
154. J. Kaulard et al. (SINDRUM II Collaboration), Phys. Lett. B 422, 334 (1998).
155. B. Yeo, Y. Kuno, M. Lee, and K. Zuber, Phys. Rev. D 96, 075027 (2017).
156. J.M. Berryman, A. de Gouvêa, K.J. Kelly, and A. Kobach, Phys. Rev. D 95, 115010 (2017).
157. T. Geib, A. Merle, and K. Zuber, Phys. Lett. B 764, 157 (2017).

CHAPTER

13

Spectrometry of the Earth Using Neutrino Oscillations

C. Rott[1], A. Taketa[2] & D. Bose[1]

[1]Department of Physics, Sungkyunkwan University, Suwon, 440-746, South Korea

[2]Earthquake Research Institute, University of Tokyo, 1-1-1 Yayoi, Bunkyo-ku, Tokyo, Japan

ABSTRACT

The unknown constituents of the interior of our home planet have provoked the human imagination and driven scientific exploration. We herein demonstrate that large neutrino detectors could be used in the near future to significantly improve our understanding of the Earth's inner chemical composition. Neutrinos, which are naturally produced in the atmosphere, traverse the Earth and undergo oscillations that depend on the Earth's electron density. The Earth's chemical composition can be determined by combining observations from large neutrino detectors with seismic measurements

Citation: Rott, C., Taketa, A. & Bose, D. "Spectrometry of the Earth using Neutrino Oscillations". Sci Rep 5, 15225 (2015). https://doi.org/10.1038/srep15225

of the Earth's matter density. We present a method that will allow us to perform a measurement that can distinguish between composition models of the outer core. We show that the next-generation large-volume neutrino detectors can provide sufficient sensitivity to reject extreme cases of outer core composition. In the future, dedicated instruments could be capable of distinguishing between specific Earth composition models and thereby reshape our understanding of the inner Earth in previously unimagined ways.

INTRODUCTION

Understanding the inner structure and composition of the Earth is fundamental to Earth science. While Earth's matter density distribution can be inferred from geophysical observations, its compositional structure is far more difficult to determine. The state and composition of the core, which constitutes 32% of Earth's mass and 16% of its volume, remains largely uncertain. The core consists of iron-nickel mixture and is divided into inner and outer regions distinguished by a large density difference at a depth of approximately 5,100 km. The inner core is solid, while the lack of s-wave propagation in the outer core and lower density indicate it to be liquid. The density deficit in the outer core, however, cannot be simply explained by a difference in state, but rather requires the presence of light elements at about 5 wt% (weight percent) to 10 wt%. There is great excitement in Earth science with regard to determining these light components in the outer core in order to understand the evolution of the Earth and the geodynamo. We introduce a new technique based on neutrino oscillations in order to remotely measure electron density and demonstrate how, in the near future, this method could be used to distinguish between different composition models of the inner Earth.

Analyses of seismic waves have resulted in the well-understood shell structure of the Earth, consisting of crust, upper mantle, lower mantle, outer core and inner core. The matter density structure of the Earth has been accurately determined by combining astronomic-geodetic parameters, free oscillation frequencies and seismic wave velocity measurements1. The composition of the crust near the surface can be measured directly. Drill core samples have resulted in composition measurements down to a depth of approximately 12 km2. The upper mantle composition can be probed through eruption entrainment sampling3. The state and composition of the Earth's core, at a depth of approximately 2,900 km remains far more uncertain with no prospects of sampling materials.

The outer core composition can be inferred to be mostly iron-nickel mixture with traces of light elements, by combining seismological velocity profiles and the composition of primitive meteorites4. Through recent progress in high-pressure experiments, hydrogen, carbon, oxygen, silicon and sulfur have been suggested as light element candidates5. However, the abundance of these light elements remains uncertain.

Obtaining reliable estimates for the abundances of light elements in the Earth's core is essential to understanding the formation and evolution of the Earth6 and to determining the origin of the geomagnetic field7, which are two of the major problems in Earth science.

Neutrinos (denoted ν) are remarkable particles that have enjoyed an ever more important role in particle physics, cosmology and astrophysics since they were predicted by theorist Wolfgang Pauli in 1930 and first observed in 19568. There exist three different types (referred to as flavours) of neutrinos, ν_e, ν_μ and ν_τ, which relate to how the neutrino was produced. However, a neutrino's flavour can change. For example, a neutrino produced as a ν_μ can be detected as a ν_e. This process, which solved the solar neutrino problem9, is known as neutrino oscillation10. At a later date, three flavour neutrino oscillation was confirmed by observing the neutrinos produced in the Earth atmosphere11. Neutrino oscillations are a quantum mechanical consequence of neutrinos having mass and as such the behaviour of these oscillations can be described precisely.

In the present study, we propose a novel technique for measuring the average chemical composition of the deep Earth using neutrinos. Due to their tiny interaction cross section, neutrinos can pass through the entire Earth without interacting. As mentioned earlier, due to neutrino oscillations, a flavour of one neutrino can convert to another flavour. Neutrino oscillations depend on the medium traversed, or, more specifically, on the electron density along the path of the neutrino through the Earth12. The compositional structure of the Earth can be obtained as the average ratio of the atomic number to the atomic weight (Z/A), by comparing the electron density distribution and the Earth's matter density distribution. This effect makes neutrinos unique messenger particles to remotely probe the Earth's interior.

Large-volume neutrino detectors have emerged as powerful tools in particle physics and astrophysics. Operating instruments have demonstrated their tremendous potential in groundbreaking discoveries, such as the observation of high-energy extra-terrestrial neutrinos by IceCube and through the observation of neutrino oscillations by Super-Kamiokande. There is a great interest in constructing the next generation of neutrino detectors with larger volumes and improved performance. This new generation of large-volume detectors could be capable of observing neutrinos at sufficiently high rates to perform the first experimental measurement of the Earth's interior. With the advent of Hyper-Kamiokande (Hyper-K)13 and the Precision IceCube Next-Generation Upgrade (PINGU)14, spectrometry using neutrino oscillations could be demonstrated and a first direct measurement of the compositional structure of the Earth performed. Even more visionary ideas, such as large ocean-going15 or ice-based detectors, could see neutrino spectrometry emerge as a precision science capable of distinguishing different compositional models.

Preceding research of geophysics using neutrinos can be divided into three categories: (1) measurement of the radioactive nuclei density in the Earth using geo-neutrinos generated through nuclear decays, (2) measurement of Earth's matter density using neutrino absorption and (3) measurement of Earth's matter density using neutrino oscillations16·17·18·19·20·21·22·23. In the present study, we introduce a new fourth category. We apply neutrino oscillations for a composition measurement, exploiting the fact that neutrino oscillations are dependent on electron density, which is the product of the matter density and the ratio of the average atomic number to the atomic weight. While the underlying physical phenomena are well understood, we focus in particular on the relevance of these effects to geophysics and discuss the prospects for an Earth composition measurement that could be performed within the next two decades.

RESULTS

Neutrino oscillations in the Earth

In geophysics, neutrinos have received attention due to the information on the inner Earth they provide, as demonstrated by the measurement of radiogenic heat generated in the Earth through the observations of neutrinos from nuclear decays of uranium and thorium24·25. The success in detecting these geoneutrinos has confirmed the feasibility of using neutrinos in Earth science. While geoneutrinos are generated through nuclear decay and carry energies of approximately 10^6 eV (one electron volt (eV) = 1.602×10^{-19} joules), the neutrinos used for the proposed method have energies of a few GeV (10^9 eV) and are naturally produced when energetic cosmic rays collide with the upper Earth's atmosphere. The majority of atmospheric neutrinos produced are type ν_μ and their flavour changes as they pass straight through the Earth. The neutrino oscillation probability depends on a set of oscillation parameters, the neutrino energy, E_ν, the distance travelled and the electron density along its path. The path length, L, is the distance that the neutrino travels from its point of origin in the atmosphere to the detector. Since all neutrinos relevant for this analysis are generated in the Earth's atmosphere, L is simply a function of the zenith angle, Θ, of the neutrino arrival direction at the detector. Figure 1(a) shows the neutrino path through the Earth.

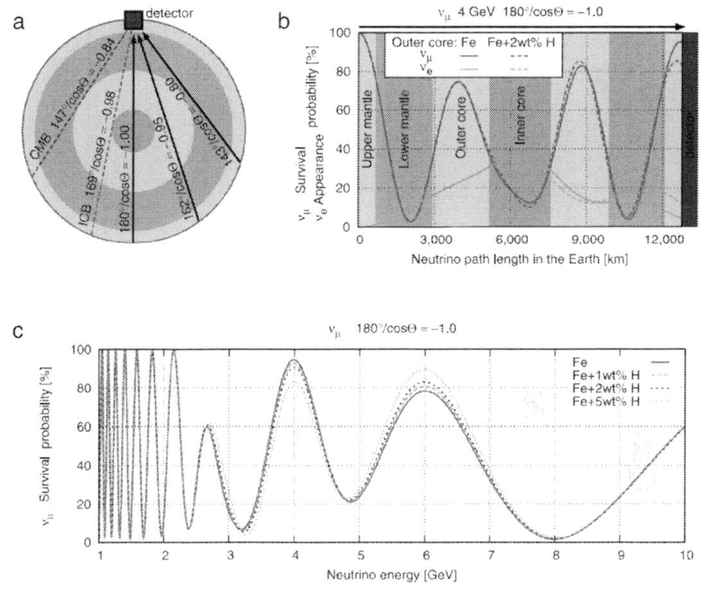

Figure 1. (a) Schematic diagram of a neutrino's path through the Earth and the corresponding zenith angles. The inner core boundary (ICB) at Θ = o 169 and the core mantle boundary (CMB) at Θ = o 147 are indicated by dashed red and blue lines, respectively. (b) ν_e appearance probability (green) and ν_μ survival probability (red) as functions of path length in the Earth. The neutrino direction is Θ = o 180 , as shown in (a). The solid/dashed line corresponds to the case in which the composition of the outer core is pure iron/a mixture of iron and 2wt% hydrogen. (c) Θ =180° − ν_μ survival probabilities as a function of neutrino energy for different outer core compositions. The solid (red), long dashed (green), short dashed (blue), and dotted (gray) lines represent iron, a mixture of iron and 1wt% hydrogen, a mixture of iron and 2wt% hydrogen, and a mixture of iron and 5wt% hydrogen, respectively.

We calculate neutrino oscillation probabilities, following the approach of Barger et al.12 and use the numerical implementation of the NuCraft software package26. The oscillation parameters, which are well measured, are taken from the global fit given by Capozzi et al.27, assuming the case of a normal mass hierarchy. We use the modified Preliminary Reference Earth Model (PREM) matter density model1,28 to describe the Earth density and structure. We fix the mantle composition to pyrolite, the hypothetical mixture of distinct minerals, which accurately represent the constituents of the Earth's mantle4. We constrain the composition of the inner core to iron and only vary that of the of the outer core.

Figure 1(b) shows the ν_μ survival probability and the ν_e appearance probability as a function of the path length for a neutrino with an energy of 4GeV(10^9eV) passing vertically through the Earth. The survival probability is the probability that a created neutrino of specific flavour is observed as such. In this case, we consider a muon neutrino observed as such $P(\nu_\mu \rightarrow \nu_\mu)$. The appearance probability is the chance that a neutrino of one flavour is observed as a neutrino of a different flavour, for example $P(\nu_\mu \rightarrow \nu_e)$. The flavour change as a function of the distance travelled in the Earth is shown. In order to visually show the impact of the outer core composition on the oscillation probability, we compare the cases of a mixture of iron and 2wt% hydrogen with iron. Figure 1(c) shows the ν_μ survival probability at the surface of the Earth, as a function of the neutrino's energy for four different core compositions. In order to visualize the difference in survival probability for different outer core compositions, we selected (1) iron, (2) a mixture of iron and 1wt% hydrogen, (3) a mixture of iron and 2wt% hydrogen, and (4) a mixture of iron and 5wt% hydrogen as extreme examples of the outer core composition.

Z/A ratios for different outer core models

Iron is the most abundant element in the outer Earth core and throughout this document we have chosen pure iron as our default composition. Models adding single or multiple elements to iron have been proposed29,30,31. In Table 1, we introduce some selected outer core composition models and characterize them according to Z/A ratio. The estimated maximal abundance of light elements5,32 for mixtures of iron are listed in Table 1. Note that nickel is thought to co-exist with iron in the outer core, with an estimated content of approximately 5%33. Since there is only a slight difference between Z/A values, using an mixture of iron and 5wt% nickel as the base composition rather than iron will result in only a marginal change in Z/A from 0.4656 to 0.4661.

Table 1. Z/A ratios for mixtures of iron and light elements and some selected composition models.

Model name	Z/A ratio	O(wt%)	C(wt%)	S(wt%)	H(wt%)	Si(wt%)
Single-light-element model (maximum abundance)						
Fe + 11 wt%O[32,56]	0.4693	11	—	—	—	—
Fe + 12 wt%C[5]	0.4697	—	12	—	—	—
Fe + 13 wt%S[5]	0.4699	—	—	13	—	—
Fe + 1 wt%H[5]	0.4709	—	—	—	1	—
Fe + 18 wt%Si[32]	0.4715	—	—	—	—	18
Multiple-light-element model						
Huang2011[31]	0.4678	0.1	—	5.7	—	—
Mc-Donough2003[30]	0.4682	0	0.2	1.9	0.06	6
Allegre2001[29]	0.4699	5	—	1.21	—	7

Oscillation probabilities for different outer core models

As neutrino oscillations simply depend on the neutrino's energy, path length and composition along the path, we can determine the probability that a neutrino will change flavours as a function of the zenith angle and the energy. We calculated the oscillation probabilities for different core models. Figure 2 shows such an oscillogram, i.e., oscillation probabilities as a function of zenith angle and neutrino energy, for two different outer core compositions. Subtle differences in the neutrino survival probability can be exploited in order to distinguish between different composition models. The most pronounced differences in survival probability are for neutrinos with energies between 2 GeV and 8 GeV that traverse the outer core, i.e., their zenith angles are larger than 147°.

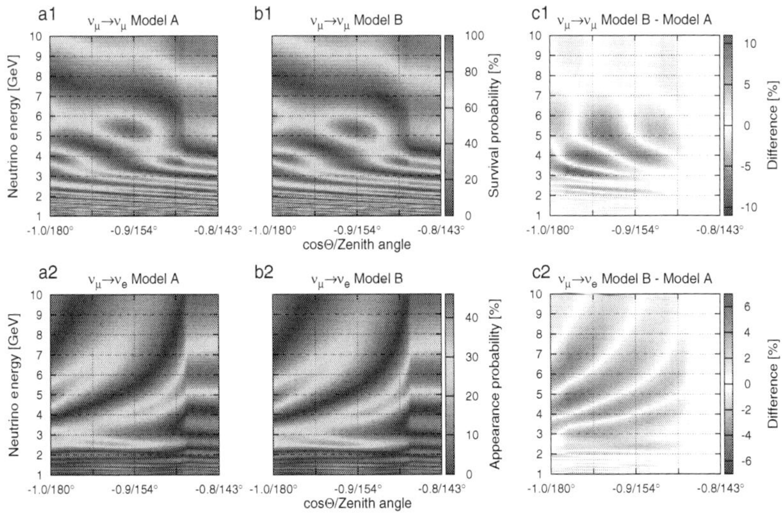

Figure 2. Comparison of oscillation probabilities for two different core compositions: Model A – iron; Model B – an mixture of iron and 2 wt% hydrogen.

a1 and b1 show the ν_μ survival probabilities as a function of neutrino energy and zenith angle for Models A and B, respectively. a2 and b2 show the appearance probability for ν_μ to ν_e for Models A and B, respectively. c1 shows the difference between a1 and b1 and c2 shows the difference between a2 and b2.

Detector requirements

The described differences in neutrino oscillation effects that depend on the Earth's composition could be detectable with a neutrino detector if the detector combines good energy and angular resolution in the relevant energy range and observes GeV neutrinos at sufficiently high rates to accumulate sufficient statistical samples. Due to the small neutrino interaction cross section, a large detector volume of megaton scale is necessary in order to acquire a sufficient number of neutrino events and not suffer from limited statistics. Good neutrino flavour identification can be beneficial.

Neutrino detectors

Neutrinos can easily be detected through Cherenkov light emissions from one or more energetic particles produced in neutrino interactions in an optical transparent medium such as ice or water. The IceCube neutrino telescope34 is the world's largest neutrino detector. It uses one gigaton (1,000 megatons) of ice at the Geographic South Pole that was instrumented with more than 5,000 photosensors. The optical sensor array relies on the ultra-pure Antarctic ice as a detection medium, in this way the giant detector size was realized in a cost-effective manner. The detector is working extremely well and the recent discovery of high-energy astrophysical neutrinos demonstrates the potential of large neutrino detectors35. Precision physics with neutrinos of a few GeV has been pioneered by Super-Kamiokande36. This neutrino detector consists of a 50 kiloton water tank surrounded by 11,000 photosensors to observe Cherenkov light, allowing neutrino energies to be determined precisely and neutrino flavours to be identified reliably. The underlying technology applied by IceCube and Super-Kamiokande is well established and is the basis for future detectors. Next-generation detectors could benefit from better photosensors with higher photon detection efficiency.

Sensitivity of benchmark detectors

We calculate the significance with which the composition of the outer core could be determined for some benchmark neutrino detectors. We use a generic neutrino detector description based on performance parameters to estimate sensitivities. Our parameterization can easily be converted into hardware and design requirements for the planning of new detectors. We compute expected event rates as a function of the neutrino energy and the zenith angle as a function of the product of detector size and the exposure time in megaton-years. In this way, we calculate the number of neutrino events for a certain energy, direction, and flavour. In our calculations we treat neutrinos (ν) and their anti-particles, the anti-neutrinos $(\overline{\nu})$, separately. As we refrain from discussing detectors with separation capabilities between ν and $(\overline{\nu})$, we use in the following the term neutrino for both ν and $(\overline{\nu})$, unless other specified. We create templates of the expected event rates for different outer core models. Event rates were calculated from the atmospheric neutrino flux, oscillation probabilities, neutrino cross section, detector volume, and exposure time. The atmospheric neutrino flux and energy spectrum are well understood for our purposes. We adopt the atmospheric neutrino flux model of Athar et al.37. For the neutrino (ν_μ) and anti-neutrino $(\overline{\nu}_\mu)$ interaction cross sections, we use the approximate values of $7.0 \times (E/GeV) \times 10^{-39}cm^2$ and $3.0 \times (E/GeV) \times 10^{-39}cm^2$, respectively38

The outcome of any experimental measurement that deals with individual events, such as the detection of neutrinos, will be subject to statistical fluctuations. We consider a large number of potential experimental outcomes, called pseudo experiments, in order to estimate the chance that models could be distinguished through an actual measurement. For each pseudo experiment, we compare the number of observed events to the number of expected events for a specific model. We calculate events for given ranges of energy and zenith angle. For each of these bins in energy and zenith angle, events follow Poisson statistics and we determine the probability for the observation. We then compute the likelihood of this experimental outcome with respect to a specific Earth model assumption. The total likelihood is then given by the product of the likelihoods for the individual bins. In order to compare the likelihood of one model with that of another, we compute the likelihood ratio. For the calculation of the expected significance, we perform a set of pseudo experiments and apply the log-likelihood ratio (LLR) method39. Figure 3 illustrates the significance calculation.

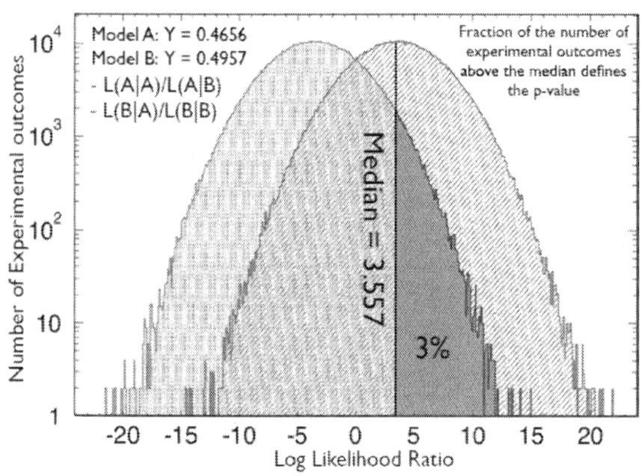

Figure 3. Method to determine significance of the expected experimental outcome using pseudo-experiments.

The significance is defined as the fraction of outcomes in which the log of the likelihood ratio is above the median of the distribution with interchanged models. For the example shown here it can be seen that Model A (Iron) can be distinguished from Model B (Pyrolite) with a 97% confidence. We assumed an exposure of 10 MTyrs and the benchmark detector with energy resolution and angular resolution, as defined by $\Delta E_\nu / E_\nu = 0.20$ and $\Delta\Theta = 0.25/\sqrt{E/\mathrm{GeV}}$.

For simplicity of the analysis, we consider only muon neutrino $(\nu_\mu + \overline{\nu}_\mu)$ events, which are the most relevant for neutrino spectrometry.

Muon neutrino events can be identified with high efficiency by proposed next-generation detectors, such as PINGU or Hyper-K. For PINGU, the resolution for the muon neutrino energy is expected to be better than 25% at 5 GeV and the zenith angle resolution for the neutrino has been reported to be approximately 13°[14]. At Hyper-K, better angular and energy resolutions compared to PINGU can be expected, in addition to a larger than 99% efficiency to identify the interaction products of a neutrino interaction[13,40].

Figure 4(a,b) show the sensitivity for rejecting outer core compositions given by their Z/A ratios with respect to iron for a generic neutrino detector. The detector is characterized by energy resolution and angular resolution, as defined by $\Delta E_\nu / E_\nu = \alpha$ and $\Delta\Theta = \beta / \sqrt{E/\mathrm{GeV}}$, respectively. We choose α=0.20 and β=0.25 as a default generic benchmark detector and show the sensitivity depending on the product of lifetime and detector volume (megaton-years) in Fig. 4(b). The specified benchmark detector parameters correspond to a 20% energy resolution independent of energy and a 6.5° zenith angle resolution at 5GeV. The angular and energy resolution dependences of the sensitivity is shown in Fig. 4(c,d).

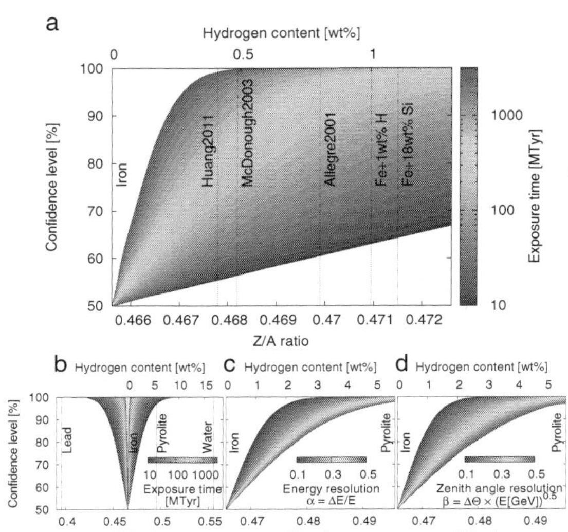

Figure 4. (**a**) Expected confidence level for rejecting a specific outer core composition with respect to iron plotted as a function of the corresponding Z/A ratio. A generic detector case with an energy resolution of 20% and an angular resolution of $0.25 \times (E/\mathrm{GeV})^{-0.5}$ is shown as an example. The colour indicates the exposure time given in megaton-years. We indicate the Z/A ratios for some selected outer core composition models (see Table 1 for details) as black dot-

ted vertical lines. (**b**) The same plot as (**a**) for a larger Z/A range. Sensitivity dependences on (**c**) energy resolution and (**d**) angular resolution for a generic detector with an exposure time of 30 megaton-years for an angular resolution of $0.25 \times (E/\mathrm{GeV})^{-0.5}$ and an energy resolution of 20%, respectively.

With an acquired dataset of 10 megaton-years, a neutrino detector could for the first time confirm an iron-like core through experimental measurements. Extreme cases of outer core composition such as lead or water could be rejected with more than 99% confidence with respect to iron. The Z/A ratios of iron, lead, mantle (pyrolite) and water are 0.4656, 0.3958, 0.4957 and 0.5556, respectively. The large hydrogen content appearing at the axis at the top of each plot could be rejected.

Figure 4(a) shows the future prospects of neutrino spectrometry. A one-gigaton-year (or 1,000 megaton-year) dataset, which is equivalent to observation for 20 year using a 50 megaton detector, would provide the ability to discriminate between established outer core composition models. Furthermore, the hydrogen content of the outer core could be measured with a precision of 0.4 wt%. Better sensitivities could be achieved if the detector exceeds the benchmark detector performance parameters selected for use in the present study as shown in Fig. 4(c,d) for the energy resolution and angular resolution, respectively.

Note that neutrino spectrometry determines the Z/A ratio. Based on this measurement Earth composition models can be distinguished by their corresponding Z/A ratios. Some Earth composition models however predict very similar values for this ratio. The origin for this lies in the fact that for example oxygen, sulfur, silicon and carbon are being nearly degenerate in their ratio of atomic number to atomic mass. Hence, a model with Fe (90 wt%)+O (10 wt%) (Z/A=0.4690) could be distinguished from an iron (Z/A=0.4656), but would be indistinguishable from a model with Fe (90 wt%)+S (10 wt%) (Z/A=0.4689) through neutrino spectrometry alone. The full impact of neutrino spectrometry can only be assessed by its complimentarily to existing methods in Earth science. For example, degeneracies between composition models could be resolved by using neutrino spectrometry in the combination with high-pressure experiments. High-pressure experiments measure the matter properties, such as density and sound velocity, of mixtures of iron with other light elements at core pressure-temperature conditions41. Measurements have resulted in limits on the content of each of the light elements in the core. Further studies, which are beyond the scope of this work, will be needed to show how a combination of neutrino spectrometry with high-pressure experiments could distinguish composition models.

UNCERTAINTIES

We discuss the feasibility of the neutrino spectrometry measurement with future neutrino detectors with respect to theoretical and experimental uncertainties.

Uncertainties in the composition measurement originate from limited knowledge of the neutrino oscillation parameters, atmospheric neutrino flux uncertainties, neutrino cross section uncertainties and uncertainties in the Earth's matter density profile. In addition to these theoretical uncertainties, detector related uncertainties must be determined. However, this is beyond the scope of the present study and would have to be carried out through experimental collaborations.

To examine the feasibility of the neutrino spectrometry measurement, we calculated the confidence level by changing the oscillation parameters and the matter density models. To calculate the confidence level, we used the same oscillation parameters and matter density model for each template. Using the current best-fit oscillation parameters and their uncertainty27, the error in the confidence level curve was approximately ±4% at 90% (see Fig. 5:Left). To calculate the uncertainty from oscillation parameters, we performed a grid scan of the oscillation parameters, each of them was varied with a 0.1σ step size for the 1σ uncertainty range and a 0.25σ step size for the 3σ range. Several experiments are planned for the near future in order to realize more precise measurements of the neutrino mixing parameters42·43·44. A better determination of the neutrino oscillation parameters will reduce uncertainties.

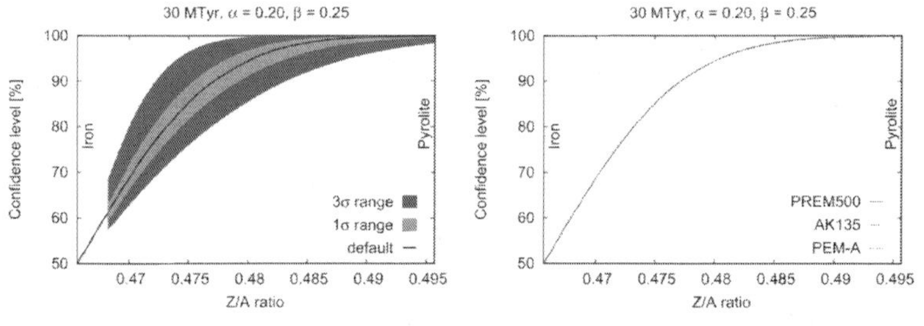

Figure 5. Systematic error of the expected confidence level as a function of Z/A.

Left: Systematic error resulting from the uncertainty in oscillation parameters. The blue line represents default mixing parameter case, the red area represents its 3σ uncertainty range and the green area represents its 1σ uncertainty range. A generic detector case with an exposure time of 30 MTyr, an energy resolution of 20% and an angular resolution of $0.25 \times (E/\mathrm{GeV})^{-0.5}$ is shown. Right: Systematic error resulting from matter density models. Expected confidence level for rejecting a specific outer core composition with respect to iron plotted as a function of the corresponding Z/A ratio. A generic detector case with an exposure time of 30 MTyr, an energy resolution of 20% and an angular resolution of $0.25 \times (E/\mathrm{GeV})^{-0.5}$ is shown. We estimated the confidence level using three different density models. The solid (red), dashed (green) and dotted (blue) lines represent the modified PREM, AK135 and PEM-A, respectively.

In order to estimate the uncertainty resulting from the matter density models, we have calculated the confidence level curves using three different density models (PREM5001·28, AK13545, PEM-A46; see Supplementary Figure 1). The systematic error resulting from the matter density in the confidence level curve was negligible compared with the systematic error resulting from the mixing parameters (see Fig. 5:Right). The expected uncertainty is sufficiently small to distinguish the models introduced in the present study. However, in order to determine the light material contents in the outer core, a more precise mixing parameter and matter density model, which may be available in the near future, are needed. We have limited our sensitivity study to the dominant systematic uncertainties. Capozzi *et al.*47 showed that systematic uncertainties related to resolution functions in energy and angle are sub-dominant to the oscillation parameter uncertainties. We note that effects related to the detector energy and angular resolution deserve further investigations, but go beyond the scope of this work.

At present, the neutrino mass hierarchy, one of the fundamental neutrino properties, remains unknown27·48. By the time that the proposed composition measurement is performed, we can safely assume that the mass hierarchy will have been determined, potentially even at the same neutrino detector considered for our measurement49·50. For the case of a normal (inverted) neutrino mass hierarchy matter induced neutrino oscillations will act on neutrinos (anti-neutrinos). A normal mass hierarchy would be favourable for our proposed measurement because the atmospheric flux of neutrinos is larger than that of anti-neutrinos and the interaction cross section of neutrinos is larger compared to that of anti-neutrinos. For our study we have assumed a normal mass hierarchy. In the case of an inverted mass hierarchy a roughly six times larger dataset would be needed to achieve the same sensitivity compared to the normal mass hierarchy.

DISCUSSIONS

Neutrino oscillations provide a way to distinguish different Earth composition models by probing the Z/A ratio. The novel method introduced here can provide a measurement of the chemical abundances of the inner Earth, which is direct, independent and complimentary to phase state measurements in high pressure experiments. A better measurement of the core composition is essential to improve scenarios of the formation of the Earth, which depend on composition models6. Further, a composition measurement is essential to understand convection in the outer liquid core. Understanding the convection effects is key to explain the gyodynamo, which is expected to be driven by thermal and compositional convection in the outer core7. Neutrino spectroscopy will lead us to a better understanding of the Earth's evolution and the origin of the geomagnetic field.

Large-volume neutrino detectors with good angular and energy resolutions at neutrino energies of 2−8 GeV are needed for our proposed measurements. While at present no instrument exists that combines the necessary capabilities discussed here, the technology is however well established and proven through large volume and high-precision detectors. In the near future however, new neutrino telescopes and upgrades to existing instruments will significantly enhance neutrino detection capabilities in the most relevant energy range for the spectroscopic measurement discussed here. The Hyper-K project will see the construction of a 0.6-megaton fiducial volume detector comprising eight compartments and approximately 100,000 photosensors. PINGU will use a few megatons of clearest ice in the centre of the IceCube detector. A detector of similar size named as ORCA is also considered as a deep-sea neutrino telescope in the Mediterranean Sea as part of the KM3NeT project51·52. Construction of a gigaton volume detector (GVD)53 has also started in lake Baikal and complementation with a PINGU-like detector could be considered. All these detectors in the planning and construction phase offer sufficient sensitivity to demonstrate neutrino spectrometry measurements and are expected to exclude extreme cases of the Earth's composition, such as water or lead. If complimented and carefully optimized for neutrino spectrometry, the first meaningful bounds on the hydrogen content in the core could be in reach. In the future, dedicated next-generation neutrino experiments could be used to distinguish between different composition models, as we have demonstrated.

We have limited our studies to muon neutrinos. The detection of neutrinos of different flavours is expected to enhance the sensitivity of the proposed method, warranting further investigation in the future.

In the present study, we focused on the composition of the outer core, but neutrino spectrometry could also be applied to the mantle, especially in order to elucidate the water content of the lower mantle. Through recent progress in diamond inclusion sampling, high-pressure experiments and dense seismic velocity measurements, it was found that the uppermost part of the lower mantle can reserve 1 wt% water54. Neutrino spectrometry has the ability to provide an upper limit for the water content of the lower mantle in the same way as the hydrogen content of the outer core.

METHODS

Flux calculation

Event distributions for a generic neutrino detector defined by volume, energy resolution and angular resolution were calculated for angle-averaged atmospheric neutrino fluxes after propagation through the Earth. The calculation proceeded in two steps: (1) first we calculated the transition probabilities for neutrinos of all flavours as function of zenith angle and energy. For this step we computed full three-flavour neutrino oscillations with the NuCraft software package26 using the 400 layers PREM Earth structural model. The results were binned in 720 bins of the cosine of the zenith angle and 400 bins of energy to obtain a transfer matrix *T*. We computed *T* for different outer core Z/A ratios and for the purpose of systematic studies, we used different Earth structural models (see Supplementary Figure 1) and oscillation parameters. (2) In a second step we obtain detector event rates. The predicted neutrino flux for muon neutrinos obtained from the Honda model37,55 was propagated through the Earth using transition tables describing the muon neutrino survival probability. Interaction rates were calculate for our generic detector and coarsely binned in 40 bins of the log of reconstructed neutrino energy and in 20 bins of the cosine of the reconstructed zenith angle for the range of $\cos\Theta = (-1...0)$ and log(E/GeV)=0…1. The transitions to this coarser binned matrix *M* was done to have bin sizes that are closer to expected detector resolutions and to insure a reasonable event statistics per bin, avoiding empty bins. To obtain reconstructed zenith angle and neutrino energies we randomly sampled from the expected distribution as defined by the detector model (assuming Gaussian distributions). For each bin of *T*, 100,000 events were generated and mapped into the reconstruction matrix, weighted by the expected neutrino flux calculation37. Rates were calculated according to the neutrino interaction cross section and the product of the detector volume and operation time (megaton-years). Each bin m_{ij} of M obtains the expectation value of an observation, were a measurement to be performed. Using a different Earth composition model and the same generic detector model, we can obtain the expectation values for this model. Let m_{ij}^{A} and m_{ij}^{B} be the templates for two different models, respectively A and B.

Log-likelihood method

Ensembles of pseudo datasets are drawn from each template m_{ij}^{A} and m_{ij}^{B} that are repeatedly varied following Poisson statistics. The log of the Poisson likelihood of the pseudo data for a specific bin is calculated with respect to the corresponding bin in m_{ij}^{A} and m_{ij}^{B}. Given an experimental outcome O, with bins o_{ij}, we take the sum of $(-2)\log\left(P\left(o_{ij}, m_{ij}^{A}\right)\right)$ to obtain the total likelihood. In this way, for each pseudo dataset, two likelihoods are calculated and are labelled $\mathcal{L}$ (pseudo data, template). The likelihoods are used to calculate LLR. We calculate two distributions for pseudo data oij drawn from m_{ij}^{A} given by $\mathcal{L}(o(M^{A})|M^{A})/\mathcal{L}(o(M^{A})|M^{B})$ and $\mathcal{L}(o(M^{B})|M^{A})/\mathcal{L}(o(M^{B})|M^{B})$total of 10,000 pseudo datasets are used to achieve adequate coverage of the probability space. We obtain expected significances from our ensemble of pseudo experiments. The probability of distinguishing model A from B is obtained by calculating the fraction of cases in which events drawn from M^{A} have a likelihood ratio that is more consistent with M^{A} than M^{B}. An illustration of this method is given in Fig. 3.

REFERENCES

1. Dziewonski, A. M. & Anderson, D. L. Preliminary reference Earth model. Phys. Earth Planet. Inter. 25, 297–356 (1981).
2. Popov, Y. A., Pevzner, S. L., Pimenov, V. P. & Romushkevich, R. A. New geothermal data from the Kola superdeep well SG-3. Tectonophysics 306, 345–366 (1999).
3. Hofmann, A. W. Mantle geochemistry: the message from oceanic volcanism. Nature 385, 219–229 (1997).
4. McDonough, W. & Sun, S. The composition of the Earth. Chem. Geol 120, 223–253 (1995).
5. Li, J. & Fei, Y. Experimental Constraints on Core Composition. In Treatise on Geochemistry vol. 3, 1–31 (Elsevier, 2007).
6. Allègre, C. J., Poirier, J., Humler, E. & Hofmann, A. W. The chemical composition of the Earth. Earth Planet. Sci. Lett 134, 515–526 (1995).
7. Fearn, D. R. & Loper, D. E. Compositional convection and stratification of Earth's core. Nature 289, 393–394 (1981).
8. Cowan, C. L., Reines, F., Harrison, F. B., Kruse, H. W. & McGuire, A. D. Detection of the Free Neutrino: a Confirmation. Science 124, 103–4 (1956).
9. Bahcall, J. N. & Pinsonneault, M. H. Standard solar models, with and without helium diffusion and the solar neutrino problem. Rev. Mod. Phys 64, 885–926 (1992).
10. Pontecorvo, B. Mesonium and anti-mesonium. Sov.Phys.JETP 6, 429 (1957).
11. Fukuda, Y. et al. Evidence for Oscillation of Atmospheric Neutrinos. Phys. Rev. Lett 81, 1562–1567 (1998).
12. Barger, V., Whisnant, K., Pakvasa, S. & Phillips, R. J. N. Matter effects on three-neutrino oscillations. Phys. Rev. D 22, 2718–2726 (1980).
13. Abe, K. et al. Letter of Intent: The Hyper-Kamiokande Experiment—Detector Design and Physics Potential—. Preprint arXiv:1109.3262 (2011).
14. The IceCube-PINGU Collaboration. Letter of Intent: The Precision IceCube Next Generation Upgrade (PINGU). Preprint arXiv:1401.2046 (2014).
15. Kistler, M. D., Yüksel, H., Ando, S., Beacom, J. F. & Suzuki, Y. Core-collapse astrophysics with a five-megaton neutrino detector. Phys. Rev.

D 83, 1–11 (2011).

16. Nicolaidis, A. Neutrinos for geophysics. Phys. Lett. B 200, 553–559 (1988).
17. Jacobsson, B., Ohlsson, T., Snellman, H. & Winter, W. The Effects of matter density uncertainties on neutrino oscillations in the earth. J.Phys. G29, 1873–1876 (2003).
18. Lindner, M. Tomography of the Earth's core using supernova neutrinos. Astropart. Phys. 19, 755–770 (2003).
19. Geller, R. J. & Hara, T. Geophysical aspects of very long baseline neutrino experiments. Nucl. Instr. Meth. A 503, 187–191 (2003).
20. Winter, W. Probing the absolute density of the Earth's core using a vertical neutrino beam. Phys. Rev. D 72, 1–4 (2005).
21. Winter, W. Neutrino tomography: Learning about the earth's interior using the propagation of neutrinos. Earth Moon Planets 99, 285–307 (2006).
22. Gonzalez-Garcia, M., Halzen, F., Maltoni, M. & Tanaka, H. Radiography of Earths Core and Mantle with Atmospheric Neutrinos. Phys. Rev. Lett 100, 061802 (2008).
23. Agarwalla, S. K., Li, T., Mena, O. & Palomares-Ruiz, S. Exploring the Earth matter effect with atmospheric neutrinos in ice. Preprint arXiv:1212.2238 (2012).
24. Araki, T. et al. Experimental investigation of geologically produced antineutrinos with KamLAND. Nature 436, 499–503 (2005).
25. Bellini, G. et al. Measurement of geo-neutrinos from 1353 days of Borexino. Phys. Lett. B 722, 295–300 (2013).
26. Wallraff, M. & Wiebusch, C. Calculation of oscillation probabilities of atmospheric neutrinos using nuCraft. Preprint arXiv:1409.1387 (2014).
27. Capozzi, F. et al. Status of three-neutrino oscillation parameters, circa 2013. Phys. Rev. D 89, 093018 (2014).
28. Durek, J. & Ekström, G. A radial model of anelasticity consistent with long-period surface-wave attenuation. Bull. Seismol. Soc. Am. 86, 144–158 (1996).
29. Allègre, C., Manhes, G. & Lewin, E. Chemical composition of the Earth and the volatility control on planetary genetics. Earth. Planet. Sci. Lett 185, 49–69 (2001).

30. Mcdonough, W. F. Compositional Model for the Earth's Core. In Treatise on Geochemistry vol. 2, 547–566 (Elsevier, 2003).
31. Huang, H. et al. Evidence for an oxygen-depleted liquid outer core of the Earth. Nature 479, 513–6 (2011).
32. Poirier, J. P. & Shankland, T. J. Dislocation melting of iron and the temperature of the inner core boundary, revisited. Geophys. J. Int 115, 147–151 (1993).
33. Wood, B. J., Walter, M. J. & Wade, J. Accretion of the Earth and segregation of its core. Nature 441, 825–833 (2006).
34. Ahrens, J. et al. Sensitivity of the IceCube detector to astrophysical sources of high energy muon neutrinos. Astropart. Phys 20, 507–532 (2004).
35. Aartsen, M. G. et al. Evidence for high-energy extraterrestrial neutrinos at the IceCube detector. Science 342, 1242856 (2013).
36. Fukuda, Y. et al. Measurement of a small atmospheric *νμ/νe* ratio. Phys. Lett. B 433, 9–18 (1998).
37. Sajjad Athar, M., Honda, M., Kajita, T., Kasahara, K. & Midorikawa, S. Atmospheric neutrino flux at INO, South Pole and Pyhäsalmi. Phys. Lett. B 718, 1375–1380 (2013).
38. Formaggio, J. A. & Zeller, G. P. From eV to EeV: Neutrino cross sections across energy scales. Rev. Mod. Phys 84, 1307–1341 (2012).
39. James, F. Statistical methods in experimental physics (World Scientific, 2006).
40. Abe, K. et al. A Long Baseline Neutrino Oscillation Experiment Using J-PARC Neutrino Beam and Hyper-Kamiokande. Preprint arXiv:1412.4673 (2014).
41. Hirose, K., Labrosse, S. & Hernlund, J. Composition and State of the Core. Annu. Rev. Earth Planet. Sci 41, 657–691 (2013).
42. Abe, K. et al. The T2K experiment. Nucl. Instr. Meth. A 659, 106–135 (2011).
43. An, F. P. et al. Observation of Electron-Antineutrino Disappearance at Daya Bay. Phys. Rev. Lett 108, 171803 (2012).
44. Li, Y. F. Overview of the Jiangmen Underground Neutrino Observatory (JUNO). Preprint arXiv:1402.6143 4 (2014).
45. Kennett, B. L. N., Engdahl, E. R. & Buland, R. Constraints on seismic velocities in the Earth from traveltimes. Geophys. J. Int 122, 108–124

(1995).

46. Dziewonski, A., Hales, A. & Lapwood, E. Parametrically simple earth models consistent with geophysical data. Phys. Earth Planet. Inter. 10, 12–48 (1975).
47. Capozzi, F., Lisi, E. & Marrone, A. PINGU and the neutrino mass hierarchy: Statistical and systematic aspects. Phys. Rev. D 91, 1–18 (2015).
48. Fogli, G. L. et al. Global analysis of neutrino masses, mixings and phases: Entering the era of leptonic CP violation searches. Phys. Rev. D 86, 1–11 (2012).
49. Akhmedov, E. K., Razzaque, S. & Smirnov, A. Y. Mass hierarchy, 2-3 mixing and CP-phase with huge atmospheric neutrino detectors. J. High Energy Phys 1302, 082 (2013).
50. Ge, S. F., Hagiwara, K. & Rott, C. A novel approach to study atmospheric neutrino oscillation. J. High Energy Phys 1406, 150 (2014).
51. Katz, U. Status of the KM3NeT project. Nucl. Instr. Meth. A 602, 40–46 (2009).
52. Katz, U. F. for the KM3NeT Collaboration. The ORCA Option for KM3NeT. Preprint arXiv:1402.1022 (2014).
53. Avrorin, A. et al. The prototyping/early construction phase of the BAIKAL-GVD project. Nucl. Instr. Meth. A 742, 82–88 (2014).
54. Schmandt, B., Jacobsen, S. D., Becker, T. W., Liu, Z. & Dueker, K. G. Dehydration melting at the top of the lower mantle. Science 344, 1265–1268 (2014).
55. Honda, M., Kajita, T., Kasahara, K. & Midorikawa, S. Improvement of low energy atmospheric neutrino flux calculation using the JAM nuclear interaction model. Phys. Rev. D 1–46 (2011).
56. Anderson, O. L. Mineral physics of iron and of the core. Rev. Geophys 33, 429–442 (1995).

SECTION 4
ATMOSPHERIC NEUTRINOS AND SOLAR NEUTRINOS

CHAPTER

14

Atmospheric Neutrino Flux Measurement

Kimihiro Okumura[1]

[1]Research Center for Cosmic Neutrino (RCCN), Institute for Cosmic Ray Research (ICRR), University of Tokyo, 5-1-5 Kashiwa-no-ha, Kashiwa, Chiba, Japan

ABSTRACT

The atmospheric neutrino flux has been measured by several experiments and its understanding has been improved by comparing with the theoretical predictions in detail. In this paper recent results of the flux measurements will be reviewed and discussed.

INTRODUCTION

Atmospheric neutrinos are the end particles produced by meson decay in air showers, initiated by cosmic ray interactions with the atmosphere. They are most abundant in nature above sub-GeV energies. The fluxes on the Earth are well calculated by several simulation studies using the inputs of the measured cosmic ray flux and the hadronic interactions. However, the

Citation: Kimihiro Okumura. "Atmospheric Neutrino Flux Measurement" EPJ Web Conf., 208 (2019) 07002 DOI: https://doi.org/10.1051/epjconf/201920807002

neutrino flux predictions are affected by several factors, such as the primary cosmic ray flux, compositions, hadron interactions, atmosphere model, seasonal variations, geomagnetic effects, etc. The flux predictions can be tested directly by atmospheric neutrino flux measurements.

Generally atmospheric neutrinos are produced by the decay of particles such as pions (π) and kaons (K), and dominate below several hundred TeV. Those are called as the "conventional" atmospheric neutrino flux. The nominal energy spectrum follows a power-law with spectral index of-3.7, $dN/dE \propto E^{-3.7}$. The flavor ratio of muon and electron neutrinos, $(\nu_\mu+\bar{\nu}_\mu)/(\nu_e+\bar{\nu}_e)$, is about two around 1 GeV where pion decay neutrinos are dominant. Above O(10) GeV the νe flux becomes smaller by the steeper spectrum since the parent muons (μ) reach ground level without decay. Instead of the μ/π-decay neutrino, K-decay neutrino becomes dominant because of its faster decay time constant. The flux prediction of K-decay neutrino largely differs depending on the K production models, causing the major flux uncertainties in this energy range.

As the energy increases further above several ten TeV, K mesons also reaches to the ground without decay and the conventional neutrino flux is suppressed. Instead the "prompt" flux produced by D meson particles begins to dominate. The understanding of the prompt neutrino is important because they could be a substantial background of the astrophysical neutrino observation. The flux predictions based on the simulation are available but its detection has not yet been successfully done

The primary cosmic rays such as protons and helium are deflected by the Earth's geomagnetic field, and suffer from the rigidity cutoff depending on the detector site below several ten GeV. Therefore the so called "East-West" anisotropy, which is the flux modulation depending on the azimuthal direction, is observed in atmospheric neutrinos as well as in cosmic rays. The anisotropy is anticipated to be increased in the GeV region and near horizontal directions. The feature of the geomagnetic effect in the flux simulation, such as the bending of the primary and secondary cosmic ray particles, can be tested with the atmospheric neutrino data. The cosmic ray flux is known to be anti-correlated with solar activity, which cycles in an 11 year period, since the cosmic rays entering the Earth are scattered out by the solar wind. Accordingly the modulation of the atmospheric neutrino flux by the solar cycle is expected, especially in the low energies below 10 GeV and in the direction of the polar regions. The understanding of the atmospheric neutrino flux is absolutely imperative because the atmospheric neutrino itself is utilized for the measurement of neutrino oscillations. Also atmospheric neutrinos could be a background of rare event search such as proton decay, dark matter, astronomical neutrinos, etc. The background level is estimated based on the flux prediction. The flux measurement can give a severe test directly to the flux predictions, and make it more reliable.

METHODOLOGY

Basically the atmospheric neutrino flux is estimated by counting neutrino events produced by charged-current (CC) interactions with detector mass. In order to measure the flavor dependent flux, the particle identification capability of the out-going lepton is required. Also neutralcurrent (NC) events could be a background, and the data quality of the flux sample depends on the rejection performance of NC backgrounds. In addition the measured flux is affected by neutrino oscillations below 100 GeV, and the Earth absorption at O(10) TeV and higher. Such effects should be taken into account.

In this paper the results of the flux measurement will be discussed in two categories, which are roughly grouped by the detector mass scale; the first category is kiloton scale detectors, such as Frejus and Super-Kamiokande (SuperK), measuring the flux below TeV, mainly GeV energies. Super-K detector is equipped with the fine granularity of the photo-sensors for observing the detailed event topology providing excellent capabilities of particle identification and NC background rejection. Another category is kilometer cubic (km3) scale detectors such as AMANDA, ANTARES, IceCube, using natural ice in the South Pole and the water of Mediterranean sea as the detector material. These detectors have sensitivities to measure smaller fluxes at energies greater than O(100) GeV, thanks to the huge detector mass. However, due to the coarser arrangement of the photo-sensors, the event identification capability is not as good as compared to that of kiloton scale detectors. As a result, though the relatively good quality of data is provided for ν_μ samples by identifying the muon track in the event, the contamination of NC background events is inevitable for ν_e measurement due to similar event topology.

FLUX MEASUREMENT ABOVE TEV ENERGIES

The flux of muon neutrino (ν_μ) above TeV energies is well studied by several experiments, as shown in Figure 1. Basically upward-going ν_μ sample are utilized for flux estimation to reduce the cosmic muon backgrounds. The measured spectra are covered from O(100) GeV up to several PeV, and consistent with each other and also with theoretical predictions. Meanwhile the recent IceCube result shows a clear excess beyond 100 TeV [1]. The excess is too large to be explained by the prompt component, and thought to be due to the external origin neutrinos produced by astronomical high energy objects in the universe.

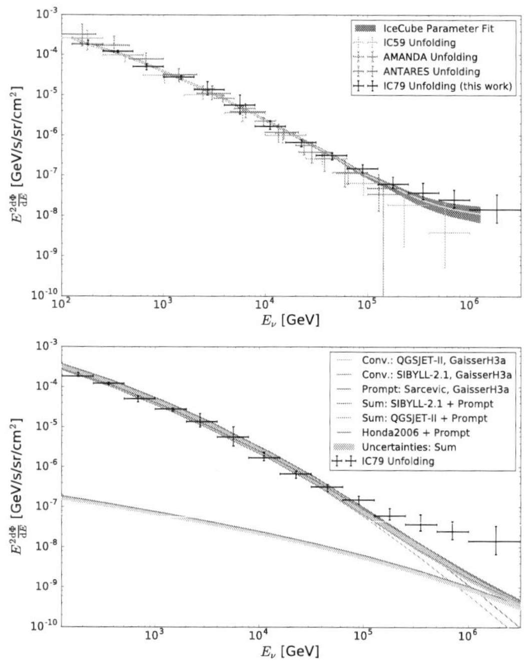

Figure 1. The atmospheric muon neutrino flux measured by several experiments from 100 GeV to several PeV (upper), and the comparison of IceCube measurements and the theoretical predictions of conventional and prompt fluxes (lower). Figures are taken from [1].

Figure 2 shows the ν_e energy spectrum measured by the IceCube experiment. The measurement of ν_e flux is important because there are still large flux uncertainties in the flux calculation above O(100) GeV due to K production cross section error in hadronic interactions. Since the contributions from K decay is dominant rather than for π decay neutrinos in this energy range, this result gives constraint on K originated flux. Also the detection of the prompt flux component is expected by measuring the ν_e spectrum in the high energy end above O(10) TeV since the prompt flux begins to exceed the conventional one. Actually, ν_e measurement in km3 detectors is not easy due to NC backgrounds. A sophisticated event selection using the likelihood method is employed to discriminate ν_e events in the IceCube experiment [2]. The observed ν_e flux shows about 30% higher than flux prediction, which indicates a larger K flux component than expected. No evidence of the prompt flux is obtained from the fitting result using the observed flux and spectrum shape

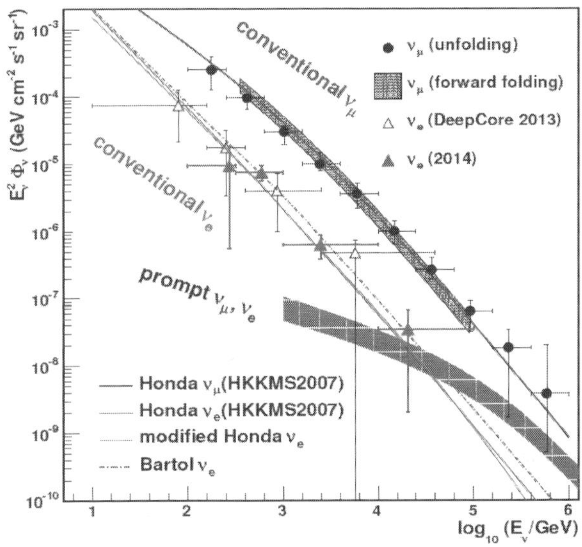

Figure 2. The atmospheric ν_e flux results by IceCube shown with the flux predictions. Figure is taken from [2].

FLUX MEASUREMENT BELOW TEV ENERGIES

Figure 3 shows the measured ν_μ and ν_e energy spectra by the Super-K experiment compared with flux predictions and other experimental results [3]. Super-K data covers the wider energy range from sub-GeV up to several TeV, providing the significantly improved flux data. At the higher energy end the energy spectrum overlaps with $\nu\mu$ measurements by km3 detectors, and the measured energy spectra are consistent and smoothly connected between lower and higher energy ranges. Figure 4 shows the detailed comparison of the measured fluxes with the flux predictions. They are consistent with each other within estimated uncertainties. The total systematic errors are estimated to be approximately 20%, which comes dominantly from the uncertainties related to neutrino interactions.

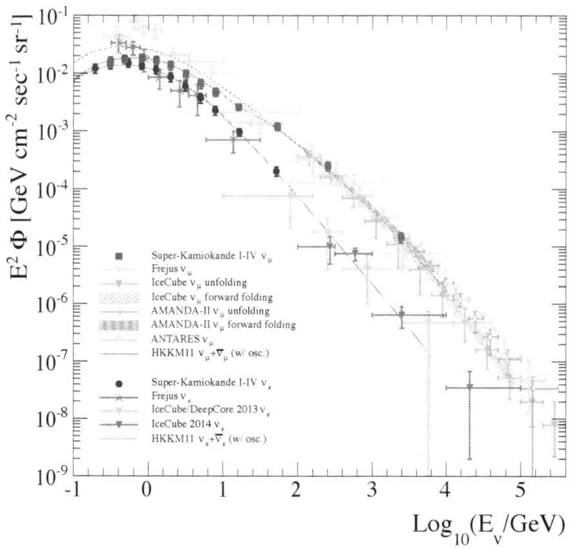

Figure 3. The measured energy spectra of the atmospheric ν_e and ν_μ fluxes by Super-Kamiokande, shown with measurements by other experiments.

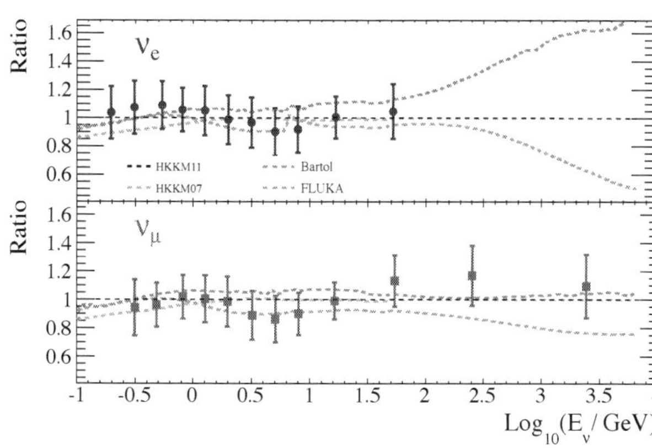

Figure 4. Measured energy spectra compared to the flux model predictions (dotted lines) from the HKKM11 (black), HKKM07 (light green), Bartol (red), and FLUKA (light blue) models. The vertical axis represents the flux ratio from the data (or each model) to HKKM11. Error bars include both statistical and systematic uncertainties.

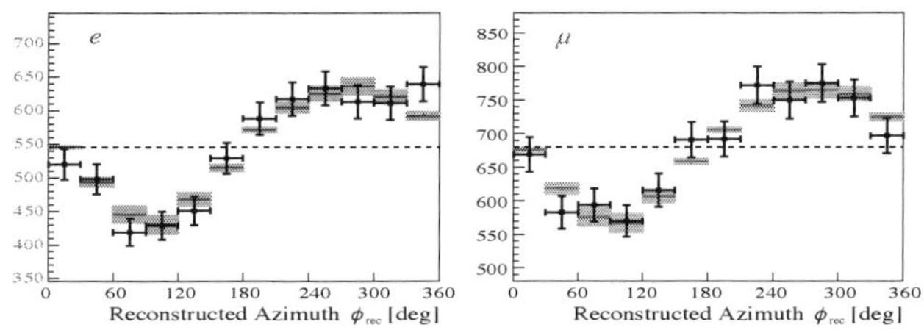

Figure 5. The azimuthal distributions of a subselection of e-like (left) and μ-like (right) events, from the Super-K data (points with statistical error) and MC simulations (boxes with systematic error). The subselection is optimized to obtain the highest significance of the final A parameters, by using only events with $0.4<E_{rec}<0.6$ and $|\cos\theta_{rec}|<0.6$.

Figure 5 shows the azimuthal spectra of Super-K ν_e (e-like) and ν_μ (μ-like) sample. Those event samples are chosen to be optimal for getting the largest significance of the anisotropy. Clear modulations depending on the azimuthal direction are seen for both e-like and μ-like data, which significances are 6.0σ and 8.0σ, respectively. The azimuthal asymmetry are also investigated quantitatively to validate the implementation of the geomagnetic effect in the flux simulation. Two parameters, A and B, are introduced for measuring the azimuthal modulation, representing the degree of the flux asymmetry between East and West directions, and the phase of the modulation curve, respectively. These parameter values are compared between data and expectation in several energies and zenith directions, as shown in Figures 6 and 7.

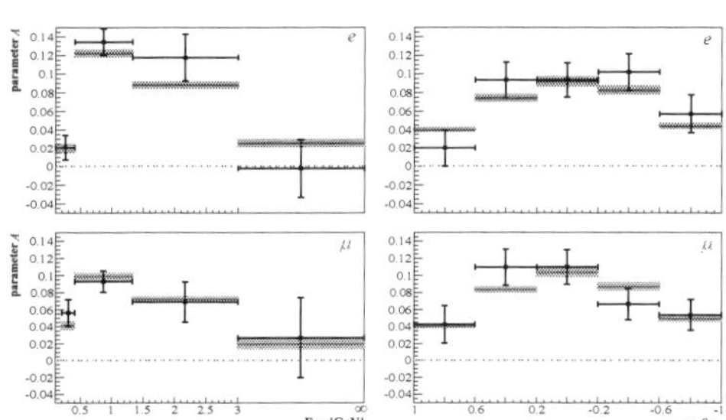

Figure 6. The parameter, A, depending on the reconstructed energy (left) and zenith angle (right), for e-like (upper) and μ-like (lower) events, from the Super-K data (points with statistical error) and MC simulations (boxes with systematic error). A grey dotted line is drawn at A=0, the case of no east-west asymmetry, as a visual guide. The highest energy bin in the left plots is unbounded from above.

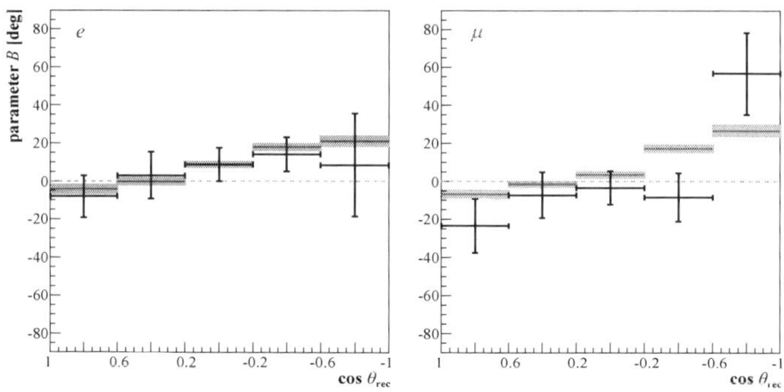

Figure 7. The parameter, B, depending on the cosine of the zenith angle, for e-like (left) and μ-like (right) events, from the Super-K data (points with statistical error) and MC simulations (boxes with systematic error).

The measured data agree well with the predictions, confirming the implementation of the geomagnetic field in the flux calculation. Super-K has performed a search for the correlation with solar activity using the atmospheric neutrino data in the period covering almost two solar cycles, as shown in Figure 8.

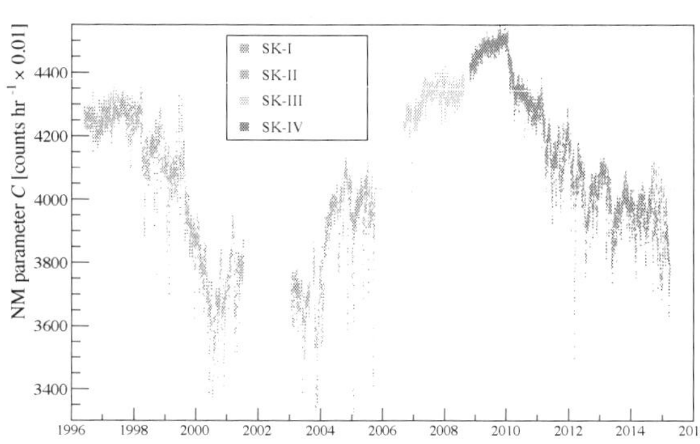

Figure 8. The neutron monitor parameter, C, over the SuperK operational period, where low values of C correspond to high solar activity, showing that almost two solar cycles are covered by the Super-K data.

The neutron monitor count, C, which is known to be well correlated with the cosmic ray flux, is utilized as an indicator of the solar activity in the observed time.

The expected flux changes due to solar modulation, as a function of neutrino energy and direction, are provided by the flux calculation group and introduced in this study. Figure 9 shows the observed event rate in the sub-GeV energy range, which is optimized for solar modulation search, as a function of neutron monitor count C.

A test is performed by introducing an α parameter to quantify the degree of flux modulation, in which α=0 and 1 correspond to no correlation and as predicted, respectively. The best fit gives $\alpha=0.62^{+0.57}_{-0.58}$, and the data weakly prefers the correlation with solar activity (1.1σ). Additional studies for the sea sonable variation of the atmospheric neutrino due to the variation of the atmosphere density in the air are investigated. No seasonable variation is seen as expected since the change is estimated to be small (

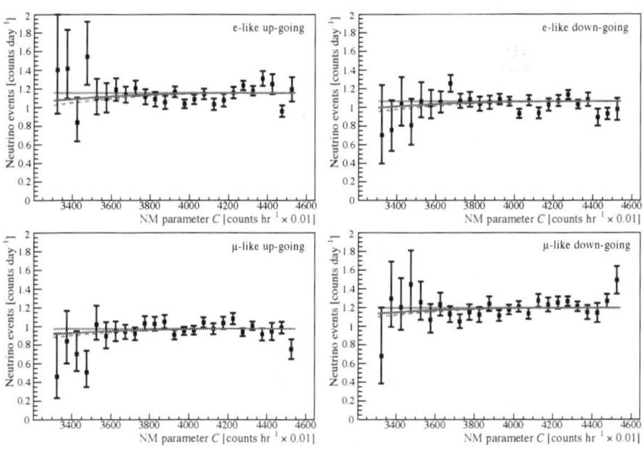

Figure 9. The test for a solar modulation correlation using the Super-K data (points). The solar correlation hypotheses are shown for no correlation (α=0, grey), best fit (α=0.62, red), and the default prediction (α=1, grey dotted) for each of the four data samples.

SUMMARY AND PROSPECT

The results of the atmospheric neutrino flux measurement have been reviewed in this paper. The ν_μ and ν_e flux spectra, ranging from sub-GeV up to PeV, are measured by several experiments using the variety types of the detectors. The measured fluxes are compared and tested to the flux predictions, and no significant deviations are seen. The flux component originating from K decay is investigated using ν_e flux measurements above several ten GeV, indicating larger contributions than prediction. The prompt flux, which is produced by the D meson decay, has not been detected yet.

Super-K data shows clear azimuthal modulation due to the geomagnetic field effect, validated in the flux simulation. The solar modulation is indicated by Super-K study but the significance is not large.

In future there are several prospects expected for the improvements of the future atmospheric neutrino flux study in the following aspects: (i) More precise flux predictions will be available using the latest cosmic ray spectrum measurement like AMS-2 and the hadronic interactions. (ii) The reduction of systematic uncertainties is necessary for further investigations of the neutrino flux, which is anticipated by future experiments using accelerator beam neutrinos. (iii) There is a possibility to measure the independent flux of neutrino and anti-neutrino, respectively, since Super-K detector is going to upgrade to SKGd by dissolving Gadolinium into the water and enhance the anti-neutrino event tagging by detecting the recoiled neutron. (iv) There are some demands for understanding of atmospheric flux below O(100) MeV from rare search experiments, such as dark matter, supernova relic neutrinos because the atmospheric neutrino will be a significant background in future experiments. (v) Future neutrino detectors with larger detector mass, such as HyperKamiokande, DUNE, IceCube-Gen2, PINGU, KM3Net, will improve the sensitivities of the flux measurement.

REFERENCES

1. M.G. Aartsen et al. (IceCube), Eur. Phys. J. C77, 692 (2017), 1705.07780
2. M.G. Aartsen et al. (IceCube), Phys. Rev. D91, 122004 (2015), 1504.03753
3. E. Richard et al. (Super-Kamiokande), Phys. Rev. D94, 052001 (2016), 1510.08127

CHAPTER

15

Improving the Prediction of the Atmospheric Neutrino Flux Using the Atmospheric Muon Flux

Morihiro Honda[1]

[1]Institute for Cosmic Ray Research, the University of Tokyo, 5-1-5 Kashiwa-no-ha, Kashiwa, Chiba 277-8582, Japan

ABSTRACT

It is well known that the correlation of atmospheric neutrinos and muons are simply correlated in the energy region of 1–10 GeV, and used for the test bench of the hadronic interaction model used for the calculation of the atmospheric neutrino flux. However, the correlation becomes unclear for neutrinos in the energy range below 1 GeV, which is important for the study of mass ordering of neutrino and CP phase of the neutrino mass. We extend the study of the correlation to the lower neutrino energies and find that the atmospheric muon flux observed at high altitude shows a good correlation

Citation: Morihiro Honda. "Improving the prediction of the Atmospheric neutrino flux using the atmospheric muon flux" EPJ Web Conf., 208 (2019) 07001 DOI: https://doi.org/10.1051/epjconf/201920807001

to the atmospheric neutrino flux, and could be used to calibrate the hadronic interaction model.

INTRODUCTION

Neutrino oscillation physics has now entered into the precision era, after it has been well established that θ13 has a finite value and clear evidence for neutrino oscillations of one flavor to another have been seen. Now experimentalists are working hard to determine very precisely the neutrino mixing parameters, mass hierarchy and CP violation in the lepton sector, and tomography of the Earth. For those purposes, new generation detectors, such as INO [1], JUNO[2], South Pole [3], Hyper [4], DUNE [5] etc, are under the construction.

In those detectors, the atmospheric neutrino is still one of the main targets, and high precision measurements are expected, especially at low energies (1GeV). However, it is difficult to calculate the atmospheric neutrino flux below 1 GeV accurately, only with the results of high energy experiments, or with the observed atmospheric muon flux as we have done in Ref [6]. In this paper we study the correlation of atmospheric muons and neutrinos at low energies. Introducing a pseudo-analytic expression for the calculation we note that the simulation and the pseudoanalytic expression for the calculation of the atmospheric muon and neutrino are similar, but there are large differences between them. The important region in the phase space of hadronic interactions for atmospheric muons are different from that of the neutrino, and varies with the observation altitude. Therefore the atmospheric muon response to the variation of the hadronic interaction model is different from that of the atmospheric neutrino depending on the observation site, especially at low energies. In the following sections, we study the correlation of atmospheric muons and neutrinos, and study the possibility to reduce the calculation error of the atmospheric neutrino flux. Let us consider the projection of the atmospheric lepton flux calculation process to the hadronic interaction phase space spanned by the projectile particle momentum and the momentum of the produced meson which creates the lepton in the decay cascade. Writing the whole process in a pseudo-analytic expression as eq. 1,

$$\Phi_L^{obs}(p_L^{obs},x^{obs}) = \sum_{N^{proj}}\sum_{M^{born}}\int\int$$
$$\Big[\int M2L(M^{born},p_M^{born},x^{int},L^{obs},p_L^{obs},x^{obs})$$
$$\times H_{int}(N^{proj},p_N^{proj},x^{int},M^{born},p_M^{born})$$
$$\times \sigma^{prod}(N^{proj},E^{proj})\cdot \rho_{air}(x^{int})$$
$$\times \Phi_{proj}(N^{proj},p_N^{proj},x^{int})dx^{int}\Big]\; dp_M^{born}dp_N^{proj}. \quad (1)$$

we may consider the integrand is the projection for each combination of projectile and meson, where $M2L(M,p_M^{born},x^{born},L,p_L,x^{obs})$ represents the probability that a meson M^{born} with momentum p^{born}_M at x^{born} decays and results in the lepton L^{obs} with momentum p^{obs}_L at x^{obs}_L without a hadronic interaction with air nuclei, $H_{int}(N^{proj},p_N^{proj},M^{born},p_M^{born})$ is the probability that the M^{born} with momentum p^{born}_M is produced in the hadronic interaction of a hadronic particle N^{proj} with momentum p^{proj}_N with air nuclei, $\sigma^{prod}(N^{proj},E^{proj})$ is the production cross section of N^{proj} particle and air nuclei, ρ_{air}(x^{int} is the nucleus density of the air at x^{int}, and $\Phi_{proj}(N^{proj},p_N^{proj},x^{int})$ the flux of N^{proj}-particle primary or secondary cosmic ray at xint with momentum p^{proj}_N, which could be the projectile particle in the hadronic interaction.

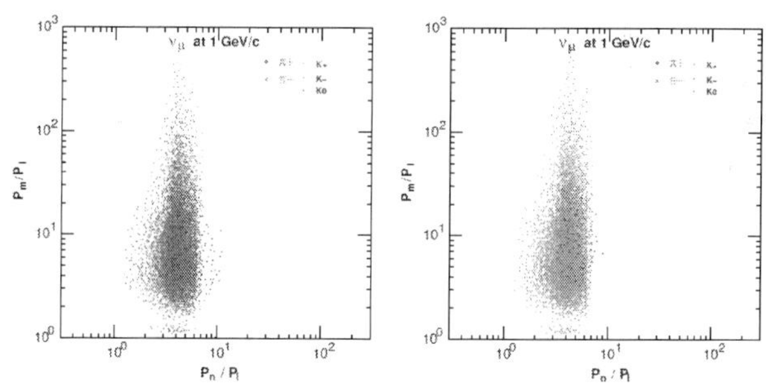

Figure 1. The density of D-function in eq.2 in projectile momentum and Meson momentum space for atmospheric muons at 1GeV/c for vertically down going direction at Kamioka (sea level)

We note we can make the pseudo analytic expression eq.1 to a realistic one, and carry out an analytic calculation in a 1-dimensional calculation scheme with some approximations. However, it is not practical to carry out the analytic calculation in a 3-dimensional scheme, and normally the Monte Carlo simulation is used. We use eq. 1 just for illustrating the study of the uncertainty of the hadronic interaction model. In the following study, it is convenient to change the integral variable and rewrite eq.1 as,

$$\Phi_L^{obs}(p_L^{obs}, x^{obs}) = \sum_{N^{proj}} \sum_{M^{born}} \int\int D(N^{proj}, lr_N^{proj}, M^{born}, lp_M^{born}, L^{obs}, lp_L^{obs}) \, dlp_M^{born} dlr_N^{proj} \, , \qquad (2)$$

and

$$\begin{aligned} D(N^{proj}, lp_N^{proj}, M^{born}, lp_M^{born}, L^{obs}, lp_L^{obs}) &= \int M2L(M^{born}, p_M^{born}, x^{int}, L^{obs}, p_L^{obs}, x^{obs}) \\ &\times H_{int}(N^{proj}, p_N^{proj}, x^{int}, M^{born}, p_M^{born}) \\ &\times \sigma^{prod}(N^{proj}, E^{proj}) \cdot \rho_{air}(x^{int}) \\ &\times \Phi_{proj}(N^{proj}, p_N^{proj}, x^{int}) \, dx^{int} \, , \qquad (3) \end{aligned}$$

The Monte Carlo simulation we use here is the same one which we have used for our calculation of atmospheric neutrinos. Tagging all the particles appearing in the simulation, we record two momenta (p_M^{born}, p_N^{proj}), one is that of the meson which created the lepton in the decay and the other is that of the projectile particle which created the meson in the hadronic interaction with the air nuclei. The set of two momenta (p_M^{born}, p_N^{proj}) is considered as a point in the phase space of the hadronic interaction. We show the projection of such points in the phase space in Fig. 1 for muon flux calculations and in Fig. 2 for neutrino flux calculations, summing all the different kinds of projectiles but for each kind of meson.

where $\Delta_{int}(N^{proj}, M^{born}, p_N^{proj}, p_M^{born})$ is a "variation function" defined for each combination of projectile particle and meson defined in the phase space of the hadronic interaction

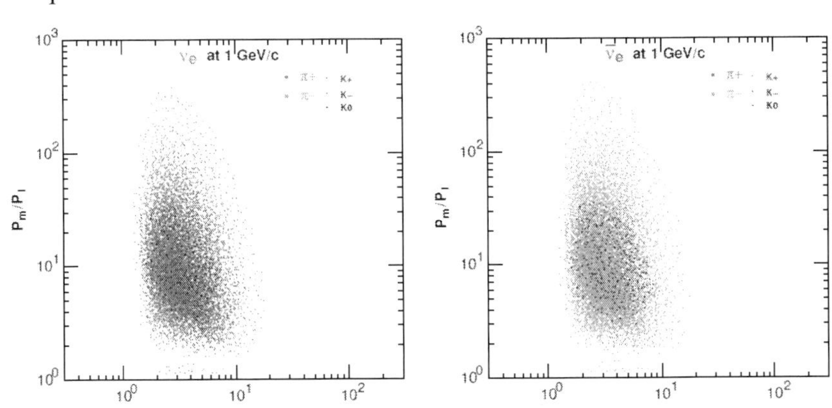

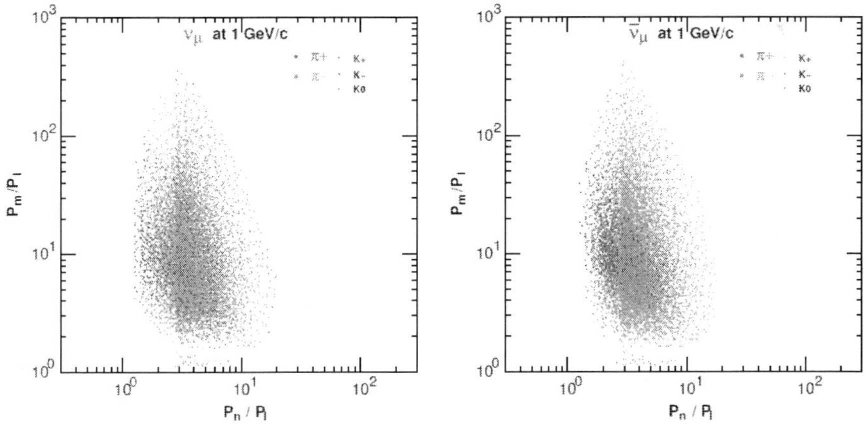

Figure 2. The density of D-function in eq.2 in projectile momentum and Meson momentum space for atmospheric neutrinos at 1GeV for vertically down going direction at Kamioka (sea level)

If we assume that the $\Phi_{proj}(N^{proj}, p_N^{proj}, x^{int})$ is not largely affected by the variation of the hadronic interaction, we can calculate the lepton flux with a small variation of the interaction model by substituting the corresponding probability function $\tilde{H}_{int}(N^{proj}, p_N^{proj}, x^{int}, M^{born}, p_M^{born})$ in the expression eq.1. It means we can study a small variation of the interaction model and resulting atmospheric lepton flux without repeating the Monte Carlo simulation. The lepton flux calculated by the variation of the interaction model may be written as;

$$
\begin{aligned}
\tilde{\Phi}_L^{obs}(p_L^{obs}, x^{obs}) = \sum_{N^{proj}} \sum_{M^{born}} \int \int \\
\Big[\int M2L(M^{born}, p_M^{born}, x^{int}, L^{obs}, p_L^{obs}, x^{obs}) \\
\times H_{int}(N^{proj}, p_N^{proj}, x^{int}, M^{born}, p_M^{born}) \\
\cdot \left(1 + \Delta_{int}(N^{proj}, M^{born}, p_N^{proj}, p_M^{born})\right) \\
\times \sigma^{prod}(N^{proj}, E^{proj}) \cdot \rho_{air}(x^{int}) \\
\times \Phi_{proj}(N^{proj}, p_N^{proj}, x^{int}) dx^{int} \Big] dp_M^{born} dp_N^{proj}
\end{aligned}
\qquad (5)
$$

$$= \sum_{N^{proj}} \sum_{M^{born}} \int \int D(N^{proj}, p_N^{proj}, M^{born}, p_M^{born}, L^{obs}, p_L^{obs}) \times \left(1 + \Delta_{int}(N^{proj}, M^{born}, p_N^{proj}, p_M^{born})\right) dp_M^{born} dp_N^{proj} . \quad (6)$$

In the following, we study the effect of the variation of hadronic interaction model, introducing an artificial variation in the interaction model, with a set of random numbers and 3rd order b-spline. The 3rd order b-spline function with constant knots separation is represented as

$$B_\Delta^i(x) = b\left(\frac{x - i \cdot \Delta}{\Delta} - x_0\right) \quad (7)$$

where

$$b(t) = \begin{cases} \frac{1}{6}(3|t|^3 - 6|t|^2 + 4) & (|t| \leq 1) \\ -\frac{1}{6}(|t| - 2)^3 & (1 \leq |t| \leq 2) \\ 0 & (|t| \geq 2) \end{cases} \quad (8)$$

where Δ is the knots separation, and x_0 is for a shift less than Δ. The linear combination of 3rd order b-spline function (eq.7) is contiguous up to 2nd order derivative, and is often used to connect the discrete data points or to fit them. We can extend it to a 2-dimensional spline function as

$$B_\Delta^{ij}(x, y) = b\left(\frac{x - i \cdot \Delta}{\Delta} - x_0\right) b\left(\frac{y - j \cdot \Delta}{\Delta} - y_0\right) \quad (9)$$

Here, we took the same knots separation for x and y. We may construct an artificial variation function in eq.4 with those 2-dimensional b-spline functions as

$$\Delta_{int}(N^{proj}, M^{born}, p_N^{proj}, p_M^{born}) = \delta \cdot \sum_i \sum_j R_N^{ij} \cdot B_\Delta^{ij}\left(\log_{10}(p_N^{proj} xs, \log_{10}(p_M^{born})\right) \quad (10)$$

where $\{R_N^{ij}\}$ is the set of random numbers with a normal distribution. Therefore, for each combination of the set of random numbers $\{R_N^{ij}\}$ and δ, we have a variation of the interaction model and a flux spectrum as,

$$\tilde{\Phi}_L^{obs}(p_L^{obs}, x^{obs}) = \sum_{N^{proj}} \sum_{M^{born}} \int\int D(N^{proj}, p_N^{proj}, M^{born}, p_M^{born}, L^{obs}, p_L^{obs}) \times \left(1 + \delta \sum_i \sum_j R_N^{ij} \cdot B_\Delta^{ij}\left(\log_{10}(p_N^{proj}), \log_{10}(p_M^{born})\right)\right) dp_M^{born} dp_N^{proj}, \quad (11)$$

Note, we assume δ has a small value ($\lesssim 1$) so that we consider a small variation from the hadronic interaction we used in our calculation of atmospheric lepton calculations. With this expression (eq.11), we can write the variation of the lepton flux as

$$\Delta\Phi_L^{obs}(p_L^{obs}, x^{obs}) \equiv \tilde{\Phi}_L^{obs}(p_L^{obs}, x^{obs}) - \Phi_L^{obs}(p_L^{obs}, x^{obs}) = \delta \sum_{N^{proj}} \sum_{M^{born}} \sum_i \sum_j \int\int D(N^{proj}, p_N^{proj}, M^{born}, p_M^{born}, L^{obs}, p_L^{obs}) \times R_N^{ij} \cdot B_\Delta^{ij}\left(\log_{10}(p_N^{proj}), \log_{10}(p_M^{born})\right) dp_M^{born} dp_N^{proj}, \quad (12)$$

Once we have calculated the atmospheric lepton flux with a Monte Carlo simulation, it is easy and quick to calculate the variation due to a variation of the hadronic interaction model with the above formalism. Applying this formalism to atmospheric neutrinos and muons, we can study the correlation of the variation of atmospheric neutrino and muon fluxes. Let us study the correlation of atmospheric neutrinos and muons for the variation of hadronic interaction made as in eq.11 using the correlation coefficient;

$$\gamma(p_\nu^{obs}, x_\nu^{obs}; p_\mu^{obs}, x_\mu^{obs}) = \frac{\sum\left(\Delta\Phi_\nu(p_\nu^{obs}, x_\nu^{obs})\Delta\Phi_\mu(p_\mu^{obs}, x_\mu^{obs})\right)}{\sqrt{\sum\left(\Delta\Phi_\nu^k(p_\nu^{obs}, x_\nu^{obs})\right)^2 \sum\left(\Delta\Phi_\mu^k(p_\mu^{obs}, x_\mu^{obs})\right)^2}} \quad (13)$$

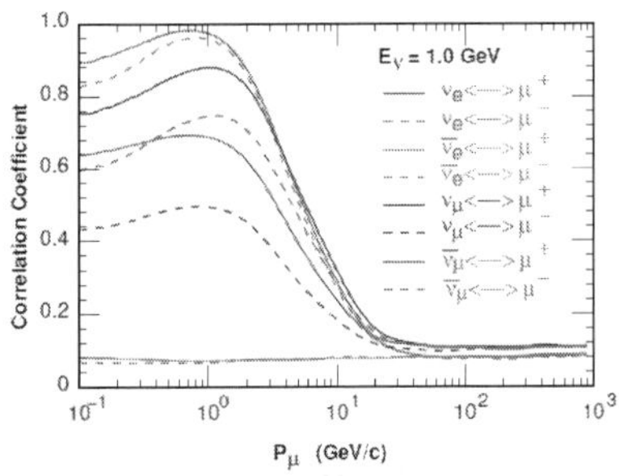

Figure 3. Correlation coefficient for each combination of neutrinos $(\nu_\mu, \bar{\nu}_\mu, \nu_e, \bar{\nu}_e)$ and muons (μ^+, μ^-) at neutrino energy of 1 GeV

In Fig.3, we show the correlation coefficient as a function of vertical down going muon momentum observed at Kamioka, and for the vertical down going 1 GeV neutrino also observed at Kamioka. We consider here all the possible combination of the kind of neutrinos $(\nu_\mu, \bar{\nu}_\mu, \nu_e, \bar{\nu}_e)$ and of muons (μ^+, μ^-) for the correlation here. In the left panel of Fig.4 we plotted the muon momenta at the maximum correlation and 90% of it for the combinations of neutrinos and muons, but when the correlation coefficient is larger than 0.3. We find that the muon momenta at the maximum correlation is about 3 times the neutrino energy above a few GeV, but it quickly decreases as the neutrino energy decreases.

Next we take $\delta = 1$ and construct 3000000 set of random numbers $\{R_N^{ij}\}$ to study the variation of muons and neutrino fluxes. In the right panel of Fig.4, we plot the variation of the neutrino flux calculating the ratio $\Delta\Phi_{\nu\mu}$ / $\Phi_{\nu\mu}$ at 1 GeV, and the same quantity when the variation of interaction model satisfies $\Delta\Phi_\mu/\Phi_\mu < 0.1$ for the calculated muon flux in the momentum range having a correlation coefficient larger than 90% of the maximum. We find that with the selection of hadronic interaction model variation with the muon flux variation, the variation of the neutrino flux is largely suppressed and the standard deviation is around 0.5.

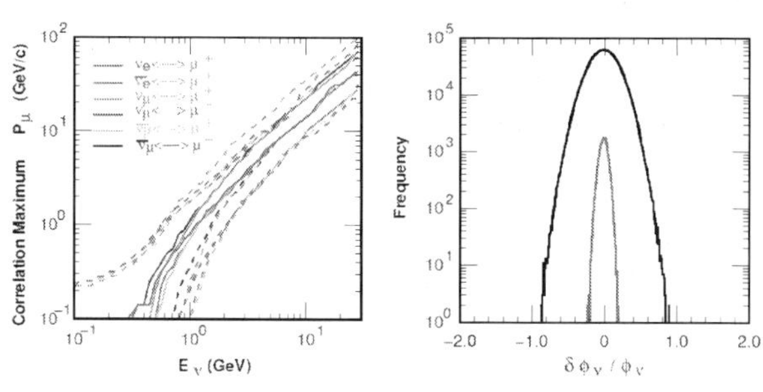

Figure 4. Left panel: Muon momentum at the correlation coefficient maximum for each combination of neutrinos $(\nu_\mu, \bar{\nu}_\mu, \nu_e, \bar{\nu}_e)$ and muons (μ^+, μ^-) both for vertical down going direction, as a function of neutrino energy. When the maximum correlation coefficient is smaller than than 0.3, the lines are not plotted. Right panel: The outside curve is the distribution of $\Delta\Phi_{\nu\mu}/\Phi_{\nu\mu}$ at 1GeV with the variation of hadronic interaction model by eq.12 for 3000000 set of random numbers and $\delta = 1$. The inside curve is the same as above, but selected for the case when the hadronic interaction model which gives $\Delta\Phi_\mu/\Phi_\mu < 0.1$ is satisfied in the muon momentum region with the correlation coefficient larger than 90% of the maximum.

With this observation, we may consider that when we reconstruct the accurately observed atmospheric muon flux accurately in the muon momentum region where the correlation coefficient of muon flux and the neutrino flux is larger than 90% of the maximum, the expected error of the neutrino calculation is also largely suppressed.

However in the practical sense, it is difficult to measure the atmospheric muon flux accurately in the low momentum region. Therefore, we consider the cases that the available minimum muon is 0.1, 0.3, 0.5, and 1.0 GeV/c for the accurate muon flux data. Then we plotted the variation of the atmospheric neutrino variation standard deviation in the ratio to flux value $\Delta\Phi_{\nu\mu}/\Phi_{\nu\mu}$ for the variation of muon flux satisfying $\Delta\Phi_{\mu}/\Phi_{\mu} < 0.1$ in Fig.5, both for muon fluxes observed at Kamioka (sea level) and Mt. Norikura (~2700m A.S.L.). We find the suppression of the neutrino flux error works better at the higher observation altitude.

SUMMARY

We have developed a quick method to study the effect of the variation of hadronic interaction model in the calculation of the atmospheric lepton flux. Using this method, we find a procedure to reduce the error in the calculation of the atmospheric neutrino flux at low energies with the accurately measured atmospheric muon flux. However, the observation site for muon flux seems important. The muon flux observed at Mt. Norikura (~ 2700m A.S.L.) gives a more stringent suppression of neutrino flux error than that at Kamioka (sea level). It is expected that the observations of muon flux on a much higher mountain will give a more stringent suppression of atmospheric neutrino calculation errors.

REFERENCES

1. S. Ahmed et al. [ICAL Collaboration], arXiv:1505.07380 [physics.ins-det].
2. F. An et al. [Juno Collaboration] arXiv:1505.95613 [physics.ins-det].
3. M.G. Aartsen et al. arXiv:1607.02671 [physics.insdet]
4. K. Abe et al. [Hyper-Kamiokande Collaboration] arXiv:1805.04163 [physics.ins-det].
5. B. Abi et al. [DUNE Collaboration] arXiv:1807.10334 [physics.ins-det].
6. T. Sanuki et al. Phys. Rev.,D75 043005, 2007.

CHAPTER

16

Sub-GeV Atmospheric Neutrinos and CP Violation in DUNE

Kevin J. Kelly[1], Pedro A. N. Machado[2], Ivan Martinez-Soler[2,3,4], Stephen J. Parke[2], and Yuber F. Perez-Gonzalez[2,3,4]

[1]Theory Department, Fermi National Accelerator Laboratory, Batavia, Illinois 60510, USA

[2]Theory Department, Fermi National Accelerator Laboratory, P.O. Box 500, Batavia, Illinois 60510, USA

[3]Department of Physics & Astronomy, Northwestern University, Evanston, Illinois 60208, USA

[4]Colegio de Física Fundamental e Interdisciplinaria de las Américas (COFI), 254 Norzagaray street, San Juan, Puerto Rico 00901, USA

ABSTRACT

We propose to use the unique event topology and reconstruction capabilities of liquid argon time projection chambers to study sub-GeV atmospheric neutrinos. The detection

Citation: Kevin J. Kelly, Pedro A. N. Machado, Ivan Martinez-Soler, Stephen J. Parke, and Yuber F. Perez-Gonzalez. "Sub-GeV Atmospheric Neutrinos and CP Violation in DUNE" Phys. Rev. Lett. 123, 081801 (2019). https://doi.org/10.1103/PhysRevLett.123.081801

of low energy recoiled protons in DUNE allows for a determination of the leptonic CP-violating phase independent from the accelerator neutrino measurement. Our findings indicate that this analysis can exclude a range of values of δ_{CP} beyond the 3σ level. Moreover, the determination of the sub-GeV atmospheric neutrino flux will have important consequences in the detection of diffuse supernova neutrinos and in dark matter experiments.

INTRODUCTION

Atmospheric neutrinos, produced by cosmic-ray interactions in Earth's atmosphere, have played a crucial role in the discovery of neutrino oscillations [1], the only evidence of nonzero neutrino masses [2,3]. Even now, atmospheric neutrinos contribute significantly to our understanding of neutrino oscillations and mixing in the lepton sector [4,5]. In this Letter, we are particularly interested in such neutrinos with energies in the 100 MeV to 1 GeV region, i.e., sub-GeV atmospheric neutrinos. The oscillation phenomenology of this sample is exceptionally rich [6–15]. The physical reason behind this is twofold. First, for baselines comparable to Earth's radius, oscillation of sub-GeV neutrinos are strongly affected by both solar and atmospheric mass splittings. Second, the broad energy spectrum and large matter effects induced by Earth's matter density profile lead to nontrivial oscillation effects, namely MSW [16,17] and parametric [18,19] resonances. Compared to long baseline accelerator neutrinos, the effects on oscillation probabilities of the leptonic CP-violating phase δCP is much more pronounced in sub-GeV atmospheric neutrinos, and, therefore, a measurement of their oscillation pattern can yield important new information on δ_{CP}.At present, only Cherenkov light detectors like Super-Kamiokande and IceCube are large enough to have significant sensitivity to the broad spectrum of atmospheric neutrinos. Nevertheless, a precise measurement of sub-GeV neutrinos is still lacking. The difficulty in studying these neutrinos is related to the event reconstruction, which is very challenging at these low energies. When a sub-GeV neutrino scatters on a nucleon via a charged current interaction, it produces a charged lepton and recoils the nucleon isospin partner, for instance $\nu_e n \rightarrow e^- p^+$. However, such low energy protons will not emit Cherenkov light in water, and thus are invisible in these detectors.In neutrino scattering events, the kinematics of the outgoing lepton bears correlation with the neutrino energy and direction. Nevertheless, at these low energies, much of such correlation is lost, and there is a large spread in outgoing lepton momentum and angle. On top of that, the lack of charge identification impedes the separation of events originated from neutrinos or antineutrinos, concealing CP-violating effects. All this results in very poor reconstruction of the neutrino energy, direction, and flavor (between ν and $\overline{\nu}$), making the use of sub-GeV atmospheric neutrinos to probe CP violation in Cherenkov detectors impractical, unless detectors are gigantic, at the multimegaton scale [20].In liquid argon time projection chambers (LArTPCs), the situation is completely different. The LArTPC technology allows for excellent event topology reconstruction by detecting the tracks of all charged particles and identifying them by topology and energy loss. Recently it was shown that protons with kinetic energy above 21 MeV can be efficiently identified in the ArgoNeut experiment [21]. Their three momenta can still be reconstructed with good resolution, which will allow for a pioneering measurement of sub-GeV neutrino energies and angles. Besides, the capability of detecting these protons enables statistical separation between neutrinos and antineutrinos, since the former are significantly more likely to kick out a single proton from Argon than the latter [21]. Together with the fact that the cross section for neutrinos is about a factor of 2 larger than the one for antineutrinos, one expects that events with one lepton, one proton, and no pions (CC-1p0π) are "neutrino rich," while events with only an outgoing lepton (CC-0p0π) are somewhat "antineutrino rich."In this Letter, we propose to use the unique event reconstruction capabilities of LArTPC to estimate how the future Deep Underground Neutrino Experiment (DUNE) [22] will be able to measure sub-GeV atmospheric neutrinos and extract information on δ_{CP} complementary to the accelerator neutrino program. We also comment on the impact of determining the sub-GeV atmospheric neutrino flux in diffuse supernova neutrino measurements and dark matter experiments.

PHYSICS WITH LOW-ENERGY ATMOSPHERIC NEUTRINOS

In general terms, neutrino oscillations are driven by a phase $\propto (\Delta m^2_{ij}/\text{eV}^2)(L/\text{km})(\text{GeV}/E)$, where L is the distance traveled between neutrino production and detection, E is the neutrino energy, and $\Delta m^2_{ij} \equiv m^2_i - m^2_j$ is the squared mass splitting. When E≳1 GeV, oscillations are induced largely by the aptly named atmospheric mass splitting $|\Delta m^2_{31}| \simeq 2.5 \times 10^{-3}\ \text{eV}^2$ [23,24], and they develop over scales L∼O(RE), the radius of Earth. Oscillations of atmospheric neutrinos with energies 100 MeV<E<1 GeV, are governed by both the atmospheric mass splitting and the smaller solar mass splitting, $\Delta m^{22}{}_1 \simeq 7.4\times10^{-5}\text{eV}^2$ [25–30]. In what follows, we will consider two aspects of major significance to our analysis, CP violation and matter effects. We adopt the usual parametrization for neutrino mixing [31]. To set convention, we define the zenith angle such that $\cos\theta_z=-1$ corresponds to neutrinos coming from directly below the detector, while $\cos\theta_z=0$ indicates the horizon direction.

First we discuss the effects of δ_{CP} in oscillations of sub-GeV neutrinos. In vacuum, for simplicity, the CP-violating term in neutrino oscillation probability is given by [32]

$$P_{CP} = -8J_r \sin\delta_{CP} \sin\Delta_{21} \sin\Delta_{31} \sin\Delta_{32}, \tag{1}$$

which includes the Jarlskog invariant [33,34] $J_r \sin\delta_{CP}$ (in our convention) and $\Delta_{ij} \equiv \Delta m^2_{ij} L/4E$ are the oscillation phases. Oscillations of beam neutrinos probe the atmospheric splitting $\Delta_{31} \sim O(1)$, while $\Delta_{21} \ll 1$. There, the CP term is suppressed by $\Delta m^2_{21}/\Delta m^2_{31} \times \pi/2 \sim 1/20$ due to the fact that oscillations driven by Δm^2_{21} do not have time to develop. This yields $P_{CP} \simeq -0.4J_r \sin\delta_{CP} \sin\Delta_{31} \sin\Delta_{32}$. Sub-GeV atmospheric neutrino oscillations, on the other hand, probe the solar splitting. In this case, the oscillations driven by $\Delta m^2_{31;32}$ are fast and average out. The resulting factor is just 1=2, leading to a much larger CP-violating term relative to beam neutrinos, namely PCP $\simeq -4J_r \sin\delta_{CP} \sin\Delta_{21}$ with $\Delta_{21} \sim O(1)$.

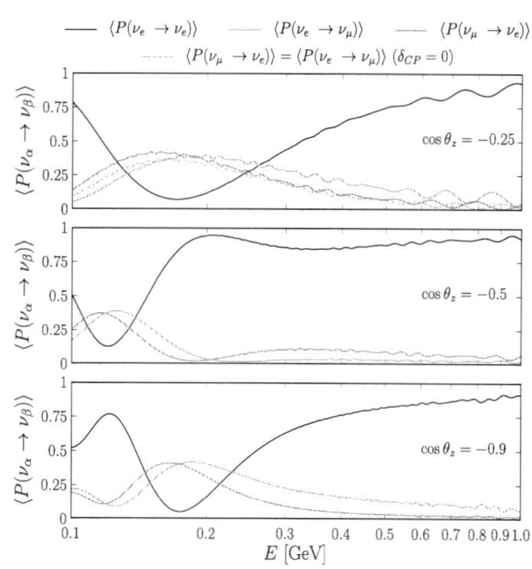

Figure 1. Oscillation probabilities for $\nu_e \rightarrow \nu_e$ (black), $\nu_e \rightarrow \nu_\mu$ (green), and $\nu_\mu \rightarrow \nu_e$ (purple) for $\delta_{CP} = 3\pi/2$ and various values of the zenith angle $\cos\theta_z$ as indicated. In the upper panel we also present $P(\nu_\mu \rightarrow \nu_e) = P(\nu_\mu \rightarrow \nu_e)$ for $\delta_{CP} = 0$ in the upper panel (gray dashed). Earth's matter profile was implemented using the PREM model [36].

In Fig. 1, we present three oscillation probability curves [35] as a function of neutrino energy for various zenith angles. A mild energy averaging of $10\%/\sqrt{E/\text{GeV}}$ was introduced which washes out unobservable fast oscillations induced by Δm^2_{31}. In black, $P(\nu_e \to \nu_e)$ is displayed which does not depend on δ_{CP}. For $\delta_{CP}=0$, the $\nu_e \to \nu_\mu$ and $\nu_\mu \to \nu_e$ transition probabilities are identical (upper panel, gray dashed line). The reason for that is because these oscillations are related by T parity and CPT is conserved. A nonzero δ_{CP} makes these probabilities distinct. In all panels we show $P(\nu_e \to \nu_\mu)$ and $P(\nu_\mu \to \nu_e)$ for $\delta_{CP}=3\pi/2$ in green and purple, respectively. If these probabilities can be distinguished, then the CP phase can be measured. The second feature that stands out is the impact of different zenith angles, which is related to matter effects. We turn our attention to them now.

The second crucial feature of sub-GeV atmospheric neutrino oscillations are matter effects. Interactions with matter in the Earth, specifically in the dense mantle and core, may significantly modify neutrino oscillations. These effects are quite rich and have been studied in great depth [6–15]. Here we restrict ourselves to review some oscillation aspects and provide a few examples.

Up-going atmospheric neutrinos that traverse the Earth may go through an MSW resonance [16,17] in the solar sector, maximizing oscillations between ν_e and $\nu_{\mu;\tau}$, when

$$\Delta m^2_{21} \cos\theta_{12} = 2\sqrt{2} E G_F n_e, \quad (2)$$

where G_F is the Fermi constant, and ne is the electron number density. In the solar sector, the MSW resonance happens only for neutrinos, not for antineutrinos, as observed in oscillation of neutrinos produced in the Sun. We will focus on the $\nu_e \to \nu_e$ oscillation dependence on the zenith angle, shown as black curves in the different panels of Fig. 1. In the crust (upper panel, $-0.44 < \cos\theta_z$), mantle (middle panel, $-0.84 < \cos\theta_z < -0.44$), and core (bottom panel, $\cos\theta_z < -0.84$), the MSW resonant energies are found to be around 180, 130, and 50 MeV, respectively. Although this energy in Earth's core is below 100 MeV, another type of resonance occurs about E $\sim$ 170 MeV, a parametric resonance [18,19,24]. A parametric resonance happens when changes to the matter density profile occur on the same scale as the neutrino oscillation length. The phenomenon is analogous to a resonant spring oscillator. Note that, due to the near-maximal value of θ_{23}, ν_e oscillates approximately equally into ν_μ and ν_τ.

The CP-violating and matter effects displayed in Fig. 1 show that the δ_{CP} effect is broad in neutrino energy, but there are large variations of oscillation curves for different zenith angles. Therefore, the precise reconstruction of the neutrino energy will not be as important as the determination of the incoming neutrino direction for the measurement of δ_{CP}. LArTPCs have excellent energy resolution and tracking reconstruction, and hence the incoming neutrino direction may be determined by considering the full event topology in charged current quasielastic events, $\nu_l n \rightarrow l^- p^+$.

SIMULATION DETAILS

To simulate the atmospheric neutrino flux at sub-GeV energies, we use Ref. [37]. The atmospheric neutrinos flux for a given flavor is parametrized by

$$\Phi_\alpha(E) = \Phi_{\alpha,0} f_\alpha(E) \left(\frac{E}{E_0}\right)^\gamma, \quad (3)$$

where $f_\alpha(E)$ gives the shape of the neutrino energy spectrum for each flavor; $\Phi_{\alpha,0}$ is the normalization of flavor $\alpha = \nu_e, \nu_\mu, \bar{\nu}_e, \bar{\nu}_\mu, E_0$ is an arbitrary reference energy; and γ accounts for spectral distortions. To account for unknowns on the meson production in the atmosphere, we consider systematic uncertainties on the following quantities (see Supplemental Material [38] for details): overall normalization (40%); the ratio r_e between ν_e and ν_μ fluxes (5%); the ratio rv between neutrinos and antineutrinos fluxes (2%); and the spectral distortion parameter γ with 0.2 absolute uncertainty.

Neutrino events in DUNE will be classified by topology. We consider events with a charged lepton (electrons or muons) and up to two outgoing protons and no pions, namely CC-Np0π (N=0, 1, 2). Neutrino interactions were modeled with the NuWro event generator [39]. This is an important step as recoiled nucleons may reinteract still inside the nucleus, a process typically referred to as final state interactions or intranuclear cascades. A pictorial representation of neutrino scattering on free nucleons and the effect of intranuclear cascades is shown in Fig. 2. To account for detector response, a cut on the minimum proton kinetic energy of 30 MeV was implemented [22]. Momentum and angular resolutions of (5%,5%,10%) and (5°,5°,10°) for electrons, muons, and protons were assumed [40,41]. It has been shown that LArTPCs have a near 100% efficiency to reconstruct tracks [42], and thus we do not take efficiencies into account here.

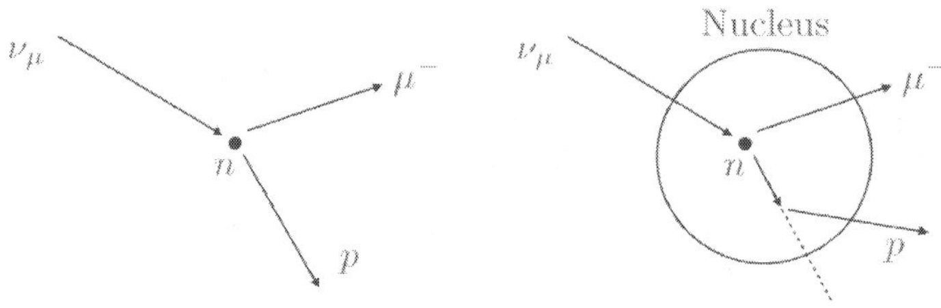

Figure 2. Pictorial representations of a neutrino scattering on a free nucleon (left) and the effect of intranuclear cascades (right).

We define two observables: the deposited energy E_{dep} (the sum of the energy of all detected particles) and deposited energy direction θ_z. For example, in a CC-2p0π event we would have $E_{\rm dep} = E_\ell + K_p^{(1)} + K_p^{(2)}$, where K_p indicates the proton kinetic energy. The direction is simply the direction of the sum all outgoing charged particles three momenta. Besides the imperfect detector response, intranuclear cascades effects and outgoing neutrons (which we consider to always go undetected) can affect distribution of E_{dep} and θ_z. We find that the largest contribution to the spread in deposited energy and direction arrives from intranuclear cascades [39]. A similar technique was proposed in Refs. [43,44] to improve the DUNE sensitivity for dark matter annihilation in the Sun using pointing.

To evaluate the experimental sensitivity to δ_{CP}, we have calculated the oscillation probabilities for $-1 \leq \cos\theta_z \leq 1$ and 100 MeV$\leq E_\nu \leq$1 GeV, assuming the PREM Earth density model [36] and marginalizing over other oscillation parameters assuming current central values and Gaussian priors on solar parameters [45] and expected priors from DUNE on atmospheric parameters [22], namely, $\sin^2\theta_{12}$=0.31±0.013, $\sin^2\theta_{13}$=0.0224±0.00066, $\sin^2\theta_{23}$=0.58±0.01, Δm^2_{21}=(7.39±0.21)×10^{-5}eV^2, and Δm^2_{31}=(+2.53±0.01)×10^{-3}eV^2. Throughout this Letter we assume an exposure of 400 kton-year. For the best fit values, we expect roughly 40 00 ν_e, 5000 ν_μ, and 1000 $\bar{\nu}_e$ and $\bar{\nu}_\mu$ events. The majority of $\nu\,(\bar{\nu})$ events are of the CC-1p 0π (CC-0p 0π) topology. The backgrounds to this analysis are very small: beam events can be vetoed with timing information; cosmic ray backgrounds are expected to be suppressed by optical filtering techniques together with cuts on track length and topology [46]; and misidentified $\pi^\pm$ produced by higher energy neutrinos should be subleading, as they are suppressed by the lower fluxes, the neutral current cross section and the misidentification rate which should be below 10% [47].

In Fig. 3, $\cos\theta_z$ event spectra for a small subset of the data are shown for $\delta_{CP}=3\pi/2$ (green) and $\delta_{CP}=3\pi/4$ (purple). Two-dimensional event spectra in the $\cos\theta_z \times E_{dep}$ plane can be found in the Supplemental Material [38]. These spectra are used to calculate a χ^2 test statistics for each distinct final-state event topology, assuming no charge identification, but perfect μ–e separation. The sensitivity to δ_{CP}, presented in the following sections, comes from combining of all these event topologies and marginalizing the test statistics over the systematic uncertainties aforementioned.

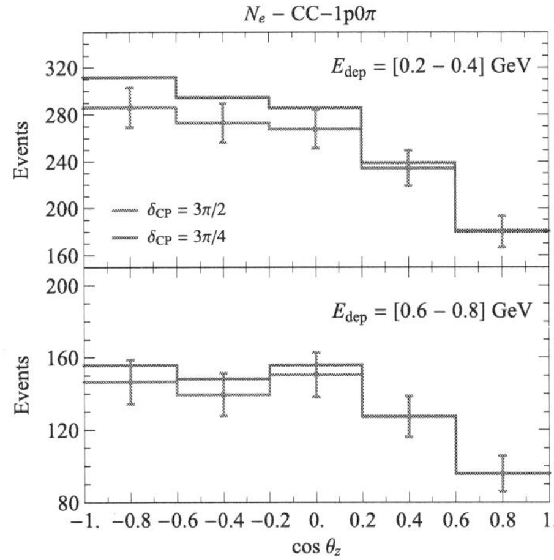

Figure 3. Event spectra as a function of the zenith direction of the deposited energy θ_z, for $\delta_{CP}=3\pi/2$ (green) and $\delta_{CP}=3\pi/4$ (purple), and for a small subset of the entire data: deposited energies between 0.2–0.4 GeV (upper panel) and 0.6–0.8 GeV (lower panel). The error bars include only statistical uncertainties.

DISCUSSION

The sensitivity to δ_{CP} for an input value of $\delta_{CP}=3\pi/2$ is shown in Fig. 4. The individual $\Delta\chi^2$ contribution for each topology is shown, as well as the combined fit. A significant sensitivity to δ_{CP} may be achieved, allowing for excluding regions of the parameter space beyond the 3σ level.

Several factors contribute to this sensitivity. As already discussed, the CP violation effect for sub-GeV atmospheric neutrinos is a sizable effect, an order of magnitude larger than the corresponding one for beam neutrinos. To observe CP violation, one should be able to independently measure oscillations of neutrinos and antineutrinos and/or the time-conjugated channels $\nu_\mu \rightarrow \nu_e$ and $\nu_e \rightarrow \nu_\mu$. At these low energies, a neutrino interaction is more likely to kick out a proton from a nucleus than an antineutrino interaction, and vice-versa for neutrons—therefore, the CC-1p0π sample is neutrino rich while CC-0p0π is antineutrino rich. Combining these two samples allows for measuring, statistically, the flux of ν and $\bar{\nu}$ from the atmosphere. Besides, θ_z has a typical spread between $\Delta\theta \sim 20°$–$30°$, mainly due to intranuclear cascades, except for the CC-0p0π topology which has $\Delta\theta \sim 50°$. This allows us to disentangle the baseline dependent oscillation effects discussed earlier fairly well. These aspects indicate a synergy between each distinct topology, as it can be seen in Fig. 4: the sum of the individual $\Delta\chi^2$ contributions for each topology is significantly below the combined sensitivity.

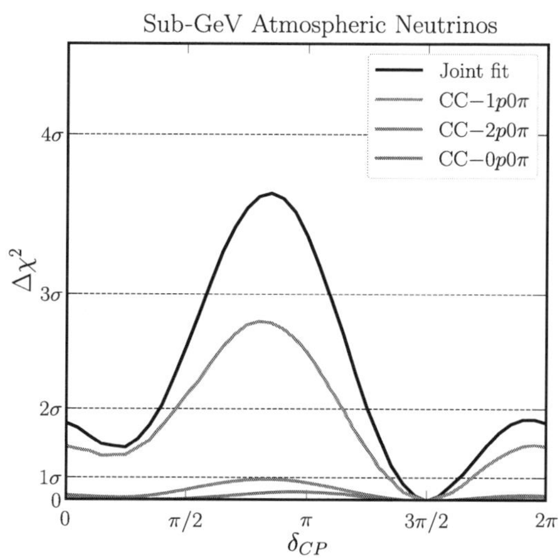

Figure 4. DUNE sensitivity to the leptonic CP-violating phase δ_{CP} using sub-GeV atmospheric neutrinos, for an input value $\delta_{CP}=3\pi/2$ and 400 kton-year exposure.

We have found that DUNE constrains the pull parameters beyond the uncertainties adopted here, namely (2%, 2%, 1%, 0.02) for $(\Phi_0, r_e, r_\nu, \gamma)$, evidencing that the experimental sensitivity is not induced by any prior uncertainty on the atmospheric fluxes, and therefore is quite robust (see Supplemental Material [38] for details). As we see in Fig. 3, the effects of δ_{CP} on the atmospheric spectra are highly nontrivial. Therefore, the available range energies and baselines (given by θ_z) helps to disentangle these effects from the several uncertainties in the sub-GeV atmospheric neutrino flux. This will have significant consequences for determining the atmospheric background in diffuse supernovae neutrino measurements [48] and dark matter experiments [49].

The sensitivity to δ_{CP} obtained here, though not as powerful as the one obtained with beam neutrinos [50], is competitive, providing an important cross check for the determination of δ_{CP} involving different energies and baselines. The sensitivity to δ_{CP} is mostly encoded by neutrinos with energies below ~0.6 GeV, while it is mainly limited by statistics and the spread in θ_z distribution. The addition of multi-GeV atmospheric neutrinos and other event topologies, should improve the sensitivity to δ_{CP} and new physics scenarios [51]. Furthermore, other oscillation parameters can be constrained with this analysis. As an example we show the sensitivity in $\delta_{CP}\times\sin^2\theta_{23}$ in the Supplemental Material [38]. A similar analysis could be performed for the JUNO experiment [52].

Finally, one could wonder about the impact of neutrino interaction uncertainties in this analysis. Overall cross section uncertainty plays no role as it is degenerate with the flux normalization. Intranuclear cascade uncertainties are expected to be more relevant and we will investigate this in detail in a forthcoming publication. The exploration of these effects is quickly evolving with the liquid argon neutrino program [21], and we expect significant advances by the time DUNE will be taking data. In fact, the short-baseline neutrino program [53] at Fermilab and DUNE itself will explore the sub-GeV region with high statistics using beam neutrinos. The DUNE-PRISM concept [54], a movable near detector to probe off-axis neutrinos, could greatly enhance our knowledge of neutrino-argon interactions at sub-GeV energies if the near detector hall allows for an off-axis distance of at least 25–30 meters. This would enable a pioneering data-driven analysis of CP violation using sub-GeV atmospheric neutrinos in DUNE, fully exploring the unique capabilities of liquid argon time projection chambers.

CONCLUSIONS

We have proposed to use the unique capabilities of LArTPCs to explore the physics of sub-GeV atmospheric neutrinos. By detecting low energy charged particles, the direction and energy of incoming neutrinos can be inferred, furnishing LArTPCs with a unique opportunity to probe CP violation with sub-GeV atmospheric neutrinos. We have shown, with a detailed simulation, that DUNE's sensitivity to the CP phase from this atmospheric sample is competitive, possibly ruling out regions of the parameter space beyond the 3σ level, and providing an important cross-check of the CP phase determination with beam neutrinos. This measurement will have significant consequences for diffuse supernovae neutrino measurements and dark matter experiments. We also highlight the possibility of performing a data-driven analysis using inputs from highly off-axis DUNE-PRISM measurements on neutrino-argon interactions.

ACKNOWLEDGEMENTS

We thank Alvaro Hernandez-Cabezudo, Shirley Li, Bryce Littlejohn, Alberto Marchionni, Ornella Palamara, and Sam Zeller for useful discussions. We are especially grateful to John Beacom for his comments on an earlier version of this manuscript which led to substantial improvement. Fermilab is operated by the Fermi Research Alliance, LLC under contract No. DE-AC02-07CH11359 with the United States Department of Energy. This project has received funding/support from the European Unions Horizon 2020 research and innovation program under the Marie Sklodowska-Curie Grant Agreement No. 690575 and No. 674896. I. M. S. and Y. P. G. acknowledge travel support from the Colegio de Fisica Fundamental e Interdisciplinaria de las Americas (COFI)

REFERENCES

1. Y. Fukuda et al. (Super-Kamiokande Collaboration), Phys. Rev. Lett. 81, 1562 (1998).
2. T. Kajita, Rev. Mod. Phys. 88, 030501 (2016).
3. A.B. McDonald, Rev. Mod. Phys. 88, 030502 (2016).
4. K. Abe et al. (Super-Kamiokande Collaboration), Phys. Rev. D 97, 072001 (2018).
5. M.G. Aartsen et al. (IceCube Collaboration), Phys. Rev. Lett. 120, 071801 (2018).
6. V.D. Barger, T.J. Weiler, and K. Whisnant, Phys. Lett. B 440, 1 (1998).
7. A. Friedland, C. Lunardini, and M. Maltoni, Phys. Rev. D 70, 111301(R) (2004).
8. P. Huber, M. Maltoni, and T. Schwetz, Phys. Rev. D 71, 053006 (2005).
9. E.K. Akhmedov, M. Maltoni, and A. Yu. Smirnov, J. High Energy Phys. 2008, 072.
10. O. Mena, I. Mocioiu, and S. Razzaque, Phys. Rev. D 78, 093003 (2008).
11. O.L.G. Peres and A. Yu. Smirnov, Phys. Rev. D 79, 113002 (2009).
12. E. Fernandez-Martinez, G. Giordano, O. Mena, and I. Mocioiu, Phys. Rev. D 82, 093011 (2010).
13. V. Barger, R. Gandhi, P. Ghoshal, S. Goswami, D. Marfatia, S. Prakash, S.K. Raut, and S.U. Sankar, Phys. Rev. Lett. 109, 091801 (2012).
14. E.K. Akhmedov, S. Razzaque, and A. Yu. Smirnov, J. High Energy Phys. 2013, 082; 2013, 26.
15. W. Winter, Phys. Rev. D 88, 013013 (2013).
16. L. Wolfenstein, Phys. Rev. D 17, 2369 (1978).
17. S.P. Mikheyev and A. Yu. Smirnov, Sov. J. Nucl. Phys. 42, 913 (1985).
18. E.K. Akhmedov, Yad. Fiz. 47, 475 (1988); [Sov. J. Nucl. Phys. 47, 301 (1988)].
19. P.I. Krastev and A. Yu. Smirnov, Phys. Lett. B 226, 341 (1989).
20. S. Razzaque and A. Yu. Smirnov, J. High Energy Phys. 2015, 139.
21. O. Palamara (ArgoNeuT Collaboration), J. Phys. Soc. Jpn. Conf. Proc. 12, 010017 (2016).
22. R. Acciarri et al. (DUNE Collaboration), arXiv:1512.06148.
23. M.A. Acero et al. (NOvA Collaboration), Phys. Rev. D 98, 032012

(2018).

24. K. Abe et al. (T2K Collaboration), Phys. Rev. Lett. 121, 171802 (2018).
25. A. Gando et al. (KamLAND Collaboration), Phys. Rev. D 88, 033001 (2013).
26. B.T. Cleveland, T. Daily, R. Davis, Jr., J.R. Distel, K. Lande, C.K. Lee, P.S. Wildenhain, and J. Ullman, Astrophys. J. 496, 505 (1998).
27. J.N. Abdurashitov et al. (SAGE Collaboration), Phys. Rev. C 80, 015807 (2009).
28. G. Bellini et al., Phys. Rev. Lett. 107, 141302 (2011).
29. J. Hosaka et al. (Super-Kamiokande Collaboration), Phys. Rev. D 73, 112001 (2006).
30. B. Aharmim et al. (SNO Collaboration), Phys. Rev. C 88, 025501 (2013).
31. M. Tanabashi et al. (Particle Data Group), Phys. Rev. D 98, 030001 (2018).
32. P.B. Denton, H. Minakata, and S.J. Parke, J. High Energy Phys. 2016, 051.
33. C. Jarlskog, Phys. Rev. Lett. 55, 1039 (1985).
34. C. Jarlskog, Z. Phys. C 29, 491 (1985).
35. See https://imgur.com/HoWUniu for an animation of how δCP changes oscillation probabilities as a function of zenith angle and neutrino energy.
36. A.M. Dziewonski and D.L. Anderson, Phys. Earth Planet. Interiors 25, 297 (1981).
37. M. Honda, M. Sajjad Athar, T. Kajita, K. Kasahara, and S. Midorikawa, Phys. Rev. D 92, 023004 (2015).
38. See Supplemental Material at http://link.aps.org/supplemental/10.1103/PhysRevLett.123.081801 for further discussions on the relevant oscillation probabilities, differential event rates per event topology, as well as further considerations regarding systematic uncertainties and the statistical treatment of the current determination of oscillation parameters.
39. T. Golan, J.T. Sobczyk, and J. Zmuda, Nucl. Phys. B, Proc. Suppl. 229–232, 499 (2012).
40. R. Acciarri et al. (ArgoNeuT Collaboration), Phys. Rev. D 90, 012008 (2014).

41. T. Alion et al. (DUNE Collaboration), arXiv:1606.09550.
42. R. Acciarri et al. (MicroBooNE Collaboration), J. Instrum. 12, P12030 (2017).
43. C. Rott, S. In, J. Kumar, and D. Yaylali, J. Cosmol. Astropart. Phys. 2017, 016.
44. C. Rott, D. Jeong, J. Kumar, and D. Yaylali, J. Cosmol. Astropart. Phys. 2019, 006.
45. I. Esteban, M.C. Gonzalez-Garcia, A. Hernandez-Cabezudo, M. Maltoni, and T. Schwetz, J. High Energy Phys. 2019, 106.
46. C. Adams et al. (MicroBooNE Collaboration), arXiv:1812.05679.
47. J.N. Esquivel, μ/π Separation Using Convolutional Neural Networks for the MicroBooNE Charged Current Inclusive Cross Section Measurement, Syracuse University, 2018.
48. J.F. Beacom, Annu. Rev. Nucl. Part. Sci. 60, 439 (2010).
49. A. Gütlein, C. Ciemniak, F. von Feilitzsch, N. Haag, M. Hofmann, C. Isaila, T. Lachenmaier, J.-C. Lanfranchi, L. Oberauer, S. Pfister, W. Potzel, S. Roth, M. von Sivers, R. Strauss, and A. Zöller, Astropart. Phys. 34, 90 (2010).
50. B. Abi et al. (DUNE Collaboration), arXiv:1807.10334.
51. K.J. Kelly, P.A.N. Machado, I. Martinez-Soler, S.J. Parke, and Y.F. Perez-Gonzalez (to be published).
52. F. An et al. (JUNO Collaboration), J. Phys. G 43, 030401 (2016).
53. P.A. Machado, O. Palamara, and D.W. Schmitz, Annu. Rev. Nucl. Part. Sci. 69, (2019).
54. T. Cai, J. Calcutt, K. Mahn, L. Pickering, S. Manly, H. Tanaka, C. Vilela, M. Wilking, and G. Yang, The dune-prism near detector, https://indico.fnal.gov/event/16205/contribution/2/material/0/0.pdf, 2018.

CHAPTER

17

DUNE as the Next-Generation Solar Neutrino Experiment

Francesco Capozzi[1,2,3], Shirley Weishi Li[1,2,4], Guanying Zhu[1,2], and John F. Beacom[1,2,5]

[1]Center for Cosmology and AstroParticle Physics (CCAPP), Ohio State University, Columbus, Ohio 43210, USA

[2]Department of Physics, Ohio State University, Columbus, Ohio 43210, USA

[3]Max-Planck-Institut für Physik (Werner-Heisenberg-Institut), 80805 München, Germany

[4]SLAC National Accelerator Laboratory, Menlo Park, California 94025, USA

[5]Department of Astronomy, Ohio State University, Columbus, Ohio 43210, USA

ABSTRACT

We show that the Deep Underground Neutrino Experiment (DUNE), with significant but feasible new efforts, has the potential to deliver world-leading results in solar

Citation: Francesco Capozzi, Shirley Weishi Li, Guanying Zhu, and John F. Beacom. "DUNE as the Next-Generation Solar Neutrino Experiment" Phys. Rev. Lett. 123, 131803 (2019). https://doi.org/10.1103/PhysRevLett.123.131803

neutrinos. With a 100 kton−yr exposure, DUNE could detect $\gtrsim 10^5$ signal events above 5 MeV electron energy. Separate precision measurements of neutrino-mixing parameters and the ^{8}B flux could be made using two detection channels ($\nu e+^{40}Ar$ and $\nu_{e,\mu,\tau}+e^-$) and the day-night effect (>10σ).

New particle physics may be revealed through the comparison of solar neutrinos (with matter effects) and reactor neutrinos (without), which is discrepant by ~2σ (and could become 5.6σ). New astrophysics may be revealed through the most precise measurement of the 8B flux (to 2.5%) and the first detection of the hep flux (to 11%). DUNE is required: No other experiment, even proposed, has been shown capable of fully realizing these discovery opportunities.

INTRODUCTION

Tremendous scientific opportunities remain in solar neutrinos. What are the particle properties of neutrinos? What are the nuclear processes that power our Sun and other stars? Although the basics are known [1–6], there are multiple unknowns and discrepancies.

To progress, we need precise measurements of all neutrino-producing processes, plus ways to isolate new physics from new astrophysics. Here we focus on high energies (>5 MeV electron energy).

For particle physics, the primary opportunity is to test for new physics through a precision comparison of neutrino-mixing parameters [7–14] measured in solar versus reactor experiments. Figures 1 and 2 preview this. There is a ~2σ discrepancy for Δm^2_{21} [6,15–17]. The reactor measurement will soon be greatly improved by the Jiangmen Underground Neutrino Observatory (JUNO) experiment [18], but testing new physics depends on improving the solar measurement, too.

The contrast in physical conditions is striking: neutrinos versus antineutrinos, matter versus vacuum mixing, plus a much larger distance, giving sensitivity to CPT violation [7,8], nonstandard neutrino interactions [9,19], neutrino decay [20,21], and more.

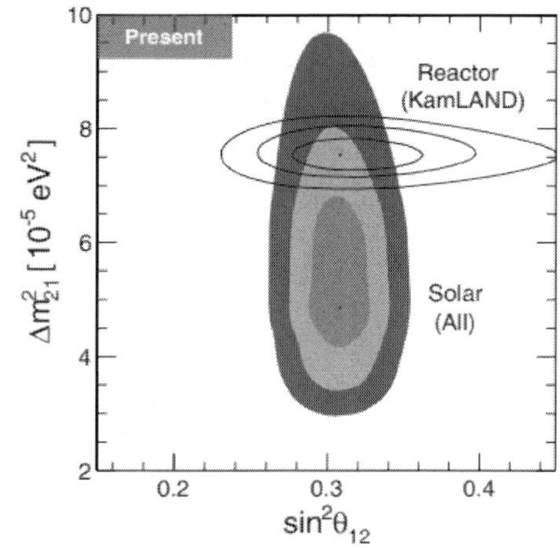

Figure 1. Present measurements (1, 2, and 3σ) of neutrino mixing with solar [1–6] and reactor [15] neutrinos.

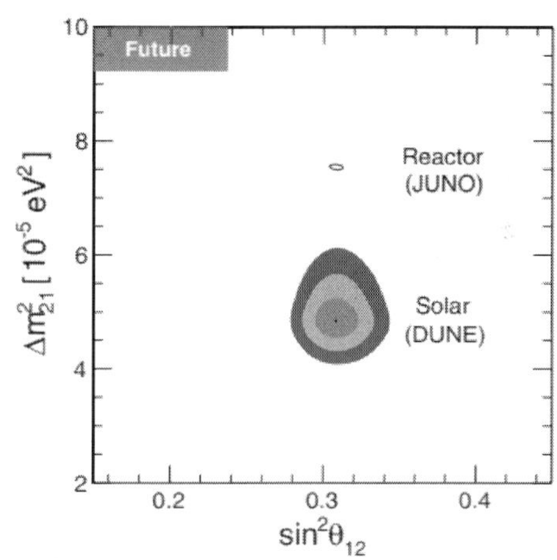

Figure 2. Future precision of neutrino mixing with solar (DUNE alone; 1, 2, and 3σ) and reactor (JUNO alone; 3σ [18,22]) neutrinos, using present best-fit points and 100 kton-yr for each.

For astrophysics, the primary opportunity is to make an independent precise measurement of the ^{8}B flux, which is extremely sensitive to the solar core temperature ($\sim T^{25}_c$ [23]) and which is an important ingredient for resolving the solar-metallicity discrepancy (requiring also progress on the 7Be and CNO fluxes) [24–28]. Discovery of the hep flux [24–30], the highest-energy neutrino process, would probe physical conditions far from the solar center while still having large matter effects.

How can these opportunities be realized, especially simultaneously? This requires a new multi-10-kton-scale experiment, plus breakthroughs in detection strategy.

We propose that the Deep Underground Neutrino Experiment (DUNE)—intended to make transformative studies of GeV long-baseline neutrino mixing, proton decay, and supernova neutrino bursts [31–33]—has the potential to do the same for solar neutrinos. The budgeted plans for DUNE provide a large active volume, a huge overburden, and excellent technical capabilities, including at MeV energies (for supernovae) [31–33]. For solar neutrinos, DUNE would need new investments, detailed below, that would also enhance its planned programs. Building on prior work on solar neutrino detection in liquid argon [34–37], our Letter goes much further.

We review the challenges in solar neutrinos, outline our proposed strategy for DUNE, calculate the signals and backgrounds, calculate the physics reach, define technical requirements, and conclude. To show what DUNE could achieve, we calculate our main results under optimistic but feasible assumptions; we also discuss the impact of varying these assumptions.

SOLAR NEUTRINOS: STATUS AND OBSTACLES

The fundamental challenge in solar neutrinos is disentangling neutrino-mixing effects and source properties. Super-Kamiokande (Super-K) and Sudbury Neutrino Observatory (SNO) measurements of ^{8}B neutrinos dominate the precision of solar determinations of $\sin^2\theta_{12}$ and Δm^2_{21}, as well as $\phi(^8B)$, the total ^{8}B flux [5,6,12,40,41]. The hep flux, $\phi(hep) \sim 10^{-3}\phi(^8B)$, has not been detected [42,43].

Super-K and SNO measurements are consistent with an energy-independent ν_e survival probability $P_{ee} \simeq \sin^2\theta_{12}$; the lack of an observed upturn in P_{ee} at low energies sets a weak upper limit on Δm^2_{21} [5,6]. Within the theoretical framework of matter-affected neutrino mixing [44–49], these results are consistent with lower-energy solar neutrino data [1–4]. Two other results were key.

(1) SNO separately measured $\phi(^8B)$ and $\sin^2\theta_{12}$ using two channels: $\nu_{e,\mu,\tau}+d \rightarrow \nu_{e,\mu,\tau}+p+n$, which is equally sensitive to all active flavors and, hence, measures the total flux, and $\nu_e+d \rightarrow e^-+p+p$, from which they can then extract the mixing angle. Progress on $\sin^2\theta_{12}$ is limited primarily by SNO's final precision for $\phi(^8B)$ of $\simeq$4% ($\simeq$3% statistical) and partially

by the 1.7% systematic uncertainty on the elastic-scattering channel in Super-K [5,6,12,40].

(2) Super-K best constrains Δm^2_{21} by measuring the day-night flux asymmetry (at $\simeq 3\sigma$) [6] with the $\nu_{e,\mu,\tau}+e^- \rightarrow \nu_{e,\mu,\tau}+e^-$ channel, where P_{ee} at night is increased by several percent due to the matter effect in Earth [49–52]. Progress is limited by the slow increase in statistics after 20 years of exposure.

UNIQUE ADVANTAGES OF DUNE

DUNE will be in the Homestake mine in South Dakota (4300 m.w.e.). Each of two liquid-argon (LAr) modules (eventually four) will have a fiducial mass of 10 kton, surrounded by ~1 m LAr shielding (details depend on single or dual phase). Readout is by the time-projection technique—here, drifting charge deposited in the volume onto wire planes at the boundaries—plus prompt detection of scintillation light [31–33,53].

DUNE can simultaneously measure neutrino-mixing parameters and solar neutrino fluxes. Here, we first state our underlying ideas and simple estimates.

(a) The degeneracy between $\sin^2\theta_{12}$ and $\phi(^8B)$ can be broken using two detection channels:

$$\nu_e + {}^{40}\text{Ar} \rightarrow e^- + {}^{40}\text{K}^*, \quad (1)$$

where the rate $R_{Ar} \propto \phi(^8B)\sin^2\theta_{12}$, and

$$\nu_{e,\mu,\tau} + e^- \rightarrow \nu_{e,\mu,\tau} + e^-, \quad (2)$$

where $R_e \propto \phi(^8B)(\sin^2\theta_{12}+1/6\cos^2\theta_{12})$. These channels can be adequately separated with a crude angular cut. Though the dependence on the $\nu\mu,\tau$ content is weak, DUNE can improve on SNO due to its huge statistics. Figure 3 illustrates this.

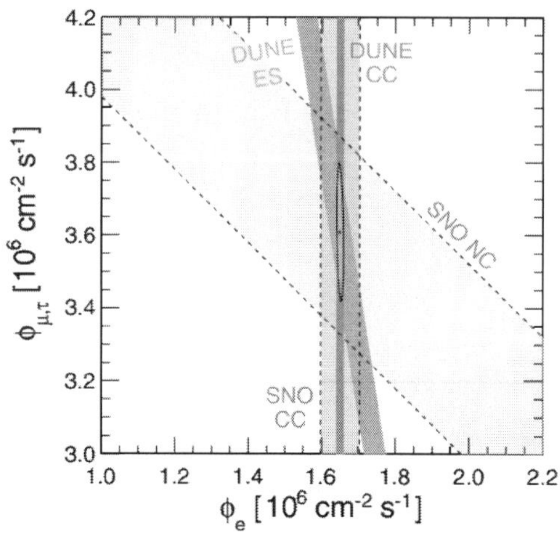

Figure 3. Estimated precision of the ν_e and $\nu_{\mu,\tau}$ content of the 8B flux, present (SNO [5,54]) and future (DUNE), with the ellipse for DUNE alone. Based on a simplified analysis, with only statistical uncertainties (1σ) but assuming 2 d.o.f., and with SNO fluxes slightly rescaled to match their global-fit 8B flux. Note the small axis ranges. Full analysis in the text.

(b) Δm^2_{21} can be isolated through the day-night flux asymmetry, $A_{D-N}=(D-N)/\frac{1}{2}(D+N)$, which scales as $\propto E_\nu/\Delta m^2_{21}$. For the solar Δm^2_{21}, an exposure of 100 kton-yr, and using only events above 6 MeV electron energy (effective threshold; see below) and outside the forward cone, we expect $D=3.04\times10^4$ and $N=3.29\times10^4$ signal $\nu_e+^{40}Ar$ events, along with 0.83×10^4 background events in total (as detailed below, and conservatively including $\nu_{e,\mu,\tau}+e-$ events). Considering for now only statistical uncertainties, we expect $A_{D-N}\simeq-(7.9\pm0.8)\%$ (~10σ). DUNE can improve on Super-K, because the $\nu_e+^{40}Ar$ channel has a larger cross section, emphasizes larger neutrino energies, and has a tighter relation between neutrino and electron energy.

SOLAR NEUTRINOS IN DUNE

The MeV-range capabilities of DUNE [31–33] are designed for detecting supernova neutrinos. Above 5 MeV, we assume electrons can be detected with high efficiency and 7% energy-independent energy resolution [33]. For solar signals, electrons lose energy dominantly by ionization, as the critical energy of LAr is 32 MeV [55–57]. The angular resolution of DUNE is uncertain; we adopt 25°, based on ICARUS simulations [35]. Below, we discuss the impact of different assumptions.

We use neutrino spectra from Refs. [58,59] and radial distributions from the BS05(OP) model of Refs. [25,26]. As nominal fluxes, we use $\phi(^8B)=5.25\times10^6\ cm^{-2}\,s^{-1}$ (4% uncertainty, from SNO [5]) and $\phi(hep)=8.25\times10^3\ cm^{-2}\,s^{-1}$ (30% uncertainty, from theory [28,30]). The end point energies are $\simeq$15 and $\simeq$19 MeV.

For the charged-current (CC) channel $\nu_e+^{40}Ar$ [60–74], the ground state threshold is $Q_{g.s}$.=1.5MeV [75], but this transition is forbidden. The cross section is dominated by transitions to nuclear excited states in $^{40}K^*$ (a superallowed Fermi transition with ΔE_i=4.4 MeV, plus several Gamow-Teller transitions), which promptly produce gamma rays by nuclear deexcitation. Because of these nuclear thresholds, DUNE is most sensitive to $E_\nu\gtrsim9$ MeV. We define the detectable energy of an event as the electron kinetic energy T_e, given by $T_e=E_\nu-Q$, where $Q=Q_{g.s.}+\Delta E_i$, conservatively neglecting the detectability of the ΔE_i in gamma rays (if these gamma rays were detectable, that would dramatically improve event identification, background rejection, energy reconstruction, and sensitivity). The electrons are emitted near isotropically. Details, including cross section uncertainties, are discussed below [38].

For the elastic-scattering (ES) channel $\nu_{e,\mu,\tau}+e^-$, there is no threshold, and the cross section is known with subpercent precision [76]. All flavors participate, but the sensitivity to the $\nu_{\mu,\tau}$ content is reduced, as these have only neutral-current couplings. DUNE is sensitive to $E_\nu\gtrsim5$ MeV, though the broad differential cross section effectively raises that. The direction of the scattered electron is well correlated to the neutrino direction, with a maximum scattering angle of about 20°. We adequately separate $\nu_{e,\mu,\tau}+e^-$ and $\nu_e+^{40}Ar$ events by defining a forward cone of half-angle 40°, maximizing the signal to background ratios for both event categories in the cone away from the Sun and its complement. Inside the cone, which includes 81% of $\nu_{e,\mu,\tau}+e^-$ events [35], they dominate; outside the cone, which includes 88% of $\nu_e+^{40}Ar$ events, they dominate.

In principle, DUNE could use the neutral-current (NC) channel $\nu_{e,\mu,\tau}+^{40}Ar\rightarrow\nu_{e,\mu,\tau}+^{40}Ar^*$, where the final state is detected through nuclear gamma rays [68,70,77]. We treat this as a background because the cross section seems small. If not, this could be an important new signal.

Backgrounds must be mitigated with standard MeV-detector techniques: defining a fiducial volume, removing U and Th from liquids and Rn from air, selecting low-background materials, applying dead time after high-energy events, etc. [1–6,15,18,78,79]. Three important backgrounds will remain. First are neutron captures on 40Ar, releasing a total of 6.1 MeV in several gamma rays [80–82]; these Compton scatter or pair-produce electrons. These neutrons, most less than a few MeV, are dominantly produced by (α,n) interactions in the rock following U- and Th-chain decays [35,83,84]; muon-induced neutrons are relatively negligible [39]. Once neutrons enter the detector, they fill the volume, due to their small cross section on argon. We assume a hermetic, passive water (or oil or plastic) shield of 40-cm thickness, reducing this background by $\sim 4\times10^3$. Below, we show that even no shielding is acceptable. Second, neutral-current $\nu_{e,\mu,\tau}+^{40}Ar$ events cause a peak near 9 MeV [77]. Third, at the highest energies, are beta-decaying radioactivities induced by muons [39,85–87], for which we apply simple cuts. The pileup rates from these and other backgrounds (e.g., ^{39}Ar and ^{42}Ar decays) are negligible [39].

Figure 4 shows the solar neutrino signal and background spectra in DUNE. Our calculations include three-flavor neutrino mixing [44–51], realistic detection effects (differential cross sections [69,88–90], energy smearing, angular cuts [35], background reduction [75,91–104]), and a 100 kton-yr exposure.

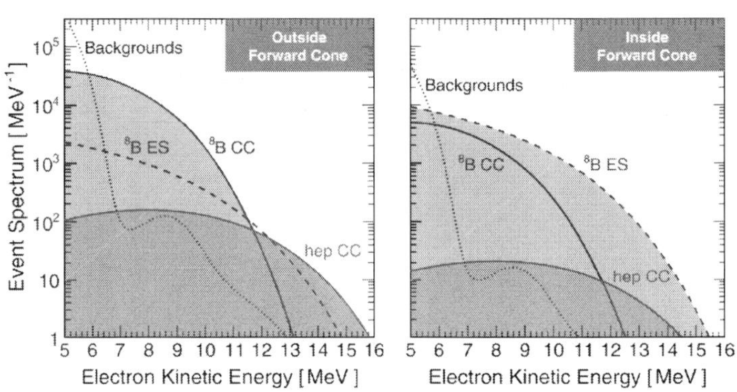

Figure 4. Predicted solar neutrino signals and backgrounds in DUNE for 100 kton-yr, using a forward cone of half-angle 40° and (here only) combining day and night data. We include all factors discussed in the text.

For ^8B events, the two channels are well separated and have superb yields above 5 MeV electron energy. For $\nu_e+^{40}Ar$, there are 9.9×10^4 events outside the forward cone. For $\nu_{e,\mu,\tau}+e^-$, there are 2.6×10^4 events inside the forward cone. This channel provides better sensitivity to lower-energy neutrinos and the only sensitivity to $\nu_{\mu,\tau}$. For hep events, the $\nu_e+^{40}Ar$ channel allows a clear separation at high electron energies, with 150 events above 11 MeV (the small $\nu_{e,\mu,\tau}$+e− channel is not shown).

DUNE PHYSICS REACH

DUNE can significantly improve the precision of solar neutrino observables. We jointly fit (without priors) four parameters: $\sin^2\theta_{12}$, Δm^2_{21}, $\phi(^8B)$, and ϕ(hep). When reporting projected uncertainties for n parameters, we marginalize over the others, adopting $\Delta\chi^2$ confidence levels for n d.o.f. We assume that new physics affecting solar neutrinos is reflected in mixing-parameter values that differ from the reactor values. We use the theoretically expected counts for signals and backgrounds, with Poisson uncertainties (below, we discuss systematics). We partition data into bins of energy, bins of Earth zenith angle (night only), and outside or inside the forward cone (as in Fig. 4). We use the total electron spectra, assuming only statistical separation of the components and not requiring neutrino-energy reconstruction.

Figure 2 shows the projected precision for DUNE's measurement of neutrino-mixing parameters, assuming the solar best-fit values ($\sin^2\theta_{12}$=0.308 and $\Delta m^2_{21}=4.85\times10^{-5}eV^2$ [6]). The uncertainties are 3.0% and 5.9%, a factor of $\simeq$1.5 and $\simeq$3 better than from all solar experiments to date, respectively, as shown in Fig. 1. The sensitivity to Δm^2_{21} comes primarily from the day-night effect (10.4σ). The ^{8}B flux (marginalizing over other parameters) can be measured to 2.5%, a factor of $\simeq$1.6 better than from SNO. DUNE can make a robust first detection of hep neutrinos, with a precision of 11%, a factor of $\simeq$3 better than the current theoretical uncertainty.

GOING FORWARD

New investments are needed to enhance the MeV capabilities of DUNE. At the trigger level, this includes enhancing data acquisition, storage, and processing for a steady rate of MeV events. Calibration at MeV energies across the detector volume will be crucial to controlling systematics. An enhanced light-detection system would enhance MeV detection.

Backgrounds must be controlled, and the biggest concern, after standard cuts [1–6,15,18,78,79], is neutron captures [39]. We see three possible strategies.

1. *40 cm of shielding, as assumed above.*—This allows a low threshold ($\simeq$5.8 MeV) to test for shape distortions in the spectrum

and to enhance particle-identification techniques.

2. *Additional run time.*—With less or no shielding, the effective analysis threshold would be higher. This can be compensated by a larger exposure than 100 kton-yr. With 30, 20, 10, or 0 cm of shielding, the effective analysis threshold is ≃6.2, 6.5, 6.9, or 7.2 MeV, and the exposure needed for comparable results increases at most of a factor of ~2.
3. *Better particle-identification techniques.*—We assume that neutrino-interaction events and neutron-capture events with the same electron energy are indistinguishable. This is too conservative, because νe,μ,τ+e−, νe+40Ar, and neutron-capture events would have one electron, an electron with γ rays, and multiple γ rays, respectively. For γ rays, the radiation length is 14 cm and Compton scattering dominates [55], so these event classes should be distinct.

Here, we summarize some key points. The sensitivity to Δm^2_{21} is very robust, because it is determined from the day-night effect, which cancels many inputs. The uncertainty on the ν_e+^{40}Ar cross section (presently 10%) and the detector systematics should be reduced to 1%. Without this, DUNE alone cannot break the degeneracy between $\sin^2\theta_{12}$ and $\phi(^8B)$, though its measurement of Δm^2_{21} would be unaffected and its measurement of ϕ(hep) be only modestly affected. Importantly, the intended precision of $\sin^2\theta_{12}$ and $\phi(^8B)$ could be largely restored by combining with existing solar data. If the energy resolution is 20%, the intended precision of ϕ(hep) would degrade to 18%, though that of $\sin^2\theta_{12}$, $\phi(^8B)$, and Δm^2_{21} could be largely restored by increasing exposure by ~2. DUNE's sensitivity would be lost if the energy resolution is poor *and* backgrounds are not reduced.

CONCLUDING PERSPECTIVES

This is the first study to detail how DUNE, with a new, challenging but feasible solar neutrino program, would open substantial discovery space in both particle physics and astrophysics. With DUNE's precision measurements of $\sin^2\theta_{12}$ and Δm^2_{21}, the comparison to JUNO's may reveal new particle physics of neutrinos. Simultaneously, with DUNE's precision measurements of $\phi(^8B)$ and ϕ(hep), it may reveal new astrophysics. Further studies are needed to evaluate our proposal and to optimize its sensitivity.

No other planned experiment has been shown potentially capable of meeting all of these goals. Of proposed experiments, Hyper-Kamiokande

(Hyper-K) [105,106] stands out, and it can nicely complement DUNE. Hyper-K would have only one channel, $\nu_{e,\mu,\tau}+e^-$, but huge statistics. DUNE and Hyper-K would measure Δm^2_{21} (from the day-night asymmetry) comparably well. Hyper-K would have a significant advantage on measuring the upturn in the νe survival probability. DUNE would measure ϕ(hep) much better. Their combined impact would be significantly enhanced by new experiments for low-energy solar neutrinos [36,107,108].

Solar neutrino studies, begun long ago, are not done. DUNE can lead the next generation of discoveries.

ACKNOWLEDGEMENTS

For helpful discussions, we are grateful to Manojeet Bhattacharya, Ed Blucher, Steve Brice, Mauricio Bustamante, Gustavo Cancelo, David Caratelli, Flavio Cavanna, Alex Friedland, Dick Furnstahl, Cristiano Galbiati, Alejandro Garcia, Mathew Graham, Jeff Hartnell, Alex Himmel, Brennan Jordan, Josh Klein, Pierre Lasorak, Eligio Lisi, Joe Lykken, Kenny Ng, Gabriel Orebi Gann, Ornella Palamara, Wendy Panero, Stephen Parke, Ryan Patterson, Simon Peeters, Georg Raffelt, Juergen Reichenbacher, Elizabeth Ricard-McCutchan, David Schmitz, Kate Scholberg, Michael Smy, Bob Svoboda, Mark Vagins, Francesco Vissani, Petr Vogel, and Kyle Wendt. The work of all authors was supported by National Science Foundation Grants No. PHY-1404311 and No. PHY-1714479 awarded to J. F. B. F. C. was later supported by Deutsche Forschungsgemeinschaft Grants No. EXC 153 and No. SFB 1258, as well as by European Union Grant No. H2020-MSCA-ITN-2015/674896. S. W. L. was also supported by an Ohio State Presidential Fellowship and later at SLAC by the Department of Energy under Contract No. DE-AC02-76SF00515.

We speak for ourselves as theorists, not on behalf of the DUNE Collaboration. This work is based on our ideas, our calculations, and publicly available information.

REFERENCES

1. B.T. Cleveland, T. Daily, R. Davis, Jr., J.R. Distel, K. Lande, C.K. Lee, P.S. Wildenhain, and J. Ullman, Astrophys. J. 496, 505 (1998).
2. M. Altmann et al. (GNO Collaboration), Phys. Lett. B 616, 174 (2005).
3. J.N. Abdurashitov, V.N. Gavrin, V.V. Gorbachev, P.P. Gurkina, T.V. Ibragimova, A.V. Kalikhov, N.G. Khairnasov, T.V. Knodel, I.N. Mirmov, and A.A. Shikhin et al. (SAGE Collaboration), Phys. Rev. C 80, 015807 (2009).
4. G. Bellini et al. (Borexino Collaboration), Phys. Rev. Lett. 107, 141302 (2011).
5. B. Aharmim et al. (SNO Collaboration), Phys. Rev. C 88, 025501 (2013).
6. K. Abe et al. (Super-Kamiokande Collaboration), Phys. Rev. D 94, 052010 (2016).
7. G. Barenboim, L. Borissov, J.D. Lykken, and A.Y. Smirnov, J. High Energy Phys. 2002, 001.
8. A. de Gouvea and C. Peña-Garay, Phys. Rev. D 71, 093002 (2005).
9. A. Friedland, C. Lunardini, and C. Peña-Garay, Phys. Lett. B 594, 347 (2004).
10. M. Cirelli, M.C. González-García, and C. Peña-Garay, Nucl. Phys. B719, 219 (2005).
11. A. Palazzo, Phys. Rev. D 83, 113013 (2011).
12. M. Maltoni and A. Yu. Smirnov, Eur. Phys. J. A 52, 87 (2016).
13. F. Capozzi, I.M. Shoemaker, and L. Vecchi, J. Cosmol. Astropart. Phys. 2017, 021.
14. S.-H. Seo and S.J. Parke, Phys. Rev. D 99, 033012 (2019).
15. A. Gando, Y. Gando, H. Hanakago, H. Ikeda, K. Inoue, K. Ishidoshiro, H. Ishikawa, M. Koga, R. Matsuda, and S. Matsuda et al. (KamLAND Collaboration), Phys. Rev. D 88, 033001 (2013).
16. I. Esteban, M.C. González-García, M. Maltoni, I. Martínez-Soler, and T. Schwetz, J. High Energy Phys. 2017, 087.
17. F. Capozzi, E. Lisi, A. Marrone, and A. Palazzo, Prog. Part. Nucl. Phys. 102, 48 (2018).
18. F. An et al. (JUNO Collaboration), J. Phys. G 43, 030401 (2016).
19. J. Liao, D. Marfatia, and K. Whisnant, Phys. Lett. B 771, 247 (2017).

20. J.F. Beacom and N.F. Bell, Phys. Rev. D 65, 113009 (2002).
21. J.M. Berryman, A. de Gouvea, and D. Hernandez, Phys. Rev. D 92, 073003 (2015).
22. F. Capozzi, E. Lisi, and A. Marrone, Phys. Rev. D 92, 093011 (2015).
23. J.N. Bahcall and A. Ulmer, Phys. Rev. D 53, 4202 (1996).
24. N. Grevesse and A.J. Sauval, Space Sci. Rev. 85, 161 (1998).
25. J.N. Bahcall, A.M. Serenelli, and S. Basu, Astrophys. J. 621, L85 (2005).
26. J.N. Bahcall, Software and data for solar neutrino research, 2005.
27. M. Asplund, N. Grevesse, A.J. Sauval, and P. Scott, Annu. Rev. Astron. Astrophys. 47, 481 (2009).
28. N. Vinyoles, A.M. Serenelli, F.L. Villante, S. Basu, J. Bergström, M.C. González-García, M. Maltoni, C. Peña-Garay, and N. Song, Astrophys. J. 835, 202 (2017).
29. L.E. Marcucci, R. Schiavilla, M. Viviani, A. Kievsky, S. Rosati, and J.F. Beacom, Phys. Rev. C 63, 015801 (2000).
30. T.S. Park, L.E. Marcucci, R. Schiavilla, M. Viviani, A. Kievsky, S. Rosati, K. Kubodera, D.P. Min, and M. Rho, Phys. Rev. C 67, 055206 (2003).
31. R. Acciarri et al. (DUNE Collaboration), arXiv:1512.06148.
32. J. Strait et al. (DUNE Collaboration), arXiv:1601.05823.
33. R. Acciarri et al. (DUNE Collaboration), arXiv:1601.02984.
34. J.N. Bahcall, M. Baldo-Ceolin, D.B. Cline, and C. Rubbia, Phys. Lett. B 178, 324 (1986).
35. F. Arneodo et al., Nucl. Instrum. Methods Phys. Res., Sect. A 455, 376 (2000).
36. D. Franco et al., J. Cosmol. Astropart. Phys. 2016, 017.
37. A.N. Ioannisian, A.Y. Smirnov, and D. Wyler, Phys. Rev. D 96, 036005 (2017).
38. See Supplemental Material at http://link.aps.org/supplemental/10.1103/PhysRevLett.123.131803 for a thorough discussion on charged-current neutrino interactions, the calculation of backgrounds, and the statistical analysis with different input choices.
39. G. Zhu, S.W. Li, and J.F. Beacom, Phys. Rev. C 99, 055810 (2019).
40. G.L. Fogli, E. Lisi, A. Marrone, and A. Palazzo, Prog. Part. Nucl.

Phys. 57, 742 (2006).

41. B. Aharmim et al. (SNO Collaboration), Phys. Rev. C 72, 055502 (2005).
42. J. Hosaka et al. (Super-Kamiokande Collaboration), Phys. Rev. D 73, 112001 (2006).
43. A.T. Mastbaum, Constraining the Hep Solar Neutrino and Diffuse Supernova Neutrino Background Fluxes with the Sudbury Neutrino Observatory, University of Pennsylvania, 2016.
44. L. Wolfenstein, Phys. Rev. D 17, 2369 (1978).
45. S.P. Mikheev and A. Yu. Smirnov, Yad. Fiz. 42, 1441 (1985); [Sov. J. Nucl. Phys. 42, 913 (1985)].
46. S.P. Mikheev and A. Yu. Smirnov, Nuovo Cimento C 9, 17 (1986).
47. W.C. Haxton, Phys. Rev. Lett. 57, 1271 (1986).
48. S.J. Parke, Phys. Rev. Lett. 57, 1275 (1986).
49. C. Giunti and C.W. Kim, Fundamentals of Neutrino Physics and Astrophysics (Oxford University Press, Oxford, 2007).
50. E. Lisi and D. Montanino, Phys. Rev. D 56, 1792 (1997).
51. M. Maris and S.T. Petcov, Phys. Rev. D 56, 7444 (1997).
52. W.C. Forsythe, E.J. Rykiel, Jr., R.S. Stahl, H.-i. Wu, and R.M. Schoolfield, Ecol. Model. 80, 87 (1995).
53. A. Marchionni, Annu. Rev. Nucl. Part. Sci. 63, 269 (2013).
54. B. Aharmim et al. (SNO Collaboration), Phys. Rev. C 87, 015502 (2013).
55. C. Patrignani et al. (Particle Data Group), Chin. Phys. C 40, 100001 (2016).
56. S. Amoruso et al. (ICARUS Collaboration), Eur. Phys. J. C 33, 233 (2004).
57. R. Acciarri et al. (MicroBooNE Collaboration), J. Instrum. 12, P09014 (2017).
58. W.T. Winter, S.J. Freedman, K.E. Rehm, and J.P. Schiffer, Phys. Rev. C 73, 025503 (2006).
59. J.N. Bahcall, Phys. Rev. C 56, 3391 (1997).
60. E.J. Konopinski, The Theory of Beta Radioactivity (Oxford University Press, Oxford, 1950).
61. W.E. Ormand, P.M. Pizzochero, P.F. Bortignon, and R.A. Broglia, Phys.

Lett. B 345, 343 (1995).
62. M. Bhattacharya et al., Phys. Rev. C 58, 3677 (1998).
63. J.F. Beacom and P. Vogel, Phys. Rev. D 60, 033007 (1999).
64. P. Vogel and J.F. Beacom, Phys. Rev. D 60, 053003 (1999).
65. J.F. Beacom and S.J. Parke, Phys. Rev. D 64, 091302(R) (2001).
66. A. Kurylov, M.J. Ramsey-Musolf, and P. Vogel, Phys. Rev. C 67, 035502 (2003).
67. I. Gil Botella and A. Rubbia, J. Cosmol. Astropart. Phys. 2003, 009.
68. E. Kolbe, K. Langanke, G. Martinez-Pinedo, and P. Vogel, J. Phys. G 29, 2569 (2003).
69. M. Bhattacharya, C.D. Goodman, and A. Garcia, Phys. Rev. C 80, 055501 (2009).
70. M.-K. Cheoun, E. Ha, and T. Kajino, Phys. Rev. C 83, 028801 (2011).
71. T. Suzuki, M. Honma, A.B. Balantekin, T. Kajino, and S. Chiba, EPJ Web Conf. 66, 07025 (2014).
72. J.C. Hardy and I.S. Towner, Phys. Rev. C 91, 025501 (2015).
73. M. Karakoç, R.G.T. Zegers, B.A. Brown, Y. Fujita, T. Adachi, I. Boztosun, H. Fujita, M. Csatlós, J.M. Deaven, and C.J. Guess et al., Phys. Rev. C 89, 064313 (2014).
74. D. Akimov et al. (COHERENT Collaboration), Science 357, 1123 (2017).
75. National Nuclear Data Center, NuDat 2.7, 2016.
76. J.N. Bahcall, M. Kamionkowski, and A. Sirlin, Phys. Rev. D 51, 6146 (1995).
77. R.S. Raghavan, S. Pakvasa, and B.A. Brown, Phys. Rev. Lett. 57, 1801 (1986).
78. Y. Takeuchi et al. (SuperKamiokade Collaboration), Phys. Lett. B 452, 418 (1999).
79. I. Blevis et al., Nucl. Instrum. Methods Phys. Res., Sect. A 517, 139 (2004).
80. R. Hardell and C. Beer, Phys. Scr. 1, 85 (1970).
81. C.D. Nesaraja and E.A. McCutchan, Nucl. Data Sheets 133, 1 (2016).
82. National Nuclear Data Center, CapGam, 2013.
83. H. Wulandari, J. Jochum, W. Rau, and F. von Feilitzsch, Astropart. Phys. 22, 313 (2004).

84. L. de Viveiros, Optimization of Signal versus Background in Liquid Xe Detectors Used for Dark Matter Direct Detection Experiments, Brown University, 2010.
85. S.W. Li and J.F. Beacom, Phys. Rev. C 89, 045801 (2014).
86. S.W. Li and J.F. Beacom, Phys. Rev. D 91, 105005 (2015).
87. S.W. Li and J.F. Beacom, Phys. Rev. D 92, 105033 (2015).
88. H. Behrens and J. Jänecke, Numerical Tables for Beta-Decay and Electron Capture (Springer, New York, 1969).
89. G.K. Schenter and P. Vogel, Nucl. Sci. Eng. 83, 393 (1983).
90. A.C. Hayes and P. Vogel, Annu. Rev. Nucl. Part. Sci. 66, 219 (2016).
91. H. Rogers, in Metallogeny of Gold in the Black Hills, South Dakota (Society of Economic Geologists, 1990), p. 204.
92. J.K. Shultis and R.E. Faw, Fundamentals of Nuclear Science and Engineering (Taylor & Francis, London, 2002).
93. A. Ferrari, P.R. Sala, A. Fasso, and J. Ranft, FLUKA: A multi-particle transport code (Program version 2005), 2005.
94. G. Battistoni, S. Muraro, P.R. Sala, F. Cerutti, A. Ferrari, S. Roesler, A. Fasso, and J. Ranft, AIP Conf. Proc. 896, 31 (2007).
95. K.T. Lesko et al., arXiv:1108.0959.
96. Y.-D. Chan, The Low-Background Construction of Laboratories at the 4850-ft Level Davis Campus, 2012.
97. B. Ricci, F. Mantovani, M. Baldoncini, J. Esposito, L. Ludhova, and S. Zavatarelli, in Proc. Sci., No. 2013, p. 077.
98. J. Heise, J. Phys. Conf. Ser. 606, 012015 (2015).
99. S. Westerdale and P.D. Meyers, Nucl. Instrum. Methods Phys. Res., Sect. A 875, 57 (2017).
100. J.F. Beacom et al. (Jinping Collaboration), Chin. Phys. C 41, 023002 (2017).
101. T.K. Gaisser and M. Honda, Annu. Rev. Nucl. Part. Sci. 52, 153 (2002).
102. K. Abe et al. (Super-Kamiokande Collaboration), Phys. Rev. D 97, 072001 (2018).
103. J.F. Beacom, Annu. Rev. Nucl. Part. Sci. 60, 439 (2010).
104. K. Bays et al. (Super-Kamiokande Collaboration), Phys. Rev. D 85, 052007 (2012).
105. K. Abe et al. (Hyper-Kamiokande Collaboration), arXiv:1109.3262.

106. K. Abe et al. (Hyper-Kamiokande Collaboration), arXiv:1805.04163.
107. S. Andringa et al. (SNO+Collaboration), Adv. High Energy Phys. 2016, 6194250 (2016).
108. J. Aalbers et al. (DARWIN Collaboration), J. Cosmol. Astropart. Phys. 2016, 017.

SECTION 5
NEUTRINOS FROM SUPERNOVAE AND ULTRA-HIGH ENERGETIC COSMIC NEUTRINOS

CHAPTER

18

Production of keV Sterile Neutrinos in Supernovae: New constraints and Gamma-Ray Observables

Carlos A. Argüelles[1,2], Vedran Brdar[3], and Joachim Kopp[3]

[1]Massachusetts Institute of Technology, Cambridge, Massachusetts 02139, USA

[2]Wisconsin IceCube Particle Astrophysics Center, Madison, Wisconsin 53703, USA

[3]PRISMA Cluster of Excellence, 55099 Mainz, Germany and Mainz Institute for Theoretical Physics, Johannes Gutenberg-Universität Mainz, 55099 Mainz, Germany

INTRODUCTION

One of the most auspicious candidate particles for the dark matter in the Universe is the sterile neutrino—an electrically neutral fermion with a mass

Citation: Carlos A. Argüelles, Vedran Brdar, and Joachim Kopp. "Production of keV sterile neutrinos in supernovae: New constraints and gamma-ray observables" Phys. Rev. D 99, 043012 (2019). https://doi.org/10.1103/PhysRevD.99.043012

on the order of keV-MeV that couples to ordinary matter only through a tiny mass mixing with Standard Model (SM) neutrinos [1,2]. In the simplest sterile neutrino scenarios, it is assumed that the abundance of sterile neutrinos ν_s is zero at the end of inflation, and they are later produced through their mixing with SM (active) neutrinos ν_a [3,4] (see also [5–7]). Experimental constraints on the mass of keV sterile neutrinos and their mixing with SM neutrinos arise from the measured DM relic density [8,9], from Pauli blocking (the Tremaine-Gunn bound) [10,11], from Lyman- α forests [12–14] and from X-ray searches for radiative decays of sterile neutrinos $\nu_s \to \nu_a + \gamma$ [15–20]. From the combination of these constraints, one concludes that the ν_s mass should be $m_s \gtrsim 4$ keV, and its mixing angle with the SM neutrinos should be $\sin^2 2\theta \lesssim 10^{-6}$ in the simple two-flavor approximation.

In this paper, we add a new limit to this inventory of constraints by considering sterile neutrino production in core-collapse supernovae (SN) [21–31]. A supernova develops when a $\gtrsim 9\ M_\odot$ star runs out of nuclear fuel. The thermal pressure that normally counteracts gravity disappears, and the core of the star collapses into a neutron star. The temperature in the nascent neutron star is $\gtrsim$MeV, so that a thermal population of (active) neutrinos is produced. These ν_a can oscillate into ν_s, which escapes the exploding star unhindered and may carry away significant amounts of energy [24,28,32]. Constraints on anomalous energy loss from SN 1987A will thus allow us to constrain the sterile neutrino parameters. As $\nu_a \to \nu_s$ conversion can be resonant thanks to the ultrahigh matter density $\sim 10^{14}$g/cm^3 in the SN core, these limits will be very strong. The flux of sterile neutrinos with O(MeV) energies escaping from a supernova leads to a flux of secondary gamma-rays when they decay, and we study this flux, as well.

Our main results are summarized in Fig. 1. The solid orange exclusion region shows that, in the mass range ms~2–80 keV, limits from energy loss in supernovae surpass previous limits by up to 2 orders of magnitude in $\sin^2 2\theta$ for the case when ν_s mixes with ν_μ or ν_τ. Note that, unlike the other limits shown in Fig. 1, our bounds would still hold if sterile neutrinos are not part of the DM in the Universe. In the following, we discuss in detail how we have obtained our new limits and sensitivity estimates.

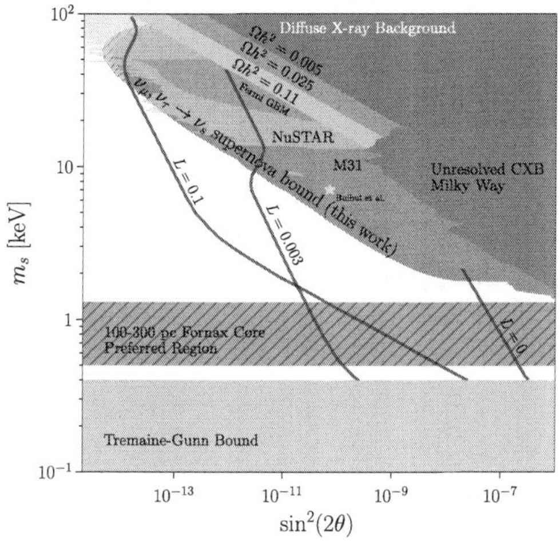

Figure 1. Supernova bounds on the sterile neutrino mass m_s and mixing $\sin^2 2\theta$ (orange, this work) between ν_x (x=μ or τ) and νs flavors compared to previous constraints [17–19] from the Tremaine-Gunn bound [10,11] (yellow), NuSTAR observation of galactic center [33] (pale orange), X-ray searches in the Andromeda Galaxy M31 [15] (green), the Fermi GBM all-sky analysis [34] (purple), and the galactic [17] (dark blue) and extragalactic [16,18,19] (gray) diffuse X-ray background. For the latter constraint, we show also how it is modified if sterile neutrinos account for only a fraction of the DM in the Universe. It is understood that the other X-ray limits would shift in a similar way if $\Omega h^2<0.11$. The hatched pink region indicates the region preferred by the properties of the core of the Fornax dwarf galaxy [35], whereas a yellow star shows the parameter point that could explain the 3.5 keV X-ray line hinted at by Refs. [19,36]. Black curves illustrate the parameter regions in which the Dodelson-Widrow mechanism [3] (L=0) or the Shi-Fuller mechanism [4] with lepton assymetry L>0 would yield the correct DM relic density $\Omega h^2=0.11$. The hatched band at large ms indicates the theoretical uncertainty in our limit stemming from our treatment of energy transport inside the supernova.

STERILE NEUTRINO PRODUCTION IN SUPERNOVAE

We consider a simplified two-flavor oscillation picture with mixing between a sterile neutrino[1] ν_s and one species of active neutrinos ν_x, x=μ or τ. We do not consider mixing between ν_s and electron neutrinos to avoid complications arising from charged current interactions between νe and electrons/positrons. The flavor basis Hamiltonian describing neutrino propagation in matter includes the vacuum oscillation term and a Mikheyev-Smirnov-Wolfenstein

(MSW) potential Vx that describes coherent forward scattering of νx on the background matter via Z exchange [37–39]:

$$H = \frac{\Delta m^2}{4E}\begin{pmatrix} -\cos 2\theta & \sin 2\theta \\ \sin 2\theta & \cos 2\theta \end{pmatrix} + \begin{pmatrix} V_x & 0 \\ 0 & 0 \end{pmatrix} \qquad (1)$$

Here, $\theta \ll 1$ is the ν_s–ν_x mixing angle in vacuum, and $\Delta m^2 \simeq m^2_s$ is the mass squared difference between the two mass eigenstates.[2] The MSW potential is $V_x = \pm\sqrt{2}G_F(-N_n/2 + N_{\nu e} - N_{\overline{\nu}e})$, where GF is the Fermi constant and N_n, $N_{\nu e}$, and $N_{\overline{\nu}e}$ are the neutron, electron neutrino, and electron antineutrino number densities, respectively. The + (−) sign corresponds to the potential experienced by neutrinos (antineutrinos). As $|N_{\nu e} - N_{\overline{\nu}e}| \ll N_n/2$, the potential is positive for antineutrinos and negative for neutrinos. Note that we neglect terms that would arise from differences between the ν_x and $\overline{\nu}x$ number densities because such differences are expected to be small in the parameter region where our limits will lie [32]. We have checked that even adding an MSW potential corresponding to a maximal asymmetry does not alter our results significantly. We also neglect the momentum and angular dependence of neutrino self-interactions [40,41] and instead restrict ourselves to the simplified formalism used in [24,25,28]. Because N_n, $N_{\nu e}$, and $N_{\overline{\nu}e}$ are extremely large in the supernova core and gradually decrease with radius, most antineutrinos will encounter an MSW resonance on their way out of the exploding star. At the resonance,

$$\cos 2\theta = 2V_x E_{\text{res}}/m_s^2, \qquad (2)$$

and the effective mixing angle θ_m in matter becomes maximal [37–39,42].

We consider two physically different mechanisms for sterile neutrino production in supernovae:

(i) *Adiabatic flavor conversion at an MSW resonance.* A $\overline{\nu}x$ of energy E streaming away from the supernova core can convert to sterile antineutrino ($\overline{\nu}s$) when the matter density, and thus the MSW potential Vx, has reached the value satisfying the resonance condition, Eq. (2) [43]. Each point (t,R) in time and space corresponds to a specific value of V_x; therefore, antineutrinos of a specific energy E_{res}(t,R) are resonantly converted at this point. We take the local antineutrino luminosities and spectra and the local matter densities from the simulation of an 8.8 $M_\odot$ supernova by the Garching group [44] (see also [45–48]).

Hard scattering processes must be rare to give antineutrinos sufficient time to convert adiabatically. We therefore require the spatial width of the MSW resonance regions,

$$R_{\rm width} = 2\sin 2\theta \left(\frac{1}{V_x} \frac{\partial V_x}{\partial R} \right)^{-1}, \tag{3}$$

to be smaller than the mean free path λ_{mfp}. The number $d\mathcal{N}_s^{\rm MSW}/dE$ of $\bar{\nu}_s$ in an energy interval [E,E+dE] produced by adiabatic flavor conversion at the MSW resonance is given by [24]

$$\frac{d\mathcal{N}_s^{\rm MSW}}{dE} = \int_0^t dt' 4\pi R_{\rm res}^2 n_\nu(t', R_{\rm res}) \times f_x(E) P_{\rm res}(E) \frac{E^2}{\bar{E}^3} \Theta(\lambda_{\rm mfp} - R_{\rm width}) \tag{4}$$

Here, the time integral runs from the time of core bounce ($t'=0$) until ~9 s later, and $R_{res}(t',E)$ is the radius at which the resonance energy is E at time t′. The quantity $n\nu(t',R_{res})$ is the active (anti)neutrino number density at time t′ and radius R_{res}, $f_x(E)$ is the energy distribution of active (anti)neutrinos, $\bar{E}$ is the average (anti)neutrino energy, $P_{res}(E)$ is the flavor conversion probability at the resonance, and the Heaviside Θ function implements the condition that antineutrinos must have enough time between collisions to convert adiabatically. It is crucial that antineutrinos do not encounter more than one MSW resonance on their way out of the supernova. (This would be different if we considered mixing between ν_s and ν_e instead of $\nu_{\mu,\tau}$ [26,27].) Note that we can make the strongly simplifying assumption of radial symmetry, and we also neglect the depletion of active antineutrinos by conversion into $\bar{\nu}s$. Moreover, we do not need to consider $\bar{\nu}x$ streaming inwards. They would convert to $\bar{\nu}s$ at the resonance, then travel through the core, and convert back to $\bar{\nu}x$ on its far side.

We parameterize fx(E) as [49]

$$f_x(E) = \frac{(1+\alpha)^{3+\alpha}}{\Gamma(3+\alpha)} \left(\frac{E}{\bar{E}}\right)^{\alpha} \exp\left[-(1+\alpha)\frac{E}{\bar{E}}\right] \tag{5}$$

with normalization $\int_0^\infty dE E^2 f_x(E) = \bar{E}^3$. (This relation defines $\bar{E}$.) The "pinching parameter" α desribes the degree to which fx(E) differs from a Maxwell-Boltzmann distribution.

The $\bar{\nu}_x \to \bar{\nu}_s$ conversion probability at the resonance is given by the Landau-Zener formula [24,25,42],

$$P_{\text{res}}(E) = 1 - \exp\left[-\frac{\pi^2}{2}\frac{R_{\text{width}}}{L_{\text{osc}}^{\text{res}}}\right] \tag{6}$$

with the oscillation length at the resonance, $L_{\text{osc}}^{\text{res}} \simeq 2\pi/(V_x \sin 2\theta)$.

We find that adiabatic flavor conversion occurs mostly at radii ~10–15 km, still inside the neutrino sphere at ~20–30 km. In Fig. 2, we compare dN_s^{MSW}/dE to the spectrum of active neutrinos. We see that the sterile antineutrino spectrum extends to higher energies because most of the flavor conversion happens in a high-temperature region from which ν_s can stream out freely, while ν_x are still trapped.

(ii) *Collisional production.* Sterile neutrinos can also be produced from a mixed ν_x–ν_s state in hard scatterings on nucleons, electrons, and positrons. We take the interaction rate Γ_x to be approximately equal to the dominant scattering rate on neutrons, with the corresponding cross section given by [50,51] $\sigma_{\nu n} \simeq G_F^2 E_\nu^2 (1 + g_A^2)/4\pi$, with $g_A \simeq 1.23$. The physical picture for collisional ν_s production is as follows: starting with an ensemble of only ν_x at t=0, each of them soon acquires a small ν_s admixture $\propto \sin^2 2\theta_m$ by oscillation. A collision causes the collapse of the resulting mixed state into either a pure ν_x or a pure ν_s. Afterwards, oscillations start anew, and ν_x are quickly replenished. After many collisions, the ν_s abundance is proportional to $\sin^2 2\theta_m$ and Γ_x.

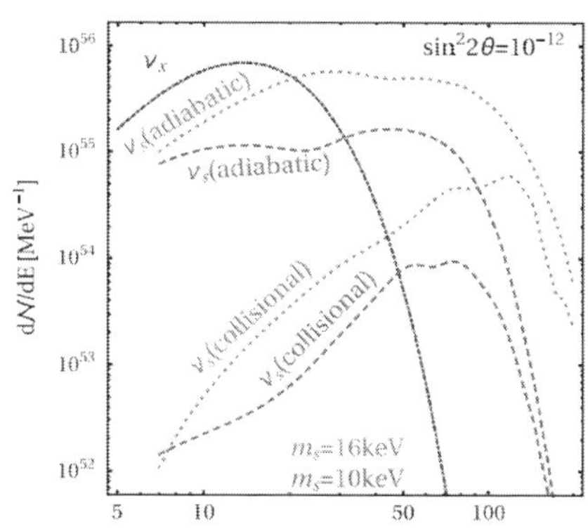

Figure 2. Time-integrated spectra of active neutrinos (blue) and sterile neutrinos (green and red) produced in an 8.8 $M_\odot$ supernova [44] during the first ~9 s after the core bounce. Contributions from adiabatic flavor conversion

and collisional production are shown separately. The ν_s spectrum extends to much higher energies than the ν_x spectrum because flavor conversion occurs in a region where temperatures are much higher than those at the last scattering surface of ν_x.

Up to a factor of 1/2, this intuitive picture leads to the correct quantum mechanical Boltzmann equation [1,3,23,52,53],

$$\frac{\partial}{\partial t}\frac{dn_s}{dE} = \frac{\Gamma_x}{2}\langle P_{\nu_x\to\nu_s}\rangle \frac{dn_x}{dE} \quad (7)$$

Here, dn_s/dE and dn_x/dE are the energy spectra of sterile and active neutrinos, respectively. At any given spacetime point, $dn_x(t,R,E)/dE$ is related to the distribution function in Eq. (5) via $(1/n_x)dn_x/dE = (E^2/\bar{E}^3)f_x(E)$. The averaged oscillation probability $\langle P_{\nu_x\to\nu_s}\rangle$ is [1]

$$\langle P_{\nu_x\to\nu_s}\rangle = \frac{1}{2}\frac{\sin^2 2\theta}{(\cos 2\theta - 2V_x E/m_s^2)^2 + \sin^2 2\theta + (\Gamma_x E/m_s^2)^2} \quad (8)$$

The extra term $(\Gamma x_E/m^2 s)^2$ in the denominator compared to the usual expression for the mixing angle in matter [42] accounts for the suppression of ν_s production when λ_{mfp} is much smaller than the oscillation length, so that oscillations do not have time to develop between collisions (quantum Zeno effect).

Integrating Eq. (7) over time and radius leads to the energy spectrum of sterile neutrinos produced collisionally,

$$\frac{dN_s^{coll}}{dE} = \int_0^t dt' \int_0^{R_{core}} dR' 4\pi R'^2 \frac{\Gamma_x}{2}\langle P_{\nu_x\to\nu_s}\rangle \frac{dn_x}{dE} \quad (9)$$

We again evaluate dN_s/dE numerically using the data from [44]. The resulting ν_s spectra, shown in Fig. 2, can be harder than the ones from adiabatic production because the collisional production rate depends on Γ_x, which grows proportional to E^2.

CONSTRAINTS FROM SUPERNOVA LUMINOSITY

We can constrain the energy output in sterile neutrinos from SN 1987A by comparing the observed energy output in active neutrinos of E_a=few$\times 10^{53}$ ergs [54,55] to the gravitational energy released in the collapse of a stellar core at the Chandrasekhar mass, which is also on the order of E_{tot}=few$\times 10^{53}$ ergs [44,56]. If a substantial fraction of E_{tot} was carried away by sterile neutrinos, the observed E_a could not be explained.[3] We therefore consider the ratio $R(\sin^2 2\theta, m_s) \equiv Es(\sin^2 2\theta, m_s)/E_{tot}$ [21]. We assume that R depends only weakly on the mass and type of the progenitor star, so that the values obtained for the supernova simulated in [44] are a good proxy for SN 1987A. Indeed, the energy emitted in neutrinos from the simulated supernova is roughly 1.5×10^{53} erg [44], which is on the same order of magnitude as in Supernova 1987A.

Our computation of ν_s production is only self-consistent for $R \ll 1$ because we neglect depletion of active neutrinos. Extrapolating it to larger values nevertheless and setting a limit by requiring $R<1$ yields the leftmost edge of the hatched uncertainty band in Fig. 1. To obtain the right edge of the band, we have also implemented a more robust treatment of active neutrino fluxes that parametrizes both depletion due to ν_s production and replenishment from diffusion.

Our limits on the ν_s parameter space are shown in Fig. 3 for adiabatically and collisionally produced ν_s separately and in Fig. 1 for the combination of both production mechanisms (requiring $R<1$). Due to MSW enhancement, our constraints become $\sin^2 2\theta \sim 10^{-14}$ at ms~10–100 keV, surpassing all other limits in this mass range. The strength of our bounds can be understood by noting that, according to Eq. (8), strong resonant conversion is possible down to $\sin^2 2\theta \sim 10^{-14}$. At smaller values, the damping term $(\Gamma_x E/m^2 s)^2$ in the denominator suppresses $\langle P_{\nu x \to \nu s} \rangle$ at resonance. Moreover, considering the smooth density profiles from [44] rather than assuming a constant density core as, e.g., in Ref. [28] implies that, in the mass range between 10–100 keV, neutrinos of any energy will experience a resonance somewhere along their trajectory. It is primarily due to this effect that the limits on sin22θ presented here are (in the most sensitive region around m_s=50 keV) about 3–4 orders of magnitude stronger than those in Ref. [28].

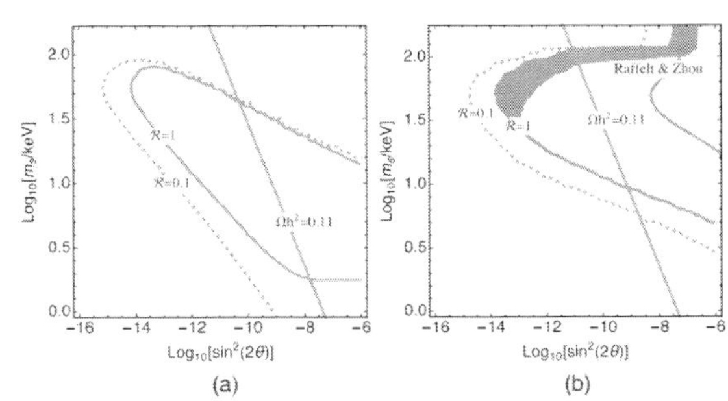

Figure 3. Constraints on the sterile neutrino parameter space from energy loss in supernovae, considering (a) adiabatic $\nu_x \to \nu_s$ conversion only or

(b) collisional ν_s production only. The solid and dotted orange curves correspond to different assumptions on the maximum allowed energy loss, expressed here in terms of the ratio R of the energy output in sterile neutrinos and the total energy output. For comparison, we also show in green the parameters for which the Dodelson-Widrow mechanism [3] predicts the correct relic abundance of ν_s. In panel (b) for R=1, the shaded region illustrates the theoretical uncertainty in our limit stemming from our treatment of energy transport inside the supernova.

The shape of the exclusion regions in Fig. 3 can be understood as follows. For adiabatic conversion at small m_s, the oscillation length at the MSW resonance L^{res}_{osc} is large, making flavor conversion nonadiabatic according to Eq. (6). At large $m_s \gtrsim 100$ keV, the resonance condition of Eq. (2) cannot be satistifed. At ms~few×10 keV, adiabatic flavor conversion is most effective at low $\sin^2 2\theta \sim 10^{-14}$. For larger $\sin^2 2\theta$, the radial width R_{width} of the resonance region becomes too large, so that neutrinos scatter before having a chance to convert. Also, collisional production is most effective when the mixing angle in matter is MSW enhanced. At $m_s \lesssim 10$ keV, the resonance condition of Eq. (2) is fulfilled only in the outer regions of the supernova, where the scattering rate is too low for effective collisional production. At $m_s \gtrsim 100$ keV, the resonance condition is never satisfied.

PHOTON FLUX FROM $N_S \rightarrow N_A \Gamma$

Sterile neutrinos, once produced, decay to photons via $\nu_s \rightarrow \nu_a \gamma$, a process mediated by loop diagrams involving charged leptons and W bosons. The decay rate is [57,58]

$$\Gamma_{\nu_s \rightarrow \nu_a \gamma} = 1.38 \times 10^{-29}\ \mathrm{sec}^{-1} \left(\frac{\sin^2 2\theta}{10^{-7}}\right)\left(\frac{m_s}{1\ \mathrm{keV}}\right)^5 \tag{10}$$

Thus, if ν_s are abundantly produced in a supernova, we expect the explosion to be accompanied by a flux of energetic [O(1–100 MeV)] secondary gamma-rays. Photons in this energy range are not normally expected from a supernova or supernova remnant, both of which emit X-rays only at energies $\lesssim 10$ keV [59,60]. The arrival times of the gamma-rays from ν_s decay are spread out over a time interval,

$$\Delta t \simeq 3.6\ \mathrm{hrs} \times \left(\frac{d}{1\ \mathrm{kpc}}\right)\left(\frac{m_s}{1\ \mathrm{keV}}\right)^2\left(\frac{1\ \mathrm{MeV}}{E_\gamma}\right)^2, \tag{11}$$

where d is the distance to the supernova and E_γ is the gamma-ray energy. Sterile neutrinos decaying immediately after their production lead to gamma-rays that reach the Earth at the same time as the active neutrino burst. Gamma-rays from νs decaying only after traveling a distance ≃d are delayed by Δt.

The photon flux from ν_s decay in units of $cm^{-2}s^{-1}MeV^{-1}$ is given by

$$\frac{d\phi_\gamma(E_\gamma)}{dE_\gamma} = \frac{2}{4\pi\Delta t d^2}\left[1 - \exp\left(-\frac{d}{\gamma}\Gamma_{\nu_s\to\nu_a\gamma}\right)\right]\frac{dN_{\nu_s}(E_\nu)}{dE_\nu}, \tag{12}$$

where $\gamma=2E_\gamma/m_s$ is the Lorentz boost factor. The factor in brackets accounts for the ν_s decay probability. Note that the sterile neutrino spectrum $dN_{\nu s}(E_\nu)/dE_\nu$ is evaluated at $E_\nu=2E_\gamma$.

We plot $d\phi_\gamma(E_\gamma)/dE_\gamma$ for two sets of benchmark parameters in Fig. 4, and the total gamma-ray flux as a function of $\sin^2 2\theta$ and m_s is shown in Fig. 5. As expected from Fig. 2, and taking into account $E_\gamma=E_\nu/2$, the photon spectrum peaks at E~10–100 MeV. Compared to observations [61–67], we find that the signal is still several orders of magnitude below the uncertainty on the background. A direct observation of the νs-induced photon signal is therefore challenging even for the next generation of Compton telescopes [68–70] and would require a factor ~10^5 improvement compared to the projected sensitivity of ComPair [69] for a supernova at 1 kpc.

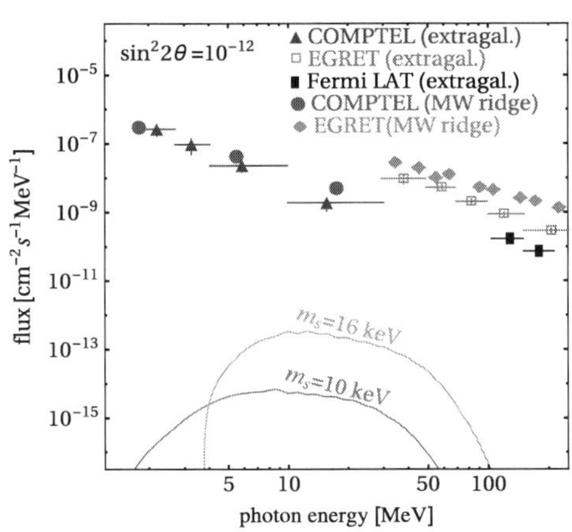

Figure 4. Photon flux from the decay of sterile neutrinos produced in a nearby (d=1 kpc) supernova for two different benchmark points in the m_s–$\sin^2 2\theta$ plane. We compare to observed spectra of diffuse gamma-rays from COMPTEL [61–63], EGRET [64–66], and Fermi-LAT [67], assuming angular bin sizes of 1 degree for COMPTEL, 5 degrees for EGRET, and 3 degrees for Fermi-LAT [71].

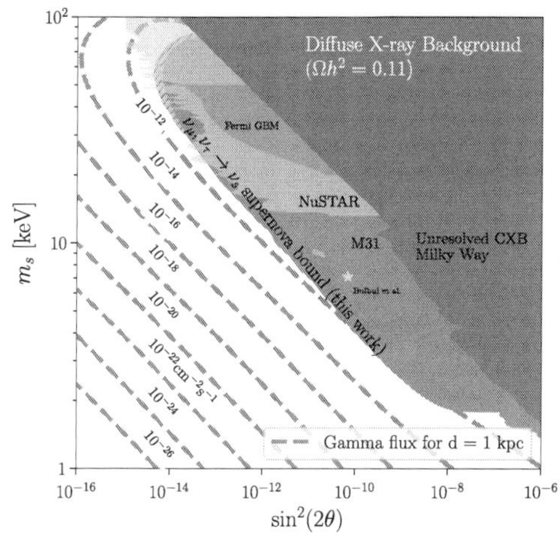

Figure 5. Expected photon flux from the decay of sterile neutrinos produced in a supernova at distance d=1 kpc. We compare to our new energy loss limits from Fig. 1 (orange region) and to constraints from X-ray searches [15–19,34] (green, dark blue, gray, and light blue regions).

DISCUSSION AND CONCLUSIONS

Notes

[1] In this work, for brevity, the term “sterile neutrino” is denoting both sterile neutrinos and antineutrinos. The only exception is when discussing the adiabatic production where we refer explicitly to “sterile antineutrinos” because the resonance appears in the antineutrino channel.

[2] In this paper, we refer to m_s as the sterile neutrino mass, even though we mean of course the mass of the heavy mass eigenstate which is mostly a ν_s, with only a tiny admixture of ν_x.

[3] A similar constraint could also be obtained by considering the observed duration of the neutrino emission from SN 1987A. Efficient production of sterile neutrinos would expedite the cooling of the supernova core, shortening the neutrino burst [28,32].

ACKNOWLEDGMENTS

We are indebted to Georg Raffelt and Ninetta Saviano for many useful comments on the manuscript, and to Thomas Janka and his collaborators for providing the data from [44] in machine-readable form. We would also like to thank Markus Ackermann, John Beacom, Alexei Smirnov, MacKenzie Warren, and Shun Zhou for very useful discussions. V. B. would like to thank the Wisconsin IceCube Particle Astrophysics Center (WIPAC) for hospitality. The work of V. B. and J. K. is supported by the German Research Foundation (DFG) under Grants No. KO 4820/1-1 and No. FOR 2239 and by the European Research Council (ERC) under the European Union's Horizon 2020 research and innovation programme (Grant Agreement No. 637506, "ν Directions"). V. B. is supported by the DFG Graduate School Symmetry Breaking in Fundamental Interactions (GRK 1581). Additional support has been provided by the Cluster of Excellence "Precision Physics, Fundamental Interactions and Structure of Matter" (PRISMA—EXC 1098), Grant No. 05H12UME of the German Federal Ministry for Education and Research (BMBF). C.A. was supported in part by the National Science Foundation (OPP-0236449, NSF-PHY-1505858, and PHY-0969061) and by the University of Wisconsin Research Committee with funds granted by the Wisconsin Alumni Research Foundation.

REFERENCES

1. K. Abazajian, G.M. Fuller, and M. Patel, Sterile neutrino hot, warm, and cold dark matter, Phys. Rev. D 64, 023501 (2001).
2. A. Kusenko, Sterile neutrinos: The dark side of the light fermions, Phys. Rep. 481, 1 (2009).
3. S. Dodelson and L.M. Widrow, Sterile-Neutrinos as Dark Matter, Phys. Rev. Lett. 72, 17 (1994).
4. X.-D. Shi and G.M. Fuller, A New Dark Matter Candidate: Nonthermal Sterile Neutrinos, Phys. Rev. Lett. 82, 2832 (1999).
5. K. Petraki and A. Kusenko, Dark-matter sterile neutrinos in models with a gauge singlet in the Higgs sector, Phys. Rev. D 77, 065014 (2008).
6. A. Merle, V. Niro, and D. Schmidt, New production mechanism for keV sterile neutrino dark matter by decays of frozen-in scalars, J. Cosmol. Astropart. Phys. 2014, 028.
7. A. Merle, A. Schneider, and M. Totzauer, Dodelson-Widrow production of sterile neutrino dark matter with non-trivial initial abundance, J. Cosmol. Astropart. Phys. 2016, 003.
8. P.A.R. Ade et al. (Planck Collaboration), Planck 2015 results. XIII. Cosmological parameters, Astron. Astrophys. 594, A13 (2016).
9. A.C. Vincent, E.F. Martinez, P. Hernandez, M. Lattanzi, and O. Mena, Revisiting cosmological bounds on sterile neutrinos, J. Cosmol. Astropart. Phys. 2015, 006.
10. S. Tremaine and J.E. Gunn, Dynamical Role of Light Neutral Leptons in Cosmology, Phys. Rev. Lett. 42, 407 (1979).
11. D. Iakubovskyi, Constraining Properties of Dark Matter Particles Using Astrophysical Data, Leiden University, 2013.
12. J. Baur, N. Palanque-Delabrouille, C. Yche, C. Magneville, and M. Viel, Lyman-alpha forests cool warm dark matter, in SDSS-IV collaboration meeting, 2015, J. Cosmol. Astropart. Phys. 2016, 012.
13. A. Merle and A. Schneider, Production of sterile neutrino dark matter and the 3.5 keV line, Phys. Lett. B 749, 283 (2015).
14. A. Schneider, Astrophysical constraints on resonantly produced sterile neutrino dark matter, J. Cosmol. Astropart. Phys. 2016, 059.
15. S. Horiuchi, P.J. Humphrey, J. Onorbe, K.N. Abazajian, M. Kaplinghat, and S. Garrison-Kimmel, Sterile neutrino dark matter bounds from

galaxies of the Local Group, Phys. Rev. D 89, 025017 (2014).

16. A. Boyarsky, A. Neronov, O. Ruchayskiy, and M. Shaposhnikov, Constraints on sterile neutrino as a dark matter candidate from the diffuse x-ray background, Mon. Not. R. Astron. Soc. 370, 213 (2006).
17. K.N. Abazajian, M. Markevitch, S.M. Koushiappas, and R.C. Hickox, Limits on the radiative decay of sterile neutrino dark matter from the unresolved cosmic and soft X-ray backgrounds, Phys. Rev. D 75, 063511 (2007).
18. K.N. Abazajian, Detection of Dark Matter Decay in the X-Ray, Detection of dark matter decay in the X-ray arXiv:0903.2040.
19. E. Bulbul, M. Markevitch, A. Foster, R.K. Smith, M. Loewenstein, and S.W. Randall, Detection of an unidentified emission line in the stacked X-ray spectrum of galaxy clusters, Astrophys. J. 789, 13 (2014).
20. N. Sekiya, N.Y. Yamasaki, and K. Mitsuda, A search for a keV signature of radiatively decaying dark matter with Suzaku XIS observations of the X-ray diffuse background, Publ. Astron. Soc. Jpn. 68, SP1 (2015).
21. S.P. Mikheev and A. Yu. Smirnov, Resonant oscillations and limitations found on neutrino parameter values from the possible observation of a neutrino burst from the gravitational collapse of a star, Pis'Ma Zh. Eksp. Teor. Fiz. 46, 11 (1987); [JETP Lett. 46, 10 (1987)].
22. K. Kainulainen, J. Maalampi, and J.T. Peltoniemi, Inert neutrinos in supernovae, Nucl. Phys. B358, 435 (1991).
23. G. Raffelt and G. Sigl, Neutrino flavor conversion in a supernova core, Astropart. Phys. 1, 165 (1993).
24. X. Shi and G. Sigl, A Type II supernovae constraint on electron-neutrino—sterile-neutrino mixing, Phys. Lett. B 323, 360 (1994); Erratum, 324, 516 (1994).
25. H. Nunokawa, J.T. Peltoniemi, A. Rossi, and J.W.F. Valle, Supernova bounds on resonant active sterile neutrino conversions, Phys. Rev. D 56, 1704 (1997).
26. J. Hidaka and G.M. Fuller, Dark matter sterile neutrinos in stellar collapse: Alteration of energy/lepton number transport and a mechanism for supernova explosion enhancement, Phys. Rev. D 74, 125015 (2006).
27. J. Hidaka and G.M. Fuller, Sterile neutrino-enhanced supernova explosions, Phys. Rev. D 76, 083516 (2007).

28. G.G. Raffelt and S. Zhou, Supernova bound on keV-mass sterile neutrinos reexamined, Phys. Rev. D 83, 093014 (2011).
29. M.-R. Wu, T. Fischer, L. Huther, G. Martnez-Pinedo, and Y.-Z. Qian, Impact of active-sterile neutrino mixing on supernova explosion and nucleosynthesis, Phys. Rev. D 89, 061303 (2014).
30. M.L. Warren, M. Meixner, G. Mathews, J. Hidaka, and T. Kajino, Sterile Neutrino Oscillations in Core-Collapse Supernova Simulations, Sterile neutrino oscillations in core-collapse supernova simulations arXiv:1405.6101.
31. M. Warren, G.J. Mathews, M. Meixner, J. Hidaka, and T. Kajino, A review of the impact of sterile neutrino dark matter on core-collapse supernovae, Int. J. Mod. Phys. A 31, 1650137 (2016).
32. G.G. Raffelt, Astrophysical methods to constrain axions and other novel particle phenomena, Phys. Rep. 198, 1 (1990).
33. K. Perez, K.C.Y. Ng, J.F. Beacom, C. Hersh, S. Horiuchi, and R. Krivonos, Almost closing the MSM sterile neutrino dark matter window with NuSTAR, Phys. Rev. D 95, 123002 (2017).
34. K.C.Y. Ng, S. Horiuchi, J.M. Gaskins, M. Smith, and R. Preece, Improved limits on sterile neutrino dark matter using full-sky fermi-GBM data, Phys. Rev. D 92, 043503 (2015).
35. L.E. Strigari, J.S. Bullock, M. Kaplinghat, A.V. Kravtsov, O.Y. Gnedin, K. Abazajian, and A.A. Klypin, A large dark matter core in the fornax dwarf spheroidal galaxy, Astrophys. J. 652, 306 (2006).
36. A. Boyarsky, O. Ruchayskiy, D. Iakubovskyi, and J. Franse, An Unidentified Line in X-Ray Spectra of the Andromeda Galaxy and Perseus Galaxy Cluster, Phys. Rev. Lett. 113, 251301 (2014).
37. L. Wolfenstein, Neutrino Oscillations in Matter, Phys. Rev. D 17, 2369 (1978).
38. S.P. Mikheev and A. Yu. Smirnov, Resonance amplification of oscillations in matter and spectroscopy of solar neutrinos, Yad. Fiz. 42, 1441 (1985); Sov. J. Nucl. Phys. 42, 913 (1985).
39. S.P. Mikheev and A. Yu. Smirnov, Resonant amplification of neutrino oscillations in matter and solar neutrino spectroscopy, Nuovo Cimento C 9, 17 (1986).
40. H. Duan, G.M. Fuller, and Y.-Z. Qian, Collective neutrino oscillations, Annu. Rev. Nucl. Part. Sci. 60, 569 (2010).
41. B. Dasgupta and A. Dighe, Collective three-flavor oscillations of

supernova neutrinos, Phys. Rev. D 77, 113002 (2008).
42. E.K. Akhmedov, Neutrino Physics, Neutrino physics arXiv:hep-ph/0001264.
43. S.P. Mikheev and A. Yu. Smirnov, Neutrino oscillations in a variable density medium and neutrino bursts due to the gravitational collapse of stars, Zh. Eksp. Teor. Fiz. 91, 7 (1986); Sov. Phys. JETP 64, 4 (1986).
44. L. Hüdepohl, B. Müller, H. Janka, A. Marek, and G. G. Raffelt, Neutrino Signal of Electron-Capture Supernovae from Core Collapse to Cooling, Phys. Rev. Lett. 104, 251101 (2010); Erratum, 105, 249901(E) (2010). machine-readable data from ; we use results for the model with full neutrino interactions (Sf).
45. M. Rampp and H.T. Janka, Radiation hydrodynamics with neutrinos: Variable Eddington factor method for core collapse supernova simulations, Astron. Astrophys. 396, 361 (2002).
46. M.Liebendoerfer,M.Rampp,H.T.Janka,andA.Mezzacappa,Supernova simulations with Boltzmann neutrino transport: A Comparison of methods, Astrophys. J. 620, 840 (2005).
47. A. Marek, H. Dimmelmeier, H.T. Janka, E. Muller, and R. Buras, Exploring the relativistic regime with Newtonian hydrodynamics: An Improved effective gravitational potential for supernova simulations, Astron. Astrophys. 445, 273 (2006).
48. R. Buras, M. Rampp, H.T. Janka, and K. Kifonidis, Two-dimensional hydrodynamic core-collapse supernova simulations with spectral neutrino transport. 1. Numerical method and results for a 15 solar mass star, Astron. Astrophys. 447, 1049 (2006).
49. M.T. Keil, G.G. Raffelt, and H.-T. Janka, Monte Carlo study of supernova neutrino spectra formation, Astrophys. J. 590, 971 (2003).
50. C. Giunti and C.W. Kim, Fundamentals of Neutrino Physics and Astrophysics (Oxford University Press, Oxford, UK, 2007).
51. J. Peltoniemi, The ultimate neutrino page, http://cupp.oulu.fi/neutrino/.
52. L. Stodolsky, On the Treatment of neutrino oscillations in a thermal environment, Phys. Rev. D 36, 2273 (1987).
53. M.J. Thomson, Damping of quantum coherence by elastic and inelastic processes, Phys. Rev. A 45, 2243 (1992).
54. T.J. Loredo and D.Q. Lamb, Bayesian analysis of neutrinos observed from supernova SN-1987A, Phys. Rev. D 65, 063002 (2002).

55. G. Pagliaroli, F. Vissani, M.L. Costantini, and A. Ianni, Improved analysis of SN1987A antineutrino events, Astropart. Phys. 31, 163 (2009).
56. K. Zuber, Neutrino Physics (Taylor & Francis, London, 2011), 2nd ed.
57. P.B. Pal and L. Wolfenstein, Radiative decays of massive neutrinos, Phys. Rev. D 25, 766 (1982).
58. Z. Xing and S. Zhou, Neutrinos in Particle Physics, Astronomy and Cosmology (Springer, Berlin, Heidelberg, 2011).
59. A.M. Soderberg et al., An extremely luminous X-ray outburst marking the birth of a normal supernova, Nature (London) 453, 469 (2008).
60. J. Vink, Supernova remnants: The X-ray perspective, Astron. Astrophys. Rev. 20, 49 (2012).
61. A.W. Strong, K. Bennett, H. Bloemen, R. Diehl, W. Hermsen, D. Morris, V. Schoenfelder, J.G. Stacy, C. de Vries, M. Varendorff, C. Winkler, and G. Youssefi, Diffuse continuum gamma rays from the Galaxy observed by COMPTEL, Astron. Astrophys. 292, 82 (1994).
62. S.C. Kappadath, J. Ryan, K. Bennett, H. Bloemen, D. Forrest, W. Hermsen, R.M. Kippen, M. McConnell, V. Schoenfelder, R. van Dijk, M. Varendorff, G. Weidenspointner, and C. Winkler, The preliminary cosmic diffuse γ-ray spectrum from 800 keV to 30 MeV measured with COMPTEL, Astron. Astrophys. Proc. Suppl. 120, 619 (1996).
63. R.L. Kinzer, W.R. Purcell, and J.D. Kurfess, Gamma-ray emission from the inner galactic ridge, Astrophys. J. 515, 215 (1999).
64. S.D. Hunter et al., EGRET observations of the diffuse gamma-ray emission from the galactic plane, Astrophys. J. 481, 205 (1997).
65. P. Sreekumar et al. (EGRET Collaboration), EGRET observations of the extragalactic gamma-ray emission, Astrophys. J. 494, 523 (1998).
66. P. Sreekumar, F.W. Stecker, and S.C. Kappadath, The extragalactic diffuse gamma-ray emission, AIP Conf. Proc. 410, 344 (1997).
67. M. Ackermann et al. (Fermi-LAT Collaboration), The spectrum of isotropic diffuse gamma-ray emission between 100 MeV and 820 GeV, Astrophys. J. 799, 86 (2015).
68. S.D. Hunter et al., A pair production telescope for medium-energy gamma-ray polarimetry, Astropart. Phys. 59, 18 (2014).
69. A.A. Moiseev et al., Compton-Pair Production Space Telescope (ComPair) for MeV Gamma-Ray Astronomy, Compton-pair

production space telescope (ComPair) for MeV gamma-ray astronomy arXiv:1508.07349.

70. M. Tavani et al. (e-ASTROGAM Collaboration), Science with e-ASTROGAM (A Space Mission for MeV-GeV Gamma-Ray Astrophysics), Science with e-ASTROGAM (A space mission for MeV-GeV gamma-ray astrophysics) arXiv:1711.01265.
71. V. Schönfelder and G. Kanbach, Observing Photons in Space: A Guide to Experimental Space Astronomy (Springer, New York, 2013), pp. 225–242.

CHAPTER

19

Multimessenger Search for the Sources of Cosmic Rays Using Cosmic Neutrinos

Francis Halzen and Ali Kheirandish*

Department of Physics & Wisconsin IceCube Particle Astrophysics Center, UW-Madison, Madison, WI, United States

ABSTRACT

After updating the status of the measurements of the cosmic neutrino flux by the IceCube experiment, we summarize the evidence for the first identified source of cosmic rays and speculate on the connection between the two observations.

DETECTING COSMIC NEUTRINOS

The highest energy radiation reaching us from the universe is not radiation at all; it is cosmic rays, high-energy nuclei that are mostly protons, except

Citation: Halzen F and Kheirandish A (2019) "Multimessenger Search for the Sources of Cosmic Rays Using Cosmic Neutrinos". Front. Astron. Space Sci. 6:32. https://doi.org/10.3389/fspas.2019.00032

possibly at the end of the spectrum. We do not know where they come from or how they are accelerated to such extreme energies. However, with the compelling evidence for a distant blazar as the source of a very high energy muon neutrino, neutrinos have provided a breakthrough in solving this puzzle (Aartsen et al., 2018a,b).

The rationale for searching for cosmic ray sources by observing neutrinos is straightforward: in the vicinity of neutron stars and black holes, the gravitational energy released in the accretion of matter accelerates protons or heavier nuclei, which subsequently interact with gas or ambient radiation to produce neutrinos originating from the decay of pions and other secondary particles[1]. When accelerated, protons interact with photons; for instance, both neutral and charged pion secondaries are produced. While charged pions decay into three high-energy neutrinos and antineutrinos, neutral pions decay and create a flux of high-energy gamma rays. These photons, referred to as pionic photons, are distinguished from photons radiated by electrons that may be accelerated along with the cosmic rays. Pionic photons are what provide the motivation for multimessenger observations.

The search for high-energy astrophysical neutrinos led to the development of large scale detectors: DUMAND in the Pacific Ocean, NT200 in Lake Baikal, AMANDA at the South Pole, and ANTARES in the Mediterranean Sea. These detectors use highly transparent natural water or ice as a Cherenkov medium that is then instrumented with photomultipliers; see (Katz and Spiering, 2012). The Cherenkov light patterns radiated by secondary charged particles produced by neutrinos interacting inside or near the instrumented volume encode the flavor, energy, and arrival direction of the neutrino. Based on the expectation that cosmic accelerators may produce similar energies in cosmic rays, gamma rays, and neutrinos, it had been anticipated that the construction of a cubic-kilometer, gigaton-scale detector would be sufficient to make a first detection of high-energy neutrinos of cosmic origin.

The IceCube Neutrino Observatory (Aartsen et al., 2017c) transformed one cubic kilometer of natural ice into a Cherenkov detector, operating as the world's premier neutrino detector above energies of approximately 10 GeV. IceCube was constructed between 2004 and 2010 near the Amundsen-Scott South Pole Station, Antarctica, at the geographic South Pole. Below a depth of 1,450 m, a cubic-kilometer of glacial ice is instrumented with 86 cables called "strings," each of which is equipped with 60 digital optical modules (DOMs). The DOM is a glass pressure vessel containing a 10-inch photomultiplier tube (PMT) and digitizing electronics which capture the PMT light signal (Aartsen et al., 2017c). These digitized waveforms are the basis of IceCube analysis: the recorded arrival time and amplitude of the photons are used to reconstruct the direction, energy, and location of particle interactions in the ice. The central strings are more densely spaced, forming the DeepCore component of IceCube, and most DOMs in this region are equipped with high-quantum-efficiency PMTs to detect lower energy particle interactions. IceCube also includes a surface cosmic ray air shower detector, IceTop, that consists of 81 stations, each comprising two light-tight tanks of (frozen) water instrumented with two DOMs each.The deep ice itself constitutes the detector, forming both support structure and Cherenkov medium. Its optical properties, scattering and absorption, are depth-dependent, predominantly due to layers of dust resulting from past glacial-maximum climate events, with scattering in the glacier being anisotropic (Aartsen et al., 2013b). Additionally, scattering results from bubbles in the re-frozen "hole ice" in which the strings have been deployed. Each DOM is equipped with LED flashers that are used as calibration light sources to determine the optical properties of the natural ice. The high statistics data sets, as well as extensive calibration data, have enabled sophisticated modeling of the optics of the ice.In principle, the particle trajectory is determined by the arrival times of photons at the optical sensors, while the number of photons observed is used to determine the energy deposited by charged secondary particles in the detector. For illustration, the Cherenkov patterns initiated by an electron (or tau) neutrino of 1 PeV energy and a neutrino induced muon losing 2.6 PeV energy while traversing the detector are contrasted in Figure 1. For the cascade (shower) event shown in the left panel of Figure 1, more than 300 digital optical modules (DOMs) report a total of more than 100,000 photoelectrons.

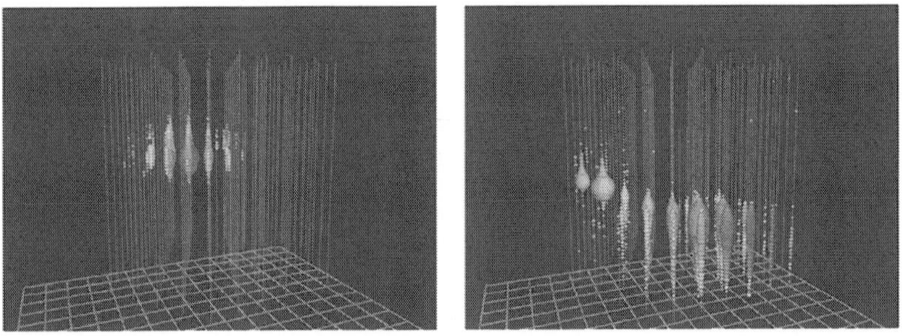

Figure 1. (Left) Light pool produced in IceCube by a shower initiated by an electron or tau neutrino. The measured energy is 1.14 PeV, which represents a lower limit on the energy of the neutrino that initiated the shower. White dots represent sensors with no signal. For the colored dots, the color indicates arrival time, from red (early) to purple (late) following the rainbow,

and size reflects the number of photons detected. (Right) An upgoing muon track traverses the detector at an angle of 11° below the horizon. The deposited energy, i.e., the energy equivalent of the total Cherenkov light of all charged secondary particles inside the detector, is 2.6 PeV. Image: IceCube Collaboration, permission: Elsevier.

Long tracks resulting from muon neutrino interactions can be pointed back to their sources with a $\leq 0.4°$ angular resolution. In contrast, the reconstruction of cascade directions, in principle currently possible to a few degrees, is still in the development stage in IceCube, where originally it was limited to within 10°~ 15° of the direction of the incident neutrino (Aartsen et al., 2014; Yuan, 2017). Determining the deposited energy from the observed light pool is, however, relatively straightforward, and a resolution of better than 15 % is possible; the same value holds for the reconstruction of the energy deposited by a muon track inside the detector.

IceCube operates continuously, and monitors the whole sky simultaneously, in the search for high-energy neutrinos of cosmic origin, collecting very high statistics of cosmic ray muons and atmospheric neutrino data sets. Detected neutrino energies cover more than six orders of magnitude, from 5 GeV in DeepCore to beyond 5 PeV. Soon after the completion of the detector, with 2 years of data, IceCube discovered an extragalactic flux of cosmic neutrinos with an energy density in the local Universe that is similar to that in gamma rays (See Ackermann et al., 2015; Fang and Murase, 2018).

Two principal methods are used to identify cosmic neutrinos. The first method has traditionally focused on the observation of muon neutrinos that interact primarily outside the detector. High-energy muon neutrinos interact and produce kilometer-long muon tracks passing through the instrumented volume. Although this allows identifying neutrinos that interact outside the detector, it relies on using the Earth as a filter in order to remove the background of cosmic-ray muons. This method limits the neutrino view to a single flavor and half the sky. IceCube has measured the atmospheric neutrino background flux over more than 5 orders of magnitude in energy with a result that is consistent with theoretical calculations. With 8 years of data, IceCube has observed an excess of events at energies beyond 100 TeV (Aartsen et al., 2015, 2016a, 2017b) that cannot be accounted for by the atmospheric flux. Even acknowledging large uncertainties in the extrapolation of the atmospheric component to higher energy, the statistical significance of the excess astrophysical flux is 6.7σ. Although the detector only records the energy of the secondary muons inside the detector, from Standard Model physics we can infer the energy spectrum of the parent neutrinos. The resulting high-energy cosmic muon neutrino flux is well-described by a power law with a spectral index of 2.19±0.10 and a normalization at 100 TeV neutrino energy of $(1.01^{+0.26}_{-0.23}) \times 10^{-18} GeV^{-1} cm^{-2} sr^{-1}$ (Aartsen et al., 2017b).

An alternative method exclusively identifies high-energy neutrinos interacting inside the detector, so-called high-energy starting events (HESE). It divides the instrumented volume of ice into an outer veto shield and a ~500-megaton inner fiducial volume. The advantage of focusing on neutrinos interacting inside the instrumented volume of ice is that the detector functions as a total absorption calorimeter (Aartsen et al., 2014). With this method, neutrinos from all directions in the sky and all flavors can be identified, which includes both muon tracks as well as secondary showers, produced by charged-current interactions of electron and tau neutrinos, and neutral current interactions of neutrinos of all flavors. Furthermore, this method can separate neutrinos of atmospheric and cosmic origin. Atmospheric neutrinos reaching us from the Southern Hemisphere are readily distinguishable because they are accompanied by particles produced in the same air shower where the neutrinos originate. They are further separable from cosmic neutrinos by their well-measured energy. A sample event with a light pool of roughly one hundred thousand photoelectrons extending over more than 500 m is shown in the left panel of Figure 1.

The advantages of these two methods of separating cosmic neutrinos from the atmospheric backgrounds are complementary. It was the HESE method that succeeded in revealing the first evidence for cosmic neutrinos (Aartsen et al., 2013a,c). Observed events with PeV energies, and no trace of accompanying muons from an atmospheric shower, are highly unlikely to be of atmospheric origin. The 7-year data set contains a total of 60 neutrino events with deposited energies ranging from 60 TeV to 10 PeV. The data are consistent with an astrophysical component with a spectrum close to E^{-2} above an energy of ~ 200 TeV.

We should mention here that there is yet another method to conclusively identify cosmic neutrinos: the observation of very high energy tau neutrinos. Tau neutrinos above about 100 TeV produce two spatially separated showers in the detector, one from the interaction of the tau neutrino and the second one from the tau decay; the mean tau lepton decay length is about 50 m/PeV. Two such candidate events have been recently identified (Taboada, 2018). In addition to the double cascade candidates, we should also mention that a first candidate event has been attributed to the Glashow resonance. This event was identified in a search for partially contained events, and represents

~ 6 PeV in energy deposited in the detector; See Figure 2 (Aartsen et al., 2017b; Lu, 2018).

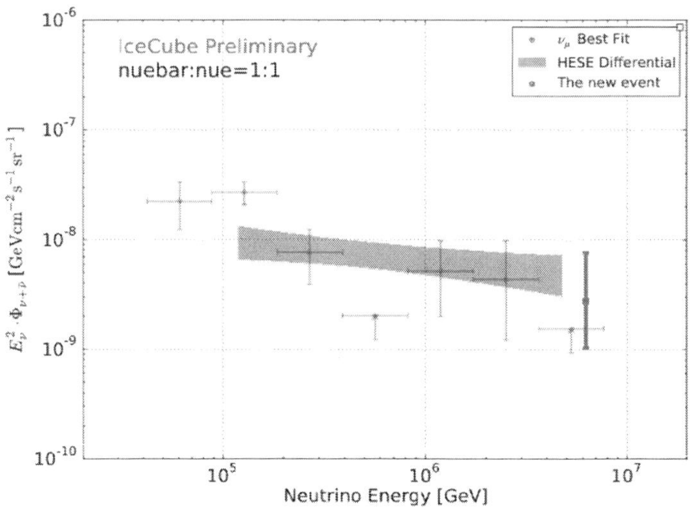

Figure 2. The IceCube diffuse cosmic neutrino spectrum. The best fit to the spectrum of neutrinos observed through the Earth is compared to the unfolded spectrum of events starting inside the detector. Also shown is the flux corresponding to a single candidate event (red) attributed to the Glashow resonance. Figure from Lu (2018).

ICECUBE NEUTRINOS AND FERMI PHOTONS

The most important message emerging from the IceCube measurements of the high-energy cosmic neutrino flux is not apparent yet: the prominent role of hadrons relative to leptons in the high-energy universe. As highlighted in the introduction, photons are inevitably produced in association with neutrinos when accelerated cosmic rays produce neutral and charged pions in interactions with target material in the vicinity of the accelerator. Targets include strong radiation fields that may be associated with the accelerator as well as concentrations of matter, such as hydrogen or molecular clouds in their vicinity. The subsequent decay of charged pions via $\pi^+ \to \mu^+ + \nu_\mu$ followed by the decay of the muon $\mu^+ \to e^+ + \nu_e + \bar{\nu}_\mu$ and the charge–conjugate processes produces a flux of high-energy neutrinos. High-energy gamma rays result from the decay of neutral pions, $\pi^0 \to \gamma + \gamma$. On average, pionic gamma rays and neutrinos carry 1/2 and 1/4 of the energy of the parent pion, respectively. Using these approximations, one can relate the production rate Q (units of

GeV^{-1}s^{-1}) of neutrinos and gamma rays to their parent pions. For neutrinos, summing over all flavors, the production rate can be related to the one for charged pions as

$$\sum_{\alpha} E_\nu Q_{\nu_\alpha}(E_\nu) \simeq 3[E_\pi Q_{\pi^\pm}(E_\pi)]_{E_\pi \simeq 4E_\nu}. \tag{1}$$

Similarly, the production rate of pionic gamma rays is related to the one for neutral pions as

$$E_\gamma Q_\gamma(E_\gamma) \simeq 2[E_\pi Q_{\pi^0}(E_\pi)]_{E_\pi \simeq 2E_\gamma}. \tag{2}$$

The relative production rates of pionic gamma rays and neutrinos only depend on the ratio of charged-to-neutral pions produced in cosmic-ray interactions, denoted by K_π. In photohadronic interactions, the production is dominated by the production of the Δ resonance: $p + \gamma \rightarrow \Delta^+ \rightarrow \pi^0 + p$ and $p + \gamma \rightarrow \Delta^+ \rightarrow \pi^+ + n$. These channels produce charged and neutral pions with probabilities 2/3 and 1/3, respectively. However, the additional contribution of non-resonant pion production modifies the ratios to approximately 1/2 and 1/2. In contrast, if neutrinos originate in the interactions of a hadronic beam with matter, e.g., hydrogen or molecular clouds, equal numbers of pions of all three charges originate: $p+p \rightarrow N_\pi[\pi^0+\pi^++\pi^-]+X$, where N_π is the pion multiplicity. Therefore, $K_\pi \simeq 2$ for hadronuclear interaction, while $K_\pi \simeq 1$ for photohadronic process (Ahlers and Halzen, 2018). We can combine Equations (1) and (2) to derive a direct relation between the pionic gamma-ray and neutrino production rates:

$$\frac{1}{3}\sum_{\alpha} E_\nu^2 Q_{\nu_\alpha}(E_\nu) \simeq \frac{K_\pi}{4}\left[E_\gamma^2 Q_\gamma(E_\gamma)\right]_{E_\gamma = 2E_\nu}. \tag{3}$$

The pre-factor 1/4 accounts for the energy ratio $\langle E_\nu\rangle/\langle E_\gamma\rangle \simeq 1/2$ and the two gamma rays produced in the neutral pion decay. Note that this relation relates pionic neutrinos and gamma rays without any reference to the cosmic ray beam that initiates the interactions; it simply reflects the fact that a neutral pion produces two gamma rays for every charged pion producing a $\nu_\mu + \bar{\nu}_\mu$ pair, which cannot be separated by current experiments.

Before applying this relation to data, one must take into account the fact that, unlike neutrinos, gamma rays would get absorbed and cascade to lower energies, as they interact with photons of the cosmic microwave background before reaching Earth. If the source is not transparent, intrinsic absorption would also result in lower energy gamma rays leaving the source. The resulting electromagnetic shower subdivides the initial photon energy, leading to multiple photons in the range of GeV to TeV energies by the time the gamma rays arrive at Earth (Protheroe and Stanev, 1993; Ahlers et al., 2010).

As described, pionic gamma rays should accompany neutrinos at the site of production. However, since they undergo EBL absorption, they will appear at lower energies. As an illustration, the gamma ray flux accompanying the IceCube diffuse cosmic neutrino flux is described by a simple power law with spectral index of –2.19, which describes the data obtained from both analyses, above energy of 100 TeV. The result is shown for $K_\pi \simeq 2$ by the blue lines in Figure 3. The high-energy cosmic neutrino flux (blue dashed line) follows this initial CR spectrum. Note, however, the observable flux of gamma rays at Earth (blue solid lines) gets strongly attenuated at energies above 100 GeV by interactions with EBL (Ahlers and Halzen, 2018).

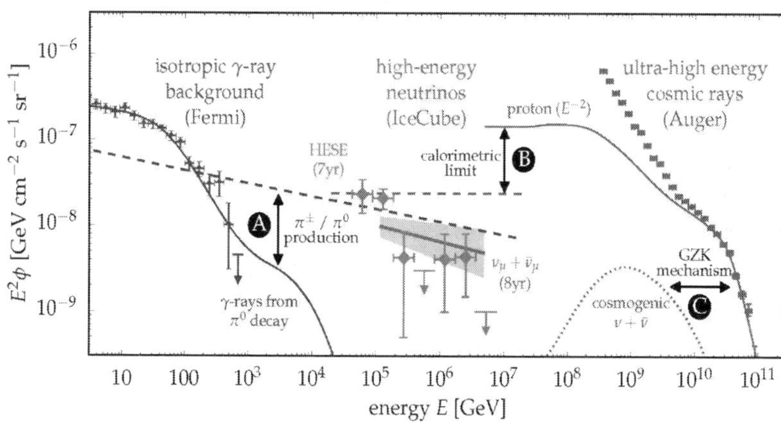

Figure 3. The spectral flux (ϕ) of neutrinos inferred from the 8-year upgoing track analysis (red solid line) with 1σ uncertainty range (shaded range) and the 7-year HESE analysis (magenta data) compared to the flux of unresolved extragalactic gamma ray sources (Ackermann et al., 2015) (blue data) and ultra-high-energy cosmic rays (Aab et al., 2015) (green data). We highlight the various multimessenger interfaces: **(A)** The joined production of charged pions ($\pi^{\pm}$) and neutral pions (π^0) in cosmic-ray interactions leads to the emission of neutrinos (dashed blue) and gamma rays (solid blue), respectively. **(B)** Cosmic ray emission models (solid green) of the most energetic cosmic rays imply a maximal flux (calorimetric limit) of neutrinos from the same sources (green dashed). **(C)** The same cosmic ray model predicts the emission of cosmogenic neutrinos from the collision with cosmic background photons (GZK mechanism). Figure from Ahlers and Halzen (2018), permission: Elsevier.

The overall normalization of the emission has been chosen not to exceed the isotropic gamma ray background observed by the Fermi satellite (blue data). Interestingly, the high-energy cosmic neutrino flux above 100 TeV shown in Figure 3 saturates this limit. Moreover, the starting event data that extends to lower energies, below 100 TeV, are only marginally consistent with the upper bound (blue dashed line). This high intensity of the neutrino flux at lower energies compared to the isotropic gamma-ray background might indicate that these sources are more efficient neutrino than gamma-ray sources (Murase et al., 2016; Bechtol et al., 2017). The similar energy densities of the extragalactic gamma-ray flux observed by Fermi and the high-energy cosmic neutrino flux suggest that, rather than detecting some exotic sources, it is more likely that IceCube to a large extent observes the same universe that conventional astronomy does. Clearly, an extreme universe modeled exclusively on the basis of electromagnetic processes is no longer realistic. We must conclude that a large fraction, possibly most, of the energy in the non-thermal universe originates in hadronic processes, indicating a larger role than previously thought.

Interestingly, the common energy density of photons and neutrinos is also comparable to that of the ultra-high-energy extragalactic cosmic rays above 10^9 GeV) (Aab et al., 2015) (green data). A scenario where the high-energy neutrinos observed at IceCube could actually originate in the same sources could be realized as follows: the cosmic ray sources can be embedded in environments that act as "storage rooms" for cosmic rays with energies far below the "ankle" ($E_{CR} \ll 1$EeV). This energy-dependent confinement is obtained by cosmic ray diffusion in magnetic fields. Gamma rays and neutrinos are produced through collisions with gas. Provided that the conditions are met, this mechanism will be so efficient that the total energy stored in cosmic rays below the ankle is converted to that of gamma rays and neutrinos. These "calorimetric" conditions can be achieved, for instance, in starburst galaxies (Loeb and Waxman, 2006) or galaxy clusters (Berezinsky et al., 1997). The qualitatively equal energy in cosmic rays and neutrinos is suggestive of a common origin within the large measurement errors. Additionally, it is further affected by the assumptions on the injected spectrum, the composition of the extragalactic cosmic rays, and, additionally, the transition energy from the Galactic to extragalactic component; see (Globus et al., 2015; Unger et al., 2015; Aab et al., 2017; Kachelrieß et al., 2017) for details.

The extragalactic gamma ray background observed by Fermi (Ackermann et al., 2015) has contributions from identified sources, mostly blazars, on top of an isotropic gamma ray background (IGRB) shown in Figure 3. Furthermore, this diffuse contribution is expected to result from the emission from the same class of gamma ray sources that are individually below Fermi's point-source detection threshold (see, e.g., Di Mauro and Donato, 2015). This speculation received further support when IceCube developed methods for performing real-time multiwavelength observations (Aartsen et al., 2016b, 2017d) that led to the evidence for identification of a distant flaring blazar as a cosmic ray accelerator in a multimessenger campaign launched by a 290 TeV energy neutrino detected from the constellation of Orion on September 22, 2017 (Aartsen et al., 2018a).

THE FIRST TRULY MULTIMESSENGER CAMPAIGN

Neutrinos only originate in environments where protons are accelerated to produce pions and other particles that decay into neutrinos. Neutrinos can thus exclusively pinpoint cosmic ray accelerators, and this is exactly what one neutrino did on September 22, 2017.

IceCube detects muon neutrinos, a type of neutrino that leaves a well-reconstructed track in the detector roughly every 5 min. Most of them are low-energy neutrinos produced in the Earth's atmosphere, which are of interest for studying the neutrinos themselves, but are a persistent background when doing neutrino astronomy. In 2016, IceCube installed an online filter that selects from this sample, in real time, very high energy neutrinos that are likely to be of cosmic origin (Aartsen et al., 2017d). We reconstruct their energy and celestial coordinates, typically in less than 1 min, and distribute the information automatically via the Gamma-ray Coordinate Network to a group of telescopes around the globe and in space for follow-up observations. These telescopes look for electromagnetic radiation from the arrival direction of the neutrino, searching for coincident emission that can reveal its origin.

The tenth such alert (Kopper and Blaufuss, 2017), IceCube-170922A, on September 22, 2017, reported a well-reconstructed muon neutrino with an energy of 290 TeV and, therefore, with a high probability of originating in an astronomical source. The Fermi telescope detected a flaring blazar aligned with the cosmic neutrino within 0.06 degrees. The source is a known blazar, a supermassive black hole spitting out high-energy particles in twin jets aligned with its rotation axis which is directed at Earth. This blazar, TXS 0506 + 056, had been relatively poorly studied until now, although it was identified by EGRET as the preeminent source for observation of two photons with energies above 40 GeV (Dingus and Bertsch, 2001). Observations triggered by the neutrino alert yielded a treasure trove of multiwavelength data. An optical telescope eventually measured its distance (Paiano et al., 2018), which was found at redshift of 0.34. Its large distance points to a special galaxy, which sets it apart from the ten-times-closer blazars, such as the Markarian sources that dominate the extreme gamma-ray sky observed by NASA's Fermi satellite. Getting all the elements of this multimessenger observation together has been challenging. Attempts to explain the broadband multiwavelength spectrum of the source indicate a subdominant hadronic emission for the coincident observations. The hadronic contribution to the high-energy emission is constrained by the X-ray flux expected in conventional models for blazars. For details, see (Gokus et al., 2018; Keivani et al., 2018; Murase et al., 2018; Sahakyan, 2018; Zhang et al., 2018; Cerruti et al., 2019; Gao et al., 2019).

TXS 0506 + 056 was originally flagged by the Fermi (Tanaka et al., 2017) and Swift (Keivani et al., 2017) satellite telescopes. Follow-up observations with the MAGIC air Cherenkov telescope (Mirzoyan, 2017) identified it as a rare very high energy blazar with the potential to produce the very high energy neutrino detected by IceCube. The source was subsequently scrutinized in X-ray, optical, and radio wavelengths. This is a first, truly multimessenger observation: none of the instruments could have made this breakthrough independently. In total, more than 20 telescopes observed the flaring blazar as a highly variable source in a high state (Aartsen et al., 2018a).

It is important to realize that nearby blazars like the Markarian sources are at a redshift that is ten times smaller, and therefore TXS 0506 + 056, with a similar flux despite the greater distance, is one of the most luminous sources in the Universe. It likely belongs to a special class of blazars that accelerate proton beams as revealed by the neutrino. This explains the fact that a variety of previous attempts to associate the arrival directions of cosmic neutrinos with the bulk of Fermi blazars (Aartsen et al., 2017a) have failed.

Informed by the multimessenger campaign, IceCube searched its archival neutrino data up to and including October 2017 in the direction of IC-170922A using the likelihood routinely used in previous searches. This revealed a spectacular burst of over a dozen high-energy neutrinos in 110 days in the 2014-2015 data with a spectral index similar to the one observed for the diffuse cosmic neutrino spectrum (Aartsen et al., 2018b).

It is worth mentioning that the AGILE collaboration, which operates an orbiting X-ray and gamma ray telescope, reported a day-long flare in the direction of a previous neutrino alert sent on July 31, 2016 (Lucarelli et al., 2017). The flare occurred more than one day before the time of the alert. In light of the rapid daily variations observed near the peak emission of the TXS 0506 + 056 flare at the time of IC-170922A, this may well be a genuine coincidence.

FLARING SOURCES AND THE HIGH-ENERGY NEUTRINO FLUX

As we will show further on, it is clear that a source that produces more than a dozen high-energy neutrinos in only 110 days cannot be a routine blazar in the Fermi sky. This single flare in 2014-2015 dominates the flux of the source over the 9.5 years of archival IceCube data, leaving IC-170922A as a less luminous second flare in the sample. With the density of photons in the target required to produce the neutrinos, the source cannot be transparent to the high-energy photons routinely observed by Fermi. This, as well as the detection of this extraordinary large neutrino flux from a source at a redshift of 0.34, when photon-bright blazars are observed at distances closer by one order of magnitude, suggests that TXS 0506+056 belongs to a special class of sources, those that produce cosmic rays. We will show next that a subset of about 5% of all blazars can accommodate the diffuse cosmic neutrino flux observed by IceCube assuming they burst once in 10 years at the level of TXS 0506+056 in 2014. We have already pointed out in the previous section that the energy density of this flux is similar to that in the highest energy cosmic rays.

The all-sky flux of high-energy neutrinos from a fraction (F) of a class of sources with density ρ, neutrino luminosity L_ν, and episodic emission of flares of duration Δt over a total observation time T is

$$\sum_{\alpha} E_\nu^2 \frac{dN_\nu}{dE_\nu} = \frac{1}{4\pi} \frac{c}{H_0} \xi_z L_\nu \rho \mathcal{F} \frac{\Delta t}{T}. \tag{4}$$

where ξ_z is a factor of order unity that parameterizes the integration over the redshift evolution of the sources (Halzen et al., 2018). Applying this relation to the 2014 TXS 0506+056 burst that dominates the flux over the 9.5 years of neutrino observations, yields

$$3 \times 10^{-11}\mathrm{TeVcm}^{-2}\mathrm{s}^{-2}\mathrm{sr}^{-1} = \frac{\mathcal{F}}{4\pi}\frac{c}{H_0}\left(\frac{\xi_z}{0.7}\right)\left(\frac{L_\nu}{1.2\times10^{47}\mathrm{erg/s}}\right) \left(\frac{\rho}{1.5\times10^{-8}\mathrm{Mpc}^{-3}}\right)\left(\frac{\Delta t}{110\mathrm{d}}\frac{10\mathrm{yr}}{T}\right), \tag{5}$$

a relation which is satisfied for F~0.05F~0.05. In summary, a special class of BL Lac blazars that undergo ~110-day duration flares like TXS 0506+056 once every 10 years accommodates the observed diffuse flux of high-energy cosmic neutrinos. The class of such neutrino-flaring sources represents 5% of the sources. The argument implies the observation of roughly 100 muon neutrinos per year. This is exactly the flux of cosmic neutrinos that corresponds to the $E^{-2.19}$ diffuse flux measured above 100 TeV. (Note that the majority of these neutrinos cannot be separated from the atmospheric background, leaving us with the reduced number of very high energy events discussed in the previous sections). As previously discussed, the energetics of the cosmic neutrinos is matched by the energy of the highest energy cosmic rays, with

$$\frac{1}{3}\sum_\alpha E_\nu^2 \frac{dN_\nu}{dE_\nu} \simeq \frac{c}{4\pi}\left(\frac{1}{2}\left(1 - e^{-f_\pi}\right)\xi_z t_H \frac{dE}{dt}\right) \tag{6}$$

The cosmic rays' injection rate dE/dt above 10^{16} eV is (1−2) × 10^{44}erg $\mathrm{Mpc}^{-3}\mathrm{yr}^{-1}$ (Ahlers and Halzen, 2012; Katz et al., 2013). From Equation (5) it follows that the energy densities match for a pion production efficiency of the neutrino source of $f_\pi \gtrsim 0.4$. This high efficiency requirement is consistent with the premise that a special class of efficient sources is responsible for producing the high-energy cosmic neutrino flux seen by IceCube. The sources must contain sufficient target density in photons, possibly protons, to generate the large value of f_π. It is clear that the emission of flares producing the large number of cosmic neutrinos detected in the 2014 burst must correspond to major accretion events onto the black hole lasting a few months. The pionic photons will lose energy in the source and the neutrino emission is not accompanied by a flare as was the case for the 2017 event (See Halzen et al., 2018 for details); the Fermi data, consistent with the scenario proposed, reveal photons with energies of tens of GeV, but no enhancement in gamma ray activity (Garrappa et al., 2018).

In the modeling of hadronic blazars, it is typically assumed that the interaction of accelerated protons with photons is responsible for generating pions that decay into neutrinos. The interaction is dominated by the Δ-resonance, with $p\gamma \rightarrow \Delta \rightarrow \pi N$. Accelerated cosmic rays may interact with photons in the jet or with stationary photons provided by the accretion disk. Assuming the former case, the efficiency for producing pions depends on the target photons energy, the Lorentz factor of the jet, the luminosity of the photons, and the duration of the flare:

$$f_\pi \simeq \frac{L_\gamma}{E_\gamma} \frac{1}{\Gamma^2 \Delta t} \frac{3\sigma_\Delta \langle x_{p\to\pi} \rangle}{4\pi c^2}. \quad (7)$$

Since TXS 0506+056 is an intermediate synchrotron-peaked blazar, we consider the target photon at the synchrotron energy peak, 10 eV, to constrain the parameter space for the pion efficiency:

$$f_\pi \gtrsim 0.4 \simeq \left(\frac{L_\gamma}{2\times 10^{46}\mathrm{erg/s}}\right)\left(\frac{10\mathrm{eV}}{E_\gamma}\right)\left(\frac{1}{\Gamma^2}\right) \left(\frac{110\mathrm{d}}{\Delta t}\right)\left(\frac{3\sigma_\Delta \langle x_{p\to\pi} \rangle}{4\pi c^2}\right) \quad (8)$$

This suggests that sources with a small Lorentz factor and UV luminosity of $O(10^{46})$ can satisfy the requirements for producing the high-energy cosmic neutrinos observed. This level of UV luminosity is compatible with the Swift-UVOT measurement reported by Padovani et al. (2018). Similar values for UV luminosity were also reported in Krauß et al. (2014). In addition, the study of radio structure of TXS 0506+056 suggests the feasibility of a small Lorentz factor (Kun et al., 2018). The case for a small Lorentz factor also follows arguments made in the context of a tentative observation of high-energy neutrinos by AMANDA in coincidence with a flare of 1ES 1959+560 in 2002 (Halzen and Hooper, 2005).

A key question is whether the neutrino and gamma-ray spectra for the 2014 neutrino burst from TXS 0506+056 satisfy the multimessenger relationship introduced in section 2. With the low statistics of the very high energy gamma-ray measurements during the burst period, the energetics is a more robust measure for evaluating the connection, especially because the source is opaque to high-energy gamma rays, as indicated by the large value of f_π, and the pionic gamma rays will lose energy inside the source before cascading in the microwave photon background; for details, see (Halzen et al., 2018).

The gamma-ray opacity can be related to f_π as

$$\tau_{\gamma\gamma} \approx \frac{\eta_{\gamma\gamma}\sigma_{\gamma\gamma}}{\eta_{p\gamma}\hat{\sigma}_{p\gamma}} f_\pi, \quad (9)$$

where $\hat{\sigma}_{p\gamma} \sim 0.7 \times 10^{-28}\mathrm{cm}^2$ and $\sigma_{\gamma\gamma} \simeq 6.65 \times 10^{-25}\mathrm{cm}^2$ (Murase et al., 2016). We assume $\eta_{\gamma\gamma} \sim 0.1$ and $\eta_{p\gamma} \simeq 1$. This implies $\tau_{\gamma\gamma} \simeq O(100)$ for $f_\pi \gtrsim 0.4$. This high opacity will make it impossible for the very high energy pionic gamma rays, with energies similar to those of the neutrinos, to leave the source.

Absorption and interactions intrinsic to the source due to the high opacity, followed by the interaction with EBL, will result in a suppressed gamma-ray flux, and hence a gamma-ray flare is not expected as the source is a highly efficient neutrino emitter (Halzen et al., 2018). This is consistent with the Fermi observation of the source during the neutrino flare in 2014 (Garrappa et al., 2018; Padovani et al., 2018). Potential hardening reported by Padovani et al. (2018) could be explained in this picture (see Halzen et al., 2018). The absence of gamma rays during a neutrino burst could also be a result of secondary gamma rays spreading in a larger opening angle compared to the original beam; see (Neronov and Semikoz, 2002) for details.

It is worth noting that this model for the diffuse neutrino flux clarifies why earlier attempts to associate it with the bulk of blazars were unsuccessful and why the limits built on source population and non-detection of a steady source do not apply. Clearly, the time-integrated studies are not applicable to time-dependent sources. Moreover, with a subclass of more energetic sources with lower density responsible for the diffuse flux, the constraints on blazars obtained from the relation between the point source limits and the diffuse flux are mitigated. The idea of a subclass of sources with strong emission is further supported by the rapid evolution of sources; see (Neronov and Semikoz, 2018) for details. These speculations are qualitative and may be premature, but the hope is that multimessenger astronomy will provide us with more clues after the breakthrough event of September 22, 2017, which generated an unmatched data sample over all wavelengths of the electromagnetic spectrum of a likely cosmic ray accelerator.

SUMMARY

Our main conclusion is that this aspect of multimessenger astronomy is likely to be subtle. The spectacular neutrino burst in 2014-2015 must originate from a source that is obscured to high-energy photons and leaves a signature in the Fermi data that is not straightforward and likely to change on an event by event basis. On the other hand, the intense photon flare in 2017 was accompanied by a single very high energy neutrino, most likely a fluctuation from a much lower flux. Getting all the elements of this puzzle to fit together will not be easy, especially because we are missing the rich spectrum of multiwavelength observations in 2014 that were the result of the extensive multimessenger campaign at all wavelengths in September 2017. We need more neutrinos from multiple and more sensitive detectors, and we need a future gamma-ray mission with uninterrupted all-sky coverage.

Unlike the previous SN1987A and GW170817 multimessenger events, this event could not have been observed with a single instrument. Without the initial coincident observation, IC-170922A would be just one more of the few hundred cosmic neutrinos detected by IceCube and the accompanying radiation just one more flaring blazar observed by Fermi-LAT. Neutrino astronomy was born with a supernova in 1987. Thirty years later, this recent event involves neutrinos that are tens of millions of times more energetic and are from a source a hundred thousand times more distant.

Author Contributions

The introductory material was mostly written by FH, the more substantive material about the multimessenger event and the interpretation of the data was a collaborative effort of AK and FH.

Conflict of Interest Statement

The authors declare that the research was conducted in the absence of any commercial or financial relationships that could be construed as a potential conflict of interest.

ACKNOWLEDGMENTS

The authors are supported in part by NSF under grants PLR-1600823 and PHY-1607644 and by the University of Wisconsin Research Council with funds granted by the Wisconsin Alumni Research Foundation.

Footnotes

It is worth mentioning that other scenarios could lead to particle acceleration, such as Poynting flux in neutron stars.

REFERENCES

1. Aab, A., Abreu, P., Aglietta, M., Ahn, E. J., Al Samarai, I., Albuquerque, I. F. M., et al. (2015). "The Pierre Auger Observatory: Contributions to the 34th International Cosmic Ray Conference (ICRC 2015)," in Proceedings, 34th International Cosmic Ray Conference (ICRC 2015) (The Hague).
2. Aab, A., Abreu, P., Aglietta, M., Al Samarai, I., Albuquerque, I. F. M., Allekotte, I., et al. (2017). Inferences on mass composition and tests of hadronic interactions from 0.3 to 100 EeV using the water-Cherenkov detectors of the Pierre Auger Observatory. Phys. Rev. D96:122003. doi: 10.1103/PhysRevD.96.122003
3. Aartsen, M. G., Abbasi, R., Abdou, Y., Ackermann, M., Adams, J., Aguila, J. A., et al. (2013a). First observation of PeV-energy neutrinos with IceCube. Phys. Rev. Lett. 111:021103. doi: 10.1103/PhysRevLett.111.021103
4. Aartsen, M. G., Abbasi, R., Abdou, Y., Ackermann, M., Adams, J., Aguila, J. A., et al. (2013b). Measurement of South Pole ice transparency with the IceCube LED calibration system. Nucl. Instrum. Methods A711, 73–89. doi: 10.1016/j.nima.2013.01.054
5. Aartsen, M. G., Abbasi, R., Abdou, Y., Ackermann, M., Adams, J., Aguilar, J. A., et al. (2013c). Evidence for high-energy extraterrestrial neutrinos at the IceCube detector. Science 342:1242856. doi: 10.1126/science.1242856
6. Aartsen, M. G., Abbasi, R., Ackermann, M., Adams, J., Aguilar, J. A., Ahlers, M., et al. (2014). Energy reconstruction methods in the IceCube neutrino telescope. JINST 9:P03009. doi: 10.1088/1748-0221/9/03/P03009
7. Aartsen, M. G., Abraham, K., Ackermann, M., Adams, J., Aguilar, J. A., Ahlers, M., et al. (2015). Evidence for astrophysical muon neutrinos from the Northern sky with IceCube. Phys. Rev. Lett. 115:081102. doi: 10.1103/PhysRevLett.115.081102
8. Aartsen, M. G., Abraham, K., Ackermann, M., Adams, J., Aguilar, J. A., Ahlers, M., et al. (2016a). Observation and characterization of a Cosmic Muon Neutrino Flux from the Northern Hemisphere using six years of IceCube data. Astrophys. J. 833:3. doi: 10.3847/0004-637X/833/1/3
9. Aartsen, M. G., Abraham, K., Ackermann, M., Adams, J., Aguilar, J.

A., Ahlers, M., et al. (2016b). Very high-energy gamma-ray follow-up program using Neutrino triggers from IceCube. JINST 11:P11009. doi: 10.1088/1748-0221/11/11/P11009

10. Aartsen, M. G., Abraham, K., Ackermann, M., Adams, J., Aguilar, J. A., Ahlers, M., et al. (2017a). The contribution of Fermi-2LAC blazars to the diffuse TeV-PeV neutrino flux. Astrophys. J. 835:45. doi: 10.3847/1538-4357/835/1/45
11. Aartsen, M. G., Ackerman, M., Adams, J., Aguilar, J. A., Ahlers, M., Ahrens, M., et al. (2018a). Multimessenger observations of a flaring blazar coincident with high-energy neutrino IceCube-170922A. Science 361:eaat1378. doi: 10.1126/science.aat1378
12. Aartsen, M. G., Ackermann, M., Adams, J., Aguilar, J. A., Ahlers, M., Ahrens, M., et al. (2017b). The IceCube neutrino observatory - Contributions to ICRC 2017 part II: properties of the atmospheric and astrophysical neutrino flux. Available online at: https://arxiv.org/abs/1710.01191
13. Aartsen, M. G., Ackermann, M., Adams, J., Aguilar, J. A., Ahlers, M., Ahrens, M., et al. (2017c). The IceCube neutrino observatory: instrumentation and online systems. JINST 12:P03012. doi: 10.1088/1748-0221/12/03/P03012
14. Aartsen, M. G., Ackermann, M., Adams, J., Aguilar, J. A., Ahlers, M., Ahrens, M., et al. (2017d). The IceCube realtime alert system. Astropart. Phys. 92, 30–41. doi: 10.1016/j.astropartphys.2017.05.002
15. Aartsen, M. G., Ackermann, M., Adams, J., Aguilar, J. A., Ahlers, M., Ahrens, M., et al. (2018b). Neutrino emission from the direction of the blazar TXS 0506+056 prior to the IceCube-170922A alert. Science 361, 147–151. doi: 10.1126/science.aat2890
16. Ackermann, M., Ajello, M., Albert, A., Atwood, W. B., Baldini, L., Ballet, J., et al. (2015). The spectrum of isotropic diffuse gamma-ray emission between 100 MeV and 820 GeV. Astrophys. J. 799:86. doi: 10.1088/0004-637X/799/1/86
17. Ahlers, M., Anchordoqui, L., Gonzalez-Garcia, M., Halzen, F., and Sarkar, S. (2010). GZK Neutrinos after the Fermi-LAT diffuse photon flux measurement. Astropart. Phys. 34, 106–115. doi: 10.1016/j.astropartphys.2010.06.003
18. Ahlers, M., and Halzen, F. (2012). Minimal cosmogenic neutrinos. Phys. Rev. D86:083010. doi: 10.1103/PhysRevD.86.083010

19. Ahlers, M., and Halzen, F. (2018). Opening a new window onto the Universe with IceCube. Prog. Part. Nucl. Phys. 102, 73–88. doi: 10.1016/j.ppnp.2018.05.001
20. Bechtol, K., Ahlers, M., Di Mauro, M., Ajello, M., and Vandenbroucke, J. (2017). Evidence against star-forming galaxies as the dominant source of IceCube neutrinos. Astrophys. J. 836:47. doi: 10.3847/1538-4357/836/1/47
21. Berezinsky, V., Blasi, P., and Ptuskin, V. (1997). Clusters of galaxies as a storage room for cosmic rays. Astrophys J. 487, 529–535. doi: 10.1086/304622
22. Cerruti, M., Zech, A., Boisson, C., Emery, G., Inoue, S., and Lenain, J. P. (2019). Lepto-hadronic single-zone models for the electromagnetic and neutrino emission of TXS 0506+056. Mon. Notices R. Astron. Soc. 483:L12. doi: 10.1093/mnrasl/sly210
23. Di Mauro, M., and Donato, F. (2015). Composition of the Fermi-LAT isotropic gamma-ray background intensity: emission from extragalactic point sources and dark matter annihilations. Phys. Rev. D91:123001. doi: 10.1103/PhysRevD.91.123001
24. Dingus, B. L., and Bertsch, D. L. (2001). The highest energy emission detected by egret from blazars. AIP Conf. Proc. 587, 251–255. doi: 10.1063/1.1419408
25. Fang, K., and Murase, K. (2018). Linking high-energy cosmic particles by black hole jets embedded in large-scale structures. Phys. Lett., 14:396. doi: 10.1038/s41567-017-0025-4
26. Gao, S., Fedynitch, A., Winter, W., and Pohl, M. (2019). Modelling the coincident observation of a high-energy neutrino and a bright blazar flare. Nat. Astron. 3, 88–92. doi: 10.1038/s41550-018-0610-1
27. Garrappa, S. (2018). Gamma ray counterparts to IceCube high-energy tracks. in 8th International Fermi Symposium. Baltimore, MD.
28. Globus, N., Allard, D., and Parizot, E. (2015). A complete model of the cosmic ray spectrum and composition across the Galactic to extragalactic transition. Phys. Rev. D92:021302. doi: 10.1103/PhysRevD.92.021302
29. Gokus, A., Richter, S., Spanier, F., Kreter, M., Kadler, M., Mannheim, K., and Wilms, J. (2018) Decomposing blazar spectra into lepto-hadronic emission components. Availble online at: https://arxiv.org/abs/1808.05540

30. Halzen, F., and Hooper, D. (2005). High energy neutrinos from the TeV blazar 1ES 1959+650. Astropart. Phys. 23, 537–542. doi: 10.1016/j.astropartphys.2005.03.007
31. Halzen, F., Kheirandish, A., Weisgarber, T., and Wakely, S. P. (2018). On the neutrino flares from the direction of TXS 0506+056. Astrophys. J. 874, L9. doi: 10.3847/2041-8213/ab0d27
32. Kachelrieß, M., Kalashev, O., Ostapchenko, S., and Semikoz, D. V. (2017). Minimal model for extragalactic cosmic rays and neutrinos. Phys. Rev. D96:083006. doi: 10.1103/PhysRevD.96.083006
33. Katz, B., Waxman, E., Thompson, T., and Loeb, A. (2013). The energy production rate density of cosmic rays in the local universe is $\sim 10^{44-45}$ergMpc^{-3}yr^{-1} *at all particle energies*. Available online at: https://arxiv.org/abs/1311.0287
34. Katz, U. F., and Spiering, C. (2012). High-Energy Neutrino Astrophysics: Status and Perspectives. Prog. Part. Nucl. Phys. 67, 651–704. doi: 10.1016/j.ppnp.2011.12.001
35. Keivani, A., Murase, K., Petropoulou, M., Fox, D. B., Cenko, S. B., Chaty, S., et al. (2018). A multimessenger picture of the flaring blazar TXS 0506+056: implications for high-energy neutrino emission and cosmic ray acceleration. Astrophys. J. 864:84. doi: 10.3847/1538-4357/aad59a
36. Keivani, P. A., Evans, A., Kennea, J. A., Fox, D. B., Cowen, D. F., Osborne, J. P., et al. (2017). "Further Swift-XRT observations of IceCube 170922A". in The Astronomer's Telegram, 10792.
37. Kopper, C., and Blaufuss, E. (2017). "IceCube-170922A - IceCube observation of a high-energy neutrino candidate event.," in GRB Coordinates Network, Circular Service, No. 21916, #1 (2017), 21916.
38. Krauß, F., Kadler, M., Mannheim, K., Schulz, R., Trüstedt, J., Wilms, J., et al. (2014). TANAMI blazars in the IceCube PeV Neutrino fields. Astron. Astrophys. 566:L7. doi: 10.1051/0004-6361/201424219
39. Kun, E., Biermann, P. L., and Gergely, L. (2018). VLBI radio structure and core-brightening of the high-energy neutrino emitter TXS 0506+056. Mon. Not. Roy. Astron. Soc. 483, L42–L46. doi: 10.1093/mnrasl/sly216
40. Loeb, A., and Waxman, E. (2006). The cumulative background of high energy neutrinos from starburst galaxies. JCAP 0605:003. doi: 10.1088/1475-7516/2006/05/003

41. Lu, L. (2018). "Recent IceCube results - evidences of neutrino emission from the blazar TXS 0506+056 and searches for Glashow resonance." Contribution to Ultra High Energy Cosmic Rays 2018. Paris, France.
42. Lucarelli, F., Pittori, C., Verrecchia, F., Donnarumma, I., Tavani, M., Bulgarelli, A., et al. (2017). AGILE detection of a candidate gamma-ray precursor to the ICECUBE-160731 neutrino event. Astrophys. J. 846:121. doi: 10.3847/1538-4357/aa81c8
43. Mirzoyan, R. (2017). "First-time detection of VHE gamma rays by MAGIC from a direction consistent with the recent EHE neutrino event IceCube-170922A," in The Astronomer's Telegram, 10817.
44. Murase, K., Guetta, D., and Ahlers, M. (2016). Hidden cosmic-ray accelerators as an origin of TeV-PeV cosmic neutrinos. Phys. Rev. Lett. 116:071101. doi: 10.1103/PhysRevLett.116.071101
45. Murase, K., Oikonomou, F., and Petropoulou, M. (2018). Blazar flares as an origin of high-energy cosmic neutrinos? Astrophys. J 865:124. doi: 10.3847/1538-4357/aada00
46. Neronov, A., and Semikoz, D. V. (2018). Self-consistent model of extragalactic neutrino flux from evolving blazar population. Available online at: https://arxiv.org/abs/1811.06356
47. Neronov, A. Y., and Semikoz, D. V. (2002). Which blazars are neutrino loud? Phys. Rev. D66:123003. doi: 10.1103/PhysRevD.66.123003
48. Padovani, P., Giommi, P., Resconi, E., Glauch, T., Arsioli, B., Sahakyan, N., et al. (2018). Dissecting the region around IceCube-170922A: the blazar TXS 0506+056 as the first cosmic neutrino source. Mon. Notices R. Astron. Soc. 480:192. doi: 10.1093/mnras/sty1852
49. Paiano, S., Falomo, R., Treves, A., and Scarpa, R. (2018). The redshift of the BL Lac object TXS 0506+056. Astrophys. J. 854:L32. doi: 10.3847/2041-8213/aaad5e
50. Protheroe, R. J., and Stanev, T. (1993). Electron-photon cascading of very high-energy gamma-rays in the infrared background. Mon. Notices R. Astron. Soc. 264, 191–200.
51. Sahakyan, N. (2018). Lepto-hadronic γ-ray and neutrino emission from the jet of TXS 0506+056. Astrophys. J. 866:109. doi: 10.3847/1538-4357/aadade
52. Taboada, I. (2018). "A View of the Universe with the IceCube and ANTARES Neutrino Telescopes." Neutrino 2018. Heidelberg,

Germany. doi: 10.5281/zenodo.1286919

53. Tanaka, Y. T., Buson, S., and Kocevski, D. (2017). "Fermi-LAT detection of increased gamma-ray activity of TXS 0506+056, located inside the IceCube-170922A error region," in The Astronomer's Telegram, 10791.
54. Unger, M., Farrar, G. R., and Anchordoqui, L. A. (2015). Origin of the ankle in the ultrahigh energy cosmic ray spectrum, and of the extragalactic protons below it. Phys. Rev. D92:123001. doi: 10.1103/PhysRevD.92.123001
55. Yuan, T. (2017). "Improving the angular resolution in icecube cascade reconstruction," in TeV Particle Astrophysics 2017 (TeVPA2017) (Columbus, OH).
56. Zhang, H., Fang, K., and Li, H. (2018). Variability and optical polarization can probe the neutrino and electromagnetic emission mechanisms of TXS 0506+056. Available online at: https://arxiv.org/abs/1807.11069

CHAPTER

20

IceCube: Neutrinos and Multimessenger Astronomy

Markus Ahlers, Francis Halzen

Progress of Theoretical and Experimental Physics, Volume 2017, Issue 12, December 2017, 12A105, https://doi.org/10.1093/ptep/ptx021

ABSTRACT

We review the status of the IceCube observations of cosmic neutrinos. We investigate model-independent constraints on the properties of the sources where they originate. Specifically, we evaluate the multimessenger relations connecting neutrino, gamma ray, and cosmic ray observations and conclude that neutrinos are ubiquitous in the nonthermal universe, suggesting a more significant role than previously anticipated. Subsequently, we study the implications of IceCube's upper limits on the flux from individual point sources, as well as on the "guaranteed" flux of cosmogenic neutrinos.

Citation: Markus Ahlers, Francis Halzen, "IceCube: Neutrinos and multimessenger astronomy, Progress of Theoretical and Experimental Physics", Volume 2017, Issue 12, December 2017, 12A105, https://doi.org/10.1093/ptep/ptx021

INTRODUCTION: MULTIMESSENGER ASTRONOMY

With the commissioning of the IceCube [1,2] and Advanced LIGO [3] facilities, we are able to observe the universe for the first time using three distinct astronomical messengers. In addition to photons, from radio waves to gamma rays, we can now simultaneously observe the sky with gravitational waves and high-energy neutrinos. This new era of multimessenger astrophysics will offer a unique view of our universe and provide powerful insights into the workings of some of the most energetic and enigmatic objects in the cosmos. In fact, initial findings have been astonishing. The first direct observation of a gravitational wave may become a footnote in history to the fact that thirty-solar-mass black holes exist to generate its energy. The mere existence of black holes with such a mass challenges our understanding of the universe.

Powered by gravity, shocks in the relativistic particle flows in the vicinity of neutron stars and black holes may transform a fraction of the gravitational energy into the acceleration of particles, mostly protons that are observed as cosmic rays. We do not know how, or even where, this happens, but we have detected cosmic rays with Joule kinetic energy that bear witness to the enigmatic processes that pack a macroscopic energy into a single elementary particle.

With IceCube's discovery that neutrinos from the cosmos can be observed with a cubic kilometer detector, cosmic neutrinos may now reveal an unobstructed view of the universe at wavelengths above tens of TeV where the universe is opaque to light. With more than one thousand times the energy of the highest-energy neutrinos produced with earthbound accelerators, cosmic neutrinos also exceed by a factor of one billion the energy of the neutrinos detected from a supernova that exploded in the Large Magellanic Cloud in February 1987, the only neutrinos that have reached us from outside the solar system prior to IceCube's breakthrough. It is therefore probably somewhat counterintuitive that the more puzzling property of the observed cosmic neutrinos is not their energy but their large intensity.

An immediate inference made about the large neutrino flux observed by IceCube, which is predominantly extragalactic in origin, is that the total energy density of neutrinos in the high-energy universe is similar to that of gamma rays. This is worthy of a closer look. Neutrinos are the decay products of pions. Protons accelerated in regions of high magnetic fields near neutron stars or black holes may interact with the radiation or dust surrounding them to produce pions and kaons that decay into neutrinos. This is the mechanism by which neutrino beams are produced at Fermilab, where the target material is arranged to be sufficiently dense so as to absorb all secondary particles created in the collisions, except for the neutrinos of course. Not so in the sky, where pions and kaons are produced in more tenuous radiation fields or in dust in the vicinity of the accelerator. In fact, the accelerator and the pion-producing target may be separated, as is the case for the abundant production of gamma rays in the Galactic plane. Elementary particle physics dictates that neutral pions, which promptly decay into two gamma rays, inevitably accompany charged pions generating neutrinos.

It seems, therefore, surprising that no gamma ray has ever been observed matching the 100 to 10,000 TeV energy range of IceCube neutrinos. However, this is just a consequence of the universe's opacity to high-energy photons. Unlike neutrinos, gamma rays interact with photons of the cosmic microwave background before reaching Earth. The resulting electromagnetic shower subdivides the initial photon energy, resulting in multiple photons in the GeV–TeV energy range by the time the shower reaches Earth. Calculating the cascaded gamma ray flux accompanying IceCube neutrinos is straightforward. It is intriguing that the the resulting flux shown in Fig. 1 matches the extragalactic high-energy gamma ray flux observed by the Fermi satellite.

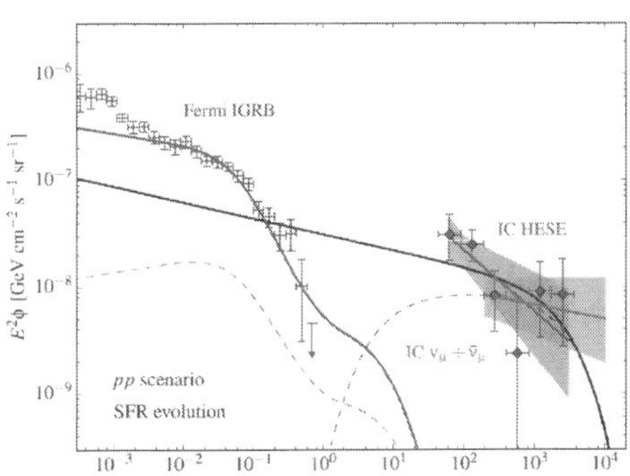

Figure 1. Two models of the astrophysical neutrino flux (black lines) observed by IceCube and the corresponding cascaded gamma ray flux (blue lines) observed by Fermi. The models assume that the decay products of neutral and charged pions from pppp interactions are responsible for the nonthermal emission in the universe [4]. The thin dashed lines represent an attempt to minimize the contribution of the pionic gamma ray flux to the Fermi observations. It assumes an injected flux of E−2E−2 with exponential cutoff at low and high energy. The green data show the binned neutrino spectrum inferred from the four-year "high-energy starting event" (HESE) analysis [5].

The green solid line and shaded band indicate the corresponding power-law fit with uncertainty range. Also shown as a red solid line and shaded band is the best fit to the flux of high-energy muon neutrinos penetrating the Earth [6].

The matching energy densities of the extragalactic gamma ray flux detected by Fermi and the high-energy neutrino flux measured by IceCube suggest that, rather than detecting some exotic sources, it is more likely that IceCube to a large extent observes the same universe astronomers do. The finding implies that a large fraction, possibly most, of the energy in the nonthermal universe originates in hadronic processes, indicating a larger role than previously thought. IceCube is developing methods, most promisingly real-time multiwavelength observations with astronomical telescopes, to identify the sources and build on the discovery of cosmic neutrinos to launch a new era in astronomy [7,8].

The outline of this paper is as follows. After briefly introducing IceCube in Sect. 2, we will review the status of its observations in Sect. 3. Subsequently, we will discuss the initial lessons learned for multimessenger astronomy in Sect. 4. The nonobservation of extragalactic neutrino point sources allows the placement of exclusion limits on the density of feasible point source populations, as we will discuss in Sect. 5. We then discuss the possibility that the sources of ultra-high-energy cosmic rays (UHE CRs) are related to the observed flux of cosmic neutrinos in Sect. 6. In Sect. 7, we review the predictions of cosmogenic neutrinos, which are considered a "guaranteed" contribution of the sources of UHE CRs. We finally consider the possibility of finding correlations of UHE CR events with neutrino events in Sect. 8 before we conclude in Sect. 9.

ICECUBE

The IceCube detector [1] transforms deep natural Antarctic ice 1450 m below the geographic South Pole into a Cherenkov detector. The instrument consists of 5160 digital optical modules that instrument a cubic kilometer of ice; see Fig. 2. Each digital optical module consists of a glass sphere that contains a 10 in photomultiplier and the electronics board that digitizes the signals locally using an onboard computer. The digitized signals are given a global time stamp with residuals accurate to 2 ns and are subsequently transmitted to the surface. Processors at the surface continuously collect the time-stamped signals from the optical modules, each of which functions independently. These signals are sorted into telltale patterns of light that reveal the direction, energy, and flavor of the incident neutrino.

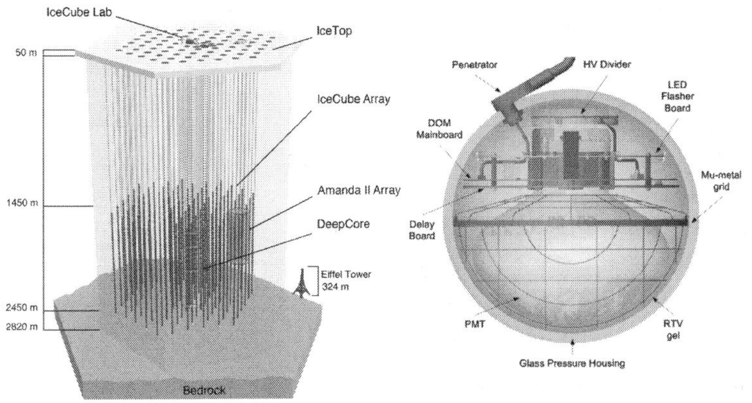

Figure 2. Architecture of the IceCube observatory (left) and the schematics of a digital optical module (right).

Even at a depth of 1450 m, IceCube detects a background of atmospheric cosmic ray muons originating in the Southern Hemisphere at a rate of 3000 per second. Two methods are used to identify neutrinos. Traditionally, neutrino searches have focused on the observation of muon neutrinos that interact primarily outside the detector to produce kilometer-long muon tracks passing through the instrumented volume. Although this allows the identification of neutrinos that interact outside the detector, it is necessary to use the Earth as a filter in order to remove the huge background of cosmic ray muons. This limits the neutrino view to a single flavor and half the sky. An alternative method exclusively identifies neutrinos interacting inside the detector [9]. It divides the instrumented volume of ice into an outer veto shield and a 500-megaton inner fiducial volume. The advantage of focusing on neutrinos interacting inside the instrumented volume of ice is that the detector functions as a total absorption calorimeter, measuring the neutrino energy with a 10%%–15%% resolution. Furthermore, with this method, neutrinos from all directions in the sky can be identified, including both muon tracks as well as secondary showers, produced by charged-current interactions of electron and tau neutrinos, and neutral current interactions of neutrinos of all flavors. The Cherenkov patterns initiated by an electron (or tau) neutrino of 1 PeV energy and a muon neutrino depositing 2.6 PeV energy while traversing the detector are contrasted in Fig. 3.

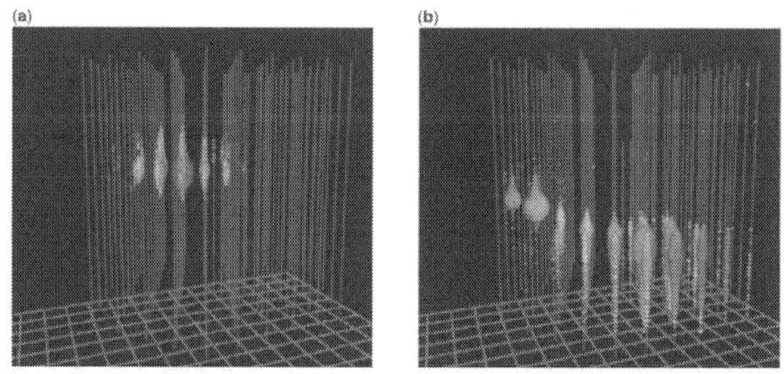

Figure 3. (a) Light pool produced in IceCube by a shower initiated by an electron or tau neutrino. The measured energy is 1.141.14 PeV, which represents a lower limit on the energy of the neutrino that initiated the shower. White dots represent sensors with no signal. For the colored dots, the color indicates arrival time, from red (early) to purple (late) following the rainbow, and size reflects the number of photons detected. (b) An upgoing muon track traverses the detector at an angle of 11∘11∘ below the horizon. The deposited energy inside the detector is 2.6 PeV.

In general, the arrival times of photons at the optical sensors determine the particle's trajectory [10], while the number of photons is a proxy for the deposited energy. The two methods for separating neutrinos from the cosmic ray muon background have complementary advantages. The long tracks produced by muon neutrinos can be pointed back to their sources with a ≤0.4° angular resolution. In contrast, the reconstruction of the direction of secondary showers, in principle possible to a few degrees, is still in the development stage in IceCube [11]. They can be pointed to within ~~10°–15° of the direction of the incident neutrino. Determining the deposited energy from the observed light pool is, however, relatively straightforward, and a resolution of better than 15% is possible; the same value holds for the reconstruction of the energy deposited by a muon track inside the detector.

COSMIC NEUTRINOS

For neutrino astronomy, the first challenge is to select a pure sample of neutrinos, roughly 100,000 per year above a threshold of 0.1 TeV for IceCube, in a background of ten billion cosmic ray muons, while the second is to identify the small fraction of these neutrinos that is astrophysical in origin, observed at the level of tens of events per year. Atmospheric neutrinos are an overwhelming background for cosmic neutrinos, at least at energies below ~~100 TeV. Above this energy, however, the atmospheric neutrino flux is too low to produce events, even in a kilometer-scale detector, and events in that energy range are cosmic in origin.

Using the Earth as a filter, a flux of neutrinos has been identified that is predominantly of atmospheric origin. IceCube has measured this flux over three orders of magnitude in energy with a result that is consistent with theoretical calculations [12,13]. However, in seven years of data, an excess of events is observed at energies beyond 100 TeV [6,14], which cannot be accommodated by the atmospheric flux; see Fig. 4. Allowing for large uncertainties on the extrapolation of the atmospheric component, the statistical significance of the excess astrophysical flux is 6σ6σ□. While IceCube measures only the energy deposited by the secondary muon inside the detector, from Standard Model physics we can infer the energy spectrum of the parent neutrinos represented in the figure. For the highest-energy event, already shown in Fig. 3, the most likely energy of the parent neutrino approaches 10 PeV. Independent of any calculation, the energy lost by the muon inside the instrumented detector volume is 2.6±0.3 PeV. The cosmic neutrino flux is well described by a power law with a spectral index γ=2.13±0.13γ=2.13±0.13 and a normalization at 100 TeV neutrino energy of $(0.90^{+0.30}_{-0.27}) \times 10^{-18}\,\mathrm{GeV}^{-1}\,\mathrm{cm}^{-2}\,\mathrm{s}^{-1}\,\mathrm{sr}^{-1}$ [6]. The error range is estimated from a profile likelihood using Wilks' theorem and includes both statistical and systematic uncertainties. The neutrino energy contributing to this flux covers the range 200 TeV to 9 PeV.

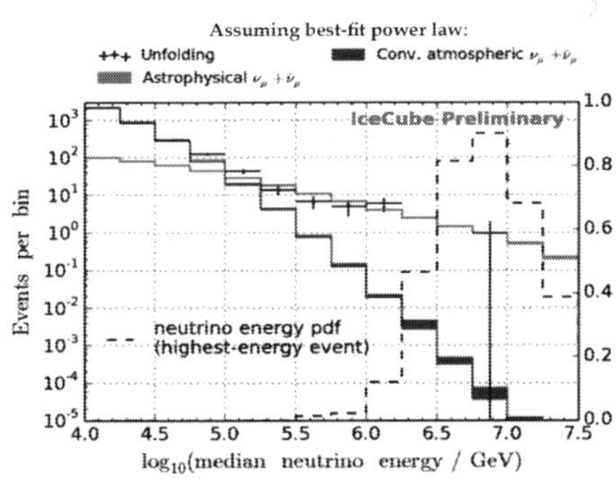

Figure 4. Spectrum of secondary muons initiated by muon neutrinos that have traversed the Earth, i.e., with zenith angle less than 5∘5∘ above the horizon, as a function of the energy they deposit inside the detector. For each reconstructed muon energy, the median neutrino energy is calculated assuming the best-fit spectrum. The colored bands (blue/red) show the expectation for the conventional and astrophysical contributions. The black crosses show the measured data. Additionally, the neutrino energy probability density function for the highest-energy event assuming the best-fit spectrum is shown (dashed line).

However, it was the alternative method, which selects isolated neutrinos interacting inside the detector, that revealed the first evidence for cosmic neutrinos [15,16]. Their isolation and well-measured energy allows for a clear separation between neutrinos of atmospheric origin and those of cosmic origin; a sample event with a light pool of roughly 100,000 photoelectrons extending over more than 500 m is shown in Fig. 3. The geometry of the veto and active signal regions has been optimized to reduce the background of atmospheric muons and neutrinos to a handful of events per year while keeping 98%% of the cosmic signal.

With PeV energy and no trace of accompanying muons from an atmospheric shower, these events are highly unlikely to be of atmospheric origin. It is indeed important to realize that the muon produced in the same pion or kaon decay as an atmospheric neutrino will reach the detector provided that the neutrino energy is sufficiently high and the zenith angle sufficiently small [9,17]. PeV atmospheric neutrinos come with their own self-veto. This self-veto is applied to IceCube cosmic neutrino candidates that exclusively consist of isolated neutrino events.

The energy dependence of the high-energy neutrinos collected in four years of data [5] is compared to that of atmospheric backgrounds in Fig. 5. It is, above an energy of 200 TeV, consistent with the flux of muon neutrinos penetrating the Earth shown in Fig. 4. A purely atmospheric explanation of the observation is excluded at 7σ.

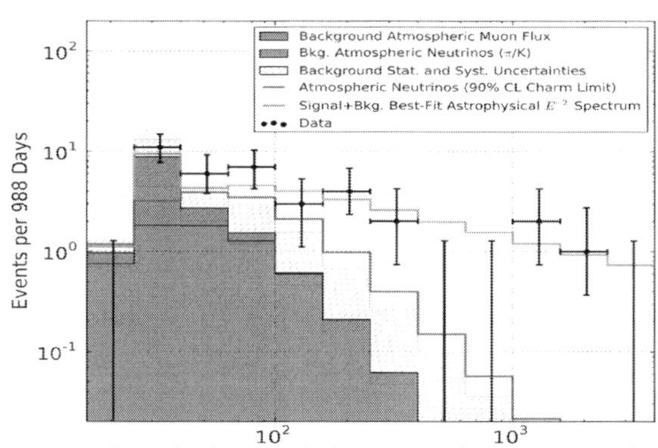

Figure 5. Deposited energies of muons observed in four years of data [5]. The hashed region shows uncertainties on the sum of all backgrounds. The atmospheric muon flux (red) and its uncertainty is computed from simulation to overcome statistical limitations in our background measurement and scaled to match the total measured background rate. The atmospheric neutrino flux is

derived from previous measurements of both the π, K, and charm components of the atmospheric spectrum [25]. Also shown are two illustrative power-law fits to the spectrum.

In summary, IceCube has observed cosmic neutrinos using both methods for rejecting background; each analysis has reached a statistical significance of more than 6σ. Based on different methods for reconstruction and energy measurement, their results agree, pointing at extragalactic sources whose flux has equilibrated in the three flavors after propagation over cosmic distances [18] with $\nu_e:\nu_\mu:\nu_\tau \sim 1:1:1$. Its total energy matches that of extragalactic photons, as seen in Fig. 1, and also that of UHE CRs, as we will discuss in one of the following sections.

The four-year data set contains a total of 54 neutrino events with deposited energies ranging from 30 to 2000 TeV. The data in both Figs. 4 and 5 support an astrophysical component with a spectrum close to E^{-2} above an energy of 200 TeV. An extrapolation of this high-energy flux to lower energy suggests an excess of events in the 30–100 TeV energy range over and above a single power-law fit. This conclusion is supported by a subsequent analysis that has lowered the threshold of the starting-event analysis [19]. The astrophysical flux measured by IceCube is not featureless; either the spectrum of cosmic accelerators cannot be described by a single power law or a second component of cosmic neutrino sources emerges in the spectrum. The events are isolated neutrinos, and it is therefore very difficult to accommodate them as a feature in the atmospheric background, of charm origin or not [20]. The excess is already hinted at in the data shown in Fig. 1 and, in the context of that discussion, the energy associated with the photons that accompany the neutrino "excess" is not seen in the Fermi data [4]. This might indicate that the neutrinos originate in hidden sources [21] or in sources with a very strong cosmological evolution resulting in a shift of the photons to sub-GeV energies [22].

In Fig. 6 we show the arrival directions of the most energetic events of the six-year upgoing $\nu_\mu + \bar{\nu}_\mu$ analysis (⊙) and the four-year HESE analysis, separated into tracks (⊗) and cascades (⊕). The median angular resolution of the cascade events is indicated by thin circles around the best-fit position. The apparent anisotropy of the arrival directions is dominated by the effective area of the analysis. The most energetic muons with energy E_μ>200 TeV in the upgoing $\nu_\mu + \bar{\nu}_\mu$ analysis accumulate just below the horizon in the Northern Hemisphere due to Earth absorption. The HESE events with deposited energy of E_{dep}>100 TeV also suffer from Earth absorption, but can also be visible in the Southern Hemisphere. Various analyses of the IceCube event distribution could not reveal a strong anisotropy from extended emission regions, which could indicate, e.g., a contribution from Galactic sources along the Galactic plane [23,24]. In fact, no correlation of the arrival directions of the highest-energy events, shown in Fig. 6, with potential sources or source classes has reached the level of 3σ [19].

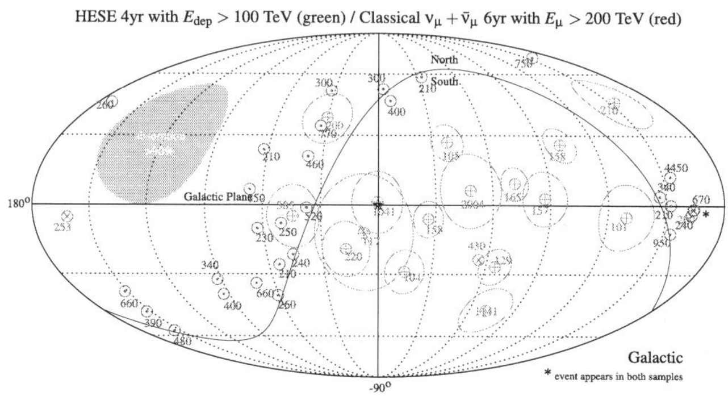

Figure 6. Mollweide projection in Galactic coordinates of the arrival direction of neutrino events. We show the results of the six-year upgoing track analysis [6] with energy proxy M_uE_x>50 (⊙). The red numbers show the most probable neutrino energy (in TeV) assuming the best-fit astrophysical flux of the analysis [6]. The events of the four-year high-energy starting event (HESE) analysis with deposited energy (green numbers) larger than 60 TeV (tracks ⊗ and cascades ⊕) are also shown [5,24]. Cascade events (⊕) are indicated together with their median angular uncertainty (thin circles). One event (*) appears in both event samples. The gray-shaded region indicates the zenith angle range where Earth absorption of 100 TeV neutrinos is larger than 90%. The star symbol (★) indicates the Galactic Center and the thin curved solid black line indicates the horizon.

Various scenarios have been invoked to explain the observed diffuse emission. The absence of strong anisotropies in the arrival direction of the data disfavors scenarios with strong Galactic emission. However, the limited event number and the low angular resolution of cascade-dominated samples can hide this type of emission. Scenarios that could have a partial contribution to the diffuse neutrino flux in the TeV–PeV energy range are the diffuse emission from Galactic CRs [32–37], the joint emission of Galactic CR sources [38–40], or very extended emission from the Fermi bubbles [32,41,42] or the Galactic halo [43,44]. More exotic scenarios consider dark matter decay [45–51] in the Galactic dark matter halo. A smoking gun of these local scenarios would be the observation of PeV gamma rays [32,52] that can only survive over Galactic distance scales.

On the other hand, an isotropic arrival direction of neutrinos is expected for extragalactic source populations. Various scenarios have been considered, including galaxies with intense star formation [4,53–59], cores of active galactic nuclei (AGNs) [60–62], low-luminosity AGNs [63,64], quasar-driven outflows [22], blazars [65–72], low-power gamma ray bursts (GRBs) [73–76], choked GRBs [77,78], cannonball GRBs [79], intergalactic shocks [80], galaxy clusters [4,81–83], tidal disruption events [84–87], or cosmogenic neutrinos [88,89].

An overview of the current information on the flux of cosmic neutrinos is shown in Fig. 7. As already mentioned, a challenge of most of these Galactic and extragalactic scenarios is the high intensity of the neutrino data at 10–100 TeV (see Fig. 7), which implies an equally high intensity of gamma rays produced via neutral pion production and decay. For extragalactic scenarios, this emission is not directly visible due to the strong absorption in the extragalactic radiation background. However, this emission induces electromagnetic cascades that contribute strongly to the Fermi gamma ray background in the GeV–TeV range. We will discuss these multimessenger relations in the next section.

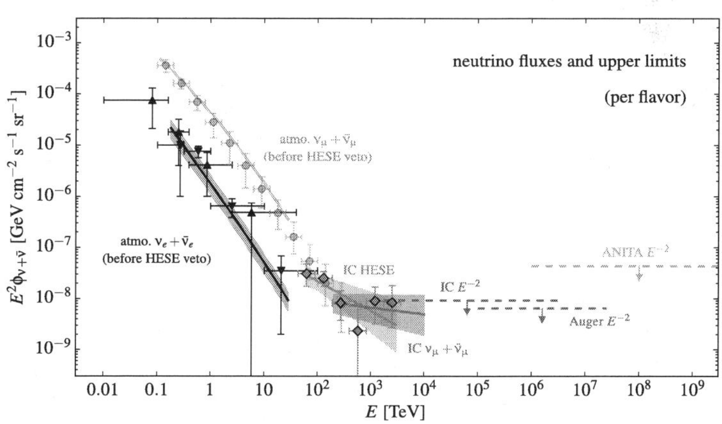

Figure 7. Summary of neutrino observations and upper limits (per flavor). The black and gray data show IceCube's measurement of the atmospheric $\nu_e + \bar{\nu}_e$ [26,27] and $\nu_\mu + \bar{\nu}_\mu$ [28] spectra. The green data show the inferred bin-wise spectrum of the four-year high-energy starting event (HESE) analysis. The green line and green-shaded area indicate the best-fit and 1σ uncertainty range of a power-law fit to the HESE data. Note that the HESE analysis ve-

toes atmospheric neutrinos, and the true background level is much lower as indicated in the plot (cf. Fig. 5). In red we show the corresponding fit to the six-year $\nu_\mu + \bar{\nu}_\mu$ analysis. The dashed lines show 90%% C.L. upper limits of an E^{-2} neutrino emission flux (dashed) at higher energies from IceCube [29] (brown), ANITA [30] (orange), and Auger [31] (blue).

MULTIMESSENGER RELATIONS

Having established a prominent role for hadronic accelerators in the nonthermal universe, we investigate how the accelerated cosmic rays may produce photons and neutrinos after the relatively brief acceleration process. The principal mechanism at work is the production of pions in interactions of high-energy cosmic rays with photons or nuclei. Targets include strong radiation fields that may be associated with the accelerator as well as any concentrations of matter or molecular clouds in their vicinity. Finally, attenuation of the cosmic rays when propagating through the interstellar or intergalactic backgrounds can lead to further production of pions. A high-energy flux of neutrinos is then produced in the subsequent decay of charged pions via $\pi^+ \rightarrow \mu^+ + \nu_\mu$ followed by $\mu^+ \rightarrow e^+ + \nu_e + \bar{\nu}_\mu$ and the charge-conjugate processes. High-energy gamma rays are produced in the decay of neutral pions, $\pi^0 \rightarrow 2\gamma$.

Pion production of cosmic rays via scattering off photons can proceed resonantly via $p+\gamma \rightarrow \Delta^+ \rightarrow \pi^0+p$ or $p+\gamma \rightarrow \Delta^+ \rightarrow \pi^++n$. These channels produce charged and neutral pions with probabilities of 2/3 and 1/3, respectively. However, the contribution of nonresonant pion production at the resonance changes this ratio to about 1/2 and 1/2. In contrast, cosmic rays interacting with hydrogen, e.g., in the Galactic disk, produce equal numbers of pions of all three charges in hadronic collisions: $p+p \rightarrow N_\pi[\pi^0+\pi^++\pi^-]+X$, where N_π is the pion multiplicity.

To evaluate the flux of neutrinos from the cosmic ray interaction region, we start from the pion production rate $Q_\pi\pm$, providing the number of charged pions per unit energy and time (units of $GeV^{-1}s^{-1}$). This quantity is proportional to the corresponding cosmic ray nucleon density Q_N by a "bolometric" proportionality factor $f_\pi \leq 1$ that parametrizes the efficiency of the conversion of cosmic ray energy into pion energy:

$$E_\pi^2 Q_{\pi^\pm}(E_\pi) \simeq f_\pi \frac{K_\pi}{1+K_\pi} \left[E_N^2 Q_N(E_N)\right]_{E_N=E_\pi/\kappa_\pi} \quad (1)$$

The factor introducing K_π accounts for the different ratio between charged and neutral pions in interactions with gas or dust ("pp") and with radiation ("p_{γ}"), with $K_{\pi} \simeq 2$ and 1, respectively. The total energy loss of the hadronic interaction (pp or pγ) is in the form of pions with average energy E_{π}, average multiplicity N_{π}, and total inelasticity κ. For pγ interactions, the total inelasticity is about κ≃0.2, whereas it is about κ≃0.5 for pp interactions. In both cases the average inelasticity per pion can be approximated as $\kappa_{\pi}=\kappa/N_{\pi} \simeq 0.2$ [90]. The average energy per pion is then $E_{\pi}=\kappa_{\pi}EN$. For a target with nucleon density nn and diameter ℓ, the efficiency factor for producing pions can be expressed as

$$f_{\pi} = 1 - \exp(-\kappa \ell \sigma n), \tag{2}$$

with cross section σ and inelasticity κ for either pγ or pp interactions.

4Subsequently, the pions decay into gamma rays and neutrinos that carry, on average, 1/2 and 1/4 of the energy of the parent pion. We here make the approximation that, on average, the four leptons in the decay of $\pi^{\pm}$ equally share the charged pion's energy. The energy of the pionic leptons relative to the proton is:

$$x_{\nu} = \frac{E_{\nu}}{E_p} = \frac{1}{4}\kappa_{\pi} \simeq \frac{1}{20} \text{ and } x_{\gamma} = \frac{E_{\gamma}}{E_p} = \frac{1}{2}\kappa_{\pi} \simeq \frac{1}{10} \tag{3}$$

With this approximation, the neutrino production rate QvαQvα can be related to the one for charged pions:

$$\frac{1}{3}\sum_{\alpha} E_{\nu} Q_{\nu_{\alpha}}(E_{\nu}) \simeq [E_{\pi} Q_{\pi^{\pm}}(E_{\pi})]_{E_{\pi} \simeq 4E_{\nu}} \tag{4}$$

Using Eqs. (1) and (4), we arrive at the final relation for neutrino production:

$$\frac{1}{3}\sum_{\alpha} E_{\nu}^2 Q_{\nu_{\alpha}}(E_{\nu}) \simeq \frac{1}{4} f_{\pi} \frac{K_{\pi}}{1+K_{\pi}} [E_N^2 Q_N(E_N)]_{E_N = 4E_{\nu}/\kappa_{\pi}} \tag{5}$$

The production rate of gamma rays from the decay of neutral pions can be obtained in exactly the same way.

From the two equations for the productions of neutrinos and gamma rays, one can eliminate Q_N to obtain a model-independent relation that is independent of the details of the cosmic ray beam, except for the relative contribution of charged-to-neutral pions,

$$\frac{1}{3}\sum_{\alpha} E_{\nu}^2 Q_{\nu_{\alpha}}(E_{\nu}) \simeq \frac{K_{\pi}}{4} \left[E_{\gamma}^2 Q_{\gamma}(E_{\gamma})\right]_{E_{\gamma} = 2E_{\nu}} \tag{6}$$

Here, the prefactor 1/41/4 accounts for the energy ratio Eν/Eγ≃1/2 and the two gamma rays produced in the neutral pion decay. The relation simply reflects the fact that a π^0 produces two γ rays for every charged pion producing a $\nu_\mu + \bar{\nu}_\mu$ pair, which cannot be separated by current experiments. This is the simple counting that was used to derive the energy density in gamma rays in the universe accompanying the flux of cosmic neutrinos observed by IceCube, shown in Fig. 1.

CONSTRAINTS ON NEUTRINO SOURCE POPULATIONS

The failure of IceCube to identify sources in the diffuse cosmic neutrino flux [91] leads to constraints on the potential origin of these neutrinos. Individual neutrino sources at redshift zz contribute a flux (units $GeV^{-1}cm^{-2}s^{-1}$ and summed over flavors)

$$\phi_\nu^{PS}(E_\nu) = \frac{(1+z)^2}{4\pi d_L^2(z)} \sum_\alpha Q_{\nu_\alpha}((1+z)E_\nu) \tag{7}$$

to the total diffuse flux, where dL is the luminosity distance that, for a flat universe, is given by

$$d_L(z) = (1+z) \int_0^z \frac{dz'}{H(z')} \tag{8}$$

with Hubble parameter H. For the standard ΛCDM cosmological model, the Hubble parameter scales as $H^2(z)=H^2_0[(1+z)^3\Omega_m+\Omega_\Lambda]$, with $\Omega_m \simeq 0.3$, $\Omega_\Lambda \simeq 0.7$, and $c/H_0 \simeq 4.4$ Gpc [92]. The diffuse neutrino flux from extragalactic sources is given by an integral over comoving volume $dV_c=4\pi(dL/(1+z))2dz/H(z)$, weighting each neutrino source by its density per comoving volume $\rho(z)$,

$$\phi_\nu(E_\nu) = \frac{c}{4\pi} \int_0^\infty \frac{dz}{H(z)} \rho(z) \sum_\alpha Q_{\nu_\alpha}((1+z)E_\nu) \tag{9}$$

In the following, we will assume that the neutrino emission rate $Q_{\nu\alpha}$ follows a power law $E^{-\Gamma}$. The flavor-averaged neutrino flux can then be written as

$$\frac{1}{3}\sum_\alpha E_\nu^2 \phi_{\nu_\alpha}(E_\nu) = \frac{c}{4\pi}\frac{\xi_z}{H_0}\rho_0 \frac{1}{3}\sum_\alpha E_\nu^2 Q_{\nu_\alpha}(E_\nu), \tag{10}$$

where we introduce the redshift factor

$$\xi_z = \int_0^\infty dz \frac{(1+z)^{-\Gamma}}{\sqrt{\Omega_\Lambda + (1+z)^3\Omega_m}} \frac{\rho(z)}{\rho(0)} \quad (11)$$

A spectral index of $\Gamma \simeq 2.0$ and no source evolution, $\rho(z)=\rho_0$, yields $\xi_z \simeq 0.6$, whereas the same spectral index and source evolution following the star formation rate yields $\rho \simeq 2.4$.

We can now investigate under what circumstances IceCube can detect the neutrino emission from individual, presumably nearby, point sources that contribute to the diffuse emission. Equation (10) relates the average luminosity of individual neutrino sources to the diffuse flux that is measured by the experiment to be $E^2_{\phi\nu} \simeq 10^{-8}\text{GeVcm}^{-2}\text{s}^{-1}\text{sr}^{-1}$ for energies in excess of $\simeq 100 \simeq 100$ TeV; see Fig. 7. From the measurement, we can infer the average emission from a single source [93]:

$$\frac{1}{3}\sum_\alpha E_\nu^2 Q_{\nu_\alpha}(E_\nu) \simeq 1.8 \times 10^{43} \left(\frac{\xi_z}{2.4}\right)^{-1} \left(\frac{\rho_0}{10^{-8}\,\text{Mpc}^{-3}}\right)^{-1} \text{erg s}^{-1} \quad (12)$$

For a homogenous distribution of sources, we expect, within the partial field of view f_{sky} of the full sky, one source within a distance d_1:

$$4\pi/3 d_1^3 \, \rho_0 f_{\text{sky}} = 1. \quad (13)$$

Defining F_1 as the point source flux of a source with emission rate (12) at distance r_1, we can write down the probability distribution of the sources:

$$F\frac{dn}{dF} = \frac{3}{2}\left(\frac{F_1}{F}\right)^{\frac{3}{2}} e^{-\left(\frac{F_1}{F}\right)^{\frac{3}{2}}} \quad (14)$$

The average flux from the closest source is then $\langle F \rangle \simeq 2.7F_1$, with median $F_{med} \simeq 1.3F_1$. Assuming the latter as the emission of the closest source, we arrive at

$$\frac{1}{3}\sum_\alpha E_\nu^2 \phi_{\nu_\alpha} \simeq 9 \times 10^{-13} \left(\frac{f_{\text{sky}}}{0.5}\right)^{\frac{2}{3}} \left(\frac{\xi_z}{2.4}\right)^{-1} \left(\frac{\rho_0}{10^{-8}\,\text{Mpc}^{-3}}\right)^{-\frac{1}{3}} \text{TeV cm}^{-2}\,\text{s}^{-1} \quad (15)$$

Interestingly, this value is comparable to IceCube's point source sensitivity at the level of $5 \times 10^{-13}\text{TeVcm}^{-2}\text{s}^{-1}$ in the Northern Hemisphere [91].

In Fig. 8, we show the inferred local density and luminosity of neutrino source candidates from Ref. [94]. The gray-shaded region is excluded by nonobservation of sources as individual point sources, assuming the aforementioned sensitivity of IceCube in the Northern Hemisphere. The green lines show the combination of density and luminosity for sources at the level of the observed IceCube flux, assuming a source density evolution following the star formation rate (solid line) or no evolution (dotted line). We conclude that IceCube is presently sensitive to source populations with local source densities lower than about 6×10^{-8}Mpc^{-3}, and much lower local densities, like flat spectrum radio quasars, are excluded by the nonobservation of individual sources. Some source classes, like Fanaroff–Riley (FR) radio galaxies, have an estimated neutrino luminosity that is too low for the observed flux. Note that these estimates depend on the evolution parameter ξz, and the exact sensitivity estimate depends on the particular redshift evolution of the source luminosity density. In addition, this simple estimate can be refined by considering not only the closest source of the population but the combined emission of known local sources; see, e.g., Ref. [93].

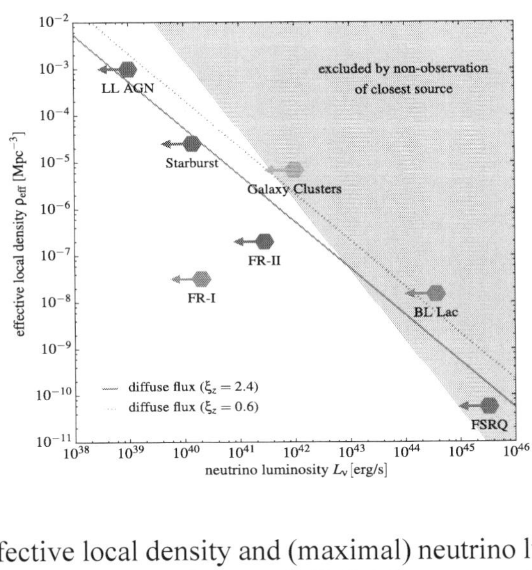

Figure 8. The effective local density and (maximal) neutrino luminosity of various neutrino source candidates from Ref. [94]. The green solid (green dotted) line shows the local density and luminosity of the population of sources responsible for the diffuse neutrino flux of $E2\phi\simeq10^{-8}$GeVcm^{-2}s^{-1}sr^{-1} observed with IceCube, assuming source evolution following the star formation rate ($\xi_z\simeq2.4$) or no source evolution ($\xi_z\simeq0.6$), respectively. The gray-shaded area indicates source populations that are excluded by the nonobservation of point sources in the Northern Hemisphere ($f_{sky}\simeq0.5$) with sensitivity $E^2\phi PS\simeq5\times10^{-13}$TeVcm^{-2}s^{-1}.

THE ENERGY DENSITY OF EXTRAGALACTIC COSMIC RAYS

Is it possible that the sources of the extragalactic cosmic rays are themselves neutrino sources? From the measured cosmic ray spectrum, one can derive that the emission rate density of nucleons is at the level of [95,96]

$$\mathcal{L}_N = \rho_0 E_N^2 Q_N(E_p) \simeq (1-2) \times 10^{44}\,\mathrm{erg\,Mpc^{-3}\,yr^{-1}} \tag{16}$$

Combining this with Eq. (5), we can derive the diffuse neutrino flux:

$$\frac{1}{3}\sum_{\alpha} E_{\nu}^2 \phi_{\nu_\alpha}(E_\nu) \simeq f_\pi \frac{\xi_z K_\pi}{1+K_\pi}(2-4) \times 10^{-8}\,\mathrm{GeV\,cm^{-2}\,s^{-1}\,sr^{-1}} \tag{17}$$

Here, ξ_z is the evolution factor previously introduced. The requirement $f_\pi \leq 1$ limits the neutrino production by the actual sources of the cosmic rays as pointed out in the seminal work by Waxman and Bahcall [12]. For optically thin sources, $f_\pi \ll 1$, neutrino production is only a small by-product of the acceleration process. The energy loss associated with pion production must not limit the sources' ability to accelerate the cosmic rays. On the other hand, optically thick sources, $f_\pi \simeq 1$, may be efficient neutrino emitters. Realistic sources of this type need different zones, one zone for the acceleration process ($f_\pi \ll 1$) and a second zone for the efficient conversion of cosmic rays to neutrinos ($f_\pi \simeq 1$). Examples of this scenario are sources embedded in starburst galaxies, where cosmic rays can be stored over sufficiently long timescales to yield significant neutrino production.

For $\xi_z \simeq 2.4$ and $K_\pi \simeq 1$–22, the upper bound resulting from Eq. (17) and $f_\pi = 1$ is at the level of the neutrino flux observed by IceCube. Therefore, it is possible that the observed extragalactic cosmic rays and neutrinos have the same origin. A plausible scenario is a "calorimeter" in which only cosmic rays with energy below a few 10 PeV interact efficiently. An energy dependence of the calorimetric environment can be introduced by energy-dependent diffusion. If D(E) is the diffusion coefficient, then the timescale of escape from the calorimeter is given by the solution to $6D(E)t=d^2$, where d is the effective size of the region. Typically, we have $D(E) \propto E^\delta$ with $\delta \simeq 0.3$–0.6. In the following, we again consider the case of protons. Taking $\sigma_{pp} \simeq 8\times10^{-26}\,\mathrm{cm}^2$ at Ep=100 PeV and the diffusion coefficient of $D(E_p) \simeq D_{\mathrm{GeV}}(E_p/1\,\mathrm{GeV})^{1/3}$, the pp thickness can be expressed as $\tau_{pp} \simeq ctn_{\mathrm{gas}}\sigma_{pp}$, or

$$\tau_{pp} \simeq 0.18 \left(\frac{d}{100\,\mathrm{pc}}\right)^2 \left(\frac{D_{\mathrm{GeV}}}{10^{26}\,\mathrm{cm}^2\,\mathrm{s}^{-1}}\right)^{-1} \left(\frac{E_p}{10\,\mathrm{PeV}}\right)^{-1/3} \left(\frac{n}{100\,\mathrm{cm}^{-3}}\right) \tag{18}$$

Here, we have used feasible parameters of starburst galaxies [4,53]. Therefore, depending on the calorimetric environment, it is possible that the flux below a few PeV is efficiently converted to neutrinos and contributes to the TeV–PeV diffuse emission observed by IceCube.

COSMOGENIC NEUTRINOS

The production of neutrinos in the sources that accelerate the high-energy cosmic rays depends on the source environment. In order to efficiently accelerate cosmic rays, any loss mechanism, including pion production in pγ and pp interactions, must be suppressed as it reduces the acceleration time. Efficient accelerators are likely to be inefficient beam dumps for producing neutrinos. High-efficiency neutrino production can be achieved by separating the sites of acceleration and neutrino production. For instance, after acceleration, extragalactic cosmic rays propagate over cosmological distances of more than 10 Mpc and can efficiently produce neutrinos on the dilute extragalactic medium.

In this section, we will discuss the production of neutrinos in the interactions of extragalactic cosmic rays with cosmic radiation backgrounds. Soon after the discovery of the cosmic microwave background (CMB), Greisen, Zatsepin, and Kuzmin [97,98] (GZK) realized that extragalactic cosmic rays are attenuated by interactions with background photons. Actually, protons interact resonantly via $p\gamma \rightarrow \Delta^+ \rightarrow \pi^+ n$ with background photons with mean energy $\epsilon \simeq 0.33$ meV at energies $E_p \simeq (m_\Delta^2 - m_p^2)/4/\epsilon \simeq 500\,\mathrm{EeV}$. The width of the Planck spectrum leads to a significant attenuation of proton fluxes after propagation over distances on the order of 200 Mpc at an energy above $E_{GZK} \simeq 50$ EeV, which is known as GZK suppression. Also, heavier nuclei are attenuated at a similar energy by photodisintegration of the nucleus by CMB photons via the giant dipole resonance.

The pions produced in GZK interactions decay, resulting in a detectable flux of cosmogenic neutrinos first estimated by Berezinsky and Zatsepin [99] in 1969. This guaranteed flux of neutrinos became one of the benchmarks for high-energy neutrino astronomy leading early on to the concept of kilometer-scale detectors. The flux of cosmogenic neutrinos peaks at EeV neutrino energy depending on the chemical composition and the evolution with redshift of the unknown sources. The largest neutrino flux results from proton-dominated models [100–102]. A particularly strong emission can be expected in such models if the proton spectrum extends below the ankle. Referred to as "dip models," the ankle results from the absorption of protons by Bethe–Heitler pair production on CMB photons. A fit to the observed cosmic ray spectrum requires relative strong source evolution with redshift [103–106] that enhances pion production. However, the corresponding electromagnetic emission via neutral pions as well as $e^{\pm}$ pairs is constrained by the isotropic gamma ray background (IGRB) observed by Fermi LAT [107,108] and limits the neutrino intensity of these proton-dominated scenarios [109–114]. Recent upper limits on cosmogenic neutrinos resulting from the failure by IceCube to observe EeV neutrinos constrains proton-dominated models [29].

In contrast, the IceCube constraint can be accommodated by introducing a heavy nuclear composition. Resonant neutrino production still proceeds via the interaction of individual nucleons with background photons, but the threshold of the production is increased to $E_{CR} \gtrsim AE_{GZK}$ for nuclei with mass number A. Therefore, efficient cosmogenic neutrino production would require an injected cosmic ray flux that extends well above E_{GZK}. Especially for heavier nuclear composition of the primary flux, the production of neutrinos on photons of the extragalactic background light (EBL) becomes relatively important [95,112,115–122]. The interaction with optical photons produces neutrino fluxes in the PeV energy range. However, the overall level is much lower because of the low intensity of the EBL photons. It is unlikely that the PeV neutrino flux observed with IceCube could be related to the neutrino production in the EBL [88] (see also Ref. [89]). The observed PeV neutrino flux level is too high to be consistent with associated electromagnetic contributions to the IGRB or upper limits on the EeV neutrino flux.

CORRELATION WITH UHE CR EVENTS

Cosmic rays above the GZK cutoff have an absorption length on the order of 200 Mpc. Therefore, the observed events have to originate in the local universe. In combination with the high rigidity, $R \propto E/Z$, of these events, the deflection in extragalactic and Galactic magnetic fields can be low enough to find anisotropies in the UHE CR arrival direction. Figure 9 shows the distribution of events observed with Auger [123] above 52 EeV (×) and Telescope Array [124] above 57 EeV (+). To highlight feasible anisotropies in the arrival direction of the combined data set, we smooth the events at best-fit position n_i over the sphere with unit vector n following a von Mises–

Fisher distribution, $f_i(n)=a \exp(an \cdot n_i)/ \sinh a/(4_\pi)$. The parameter a is fixed to 11.5 such that 50% of the distribution is contained within an opening angle of 20°. From the smoothed event distribution $f(n) = i\ f_i(n)$, we define $\bar{f}(\delta)$ as the average of the distribution f in each declination bin δ and define the anisotropy as

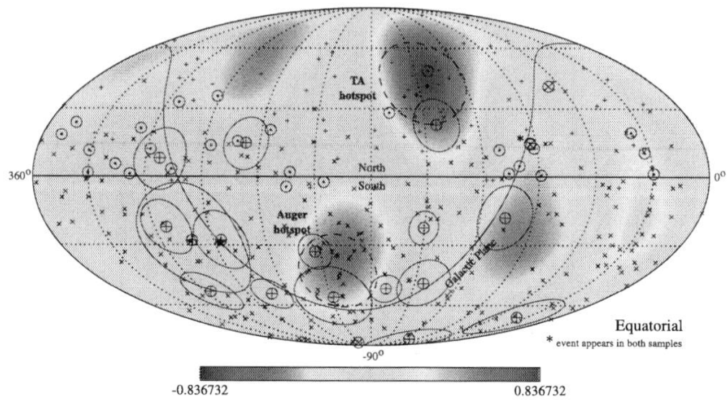

Figure 9. Mollweide projection of the arrival direction of neutrinos and UHE CRs. The neutrino sample is identical to the one shown in Fig. 6. We show events from Auger [123] above 54 EeV (×) and from Telescope Array [124] above 57 EeV (+). The background shows the anisotropy of the combined UHE CR map derived with the method described in the main text and smoothed with with $\theta_{50\%}=20°$. We highlight the excess regions found by Auger (sampling radius of 15°; post-trial p-value of 1.4×10^{-2}) and Telescope Array (sampling radius of 20°; post-trial pp-value of 3.7×10^{-4}).

$$\delta I(\mathbf{n}) = f(\mathbf{n})/\bar{f}(\delta(\mathbf{n})) - 1 \qquad (19)$$

This simple procedure ensures that spurious anisotropies coming from the detector exposures that depend mostly on declination are corrected from the map. The anisotropy map recovers two small-scale excess regions in the Northern and Southern Hemisphere that coincide with the excess regions reported by Telescope Array [124] (sampling radius of 20°; post-trial p-value of 3.7×10^{-4}) and Auger [123] in the direction of Centaurus A (sampling radius of 15°; post-trial pp-value of 1.4×10^{-2}), respectively. These are indicated as dashed circles of different sizes.

Figure 9 also shows the same neutrino event candidates that were shown in Fig. 6. It is apparent that there is no noticeable clustering of high-energy neutrino events in the direction of these hot spots. Indeed, a dedicated analysis [125] by Telescope Array, Auger, and IceCube did not identify significant cross-correlation of neutrino and UHE CR events (below 3.3σ).

However, this does not necessarily rule out the possibility that the events emerge from the same sources. Neutrino events can be observed from all UHE CR sources up to the Hubble horizon c/H0≃4.4. On the other hand, UHE CRs above the energy shown in Fig. 9 have to emerge from local sources up to 200 Mpc. Therefore, we can estimate that only a fraction of 200Mpc/4.4Gpc≃5 % of astrophysical neutrinos should correlate with UHE CRs. The total number of neutrino events shown in Fig. 9 is only 45, so maybe two events are expected to correlate with the anisotropy structure suggested by UHE CRs.

CONCLUSIONS

IceCube has discovered a flux of extragalactic cosmic neutrinos with an energy density that matches that of extragalactic high-energy photons and ultra-high-energy cosmic rays. This may suggest that neutrinos and high-energy cosmic rays share a common origin. They may originate in calorimetric environments like starburst galaxies or galaxy clusters hosting the cosmic ray accelerators. Identification of the sources by observation of multiple neutrino events from these sources with IceCube will be challenging. However, the possibility exists for revealing the sources by the comprehensive IceCube multimessenger program.

Further progress requires larger instruments. We therefore propose as a next step the extraordinary opportunity of instrumenting 10km310km3 of glacial ice at the South Pole and thereby improving on IceCube's sensitive volume by an order of magnitude [126]. This large gain is made possible by the unique optical properties of the Antarctic glacier revealed by the construction of IceCube. As a consequence of the extremely long photon absorption lengths in the deep Antarctic ice, the spacing between strings of light sensors can be increased from 125 m to over 250 m without loss of performance of the instrument. The instrumented volume can therefore grow by one order of magnitude while keeping the construction budget of a next-generation instrument at the level of the cost of the current IceCube detector. The new facility will increase the event rates of cosmic events from hundreds to thousands over several years.

ACKNOWLEDGMENTS

Discussion with collaborators inside and outside the IceCube Collaboration, too many to be listed, have greatly shaped this presentation. Thanks. This research was supported in part by the U.S. National Science Foundation under grants PLR-1600823 and PHY-1607644, and by the University of Wisconsin Research Committee with funds granted by the Wisconsin Alumni Research Foundation.

REFERENCES

1. Aartsen, M. G. et al. , arXiv:1612.05093 [astro-ph.IM] [Search inSPIRE].
2. Ahlers M. and Halzen, F. Rept. Prog. Phys. 78, 126901 (2015). (http://dx.doi.org/10.1088/0034-4885/78/12/126901)
3. Abbott, B. P. et al. , Phys. Rev. Lett. 116, 131103 (2016) [arXiv:1602.03838 [gr-qc]] [Search inSPIRE]. (http://dx.doi.org/10.1103/PhysRevLett.116.131103)
4. Murase, K. Ahlers, M. and Lacki, B. C. Phys. Rev. D 88, 121301 (2013) [arXiv:1306.3417 [astro-ph.HE]] [Search inSPIRE]. (http://dx.doi.org/10.1103/PhysRevD.88.121301)
5. Aartsen, M. G. et al. , Phys. Rev. Lett. 113, 101101 (2014) [arXiv:1405.5303 [astro-ph.HE]] [Search inSPIRE]. (http://dx.doi.org/10.1103/PhysRevLett.113.101101)
6. Aartsen, M. G. et al. , Astrophys. J. 833, 3 (2016) [arXiv:1607.08006 [astro-ph.HE]] [Search inSPIRE]. (http://dx.doi.org/10.3847/0004-637X/833/1/3)
7. Aartsen, M. G. et al. , JINST 11, P11009 (2016) [arXiv:1610.01814 [hep-ex]] [Search inSPIRE]. (https://doi.org/10.1088/1748-0221/11/11/P11009)
8. Aartsen, M. G. et al. , arXiv:1612.06028 [astro-ph.HE] [Search inSPIRE].
9. Schonert, S. Gaisser, T. K. Resconi, E. and Schulz, O. Phys. Rev. D 79, 043009 (2009) [arXiv:0812.4308 [astro-ph]] [Search inSPIRE]. (http://dx.doi.org/10.1103/PhysRevD.79.043009)
10. Ahrens J. et al. [AMANDA Collaboration], Nucl. Instrum. Meth. A 524, 169 (2004) [arXiv:astro-ph/0407044] [Search inSPIRE]. (http://dx.doi.org/10.1016/j.nima.2004.01.065)
11. Aartsen, M. G. et al. , JINST 9, P03009 (2014) [arXiv:1311.4767 [physics.ins-det]] [Search inSPIRE]. (http://dx.doi.org/10.1088/1748-0221/9/03/P03009)
12. Waxman E. and Bahcall, J. N. Phys. Rev. D 59, 023002 (1999) [arXiv:hep-ph/9807282] [Search inSPIRE]. (http://dx.doi.org/10.1103/PhysRevD.59.023002)
13. Bahcall J. N. and Waxman, E. Phys. Rev. D 64, 023002 (2001) [arXiv:hep-ph/9902383] [Search inSPIRE]. (http://dx.doi.org/10.1103/

PhysRevD.64.023002)

14. Aartsen, M. G. et al. , Phys. Rev. Lett. 115, 081102 (2015) [arXiv:1507.04005 [astro-ph.HE]] [Search inSPIRE]. (http://dx.doi.org/10.1103/PhysRevLett.115.081102)
15. Aartsen, M. G. et al. , Phys. Rev. Lett. 111, 021103 (2013) [arXiv:1304.5356 [astro-ph.HE]] [Search inSPIRE]. (http://dx.doi.org/10.1103/PhysRevLett.111.021103)
16. Aartsen, M. G. et al. , Science 342, 1242856 (2013) [arXiv:1311.5238 [astro-ph.HE]] [Search inSPIRE]. (http://dx.doi.org/10.1126/science.1242856)
17. Gaisser, T. K. Jero, K. Karle, A. and van Santen, J. Phys. Rev. D 90, 023009 (2014) [arXiv:1405.0525 [astro-ph.HE]] [Search inSPIRE]. (http://dx.doi.org/10.1103/PhysRevD.90.023009)
18. Aartsen, M. G. et al. , Phys. Rev. Lett. 114, 171102 (2015) [arXiv:1502.03376 [astro-ph.HE]] [Search inSPIRE]. (http://dx.doi.org/10.1103/PhysRevLett.114.171102)
19. Aartsen, M. G. et al. , Astrophys. J. 824, L28 (2016) [arXiv:1605.00163 [astro-ph.HE]] [Search inSPIRE]. (http://dx.doi.org/10.3847/2041-8205/824/2/L28)
20. Halzen F. and Wille, L. Phys. Rev. D 94, 014014 (2016) [arXiv:1605.01409 [hep-ph]] [Search inSPIRE]. (http://dx.doi.org/10.1103/PhysRevD.94.014014)
21. Murase, K. Guetta, D. and Ahlers, M. Phys. Rev. Lett. 116, 071101 (2016) [arXiv:1509.00805 [astro-ph.HE]] [Search inSPIRE]. (http://dx.doi.org/10.1103/PhysRevLett.116.071101)
22. Wang X. and Loeb, A. J. Cosmol. Astropart. Phys. 1612, 012 (2016) [arXiv:1607.06476 [astro-ph.HE]] [Search inSPIRE]. (http://dx.doi.org/10.1088/1475-7516/2016/12/012)
23. Ahlers, M. Bai, Y. Barger, V. and Lu, R. Phys. Rev. D 93, 013009 (2016) [arXiv:1505.03156 [hep-ph]] [Search inSPIRE]. (http://dx.doi.org/10.1103/PhysRevD.93.013009)
24. Aartsen, M. G. et al. , Proc. 34th Int. Cosmic Ray Conf. (ICRC 2015) (2015) [arXiv:1510.05223 [astro-ph.HE]] [Search inSPIRE].
25. Aartsen, M. G. et al. , Phys. Rev. D 89, 102001 (2014) [arXiv:1312.0104 [astro-ph.HE]] [Search inSPIRE]. (http://dx.doi.org/10.1103/PhysRevD.89.102001)
26. Aartsen, M. G. et al. , Phys. Rev. Lett. 110, 151105 (2013)

[arXiv:1212.4760 [hep-ex]] [Search inSPIRE]. (http://dx.doi.org/10.1103/PhysRevLett.110.151105)

27. Aartsen, M. G. et al., Phys. Rev. D 91, 122004 (2015) [arXiv:1504.03753 [astro-ph.HE]] [Search inSPIRE]. (http://dx.doi.org/10.1103/PhysRevD.91.122004)
28. Abbasi R. et al. , Phys. Rev. D 83, 012001 (2011) [arXiv:1010.3980 [astro-ph.HE]] [Search inSPIRE]. (http://dx.doi.org/10.1103/PhysRevD.83.012001)
29. Aartsen, M. G. et al. , Phys. Rev. Lett. 117, 241101 (2016) [arXiv:1607.05886 [astro-ph.HE]] [Search inSPIRE]. (http://dx.doi.org/10.1103/PhysRevLett.117.241101)
30. Gorham, P. W. et al. , Phys. Rev. D 82, 022004 (2010); (http://dx.doi.org/10.1103/PhysRevD.85.049901)85, 049901 (2012) [erratum] [arXiv:1011.5004 [astro-ph.HE]] [Search inSPIRE]. (http://dx.doi.org/10.1103/PhysRevD.82.022004)
31. Aab A. et al. , Phys. Rev. D 91, 092008 (2015) [arXiv:1504.05397 [astro-ph.HE]] [Search inSPIRE]. (http://dx.doi.org/10.1103/PhysRevD.91.092008)
32. Ahlers M. and Murase, K. Phys. Rev. D 90, 023010 (2014) [arXiv:1309.4077 [astro-ph.HE]] [Search inSPIRE]. (http://dx.doi.org/10.1103/PhysRevD.90.023010)
33. Joshi, J. C. Winter, W. and Gupta, N. Mon. Not. Roy. Astron. Soc. 439, 3414 (2014); (http://dx.doi.org/10.1093/mnras/stu2132)446, 892 (2014) [erratum] [arXiv:1310.5123 [astro-ph.HE]] [Search inSPIRE]. (http://dx.doi.org/10.1093/mnras/stu189)
34. Neronov, A. Semikoz, D. V. and Tchernin, C. Phys. Rev. D 89, 103002 (2014) [arXiv:1307.2158 [astro-ph.HE]] [Search inSPIRE]. (http://dx.doi.org/10.1103/PhysRevD.89.103002)
35. Kachelriess M. and Ostapchenko, S. Phys. Rev. D 90, 083002 (2014) [arXiv:1405.3797 [astro-ph.HE]] [Search inSPIRE]. (http://dx.doi.org/10.1103/PhysRevD.90.083002)
36. Gaggero, D. Grasso, D. Marinelli, A. Urbano, A. and Valli, M. Astrophys. J. 815, L25 (2015) [arXiv:1504.00227 [astro-ph.HE]] [Search inSPIRE]. (http://dx.doi.org/10.1088/2041-8205/815/2/L25)
37. Neronov A. and Semikoz, D. V. Astropart. Phys. 75, 60 (2016) [arXiv:1509.03522 [astro-ph.HE]] [Search inSPIRE]. (http://dx.doi.org/10.1016/j.astropartphys.2015.11.002)

38. Fox, D. B. Kashiyama, K. and Mészáros, P. Astrophys. J. 774, 74 (2013) [arXiv:1305.6606 [astro-ph.HE]] [Search inSPIRE]. (http://dx.doi.org/10.1088/0004-637X/774/1/74)
39. Gonzalez-Garcia, M. C. Halzen, F. and Niro, V. Astropart. Phys. 57–58, 39 (2014) [arXiv:1310.7194 [astro-ph.HE]] [Search inSPIRE]. (http://dx.doi.org/10.1016/j.astropartphys.2014.04.001)
40. Anchordoqui, L. A. Goldberg, H. Paul, T. C. da Silva, L. H. M. and Vlcek, B. J. Phys. Rev. D 90, 123010 (2014) [arXiv:1410.0348 [astro-ph.HE]] [Search inSPIRE]. (http://dx.doi.org/10.1103/PhysRevD.90.123010)
41. Razzaque, S. Phys. Rev. D 88, 081302 (2013) [arXiv:1309.2756 [astro-ph.HE]] [Search inSPIRE]. (http://dx.doi.org/10.1103/PhysRevD.88.081302)
42. Lunardini, C. Razzaque, S. Theodoseau, K. T. and Yang, L. Phys. Rev. D 90, 023016 (2014) [arXiv:1311.7188 [astro-ph.HE]] [Search inSPIRE]. (http://dx.doi.org/10.1103/PhysRevD.90.023016)
43. Taylor, A. M. Gabici, S. and Aharonian, F. Phys. Rev. D 89, 103003 (2014) [arXiv:1403.3206 [astro-ph.HE]] [Search inSPIRE]. (http://dx.doi.org/10.1103/PhysRevD.89.103003)
44. Kalashev O. and Troitsky, S. Phys. Rev. D 94, 063013 (2016) [arXiv:1608.07421 [astro-ph.HE]] [Search inSPIRE]. (http://dx.doi.org/10.1103/PhysRevD.94.063013)
45. Feldstein, B. Kusenko, A. Matsumoto, S. and Yanagida, T. T. Phys. Rev. D 88, 015004 (2013) [arXiv:1303.7320 [hep-ph]] [Search inSPIRE]. (http://dx.doi.org/10.1103/PhysRevD.88.015004)
46. Esmaili A. and Serpico, P. D. J. Cosmol. Astropart. Phys. 1311, 054 (2013) [arXiv:1308.1105 [hep-ph]] [Search inSPIRE]. (http://dx.doi.org/10.1088/1475-7516/2013/11/054)
47. Bai, Y. Lu, R. and Salvado, J. J. High Energy Phys. 01, 161 (2016) [arXiv:1311.5864 [hep-ph]] [Search inSPIRE]. (http://dx.doi.org/10.1007/JHEP01(2016)161)
48. Murase, K. Laha, R. Ando, S. and Ahlers, M. Phys. Rev. Lett. 115, 071301 (2015) [arXiv:1503.04663 [hep-ph]] [Search inSPIRE]. (http://dx.doi.org/10.1103/PhysRevLett.115.071301)
49. Boucenna, S. M. Chianese, M. Mangano, G. Miele, G. Morisi, S. Pisanti, O. and Vitagliano, E. J. Cosmol. Astropart. Phys. 1512, 055 (2015) [arXiv:1507.01000 [hep-ph]] [Search inSPIRE]. (http://dx.doi.org/10.1088/1475-7516/2015/12/055)

50. Chianese, M. Miele, G. and Morisi, S. J. Cosmol. Astropart. Phys. 1701, 007 (2017) [arXiv:1610.04612 [hep-ph]] [Search inSPIRE]. (http://dx.doi.org/10.1088/1475-7516/2017/01/007)
51. Cohen, T. Murase, K. Rodd, N. L. Safdi, B. R. and Soreq, Y. arXiv:1612.05638 [hep-ph] [Search inSPIRE].
52. Gupta, N. Astropart. Phys. 48, 75 (2013) [arXiv:1305.4123 [astro-ph.HE]] [Search inSPIRE]. (http://dx.doi.org/10.1016/j.astropartphys.2013.07.003)
53. Loeb A. and Waxman, E. J. Cosmol. Astropart. Phys. 0605, 003 (2006) [arXiv:astro-ph/0601695] [Search inSPIRE]. (http://dx.doi.org/10.1088/1475-7516/2006/05/003)
54. He, H. -N. Wang, T. Fan, Y. -Z. Liu, S. -M. and Wei, D. -M. Phys. Rev. D 87, 063011 (2013) [arXiv:1303.1253 [astro-ph.HE]] [Search inSPIRE]. (http://dx.doi.org/10.1103/PhysRevD.87.063011)
55. Anchordoqui, L. A. Paul, T. C. da Silva, L. H. M. Torres, D. F. and Vlcek, B. J. Phys. Rev. D 89, 127304 (2014) [arXiv:1405.7648 [astro-ph.HE]] [Search inSPIRE]. (http://dx.doi.org/10.1103/PhysRevD.89.127304)
56. Chang X. -C. and Wang, X. -Y. Astrophys. J. 793, 131 (2014) [arXiv:1406.1099 [astro-ph.HE]] [Search inSPIRE]. (http://dx.doi.org/10.1088/0004-637X/793/2/131)
57. Tamborra, I. Ando, S. and Murase, K. J. Cosmol. Astropart. Phys. 1409, 043 (2014) [arXiv:1404.1189 [astro-ph.HE]] [Search inSPIRE]. (http://dx.doi.org/10.1088/1475-7516/2014/09/043)
58. Bechtol, K. Ahlers, M. Di Mauro, M. Ajello, M. and Vandenbroucke, J. arXiv:1511.00688 [astro-ph.HE] [Search inSPIRE].
59. Senno, N. P. Mészáros, Murase, K. Baerwald, P. and Rees, M. J. Astrophys. J. 806, 24 (2015) [arXiv:1501.04934 [astro-ph.HE]] [Search inSPIRE]. (http://dx.doi.org/10.1088/0004-637X/806/1/24)
60. Stecker, F. W. Done, C. Salamon, M. H. and Sommers, P. Phys. Rev. Lett. 66, 2697 (1991). (http://dx.doi.org/10.1103/PhysRevLett.66.2697)
61. Stecker, F. W. Phys. Rev. D 88, 047301 (2013) [arXiv:1305.7404 [astro-ph.HE]] [Search inSPIRE]. (http://dx.doi.org/10.1103/PhysRevD.88.047301)
62. Kalashev, O. Semikoz, D. and Tkachev, I. J. Exp. Theor. Phys. 120, 541 (2015) [arXiv:1410.8124 [astro-ph.HE]] [Search inSPIRE]. (http://dx.doi.org/10.1134/S106377611503022X)
63. Bai, Y. Barger, A. J. Barger, V. Lu, R. Peterson, A. D. and Salvado,

J. Phys. Rev. D 90, 063012 (2014) [arXiv:1407.2243 [astro-ph.HE]] [Search inSPIRE]. (http://dx.doi.org/10.1103/PhysRevD.90.063012)

64. Kimura, S. S. Murase, K. and Toma, K. Astrophys. J. 806, 159 (2015) [arXiv:1411.3588 [astro-ph.HE]] [Search inSPIRE]. (http://dx.doi.org/10.1088/0004-637X/806/2/159)

65. Tavecchio F. and Ghisellini, G. Mon. Not. Roy. Astron. Soc. 451, 1502 (2015) [arXiv:1411.2783 [astro-ph.HE]] [Search inSPIRE]. (http://dx.doi.org/10.1093/mnras/stv1023)

66. Padovani P. and Resconi, E. Mon. Not. Roy. Astron. Soc. 443, 474 (2014) [arXiv:1406.0376 [astro-ph.HE]] [Search inSPIRE]. (http://dx.doi.org/10.1093/mnras/stu1166)

67. Dermer, C. D. Murase, K. and Inoue, Y. J. High Energy Astrophys. 3–4, 29 (2014) [arXiv:1406.2633 [astro-ph.HE]] [Search inSPIRE]. (http://dx.doi.org/10.1016/j.jheap.2014.09.001)

68. Petropoulou, M. Dimitrakoudis, S. Padovani, P. Mastichiadis, A. and Resconi, E. Mon. Not. Roy. Astron. Soc. 448, 2412 (2015) [arXiv:1501.07115 [astro-ph.HE]] [Search inSPIRE]. (http://dx.doi.org/10.1093/mnras/stv179)

69. Padovani, P. Petropoulou, M. Giommi, P. and Resconi, E. Mon. Not. Roy. Astron. Soc. 452, 1877 (2015) [arXiv:1506.09135 [astro-ph.HE]] [Search inSPIRE]. (http://dx.doi.org/10.1093/mnras/stv1467)

70. Kadler M. et al., Nature Phys. 12, 807 (2016) [arXiv:1602.02012 [astro-ph.HE]] [Search inSPIRE]. (http://dx.doi.org/10.1038/nphys3715)

71. Padovani, P. Resconi, E. Giommi, P. Arsioli, B. and Chang, Y. L. Mon. Not. Roy. Astron. Soc. 457, 3582 (2016) [arXiv:1601.06550 [astro-ph.HE]] [Search inSPIRE]. (http://dx.doi.org/10.1093/mnras/stw228)

72. Neronov, A. Semikoz, D. V. and Ptitsyna, K. arXiv:1611.06338 [astro-ph.HE] [Search inSPIRE].

73. Waxman E. and Bahcall, J. N. Phys. Rev. Lett. 78, 2292 (1997) [arXiv:astro-ph/9701231] [Search inSPIRE]. (http://dx.doi.org/10.1103/PhysRevLett.78.2292)

74. Ando S. and Beacom, J. F. Phys. Rev. Lett. 95, 061103 (2005) [arXiv:astro-ph/0502521] [Search inSPIRE]. (http://dx.doi.org/10.1103/PhysRevLett.95.061103)

75. Murase K. and Ioka, K. Phys. Rev. Lett. 111, 121102 (2013) [arXiv:1306.2274 [astro-ph.HE]] [Search inSPIRE]. (http://dx.doi.org/10.1103/PhysRevLett.111.121102)

76. Tamborra I. and Ando, S. Phys. Rev. D 93, 053010 (2016) [arXiv:1512.01559 [astro-ph.HE]] [Search inSPIRE]. (http://dx.doi.org/10.1103/PhysRevD.93.053010)
77. Meszaros P. and Waxman, E. Phys. Rev. Lett. 87, 171102 (2001) [arXiv:astro-ph/0103275] [Search inSPIRE]. (http://dx.doi.org/10.1103/PhysRevLett.87.171102)
78. Senno, N. Murase, K. and Meszaros, P. Phys. Rev. D 93, 083003 (2016) [arXiv:1512.08513 [astro-ph.HE]] [Search inSPIRE]. (http://dx.doi.org/10.1103/PhysRevD.93.083003)
79. Dado S. and Dar, A. Phys. Rev. Lett. 113, 191102 (2014) [arXiv:1405.5487 [astro-ph.HE]] [Search inSPIRE]. (http://dx.doi.org/10.1103/PhysRevLett.113.191102)
80. Kashiyama K. and Mészáros, P. Astrophys. J. 790, L14 (2014) [arXiv:1405.3262 [astro-ph.HE]] [Search inSPIRE]. (http://dx.doi.org/10.1088/2041-8205/790/1/L14)
81. Berezinsky, V. S. Blasi, P. and Ptuskin, V. S. Astrophys J. 487, 529 (1997) [arXiv:astro-ph/9609048] [Search inSPIRE]. (http://dx.doi.org/10.1086/304622)
82. Murase, K. Inoue, S. and Nagataki, S. Astrophys. J. 689, L105 (2008) [arXiv:0805.0104 [astro-ph]] [Search inSPIRE]. (http://dx.doi.org/10.1086/595882)
83. Zandanel, F. Tamborra, I. Gabici, S. and Ando, S. Astron. Astrophys. 578, A32 (2015) [arXiv:1410.8697 [astro-ph.HE]] [Search inSPIRE]. (http://dx.doi.org/10.1051/0004-6361/201425249)
84. Wang X. -Yu and Liu, R. -Yu Phys. Rev. D 93, 083005 (2016) [arXiv:1512.08596 [astro-ph.HE]] [Search inSPIRE]. (http://dx.doi.org/10.1103/PhysRevD.93.083005)
85. Senno, N. Murase, K. and Meszaros, P. arXiv:1612.00918 [astro-ph.HE] [Search inSPIRE].
86. Dai L. and Fang, K. arXiv:1612.00011 [astro-ph.HE] [Search inSPIRE].
87. Lunardini C. and Winter, W. arXiv:1612.03160 [astro-ph.HE] [Search inSPIRE].
88. Roulet, E. Sigl, G. van Vliet, A. and Mollerach, S. J. Cosmol. Astropart. Phys. 1301, 028 (2013) [arXiv:1209.4033 [astro-ph.HE]] [Search inSPIRE]. (http://dx.doi.org/10.1088/1475-7516/2013/01/028)
89. Yacobi, L. Guetta, D. and Behar, E. Astrophys. J. 823, 89 (2016) [arXiv:1510.01244 [astro-ph.HE]] [Search inSPIRE]. (http://dx.doi.

org/10.3847/0004-637X/823/2/89)

90. Kelner, S. R. Aharonian, F. A. and Bugayov, V. V. Phys. Rev. D 74, 034018 (2006); (http://dx.doi.org/10.1103/PhysRevD.79.039901)79, 039901 (2009) [erratum] [arXiv:astro-ph/0606058] [Search inSPIRE]. (http://dx.doi.org/10.1103/PhysRevD.74.034018)
91. Aartsen, M. G. et al. , arXiv:1609.04981 [astro-ph.HE] [Search inSPIRE].
92. Patrignani C. et al. , Chin. Phys. C 40, 100001 (2016). (http://dx.doi.org/10.1088/1674-1137/40/10/100001)
93. Ahlers M. and Halzen, F. Phys. Rev. D 90, 043005 (2014) [arXiv:1406.2160 [astro-ph.HE]] [Search inSPIRE]. (http://dx.doi.org/10.1103/PhysRevD.90.043005)
94. Mertsch, P. Rameez, M. and Tamborra, I. arXiv:1612.07311 [astro-ph.HE] [Search inSPIRE].
95. Ahlers M. and Halzen, F. Phys. Rev. D 86, 083010 (2012) [arXiv:1208.4181 [astro-ph.HE]] [Search inSPIRE]. (http://dx.doi.org/10.1103/PhysRevD.86.083010)
96. Katz, B. Waxman, E. Thompson, T. and Loeb, A. arXiv:1311.0287 [astro-ph.HE] [Search inSPIRE].
97. Greisen, K. Phys. Rev. Lett. 16, 748 (1966). (http://dx.doi.org/10.1103/PhysRevLett.16.748)
98. Zatsepin G. T. and Kuzmin, V. A. JETP Lett. 4, 78 (1966).
99. Berezinsky V. S. and Zatsepin, G. T. Phys. Lett. B 28, 423 (1969). (http://dx.doi.org/10.1016/0370-2693(69)90341-4)
100. Yoshida S. and Teshima, M. Prog. Theor. Phys. 89, 833 (1993). (http://dx.doi.org/10.1143/ptp/89.4.833)
101. Protheroe R. J. and Johnson, P. A. Astropart. Phys. 4, 253 (1996) [arXiv:astro-ph/9506119] [Search inSPIRE]. (http://dx.doi.org/10.1016/0927-6505(95)00039-9)
102. Engel, R. Seckel, D. and Stanev, T. Phys. Rev. D 64, 093010 (2001) [arXiv:astro-ph/0101216] [Search inSPIRE]. (http://dx.doi.org/10.1103/PhysRevD.64.093010)
103. Berezinsky, V. Gazizov, A. Z. and Grigorieva, S. I. Phys. Rev. D 74, 043005 (2006) [arXiv:hep-ph/0204357] [Search inSPIRE]. (http://dx.doi.org/10.1103/PhysRevD.74.043005)
104. Fodor, Z. Katz, S. D. Ringwald, A. and Tu, H. J. Cosmol. Astropart.

Phys. 0311, 015 (2003) [arXiv:hep-ph/0309171] [Search inSPIRE]. (http://dx.doi.org/10.1088/1475-7516/2003/11/015)

105. Yuksel H. and Kistler, M. D. Phys. Rev. D 75, 083004 (2007) [arXiv:astro-ph/0610481] [Search inSPIRE]. (http://dx.doi.org/10.1103/PhysRevD.75.083004)
106. Takami, H. Murase, K. Nagataki, S. and Sato, K. Astropart. Phys. 31, 201 (2009) [arXiv:0704.0979 [astro-ph]] [Search inSPIRE]. (http://dx.doi.org/10.1016/j.astropartphys.2009.01.006)
107. Abdo, A. A. et al., Phys. Rev. Lett. 104, 101101 (2010) [arXiv:1002.3603 [astro-ph.HE]] [Search inSPIRE]. (http://dx.doi.org/10.1103/PhysRevLett.104.101101)
108. Ackermann M. et al. , Astrophys. J. 799, 86 (2015) [arXiv:1410.3696 [astro-ph.HE]] [Search inSPIRE]. (http://dx.doi.org/10.1088/0004-637X/799/1/86)
109. Berezinsky, V. Gazizov, A. Kachelriess, M. and Ostapchenko, S. Phys. Lett. B 695, 13 (2011) [arXiv:1003.1496 [astro-ph.HE]] [Search inSPIRE]. (http://dx.doi.org/10.1016/j.physletb.2010.11.019)
110. Ahlers, M. Anchordoqui, L. A. Gonzalez-Garcia, M. C. Halzen, F. and Sarkar, S. Astropart. Phys. 34, 106 (2010) [arXiv:1005.2620 [astro-ph.HE]] [Search inSPIRE]. (http://dx.doi.org/10.1016/j.astropartphys.2010.06.003)
111. Gelmini, G. B. Kalashev, O. and Semikoz, D. V. J. Cosmol. Astropart. Phys. 1201, 044 (2012) [arXiv:1107.1672 [astro-ph.CO]] [Search inSPIRE]. (http://dx.doi.org/10.1088/1475-7516/2012/01/044)
112. Decerprit G. and Allard, D. Astron. Astrophys. 535, A66 (2011) [arXiv:1107.3722 [astro-ph.HE]] [Search inSPIRE]. (http://dx.doi.org/10.1051/0004-6361/201117673)
113. Heinze, J. Boncioli, D. Bustamante, M. and Winter, W. Astrophys. J. 825, 122 (2016) [arXiv:1512.05988 [astro-ph.HE]] [Search inSPIRE]. (http://dx.doi.org/10.3847/0004-637X/825/2/122)
114. Supanitsky, A. D. Phys. Rev. D 94, 063002 (2016) [arXiv:1607.00290 [astro-ph.HE]] [Search inSPIRE]. (http://dx.doi.org/10.1103/PhysRevD.94.063002)
115. Hooper, D. Taylor, A. and Sarkar, S. Astropart. Phys. 23, 11 (2005) [arXiv:astro-ph/0407618] [Search inSPIRE]. (http://dx.doi.org/10.1016/j.astropartphys.2004.11.002)
116. Ave, M. Busca, N. Olinto, A. V. Watson, A. A. and Yamamoto, T.

Astropart. Phys. 23, 19 (2005) [arXiv:astro-ph/0409316] [Search inSPIRE]. (http://dx.doi.org/10.1016/j.astropartphys.2004.11.001)

117. Hooper, D. Sarkar, S. and Taylor, A. M. Astropart. Phys. 27, 199 (2007) [arXiv:astro-ph/0608085] [Search inSPIRE]. (http://dx.doi.org/10.1016/j.astropartphys.2006.10.008)
118. Allard, D. Ave, M. Busca, N. Malkan, M. A. Olinto, A. V. Parizot, E. Stecker, F. W. and Yamamoto, T. J. Cosmol. Astropart. Phys. 0609, 005 (2006) [arXiv:astro-ph/0605327]. (http://dx.doi.org/10.1088/1475-7516/2006/09/005)
119. Anchordoqui, L. A. Goldberg, H. Hooper, D. Sarkar, S. and Taylor, A. M. Phys. Rev. D 76, 123008 (2007) [arXiv:0709.0734 [astro-ph]] [Search inSPIRE]. (http://dx.doi.org/10.1103/PhysRevD.76.123008)
120. Aloisio, R. Berezinsky, V. and Gazizov, A. Astropart. Phys. 34, 620 (2011) [arXiv:0907.5194 [astro-ph.HE]] [Search inSPIRE]. (http://dx.doi.org/10.1016/j.astropartphys.2010.12.008)
121. Kotera, K. Allard, D. and Olinto, A. V. J. Cosmol. Astropart. Phys. 1010, 013 (2010) [arXiv:1009.1382 [astro-ph.HE]] [Search inSPIRE]. (http://dx.doi.org/10.1088/1475-7516/2010/10/013)
122. Ahlers M. and Salvado, J. Phys. Rev. D 84, 085019 (2011) [arXiv:1105.5113 [astro-ph.HE]] [Search inSPIRE]. (http://dx.doi.org/10.1103/PhysRevD.84.085019)
123. Aab A. et al. , Astrophys. J. 804, 15 (2015) [arXiv:1411.6111 [astro-ph.HE]] [Search inSPIRE]. (http://dx.doi.org/10.1088/0004-637X/804/1/15)
124. Abbasi, R. U. et al. , Astrophys. J. 790, L21 (2014) [arXiv:1404.5890 [astro-ph.HE]] [Search inSPIRE]. (http://dx.doi.org/10.1088/2041-8205/790/2/L21)
125. Aartsen, M. G. et al. , J. Cosmol. Astropart. Phys. 1601, 037 (2016) [arXiv:1511.09408 [astro-ph.HE]] [Search inSPIRE]. (http://dx.doi.org/10.1088/1475-7516/2016/01/037)
126. Aartsen, M. G. et al. , arXiv:1510.05228 [astro-ph.IM] [Search inSPIRE].

SECTION 6
NEUTRINOS IN COSMOLOGY

CHAPTER 21

Constraining Sterile Neutrino Cosmology With Terrestrial Oscillation Experiments

Jeffrey M. Berryman

Center for Neutrino Physics, Department of Physics, Virginia Tech, Blacksburg, Virginia 24061, USA

INTRODUCTION

As the precision of neutrino experiments continues to improve, the importance of interdisciplinary studies grows—it becomes increasingly possible and imperative to determine the extent to which the three-neutrino oscillation framework can simultaneously describe a broad set of experimental results. In many ways, neutrino oscillations are already an interdisciplinary endeavor: neutrinos of different origins (solar, reactor, accelerator, atmospheric, cosmogenic, etc.) over orders of magnitude of energy have been studied in a common framework. As a result, synergies and tensions [1] have arisen between different sectors and between experiments within the same sector. Understanding the origins of these tensions is a central issue.

Citation: Jeffrey M. Berryman. "Constraining sterile neutrino cosmology with terrestrial oscillation experiments" Phys. Rev. D 100, 023540 (2019). https://doi.org/10.1103/PhysRevD.100.023540

One proposed solution to some (but not all) of these tensions is that additional species of neutrinos with eV-scale masses exist. The evidence driving this derives from anomalous measurements of electron-neutrino disappearance [2–5] and electron-neutrino appearance [6,7] (see, for instance, Refs. [8–14] for more details). It is important to note that this is not a silver-bullet solution to these anomalies; even among the subset of anomalous experimental results, a consistent description in terms of additional neutrinos is lacking, as has been explored in, e.g., Ref. [12].

In addition to the overall consistency of dedicated oscillation experiments, astrophysics and cosmology present stiff challenges to the proposed existence of additional neutrinos. The literature concerning the role of sterile neutrinos[1] in cosmology is vast; see Refs. [15–25] and references therein. Introducing these additional, light degrees of freedom to resolve oscillation anomalies necessarily invokes constraints from an ostensibly disconnected field of study. The objectives are to (a) understand how different sets of constraints complement each other, and (b) determine the essential characteristics of a solution to all sets of constraints.

In this work, we focus on the first of these two objectives. Our goals are to determine how constraints on sterile neutrinos derived from terrestrial oscillation experiments translate into constraints on their cosmological properties and whether these improve on constraints from dedicated astrophysics/cosmology experiments. The focus will be on accelerator and reactor (anti)neutrino experiments. Additionally, the observation of coherent elastic neutrino-nucleus scattering (CEνNS) by the COHERENT Collaboration [26] has opened a new means by which to study neutrinos and their interactions [27–31]. We will consider the sensitivities of several proposed experiments that rely on this process which exist to test the sterile-neutrino interpretation of reactor antineutrino anomalies.

This manuscript is organized as follows. In Sec. II, we review oscillations involving a sterile neutrino and the terrestrial constraints considered are discussed in Sec. III. The framework with which we study the cosmology of a sterile neutrino is presented in Sec. IV, and our results are given in Sec. V. We offer concluding thoughts in Sec. VI.

OSCILLATIONS WITH STERILE NEUTRINOS

Oscillations with three neutrinos can be extended to include a fourth neutrino by generalizing the 3×3 Pontecorvo-Maki-Nakagawa-Sakata (PMNS) matrix to a 4×4, unitary matrix, U, whose matrix elements we denote $U_{\alpha i}$ with α=e, μ, τ, s and i=1, 2, 3, 4. We need not concern ourselves with the parametrization of the entire matrix (see, for instance, Ref. [32] for more details); the relevant elements of this matrix, for this study, are

$$|U_{e4}|^2 = s_{14}^2 \tag{2.1}$$

$$|U_{\mu 4}|^2 = s_{24}^2 c_{14}^2 \tag{2.2}$$

where we define $s_{ij} \equiv \sin\phi_{ij}$ and $c_{ij} \equiv \cos\phi_{ij}$, where ϕ_{ij} are the mixing angles that parametrize U.

The probability that a neutrino of initial flavor α (=e, μ, τ, s) propagating in vacuum will be detected[2] with flavor β(=e,μ,τ,s) is

$$P_{\alpha\beta} = |\delta_{\alpha\beta} - U_{\alpha 2}U_{\beta 2}^*(1 - e^{-i\Delta_{21}}) - U_{\alpha 3}U_{\beta 3}^*(1 - e^{-i\Delta_{31}}) - U_{\alpha 4}U_{\beta 4}^*(1 - e^{-i\Delta_{41}})|^2, \tag{2.3}$$

with $\Delta_{ij} \equiv 2.54(\Delta m^2_{ij}/\text{eV}^2)(L/\text{km})(\text{GeV}/E\nu)$, where $\Delta m_{ij}^2 \equiv m_i^2 - m_j^2$ is the difference in neutrino masses squared, L is the distance traveled, and Eν is the energy. In the limit $\Delta_{31}, \Delta_{21} \ll 1$, Eq. (2.3) simplifies to

$$P_{\alpha\beta} \approx |\delta_{\alpha\beta} - U_{\alpha 4}U_{\beta 4}^*(1 - e^{-i\Delta_{41}})|^2 = \begin{cases} 1 - \sin^2 2\theta_{\alpha\alpha}\sin^2\frac{\Delta_{41}}{2} & (\alpha = \beta) \\ \sin^2 2\theta_{\alpha\beta}\sin^2\frac{\Delta_{41}}{2} & (\alpha \neq \beta) \end{cases},$$

where have employed the effective mixing angles

$$\sin^2 2\theta_{\alpha\alpha} \equiv 4|U_{\alpha 4}|^2(1 - |U_{\alpha 4}|^2), \tag{2.4}$$

$$\sin^2 2\theta_{\alpha\beta} \equiv 4|U_{\alpha 4}|^2|U_{\beta 4}|^2 \tag{2.5}$$

The quantities $\sin^2 2\theta_{ee}$ and $\sin^2 2\theta_{\mu\mu}$, as well as the mass-squared splitting Δm^2_{41}, are the primary subjects of this work.

We are interested in scenarios in which only one of the active-sterile mixing angles is nonzero; in particular, we are concerned with either nonzero $\phi_{14}(=\theta_{ee})$ or $\phi_{24}(=\theta_{\mu\mu})$.[3] The reason for this is that the formalism we introduce in Sec. IV relies on the two-flavor-oscillations approximation in the early universe. This approximation is violated if two active-sterile mixing angles are simultaneously nonzero; our framework allows for us to translate bounds on $\sin^2 2\theta_{ee}$ and $\sin^2 2\theta_{\mu\mu}$ into cosmology parameter space, but not bounds on $\sin^2 2\theta_{e\mu}$. In other words, we study neutrino *disappearance* anomalies, but not neutrino *appearance* anomalies.

When neutrinos propagate through matter, their propagation is altered by a background matter potential [33]. The propagation Hamiltonian receives an additional contribution that is diagonal in the flavor basis, $(\delta H)_{\alpha\beta}=\sqrt{2}GF\times diag(n_e,0,0,12nn)$, where ne and n are, respectively, the electron and neutron number densities along the path of propagation. In the mass basis, this potential becomes $(\delta H)_{ij}=U*\alpha_i(\delta H)_{\alpha\beta}U_{\beta j}$; it is no longer diagonal, rendering the mass-basis propagation Hamiltonian similarly nondiagonal. For antineutrinos, we must replace $(\delta H)_{\alpha\beta}\rightarrow-(\delta H)_{\alpha\beta}$, and $U\leftrightarrow U*$.

Two key assumptions in writing down this matter potential are (i) that interactions of neutrinos with the background matter are forward, coherent, and elastic, and (ii) that neutrinos do not interact with the potential produced by other neutrinos. There are physical systems in which these assumptions do not hold—for instance, neutrinos propagating out of supernovae [34]—and the resulting phenomenon is referred to as collective oscillations [35]. In Sec. IV, we discuss another such system: neutrinos in the early universe.

TERRESTRIAL CONSTRAINTS ON A FOURTH NEUTRINO

We discuss the neutrino experiments that we use to constrain sterile neutrino cosmology. There is a great deal of activity in the literature regarding searches for a fourth neutrino (see, for instance, Refs. [8,12,36,37], and references therein), but we have chosen a subset that we believe provides the most value in terms of illuminating the connection between terrestrial neutrino oscillation experiments and early universe cosmology.

Accelerator Neutrino Experiments

Neutrinos for the MINOS experiment [38,39] are produced at the Fermilab Main Injector by protons incident on a graphite target; they are directed toward a near detector and a far detector that are 1.4 km and 735 km away, respectively. We show in red the 90% C.L. constraint derived in Ref. [40].

We also show the sensitivities of two proposed, long-baseline accelerator neutrino experiments, estimated using Monte Carlo simulations. The Deep Underground Neutrino Experiment (DUNE) [41,42] is a proposed 34-kton liquid argon detector located 1300 km from the Fermilab main injector. The

Hyper-Kamiokande (Hyper-K) experiment [43–45] consists of two water Cerenkov detectors with a combined fiducial mass of 0.56 Mton located 295 km from the Japan Proton Accelerator Research Complex (J-PARC). We discuss our simulations of these experiments in the Appendix A. The resulting 95% C.L. sensitivities for DUNE (long-dashed green curve) and Hyper-K (short-dashed purple curve) in the $\sin^2 2\theta_{\mu\mu} - \Delta m^2_{41}$ plane are shown in Fig. 1.

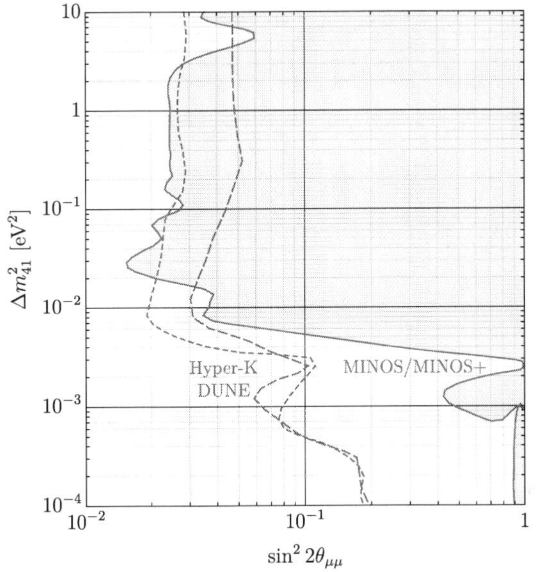

Figure 1. Constraints in the $\sin^2 2\theta_{\mu\mu}$-Δm^2_{41} plane from accelerator neutrino experiments. The red, shaded region is excluded at 90% C.L. by MINOS/MINOS+. Also shown are the expected 95% C.L. sensitivities of DUNE (long-dashed green curve) and Hyper-K (short-dashed purple curve).

Reactor Antineutrino Experiments

The Detector of AntiNeutrino based on Solid Scintillator (DANSS) project [46–48] consists of an array of gadolinium-coated plastic scintillators in an experimental hall beneath a reactor at the Kalinin Nuclear Power Plant. The detector is placed in a lifting system that allows the distance between the reactor and the detector to vary between 9.7 m and 12.2 m. The cyan curve in Fig. 2 shows the sensitivity (95% C.L.) of DANSS to the presence of a sterile neutrino as presented in Ref. [48].

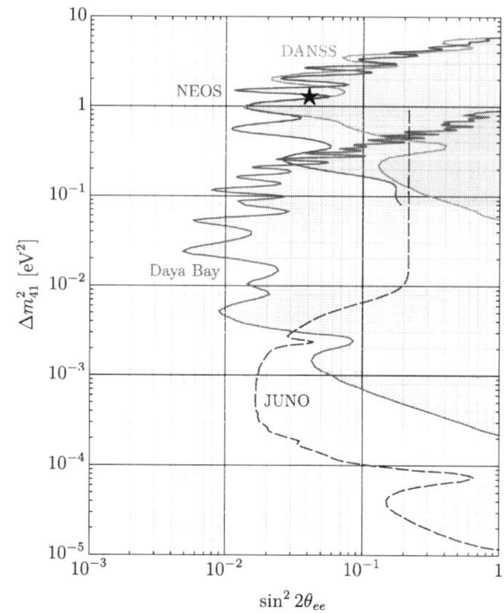

Figure 2. Constraints in the $\sin^2 2\theta_{ee}$-Δm^2_{41} plane from the reactor antineutrino experiments. The shaded regions are currently excluded by DANSS (cyan, 95% C.L.), Daya Bay (red, 95% C.L.), and NEOS (purple, 90% C.L.). Also shown is the sensitivity reach of JUNO (long-dashed blue curve, 95% C.L.). The black, five-pointed start represents the best-fit point from the global analysis of Ref. [12].

The Daya Bay experiment [49] is a collection of eight antineutrino gadolinium-doped liquid scintillator antineutrino detectors observing the antineutrinos produced by six reactor cores located in southern China. The sensitivity of Daya Bay to a sterile neutrino was investigated in Ref. [50], and the resulting exclusion limit (95% C.L.) in the $\sin^2 2\theta ee$-Δm^2_{41} plane is reproduced in red in Fig. 2. An updated dataset has recently been released [51,52], but a sterile-neutrino analysis has not yet been performed.

The NEOS experiment [53] consists of a single gadolinium-doped liquid scintillator detector located 24 m from a reactor core in southwestern South Korea. A sterile neutrino search was performed in Ref. [53], and the resulting limit (90% C.L.) is shown in purple in Fig. 2.

The Jiangmen Underground Neutrino Observatory (JUNO) experiment [54] is a proposed 20-kton liquid scintillator detector whose primary physics goal is to determine the neutrino mass hierarchy. The sensitivity of JUNO to the presence of sterile neutrinos was also studied in Ref. [54]; the result (95% C.L.) is shown in long-dashed blue in

Fig. 2 assuming a normal hierarchy of both active and sterile neutrino masses, i.e., assuming the neutrino masses are ordered $m_1<m_3$ and $m_1<m_4$.

The next few years will see a number of new short-baseline results from STEREO [55], SoLid [56], and PROSPECT [57], the first and last of which have already started collecting data. However, to avoid clutter in Fig. 2 and in our results in Sec. V, we do not consider these in our analysis. While they will ultimately be competitive in the $\Delta m^2_{41}\sim O(1-10)$ eV^2 region, they have not yet produced world-leading limits. Constraints on sterile neutrinos mixing with $\nu e/\bar{\nu}e$ from radioactive source [58–60], solar [61–72], and carbon-scattering [73,74] experiments can be nontrivial, but we do not consider these.

A global fit of the 3+1 neutrino oscillation framework to short-baseline oscillation data was recently performed in Ref. [12]. The authors find that the best-fit point to global $\bar{\nu}e$ disappearance data—including reactor, gallium, solar, and β-decay experiments—with unfixed reactor fluxes $\{\Delta m^2_{41},\sin^2 2\theta_{ee}\}=\{1.3\ \text{eV}^2, 4.04\times10^{-2}\}$. This point is indicated by the black star in Fig. 2.

Low-Threshold Experiments

The possibility of observing active-sterile oscillations through coherent elastic neutrino-nucleus scattering (CEνNS) has previously been discussed in Refs. [27–31]; these provide a complementary probe of the $\sin^2 2\theta_{ee}$-Δm^2_{41} space to reactor antineutrino experiments. In fact, the experiments that we will consider are also based at nuclear reactors, but we separate these from those of the previous subsection because the underlying signal process— $CE_\nu NS$—is distinct from inverse beta decay (IBD).

Reference [30] reports a constraint in the $\sin^2 2\theta_{ee}$-Δm^2_{41} plane from a combined analysis of $\bar{\nu}_e$ scattering data from the Krasnoyarsk [75], Rovno [76], MUNU [77], and TEXONO [78] experiments.[4] We reproduce the resulting exclusion in purple in Fig. 3. This constraint is quite weak relative to the other experiments in the figure, but it is the only such analysis that can currently exclude any portion of this parameter space.

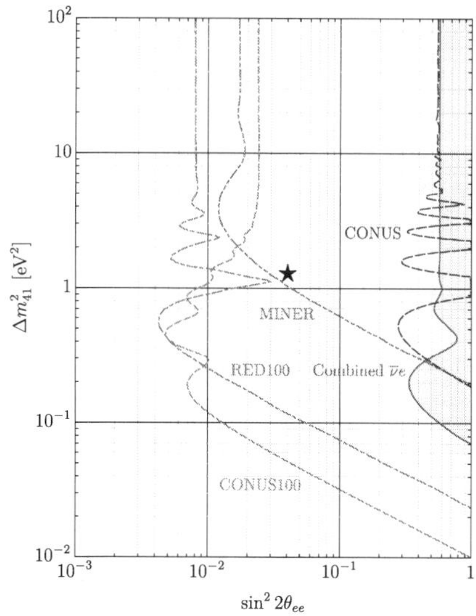

Figure 3. Constraints in the $\sin^2 2\theta_{ee}$-Δm^2_{41} plane from low-threshold experiments.

The purple, shaded region is excluded at 90% C.L. by $\bar{\nu}_e$ scattering [30]. Also shown are the expected sensitivities of RED100 (dot-dashed red curve, 90% C.L.), MINER (double-dot–dashed green curve, 90% C.L.), CONUS (long-dashed blue curve, 95% C.L.), and CONUS100 (short-dashed light blue curve, 95% C.L.). The black, five-pointed start represents the best-fit point from the global analysis of Ref. [12].

We focus on a subset the many (existing and proposed) CEνNS experiments [79–86], starting with the RED100 [87] and MINER [88] proposals. The sensitivities of these experiments to sterile neutrinos were studied in Ref. [30]. Several benchmark scenarios were considered for each of these experiments; we consider the most aggressive scenarios to assess how these experiments fare under the most optimistic assumptions. We take baselines of 15 m for RED100 and 1 m for MINER and assume a 100% efficiency for each. The 90% C.L. sensitivities for RED100 and MINER are reproduced in dot-dashed red and double-dot–dashed green curves, respectively, in Fig. 3.

The sensitivity of the COHERENT experiment [26] to sterile neutrinos has been studied in Refs. [31,86]. However, COHERENT is more sensitive to a sterile neutrino mixing with $\nu\mu/\bar{\nu}_e$ than it is to mixing with $\nu e/\bar{\nu}_e$. Consequently, we do not include COHERENT in our analysis. We remark, however, that near-term expansions to the neutrino program at the Spallation

Neutron Source could yield meaningful new constraints in this parameter space [89].

Last, we consider the CONUS experiment [90], for which an official exclusion result does not yet exist. We estimate the sensitivity of the CONUS experiment to oscillations involving a sterile neutrino following the procedures employed in Refs. [91,92]; we present this analysis fully in Appendix B, but provide some relevant details here. We consider two benchmark configurations for CONUS. The first is the nominal CONUS configuration, consisting of 4 kg of natural germanium and a recoil threshold of 1.2 keV taking data over one year; we call this configuration "CONUS." The second is a more aggressive configuration, consisting of 100 kg of 88% enriched germanium with a threshold of 0.1 keV, taking data for five years; we call this configuration "CONUS100." We also assume that systematic uncertainties will improve from O(1%) down to O(0.1%). Other relevant details are summarized in Appendix B.

The resulting sensitivities for the CONUS and CONUS100 scenarios are shown in long-dashed blue and short-dashed light blue curves, respectively, in Fig. 3. The sensitivity of the default CONUS configuration is relatively weak—it is comparable to existing bounds from $\overline{\nu}_{ee}$ scattering. The CONUS Collaboration has completed their data taking and is expected to be releasing their results in the near future, at which point it will be possible to assess the quality of the simulation performed here. However, it seems unlikely that the CONUS experiment will be able to probe the best-fit region to current short-baseline oscillation data [12], represented by a black star in Fig. 3.

The projection for CONUS100, however, is much more optimistic. Our analysis suggests that this experiment would be able to conclusively probe the parameter space preferred by short-baseline anomalies and would be competitive with other next-generation CE$_\nu$NS proposals. We remind the reader that this configuration is intended to be more optimistic than could likely be achieved, in order to demonstrate the extent to which these kinds of experiments could ultimately probe the cosmology of a sterile neutrino.

COSMOLOGY WITH STERILE NEUTRINOS

We outline the formalism for studying neutrino oscillations in the early universe in order to keep this manuscript reasonably self-contained. For a more detailed account, see, e.g., Refs. [93–100].

In the early universe, neutrino interactions with the matter background and with other neutrinos render the description of neutrino oscillations in Sec. II insufficient. Instead, one must solve for the evolution of the density matrix $\rho_{\vec{p}}$ $(\bar{\rho}_{\vec{p}})$ that describes an ensemble of neutrino (antineutrino) states with momentum $\vec{p}$. The equation of motion for $\rho_{\vec{p}}$ is given by [95–97]

$$i\frac{d\rho_{\vec{p}}}{dt} = [\Omega^0_{\vec{p}}, \rho_{\vec{p}}] + [\Omega^{\rm int}_{\vec{p}}, \rho_{\vec{p}}] + \mathbf{C}[\rho_{\vec{p}}, \bar{\rho}_{\vec{p}}]; \tag{4.1}$$

a similar equation exists for $\bar{\rho}_{\vec{p}}$. The $\Omega^0_{\vec{p}}$ term describes vacuum oscillations, while $\Omega^{\rm int}_{\vec{p}}$ describes the matter potential, including contributions from other neutrinos. The collision term $\mathbf{C}[\rho_{\vec{p}}, \bar{\rho}_{\vec{p}}]$ describes incoherent, inelastic interactions of neutrinos with their environment and with each other; it contains integrals involving $\rho_{\vec{p}}$ and $\bar{\rho}_{\vec{p}}$, and is the source of much of the technical difficulty in solving the evolution of the (anti)neutrino fluid.

To simplify the problem, we consider mixing between only two neutrinos, which we call active, a, and sterile, s,

$$\begin{pmatrix} \nu_a \\ \nu_s \end{pmatrix} = \begin{pmatrix} \cos\theta & -\sin\theta \\ \sin\theta & \cos\theta \end{pmatrix} \begin{pmatrix} \nu_1 \\ \nu_2 \end{pmatrix}, \tag{4.2}$$

where θ is the active-sterile mixing angle and ν1 and ν2 are neutrino mass eigenstates with masses m1 and m2>m1, respectively. It is for this reason that we consider only one of the ϕ_{i4} (i=1, 2) to be nonzero, and we identify $\theta=\phi_{i4}$ above, depending on which active species is being considered. Suppressing the subscript $\vec{p}$, the two-by-two matrix ρ can be decomposed as

$$\rho = \frac{1}{2} f_0 (P_0 + \vec{\sigma}\cdot\vec{P}), \tag{4.3}$$

where $\vec{\sigma}$ is the vector of the Pauli matrices and f_0 is the Fermi-Dirac distribution with vanishing chemical potential. The evolution of P0 and the Bloch vector $\vec{P}$ are governed by

$$\frac{dP_0}{dt} = R^{(a)}, \tag{4.4}$$

$$\frac{d\vec{P}}{dt} = (\vec{B} + \vec{V}^{(a)}) \times \vec{P} - D^{(a)}(P_x\hat{x} + P_y\hat{y}) + R^{(a)}\hat{z}. \tag{4.5}$$

where $\vec{B} \equiv (\frac{\Delta m^2}{2p})(\sin 2\theta, 0, -\cos 2\theta)$ and $\Delta m^2 \equiv m_2^2 - m_1^2$.

A similar decomposition exists for $\bar{\rho}$, into $\bar{P}_0$ and $\vec{\bar{P}}$.

The potential $\vec{V}^{(a)}$ depends on whether ν_a is electron-type or muon-/tau-type, as ν_e has charged-current interactions with a background of electrons that ν_μ and ν_τ do not. The matter potential is [98,100]

$$\vec{V}^{(a)} = (V_1^{(a)} + V_L^{(a)})\hat{z}, \tag{4.6}$$

$$V_1^{(a)} = -\frac{7\pi^2 G_F}{45\sqrt{2}M_Z^2} pT^4(n_{\nu_a} + n_{\bar{\nu}_a})g_a, \tag{4.7}$$

$$V_L^{(a)} = \frac{2\sqrt{2}\zeta(3)}{\pi^2} G_F T^3 L^{(a)}, \tag{4.8}$$

where n_f is the number density of species f. The constant g_a is either $g_{\mu,\tau}=1$ or $g_e = 1 + 4\sec^2\theta_W/(n_{\nu_e} + n_{\bar{\nu}_e})$, and the lepton asymmetries L(a) are given by

$$L^{(e)} = \left(\frac{1}{2} + 2\sin^2\theta_W\right)L_e + \left(\frac{1}{2} - 2\sin^2\theta_W\right)L_p - \frac{1}{2}L_n + 2L_{\nu_e} + L_{\nu_\mu} + L_{\nu_\tau}, \tag{4.9}$$

$$L^{(\mu,\tau)} = L^{(e)} - L_e - L_{\nu_e} + L_{\nu_\mu,\nu_\tau}, \tag{4.10}$$

with $L_f \equiv (n_f - n_{\bar{f}})n_f^{\mathrm{eq}}/n_\gamma^{\mathrm{eq}}$, where n_f^{eq} is the equilibrium number density of f. The damping function D(a) characterizes the loss of quantum coherence from interactions with the background, and, assuming thermal equilibrium, is approximately given by

$$D^{(a)} \approx \frac{1}{2}\Gamma^{(a)}, \quad \text{where } \Gamma^{(a)} = C^{(a)} G_F^2 T^4 p; \tag{4.11}$$

$C^{(e)} \approx 1.27$ is used for ν_e and $C^{(\mu,\tau)} \approx 0.92$ is used with $\nu_{\mu,\tau}$ [94]. Last, the repopulation function $R^{(a)}$ is given approximately by

$$R^{(a)} \approx \Gamma^{(a)}\left[\frac{f_{\mathrm{eq}}(p,\mu_{\nu_a})}{f_0} - \frac{1}{2}(P_0 + P_z)\right] \tag{4.12}$$

where $f_{eq}(p,\mu_{va})$ is the equilibrium Fermi-Dirac distribution with chemical potential μ_{va}.

It is convenient for numerical evaluation to define the quantities

$$P_i^\pm = P_i \pm \bar{P}_i, \tag{4.13}$$

$$P_a^\pm = P_0^\pm + P_z^\pm, \tag{4.14}$$

$$P_s^\pm = P_0^\pm - P_z^\pm, \tag{4.15}$$

for i=0,x,y,z. Their evolution is given by

$$\frac{dP_a^{\pm}}{dt} = B_x P_y^{\pm} + \Gamma_a (2 f_{\text{eq}}^{\pm}/f_0 - P_a^{\pm}), \tag{4.16}$$

$$\frac{dP_s^{\pm}}{dt} = \quad - B_x P_y^{\pm}, \tag{4.17}$$

$$\frac{dP_x^{\pm}}{dt} = -(B_z + V_1^{(a)}) P_y^{\pm} - V_L^{(a)} P_y^{\mp} - D^{(a)} P_x^{\pm}, \tag{4.18}$$

$$\text{where } f_{\text{eq}}^{\pm} \equiv f_{\text{eq}}(p, \mu_{\nu_a}) \pm f_{\text{eq}}(p, -\mu_{\nu_a}). \tag{4.19}$$

where $f_{\text{eq}}^{\pm} \equiv f_{\text{eq}}(p, \mu_{\nu_a}) \pm f_{\text{eq}}(p, -\mu_{\nu_a})$.

The standard ΛCDM cosmology can be augmented with a sterile neutrino through the introduction of two parameters: $\Delta N_{\text{eff}} \equiv N_{\text{eff}} - 3.046$ and $m_{\text{sterile}}^{\text{eff}} \equiv (94.1\ \text{eV})\Omega_{\text{sterile}} h^2$. In Fig. 4, we show the CMB TT power spectrum for the ΛCDM cosmology, as well as for two benchmark values of ΔN_{eff} and $m_{\text{sterile}}^{\text{eff}}$; the best-fit points for the six parameters of the base ΛCDM cosmology are shown in Table I. In Fig. 5, we show the percent deviation of the two latter scenarios relative to the former. In both figures, the gray squares represent the Planck 2015 measurements of the temperature spectrum and their uncertainties [101].

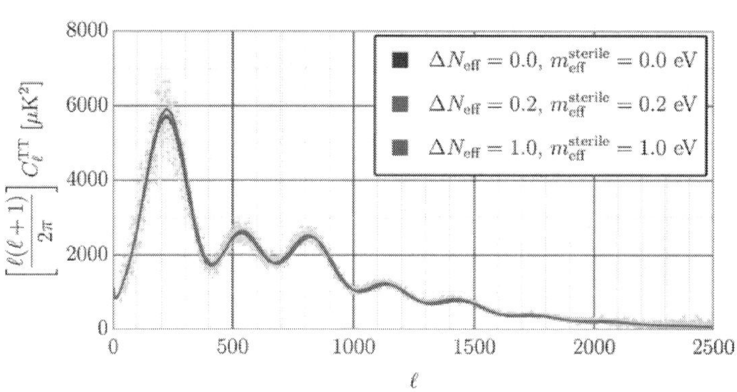

Figure 4. The CMB TT power spectrum for the base ΛCDM cosmology and for two benchmark sterile neutrino scenarios. Curves produced using the camb module [102]. The gray squares represent the Planck 2015 measurements and their uncertainties [101].

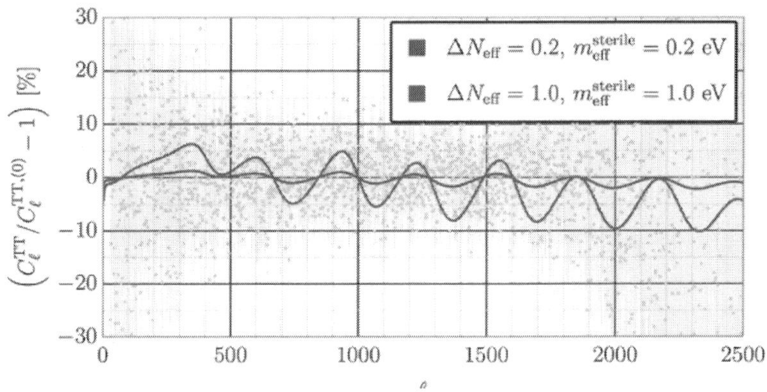

Figure 5. The percent differences between our benchmark sterile neutrino scenarios in Fig. 4 relative to the base ΛCDM cosmology. The gray squares represent the Planck 2015 measurements and their uncertainties [101].

Table 1. Best-fit parameters to the base ΛCDM cosmology used to make Figs. 4 and 5, from the " TT,TE,EE+lowP" fit to Planck data [101].

Parameter	Value
$\Omega_b h^2$	0.02225
$\Omega_c h^2$	0.1198
$100\theta_{\rm MC}$	1.04077
τ	0.079
$\ln(10^{10} A_s)$	3.094
n_s	0.9645

One can calculate ΔNeff at any point in the evolution of the Universe via

$$\Delta N_{\rm eff} = \frac{\int dx x^3 f_{\rm eq}(x, \mu = 0) P_s^+(x)}{4 \int dx x^3 f_{\rm eq}(x, \mu = 0)}, \qquad (4.20)$$

where x≡p/T. Reference [21] discusses two ways in which the mass of the fourth neutrino, m4, can be related to $m_{\rm sterile}^{\rm eff}$. The first is the Dodelson-Widrow mechanism [103], wherein the two are related via $m_{\rm sterile}^{\rm eff} = \Delta N_{\rm eff} m_4$. In this work, however, we will instead follow Ref. [21] and use the relation

$$m_{\text{sterile}}^{\text{eff}} = (\Delta N_{\text{eff}})^{3/4} m_4, \quad (4.21)$$

which assumes that sterile neutrinos are produced thermally in the early universe. We have verified that our results do not depend strongly on this assumption.

We consider two bounds in the ΔN_{eff}-$m_{\text{sterile}}^{\text{eff}}$ parameter space from cosmological data, calculated and presented in Ref. [104]. The first is a combined analysis of the Planck 2015 CMB temperature and polarization (TT, TE, and EE) spectra [101], for both high and low ℓ, and baryon acoustic oscillation (BAO) data [105–107]. Following Ref. [104], we call this " CMB+BAO," shown in dotted gray in Fig. 6. This limit is similar to, but stronger than, a similar limit derived in Ref. [21]. The second adds to this the value of the Hubble constant measured in Ref. [108] as a prior, as well as Planck cluster [109] and lensing data [110] and weak lensing data from Ref. [111], which we call " CMB+BAO+other." This is shown in black in Fig. 6.

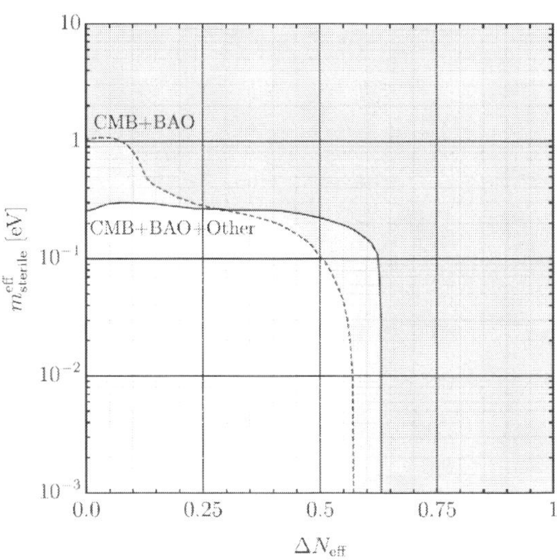

Figure 6. The 2σ C.L. exclusions in the ΔNeff-meffsterile plane derived in Ref. [104] for the CMB+BAO (dotted gray curve) and CMB+BAO+other (black curve) datasets; see text for details.

The shift in the astrophysical/cosmological limit with the inclusion of the "other" datasets stems, in part, from tension in measurements of the Hubble constant; Planck measures a smaller value [101] than local measurements indicate [108]. The origin of this discrepancy is as of yet unclear, but its effect is to allow for a modestly larger value of ΔN_{eff} at the expense of a more stringent limit on meffsterile. This is to be expected from, for instance, Figs. 29–31 in Ref. [101]—larger values of the Hubble constant result in a preference for larger ΔN_{eff}.

While the most recent Planck data release [112] assuredly implies a more stringent constraint in this space, we do not consider it here. In the absence of the Planck 2018 likelihood function, it is not possible to quantitatively determine how updated measurements of the local Hubble constant [113,114] will shift the limit in the $\Delta N_{\rm eff}$-$m^{\rm eff}_{\rm sterile}$ plane. We note, however, that the discrepancy between local and cosmological determinations of the Hubble constant persists in these recent measurements (see Ref. [115]); the resolution of this puzzle has important implications for constraints on additional neutrinos [116,117].

We assume throughout that the lightest neutrino mass vanishes; for a normal neutrino mass ordering, this means m1=0. Furthermore, we have $m_2 = \sqrt{\Delta m^2_{21}} \approx 0$, $m_3 = \sqrt{\Delta m^2_{31}}$, and $m_4 = \sqrt{\Delta m^2_{41}}$. The Planck analysis assumes that any excess in neutrino mass is attributable to a single additional state—in our case, a sterile neutrino. If m1>0 or if neutrino masses were arranged in an inverted hierarchy, then the constraints in Fig. 6 strengthen.

The primary objective of this work is to translate limits that oscillation experiments place in the $\sin^2 2\theta_{\alpha\alpha}$-$\Delta m^2_{41}$ plane (α=e, μ) into limits in the $\Delta N_{\rm eff}$-$m^{\rm eff}_{\rm sterile}$ plane. We do so using the code LASAGNA [100] to solve for the evolution of the neutrino fluid in the early universe. We use as inputs the points in the $\sin^2 2\theta_{\alpha\alpha}$-$\Delta m^2_{41}$ plane that comprise the experimental exclusions and sensitivities. For each point, LASAGNA calculates the evolution of the Bloch vector from T=40 MeV to T=1 MeV—roughly the temperature of the Universe at big bang nucleosynthesis—using the equations of motion in Eqs. (4.16)–(4.19). We determine ΔN_{eff} via Eq. (4.20) and use this to calculate $m^{\rm eff}_{\rm sterile}$ using Eq. (4.21). We assume throughout that the initial lepton asymmetry of the Universe is zero.

We are careful to distinguish between experiments that bound $\sin^2 2\theta_{\mu\mu}$ and those that bound $\sin^2 2\theta_{ee}$ because of differences in ν_e-ν_s oscillations and ν_μ-ν_s oscillations in the early universe. The effective lepton asymmetry electron neutrinos experience is different than for muon neutrinos—see Eqs. (4.9) and (4.10)—resulting in a different matter potential. We consider bounds on $\sin^2 2\theta_{\mu\mu}$ from accelerator neutrino experiments and bounds on $\sin^2 2\theta_{ee}$ coming from reactor and other low-threshold neutrino experiments; the cosmological parameter space is the same in each case, but we underscore that these constitute distinct hypotheses.

RESULTS

Accelerator Neutrino Experiments

In Fig. 7, we show the resulting exclusions/sensitivities in the $\Delta N_{\rm eff}$-$m^{\rm eff}_{\rm sterile}$ plane for accelerator neutrino experiments. Evidently, MINOS/MINOS+ can probe new regions of the cosmological parameter space—particularly compared to the CMB+BAO+other analysis—but the excursion into untouched parameter space is not quite as strong as previously reported in Ref. [21]. There, a bound in the $\Delta N_{\rm eff}$-$m^{\rm eff}_{\rm sterile}$ plane from MINOS was derived from the analysis of Ref. [118]. This analysis, however, takes some oscillation parameters to be fixed, including the solar parameters (Δm^2_{21} and θ_{12}), the reactor angle (θ_{13}), and the CP-odd phase δ_{CP}. This bound is overly optimistic; the uncertainties on the fixed parameters are nontrivial, and ignoring them artificially enhances the confidence of the final result. Therefore, we advocate that the red curve in Fig. 1 is a more accurate representation of the capabilities of MINOS/MINOS+ as a probe of neutrino cosmology.

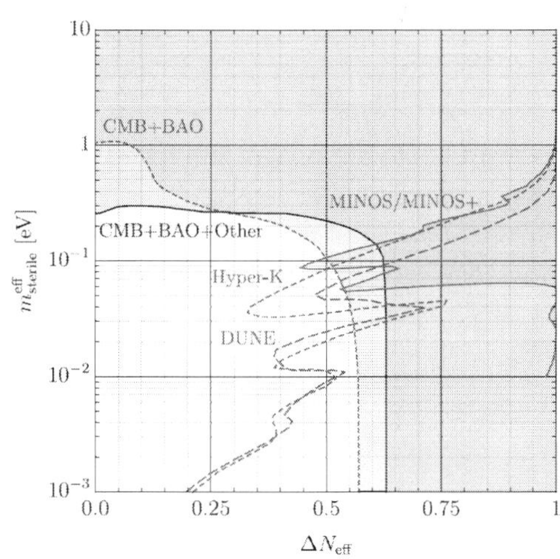

Figure 7. Constraints in the $\Delta N_{\rm eff}$-$m^{\rm eff}_{\rm sterile}$ plane derived from accelerator experiments: MINOS/MINOS+ (red curve, 90% C.L.), DUNE (long-dashed green curve, 95% C.L.), and Hyper-K (short-dashed purple curve, 95% C.L.). Note that these results rely on the assumption that sterile neutrinos are only produced via nonzero $\sin^2 2\theta_{\mu\mu}$.

Also shown in Fig. 7 are our projected 95% C.L. sensitivity limits for DUNE and Hyper-K. These long-baseline experiments have comparable sensitivity to the meffsterile≲10−2 eV portion of the ΔN_{eff}-$m^{\text{eff}}_{\text{sterile}}$ plane that is unconstrained by the CMB+BAO and CMB+BAO+other analyses, while Hyper-K has slightly more sensitivity than DUNE in the region meffsterile~10^{-1}eV. The upshot is that these experiments (will) have a capability to cut into a modest portion of the cosmological parameter space that is not currently probed by Planck and other astrophysical/cosmological experiments. Several comments are in order:

(1) These experiments are able to probe new parts of this parameter space primarily because of their sensitivities to oscillations with $\Delta m^2_{41}\sim10^{-2}$–$10^{-1}$ eV^2 and $\sin^2 2\theta_{\mu\mu}\sim$few$\times10^{-2}$. Moreover, the modest sensitivity that DUNE and Hyper-K will have in the $\Delta m^2_{41}\sim10^{-4}$–$10^{-3}$eV^2 regime translates into a substantial sensitivity to $m^{\text{eff}}_{\text{sterile}} \lesssim \mathcal{O}(10^{-1})$ eV^2.

(2) The inclusion of "other" data with the CMB+BAO dataset relaxes the bound on ΔN_{eff}. However, terrestrial oscillation experiments (will) provide some coverage in the region in which the cosmological constraints are relaxed.

(3) While the improved sensitivity is interesting, DUNE and Hyper-K will not start collecting data until at least the late 2020s. In the interim, next-generation projects such as CMB-S4 [119] will continue to whittle away at the available parameter space; CMB-S4 is expected to be completed by the mid 2020s. Even if the latter is delayed, by the time DUNE and Hyper-K have data to analyze, they will not be able to substantively probe new parameter space. On the other hand, should CMB-S4 be able to exclude, say, nonzero N_{eff} at high confidence, then these terrestrial experiments may be able to independently probe whether the relic is a sterile neutrino.

Reactor Antineutrino Experiments

Limits in the ΔN_{eff}-$m^{\text{eff}}_{\text{sterile}}$ plane from reactor antineutrino experiments are shown in Fig. 8, from which we deduce several important features. The first is that Daya Bay is (and JUNO will be) able to probe parts of this space not currently constrained by either cosmological dataset. However, while DUNE, Hyper-K, and JUNO are still years away from taking data, *Daya Bay can already rule out a significant portion of the* ΔN_{eff}-$m^{\text{eff}}_{\text{sterile}}$ *plane to which astrophysical/cosmological data are currently insensitive*. This is one of the key conclusions of this work. On the basis of Fig. 37 in Ref. [112], it is likely that this persists in the most recent data release from Planck, especially given that Daya Bay, too, possesses an updated dataset [51,52]. This is consistent with findings in Ref. [24], where a different set of astrophysical/ cosmological constraints was considered.

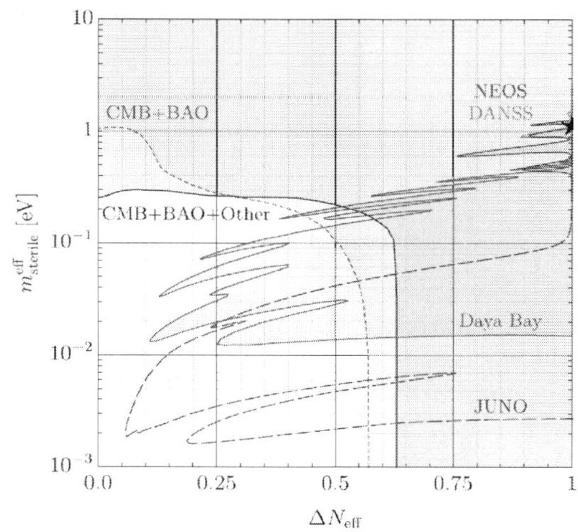

Figure 8. Constraints in the $\Delta N_{\rm eff}$-$m^{\rm eff}_{\rm sterile}$ plane derived from reactor experiments: DANSS (cyan curve, 95% C.L.), Daya Bay (red curve, 95% C.L.), NEOS (purple curve, 90% C.L.), and JUNO (long-dashed blue curve, 95% C.L.). The black, five-pointed start represents the best-fit point from Ref. [12]. Note that these results rely on the assumption that sterile neutrinos are only produced via nonzero $\sin^2 2\theta_{ee}$.

The second feature is that the bounds from DANSS and NEOS are scarcely visible in this plot; they are compressed against the boundary at $\Delta N_{eff}=1$, excluded at high significance by both the CMB+BAO and CMB+BAO+other constraints. These experiments dominate the fit to reactor data that results in the best-fit point shown, again, by the black, five-pointed star in Fig. 8. Taking this at face value implies that astrophysics and cosmology already rule out a sterile neutrino with $\Delta m^2_{41} \sim O(1)\,eV^2$.

Low-Threshold Experiments

Figure 9 depicts the limits from the low-threshold experiments we have considered. The sensitivity of MINER is scarcely visible on this plot; the sensitivity of RED100 is visible, but improves on neither the CMB+BAO nor the CMB+BAO+other exclusions. Recall that we selected these specific incarnations of the RED100 and MINER experiments because they were the most optimistic proposals for probing the $\sin^2 2\theta_{ee}$-Δm^2_{41} parameter space. However, on the basis of Figure. 9, it seems that these experiments will not offer a fresh perspective on cosmology.

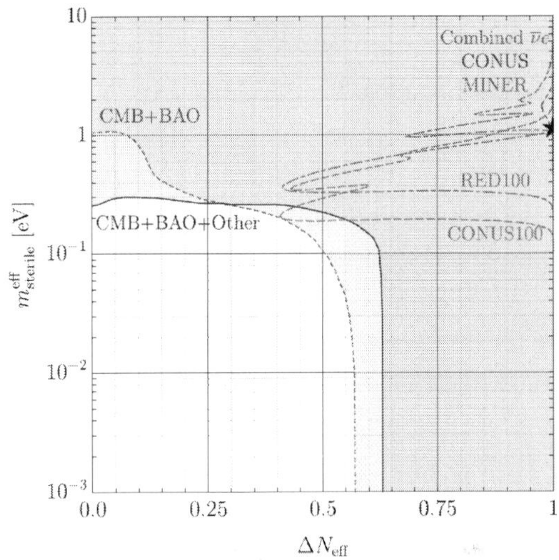

Figure 9. Constraints in the $\Delta N_{\rm eff}$-$m^{\rm eff}_{\rm sterile}$ plane derived from low-threshold experiments: $\bar{\nu}_e$ scattering (purple curve, 90% C.L.), RED100 (dot-dashed red curve, 90% C.L.), MINER (double-dot–dashed green curve, 90% C.L.), CONUS (long-dashed blue curve, 95% C.L.), and CONUS100 (short-dashed light blue curve, 95% C.L.). The black, five-pointed start represents the best-fit point from the global analysis of Ref. [12]. Note that these results rely on the assumption that sterile neutrinos are only produced via nonzero sin22θee.

Also included in Fig. 9 are the exclusion from $\bar{\nu}_e$ scattering and the sensitivity of CONUS. However, neither is visible on this plot; these constraints deviate from ΔNeff=1 by at most ~few×10^{-5}. On the other hand, CONUS100 is capable of reaching a sliver of the parameter space to which the CMB+BAO+other constraint is insensitive. However, this configuration has been chosen to be extremely optimistic—arguably unrealistically so—and even under these conditions, it is barely able to improve limits on the cosmology of a sterile neutrino. Moreover, the most recent Planck data and CMB-S4 would likely render these experiments totally impotent when it comes to excluding cosmological parameter space.

These results and those of the previous subsection present a problem. As we have mentioned, the best-fit point to the reactor antineutrino anomaly—the black star in Figs. 8 and 9—is strongly disfavored by the CMB+BAO(+others) datasets. Indeed, we have seen how experiments that either provide evidence for the anomaly (DANSS, NEOS, etc.) or are designed to probe the anomaly (RED100, MINER, CONUS100, etc.) are powerless to challenge the astrophysical/cosmological experiments in this parameter space. We are then left to question if (and how) these datasets can be rendered consistent. We enumerate a few possible resolutions.

(1) The reactor antineutrino anomaly is an aberration. Determining the reactor antineutrino flux is a complicated business; there is no shortage of ways in which theoretical calculations and experimental measurements could go awry. See, for instance, Refs. [120–126] for more details. We note that, of the reactor experiments we have considered, DANSS, Daya Bay, and NEOS[5] employ a near detector to reduce flux-related systematics; the rest, however, depend on theoretical predictions of the flux.

(2) Our understanding of cosmology is incomplete. This seems unlikely, but it is not altogether impossible that something dramatic could have happened in the early universe that the standard ΛCDM cosmology does not capture.

(3) The framework we have employed is insufficient. The two-neutrino approximation is useful to solve for the evolution of the neutrino fluid, but it may be missing some important physics. A more complete calculation would be more intensive, but potentially worth the effort.

(4) Neutrinos have extra interactions—affecting either the active or the sterile flavors—that may lead to a qualitatively different evolution of the neutrino fluid. This may be due to a new matter potential or because new degrees of freedom are relevant later in the evolution of the Universe (see, for instance, Refs. [127–130]).

(5) The initial lepton asymmetry of the Universe, L^0, is large, ~$O(10^{-3}$–$10^{-2})$. This may suppress transitions to the sterile flavor, diminishing its contribution to $m^{eff}_{sterile}$ and ΔN_{eff}. The LASAGNA module is well suited to study the evolution of the (anti)neutrino fluid in the presence of an initial lepton asymmetry, and this has been studied in Ref. [21]; see also Ref. [131]. We relegate a detailed study to future work.

Whatever its cause, the observation of this tension is the other central conclusion of this work. *There can be no satisfactory sterile neutrino solution to the reactor antineutrino anomaly that does not address this tension with astrophysical and cosmological measurements.*

CONCLUSIONS

We have considered the complementarity between terrestrial neutrino oscillation experiments—specifically, accelerator and reactor experiments—and cosmological experiments in probing the cosmological properties of a proposed sterile neutrino. Accelerator neutrino experiments have the potential to improve on current constraints, but given that these are years away from taking data, it is unlikely that they will improve on the constraints that will exist by then. Meanwhile, reactor experiments—Daya Bay, specifically—are already able to probe parameter space beyond the reach of cosmological experiments, a feature that may persist with updated measurements.

We have emphasized the tension between cosmological constraints and the mild preference for a sterile neutrino from reactor antineutrino experiments. This is highlighted by (1) the best-fit point from the global reactor analysis of Ref. [12] already being strongly disfavored by Planck, and (2) the inability of upcoming low-threshold neutrino scattering experiments to probe new parts of the cosmological parameter space. Several possibilities for resolving this tension have been enumerated, but we conclude that no solution to the reactor anomalies can be truly compelling if it does not address this tension with cosmology.

This analysis presented here is not, strictly speaking, entirely consistent. Some of the terrestrial experiments we have considered have been analyzed under the four-neutrino hypothesis (with only one active-sterile mixing angle allowed to be nonzero), whereas these bounds on sterile neutrino oscillation parameters have been translated into cosmological bounds using two-flavor oscillations. This could well be a fatal inconsistency—it is logically possible that the reduction to two-flavor oscillations has oversimplified the system, so that crucial physics is being missed. In particular, this framework offers no opportunities to study the LSND [6] and MiniBooNE [7] anomalies, since electron-neutrino appearance requires that two active-sterile mixing angles be nonzero. A more complete analysis is required, especially given the tension present in the sterile-neutrino interpretation of neutrino appearance and disappearance data [12].

The results of this work are meaningful heuristics, but that the hypotheses to which they apply—that a sterile neutrino exists, but that only one possible active-sterile mixing angle is nonzero—may be too simple to be physical. Moreover, the separation between electron-type and muon-type oscillations in our treatment of the neutrino fluid in the early universe is awkward. A more comfortable arrangement would be to use bounds on the complete four-neutrino hypothesis from a variety of experiments simultaneously to derive bounds on sterile-neutrino cosmology. However, the primary limitation of this scheme is that solving for the evolution of three active and one sterile neutrino species (and their corresponding antineutrinos) in the early universe is a technically daunting task.

This work is meant to illustrate the importance of interdisciplinary studies in neutrino physics. It is not entirely obvious, *a priori*, that an experiment such as Daya Bay can improve constraints in some part of the ΔN_{eff}-$m^{\text{eff}}_{\text{sterile}}$ plane, yet we have found this to be the case. In order to keep this work focused, we have considered only accelerator and reactor (anti)neutrino experiments and their interplay with astrophysics and cosmology. Even with the restriction that at most one active-sterile mixing angle is nonzero, similar analyses could be performed using (anti)neutrino disappearance results from solar [61–72] or atmospheric [132–134] experiments, as well as accelerator experiments that we have not considered [135–139]. These bounds, however, are at best comparable to, and are generally weaker than, the limits we have presented.

NOTES

[1] There is no reason to suspect, *a priori*, that additional neutrinos do not possess separate interactions among themselves or with other particle species. We will assume that additional neutrinos have no interactions, but use the term "sterile" to refer to any species that does not have weak interactions.

[2] It is, by hypothesis, not possible to directly detect sterile neutrino states, as they do not interact with detectors. Instead, one must infer that neutrinos have converted into the sterile flavor by observing a reduced interaction rate.

[3] The first relation is always true, while the second is true only if ϕ_{14} vanishes.

[4] This is, of course, not a CE$_\nu$NS process, which is why we have opted to call this class of bounds "low threshold."

[5] Note that NEOS uses the Daya Bay spectrum to normalize their antineutrino spectrum. The different effective fuel fractions at these

experiments introduces a small amount of dependence on the flux model.

[6] We note that 76Ge, while technically unstable, has a long enough half-life to be effectively stable.

ACKNOWLEDGMENTS

Since this work first appeared, a calculation of the evolution of the neutrino fluid using the full 3+1 mixing formalism was presented in Ref. [140]. In that work, the authors find that a sterile neutrino with the properties suggested by the reactor anomaly would still be fully thermalized: the mass splittings Δm221 and Δm231 do not dramatically affect the production of the sterile neutrino, and allowing multiple active-sterile mixing angles to be nonzero does nothing to alleviate this tension (see Figs. 9 and 10 of Ref. [140]). Consequently, the two-neutrino formalism employed in this work is not responsible for the tension between terrestrial and astrophysical experiments.

REFERENCES

1. I. Esteban, M.C. Gonzalez-Garcia, A. Hernández-Cabezudo, M. Maltoni, and T. Schwetz, Global analysis of three-flavour neutrino oscillations: Synergies and tensions in the determination of θ23, δCP, and the mass ordering, J. High Energy Phys. 2019, 106.
2. M.A. Acero, C. Giunti, and M. Laveder, Limits on nu(e) and anti-nu(e) disappearance from Gallium and reactor experiments, Phys. Rev. D 78, 073009 (2008).
3. C. Giunti and M. Laveder, Statistical significance of the gallium anomaly, Phys. Rev. C 83, 065504 (2011).
4. G. Mention, M. Fechner, T. Lasserre, T.A. Mueller, D. Lhuillier, M. Cribier, and A. Letourneau, The reactor antineutrino anomaly, Phys. Rev. D 83, 073006 (2011).
5. A.C. Hayes, J.L. Friar, G.T. Garvey, G. Jungman, and G. Jonkmans, Systematic Uncertainties in the Analysis of the Reactor Neutrino Anomaly, Phys. Rev. Lett. 112, 202501 (2014).
6. A. Aguilar-Arevalo et al. (LSND Collaboration), Evidence for neutrino oscillations from the observation of anti-neutrino(electron) appearance in a anti-neutrino(muon) beam, Phys. Rev. D 64, 112007 (2001).
7. A.A. Aguilar-Arevalo et al. (MiniBooNE Collaboration), Improved Search for ¯νμ→¯νe *Oscillations in the* MiniBooNE Experiment, Phys. Rev. Lett. 110, 161801 (2013).
8. S. Gariazzo, C. Giunti, M. Laveder, and Y.F. Li, Updated global 3+1 *analysis of short-baseline neutrino oscillations*, J. High Energy Phys. 2017, 135.
9. C. Giunti, X.P. Ji, M. Laveder, Y.F. Li, and B.R. Littlejohn, Reactor fuel fraction information on the antineutrino anomaly, J. High Energy Phys. 2017, 143.
10. M. Dentler, A. Hernández-Cabezudo, J. Kopp, M. Maltoni, and T. Schwetz, Sterile neutrinos or flux uncertainties—Status of the reactor anti-neutrino anomaly, J. High Energy Phys. 2017, 099.
11. S. Gariazzo, C. Giunti, M. Laveder, and Y.F. Li, Model-independent ¯νe *short-baseline oscillations from reactor spectral ratios*, Phys. Lett. B 782, 13 (2018).
12. M. Dentler, A. Hernández-Cabezudo, J. Kopp, P.A.N. Machado, M. Maltoni, I. Martinez-Soler, and T. Schwetz, Updated global analysis of neutrino oscillations in the presence of eV-scale sterile neutrinos, J.

High Energy Phys. 2018, 010.

13. C. Giunti, Y.F. Li, B.R. Littlejohn, and P.T. Surukuchi, Diagnosing the reactor antineutrino anomaly with global antineutrino flux data, Phys. Rev. D 99, 073005 (2019).
14. C. Giunti and T. Lasserre, EV-Scale Sterile Neutrinos, eV-scale sterile neutrinos arXiv:1901.08330.
15. A. Melchiorri, O. Mena, S. Palomares-Ruiz, S. Pascoli, A. Slosar, and M. Sorel, Sterile neutrinos in light of recent cosmological and oscillation data: A multi-flavor scheme approach, J. Cosmol. Astropart. Phys. 2009, 036.
16. M. Archidiacono, N. Fornengo, C. Giunti, and A. Melchiorri, Testing 3+1 *and* 3+2 *neutrino mass models with cosmology and short baseline experiments*, Phys. Rev. D 86, 065028 (2012).
17. M. Archidiacono, N. Fornengo, C. Giunti, S. Hannestad, and A. Melchiorri, Sterile neutrinos: Cosmology versus short-baseline experiments, Phys. Rev. D 87, 125034 (2013).
18. A. Mirizzi, G. Mangano, N. Saviano, E. Borriello, C. Giunti, G. Miele, and O. Pisanti, The strongest bounds on active-sterile neutrino mixing after Planck data, Phys. Lett. B 726, 8 (2013).
19. S. Gariazzo, C. Giunti, and M. Laveder, Light sterile neutrinos in cosmology and short-baseline oscillation experiments, J. High Energy Phys. 2013, 211.
20. M. Archidiacono, N. Fornengo, S. Gariazzo, C. Giunti, S. Hannestad, and M. Laveder, Light sterile neutrinos after BICEP-2, J. Cosmol. Astropart. Phys. 2014, 031.
21. S. Bridle, J. Elvin-Poole, J. Evans, S. Fernandez, P. Guzowski, and S. Soldner-Rembold, A combined view of sterile-neutrino Constraints from CMB and neutrino oscillation measurements, Phys. Lett. B 764, 322 (2017).
22. A. Ghalsasi, D. McKeen, and A.E. Nelson, Probing nonstandard neutrino cosmology with terrestrial neutrino experiments, Phys. Rev. D 95, 115039 (2017).
23. S. Roy Choudhury and S. Choubey, Constraining light sterile neutrino mass with the BICEP2/Keck Array 2014 B-mode polarization data, Eur. Phys. J. C 79, 557 (2019).
24. A.M. Knee, D. Contreras, and D. Scott, Cosmological Constraints on Sterile Neutrino Oscillations from Planck, Cosmological constraints

on sterile neutrino oscillations from Planck arXiv:1812.02102.

25. S.K. Kang, Roles of sterile neutrinos in particle physics and cosmology, Int. J. Mod. Phys. A 34, 1930005 (2019).
26. D. Akimov et al. (COHERENT Collaboration), Observation of coherent elastic neutrino-nucleus scattering, Science 357, 1123 (2017).
27. A.J. Anderson, J.M. Conrad, E. Figueroa-Feliciano, C. Ignarra, G. Karagiorgi, K. Scholberg, M.H. Shaevitz, and J. Spitz, Measuring active-to-sterile neutrino oscillations with neutral current coherent neutrino-nucleus Scattering, Phys. Rev. D 86, 013004 (2012).
28. B. Dutta, Y. Gao, R. Mahapatra, N. Mirabolfathi, L.E. Strigari, and J.W. Walker, Sensitivity to oscillation with a sterile fourth generation neutrino from ultra-low threshold neutrino-nucleus coherent scattering, Phys. Rev. D 94, 093002 (2016).
29. S. Kerman, V. Sharma, M. Deniz, H.T. Wong, J.-W. Chen, H.B. Li, S.T. Lin, C.-P. Liu, and Q. Yue (TEXONO Collaboration), Coherency in neutrino-nucleus elastic scattering, Phys. Rev. D 93, 113006 (2016).
30. B.C. Cañas, E.A. Garcés, O.G. Miranda, and A. Parada, The reactor antineutrino anomaly and low energy threshold neutrino experiments, Phys. Lett. B 776, 451 (2018).
31. D.K. Papoulias and T.S. Kosmas, COHERENT constraints to conventional and exotic neutrino physics, Phys. Rev. D 97, 033003 (2018).
32. J.M. Berryman, A. de Gouvêa, K.J. Kelly, and A. Kobach, Sterile neutrino at the deep underground neutrino experiment, Phys. Rev. D 92, 073012 (2015).
33. L. Wolfenstein, Neutrino oscillations in matter, Phys. Rev. D 17, 2369 (1978).
34. A. Mirizzi, I. Tamborra, H.-T. Janka, N. Saviano, K. Scholberg, R. Bollig, L. Hudepohl, and S. Chakraborty, Supernova neutrinos: Production, oscillations and detection, Riv. Nuovo Cimento 39, 1 (2016).
35. H. Duan, G.M. Fuller, and Y.-Z. Qian, Collective Neutrino Oscillations, Annu. Rev. Nucl. Part. Sci. 60, 569 (2010).
36. K.N. Abazajian et al., Light Sterile Neutrinos: A White Paper, Light sterile neutrinos: A white paper arXiv:1204.5379.
37. J. Kopp, P.A.N. Machado, M. Maltoni, and T. Schwetz, Sterile neutrino oscillations: The global picture, J. High Energy Phys. 2013, 050.

38. P. Adamson et al. (MINOS Collaboration), Measurement of Neutrino and Antineutrino Oscillations Using Beam and Atmospheric Data in MINOS, Phys. Rev. Lett. 110, 251801 (2013).
39. P. Adamson et al. (MINOS Collaboration), Combined Analysis of $\nu\mu$ *Disappearance and* $\nu\mu \rightarrow \nu e$ *Appearance in MINOS Using Accelerator and Atmospheric Neutrinos*, Phys. Rev. Lett. 112, 191801 (2014).
40. P. Adamson et al. (MINOS+ Collaboration), Search for Sterile Neutrinos in MINOS and MINOS+ *Using a Two-Detector Fit*, Phys. Rev. Lett. 122, 091803 (2019).
41. C. Adams et al. (LBNE Collaboration), The Long-Baseline Neutrino Experiment: Exploring Fundamental Symmetries of the Universe, The long-baseline neutrino experiment: Exploring fundamental symmetries of the universe arXiv:1307.7335.
42. R. Acciarri et al. (DUNE Collaboration), Long-Baseline Neutrino Facility (LBNF) and Deep Underground Neutrino Experiment (DUNE), Long-baseline neutrino facility (LBNF) and deep underground neutrino experiment (DUNE) arXiv:1512.06148.
43. K. Abe et al. (Hyper-Kamiokande Proto-Collaboration), Physics potential of a long-baseline neutrino oscillation experiment using a J-PARC neutrino beam and Hyper-Kamiokande, Prog. Theor. Exp. Phys. 2015, 053C02 (2015).
44. K. Abe et al. (Hyper-Kamiokande Collaboration), Physics potentials with the second Hyper-Kamiokande detector in Korea, Prog. Theor. Exp. Phys. 2018, 063C01 (2018).
45. K. Abe et al. (Hyper-Kamiokande Collaboration), Hyper-Kamiokande Design Report, Hyper-Kamiokande Design Report arXiv:1805.04163.
46. M. Danilov (DANSS Collaboration), Sensitivity of DANSS Detector to Short Range Neutrino Oscillations, Sensitivity of DANSS detector to short range neutrino oscillations arXiv:1412.0817.
47. I. Alekseev et al., DANSS: Detector of the reactor AntiNeutrino based on solid scintillator, J. Instrum. 11, P11011 (2016).
48. I. Alekseev et al. (DANSS Collaboration), Search for sterile neutrinos at the DANSS experiment, Phys. Lett. B 787, 56 (2018).
49. F.P. An et al. (Daya Bay Collaboration), Measurement of electron antineutrino oscillation based on 1230 days of operation of the Daya Bay experiment, Phys. Rev. D 95, 072006 (2017).

50. F.P. An et al. (Daya Bay Collaboration), Improved Search for a Light Sterile Neutrino with the Full Configuration of the Daya Bay Experiment, Phys. Rev. Lett. 117, 151802 (2016).
51. D. Adey et al. (Daya Bay Collaboration), Measurement of the Electron Antineutrino Oscillation with 1958 Days of Operation at Daya Bay, Phys. Rev. Lett. 121, 241805 (2018).
52. D. Adey et al. (Daya Bay Collaboration), Measurement of Individual Antineutrino Spectra from 235U *and* 239Pu *at* Daya Bay, Measurement of individual antineutrino spectra from 235U and 239Pu at Daya Bay arXiv:1904.07812.
53. Y. Ko et al. (NEOS Collaboration), Sterile Neutrino Search at the NEOS Experiment, Phys. Rev. Lett. 118, 121802 (2017).
54. F. An et al. (JUNO Collaboration), Neutrino physics with JUNO, J. Phys. G 43, 030401 (2016).
55. H. Almazán et al. (STEREO Collaboration), Sterile Neutrino Constraints from the STEREO Experiment with 66 Days of Reactor-On Data, Phys. Rev. Lett. 121, 161801 (2018).
56. Y. Abreu et al. (SoLid Collaboration), Performance of a full scale prototype detector at the BR2 reactor for the Solid experiment, J. Instrum. 13, P05005 (2018).
57. J. Ashenfelter et al. (PROSPECT Collaboration), First Search for Short-Baseline Neutrino Oscillations at HFIR with Prospect, Phys. Rev. Lett. 121, 251802 (2018).
58. W. Hampel et al. (GALLEX Collaboration), Final results of the Cr-51 neutrino source experiments in GALLEX, Phys. Lett. B 420, 114 (1998).
59. J.N. Abdurashitov et al. (SAGE Collaboration), Measurement of the response of the Russian-American gallium experiment to neutrinos from a Cr-51 source, Phys. Rev. C 59, 2246 (1999).
60. J.N. Abdurashitov et al., Measurement of the response of a Ga solar neutrino experiment to neutrinos from an Ar-37 source, Phys. Rev. C 73, 045805 (2006).
61. B.T. Cleveland, T. Daily, R. Davis, Jr., J.R. Distel, K. Lande, C.K. Lee, P.S. Wildenhain, and J. Ullman, Measurement of the solar electron neutrino flux with the Homestake chlorine detector, Astrophys. J. 496, 505 (1998).
62. B. Aharmim et al. (SNO Collaboration), Electron energy spectra,

fluxes, and day-night asymmetries of B-8 solar neutrinos from measurements with NaCl dissolved in the heavy-water detector at the Sudbury Neutrino Observatory, Phys. Rev. C 72, 055502 (2005).
63. J. Hosaka et al. (Super-Kamiokande Collaboration), Solar neutrino measurements in super-Kamiokande-I, Phys. Rev. D 73, 112001 (2006).
64. B. Aharmim et al. (SNO Collaboration), Determination of the νe *and total* 8B *solar neutrino fluxes with the Sudbury neutrino observatory phase I data set*, Phys. Rev. C 75, 045502 (2007).
65. J.P. Cravens et al. (Super-Kamiokande Collaboration), Solar neutrino measurements in Super-Kamiokande-II, Phys. Rev. D 78, 032002 (2008).
66. B. Aharmim et al. (SNO Collaboration), An Independent Measurement of the Total Active B-8 Solar Neutrino Flux Using an Array of He-3 Proportional Counters at the Sudbury Neutrino Observatory, Phys. Rev. Lett. 101, 111301 (2008).
67. G. Bellini et al. (Borexino Collaboration), Measurement of the solar 8B neutrino rate with a liquid scintillator target and 3 MeV energy threshold in the Borexino detector, Phys. Rev. D 82, 033006 (2010).
68. J.N. Abdurashitov et al. (SAGE Collaboration), Measurement of the solar neutrino capture rate with gallium metal. III: Results for the 2002–2007 data-taking period, Phys. Rev. C 80, 015807 (2009).
69. F. Kaether, W. Hampel, G. Heusser, J. Kiko, and T. Kirsten, Reanalysis of the GALLEX solar neutrino flux and source experiments, Phys. Lett. B 685, 47 (2010).
70. K. Abe et al. (Super-Kamiokande Collaboration), Solar neutrino results in Super-Kamiokande-III, Phys. Rev. D 83, 052010 (2011).
71. G. Bellini et al., Precision Measurement of the 7Be *Solar Neutrino Interaction Rate in* Borexino, Phys. Rev. Lett. 107, 141302 (2011).
72. A. Gando et al. (KamLAND Collaboration), 7Be *solar neutrino measurement with* KamLAND, Phys. Rev. C 92, 055808 (2015).
73. L.B. Auerbach et al. (LSND Collaboration), Measurements of charged current reactions of νe *on* 12C , Phys. Rev. C 64, 065501 (2001).
74. J.M. Conrad and M.H. Shaevitz, Limits on electron neutrino disappearance from the KARMEN and LSND νe—carbon cross section data, Phys. Rev. D 85, 013017 (2012).
75. G.S. Vidyakin, V.N. Vyrodov, I.I. Gurevich, Yu.V. Kozlov, V.P.

Martemyanov, S. V. Sukhotin, V. G. Tarasenkov, E. V. Turbin, and S. K. Khakhimov, Limitations on the magnetic moment and charge radius of the electron-anti-neutrino, Pis'Ma Zh. Eksp. Teor. Fiz. 55, 212 (1992); [JETP Lett. 55, 206 (1992)].

76. A.I. Derbin, A.V. Chernyi, L.A. Popeko, V.N. Muratova, G.A. Shishkina, and S.I. Bakhlanov, Experiment on anti-neutrino scattering by electrons at a reactor of the Rovno nuclear power plant, Pis'Ma Zh. Eksp. Teor. Fiz. 57, 755 (1993); [JETP Lett. 57, 768 (1993)].

77. C. Amsler et al. (MUNU Collaboration), The MUNU experiment, general description, Nucl. Instrum. Methods Phys. Res., Sect. A 396, 115 (1997).

78. M. Deniz et al. (TEXONO Collaboration), Measurement of nu(e)-bar -electron scattering cross-section with a CsI(Tl) scintillating crystal array at the Kuo-Sheng nuclear power reactor, Phys. Rev. D 81, 072001 (2010).

79. H.T. Wong, Ultra-low-energy germanium detector for neutrino-nucleus coherent scattering and dark matter searches, Mod. Phys. Lett. A 23, 1431 (2008).

80. A.G. Beda, V.B. Brudanin, V.G. Egorov, D.V. Medvedev, V.S. Pogosov, E.A. Shevchik, M.V. Shirchenko, A.S. Starostin, and I.V. Zhitnikov, Gemma experiment: The results of neutrino magnetic moment search, Phys. Part. Nucl. Lett. 10, 139 (2013).

81. A. Gútlein et al., Impact of coherent neutrino nucleus scattering on direct dark matter searches based on CaWO4 *crystals*, Astropart. Phys. 69, 44 (2015).

82. V. Belov et al., The νGeN *experiment at the Kalinin Nuclear Power Plant*, J. Instrum. 10, P12011 (2015).

83. J. Billard et al., Coherent neutrino scattering with low temperature bolometers at Chooz reactor complex, J. Phys. G 44, 105101 (2017).

84. A. Aguilar-Arevalo et al. (CONNIE Collaboration), Results of the engineering run of the coherent neutrino nucleus interaction experiment (CONNIE Collaboration), J. Instrum. 11, P07024 (2016).

85. R. Strauss et al., The ν-cleus experiment: A gram-scale fiducial-volume cryogenic detector for the first detection of coherent neutrino-nucleus scattering, Eur. Phys. J. C 77, 506 (2017).

86. T.S. Kosmas, D.K. Papoulias, M. Tortola, and J.W.F. Valle, Probing light sterile neutrino signatures at reactor and spallation neutron source

neutrino experiments, Phys. Rev. D 96, 063013 (2017).

87. D.Yu. Akimov et al. (RED Collaboration), Prospects for observation of neutrino-nuclear neutral current coherent scattering with two-phase Xenon emission detector, J. Instrum. 8, P10023 (2013).
88. G. Agnolet et al. (MINER Collaboration), Background studies for the MINER coherent neutrino scattering reactor experiment, Nucl. Instrum. Methods Phys. Res., Sect. A 853, 53 (2017).
89. D. Akimov et al. (COHERENT Collaboration), COHERENT 2018 at the Spallation Neutron Source, COHERENT 2018 at the spallation neutron source arXiv:1803.09183.
90. J. Hakenmüller, The CONUS experiment, https://indico.cern.ch/event/606690/contributions/2591545/attachments/1499330/2336272/Taup2017_CONUS_talk_JHakenmueller.pdf.
91. Y. Farzan, M. Lindner, W. Rodejohann, and X.-J. Xu, Probing neutrino coupling to a light scalar with coherent neutrino scattering, J. High Energy Phys. 2018, 066.
92. J.M. Berryman, V. Brdar, and P. Huber, Particle physics origin of the 5 MeV bump in the reactor antineutrino spectrum, Phys. Rev. D 99, 055045 (2019).
93. L. Stodolsky, On the treatment of neutrino oscillations in a thermal environment, Phys. Rev. D 36, 2273 (1987).
94. K. Enqvist, K. Kainulainen, and M.J. Thomson, Stringent cosmological bounds on inert neutrino mixing, Nucl. Phys. B373, 498 (1992).
95. G. Raffelt, G. Sigl, and L. Stodolsky, NonAbelian Boltzmann Equation for Mixing and Decoherence, Phys. Rev. Lett. 70, 2363 (1993); Erratum, 98, 069902 (2007).
96. B.H.J. McKellar and M.J. Thomson, Oscillating doublet neutrinos in the early universe, Phys. Rev. D 49, 2710 (1994).
97. J. Lesgourgues, G. Mangano, G. Miele, and S. Pastor, Neutrino Cosmology (Cambridge University Press, Cambridge, England, 2012).
98. S. Hannestad, I. Tamborra, and T. Tram, Thermalisation of light sterile neutrinos in the early universe, J. Cosmol. Astropart. Phys. 2012, 025.
99. C. Volpe, D. Väänänen, and C. Espinoza, Extended evolution equations for neutrino propagation in astrophysical and cosmological environments, Phys. Rev. D 87, 113010 (2013).
100. S. Hannestad, R.S. Hansen, and T. Tram, Can active-sterile neutrino

oscillations lead to chaotic behavior of the cosmological lepton asymmetry, J. Cosmol. Astropart. Phys. 2013, 032.

101. P.A.R. Ade et al. (Planck Collaboration), Planck 2015 results. XIII. Cosmological parameters, Astron. Astrophys. 594, A13 (2016).
102. A. Lewis, A. Challinor, and A. Lasenby, Efficient computation of CMB anisotropies in closed FRW models, Astrophys. J. 538, 473 (2000).
103. S. Dodelson and L.M. Widrow, Sterile-Neutrinos as Dark Matter, Phys. Rev. Lett. 72, 17 (1994).
104. L. Feng, J.-F. Zhang, and X. Zhang, A search for sterile neutrinos with the latest cosmological observations, Eur. Phys. J. C 77, 418 (2017).
105. F. Beutler, C. Blake, M. Colless, D.H. Jones, L. Staveley-Smith, L. Campbell, Q. Parker, W. Saunders, and F. Watson, The 6dF galaxy survey: Baryon acoustic oscillations and the local Hubble constant, Mon. Not. R. Astron. Soc. 416, 3017 (2011).
106. A.J. Ross, L. Samushia, C. Howlett, W.J. Percival, A. Burden, and M. Manera, The clustering of the SDSS DR7 main Galaxy sample—I. A 4 per cent distance measure at z=0.15 , Mon. Not. R. Astron. Soc. 449, 835 (2015).
107. A.J. Cuesta et al., The clustering of galaxies in the SDSS-III baryon oscillation spectroscopic survey: Baryon acoustic oscillations in the correlation function of LOWZ and CMASS galaxies in data release 12, Mon. Not. R. Astron. Soc. 457, 1770 (2016).
108. A.G. Riess et al., A 2.4% determination of the local value of the Hubble constant, Astrophys. J. 826, 56 (2016).
109. P.A.R. Ade et al. (Planck Collaboration), Planck 2015 results. XXIV. Cosmology from Sunyaev-Zeldovich cluster counts, Astron. Astrophys. 594, A24 (2016).
110. P.A.R. Ade et al. (Planck), Planck 2015 results. XV. Gravitational lensing, Astron. Astrophys. 594, A15 (2016).
111. C. Heymans et al., CFHTLenS tomographic weak lensing cosmological parameter constraints: Mitigating the impact of intrinsic galaxy alignments, Mon. Not. R. Astron. Soc. 432, 2433 (2013).
112. N. Aghanim et al. (Planck Collaboration), Planck 2018 Results. VI. Cosmological Parameters, Planck 2018 results. VI. Cosmological parameters arXiv:1807.06209.
113. A.G. Riess, S. Casertano, W. Yuan, L. Macri, J. Anderson, J.W. MacKenty, J.B. Bowers, K.I. Clubb, A.V. Filippenko, and D.O.

Jones et al., New parallaxes of galactic Cepheids from spatially scanning the Hubble space telescope: Implications for the Hubble constant, Astrophys. J. 855, 136 (2018).

114. A.G. Riess et al., Milky way Cepheid standards for measuring cosmic distances and application to Gaia DR2: Implications for the Hubble constant, Astrophys. J. 861, 126 (2018).

115. J.L. Bernal, L. Verde, and A.G. Riess, The trouble with H0 , J. Cosmol. Astropart. Phys. 2016, 019.

116. R.-Y. Guo, J.-F. Zhang, and X. Zhang, Can the H0 *tension be resolved in extensions to* ΛCDM *cosmology*, J. Cosmol. Astropart. Phys. 2019, 054.

117. S. Carneiro, P.C. de Holanda, C. Pigozzo, and F. Sobreira, Is the H0 *tension suggesting a 4th neutrino's generation*, Phys. Rev. D 100, 023505 (2019).

118. J. Huang, Ph.D. thesis, University of Texas, 2015, http://lss.fnal.gov/archive/thesis/2000/fermilab-thesis-2015-06.pdf.

119. K.N. Abazajian et al. (CMB-S4 Collaboration), CMB-S4 Science Book, First Edition, CMB-S4 Science Book, First Edition arXiv:1610.02743.

120. T.A. Mueller et al., Improved predictions of reactor antineutrino spectra, Phys. Rev. C 83, 054615 (2011).

121. P. Huber, On the determination of anti-neutrino spectra from nuclear reactors, Phys. Rev. C 84, 024617 (2011); Erratum, 85, 029901(E) (2012).

122. A.C. Hayes, J.L. Friar, G.T. Garvey, D. Ibeling, G. Jungman, T. Kawano, and R.W. Mills, Possible origins and implications of the shoulder in reactor neutrino spectra, Phys. Rev. D 92, 033015 (2015).

123. P. Huber, Reactor antineutrino fluxes—Status and challenges, Nucl. Phys. B908, 268 (2016).

124. A.C. Hayes and P. Vogel, Reactor neutrino spectra, Annu. Rev. Nucl. Part. Sci. 66, 219 (2016).

125. C. Giunti, Precise determination of the 235U *reactor antineutrino cross section per fission*, Phys. Lett. B 764, 145 (2017).

126. P. Huber, NEOS Data and the Origin of the 5 MeV Bump in the Reactor Antineutrino Spectrum, Phys. Rev. Lett. 118, 042502 (2017).

127. J.F. Cherry, A. Friedland, and I.M. Shoemaker, Short-Baseline Neutrino Oscillations, Planck, and IceCube, Short-baseline neutrino

oscillations, Planck, and IceCube arXiv:1605.06506.

128. M. Archidiacono, S. Gariazzo, C. Giunti, S. Hannestad, R. Hansen, M. Laveder, and T. Tram, Pseudoscalar—sterile neutrino interactions: Reconciling the cosmos with neutrino oscillations, J. Cosmol. Astropart. Phys. 2016, 067.
129. F. Forastieri, M. Lattanzi, G. Mangano, A. Mirizzi, P. Natoli, and N. Saviano, Cosmic microwave background constraints on secret interactions among sterile neutrinos, J. Cosmol. Astropart. Phys. 2017, 038.
130. N. Song, M.C. Gonzalez-Garcia, and J. Salvado, Cosmological constraints with self-interacting sterile neutrinos, J. Cosmol. Astropart. Phys. 2018, 055.
131. L. Johns, M. Mina, V. Cirigliano, M. W. Paris, and G. M. Fuller, Neutrino flavor transformation in the lepton-asymmetric universe, Phys. Rev. D 94, 083505 (2016).
132. M.G. Aartsen et al. (IceCube Collaboration), Determining neutrino oscillation parameters from atmospheric muon neutrino disappearance with three years of IceCube DeepCore data, Phys. Rev. D 91, 072004 (2015).
133. R. Wendell (Super-Kamiokande Collaboration), Atmospheric results from super-Kamiokande, AIP Conf. Proc. 1666, 100001 (2015).
134. M.G. Aartsen et al. (IceCube Collaboration), Searches for Sterile Neutrinos with the IceCube Detector, Phys. Rev. Lett. 117, 071801 (2016).
135. F. Dydak et al., A search for muon-neutrino oscillations in the Δm2 *range* 0.3–0.9 eV2 , Phys. Lett. 134B, 281 (1984).
136. I.E. Stockdale et al., Limits on Muon-Neutrino Oscillations in the Mass Range 30<Δm2<1000 eV2/c4 , Phys. Rev. Lett. 52, 1384 (1984).
137. G. Cheng et al. (MiniBooNE and SciBooNE Collaborations), Dual baseline search for muon antineutrino disappearance at 0.1 eV2<Δm2<100 eV2 , Phys. Rev. D 86, 052009 (2012).
138. P. Adamson et al. (NOvA Collaboration), Search for active-sterile neutrino mixing using neutral-current interactions in NOvA, Phys. Rev. D 96, 072006 (2017).
139. K. Abe et al. (T2K Collaboration), Search for light sterile neutrinos with the T2K far detector Super-Kamiokande at a baseline of 295 km, Phys. Rev. D 99, 071103 (2019).

140. S. Gariazzo, P.F. de Salas, and S. Pastor, Thermalisation of sterile neutrinos in the early Universe in the 3+1 *scheme with full mixing matrix*, J. Cosmol. Astropart. Phys. 2019, 014.

141. K.J. Kelly, Searches for new physics at the Hyper-Kamiokande experiment, Phys. Rev. D 95, 115009 (2017).

142. J.A. Formaggio and G.P. Zeller, From eV to EeV: Neutrino cross sections across energy scales, Rev. Mod. Phys. 84, 1307 (2012).

143. I. Esteban, M.C. Gonzalez-Garcia, M. Maltoni, I. Martinez-Soler, and T. Schwetz, Updated fit to three neutrino mixing: Exploring the accelerator-reactor complementarity, J. High Energy Phys. 2017, 087.

144. D. Foreman-Mackey, D.W. Hogg, D. Lang, and J. Goodman, emcee: The MCMC hammer, Publ. Astron. Soc. Pac. 125, 306 (2013).

145. D.Z. Freedman, Coherent neutrino nucleus scattering as a probe of the weak neutral current, Phys. Rev. D 9, 1389 (1974).

146. J. Erler and M.J. Ramsey-Musolf, The weak mixing angle at low energies, Phys. Rev. D 72, 073003 (2005).

147. J. Erler and R. Ferro-Hernández, Weak mixing angle in the Thomson limit, J. High Energy Phys. 2018, 196.

148. B.C. Cañas, E.A. Garcés, O.G. Miranda, and A. Parada, Future perspectives for a weak mixing angle measurement in coherent elastic neutrino nucleus scattering experiments, Phys. Lett. B 784, 159 (2018).

149. R.H. Helm, Inelastic and elastic scattering of 187-MeV electrons from selected even-even nuclei, Phys. Rev. 104, 1466 (1956).

150. V.I. Kopeikin, Flux and spectrum of reactor antineutrinos, Yad. Fiz. 75, 165 (2012); [Phys. At. Nucl. 75, 143 (2012)].

151. N. Raper, Ph.D. thesis, Rensselaer Polytechnic Institute, 2016.

CHAPTER

22

Neutrino Portals to Dark Matter

M. Blennow[1,2], E. Fernandez-Martinez[1], A. Olivares-Del Campo[3], S. Pascoli[3], S. Rosauro-Alcaraz[1], A. V. Titov[3]

[1] Departamento de Física Teórica and Instituto de Física Teórica, IFT-UAM/CSIC, Universidad Autónoma de Madrid, Cantoblanco, 28049 Madrid, Spain

[2] Department of Physics, School of Engineering Sciences, KTH Royal Institute of Technology, AlbaNova University Center, Roslagstullsbacken 21, 106 91 Stockholm, Sweden

[3] Institute for Particle Physics Phenomenology, Department of Physics, Durham University, South Road, Durham DH1 3LE, UK

ABSTRACT

We explore the possibility that dark matter interactions with Standard Model particles are dominated by interactions with neutrinos. We examine whether it is possible to construct such a scenario in a gauge invariant manner. We first study the coupling of dark matter to the full lepton doublet and

Citation: Blennow, M., Fernandez-Martinez, E., Olivares-Del Campo, A. et al. "Neutrino portals to dark matter". Eur. Phys. J. C 79, 555 (2019). https://doi.org/10.1140/epjc/s10052-019-7060-5

confirm that this generally leads to the dark matter phenomenology being dominated by interactions with charged leptons. We then explore two different implementations of the neutrino portal in which neutrinos mix with a Standard Model singlet fermion that interacts directly with dark matter through either a scalar or vector mediator. In the latter cases we find that the neutrino interactions can dominate the dark matter phenomenology. Present neutrino detectors can probe dark matter annihilations into neutrinos and already set the strongest constraints on these realisations. Future experiments such as Hyper-Kamiokande, MEMPHYS, DUNE, or DARWIN could allow to probe dark matter-neutrino cross sections down to the value required to obtain the correct thermal relic abundance.

INTRODUCTION

The unknown origin of neutrino masses and mixing together with the existence of the dark matter (DM) component of the Universe constitute our most significant experimental evidence for physics beyond the Standard Model (SM) and therefore the best windows to explore new physics. Neutrinos and DM also share an elusive nature with very weak interactions with the other SM particles. Indeed, neutrinos only participate in the weak interactions of the SM while all direct and indirect searches for DM interactions with the SM, other than gravity, are so far negative or inconclusive. A tantalising avenue of investigation is the possibility of a stronger connection between these two sectors. In this case, the best way to probe DM would be through the neutrino sector.

Several works have investigated the phenomenology of a dominant interaction between the neutrino and DM sectors and the possibility to probe DM through neutrinos both via its cosmological implications [1,2,3,4,5,6,7,8,9,10,11,12,13,14] as well as through indirect searches [14,15,16,17,18]. In the presence of this interaction, DM would no longer be collisionless, but able to scatter with neutrinos in the Early Universe, affecting matter density fluctuations. Moreover, the power spectrum would show a suppression at small scales [9, 10, 14] or even an oscillatory pattern [3,4,5, 8]. Indirect detection searches for DM annihilating to neutrinos in the galactic centre have also been performed at neutrino detectors and used to constrain DM–neutrino interactions [14,15,16]. The propagation of neutrinos through DM halos could be modified as well, leading to dips in supernova neutrino spectra due to resonant interactions with DM [19, 20], or affect the spectrum or isotropy of the high energy cosmic neutrinos observed by IceCube [21,22,23].

However, it is not straightforward to envision a scenario in which the neutrino–DM interactions dominate the DM phenomenology. Naively, gauge invariance dictates that the interactions of the left-handed (LH) SM neutrinos with DM will be equal to those of their charged lepton counterparts in the *SU*(2) doublets. In this case, the best windows to DM would instead be the charged leptons rather than the more elusive neutrinos.

In this work, we will investigate some gauge-invariant SM extensions that lead to sizeable neutrino–DM interactions, exploring if neutrino probes could dominate our sensitivity to the dark sector. This is actually a rather natural possibility. In fact, if DM does not participate in any of the SM gauge interactions, the natural expectation is that the strongest connection to DM will be via singlets of the SM gauge group. Indeed, if non-singlet fields were involved instead, the dimensionality of the operators linking the two sectors would have to increase in order to comply with gauge invariance. This reasoning leads to the three well-known SM portals to the dark sector: the "gauge boson portal" [24], the "Higgs portal" [25], and the "neutrino portal" [17, 26, 27]. The neutrino portal includes the addition of right-handed (RH) neutrinos N_R, which makes this option particularly appealing in connection to the evidence of neutrino masses and mixing from neutrino oscillations.

Since the neutrino portal relies on the mixing between N_R and the light SM neutrinos to connect the neutrino and DM sectors, this mixing needs to be sizeable. In the "canonical" seesaw mechanism [28,29,30,31,32], the smallness of neutrino masses is explained through a large Majorana mass for N_R and the mixings are then similarly suppressed by the large scale. This option, which is rather natural from the point of view of neutrino masses, worsens the Higgs hierarchy problem [33]. An interesting alternative is to explain the smallness of neutrino masses via a symmetry argument instead [34,35,36,37,38,39]. Indeed, in models with an approximate lepton number (L) symmetry such as the linear [34, 35] or inverse [40] seesaw mechanisms, neutrino masses are suppressed by the small L-breaking parameters while light neutrino mixing with N_R is unsuppressed. In the present study, we will assume relatively large mixing angles noting that they can be compatible with neutrino masses, but we will not specify a concrete neutrino mass generation mechanism, since these small lepton number violating parameters, and hence light neutrino masses, will have no significant impact on the DM-related phenomenology.

We will consider fermionic DM and, more specifically, Dirac DM, which has the richest phenomenology when interacting with SM neutrinos. Indeed, the dominant term in the annihilation cross section to neutrinos is not velocity suppressed, and DM annihilations therefore lead to interesting signatures in indirect searches. Alternative scenarios with a Majorana, scalar, or vector DM candidate will lead to a velocity-dependent annihilation cross section to neutrinos [14]. While such possibilities are viable, they are difficult to probe experimentally at neutrino detectors. This is due to the fact that the DM velocity in the halo today is $v_{halo}=10^{-3}c$ [41], which significantly reduces the annihilation rate to neutrinos.

The article is organised as follows. In Sect. 2, we summarise relevant experimental searches for DM and constraints coming from cosmology. In Sect. 3, we consider the simplest gauge-invariant scenario, in which DM is coupled directly to the full SM lepton doublet. In this case, as expected, the charged lepton probes tend to dominate the constraints on the DM parameter space. Further, in Sect. 4, we introduce the neutrino portal involving one new Dirac sterile neutrino *N*, which will communicate with the dark sector. We present two realisations of the neutrino portal, for scalar [42,43,44,45] and vector [46] interactions between the DM and *N* in Sects. 5 and 6, respectively. For both of them, we investigate the parameter space, demonstrating that current and future neutrino experiments have the dominant role in constraining it. Finally, we conclude in Sect. 7.

CONSTRAINTS ON INTERACTIONS OF DM WITH SM PARTICLES

In the next sections, we will explore the parameter space of different possible gauge-invariant ways to realise interactions of neutrinos with DM. For each realisation, we will investigate whether it is possible for these DM–neutrino interactions to play a dominant role in the DM phenomenology. In particular, we will address whether or not the DM relic abundance can be achieved via the DM–neutrino interactions and/or if indirect DM searches via its annihilation into neutrinos (probed at neutrino detectors) can be the dominant test of the model parameter space. We will use the observables presented in this section to place constraints on the parameter space of each scenario.

Indirect Detection Searches for DM Annihilation to Neutrinos

DM annihilating in high density regions such as the Milky Way can generate a significant monochromatic flux of neutrinos with energy $E_\nu=m_\chi$, where m_χ is the DM mass. This flux is proportional to the integral of the DM density squared along the line of sight and can be searched for in neutrino detectors such as Super-Kamiokande (SK) [47] or Borexino [48].

Several analyses that use neutrino detectors to probe the DM parameter space have been performed in the literature [14,15,16, 18, 49,50,51,52]. For small DM masses in the range 2–17 MeV, we can exploit the upper bound on the monochromatic antineutrino flux set by Borexino [53] and convert it to a conservative upper bound of $\langle \sigma v_r \rangle \lesssim 10^{-22}$–$10^{-20}$ cm^3/s on the thermally averaged annihilation cross section σ multiplied by the relative velocity v_r of DM particles, as discussed in Ref. [14]. Likewise, between 10 and 200 MeV, SK can place an upper bound of $\langle \sigma v_r \rangle \lesssim 10^{-25}$–$10^{-23}$ cm^3/s (depending on the DM mass) [14]. For DM with a mass between 1 GeV and 10 TeV annihilating in the galactic centre, the SK collaboration has performed a dedicated analysis and set an upper bound of $\langle \sigma v_r \rangle \sim 10^{-24}$–$10^{-22}$ cm^3/s [50]. We will also consider the general upper bound on ⟨σvr⟩ derived in Ref. [15] by calculating the cosmic diffuse neutrino signal from DM annihilations in all halos in the Universe and comparing it to the measured atmospheric neutrino background by Fréjus [54], AMANDA [55], and SK. This bound applies to m_χ in the range between 100 MeV and 100 TeV and excludes $\langle \sigma v_r \rangle \gtrsim 10^{-23}$–$10^{-21}$ cm^3/s (depending on m_χ). As argued in Ref. [15], this bound could be improved by one or even two orders of magnitude with dedicated analyses by existing neutrino experiments such as SK.

The next generation experiment Hyper-Kamiokande (HK) [56] will be sensitive to approximately one order of magnitude smaller cross sections in this mass range. Indeed, with a 187 kton fiducial mass and an exposure time of 10 years, HK could probe the parameter space almost down to the relic density cross section ($\langle \sigma v_r \rangle = 3 \times 10^{-26}$ cm^3/s [57]). Possible improvements such as additional mass from a second tank together with Gd doping for background reduction would allow to probe beyond this value [51]. Similarly, the ESSνSB project [58] envisions a 500 kton fiducial water detector, MEMPHYS [59], that would have slightly better sensitivity than HK from the additional fiducial mass. Similarly, future DM and neutrino detectors such as DARWIN [60] and DUNE [61] will be able to further constrain the DM annihilation cross section to neutrinos. DARWIN will set stronger bounds for DM masses between 100 MeV and 1 GeV [62], while DUNE will be able to exclude thermal DM masses between 25 and 100 MeV [52].

Competitive constraints from DM annihilations in the Sun to neutrinos, or other SM particles that decay to neutrinos, have also been derived by neutrino detectors such as SK [63] and IceCube [64]. These exploit the higher DM concentration expected in the solar interior since it could capture DM particles from the halo via scatterings. In all the realisations under study we explore the connection between the DM and neutrino sectors with very suppressed interactions with the rest of the SM, in particular with quarks. Thus, in these scenarios, the Sun does not accrete DM particles effectively and the constraints from these searches do not apply.

Indirect Detection Searches for DM Annihilation to Charged Leptons

DM interactions with charged leptons will always be present either at tree level, if DM couples to the full doublet, or at loop level in the neutrino portal scenarios. Therefore, we will take into account indirect detection searches for DM annihilations to charged leptons from the Fermi satellite [65], as well as from their imprint in the cosmic microwave background (CMB) as observed by Planck [66, 67].

Direct Detection Searches

DM will not couple directly to the quarks in any of the scenarios that we will discuss. Nevertheless, such couplings will arise at loop level in a similar way to the DM–charged lepton interactions. As we will see, bounds from direct detection experiments, such as XENON1T [68], are so stringent that they will still constrain the parameter space for large DM masses. Recently, direct detection of sub-GeV DM via scattering off electrons has gained significant attention [69,70,71,72]. We have also considered this process and found it to be sub-leading with respect to other relevant constraints.

Constraints from Cosmology

If DM remains in thermal equilibrium with neutrinos during Big Bang nucleosynthesis (BBN), it can spoil its predictions [73, 74]. Similarly, the effective number of neutrinos, as constrained by CMB measurements, would be affected if DM remained in equilibrium after neutrinos decoupled from the photon plasma [75,76,77]. Thus, to avoid these two effects, we will not consider DM masses m_χ<10 MeV. Moreover, DM–neutrino interactions can also have an effect in the formation of large scale structures (LSS) since, as DM particles scatter off neutrinos, they diffuse out and erase small scale perturbations. This effect leads to a suppression of the amount of small scale structures today. By comparing LSS predictions to observations, one can set an upper bound on the strength of the elastic scattering between DM and neutrinos [8, 78]. Nevertheless, for the models we are presenting in this work, the mixing between the sterile and SM neutrino suppresses the neutrino-DM elastic scattering and, consequently, its effect on LSS constrains regions of the parameter space already ruled out by CMB and BBN constraints [14].

COUPLING TO THE FULL LEPTON DOUBLET

In this section, we will study the simplest scenario, in which the neutrino–DM interaction arises from a direct coupling to the full SM *SU*(2) lepton doublet. In order to avoid specifying the nature of the mediator, we will adopt an effective field theory approach, simply adding a d=6, 4-fermion interaction.

Model

Since the 4-fermion operator needs to involve two LH SM lepton doublets $L_\alpha = (\nu_{\alpha L}, \ell_{\alpha L})^T, \alpha = e, \mu, \tau$, its Lorentz structure is fixed to be $\overline{L_\alpha}\gamma^\mu L_\alpha$. For definiteness we will assume a vector structure for the DM part. An axial coupling would instead lead to a velocity-suppressed DM annihilation cross section to neutrinos for both DM relic abundance and indirect searches. The cross section for DM annihilation to charged leptons would however have an additional term only suppressed by the lepton mass, and thus, it would tend to dominate over the annihilation cross section to neutrinos. Therefore, we will not consider this option in what follows.

The Lagrangian describing the neutrino–DM interaction is thus given by

$$\mathcal{L} = \mathcal{L}_{\text{SM}} + \overline{\chi}\left(i\partial\!\!\!/ - m_\chi\right)\chi + \frac{c_\alpha}{\Lambda^2}\overline{\chi}\gamma_\mu\chi\,\overline{L_\alpha}\gamma^\mu L_\alpha, \tag{3.1}$$

where χ is a Dirac fermion DM particle, and flavour diagonal couplings c_α/Λ^2 between DM and the lepton doublets have been assumed in order to avoid new sources of flavour violation. For the effective description to be consistent we will require that $\Lambda^2/c_\alpha \gg m_\chi^2$. The simplest UV completion which leads to the d=6 operator in Eq. (3.1) is via the exchange of a new heavy vector boson that couples both to χ and L_α.

The Lagrangian in Eq. (3.1) implies that, in this naive gauge-invariant scenario, the coupling between the SM neutrinos and DM will be accompanied by a DM–charged lepton coupling of the same strength. Therefore, the strongest constraints on this model will typically come from indirect searches for DM annihilations to charged leptons. The DM relic abundance will also be set by its annihilation into leptons, either neutrinos or charged leptons, with the annihilation cross section given by

$$\langle \sigma v_r \rangle \approx \frac{c_\alpha^2 m_\chi^2}{2\pi \Lambda^4} \left(1 - \frac{m_\alpha^2}{4m_\chi^2}\right) \sqrt{1 - \frac{m_\alpha^2}{m_\chi^2}}, \tag{3.2}$$

where m_α is the lepton mass for the different α flavour.

Results

In Fig. 1, we show regions in the parameter space of the DM mass m_χ and the new physics scale Λ excluded by different experiments. The blue line corresponds to the correct DM relic density $\Omega_{\rm DM} h^2 = 0.1193 \pm 0.0009$ [66] obtained through the thermal freeze-out mechanism. This line has been computed with micrOMEGAs [79]. In the upper hatched region, the DM–lepton interaction would be too weak, leading to overclosure of the Universe ($\Omega_{\rm DM} h^2 > 0.12$). In the region below the blue line, the relic density is smaller than the observed DM abundance. If there are additional production mechanisms contributing to the DM density, this region is also viable.

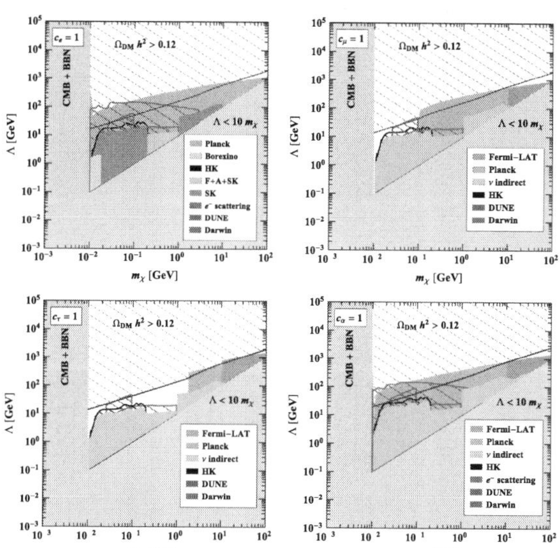

Figure 1: Constraints on the DM mass m_χ and the new physics scale Λ. The upper and bottom-left panels correspond to couplings to only one of the

lepton doublets (electron, muon, or tau), while the bottom-right panel corresponds to all three couplings being of equal strength. Along the blue line we recover the correct DM relic abundance from thermal freeze-out. The coloured shaded regions are excluded by different experiments, while the hatched areas correspond to prospective sensitivities of future experiments. The lower bound $m_\chi \gtrsim 10$ MeV is set by observations of the CMB and BBN. See text for further details

The constraints from indirect DM searches outlined in Sect. 2 are shown as different shaded regions. The light green (Planck [66, 67]) and orange (Fermi satellite [65]) regions correspond to the bounds from DM annihilation to charged leptons described in Sect. 2.2. The remaining shaded regions correspond to the constraints from DM annihilation to neutrinos as searched for in neutrino detectors and summarised in Sect. 2.1. In the upper-left panel of Fig. 1, we show in different colours the bounds coming from different neutrino experiments. The SK analyses [14, 50] are shown in red while the Borexino bounds [53] are displayed in yellow. The pink colour corresponds to the bounds from [49] obtained by combining the atmospheric neutrino data.[Footnote1] The dark red hatched region corresponds to prospective sensitivity of experiments on DM-electron scattering [71], while the blue, black, and green hatched regions correspond to prospects from different neutrino experiments as described in Sect. 2.1. In the following panels and in the rest of the paper we show all present indirect detection constraints from neutrino experiments in pink colour.

As can be seen in Fig. 1, the strongest constraints come from DM annihilation to charged leptons as probed by Fermi-LAT [65] for $\chi\overline{\chi} \to \tau^+\tau^-, \mu^+\mu^-$ and from Planck [66, 67] for $\chi\overline{\chi} \to \ell^+\ell^-, \ell = e, \mu, \tau$. The latter are in agreement with the results of Ref. [80], where, in particular, the dimension 6 operator given in Eq. (3.1) has been analysed. Indirect searches at neutrino detectors will always play a sub-leading role as long as annihilation to charged leptons is possible. Indeed, present constraints from DM annihilation to charged leptons are strong enough to rule out the entire allowed region of the parameter space that could lead to the correct DM relic density as long as the coupling to electrons is sizeable. However, if DM dominantly couples to the heavier lepton generations, allowed windows open up for $m_\chi < m_\mu \ (m_\tau)$ (see the upper-right and bottom-left panels of Fig. 1). In this case, the DM relic density would be set by its annihilation to neutrinos, and the most relevant present constraints come from the results of SK and Borexino. The prospects for HK and DUNE would be very promising in these scenarios, allowing to probe most of the parameter space up to and beyond where the relic density is entirely explained by freeze-out based on neutrino interactions.

Regarding the constraints that could be set by the DM effects in the spectrum or isotropy of high energy cosmic neutrinos as observed by IceCube [21], these would lie in the region of the parameter space already excluded by the number of relativistic degrees of freedom in the early Universe [75,76,77].

From Fig. 1 it is clear that, as long as light DM couples to the electron doublet, this option for a neutrino–DM coupling is mostly ruled out by DM–electron interactions. However, if the DM coupling to L_e is negligible and DM dominantly couples to L_μ and/or L_τ, the viable part of parameter space with $m_\chi < m_\mu$ (m_τ) can be probed by the neutrino experiments.

COUPLING VIA THE NEUTRINO PORTAL

Given the results of the previous section, we will now explore whether the neutrino portal option is able to lead to a rich DM-neutrino phenomenology without being in conflict with indirect searches involving charged leptons. The first necessary ingredient is to have sizeable mixing between the SM neutrinos and the new sterile neutrinos that will mediate the DM interaction. Therefore, the sterile-light neutrino mixing should not scale with the light neutrino masses, unlike in the canonical seesaw mechanism. Therefore, we will instead attribute the smallness of neutrino masses to an approximate lepton number (or B−L) symmetry rather than to a hierarchy of scales between the Dirac and Majorana masses. The new singlets will thus form pseudo-Dirac pairs since lepton number violation will necessarily be very small to account for the lightness of SM neutrinos. This is the case for instance in the popular "inverse" [40] and "linear" [34, 35] seesaw mechanisms based on such a symmetry.

As a simplifying assumption we will here consider the addition of only one (pseudo-)Dirac sterile neutrino that will serve as portal between the SM neutrinos and DM. Neglecting this small lepton number violation, the couplings between the SM and the new Dirac singlet neutrino are given by

$$\mathcal{L} = \mathcal{L}_{\text{SM}} + \overline{N}\left(i\partial\!\!\!/ - m_N\right)N - \lambda_\alpha \overline{L_\alpha}\tilde{H}N_R, \tag{4.1}$$

where N is the Dirac sterile neutrino $\tilde{H} = i\sigma_2 H^*$, with H being the Higgs doublet.

Electroweak symmetry breaking gives rise to the neutrino Dirac mass term

$$\left(\overline{\nu_{\alpha L}},\ \overline{N_L}\right) M_\nu N_R + \text{h.c.}, \tag{4.2}$$

where $M_\nu = (\lambda_\alpha v,\ m_N)^T$ is the neutrino mass matrix and $v = \langle H^0 \rangle = 174\,\text{GeV}$ is the Higgs vacuum expectation value (vev). Diagonalising $M_\nu M_\nu^\dagger$ with a 4×44×4 unitary matrix U,

$$U^\dagger M_\nu M_\nu^\dagger U = \text{diag}\left(m_1^2, m_2^2, m_3^2, m_4^2\right), \tag{4.3}$$

we find the mass of the heavy neutrino to be

$$m_4 = \sqrt{m_N^2 + \sum_\alpha |\lambda_\alpha|^2 v^2}. \tag{4.4}$$

As expected, the lepton number symmetry forbids light neutrino masses. In order to account for neutrino masses, small breaking of this symmetry via terms such as $\mu\, \overline{N_L} N_L^c$ (inverse seesaw), or $\lambda'_\alpha \overline{L_\alpha} \tilde{H} N_L^c$ (linear seesaw) can be added. Since these small parameters would have negligible impact in the phenomenology of neutrino–DM interactions, we will not consider them in what follows.

The neutrino mixing matrix U, which relates LH flavour neutrino fields and the neutrino fields with definite masses as

$$\begin{pmatrix} \nu_{\alpha L} \\ N_L \end{pmatrix} = U \begin{pmatrix} \nu_{iL} \\ \nu_{4L} \end{pmatrix}, \quad \alpha = e, \mu, \tau, \quad i = 1, 2, 3, \tag{4.5}$$

has the form

$$U = \begin{pmatrix} U_{\alpha i} & U_{\alpha 4} \\ U_{si} & U_{s4} \end{pmatrix}. \tag{4.6}$$

The upper-left 3×3 block $U_{\alpha i}$ would correspond to the Pontecorvo–Maki–Nakagawa–Sakata (PMNS) matrix once the small lepton number-breaking terms that induce neutrino masses are taken into account. Note that this matrix, being a 3×3 sub-block of a larger unitary matrix will, in general, not be unitary. The upper-right 3×1 block $U_{\alpha 4}$ describes the mixing between the active flavour neutrinos and the LH component of the heavy neutrino with mass m4m4. The last row of the matrix U specifies the admixture of each ν_{jL}, j=1,2,3,4, in the LH sterile neutrino NL. As we will see in what follows, the DM-related phenomenology is driven by the mixing of active-heavy mixing matrix elements $U_{\alpha 4}$. We will use the unitarity deviations of the PMNS matrix to constrain these mixings [81]. The mixing elements of interest are given by

$$U_{\alpha 4} = \frac{\theta_\alpha}{\sqrt{1 + \sum_\alpha |\theta_\alpha|^2}}, \qquad U_{s4} = \frac{1}{\sqrt{1 + \sum_\alpha |\theta_\alpha|^2}},$$
$$\sum_{i=1}^{3} |U_{si}|^2 = \sum_{\alpha=e}^{\tau} |U_{\alpha 4}|^2, \tag{4.7}$$

with $\theta_\alpha = \lambda_\alpha v/m_N$. Note that, even though the SM neutrino masses have been neglected, the mixing with the extra singlet neutrino that will act as portal can still be sizeable. For definiteness we will fix the mixing to the different flavours to their 1σ limit from Ref. [81], namely:

$$|\theta_e| = 0.031, \quad |\theta_\mu| = 0.011, \quad |\theta_\tau| = 0.044. \tag{4.8}$$

In the following sections, we will explore two possible ways in which these Dirac neutrinos could couple to the dark sector and become portals between it and the SM neutrinos.

NEUTRINO PORTAL WITH A SCALAR MEDIATOR

In this first example, we will assume that DM is composed of a new fermion, singlet under the SM gauge group, and that a new scalar mediates the Dirac neutrino–DM interactions.

Model

The Lagrangian of the model we will consider is given by

$$\mathcal{L} = \mathcal{L}_{\rm SM} + \overline{\chi}\left(i\partial\!\!\!/ - m_\chi\right)\chi + \overline{N}\left(i\partial\!\!\!/ - m_N\right)N + \partial_\mu S^* \partial^\mu S$$
$$- \left[\lambda_\alpha \overline{L_\alpha}\tilde{H} N_R + \overline{\chi}\left(y_L N_L + y_R N_R\right) S + \text{h.c.}\right]$$
$$- \mu_S^2 |S|^2 - \lambda_S |S|^4 - \lambda_{SH} |S|^2 H^\dagger H, \tag{5.1}$$

where χ is a Dirac fermion DM candidate and S is a complex scalar. The fields χ and S form the dark sector of the model (they are SM singlets), while N serves as a mediator between the dark sector and SM. The Lagrangian in Eq. (5.1) respects a global $U(1)_L$ lepton number symmetry under which L_α, N, and S_* have the same charge and which protects the

SM neutrino masses. Moreover, the Lagrangian respects a global $U(1)_D$ dark symmetry, under which χ and S have equal charges. This preserved symmetry ensures the stability of χ, if $m_\chi < m_S$, where $m_S^2 = \mu_S^2 + \lambda_{SH} v^2$ is the mass squared of the scalar S. For $m_\chi > m_S$, the roles of χ and S would change, and S would be a DM candidate. While this possibility is perfectly viable, it is more difficult to probe at neutrino detectors, as the DM annihilation cross section to neutrinos is velocity-suppressed. In what follows we assume that $m_\chi < m_S$ and focus on fermionic DM.

This model was previously considered in Refs. [43, 45]. However, we will go beyond these works by performing a comprehensive analysis of the sensitivity of neutrino experiments to the parameter space of this model.

We will limit ourselves to the case in which DM is lighter than the heavy neutrino,[Footnote2] i.e., $m_\chi < m_4$. This is the so-called direct annihilation regime [82], since DM annihilates through the mediator directly to SM particles. As intended, the only channel for DM annihilation at tree level is the one into light neutrinos. This process occurs via a diagram involving a *t*-channel exchange of the scalar mediator *S*. In the opposite regime, which is usually referred to as secluded [82], DM annihilates to heavy neutrinos, which subsequently decay. The phenomenology of this regime has been studied in Refs. [83,84,85,86].

Neglecting velocity-suppressed terms, we find the following thermally averaged cross section for DM annihilation to neutrinos:

$$\begin{aligned}\langle \sigma v_r \rangle &\approx \frac{y_L^4}{32\pi}\left(\sum_{i=1}^{3} |U_{si}|^2\right)^2 \frac{m_\chi^2}{\left(m_\chi^2 + m_S^2\right)^2} \\ &\approx \frac{y_L^4}{32\pi}\left(\sum_{\alpha=e,\mu,\tau} |\theta_\alpha|^2\right)^2 \frac{m_\chi^2}{\left(m_\chi^2 + m_S^2\right)^2}.\end{aligned} \tag{5.2}$$

The product $y_L\sqrt{\sum_\alpha |\theta_\alpha|^2}$ controls $\langle \sigma v_r \rangle$ and, in order to allow for sufficient annihilation to reproduce the observed relic density, it cannot be too small. The value of the coupling y_L is limited by the requirement of perturbativity. We will restrict ourselves to $y_L < 4\pi$. Since the coupling y_R does not enter Eq. (5.2), and thus, does not affect the tree-level DM–neutrino interactions, in what follows we set it to zero for simplicity. Regarding the mixing parameters θ_α, the bounds on them depend on the mass of the heavy neutrino. For definiteness we will assume that the heavy neutrino has a mass above the electroweak scale. At this scale the bounds on heavy neutrino mixing derived in the global analysis of flavour and electroweak precision data performed in Ref. [81] apply. If smaller masses were instead considered, more stringent constraints from collider and beam-dump searches and, eventually, production in meson and beta decays could potentially apply [87] (see discussion in Sect. 6.3). In any case, all the observables relevant to DM phenomenology have a sub-leading dependence on m4m4. We also consider the case where the coupling λ_{SH}=0, ensuring the neutrino portal regime. In Refs. [43, 45], the radiative generation of the $|S|^2 H^\dagger H$ operator was considered and its effects on m_S as well as on the invisible width of the Higgs boson were found to be negligible.

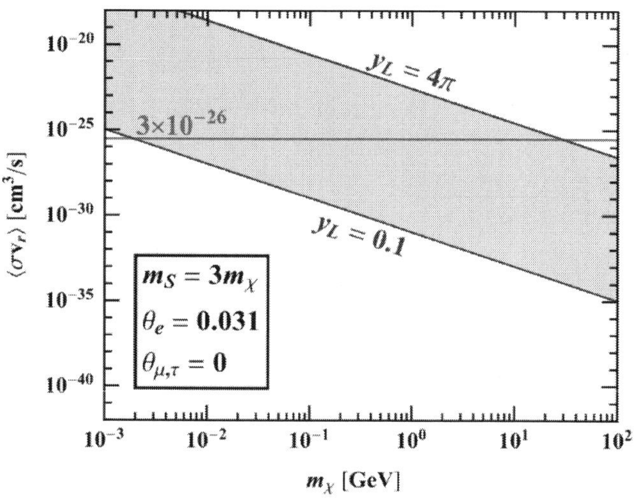

Figure 2: Thermally averaged annihilation cross section multiplied by the relative velocity for $\chi\bar{\chi} \to \nu\bar{\nu}$. We have fixed $m_S = 3m_\chi, \theta_e = 0.031,$ $\theta_\mu = \theta_\tau = 0$, and varied y_L between 0.1 and 4π.

In Fig. 2, we show the region of the parameter space for which the correct thermal relic abundance is obtained. This region spans DM masses up to 100 GeV for $|\theta_e|$=0.031, $\theta_\mu=\theta_\tau$=0, and y_L between 0.1 and 4π while keeping $m_S=3m_\chi$ as a benchmark.

Annihilation of DM into charged lepton-antilepton pairs $\ell^+\ell^-$ (ℓ=e,μ,τ) proceeds via the one-loop diagrams[Footnote3] shown in Fig. 3 (in unitary gauge).

The dominant contribution comes from the first and second diagrams, while the contribution from the last diagram is suppressed by the small Yukawa couplings of the charged leptons. The first diagram leads to the following effective operator:

$$\mathcal{L} \supset -a_{SW} \frac{g^2}{m_W^2} \overline{\chi} \gamma^\mu P_R \chi \, \overline{\ell_\alpha} \gamma_\mu P_L \ell_\beta, \tag{5.3}$$

where g is the weak coupling constant. Neglecting external momenta, the effective coupling a_{SW} is given by

$$a_{SW} = |U_{s4}|^2 U_{\alpha 4} U_{\beta 4}^* \frac{y_L^2}{(4\pi)^2} G\left(\frac{m_S^2}{m_4^2}\right), \tag{5.4}$$

where the loop function $G(x)$ reads

$$G(x) = \frac{x - 1 - \log x}{4(1-x)^2}. \tag{5.5}$$

The second diagram in Fig. 3 leads to the following effective interaction of DM with the Z boson:

$$\mathcal{L} \supset -a_Z \frac{g}{\cos\theta_W} \overline{\chi} \gamma^\mu P_R \chi Z_\mu, \tag{5.6}$$

where θW is the Weinberg angle and a_Z is the effective coupling, which in the limit of zero external momenta is given by

$$a_Z = |U_{s4}|^2 \left(1 - |U_{s4}|^2\right) \frac{y_L^2}{(4\pi)^2} G\left(\frac{m_S^2}{m_4^2}\right). \tag{5.7}$$

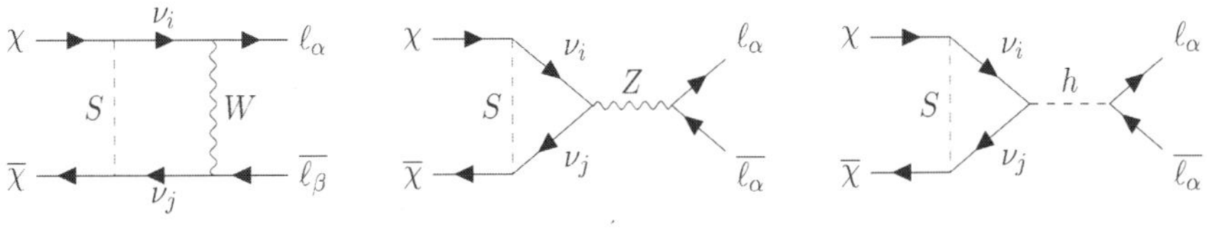

Figure 3: One-loop diagrams (in unitary gauge) contributing to annihilation of DM into charged lepton-antilepton pairs $\ell_\alpha \overline{\ell_\beta}, \alpha, \beta = e, \mu, \tau$. The indices *i* and *j* run from 1 to 4.

These contributions have been also computed using a combination of packages: FeynRules [89, 90] to produce a model file, FeynArts [91] for generating the diagrams and FormCalc [92] for computing their numerical contributions. For numerical evaluation of the Passarino-Veltman functions we have used LoopTools [92]. We have also considered the limit of zero external momenta, which effectively corresponds to the limit of small DM and charged lepton masses, and confronted the analytical results obtained in this approximation using the package ANT [93] with the LoopTools results. For DM masses between 1 MeV and 100 GeV that we are interested in, the approximation works very well. The availability of analytical expressions allows for an easier exploration of the parameter space.

In Fig. 4, we present the cross sections for annihilation of DM into e^+e^-, $\mu^+\mu^-$, and $\tau^+\tau^-$ for benchmark values of the model parameters. We fix $m_S=3m_\chi$, m4=400 GeV, $y_L=1$, $\theta_e=0.031$, and θμ,τ=0. As can be seen from the left panel, the annihilation cross sections to charged leptons are several orders of magnitude smaller than the cross section for DM annihilation into neutrinos. The difference in the cross sections becomes smaller when the DM mass approaches $m_Z/2$, and the cross sections for $\chi\overline{\chi} \to \ell^+\ell^-$ exhibit a resonant behaviour due to the second diagram in Fig. 3. In the right panel, we show the indirect detection constraints from Planck [66, 67] and Fermi-LAT [65]. Note that those constraints assume a 100% annihilation rate into a single SM channel. Even for $y_L=4\pi$ the resulting annihilation cross sections into charged leptons are well below the experimental constraints. Thus, the considered realisation of the neutrino portal does provide an example of a gauge-invariant model in which the neutrino–DM interactions dominate DM phenomenology.

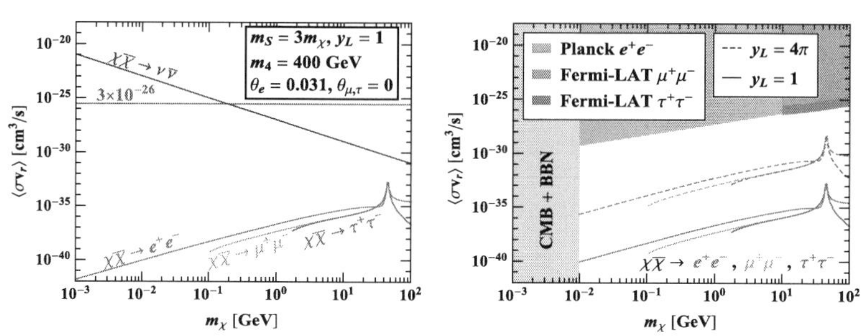

Figure 4: Thermally averaged annihilation cross section multiplied by the relative velocity for DM annihilation into $e^+e^-, \mu^+\mu^-$, and $\tau^+\tau^-$. We have fixed $m_S = 3m_\chi$, $m_4 = 400$ GeV, $y_L = 1$, $\theta_e = 0.031$ and $\theta_{\mu,\tau} = 0$. The *left* panel provides comparison with ⟨σvr⟩ for DM annihilation into neutrinos assuming the same set of model parameters. The *right* panel displays the indirect detection constraints coming from Planck and Fermi-LAT. The lower bound $m_\chi \gtrsim 10$ MeV is set by observations of the CMB and BBN. See text for further details.

At one-loop level DM also interacts with quarks via diagrams involving Z and h, which are analogous to those in Fig. 3. The corresponding effective DM-nucleon spin-independent scattering cross section reads [45]

$$\sigma_n = \frac{\mu_n^2}{\pi}\frac{\left(Zf_p + (A-Z)\, f_n\right)^2}{A^2}, \tag{5.8}$$

where μnμn is the reduced mass of the nucleon, A is the total number of nucleons in a nuclei, Z is the number of protons,

$$f_p = \left(4\sin^2\theta_W - 1\right)\frac{G_F a_Z}{\sqrt{2}}, \qquad f_n = \frac{G_F a_Z}{\sqrt{2}}, \tag{5.9}$$

with a_Z given in Eq. (5.7), and GF being the Fermi constant. The radiative coupling of DM to the Higgs, $\overline{\chi}\chi h$, would also give a contribution to direct detection searches. This contribution is however suppressed by the small quark Yukawa couplings. Direct detection of a SM singlet fermion DM candidate at one loop has been recently studied in detail in [94]. Moreover, an interesting example, which also provides radiative generation of neutrino masses, has been presented in [95].

The most stringent constraint on DM-nucleon spin-independent cross section for $m_\chi \gtrsim 10$ GeV comes from XENON1T [68]. As we will see in the next subsection, this constraint is strong enough to probe the loop-suppressed scattering process if the value of the coupling y_L is sufficiently large. We have also considered DM scattering off electrons and found that the corresponding cross section is much smaller than the projected sensitivities of silicon, germanium, and xenon experiments derived in Ref. [71]. Thus, DM-electron scattering cannot provide an additional probe of the considered neutrino portal model.

Results

In this subsection, we explore the parameter space to find regions that satisfy all direct and indirect detection constraints and in which the DM phenomenology could be dominated by its interactions with SM neutrinos. We show our results in the m_χ–m_S plane to determine the masses of the DM and the dark scalar that are presently allowed and could lead to the correct relic abundance (see Fig. 5).

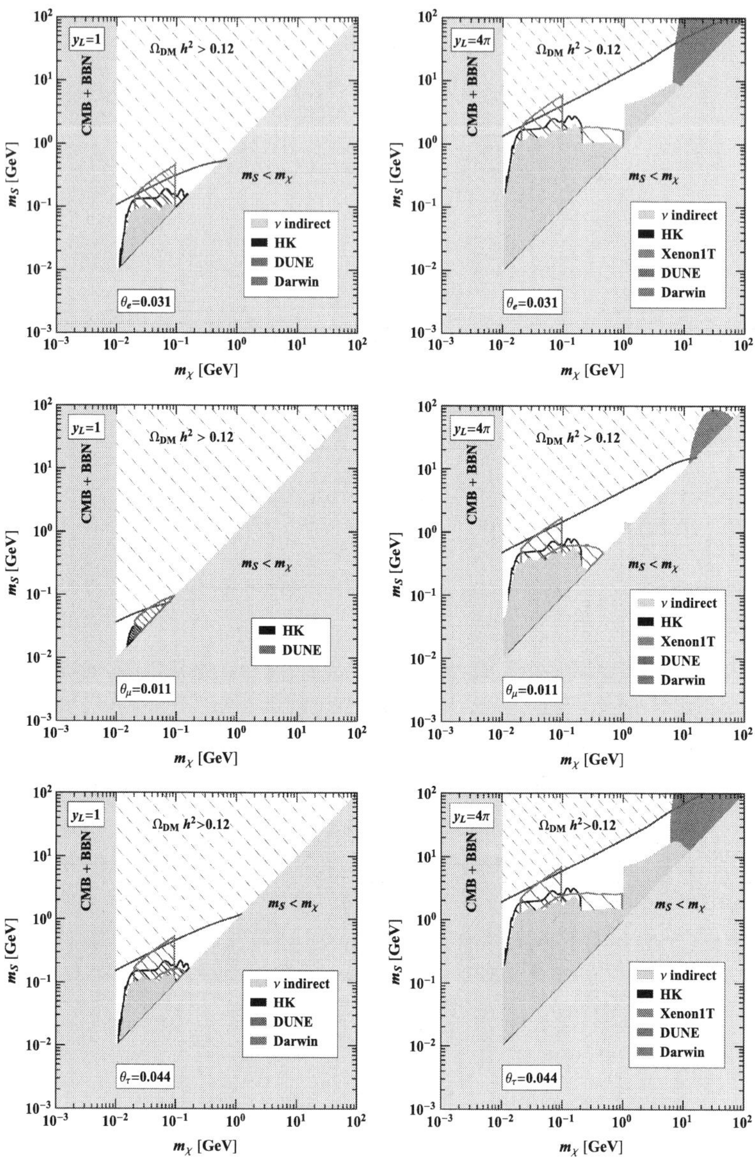

Figure 5: Constraints on the DM mass m_χ and the dark scalar mass m_S. We have fixed θ_e=0.031, $\theta_{\mu,\tau}$=0; θ_μ=0.011, $\theta_{e,\tau}$=0; and θ_τ=0.044, $\theta_{e,\mu}$=0 (from top to bottom), considering y_L=1 and 4π. Along the blue line the DM relic density matches the observed value. The coloured shaded regions are excluded by different experiments, while the hatched areas correspond to prospective sensitivities of future experiments. The lower bound $m_\chi \gtrsim 10$ MeV is set by observations of the CMB and BBN. See text for further details.

In Fig. 5, the triangular region $m_S<m_\chi$ is forbidden by DM stability. Along the blue line(s) computed with micrOMEGAs,[Footnote4] the DM relic density matches the observed value $\Omega_{\rm DM}h^2 = 0.1193 \pm 0.0009$ [66]. Above this line (the upper hatched region), the DM relic density is bigger than the measured value, i.e., DM overcloses the Universe. Below this line, the relic abundance would be smaller than the observed value. However, if there is an additional production mechanism, the relic abundance could also be compatible with this region.

As can be seen in the figure, indirect searches for annihilation to neutrinos, together with direct detection bounds by XENON1T for large DM masses, are the only probes that are presently constraining the allowed parameter space. The prospects to explore the remaining allowed regions through annihilation to neutrinos are very promising. In particular DUNE would be able to detect the neutrino signal in the range 25–100 MeV if the DM abundance is entirely due to this process.

In Fig. 6, we fix m_S to several representative values, namely m_S=0.04, 0.2, 1, and 5 GeV, and show the lines corresponding to the correct relic abundance in the m_χ–y_L plane. These results have been obtained with micrOMEGAs. Small values of y_L are ruled out since they do not lead to efficient DM annihilation. As can be seen, a lighter dark scalar allows for smaller values of y_L. For $m_S \gtrsim 500$ MeV, the values of $y_L \gtrsim 1$ are required to yield the observed relic density.

Overall, the cosmologically allowed parameter space of the model is already constrained by the current neutrino detectors as well as XENON1T.[Footnote5] Moreover, the next generation of neutrino experiments, in particular DUNE, will be able to probe thermal MeV fermion DM in the considered scenario.

NEUTRINO PORTAL WITH A VECTOR MEDIATOR

In this second example, we will again assume that DM is composed of a new Dirac fermion, this time coupled to a new massive vector boson. The Dirac singlet neutrino will also interact with this boson so as to provide the neutrino-DM interaction.

Model

The Lagrangian of the model is given by

$$\begin{aligned}\mathcal{L} = \mathcal{L}_{\rm SM} &+ \overline{\chi}\left(i\partial\!\!\!/ - m_\chi\right)\chi + \overline{N}\left(i\partial\!\!\!/ - m_N\right)N \\ &+ g'\overline{\chi_R}\gamma^\mu\chi_R Z'_\mu + g'\overline{N_L}\gamma^\mu N_L Z'_\mu \\ &- \left[\lambda_\alpha\overline{L_\alpha}\tilde{H}N_R + \text{h.c.}\right] \\ &- \frac{1}{4}Z'_{\mu\nu}Z'^{\mu\nu} + \frac{1}{2}m^2_{Z'}Z'_\mu Z'^\mu,\end{aligned} \tag{6.1}$$

where χ is a Dirac fermion DM candidate, Z' is a new vector boson mediating the interaction between neutrinos and DM, and N is the Dirac sterile neutrino connecting the dark and visible sectors through its mixing with the active neutrinos. This Lagrangian could for instance describe a new U(1)′ gauge symmetry spontaneously broken by the vev of a scalar SM singlet charged under it, that would induce masses for the Z′ as well as for the heavy neutrino N and the DM. The particular mechanism is not relevant for the rest of the discussion and will not be elaborated further. We will also assume there is an additional conserved charge (e.g., a $\mathbb{Z}_2$ symmetry) not shared between the neutrino and the DM that prevents their mixing. Note that in order to keep the Lagrangian in Eq. (6.1) anomaly free without introducing new fields, the simplest option is to couple the LH part of the Dirac sterile neutrino and the RH part of the DM to the new gauge boson with the same coupling g′.

As in the previous scenario, we will assume that the DM mass $m_\chi < m_4$ so that the dominant DM annihilation channel is to the three light SM neutrinos. This is a tree-level process and its cross section is given by

$$\begin{aligned}\langle\sigma v_r\rangle &\approx \frac{g'^4}{8\pi}\left(\sum_{i=1}^{3}|U_{si}|^2\right)^2 \frac{m_\chi^2}{(4m_\chi^2 - m_{Z'}^2)^2} \\ &\approx \frac{g'^4}{8\pi}\left(\sum_{\alpha=e,\mu,\tau}|\theta_\alpha|^2\right)^2 \frac{m_\chi^2}{(4m_\chi^2 - m_{Z'}^2)^2}.\end{aligned} \tag{6.2}$$

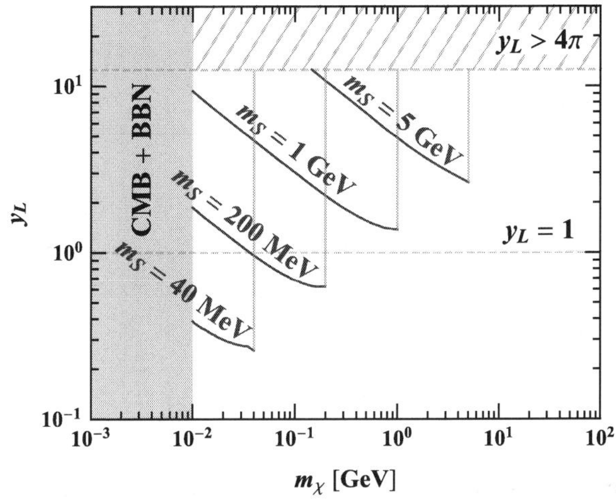

Figure 6: Values of the DM mass m_χ and the coupling y_L required to reproduce the observed relic abundance. We have fixed m_S=0.04m, 0.2, 1, and 5 GeV, and have considered the representative case of θ_e=0.031, while keeping $\theta_{\mu,\tau}$=0. Along (above) the blue lines the DM relic density matches (is less than) the observed value. The lower bound $m_\chi \gtrsim 10$ MeV is set by observations of the CMB and BBN.

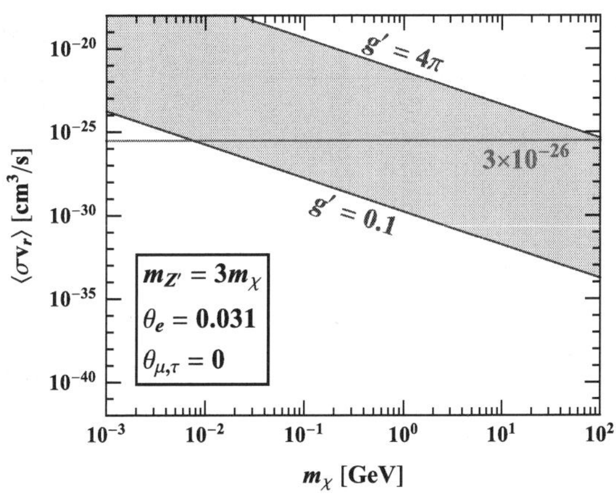

Figure 7: Thermally averaged annihilation cross section multiplied by the relative velocity for $\chi\overline{\chi} \rightarrow \nu\overline{\nu}$. We have fixed $m_{Z'} = 3m_\chi, \theta_e = 0.031$, θ_μ=θ_τ=0, and varied g′ between 0.1 and 4π.

Note however that, for $m_Z' \lesssim m_\chi$, the tree-level DM annihilation to a pair of Z′ bosons is allowed. When this channel is open, it will dominate over the direct annihilation into neutrinos, since the latter is suppressed by neutrino mixing. This is the so-called secluded annihilation regime [82], which we do not consider in the present study.

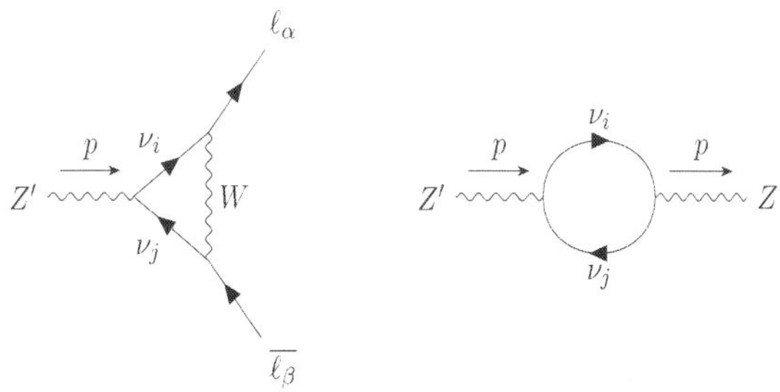

Figure 8: One-loop diagrams contributing to the coupling of the Z' boson to charged leptons (left) and to kinetic and mass mixing between the Z' and Z bosons (right).

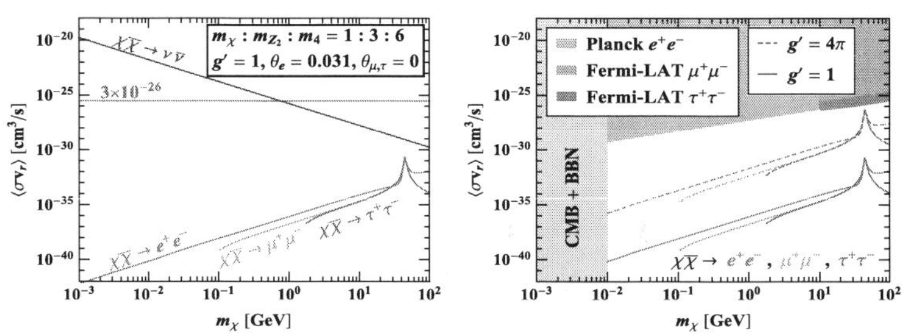

Figure 9: Thermally averaged annihilation cross section multiplied by the relative velocity for DM annihilation into $e^+e^-, \mu^+\mu^-$, and $\tau^+\tau^-$. We have fixed $m_\chi{:}m_{Z_2}{:}m_4 = 1{:}3{:}6$, $g' = 1$, $\theta_e = 0.031$, and $\theta_{\mu,\tau} = 0$. The *left* panel provides comparison with ⟨σvr⟩ for DM annihilation into neutrinos assuming the same set of model parameters. The *right* panel displays the indirect detection constraints coming from Planck and Fermi-LAT. The lower bound $m_\chi \gtrsim 10$ MeV is set by observations of the CMB and BBN. See text for further details.

In this scenario, as can be seen from Fig. 7, the correct relic abundance can be obtained purely from annihilation to the SM neutrinos for values of the new gauge coupling g′g′ between 0.1 and 4π4π, and DM masses in the 0.01–100 GeV range. In this figure, we have fixed $m_Z'=3m\chi$, $|\theta_e|=0.031$, and $\theta_\mu=\theta_\tau=0$ as benchmark values.

A direct coupling between the Z' boson and the charged leptons will also be induced through the loop diagrams in Fig. 8. Neglecting external momenta for the charged leptons, the effective vertex from the first loop diagram is given by

$$\mathcal{L} \supset -a_W g' \overline{\ell_\alpha} \gamma^\mu P_L \ell_\beta Z'_\mu, \tag{6.3}$$

where

$$a_W = |U_{s4}|^2 U_{\alpha 4} U^*_{\beta 4} \frac{g^2}{(4\pi)^2} \frac{m_4^2}{2m_W^2}. \tag{6.4}$$

Mixing with the *Z* Boson

Since the neutrino mass eigenstates have components that couple both to the Z and the Z', mixing between the two gauge bosons will be induced at loop level [24] through the second diagram in Fig. 8. The kinetic and mass mixings are described by the effective Lagrangian

$$\mathcal{L}_{Z'Z} = -\frac{\sin\epsilon}{2} Z'_{\mu\nu} Z^{\mu\nu} + \delta m^2 Z'_\mu Z^\mu. \tag{6.5}$$

Notice that these two terms could be present already at the Lagrangian level after gauge symmetry breaking. These would represent additional free parameters of the Lagrangian. However, these parameters do not contribute to the neutrino portal of interest here. Conversely, the neutrino mixing required for the neutrino portal does induce the Z–Z' mixing at the loop level. Barring fine-tuned cancellations between the allowed free parameters at the Lagrangian level and the loop-induced contributions from neutrino mixing, the minimum contribution present in our set-up will be the latter. We will therefore set the tree-level parameters to zero and require that the loop-induced contributions are below the present experimental constraints on Z–Z' mixing. We find the following results for the mixing parameters:

$$\delta m^2 = \frac{2}{(4\pi)^2} g' \frac{g}{\cos\theta_W} |U_{s4}|^2 \left(1 - |U_{s4}|^2\right) m_4^2 f_1, \tag{6.6}$$

$$\sin\epsilon = \frac{2}{(4\pi)^2} g' \frac{g}{\cos\theta_W} |U_{s4}|^2 \left(1 - |U_{s4}|^2\right) f_2, \tag{6.7}$$

where f_1 and f_2 are functions of $x \equiv m_4^2/p^2$, namely,

$$f_1(x) = \frac{1}{12}\left\{4x^2\left(1 - x^{-1}\right)^3\right.$$

$$\coth^{-1}(1-2x) + 2x - x^{-1}\log(x) - 2\sqrt{x\left(4 - x^{-1}\right)^3}$$

$$\arctan\left((4x-1)^{-1/2}\right)\Bigg\}. \tag{6.8}$$

$$f_2(x) = -\frac{x^2}{6}\Bigg\{4\left(2x-3+x^{-2}\right)\coth^{-1}(1-2x) + 4 + x^{-2}\log(x) - 2\sqrt{x^{-1}(4-x^{-1})}\left(2+x^{-1}\right) \arctan\left((4x-1)^{-1/2}\right)\Bigg\}. \tag{6.9}$$

For the purposes of this work $p^2 \sim m_X^2$, and thus, f_1 and f_2 will only depend on the ratio of the masses of the heavy neutrino and the DM particle. Following Ref. [96], we first diagonalise the kinetic term through a non-unitary transformation and then perform a rotation to diagonalise the mass term. The mass eigenstates Z_1 and Z_2 have masses given by

$$m_{Z_{1,2}}^2 = \frac{\sec^2\epsilon}{2}\left(m_Z^2 + m_{Z'}^2 - 2\delta m^2\sin\epsilon \mp \Delta\right), \tag{6.10}$$

where

$$\Delta = \mathrm{sgn}\left(m_{Z'}^2 - m_Z^2\left(1-2\sin^2\epsilon\right) - 2\delta m^2\sin\epsilon\right) \times \sqrt{m_Z^4 + m_{Z'}^4 + 4\delta m^4 - 4\left(m_Z^2+m_{Z'}^2\right)\delta m^2\sin\epsilon - 2m_Z^2 m_{Z'}^2\left(1-2\sin^2\epsilon\right)}. \tag{6.11}$$

From Eq. (6.10), one can easily verify that in the limit of small mass and kinetic mixing, i.e., $\delta m^2 \to 0$ and $\sin\epsilon \to 0$, the masses $m_{Z_1} \to m_Z$ and $m_{Z_2} \to m_{Z'}$. After the full diagonalisation, we can write the Z and Z' in terms of the mass eigenstates Z_1 and Z_2 as follows:

$$Z_\mu = (\cos\xi - \tan\epsilon\sin\xi)\, Z_{1\mu} - (\sin\xi + \tan\epsilon\cos\xi)\, Z_{2\mu}, \tag{6.12}$$

$$Z'_\mu = \sec\epsilon\left(\sin\xi\, Z_{1\mu} + \cos\xi\, Z_{2\mu}\right), \tag{6.13}$$

where ξ is the angle related to the mass diagonalisation, which is defined through

$$\tan(2\xi) = \frac{2\cos\epsilon\left(m_Z^2\sin\epsilon - \delta m^2\right)}{m_{Z'}^2 - m_Z^2\left(1-2\sin^2\epsilon\right) - 2\delta m^2\sin\epsilon}. \tag{6.14}$$

The two angles ξ and ϵ will control the phenomenology associated to the Z-Z' mixing and consequently, the possible Z' couplings to fermions.

The loop-induced kinetic mixing parameter ϵ depends solely on the ratio $x \approx m_4^2/m_X^2$, providing the coupling g' and the element U_{s4} of the neutrino mixing matrix are fixed (see Eqs. (6.7) and (6.9)), and increases with it. Fixing $|\theta_e| = 0.031$ and $\theta_{\mu,\tau} = 0$, we find that for x=4, which is the lowest value preventing the $\chi\overline{\chi} \to \nu_i\overline{\nu_4}, i = 1, 2, 3$, channels, and g′=1 (4π), the mixing parameter |sinϵ| is of order of 10^{-6} (10^{-5}). For values of *x* as large as 10^4 and g′=1 (4π), the value of |sinϵ| does not exceed approximately 10^{-5} (10^{-4}).

Generally, these values can be probed in beam dump and fixed target experiments searching for visible decay products (electrons and muons) of the Z_2 boson with mass between approximately 1 MeV and 1 GeV (see, e.g., [97, 98]). However, in the considered model the Z_2 decays mostly invisibly, either to a pair of the SM neutrinos or, if it is heavy enough, to a pair of DM particles, while its decays to charged leptons are suppressed. Thus, the bounds from fixed target experiments will not apply in this case. The supernova constraints cover nearly the same Z_2 masses, but a different range of $\epsilon \sim 10^{-10}$–10^{-7} [97], which thus are also avoided. For larger Z_2 masses, up to 100 GeV, collider experiments place the best constraints on $\epsilon \sim 10^{-4}$–10^{-3} (see, e.g., Ref. [98]). There exist also collider searchers for Z_2 decaying invisibly, which constrain $\epsilon \lesssim 10^{-3}$ for $m_{Z_2} < 8$ GeV [99]. These collider constraints are above the values of the loop-induced kinetic mixing parameter in our model. Finally, the much weaker constraint from the invisible Z_1 width, $\epsilon \lesssim 0.03$ [100], is also evaded.

Together with the first diagram in Fig. 8, the size of ξ and ϵ will determine how relevant the DM annihilation to a pair of charged leptons is. We find that the tree-level annihilation to neutrinos dominates over that to charged leptons. In Fig. 9, we show a particular example of this behaviour for $m_4 = 2m_{Z_2}$, $m_{Z_2} = 3m_\chi$, $g' = 1$, $|\theta_e| = 0.031$, and $\theta_\mu = \theta_\tau = 0$. It is clear from this figure that the annihilation to charged leptons is unconstrained by current experimental searches. Note that the Planck and Fermi-LAT constraints shown in the right panel of Fig. 9 assume a 100%100% annihilation rate into a single SM channel.

Results

The allowed regions of the parameter space in the m_χ–m_{Z2} plane that satisfy cosmological, indirect and direct detection constraints for this model are presented in Fig. 10 for g′=1 and 4π, setting $\theta_\alpha \neq 0$ one at a time and keeping two other mixing angles fixed to zero. For definiteness, in the figure we set $m_4 = 2m_{Z2}$. Notice that this choice is not relevant for the interaction between the SM neutrinos and DM and only plays a role in the loop-induced processes that are sub-dominant. Nevertheless, if the Z_2 originates from a new U(1)′ gauge group, its mass m_{Z2}, as well as that of the Dirac neutrino m_4, are generated after the breaking of the symmetry. Thus, the natural expectation is that m4m4 is not much heavier than m_{Z2} as long as the new gauge coupling g′ is O(1). Hence, unlike for the scalar example, it is not appropriate to set m4m4 to a value above the electroweak scale while exploring (sub-)GeV Z_2 boson masses.

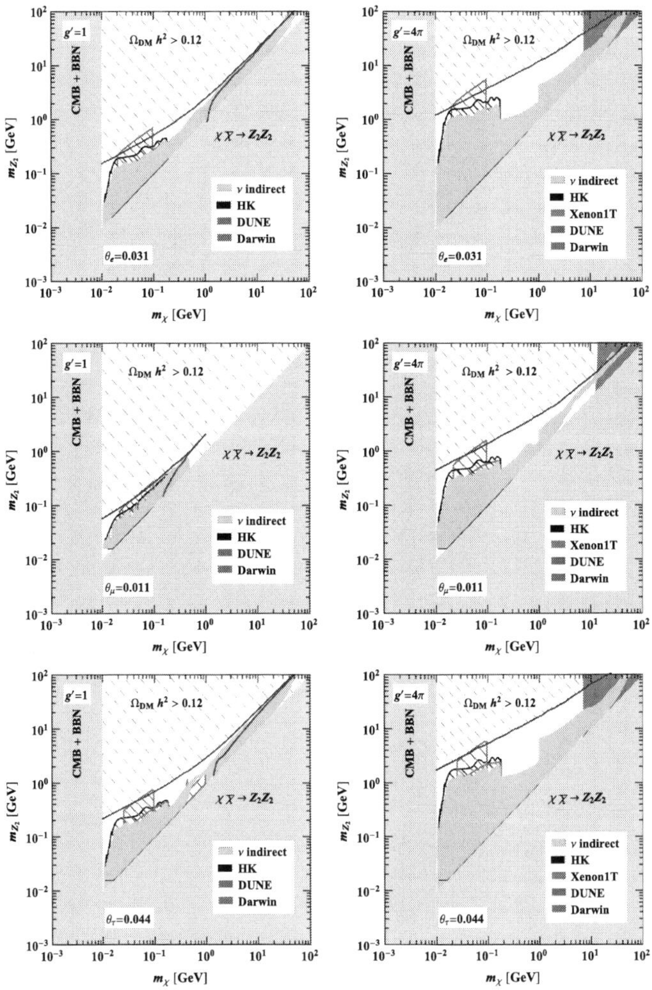

Figure 10: Constraints on the DM mass m_χ and m_{Z2}. Along the blue lines, computed with micrOMEGAs, the DM relic density matches the observed value. The coloured shaded regions are excluded by different experiments. The lower

bound $m_\chi \gtrsim 10$ MeV is set by observations of the CMB and BBN. See text for further details.

Below the electroweak scale constraints on the neutrino mixing parameters θ_α are *a priori* much more stringent [87]. However, in the model under investigation the heavy neutrino decays mostly invisibly to either a SM neutrino and the Z_2 (if $m_4>m_{Z2}$), or a SM neutrino and a pair of the DM particles (if $m_4<m_{Z2}$), assuming $g' \gtrsim 1$. This implies that the existing collider and beam dump constraints[Footnote6] should be rescaled with the corresponding branching ratios and become even weaker than the non-unitarity constraints imposed previously for the scalar realisation. The bounds from peak searches in leptonic decays of pions and kaons will however apply, since they rely entirely on the kinematics of a two-body decay. Thus, the non-unitarity constraints actually dominate down to $m_4 \approx m_K \approx 0.5$ GeV, where m_K is the kaon mass. In the region m4~0.01 GeV, the bounds on U_{e4} and $U_{\mu 4}$ from peak searches are very stringent. We do not display them explicitly in Fig. 10, because they are m_4-dependent, while all the constraints shown in the figures have an extremely sub-leading dependence on m_4, as outlined above. Thus, Fig. 10 is to be interpreted as generally valid for any neutrino mass $m_4>m_K$.

The blue line was calculated with micrOMEGAs and represents the DM and vector boson masses that will produce the correct relic abundance in a thermal scenario, while the masses in the upper hatched area would generate too much DM. A key difference with respect to the previous model is that here the DM annihilation cross section to neutrinos proceeds via an *s*-channel and thus is enhanced for $m_{Z2} \sim 2m_\chi$, as can be seen from Eq. (6.2). This explains the second branch of the blue line below the resonant condition in the panels with $g'=1$. A line where the relic abundance can be obtained below $m_{Z2}=2m_\chi$ also occurs for $g'=4\pi$ but, since the cross section is larger, the relic abundance is achieved for $m_\chi>100$ GeV, which is ruled out by XENON1T. This resonant effect also explains the shape of the indirect detection constraints which follow the same trend.

Similar to the previous model in Sect. 5, the direct detection constraints from XENON1T become relevant at large DM masses for $g'=4\pi$. However, even for values of the gauge coupling this large, we have checked that direct detection constraints from the elastic DM scattering off electrons are negligible.

The complementarity between cosmological observables, DM, and neutrino experiments allows us to set very strong bounds on the DM and Z_2 masses for this particular realisation, ruling out significant portions of the parameter space. There are still allowed regions for larger values of the gauge coupling consistent with a thermal DM candidate that yields the observed DM relic abundance. However, future neutrino experiments such as DUNE will be able to probe down to the value for which the correct relic abundance is obtained in some parts of the parameter space.

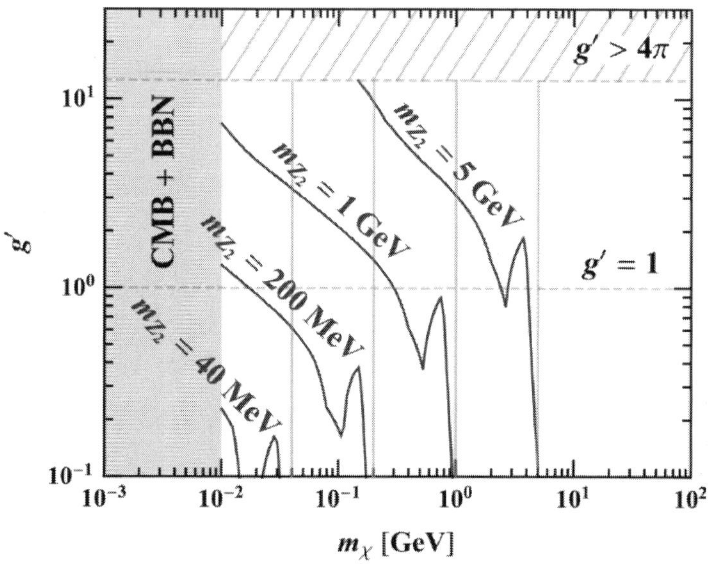

Figure 11: Values of the DM mass m_χ and the coupling g′ required to reproduce the observed relic abundance. We have fixed mZ2=0.04mZ2=0.04, 0.2, 1, and 5 GeV, and have considered the representative case of θ_e=0.031, while keeping $\theta_{\mu,\tau}$=0. Along (above) the blue lines the DM relic density matches (is less than) the observed value. We do not consider $m_\chi > m_{Z2}$ to ensure the neutrino portal regime. The lower bound $m_\chi \gtrsim 10$ MeV is set by observations of the CMB and BBN.

It is worth noticing that the sensitivity of present and future neutrino detectors to DM annihilations into neutrinos is largely independent of the flavour to which the sterile neutrino dominantly couples. Indeed, regardless of the original flavour composition produced by the DM annihilations, neutrino oscillations will tend to populate all flavours with similar fractions when the flux arrives to the detector. The main differences between the three rows in Fig. 10 are due to the different magnitude of the mixing allowed to the different flavours, with more stringent constraints applying for the mixing with muon neutrinos.

Finally, in Fig. 11, we fix m_{Z2} to several values, namely, m_{Z2}=0.04, 0.2, 1, and 5 GeV, and show the lines corresponding to the correct relic abundance in the m_χ–g' plane. These results were obtained using micrOMEGAs. Small values of $g'g'$ are ruled out except for DM masses in the proximity of the resonance, i.e., when $m_\chi \approx m_{Z2}/2$. As can be seen from this figure, a lighter dark vector boson allows for smaller values of g'. For $m_{Z2} \gtrsim 1$ GeV, values of $g' \gtrsim 1$ are required to yield the observed relic density, except for the resonance region. The dip towards $m_\chi \approx m_{Z2}$ corresponds to opening of new DM annihilation channels at tree level.

CONCLUSIONS

Despite the tremendous improvement over the last years in the sensitivity of direct, indirect and collider searches for dark matter, its discovery still eludes us. An interesting possibility is that its interactions with SM particles happen dominantly with the neutrino sector. This option would not only explain our failure to detect any DM interactions (except gravitational) so far, it would also connect our two present experimental signals of physics beyond the SM. Indeed, a rich phenomenology that would stem from the connection of these two sectors has been explored and discussed in the literature. *SU*(2) gauge invariance would naively dictate that neutrinos share all their interactions with their charged lepton counterparts, which are much easier to detect. We have therefore explored whether a dominant neutrino–DM interaction is allowed in simple gauge-invariant models without conflicting with searches through charged leptons.

We first explored the simplest scenario, in which DM couples to the full lepton doublet. We verified that, as long as the DM is heavier than the charged lepton(s) it couples to, the bounds from DM annihilation to charged leptons preclude DM–neutrino couplings sizeable enough to be probed, even ruling out all of the parameter space that would not lead to overclosure of the Universe. Alternatively, if DM couples to τ (μ) and is lighter than the charged lepton, its phenomenology is dominated by the interaction with neutrinos. This region is constrained by present neutrino detectors and will be fully probed for certain DM masses by future experiments.

We have then explored the option of the neutrino portal to DM and showed, as an example, two specific realisations with scalar and vector couplings, respectively. In the neutrino portal DM couples directly to new heavy neutrinos. Indeed, their singlet nature makes them natural candidates to probe the dark sector since they are allowed to interact with it via relevant or marginal operators. These right-handed neutrinos are also a natural addition to the SM particle content so as to account for the evidence for neutrino masses and mixings. The mixing between the SM neutrinos and the new singlets will induce DM–neutrino interactions at tree level, but DM-charged lepton couplings only at loop level.In the two realisations explored we find that it is indeed possible for neutrino detectors to place the most stringent and competitive bounds through searches for DM annihilations to neutrinos. Present searches at Super-Kamiokande, Fréjus, or Borexino are ruling out large areas of the parameter space. Interestingly, future projects such as Hyper-Kamiokande, MEMPHYS, DARWIN, or DUNE will be able to probe the cross section very close and beyond the value required to explain the DM abundance solely by annihilation to SM neutrinos. These new searches will effectively cover most of the parameter space, probing if the right-handed singlet fermions that can explain the origin of neutrino masses also represent our best window to the discovery of the dark matter sector.

ACKNOWLEDGEMENTS

We warmly thank B. Zaldivar for extremely useful discussions. We would also like to acknowledge discussions with M. Bauer, M. Chala, M. Hostert, A. Plascencia, A. Vincent, and S. Witte. This work made extensive use of the HPC-Hydra cluster at IFT. This work is supported in part by the European Union's Horizon 2020 research and innovation programme under the Marie Sklodowska-Curie Grant agreements 674896-Elusives, 690575-InvisiblesPlus, and 777419-ESSnuSB, as well as by the COST Action CA15139 EuroNuNet. MB, EFM, and SR acknowledge support from the "Spanish Agencia Estatal de Investigación" (AEI) and the EU "Fondo Europeo de Desarrollo Regional" (FEDER) through the project FPA2016-78645-P; and the Spanish MINECO through the "Ramón y Cajal" programme and through the Centro de Excelencia Severo Ochoa Program under Grant SEV-2016-0597. MB also acknowledges support from the Göran Gustafsson foundation. SP and AOD are also (partially) supported by the European Research Council under ERC Grant "NuMass" (FP7-IDEAS-ERC ERC-CG 617143). SP would like to acknowledge partial support from the Wolfson Foundation and the Royal Society. SP and AOD would also like to thank the Instituto de Física Teórica for kind hospitality during the completion of this work.

REFERENCES

1. C. Boehm, P. Fayet, R. Schaeffer, Constraining dark matter candidates from structure formation. Phys. Lett. B 518, 8–14 (2001). arXiv:astro-ph/0012504
2. C. Boehm, R. Schaeffer, Constraints on dark matter interactions from structure formation: damping lengths. Astron. Astrophys. 438, 419–442 (2005). arXiv:astro-ph/0410591
3. G. Mangano, A. Melchiorri, P. Serra, A. Cooray, M. Kamionkowski, Cosmological bounds on dark matter–neutrino interactions. Phys. Rev. D 74, 043517 (2006). arXiv:astro-ph/0606190
4. E. Bertschinger, The effects of cold dark matter decoupling and pair annihilation on cosmological perturbations. Phys. Rev. D 74, 063509 (2006). arXiv:astro-ph/0607319
5. P. Serra, F. Zalamea, A. Cooray, G. Mangano, A. Melchiorri, Constraints on neutrino–dark matter interactions from cosmic microwave background and large scale structure data. Phys. Rev. D 81, 043507 (2010). arXiv:0911.4411
6. L.G. van den Aarssen, T. Bringmann, C. Pfrommer, Is dark matter with long-range interactions a solution to all small-scale problems of ΛΛ CDM cosmology? Phys. Rev. Lett. 109, 231301 (2012). arXiv:1205.5809
7. T. Bringmann, J. Hasenkamp, J. Kersten, Tight bonds between sterile neutrinos and dark matter. JCAP 1407, 042 (2014). arXiv:1312.4947
8. R.J. Wilkinson, C. Boehm, J. Lesgourgues, Constraining dark matter–neutrino interactions using the CMB and large-scale structure. JCAP 1405, 011 (2014). arXiv:1401.7597
9. C. Boehm, J.A. Schewtschenko, R.J. Wilkinson, C.M. Baugh, S. Pascoli, Using the Milky Way satellites to study interactions between cold dark matter and radiation. Mon. Not. R. Astron. Soc. 445, L31–L35 (2014). arXiv:1404.7012
10. J.A. Schewtschenko, R.J. Wilkinson, C.M. Baugh, C. Bœhm, S. Pascoli, Dark matter-radiation interactions: the impact on dark matter haloes. Mon. Not. R. Astron. Soc. 449, 3587–3596 (2015). arXiv:1412.4905
11. F.-Y. Cyr-Racine, K. Sigurdson, J. Zavala, T. Bringmann, M. Vogelsberger, C. Pfrommer, ETHOS—an effective theory of structure formation: from dark particle physics to the matter distribution of the Universe. Phys. Rev. D 93, 123527 (2016). arXiv:1512.05344

12. M. Vogelsberger, J. Zavala, F.-Y. Cyr-Racine, C. Pfrommer, T. Bringmann, K. Sigurdson, ETHOS–an effective theory of structure formation: dark matter physics as a possible explanation of the small-scale CDM problems. Mon. Not. R. Astron. Soc. 460, 1399–1416 (2016). arXiv:1512.05349
13. O. Balducci, S. Hofmann, A. Kassiteridis, Cosmological singlet diagnostics of neutrinophilic dark matter. Phys. Rev. D 98, 023003 (2018). arXiv:1710.09846
14. Olivares-Del Campo, C. Bœhm, S. Palomares-Ruiz, S. Pascoli, Dark matter–neutrino interactions through the lens of their cosmological implications. Phys. Rev. D 97, 075039 (2018). arXiv:1711.05283
15. J.F. Beacom, N.F. Bell, G.D. Mack, General upper bound on the dark matter total annihilation cross section. Phys. Rev. Lett. 99, 231301 (2007). arXiv:astro-ph/0608090
16. S. Palomares-Ruiz, S. Pascoli, Testing MeV dark matter with neutrino detectors. Phys. Rev. D 77, 025025 (2008). arXiv:0710.5420
17. M. Lindner, A. Merle, V. Niro, Enhancing dark matter annihilation into neutrinos. Phys. Rev. D 82, 123529 (2010). arXiv:1005.3116
18. C. El Aisati, C. Garcia-Cely, T. Hambye, L. Vanderheyden, Prospects for discovering a neutrino line induced by dark matter annihilation. JCAP 1710, 021 (2017). arXiv:1706.06600
19. Y. Farzan, S. Palomares-Ruiz, Dips in the diffuse supernova neutrino background. JCAP 1406, 014 (2014). arXiv:1401.7019
20. T. Franarin, M. Fairbairn, J.H. Davis, JUNO sensitivity to resonant absorption of galactic supernova neutrinos by dark matter. arXiv:1806.05015
21. C.A. Argüelles, A. Kheirandish, A.C. Vincent, Imaging galactic dark matter with high-energy cosmic neutrinos. Phys. Rev. Lett. 119, 201801 (2017). arXiv:1703.00451
22. S. Pandey, S. Karmakar, S. Rakshit, Interactions of astrophysical neutrinos with dark matter: a model building perspective. JHEP 01, 095 (2019). arXiv:1810.04203
23. S. Karmakar, S. Pandey, S. Rakshit, Are we looking at neutrino absorption spectra at IceCube? arXiv:1810.04192
24. B. Holdom, Two U(1)'s and epsilon charge shifts. Phys. Lett. 166B, 196–198 (1986)
25. B. Patt, F. Wilczek, Higgs-field portal into hidden sectors. arXiv:hep-

ph/0605188

26. Falkowski, J. Juknevich, J. Shelton, Dark matter through the neutrino portal. arXiv:0908.1790
27. V. Gonzalez Macias, J. Wudka, Effective theories for dark matter interactions and the neutrino portal paradigm. JHEP 07, 161 (2015). arXiv:1506.03825
28. P. Minkowski, mu →→ e gamma at a rate of one out of 1-billion muon decays? Phys. Lett. B 67, 421 (1977)
29. T. Yanagida, Horizontal symmetry and masses of neutrinos, in Proceedings of Workshop on the Unified Theories and the Baryon Number in the Universe, Tsukuba, Japan, February 13–14, 1979, pp. 95–99 (1979)
30. M. Gell-Mann, P. Ramond, R. Slansky, Complex spinors and unified theories, in Proceedings of Supergravity Workshop, Stony Brook, New York, September 27–28, 1979, pp. 315–321 (1979). arXiv:1306.4669
31. S.L. Glashow, The future of elementary particle physics. NATO Sci. Ser. B 61, 687 (1980)
32. R.N. Mohapatra, G. Senjanovic, Neutrino mass and spontaneous parity nonconservation. Phys. Rev. Lett. 44, 912 (1980)
33. J.A. Casas, J.R. Espinosa, I. Hidalgo, Implications for new physics from fine-tuning arguments. 1. Application to SUSY and seesaw cases. JHEP 11, 057 (2004). arXiv:hep-ph/0410298
34. R. Mohapatra, J. Valle, Neutrino mass and baryon number nonconservation in superstring models. Phys. Rev. D 34, 1642 (1986)
35. J. Bernabeu, A. Santamaria, J. Vidal, A. Mendez, J. Valle, Lepton flavor nonconservation at high-energies in a superstring inspired standard model. Phys. Lett. B 187, 303 (1987)
36. G.C. Branco, W. Grimus, L. Lavoura, The seesaw mechanism in the presence of a conserved lepton number. Nucl. Phys. B 312, 492–508 (1989)
37. W. Buchmuller, D. Wyler, Dilatons and majorana neutrinos. Phys. Lett. B 249, 458–462 (1990)
38. Pilaftsis, Radiatively induced neutrino masses and large Higgs neutrino couplings in the standard model with Majorana fields. Z. Phys. C 55, 275–282 (1992). arXiv:hep-ph/9901206
39. J. Kersten, A.Y. Smirnov, Right-handed neutrinos at CERN LHC and

the mechanism of neutrino mass generation. Phys. Rev. D 76, 073005 (2007). arXiv:0705.3221

40. M. Malinsky, J. Romao, J. Valle, Novel supersymmetric SO(10) seesaw mechanism. Phys. Rev. Lett. 95, 161801 (2005). arXiv:hep-ph/0506296
41. G. Bertone (ed.), *Particle Dark Matter: Observations, Models and Searches* (Cambridge University Press, Cambridge, 2010)
42. Falkowski, J.T. Ruderman, T. Volansky, Asymmetric dark matter from leptogenesis. JHEP 05, 106 (2011). arXiv:1101.4936
43. Bertoni, S. Ipek, D. McKeen, A.E. Nelson, Constraints and consequences of reducing small scale structure via large dark matter–neutrino interactions. JHEP 04, 170 (2015). arXiv:1412.3113
44. V. González-Macías, J.I. Illana, J. Wudka, A realistic model for dark matter interactions in the neutrino portal paradigm. JHEP 05, 171 (2016). arXiv:1601.05051
45. B. Batell, T. Han, D. McKeen, B.Shams Es Haghi, Thermal dark matter through the Dirac neutrino portal. Phys. Rev. D 97, 075016 (2018). arXiv:1709.07001
46. J.F. Cherry, A. Friedland, I.M. Shoemaker, Neutrino portal dark matter: from dwarf galaxies to IceCube. arXiv:1411.1071
47. Super-Kamiokande Collaboration, Y. Fukuda et al., The Super-Kamiokande detector. Nucl. Instrum. Methods A 501, 418–462 (2003)
48. Borexino Collaboration, G. Alimonti et al., Science and technology of BOREXINO: a real time detector for low-energy solar neutrinos. Astropart. Phys. 16, 205–234 (2002). arXiv:hep-ex/0012030
49. H. Yuksel, S. Horiuchi, J.F. Beacom, S. Ando, Neutrino constraints on the dark matter total annihilation cross section. Phys. Rev. D 76, 123506 (2007). arXiv:0707.0196
50. Super-Kamiokande Collaboration, K. Frankiewicz, Searching for dark matter annihilation into neutrinos with Super-Kamiokande, in Proceedings, Meeting of the APS Division of Particles and Fields (DPF 2015): Ann Arbor, Michigan, USA, 4–8 Aug 2015 (2015). arXiv:1510.07999
51. Olivares-Del Campo, S. Palomares-Ruiz, S. Pascoli, Implications of a dark matter-neutrino coupling at Hyper-Kamiokande, in 53rd Rencontres de Moriond on Electroweak Interactions and Unified Theories (Moriond EW 2018) La Thuile, Italy, March 10–17, 2018

(2018). arXiv:1805.09830

52. N. Klop, S. Ando, Constraints on MeV dark matter using neutrino detectors and their implication for the 21-cm results. Phys. Rev. D 98, 103004 (2018). arXiv:1809.00671
53. Borexino Collaboration, G. Bellini et al., Study of solar and other unknown anti-neutrino fluxes with Borexino at LNGS. Phys. Lett. B 696,191–196 (2011). arXiv:1010.0029
54. FREJUS Collaboration, C. Berger et al., The Frejus nucleon decay detector. Nucl. Instrum. Methods A 262, 463 (1987)
55. AMANDA Collaboration, F. Halzen et al., The AMANDA neutrino telescope. Nucl. Phys. Proc. Suppl. 77, 474–485 (1999). arXiv:hep-ex/9809025
56. Hyper-Kamiokande Collaboration, K. Abe et al., Hyper-Kamiokande design report. arXiv:1805.04163
57. G. Jungman, M. Kamionkowski, K. Griest, Supersymmetric dark matter. Phys. Rep. 267, 195–373 (1996). arXiv:hep-ph/9506380
58. ESSnuSB Collaboration, E. Baussan et al., A very intense neutrino super beam experiment for leptonic CP violation discovery based on the European spallation source linac. Nucl. Phys. B 885, 127–149 (2014). arXiv:1309.7022
59. de Bellefon et al., MEMPHYS: a large scale water Cerenkov detector at Frejus. arXiv:hep-ex/0607026
60. DARWIN Collaboration, J. Aalbers et al., DARWIN: towards the ultimate dark matter detector. JCAP 1611, 017 (2016). arXiv:1606.07001
61. DUNE Collaboration, R. Acciarri et al., Long-baseline neutrino facility (LBNF) and deep underground neutrino experiment (DUNE). arXiv:1512.06148
62. D. McKeen, N. Raj, Monochromatic dark neutrinos and boosted dark matter in noble liquid direct detection. Phys. Rev. D 99(10), 103003 (2019). https://doi.org/10.1103/PhysRevD.99.103003 arXiv:1812.05102
63. Super-Kamiokande Collaboration, K. Choi et al., Search for neutrinos from annihilation of captured low-mass dark matter particles in the Sun by Super-Kamiokande. Phys. Rev. Lett. 114, 141301 (2015). arXiv:1503.04858
64. IceCube Collaboration, M.G. Aartsen et al., Search for annihilating dark matter in the Sun with 3 years of IceCube data. Eur. Phys. J. C 77,

146 (2017). arXiv:1612.05949

65. MAGIC and Fermi-LAT Collaboration, M.L. Ahnen et al., Limits to dark matter annihilation cross-section from a combined analysis of MAGIC and Fermi-LAT observations of dwarf satellite galaxies. JCAP 1602, 039 (2016). arXiv:1601.06590
66. Planck Collaboration, N. Aghanim et al., Planck 2018 results. VI. Cosmological parameters. arXiv:1807.06209
67. T.R. Slatyer, Indirect dark matter signatures in the cosmic dark ages. I. Generalizing the bound on s-wave dark matter annihilation from Planck results. Phys. Rev. D 93, 023527 (2016). arXiv:1506.03811
68. XENON Collaboration, E. Aprile et al., Dark matter search results from a one ton-year exposure of XENON1T. Phys. Rev. Lett. 121, 111302 (2018). arXiv:1805.12562
69. M. Battaglieri et al., US cosmic visions: new ideas in dark matter 2017: community report. arXiv:1707.04591
70. R. Essig, T. Volansky, T.-T. Yu, New constraints and prospects for sub-GeV dark matter scattering off electrons in xenon. Phys. Rev. D 96, 043017 (2017). arXiv:1703.00910
71. R. Essig, M. Sholapurkar, T.-T. Yu, Solar neutrinos as a signal and background in direct-detection experiments searching for sub-GeV dark matter with electron recoils. Phys. Rev. D 97, 095029 (2018). arXiv:1801.10159
72. M.J. Dolan, F. Kahlhoefer, C. McCabe, Directly detecting sub-GeV dark matter with electrons from nuclear scattering. Phys. Rev. Lett. 121, 101801 (2018). arXiv:1711.09906
73. P.D. Serpico, G.G. Raffelt, MeV-mass dark matter and primordial nucleosynthesis. Phys. Rev. D 70, 043526 (2004). arXiv:astro-ph/0403417
74. F. Iocco, G. Mangano, G. Miele, O. Pisanti, P.D. Serpico, Primordial nucleosynthesis: from precision cosmology to fundamental physics. Phys. Rep. 472, 1–76 (2009). arXiv:0809.0631
75. C. Boehm, M.J. Dolan, C. McCabe, A lower bound on the mass of cold thermal dark matter from Planck. JCAP 1308, 041 (2013). arXiv:1303.6270
76. K.M. Nollett, G. Steigman, BBN and the CMB constrain neutrino coupled light WIMPs. Phys. Rev. D 91, 083505 (2015). arXiv:1411.6005
77. M. Escudero, Neutrino decoupling beyond the standard model:

CMB constraints on the dark matter mass with a fast and precise NeffNeff evaluation. JCAP 1902, 007 (2019). arXiv:1812.05605

78. M. Escudero, L. Lopez-Honorez, O. Mena, S. Palomares-Ruiz, P. Villanueva-Domingo, A fresh look into the interacting dark matter scenario. JCAP 1806, 007 (2018). arXiv:1803.08427
79. G. Belanger, F. Boudjema, A. Pukhov, A. Semenov, MicrOMEGAs: a program for calculating the relic density in the MSSM. Comput. Phys. Commun. 149, 103–120 (2002). arXiv:hep-ph/0112278
80. E. Bertuzzo, C.J. Caniu Barros, G. Grilli di Cortona, MeV dark matter: model independent bounds. JHEP 09, 116 (2017). arXiv:1707.00725
81. E. Fernandez-Martinez, J. Hernandez-Garcia, J. Lopez-Pavon, Global constraints on heavy neutrino mixing. JHEP 08, 033 (2016). arXiv:1605.08774
82. M. Pospelov, A. Ritz, M.B. Voloshin, Secluded WIMP dark matter. Phys. Lett. B 662, 53–61 (2008). arXiv:0711.4866
83. M. Escudero, N. Rius, V. Sanz, Sterile neutrino portal to dark matter I: the U(1)B−LU(1)B−L case. JHEP 02, 045 (2017). arXiv:1606.01258
84. M. Escudero, N. Rius, V. Sanz, Sterile neutrino portal to dark matter II: exact dark symmetry. Eur. Phys. J. C 77, 397 (2017). arXiv:1607.02373
85. M.G. Folgado, G.A. Gómez-Vargas, N. Rius, R. Ruiz De Austri, Probing the sterile neutrino portal to dark matter with $\gamma\gamma$ rays. JCAP 1808, 002 (2018). arXiv:1803.08934
86. P. Bandyopadhyay, E.J. Chun, R. Mandal, F.S. Queiroz, Scrutinizing right-handed neutrino portal dark matter with Yukawa effect. Phys. Lett. B 788, 530–534 (2019). arXiv:1807.05122
87. Atre, T. Han, S. Pascoli, B. Zhang, The search for heavy Majorana neutrinos. JHEP 05, 030 (2009). arXiv:0901.3589
88. J. Ellis, TikZ-Feynman: Feynman diagrams with TikZ. Comput. Phys. Commun. 210, 103–123 (2017). arXiv:1601.05437
89. N.D. Christensen, C. Duhr, FeynRules—Feynman rules made easy. Comput. Phys. Commun. 180, 1614–1641 (2009). arXiv:0806.4194
90. Alloul, N.D. Christensen, C. Degrande, C. Duhr, B. Fuks, FeynRules 2.0—a complete toolbox for tree-level phenomenology. Comput. Phys. Commun. 185, 2250–2300 (2014). arXiv:1310.1921
91. T. Hahn, Generating Feynman diagrams and amplitudes with FeynArts 3. Comput. Phys. Commun. 140, 418–431 (2001). arXiv:hep-

ph/0012260

92. T. Hahn, M. Perez-Victoria, Automatized one loop calculations in four-dimensions and D-dimensions. Comput. Phys. Commun. 118, 153–165 (1999). arXiv:hep-ph/9807565
93. P.W. Angel, Y. Cai, N.L. Rodd, M.A. Schmidt, R.R. Volkas, Testable two-loop radiative neutrino mass model based on an LLQdcQdcLLQdcQdc effective operator. JHEP 10, 118 (2013). [Erratum: ibid. 11, 092 (2014)]. arXiv:1308.0463
94. J. Herrero-Garcia, E. Molinaro, M.A. Schmidt, Dark matter direct detection of a fermionic singlet at one loop. Eur. Phys. J. C 78, 471 (2018). arXiv:1803.05660
95. C. Hagedorn, J. Herrero-García, E. Molinaro, M.A. Schmidt, Phenomenology of the generalised scotogenic model with fermionic dark matter. JHEP 11, 103 (2018). arXiv:1804.04117
96. K.S. Babu, C.F. Kolda, J. March-Russell, Implications of generalized Z–Z-prime mixing. Phys. Rev. D 57, 6788–6792 (1998). arXiv:hep-ph/9710441
97. R. Harnik, J. Kopp, P.A.N. Machado, Exploring nu signals in dark matter detectors. JCAP 1207, 026 (2012). arXiv:1202.6073
98. M. Bauer, P. Foldenauer, J. Jaeckel, Hunting all the hidden photons. JHEP 07, 094 (2018). arXiv:1803.05466
99. BaBar Collaboration, J.P. Lees et al., Search for invisible decays of a dark photon produced in e+e−e+e− collisions at BaBar. Phys. Rev. Lett. 119, 131804 (2017). arXiv:1702.03327
100. Hook, E. Izaguirre, J.G. Wacker, Model independent bounds on kinetic mixing. Adv. High Energy Phys. 2011, 859762 (2011). arXiv:1006.0973

INDEX

E

F

G

H

Q

R

S